CELL BIOLOGY AND GENETICS

CECIE STARR / RALPH TAGGART

BIOLOGY

The Unity and Diversity of Life

NINTH EDITION

LISA STARR
Biological Illustrator

Australia • Canada • Mexico • Singapore • Spain • United Kingdom • United States

BIOLOGY PUBLISHER: Jack C. Carey
PROJECT DEVELOPMENT EDITOR: Kristin Milotich
MEDIA PROJECT MANAGER: Pat Waldo
EDITORIAL ASSISTANT: Daniel Lombardino
DEVELOPMENTAL EDITOR: Mary Arbogast
MARKETING TEAM: Rachel Alvelais, Mandie Houghan, Laura Hubrich, Carla Martin-Falcone
PRODUCTION EDITOR: Jamie Sue Brooks
PRODUCTION DIRECTOR: Mary Douglas, Rogue Valley Publications
EDITORIAL PRODUCTION: Karen Stough, Jamee Rae
PERMISSIONS EDITOR: Mary Kay Hancharick
TEXT AND COVER DESIGN, ART DIRECTION: Gary Head, Gary Head Design
COVER PHOTO: *Snow leopard, a currently endangered species* (*Section 28.2*). © Art Wolfe/PhotoResearchers
ART EDITOR: Mary Douglas, Jamie Sue Brooks
INTERIOR ILLUSTRATION: Lisa Starr, Raychel Ciemma, Precision Graphics, Preface Inc., Robert Demarest, Jan Flessner, Darwen Hennings, Vally Hennings, Betsy Palay, Nadine Sokol, Kevin Somerville, Lloyd Townsend
PHOTO RESEARCHER: Myrna Engler
PRINT BUYER: Vena M. Dyer
COMPOSITION: Preface, Inc. (Angela Harris)
COLOR PROCESSING: H&S Graphics (Tom Anderson, Michelle Kessel, Laurie Riggle, and John Deady)
COVER PRINTING, PRINTING AND BINDING: Von Hoffmann Press

BOOKS IN THE BROOKS/COLE BIOLOGY SERIES

Biology: The Unity and Diversity of Life, Ninth, Starr/Taggart
Biology: Concepts and Applications, Fourth, Starr
Laboratory Manual for Biology, Perry/Morton/Perry
Human Biology, Fourth, Starr/McMillan
Perspectives in Human Biology, Knapp
Human Physiology, Fourth, Sherwood
Fundamentals of Physiology, Second, Sherwood
Psychobiology: The Neuron and Behavior, Hoyenga/Hoyenga

Introduction to Cell and Molecular Biology, Wolfe
Molecular and Cellular Biology, Wolfe

Introduction to Microbiology, Second, Ingraham/Ingraham

Genetics: The Continuity of Life, Fairbanks
Human Heredity, Fifth, Cummings
Introduction to Biotechnology, Barnum
Sex, Evolution, and Behavior, Second, Daly/Wilson

Plant Biology, Rost et al.
Plant Physiology, Fourth, Salisbury/Ross
Plant Physiology Laboratory Manual, Ross

General Ecology, Second, Krohne
Living in the Environment, Eleventh, Miller
Environmental Science, Eighth, Miller
Sustaining the Earth, Fourth, Miller
Environment: Problems and Solutions, Miller
Environmental Science, Fifth, Chiras

Oceanography: An Invitation to Marine Science, Third, Garrison
Essentials of Oceanography, Second, Garrison
Introduction to Ocean Sciences, Segar
Oceanography: An Introduction, Fifth, Ingmanson/Wallace
Marine Life and the Sea, Milne

Printed in United States of America
10 9 8 7 6 5 4

For permission to use material from this work, contact us by
www.thomsonrights.com
Fax: 1-800-730-2215
Phone: 1-800-730-2214

ISBN 0-534-37940-0

For more information about this or any other Brooks/Cole products, contact:
BROOKS/COLE
511 Forest Lodge Road
Pacific Grove, CA 93950 USA
www.brookscole.com
1-800-423-0563 (Thomson Learning Academic Resource Center)

CONTENTS IN BRIEF

Highlighted chapters are included in CELL BIOLOGY AND GENETICS.

DETAILED CONTENTS

Preface

Teachers of introductory biology know all about the Red Queen effect, whereby one runs as fast as one can to stay in the same place. New and modified information from hundreds of fields of inquiry piles up daily, and somehow teachers are expected to distill it into Biology Lite, a one-course zip through the high points that still manages to help students deepen their understanding of a world of unbelievable richness.

Restricting textbook content runs the risk of splintering understanding of that world, as when an emphasis on human biology inadvertently reinforces archaic notions that everything on Earth is here in the service of Us; or when a molecular focus excludes knowledge of whole organisms. (Here I am reminded of the well intentioned but lethal blanketing of habitats with DDT and of Jacques Monod's first clueless encounters with *E. coli*.)

We offer this book as a coherent account of the sweep of life's diversity and its underlying unity. Through its examples of problem solving and experiments, it shows the power of thinking critically about the natural world. It highlights key concepts, current understandings, and research trends for major fields of biological inquiry. It explains the structure and function of a broad sampling of organisms in enough detail so students can develop a working vocabulary about life's parts and processes.

The book starts with an overview of the basic concepts and scientific methods. Three units on the principles of biochemistry, inheritance, and evolution follow. They are the conceptual framework for exploring life's unity and diversity, starting with an evolutionary survey of each kingdom. Units on comparative anatomy and physiology of plants, then animals, are next. The last unit focuses on patterns and consequences of organisms interacting with one another and with the environment. This conceptual organization parallels the levels of biological organization.

TOPIC SPREADS Ongoing feedback from teachers of more than three million students helped us refine our approach to writing. We keep the story line in focus for students by subscribing to the question "How do you eat an elephant?" and its answer, "One bite at a time." We organized descriptions, art, and supporting evidence for each topic on two facing pages, at most. Each *topic spread* starts with a numbered tab and concludes with boldface, summary statements of key points (see below). Students can use the statements to check whether they understand one topic spread's concepts before starting another.

By clearly organizing topics within chapters, we offer teachers flexibility in assigning text material to fit course requirements. For example, those who spend little time on photosynthesis might choose to bypass topic spreads with details of the properties of light or the chemiosmotic theory of ATP formation. They might or may not assign the Focus essays (one on the impact of photosynthesis on

Numbered tabs indicate the start of a new concept as the chapter's story unfolds. The green tabs identify basic chapter concepts. Blue tabs identify Focus essays that enrich the basics with examples of experiments (to demonstrate the power of critical thinking), of the nature of life, and of applying the basics to issues of human interest.

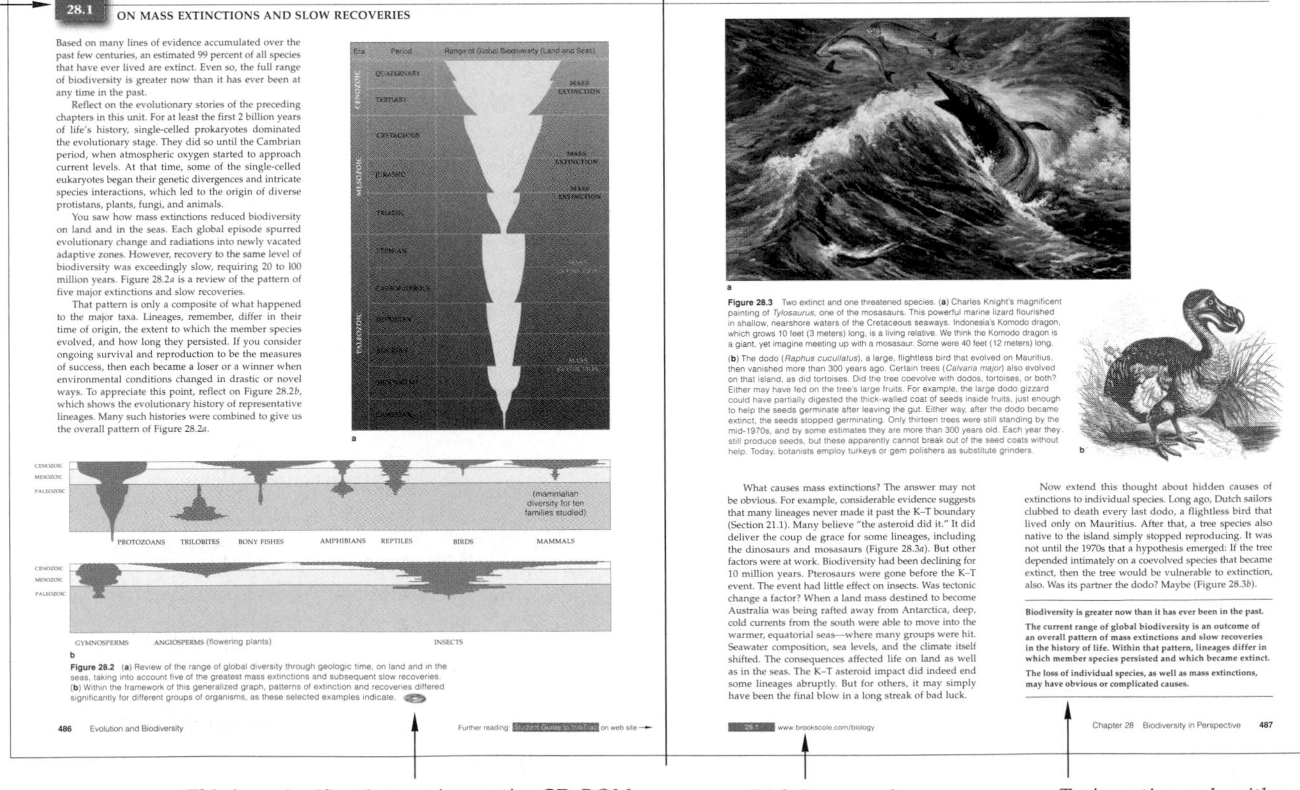

28.1 ON MASS EXTINCTIONS AND SLOW RECOVERIES

Based on many lines of evidence accumulated over the past few centuries, an estimated 99 percent of all species that have ever lived are extinct. Even so, the full range of biodiversity is greater now than it has ever been at any time in the past.

Reflect on the evolutionary stories of the preceding chapters in this unit. For at least the first 2 billion years of life's history, single-celled prokaryotes dominated the evolutionary stage. They did so until the Cambrian period, when atmospheric oxygen started to approach current levels. At that time, some of the single-celled eukaryotes began their genetic divergences and intricate species interactions, which led to the origin of diverse protistans, plants, fungi, and animals.

You saw how mass extinctions reduced biodiversity on land and in the seas. Each global episode spurred evolutionary change and radiations into newly vacated adaptive zones. However, recovery to the same level of biodiversity was exceedingly slow, requiring 20 to 100 million years. Figure 28.2*a* is a review of the pattern of five major extinctions and slow recoveries.

That pattern is only a composite of what happened to the major taxa. Lineages, remember, differ in their time of origin, the extent to which the member species evolved, and how long they persisted. If you consider ongoing survival and reproduction to be the measures of success, then each became a loser or a winner when environmental conditions changed in drastic or novel ways. To appreciate this point, reflect on Figure 28.2*b*, which shows the evolutionary history of representative lineages. Many such histories were combined to give us the overall pattern of Figure 28.2*a*.

Figure 28.2 (**a**) Review of the range of global diversity through geologic time, on land and in the seas, taking into account five of the greatest mass extinctions and subsequent slow recoveries. (**b**) Within the framework of this generalized graph, patterns of extinction and recoveries differed significantly for different groups of organisms, as these selected examples indicate.

486 Evolution and Biodiversity

Further reading: Student Guide to InfoTrac on web site

Figure 28.3 Two extinct and one threatened species. (**a**) Charles Knight's magnificent painting of *Tylosaurus*, one of the mosasaurs. This powerful marine lizard flourished in shallow, nearshore waters of the Cretaceous seaways. Indonesia's Komodo dragon, which grows 10 feet (3 meters) long, is a living relative. We think the Komodo dragon is a giant, yet imagine meeting up with a mosasaur. Some were 40 feet (12 meters) long.

(**b**) The dodo (*Raphus cucullatus*), a large, flightless bird that evolved on Mauritius, then vanished more than 300 years ago. Certain trees (*Calvaria major*) also evolved on that island, as did tortoises. Did the tree coevolve with dodos, tortoises, or both? Either may have fed on the tree's large fruits. For example, the large dodo gizzard could have partially digested the thick-walled coat of seeds inside fruits, just enough to help the seeds germinate after leaving the gut. Either way, after the dodo became extinct, the seeds stopped germinating. Only thirteen trees were still standing by the mid-1970s, and by some estimates they are more than 300 years old. Each year they still produce seeds, but these apparently cannot break out of the seed coats without help. Today, botanists employ turkeys or gem polishers as substitute grinders.

What causes mass extinctions? The answer may not be obvious. For example, considerable evidence suggests that many lineages never made it past the K–T boundary (Section 21.1). Many believe "the asteroid did it." It did deliver the coup de grace for some lineages, including the dinosaurs and mosasaurs (Figure 28.3*a*). But other factors were at work. Biodiversity had been declining for 10 million years. Pterosaurs were gone before the K–T event. The event had little effect on insects. Was tectonic change a factor? When a land mass destined to become Australia was being rafted away from Antarctica, deep, cold currents from the south were able to move into the warmer, equatorial seas—where many groups were hit. Seawater composition, sea levels, and the climate itself shifted. The consequences affected life on land as well as in the seas. The K–T asteroid impact did indeed end some lineages abruptly. But for others, it may simply have been the final blow in a long streak of bad luck.

Now extend this thought about hidden causes of extinctions to individual species. Long ago, Dutch sailors clubbed to death every last dodo, a flightless bird that lived only on Mauritius. After that, a tree species also native to the island simply stopped reproducing. It was not until the 1970s that a hypothesis emerged: If the tree depended intimately on a coevolved species that became extinct, then the tree would be vulnerable to extinction, also. Was its partner the dodo? Maybe (Figure 28.3*b*).

Biodiversity is greater now than it has ever been in the past.

The current range of global biodiversity is an outcome of an overall pattern of mass extinctions and slow recoveries in the history of life. Within that pattern, lineages differ in which member species persisted and which became extinct.

The loss of individual species, as well as mass extinctions, may have obvious or complicated causes.

28.1 www.brookscole.com/biology

Chapter 28 Biodiversity in Perspective 487

This icon signifies that our interactive CD-ROM further explores the concept being illustrated.

Website expands on the section's topic.

Topic section ends with a summary of key concepts.

the biosphere; the other on a recent, novel idea about how photosynthesis got started in the first place). All topic spreads of the chapter flow as parts of the same story, but some clearly offer depth that may be treated as optional.

The topic spreads are not gimmicks. Ongoing feedback guided decisions about when to add depth and when to loosen core material with applications. Within the spreads, headings and subheadings help students keep track of the hierarchy of information. Transitions between spreads help them keep the greater story in focus and discourage memorization for its own sake. To avoid disrupting the basic story line while still attending to interested students, we include some enriching details in optional illustrations.

The clear organization helps students find assigned topics easily and has translated into improved test scores. This is a tangible outcome, but we are more pleased that the clarity helps give students enough confidence to dig deeper into biological science. We also are happy to hear that the story has lured many of them into reading far more than they planned to do.

BALANCING CONCEPTS WITH APPLICATIONS Each chapter starts with a lively or sobering application and an adjoining list of key concepts, the chapter's advance organizer. Strategically placed examples of applications parallel core material, not so many as to be distracting but enough to keep students interested in continuing with the basics. Brief applications are integrated in the text. Focus essays afford more depth on many medical, environmental, and social issues without interrupting the conceptual flow. The book's last four pages index all applications separately for fast reference.

FOUNDATIONS FOR CRITICAL THINKING To help students increase their capacity for critical thinking, we walk them through experiments that yielded evidence in favor of or against hypotheses. The main index lists all of the selected experimental tests and observational tests (index entries *Experiment* and *Test, observational*).

We use certain chapter introductions as well as entire chapters to show students some productive results of critical thinking. The introductions to Mendelian genetics (Chapter 11), DNA structure and function (13), speciation (19), immunology (40), and behavior (47) are examples. Also, each chapter has a set of *Critical Thinking* questions. Katherine Denniston developed most of these thought-provoking questions. Daniel Fairbanks developed many of the *Genetics Problems*, which help students grasp the principles of inheritance (Chapters 11 and 12).

VISUAL OVERVIEWS OF CONCEPTS We simultaneously develop text and art as inseparable parts of the same story. We give visual learners a means to work their way through a visual overview of major processes before reading the corresponding (and possibly intimidating) text. Students repeatedly let us know how much they appreciate this art. Overview illustrations have step-by-step descriptions of biological parts and processes. Instead of "wordless" diagrams, we break down information into a series of illustrated callouts. For example, in Figure 14.14, callouts integrated with the art walk students through the stages by which a mature mRNA transcript becomes translated.

Many anatomical drawings are integrated overviews of structure/function. Students need not jump back and forth from text, to tables, to illustrations, and back again to see how an organ system is put together and what its parts do. We hierarchically arrange descriptions of parts to reflect a system's structural and functional organization.

ZOOM SEQUENCES Many illustrations progress from macroscopic to microscopic views of a system or process. Figure 7.3, for example, starts with a plant leaf and ends with reaction sites in the chloroplast. Figures 38.20 and 38.21 start with a ballerina's biceps and move down through levels of skeletal muscle contraction.

COLOR CODES Consistent use of colors for molecules, cell structures, and processes helps students track what is going on. We use these colors throughout the book:

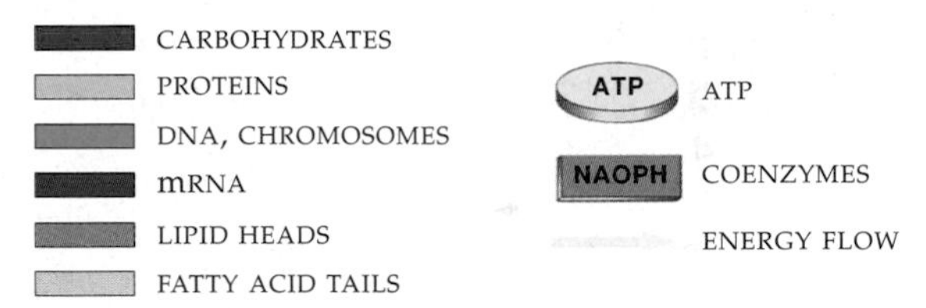

ICONS Small, simple diagrams next to an illustration help students relate a topic to the big picture. For instance, a simple diagram of a cell reminds them of the location of the plasma membrane relative to the cytoplasm. Other icons relate reactions and processes to certain locations and to how they tie in with one another. Others remind students of evolutionary relationships among organisms, as in Chapters 26 and 27. A multimedia icon directs them to art in the CD-ROM packaged in its own envelope at the back of their book. Another directs them to supplemental material on the Web and a third, to InfoTrac.

A COMMUNITY EFFORT

Each new edition starts the same way. I go back through banks of past reviews from a network of more than 2,000 teachers and reviewers. The cumulative wisdom about educational pitfalls and promises is just astounding, and it keeps me humble. Rereading pivotal journal articles follows, as do phone calls and e-mail flurries with my special advisors, contributors, and new reviewers for the next two years. Each edition, I again acknowledge those individuals whose contributions continue to shape our collective thinking. There is no way to describe their thoughtful assistance. I can only salute their commitment to quality in education.

Current configurations of Earth's oceans and land masses—the geologic stage upon which life's drama continues to unfold. Thousands of separate images were pieced together to create this remarkable, true-color image of our planet.

INTRODUCTION

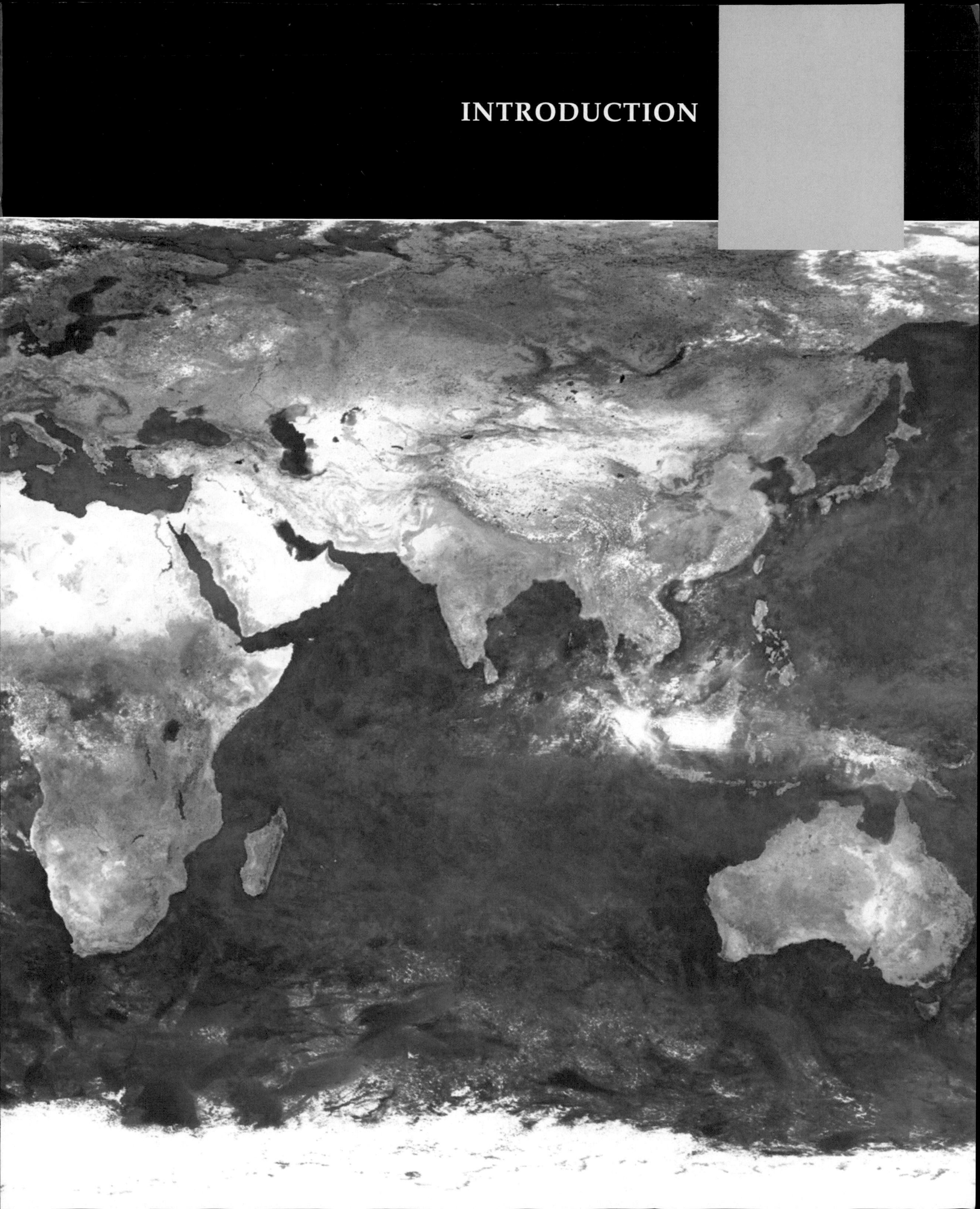

1

CONCEPTS AND METHODS IN BIOLOGY

Biology Revisited

Buried somewhere in that mass of tissue just above and behind your eyes are memories of first encounters with the living world. In that brain are early memories of discovering your hands and feet, your family, friends, the change of seasons, the scent of rain-drenched earth and grass. Still in residence are memories of your early introductions to a great disorganized parade of plants and animals, mostly living, sometimes dead. There, too, are memories of questions—*"What is life?"* and, inevitably, *"What is death?"* There also are memories of answers, some satisfying, others less so.

By making observations, asking questions, and accumulating answers, you have gradually built up a store of knowledge about life. Experience and education have been refining the questions, and no doubt some answers are difficult to come by.

Think of the world's forests—once vast, now astonishingly diminished by logging, conversion for agriculture, urban expansion, and other activities that help keep many people sheltered, warm, fed, and alive. In 1999, the human population surpassed 6 billion; it is still growing. With few forests remaining, where will we find more usable land and forest products?

Think of a college student, twenty years old, whose motorcycle skidded into a truck. Now he is comatose, and doctors say his brain is functionally dead. His breathing, heart rate, and other basic functions will continue only as long as he stays hooked up to

Figure 1.1 Think back on all you have ever known and seen. This is a foundation for your deeper probes into life.

a respirator and other artificial support systems. Would you say the unfortunate student is still "alive"?

Or think of a human egg, recently penetrated by a sperm inside a woman's body. At first the fertilized egg does not get bigger, but a series of programmed cuts divides it into a cluster of a few dozen tiny cells. Would you call the microscopically small mass a *human* life? If you learn more about how embryos actually develop, will the knowledge influence your thoughts about such incendiary issues as birth control and abortion?

If questions like these have ever crossed your mind, your thoughts about life obviously run deep. And you can approach this course in **biology**—the scientific study of life—with confidence, because you have been studying life ever since information started penetrating your brain. You simply are *revisiting* biology, in ways that might help carry your thoughts to deeper, more organized levels of understanding.

Return to the question, *What is life?* Offhandedly, you might respond that you know it when you see it. However, as you will see later in the book, the question opens up a story that has been unfolding in countless directions for about 3.8 billion years!

From the biological perspective, "life" is an outcome of ancient events by which nonliving matter—atoms and molecules—became assembled into the first living cells. "Life" is a way of capturing and using energy and raw materials. "Life" is a way of sensing and responding to changes in the environment. "Life" is a capacity to reproduce, grow, and develop. And "life" evolves, meaning that the traits characterizing the individuals of a population can change over the generations. Even so, this short list only hints at the meaning of life. Deeper insight requires wide-ranging study of life's characteristics.

Throughout this book, you will come across many diverse examples of how organisms are constructed, how they function, where they live, and what they do. The examples support certain concepts which, when taken together, will give you a sense of what "life" is.

This chapter introduces the basic concepts. It also sets the stage for forthcoming descriptions of scientific observations, experiments, and tests that help show how you can develop, modify, and refine your views of life. As you continue with your reading, you may find it useful to return occasionally to this simple overview as a way to reinforce your grasp of the details.

KEY CONCEPTS

1. Unity underlies the world of life, for all organisms are alike in key respects. They consist of one or more cells made of the same kinds of substances, put together in the same basic ways. Their activities require inputs of energy, which they must get from their surroundings. All organisms sense and respond to changing conditions in their environment. They all have a capacity to grow and reproduce, based on instructions contained in DNA.

2. The world of life shows immense diversity. Many millions of different kinds of organisms, or species, now inhabit the Earth, and many millions more lived in the past. Each one of those species is unique in some of its traits—that is, in some aspects of its body plan, body functioning, and behavior.

3. Theories of evolution, especially a theory of evolution by natural selection as formulated by Charles Darwin, help explain the meaning of life's diversity.

4. Biology, like other branches of science, is based on systematic observations, hypotheses, predictions, and observational and experimental tests. The external world, not internal conviction, is the testing ground for scientific theories.

1.1 DNA, ENERGY, AND LIFE

Nothing Lives Without DNA

DNA AND THE MOLECULES OF LIFE Picture a frog on a rock, busily croaking. Without even thinking about it, you know the frog is alive and the rock is not. Would you be able to explain why? At a fundamental level, both are no more than concentrations of the same units of matter, called protons, electrons, and neutrons. The units are building blocks of atoms, which are building blocks of larger bits of matter called molecules. And it is at the molecular level that differences between living and nonliving things start to emerge.

You will never, ever find a rock made of nucleic acids, proteins, carbohydrates, and lipids. In nature, only cells build these molecules, which you will read about later. All living things consist of one or more of cells, which are the smallest units of matter having a capacity for life. The signature molecule of cells is a nucleic acid known as **DNA**. No chunk of granite or quartz has it.

Encoded in DNA's structure are the instructions for assembling a dazzling array of proteins from a limited number of smaller building blocks, the amino acids. By analogy, if you follow suitable instructions and invest energy in the task, you might organize a heap of a few kinds of ceramic tiles (representing amino acids) into diverse patterns (representing proteins), as in Figure 1.2.

Among the proteins are enzymes. When these worker molecules get an energy boost, they swiftly build, split, and rearrange the molecules of life. Some enzymes work with a class of nucleic acids called RNAs in carrying out DNA's protein-building instructions. Think of this as a flow of information, from *DNA to RNA to protein*. As you will see from Chapter 14, this molecular trinity is central to our understanding of life.

Figure 1.2 Examples of objects built from the same materials but with different assembly instructions.

THE HERITABILITY OF DNA We humans tend to think we enter the world abruptly and leave it the same way. But we are much more than this. *We and all other organisms are part of a journey that began about 3.8 billion years ago, starting with the origin of the first living cells.*

Under present-day conditions on Earth, new cells arise only from cells that already exist. They do so by **inheritance**, which is the acquisition of traits by way of transmission of DNA from parent to offspring. Why do baby storks look like storks and not pelicans? Because they inherited stork DNA, which isn't exactly the same as pelican DNA in its molecular details.

Reproduction refers to actual mechanisms by which a cell or an organism produces offspring. Often it starts when a sperm fertilizes an egg. But the fertilized egg never could form if the sperm and egg had not formed earlier, according to DNA instructions passed on from cell to cell through countless generations according to the principles of inheritance and reproduction.

For frogs and humans and other large organisms, DNA also guides **development**, the transformation of a fertilized egg into a multicelled adult with cells, tissues, and organs specialized for certain tasks. Development proceeds through a series of stages. As one example, a moth is only the adult stage of a winged insect (Figure 1.3). First, a fertilized egg develops into an immature larval stage—a caterpillar that eats leaves and grows

Figure 1.3 "The insect"—a series of stages of development. Different adaptive properties emerge at each stage. Shown here, a silkworm moth, from the egg (**a**), to a larval stage (**b**), to a pupal stage (**c**), and on to the winged form of the adult (**d**,**e**).

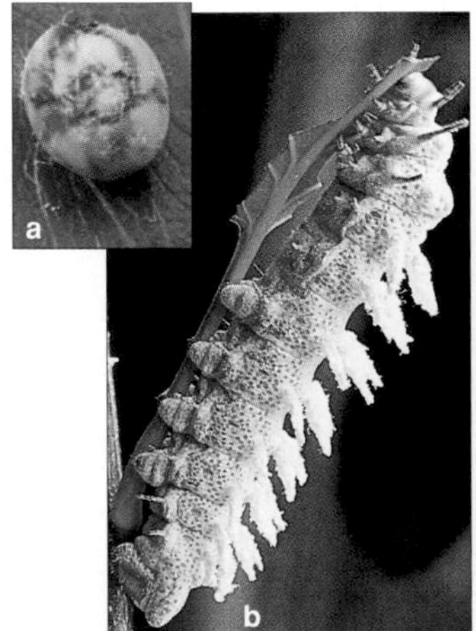

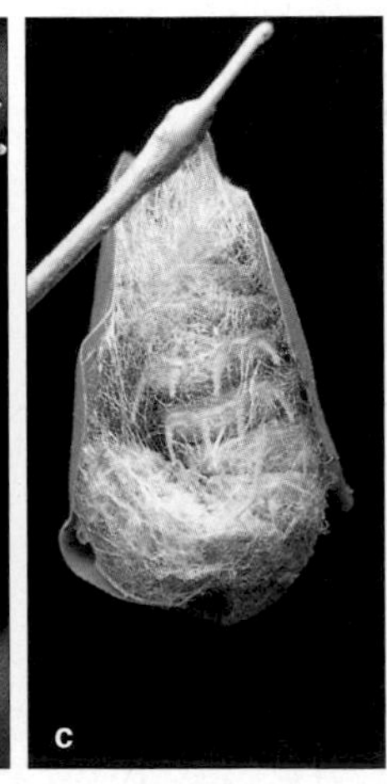

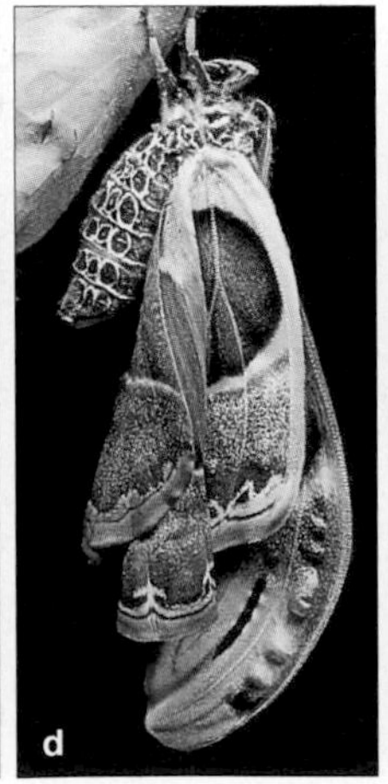

Further reading: Student Guide to InfoTrac on web site →

Figure 1.4 Response to signals from pain receptors, activated by a lion cub flirting with disaster.

rapidly until an internal alarm clock goes off. Then its tissues are remodeled into a different stage—a pupa. In time, an adult emerges that is adapted to reproduce. It produces sperm or eggs. Its wing color, patterns, and fluttering frequency are adapted for attracting a mate.

And so "the insect" is a series of organized stages. Its development from egg to adult will not proceed properly unless each stage is completed before the next begins. Instructions for each stage were written into moth DNA long before each moment of reproduction—and so the ancient moth story continues.

Nothing Lives Without Energy

ENERGY DEFINED Everything in the entire universe has some amount of **energy**, which is most simply defined as the capacity to do work. And nothing—absolutely nothing—happens in the universe without a *transfer* of energy. For instance, a single, undisturbed atom can do nothing except vibrate incessantly with its own energy. Suppose it absorbs extra energy from the sun and starts vibrating faster. Now some energy is on the move. That energy can do work by getting transferred elsewhere. If by chance the atom collides with a neighboring atom, one may give up, grab, or share energy with the other. Molecules form, become rearranged, and are split apart by such energy transfers. When this kind of molecular work is done, cells stay alive, grow, and reproduce.

METABOLISM DEFINED Each living cell has the capacity to (1) obtain and convert energy from its surroundings and (2) use energy to maintain itself, grow, and make more cells. We call this capacity **metabolism**. Think of a cell in a leaf that produces food by photosynthesis. It intercepts sunlight energy and converts it to chemical energy, in the form of ATP molecules. ATP is an energy carrier that helps drive hundreds of activities. It easily transfers some energy to metabolic workers—in this case, enzymes that assemble sugar molecules. ATP also forms by aerobic respiration. This process can release energy that cells have tucked away in sugars and other kinds of molecules.

SENSING AND RESPONDING TO ENERGY It is often said that only organisms respond to the environment. Yet even a rock shows responsiveness, as when it yields to the force of gravity and tumbles down a hill or changes its shape slowly under the repeated battering of wind, rain, or tides. The difference is this: *Organisms sense changes in their surroundings, then they make controlled, compensatory responses to them*. How? Every organism has **receptors**, which are molecules and structures that detect stimuli. A **stimulus** is a specific form of energy detected by receptors. Examples are sunlight energy, heat energy, a hormone molecule's chemical energy, and the mechanical energy of a bite (Figure 1.4).

Cells adjust metabolic activities in response to signals from receptors. Each cell (and organism) can withstand only so much heat or cold. It must rid itself of harmful substances. It requires certain foods, in certain amounts. Yet temperatures do shift, harmful substances might be encountered, and food is sometimes plentiful or scarce.

For example, after you finish a snack, simple sugars leave your gut and enter your blood. Blood is part of your *internal* environment (the other part is tissue fluid that bathes your cells). Over the long term, too much or too little sugar in the blood can cause problems, such as diabetes. When the sugar level rises, a glandular organ, the pancreas, normally steps up its secretion of insulin. Most of your cells have receptors for this hormone, which stimulates cells to take up sugar. When enough cells do so, the sugar level in blood returns to normal.

Organisms respond so exquisitely to energy changes that their internal operating conditions remain within tolerable limits. We call this a state of **homeostasis**. It is one of the key defining features of life.

All organisms consist of one or more cells, the smallest units of life. Under present-day conditions, new cells form only through the reproduction of cells that already exist.

DNA, the molecule of inheritance, encodes protein-building instructions, which RNAs help carry out. Many proteins are enzymes, the metabolic workers necessary to construct DNA and all other complex molecules of life.

Cells live only for as long as they engage in metabolism. They acquire and transfer energy that is used to assemble, break down, stockpile, and dispose of materials in ways that promote survival and reproduction.

Single cells and multicelled organisms sense and respond to environmental conditions in ways that help maintain their internal operating conditions.

1.2 ENERGY AND LIFE'S ORGANIZATION

Levels of Biological Organization

Taken as a whole, the metabolic activities of single cells and multicelled organisms maintain the great pattern of organization in nature, as sketched out in Figure 1.5. Consider the hierarchy. Life's properties emerge when DNA and other molecules become organized into cells. The **cell** is the smallest unit of organization having a capacity to survive and reproduce on its own, given DNA instructions, suitable conditions, building blocks, and energy inputs. Free-living, single cells such as an amoeba fit the definition. Does the definition of cells hold for **multicelled organisms**, which typically consist of specialized, interdependent cells organized as tissues and organs? Yes. You might find this a strange answer. After all, your own cells could never live alone in nature, because body fluids must continually bathe them. Yet even isolated human cells stay alive under controlled conditions in laboratories around the world. Investigators routinely maintain isolated human cells for use in important experiments, as in cancer studies.

BIOSPHERE
All those regions of Earth's waters, crust, and atmosphere in which organisms can exist

ECOSYSTEM
A community and its physical environment

COMMUNITY
The populations of all species occupying the same area

POPULATION
A group of individuals of the same kind (that is, the same species) occupying a given area

MULTICELLED ORGANISM
An individual composed of specialized, interdependent cells most often organized in tissues, organs, and organ systems

ORGAN SYSTEM
Two or more organs interacting chemically, physically, or both in ways that contribute to survival of the whole organism

ORGAN
A structural unit in which a number of tissues, combined in specific amounts and patterns, perform a common task

TISSUE
An organized group of cells and surrounding substances functioning together in a specialized activity

CELL
Smallest unit having the capacity to live and reproduce, independently or as part of a multicelled organism

ORGANELLE
Inside all cells except bacteria, a membrane-bound sac or compartment for a separate, specialized task

MOLECULE
A unit in which two or more atoms of the same element or different ones are bonded together

ATOM
Smallest unit of an element (a fundamental substance) that still retains the properties of that element

SUBATOMIC PARTICLE
An electron, proton, or neutron; one of the three major particles of which atoms are composed

Figure 1.5 Levels of organization in nature.

You typically find cells and multicelled organisms as part of a **population**, defined as a group of organisms of the same kind (such as a herd of zebras). The next level of organization is the **community**—all the populations of all species living in the same area (such as the African savanna's bacteria, grasses, trees, zebras, lions, and so on). The next level, the **ecosystem**, is the community *and*

Further reading: Student Guide to InfoTrac on web site →

Figure 1.6 Example of the one-way flow of energy and the cycling of materials through the biosphere.

(**a**) Plants of a warm, dry grassland called the African savanna capture energy from the sun and use it to build plant parts. Some of the energy ends up inside plant-eating organisms, including this adult male elephant. He eats huge quantities of plants to maintain his eight-ton self and produces great piles of solid wastes—dung—that still contain some unused nutrients. Although most organisms might not recognize it as such, elephant dung is an exploitable food source.

(**b**) And so we next have little dung beetles rushing to the scene almost simultaneously with the uplifting of an elephant tail. Working rapidly, they carve fragments of moist dung into round balls, which they roll off and bury in burrows. In those balls the beetles lay eggs—a reproductive behavior that will help assure their forthcoming offspring (**c**) of a compact food supply.

Thanks to beetles, dung does not pile up and dry out into rock-hard mounds in the intense heat of the day. Instead, the surface of the land is tidied up, the beetle offspring get fed, and the leftover dung accumulates in beetle burrows—there to enrich the soil that nourishes the plants that sustain (among others) the elephants.

its physical and chemical environment. The **biosphere** includes all regions of the Earth's atmosphere, waters, and crust in which organisms live. Astoundingly, *this globe-spanning organization begins with the convergence of energy, certain materials, and DNA in tiny, individual cells.*

Interdependencies Among Organisms

A great flow of energy into the world of life starts with **producers**, which are plants and other organisms that make their own food. Animals are **consumers**. Directly or indirectly, they depend on energy that became stored in the tissues of producers. For example, some energy is transferred to zebras after they browse on plants. It gets transferred again when lions devour a baby zebra that wandered away from its herd. And it is transferred again when fungal and bacterial decomposers feed on the tissues and remains of lions, elephants, or any other organism. **Decomposers** break down sugars and other biological molecules to simpler materials—which may be cycled back to producers. In time, all of the energy that the plants captured from the sun's rays returns to the environment, but that's another story.

For now, keep in mind that organisms connect with one another by a one-way flow of energy *through* them and a cycling of materials *among* them, as in Figure 1.6. Their interconnectedness affects the structure, size, and composition of populations and communities. It affects ecosystems, even the biosphere. Understand the extent of their interactions and you will gain insight into the environmental effects of acid rain, amplification of the greenhouse effect, and other modern-day problems.

Nature shows levels of organization. The characteristics of life emerge at the level of single cells and extend through populations, communities, ecosystems, and the biosphere.

A one-way flow of energy through organisms and a cycling of materials among them organizes life in the biosphere. In nearly all cases, energy flow starts with energy from the sun.

1.3 IF SO MUCH UNITY, WHY SO MANY SPECIES?

So far, we have focused on life's unity, on the characteristics that all living things have in common. Think of it! They are put together from the same "lifeless" materials. They remain alive by metabolism —by ongoing energy transfers at the cellular level. They interact in their requirements for energy and for raw materials. They have the capacity to sense and respond to the environment in highly specific ways. They all have a capacity to reproduce, based on instructions encoded in DNA. And they all inherited their molecules of DNA from individuals of a preceding generation.

Superimposed on the common heritage is immense diversity. You share the planet with many millions of different kinds of organisms, or **species**. Many millions more preceded you during the past 3.8 billion years, but their lineages vanished; they are extinct. Centuries ago, scholars tried to make sense of the confounding diversity. One of them, Carolus Linneaus, came up with a classification scheme that assigns a two-part name to each newly identified species. The first part designates the **genus** (plural, genera). Each genus encompasses all species that seem closely related by way of their recent descent from a common ancestor. The second part of the name designates a particular species within that genus.

For example, *Quercus alba* is the scientific name for the white oak, and *Q. rubra* is the name for the red oak. As this example suggests, once you spell out the name of the genus in a document, you may abbreviate the name wherever else it appears in that document.

Biologists also classify life's diversity by assigning species to groups at more encompassing levels. Among other things, they group genera that apparently share a common ancestor into the same *family*, related families into the same *order*, related orders into the same *class*, and related classes into the same *phylum* (plural, phyla). At a higher level, they assign related phyla to the same *kingdom*. They are still refining the groups. For instance, scholars once recognized only two kingdoms (animals and plants). Later, most biologists came to accept five kingdoms. Today, compelling evidence suggests there should be at least six: the **Archaebacteria**, **Eubacteria**, **Protista**, **Fungi**, **Plantae**, and **Animalia** (Figure 1.7). We use a six-kingdom scheme for this book. At this point in your reading, it is enough simply to become familiar with a few of the defining features of their members.

All of the world's bacteria (singular, bacterium) are single cells, of a type known as *prokaryotic*. The word means they don't have a nucleus, a membrane-bound sac that otherwise might keep their DNA separate from the

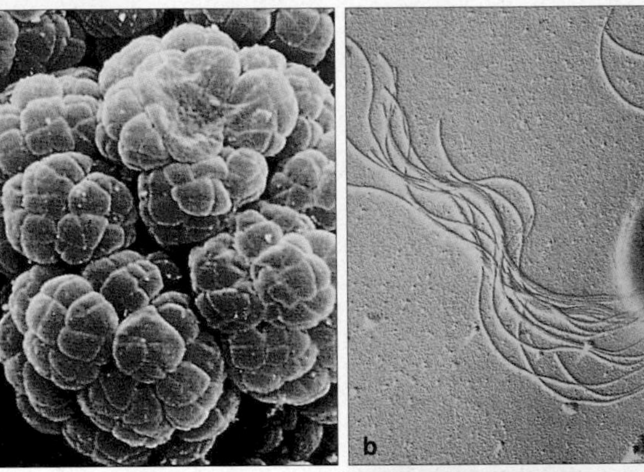

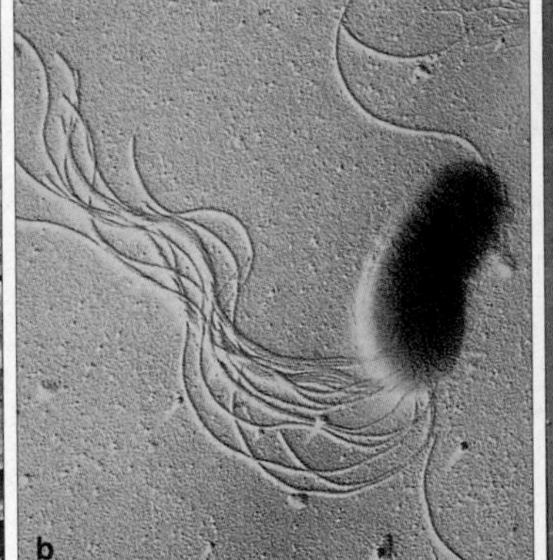

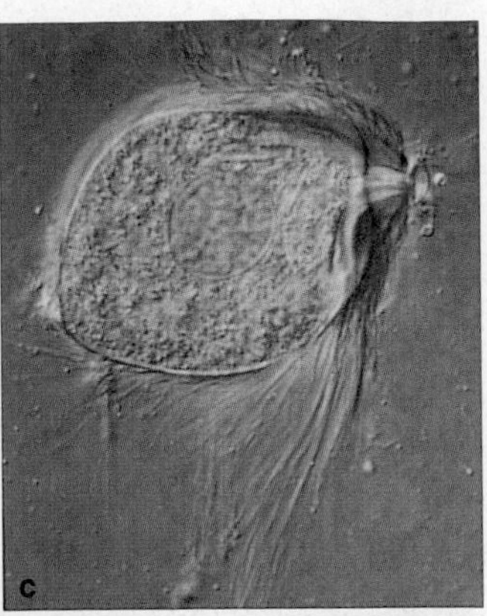

Figure 1.7 A few representatives of life's diversity.

KINGDOM ARCHAEBACTERIA. (**a**) From the muck of an anaerobic (oxygen-free) habitat, a colony of cells (*Methanosarcina*).

KINGDOM EUBACTERIA. (**b**) A eubacterium sporting a number of bacterial flagella. Flagella are used for motility.

KINGDOM PROTISTA. (**c**) A trichomonad that lives in a termite's gut. Compared to bacteria, most protistans are much larger and have greater internal complexity. They range from microscopically small single cells, such as this one, to giant seaweeds.

KINGDOM FUNGI. (**d**) A stinkhorn fungus. Some species of fungi are parasites and some cause diseases, but the vast majority are decomposers. Without decomposers, communities would gradually become buried in their own wastes.

KINGDOM PLANTAE. (**e**) Trunk of a redwood growing near the coast of California. Like nearly all plants, redwoods produce their own food by photosynthesis. (**f**) Flower of a plant from the family of composites. Its colors and patterning guide bees to nectar. The bees get food. The plant gets help reproducing. Like many other organisms, they interact in a mutually beneficial way.

KINGDOM ANIMALIA. (**g**,**h**) Male bighorn sheep competing for females and so displaying a characteristic of the kingdom—they actively move about in their environment.

Further reading: Student Guide to InfoTrac on web site →

rest of the cell interior. Different bacterial species are producers, consumers, and decomposers. Of all kingdoms, theirs shows the greatest metabolic diversity.

We find diverse archaebacteria in extreme environments, much like the forms that are thought to have prevailed soon after life originated. Eubacteria are much more successful in their world distribution. Different kinds live in about every place you might imagine. Eubacteria inside your gut and on your skin outnumber the trillions of cells making up your body.

Generalizing about protistans is not easy. Like plants, fungi, and animals, they are *eukaryotic*, meaning their DNA is located inside a nucleus. Most types are larger and show far more internal complexity than bacteria. A spectacular variety of microscopically small, single-celled producers and consumers are protistans. So are multicelled brown algae and other "seaweeds," some of which are giants of the underwater world.

Most fungi, including the common field mushrooms sold in grocery stores, are multicelled. These eukaryotic decomposers and consumers feed in a distinctive way. They secrete enzymes that digest food outside of the fungal body, then their cells absorb the digested bits.

Both plants and animals are multicelled, eukaryotic organisms. Nearly all plants are photosynthetic. Along with other producers, they make up the all-important food base for communities, especially on land. Animals are a type of consumer that ingests other organisms or their tissues. They include herbivores (which graze on photosynthesizers), carnivores (meat eaters), parasites, and scavengers. Unlike plants, animals actively move about during at least some stage of their life.

Pulling this all together, you start to get a sense of what it means when someone says life shows unity *and* diversity.

Unity threads through the world of life, for all organisms are alike in important ways. They are composed of the same substances, which are assembled in the same basic ways. They engage in metabolism, and they sense and respond to their environment. They all have a capacity to reproduce, based on heritable instructions encoded in their DNA.

Immense diversity threads through the world of life, for organisms differ enormously in body form, in the functions of their body parts, and in their behavior.

To make the study of life's diversity more manageable, we group organisms into six kingdoms—the archaebacteria, eubacteria, protistans, fungi, plants, and animals.

1.4 AN EVOLUTIONARY VIEW OF DIVERSITY

Given that organisms are so much alike, what could account for their great diversity? One key explanation is called evolution by means of natural selection. A few simple examples will be enough to introduce you to its premises, which build on the simple observation that the individuals of a population vary in their traits.

Mutation—Original Source of Variation

DNA has two striking qualities. Its instructions work to ensure that offspring will resemble their parents, yet they also permit variations in the details of most traits. As an example, having five fingers on each hand is a human trait. Yet some humans are born with six fingers on each hand instead of five. This is an outcome of a **mutation**, a molecular change in the DNA. Mutations are the original source of variations in heritable traits.

Many mutations are harmful. A change in even a bit of DNA may be enough to sabotage the body's growth, development, or functioning. One such mutation causes *hemophilia A*. If a person affected by this blood-clotting disorder gets even a small cut or bruise, an abnormally lengthy time passes before a clot forms and stops the bleeding. Yet some variations are harmless or beneficial. A classic case is a mutation in light-colored moths that results in dark offspring. Moths fly at night and rest in the day, when birds that eat them are active. Birds tend to miss light-colored moths that rest on light tree trunks. Those moths are camouflaged; they are "hiding in the open," as shown in Figure 1.8.

Suppose that people build coal-burning factories nearby. In time, smoke laden with soot darkens the tree trunks. Now the dark moths are much less conspicuous to bird predators, so they have a better chance to survive and reproduce. Where soot prevails, the variant (darker) form of the trait is more adaptive than the common form. An **adaptive trait** is any form of a trait that gives the individual an advantage, in terms of surviving and reproducing, under a given set of environmental conditions.

Figure 1.8 Example of different forms of the same trait (body surface coloration), adaptive to two different environmental conditions. (**a**) On a light-colored tree trunk, light-colored moths (*Biston betularia*) are hidden from predators, but dark ones stand out. (**b**) The dark color is more adaptive in places where tree trunks are darkened with soot.

Figure 1.9 *Facing page*: Some of the 300+ varieties of domesticated pigeons, an outcome of artificial selection practices. Breeders began with variant forms of traits in captive populations of wild rock doves.

Evolution Defined

Now imagine a population of light-colored moths in a sooty forest. At some point, a DNA mutation arose in the population, and it resulted in a moth of a darker color. When the mutated individual reproduced, some of its offspring inherited the trait. Birds saw and ate many light-colored moths, but most of the dark ones escaped detection and lived long enough to reproduce. So did their dark offspring, and so did *their* offspring. Over the generations, the frequency of the dark form of the trait increased and that of the light form decreased. As more time passes, the dark form of the trait might even become the more common, and people might end up referring to "the population of dark-colored moths." **Evolution** is under way. In biology, the word means genetically based change in a line of descent over time. Like moths, individuals of most populations typically show different forms of many (or most) of their traits. And the frequencies of those different forms relative to one another can change over successive generations.

Natural Selection Defined

Long ago, the naturalist Charles Darwin used pigeons to explain a conceptual connection between evolution and variation in traits. Domesticated pigeons display great variation in size, feather color, and other traits (Figure 1.9). As Darwin knew, pigeon breeders select certain forms of traits. For example, if breeders prefer tail feathers that are black with curly edges, they will allow only those individual pigeons having the most black and the most curl in their tail feathers to mate and produce offspring. Over time, "black" and "curly" will become the most common forms of tail feathers in the captive population, and different forms of the two traits will become less common or will be eliminated.

Pigeon breeding is a case of **artificial selection**, for the selection among different forms of a trait is taking place in an artificial environment—under contrived, manipulated conditions. Yet Darwin saw the practice as a simple model for *natural* selection, a favoring of some forms of traits over others in nature. Whereas breeders are "selective agents" that promote the reproduction of some individuals over others in captive populations, a pigeon-eating peregrine falcon is one of many selective agents that operate across the range of variation among

Further reading: Student Guide to InfoTrac on web site →

pigeons in the wild. Generation after generation, the swifter or more effectively camouflaged pigeons have a better chance of living long enough to reproduce than the not-so-swift or too-conspicuous ones among them. What Darwin identified as **natural selection** simply is a difference in *which* individuals of each generation survive and reproduce, the difference being an outcome of which have more adaptive forms of traits.

Unless you happen to be a pigeon breeder, Darwin's pigeon example may not be firing rockets through your imagination. So think about an example closer to home. Certain bacteria and fungi make *antibiotics*, metabolic products that kill bacterial competitors for nutrients in soil. Starting in the 1940s, we learned to use antibiotics to control diseases that result after bacteria invade the body and use its tissues for nutrients. Doctors routinely prescribed these "wonder drugs" for mild infections as well as serious ones. Some manufacturers even added an antibiotic to toothpaste and chewing gum.

As it turned out, antibiotics are powerful agents of natural selection. Over time, some bacteria have mutated in ways that help them resist the antibiotics produced by their bacterial neighbors. For example, streptomycin is an antibiotic that binds with some essential bacterial proteins and inhibits their activity. In certain variant strains of bacteria, mutations slightly changed the form of the proteins, so streptomycin is unable to bind with them. The mutant bacteria escape streptomycin's effects.

In infected patients, an antibiotic acts against bacteria that are susceptible to its action—but it actually favors variant strains that have resistance to it! Presently, such antibiotic-resistant strains are making it difficult to treat typhoid, tuberculosis, gonorrhea, staph (*Staphylococcus*) infections, and some other bacterial diseases. In a few patients, the "superbugs" responsible for tuberculosis cannot be successfully eliminated.

When resistance to antibiotics evolves by selection processes, the antibiotics must also evolve if they are to overcome the defenses. For example, drug companies have now modified parts of the streptomycin molecule. Such molecular changes in the laboratory produce more effective antibiotics—until new generations of more resistant superbugs enter the deadly evolutionary competition for nutrients.

Later in the book, we will consider the mechanisms by which populations of moths, pigeons, bacteria, and all other organisms evolve. Meanwhile, keep in mind the following points about natural selection. They are central to biological inquiry, for they have consistently proved useful in explaining a great deal about nature.

1. Individuals of a population vary in form, function, and behavior. Much of the variation is heritable; it can be transmitted from parents to offspring.

2. Some forms of heritable traits are more adaptive to prevailing environmental conditions. They improve an individual's chance of surviving and reproducing, as by helping it secure food, a mate, hiding places, and so on.

3. Natural selection is the outcome of differences in survival and reproduction among individuals that show variation in one or more traits.

4. Natural selection leads to a better fit with prevailing environmental conditions. Adaptive forms of traits tend to become more common and other forms less so. The population changes in its characteristics; it evolves.

In short, in the evolutionary view, *life's diversity is the sum total of variations in traits that have accumulated in different lines of descent generation after generation, as by natural selection and other processes of change.*

Mutations in DNA introduce variations in heritable traits.

Although many mutations are harmful, some give rise to variations in form, function, or behavior that are adaptive under prevailing environmental conditions.

Natural selection is a result of differences in survival and reproduction among individuals of a population that vary in one or more heritable traits. The process helps explain evolution—changes in lines of descent over the generations.

1.5 THE NATURE OF BIOLOGICAL INQUIRY

The preceding sections sketched out major concepts in biology. Now consider approaching this or any other collection of "facts" with a critical attitude. *"Why should I accept that they have merit?"* The answer requires insight into how biologists make inferences about observations and then test the predictive power of their inferences against actual experiences in nature or the laboratory.

Observations, Hypotheses, and Tests

To get a sense of "how to do science," start by following some practices that are pervasive in scientific research:

1. Observe some aspect of nature, carefully check what others have found out about it, and then frame a question or identify a problem related to your observation.

2. Develop **hypotheses**, or educated guesses, about possible answers to questions or solutions to problems.

3. Using hypotheses as a guide, make a **prediction**—that is, a statement of what you should observe in the natural world if you were to go looking for it. This is often called the "if–then" process. (*If* gravity does not pull objects toward Earth, *then* it should be possible to observe apples falling up, not down, from a tree.)

4. Devise ways to **test** the accuracy of your predictions, as by making systematic observations, building models, and conducting experiments. **Models** are theoretical, detailed descriptions or analogies that help us visualize something that has not yet been directly observed.

5. If the tests do not confirm the prediction, check to see what might have gone wrong. For example, maybe you overlooked a factor that influenced the test results. Or maybe the hypothesis is not a good one.

6. Repeat the tests or devise new ones—the more the better, for hypotheses that withstand many tests are likely to have a higher probability of being useful.

7. Objectively analyze and report the test results and the conclusions you have drawn from them.

You might hear someone refer to these practices as "the scientific method," as if all scientists march to the drumbeat of an absolute, fixed procedure. They do not. Many observe, describe, and report on some subject, then leave it to others to hypothesize about it. Some are lucky; they stumble onto information they are not even looking for, although chance does favor the prepared mind. It is not one single method they have in common. It is a critical attitude about being shown rather than told, and taking a logical approach to problem solving.

Logic encompasses thought patterns by which an individual draws a conclusion that does not contradict the evidence used to support it. Lick a cut lemon and you notice it is mouth-puckeringly sour. Lick ten more.

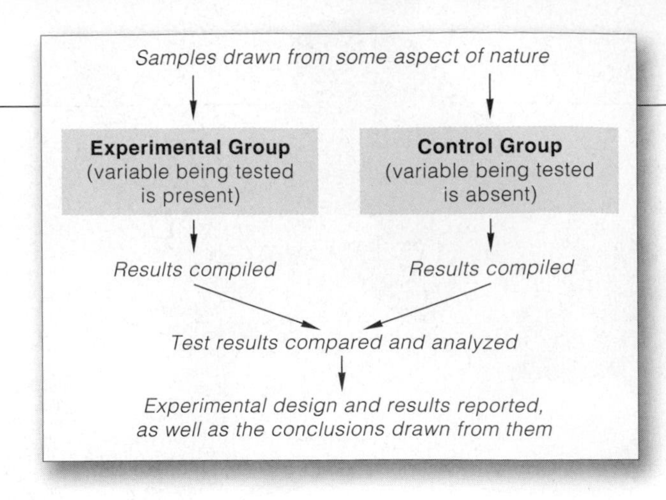

Figure 1.10 Generalized sequence of steps involved in an experimental test of a prediction based on a hypothesis.

You notice the same thing each time, so you conclude all lemons are mouth-puckeringly sour. You correlated one specific (lemon) with another (sour). By this pattern of thinking, called **inductive logic**, an individual derives a general statement from specific observations.

Express the generalization in "if–then" terms, and you have a hypothesis: "If you lick any lemon, then you will get an extremely sour taste in your mouth." By this pattern of thinking, called **deductive logic**, an individual makes inferences about specific consequences or specific predictions that must follow from a hypothesis.

You decide to test the hypothesis by tracking down and sampling all the varieties of lemons in the vicinity. One variety, the Meyer lemon, is actually mellow, for a lemon. You also discover that some people cannot taste anything. So you must modify the original hypothesis: "If most people lick any lemon *except* the Meyer lemon, they will get an extremely sour taste in their mouth." Suppose, after sampling all the known lemon varieties in the world, you conclude the modified hypothesis is a good one. You can never prove it beyond all shadow of a doubt, because there might be lemon trees growing in places people don't even know about. You *can* say the hypothesis has a high probability of not being wrong.

Comprehensive observations are a logical means to test the predictions that flow from hypotheses. So are **experiments**. These tests simplify observation in nature or the laboratory by manipulating and controlling the conditions under which observations are made. When suitably designed, observational and experimental tests allow you to predict that something will happen if a hypothesis isn't wrong (or won't happen if it *is* wrong). Figure 1.10 gives a general idea of the steps involved.

AN ASSUMPTION OF CAUSE AND EFFECT Experiments start from the premise that any aspect of nature has one or more underlying causes. With this premise, science is distinct from faith in the supernatural (meaning "beyond nature"). Experiments deal with potentially falsifiable hypotheses. Hypotheses of this sort can be tested in the natural world in ways that might disprove them.

Further reading: Student Guide to InfoTrac on web site

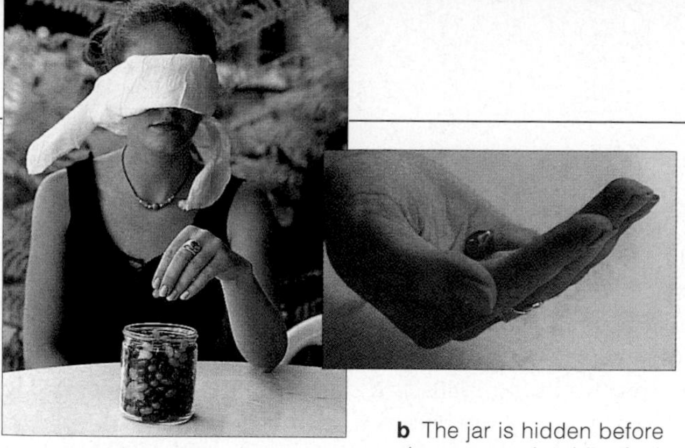

a Natalie, blindfolded, randomly plucks a jellybean from a jar of 120 green and 280 black jellybeans. That is a ratio of 30 to 70 percent.

b The jar is hidden before she removes her blindfold. She observes only a single green jellybean in her hand and assumes the jar holds only green jellybeans.

c Still blindfolded, Natalie randomly plucks 50 jellybeans from the jar and ends up with 10 green and 40 black ones.

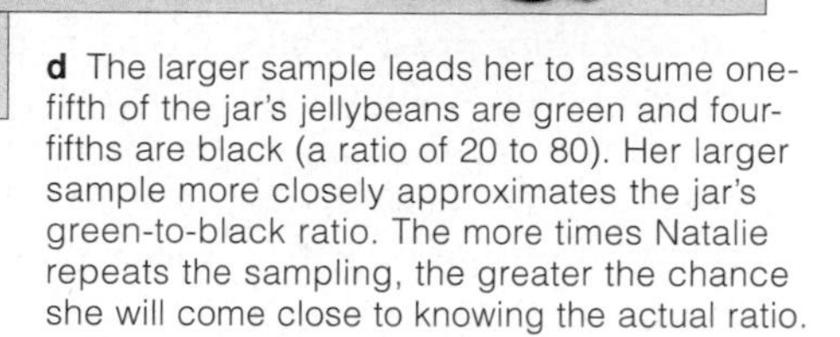

d The larger sample leads her to assume one-fifth of the jar's jellybeans are green and four-fifths are black (a ratio of 20 to 80). Her larger sample more closely approximates the jar's green-to-black ratio. The more times Natalie repeats the sampling, the greater the chance she will come close to knowing the actual ratio.

Figure 1.11 A simple demonstration of sampling error. 

EXPERIMENTAL DESIGN To get conclusive test results, experimenters rely on certain practices. They refine test designs by searching the literature for information that may relate to their inquiry. They design experiments to test one prediction of a hypothesis at a time. Each time, they set up a **control group**: a standard for comparison with one or more experimental groups. Ideally, their control group is identical with an experimental group in all respects *except* for the one variable being studied. **Variables** are specific aspects of objects or events that may differ or change over time and among individuals.

Section 1.6 shows the design of a recent experiment. As you will see, the experimenters directly manipulated a single variable in an attempt to support or disprove a prediction. They also tried to hold constant any other variables that could influence the results.

SAMPLING ERROR Rarely can experimenters observe *all* individuals of a group. Rather, they use large-enough samples to avoid risking tests with groups that are not representative of the whole. They usually must rely on samples (or subsets) of populations, events, and other aspects of nature. In general, the larger their sampling, the less likely it will be that any differences among the individuals will distort results (Figure 1.11).

About the Word "Theory"

Suppose no one has disproved a hypothesis after years of rigorous tests. Suppose scientists use it to explain more data or observations, which could involve more hypotheses. When a hypothesis meets these criteria, it may become accepted as a **scientific theory**.

You may hear someone apply the word "theory" to a speculative idea, as in the expression "It's only a theory." However, a scientific theory differs from speculation for a simple reason: *Many researchers have tested its predictive power many times, in many ways, and have yet to find evidence that disproves it*. This is why Darwin's view of natural selection is a respected theory. We use it successfully to explain many diverse issues, such as the origin of life, the relationship between plant toxins and plant-eating animals, the sexual advantages of strongly colored or patterned wings or feathers, the reason that certain cancers run in families, or why antibiotics that doctors often prescribe may no longer be effective. By yielding reasoned evidence that life evolved in the past, the theory even influenced views of Earth history.

An exhaustively tested theory might be as close to the truth as scientists can get with the evidence at hand. For example, Darwin's theory stands, with only minor modification, after more than a century's worth of many thousands of different tests. We cannot show that the theory holds under all possible conditions; an infinite number of tests would be required to do so. As for any theory, we can only say it has a *high or low probability* of being a good one. So far, biologists have not found any evidence that refutes Darwin's theory. Yet they still keep their eyes open for any new information and new ways of testing that might disprove its premises.

And this point gets us back to the value of thinking critically. Scientists must keep asking themselves: *Will observations or experiments show that a hypothesis is false?* They expect one another to put aside pride or bias by testing ideas, even in ways that may prove them wrong. Even if an individual doesn't or won't do this, others will—for science proceeds as a community that is both cooperative and competitive. Ideally, its practitioners share their ideas, with the understanding that it is just as important to expose errors as it is to applaud insights. Individuals can and often do change their minds when presented with contradictory evidence. As you will see, this is a strength of science, not a weakness.

A scientific approach to studying nature is based on asking questions, formulating hypotheses, making predictions, devising tests, and objectively reporting the results.

A scientific theory is a testable explanation about the cause or causes of a broad range of related phenomena. It remains open to tests, revision, and tentative acceptance or rejection.

1.6 THE POWER OF EXPERIMENTAL TESTS

BIOLOGICAL THERAPY EXPERIMENTS If you worry about the increasing frequencies of strains of pathogenic (disease-causing) bacteria that resist antibiotics, you are not alone. Bruce Levin of Emory University and Jim Bull of the University of Texas are among the investigators who search for possible alternatives to antibiotic therapy. They had been studying literature on a *biological therapy* that enlists bacteriophages—the "bacteria eaters"—to fight infections. Bacteriophages, a class of viruses, attack a narrow range of bacterial strains. When they contact a target, they typically inject a few enzymes and genetic material into it. What happens next is a hostile takeover of the cell's metabolic machinery, and the cell itself makes many new viral particles. The cell dies after viral enzymes rupture its outer membrane. Virus particles slip through the ruptured membrane and typically infect new cells.

The biologists wondered, as others had decades ago, whether injections of bacteriophages could help people resist or fight off bacterial infections. The discovery of antibiotics in the 1940s had diverted attention away from that idea, at least in Western countries. Now, with so many lethal pathogens breaching the antibiotic arsenal, the bacteria eaters were starting to look good again.

Levin and Bull focused on promising phage therapy experiments that were conducted in 1982 by two British researchers, H. Williams Smith and Michael Huggins. They started their research by testing whether Smith and Huggins's results could be duplicated.

They selected 018:K1:H7, a strain of *Escherichia coli* originally isolated from a human patient with meningitis. Like a harmless *E. coli* strain that lives in the intestines of humans and other mammals (Figure 1.12), this one departs from the body in feces. Bacteriophages used in laboratory work typically ignore 018:K1:H7, so Levin and Bull looked for some *E. coli* killers in samples from an Atlanta sewage treatment plant. They successfully isolated two kinds of bacteriophages and named them *H* (for Hero, an effective killer) and *W* (for the less effective Wimp).

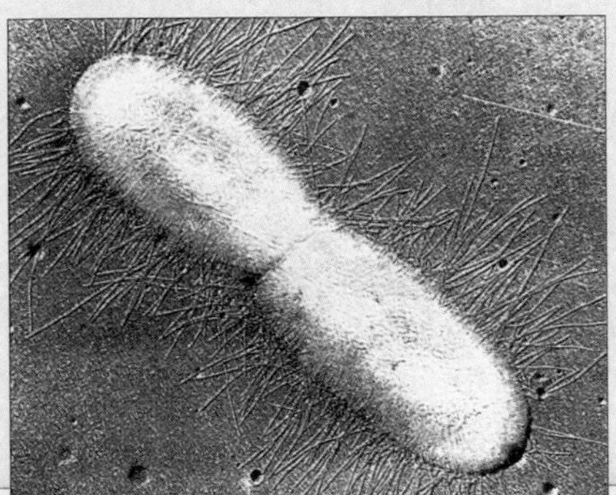

Figure 1.12 Two experiments to compare the effectiveness of bacteriophage injections against antibiotics for treating a bacterial infection. To the left, a micrograph shows a harmless *Escherichia coli* cell in the process of dividing in two.

Hypothesis: If bacteriophages specifically target and destroy cells of *E. coli* 018:K1:H7 in petri dishes, then they will do the same in laboratory mice that have been infected by that strain.

Prediction: Laboratory mice injected with a preparation that contains more than 10^7 particles of *H* bacteriophage will not die following an injection of *E. coli* strain 018:K1:H7.

Experimental test of the prediction:

Researchers establish large populations of H bacteriophage and of E. coli 018:K1:H7 from which to draw their samples, and select a specific strain of laboratory mice.

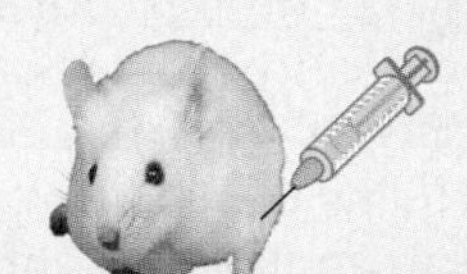

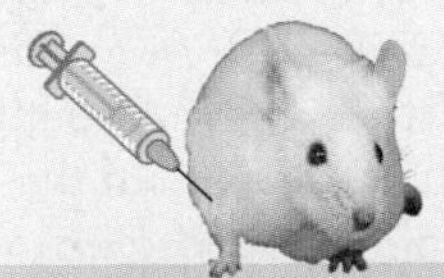

EXPERIMENTAL GROUP
E. coli injected into right thigh of 15 of the mice; bacteriophage injected into their left thigh.

CONTROL GROUP
E. coli injected into 15 other mice; no bacteriophage injected into this sampling.

Test results:

All mice survive.

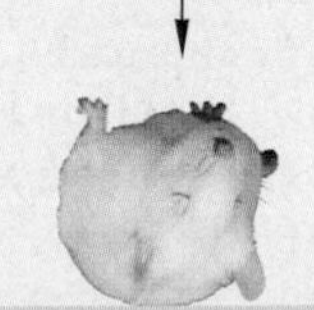

All mice die within 32 hours.

a

Another prediction derived from the same hypothesis: *H* bacteriophage that target and destroy 018:K1:H7 will be more effective than single doses of streptomycin in treating mice infected with 018:K1:H7.

Experimental test of the new prediction:

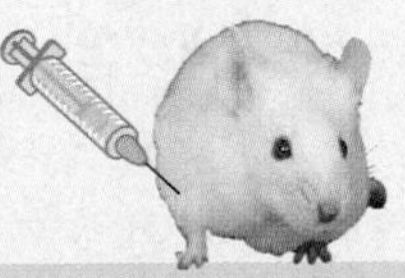

Researchers inject 48 laboratory mice with 018:K1:H7, then divide them into four groups of 12 each. Eight hours later...

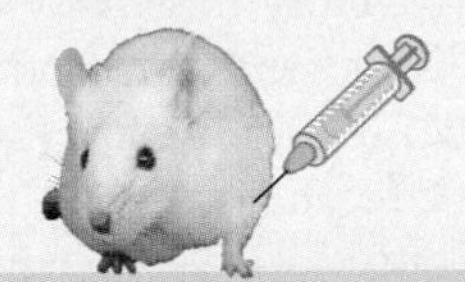

	Test results:
EXPERIMENTAL GROUP 1 *...12 mice receive a single injection of H bacteriophage.* →	*11 of 12 mice survive.*
EXPERIMENTAL GROUP 2 *...12 mice receive a single dose of 60 micrograms/gram streptomycin.* →	*5 of 12 mice survive.*
EXPERIMENTAL GROUP 3 *...12 mice receive a single dose of 100 micrograms/gram streptomycin.* →	*3 of 12 mice survive.*
CONTROL GROUP *...12 mice receive an injection of saline solution only.* →	*All control mice die.*

b

Further reading: Student Guide to InfoTrac on web site →

With graduate student Terry DeRouin and laboratory technician Nina Moore Walker, the biologists grew a large population of 018:K1:H7 in a culture flask. They selected a specific strain of laboratory mice, all females of the same age. Fifteen mice of one experimental group were each injected with 018:K1:H7 in one thigh and more than 10^7 particles of *H* bacteriophage in the other thigh. All mice survived. Fifteen mice of a control group were injected with 018:K1:H7 only. None survived. Figure 1.12*a* outlines this experimental test. Figure 1.12*b* outlines a second test of another prediction derived from the same hypothesis. The results reinforced Smith and Huggins's conclusion that certain bacteriophages can be as good as or better than antibiotics at stopping specific bacterial infections.

Each type of bacteriophage targets specific strains of one or at most a few bacterial species. At present, it takes too long for clinicians to discover which specific bacterium is infecting a patient, so the right bacteriophage might not be enlisted until it is too late to do the patient any good. On the bright side, procedures are now being developed that can dramatically accelerate the identification process.

IDENTIFYING IMPORTANT VARIABLES In nature, many factors can influence the outcome of an infection. For example, genetic differences among infected individuals lead to differences in how their immune system responds to the invasion. Age, nutrition, and health at the time of infection influence the outcome. Some pathogens are deadlier than others. And so on. That is why researchers try to simplify and control variables in their experiments. Variables, recall, are specific aspects of objects or events that may differ over time and among individuals. All of the *E. coli* cells in Levin and Bull's experiments were descended from the same parent cell and raised on the same nutrients at the same temperature in a flask. They could be expected to respond in the same way to the bacteriophage attack. All the mice were the same age and sex, and were raised under identical laboratory conditions. Each mouse in each experimental group received the same amount of bacteriophages or streptomycin. And each mouse in a given control group received the same injection of saline solution. Thus the focus was on *one variable*—a specific bacteriophage versus a specific antibiotic—in a simple, controlled, artificial situation.

BIAS IN REPORTING THE RESULTS Whether intentional or not, experimenters run the risk of interpreting data in terms of what they want to prove or dismiss. A few have even been known to fake measurements or nudge findings to reinforce their bias. That is why science emphasizes presenting test results in quantitative terms—that is, with actual counts or some other precise form. Doing so allows other experimenters to check or test the results readily and systematically, as Levin and Bull did. At this writing, they are assembling a detailed report of their own experiments, for publication in a science journal.

1.7 THE LIMITS OF SCIENCE

The call for objective testing strengthens the theories that emerge from scientific studies. It also puts limits on the kinds of studies that can be carried out. Beyond the realm of science, some events remain unexplained. Why do we exist, for what purpose? Why does any one of us have to die at a particular moment? Such questions lead to *subjective* answers, which come from within, as an outcome of all the experiences and mental connections shaping human consciousness. Because people differ vastly in this regard, subjective answers do not readily lend themselves to scientific analysis and experiments.

This is not to say subjective answers are without value. No human society can function for long unless its members share a commitment to certain standards for making judgments, even subjective ones. The moral, aesthetic, philosophical, and economic standards vary from one society to the next. But they all guide their members in deciding what is important and good, and what is not. All attempt to give meaning to what we do.

Every so often, scientists stir up controversy when they happen to explain some part of the world that was considered to be beyond natural explanation—that is, as belonging to the "supernatural." This is often the case when a society's moral codes are interwoven with religious narratives. Exploring some longstanding view of the world from the perspective of science might be misinterpreted as questioning morality even though the two are not the same thing.

As one example, centuries ago in Europe, Nicolaus Copernicus studied the planets and concluded the Earth circles the sun. Today this seems obvious enough. Back then it was heresy. The prevailing belief was that the Creator made the Earth (and, by extension, humans) the immovable center of the universe. Later a respected scholar, Galileo Galilei, studied the Copernican model of the solar system, thought it was a good one, and said so. He was forced to retract his statement publicly, on his knees, and put the Earth back as the fixed center of things. (Word has it that when he stood up he muttered, "Even so, it *does* move.") Later still, Darwin's theory of evolution ran up against the same prevailing belief.

Today, as then, society has sets of standards. Those standards might be questioned when some new, natural explanation runs counter to supernatural beliefs. This doesn't mean that the scientists who raise questions are less moral, less lawful, less sensitive, or less caring than anyone else. It simply means one more standard guides their work: *The external world, not internal conviction, must be the testing ground for scientific beliefs.*

Systematic observations, hypotheses, predictions, tests—in all these ways, science differs from systems of belief that are based on faith, force, or simple consensus.

1.8 SUMMARY

1. There is unity in the living world, for all organisms have these characteristics in common: They all consist of one or more cells. They are assembled from the same kinds of atoms and molecules according to the same laws of energy. They survive by metabolism, and by sensing and responding to specific conditions in the environment. Organisms have the capacity to survive and reproduce, based upon the heritable instructions encoded in the molecular structure of their DNA. Table 1.1 lists these characteristics.

2. The characteristics of life extend from cells, through multicelled organisms, then populations, communities, ecosystems, and the biosphere.

3. Many millions of species (kinds of organisms) exist. Many millions more were alive in the past and became extinct. Classification schemes place all known species in ever more inclusive groupings, from genus on up through family, order, class, phylum, and kingdom.

4. Life's diversity arises through mutations (changes in the structure of DNA molecules). These molecular changes are the basis for variation in heritable traits. These are traits that parents bestow on their offspring, including most details of body form and functioning.

5. Darwin's theory of evolution by natural selection is a cornerstone of biological inquiry. Its key premises are:

a. Individuals of a population differ in the details of their shared heritable traits. Variant forms of traits may affect the ability to survive and reproduce.

b. Natural selection is the outcome of differences in survival and reproduction among individuals that differ in one or more traits. Adaptive forms of a trait tend to become more common and less adaptive ones become less common or disappear. Thus a population's defining traits may change over successive generations; the population can evolve.

6. There are diverse methods of scientific inquiry. The following terms are important aspects of those methods:

a. Theory: Explanation of a broad range of related phenomena, supported by many tests. An example is Darwin's theory of evolution by natural selection.

b. Hypothesis: A proposed explanation of a specific phenomenon. Sometimes called an educated guess.

c. Prediction: A claim about what can be expected in nature, based on premises of a theory or hypothesis.

d. Test: An attempt to produce actual observations that match predicted or expected observations.

e. Conclusion: A statement about whether a theory or hypothesis should be accepted, modified, or rejected, based on tests of the predictions derived from it.

7. Logic is a pattern of thought by which an individual draws a conclusion that does not contradict evidence used to support the conclusion. Inductive logic means an individual derives a general statement from specific observations. Deductive logic means individuals make inferences about particular consequences or predictions that must follow from a hypothesis. Such a pattern of thinking is often expressed in "if–then" terms.

8. Predictions that flow from hypotheses can be tested by comprehensive observations or by experiments in nature or in the laboratory.

9. Experimental tests simplify observations in nature or the laboratory because conditions under which the observations are made are manipulated and controlled. They are based on the premise that any aspect of nature has one or more underlying causes, whether obvious or not. With such an assumption of cause and effect, only those hypotheses that can be tested in ways that might disprove them are scientific.

10. A control group is a standard against which one or more experimental groups (test groups) are compared. Ideally, it is the same as each experimental group in all variables except the one variable being investigated.

11. A variable is a specific aspect of an object or event that might differ over time and between individuals. Experimenters directly manipulate the variable they are studying to support or disprove their prediction.

12. Test results might be distorted by sampling error —chance differences between a population, event, or some other aspect of nature and the samples chosen to represent it. Test results are less likely to be distorted when samplings are large and when they are repeated.

13. Systematic observations, hypotheses, predictions, and experimental tests are the foundation of scientific theories. The external world, not internal conviction, is the testing ground for those theories.

Table 1.1 Summary of Life's Key Characteristics

SHARED CHARACTERISTICS THAT REFLECT LIFE'S UNITY
1. Organisms consist of one or more cells.
2. Organisms are constructed of the same kinds of atoms and molecules according to the same laws of energy.
3. Organisms engage in metabolism; they acquire and use energy and materials to grow, maintain themselves, and reproduce.
4. Organisms sense and make controlled responses to internal and external conditions.
5. Heritable instructions encoded in DNA give organisms their capacity to grow and reproduce. DNA instructions also guide the development of complex multicelled organisms.
FOUNDATIONS FOR LIFE'S DIVERSITY
1. Mutations (changes in the molecular structure of DNA) give rise to variation in heritable traits, including most details of body form, functioning, and behavior.
2. Diversity is the sum total of variations that accumulated in different lines of descent over the past 3.8 billion years, as by natural selection and other processes of change.

Review Questions

For this chapter and subsequent chapters, *italics* after a review question identify the section where you can find answers. They include section numbers and *CI* (for Chapter Introduction).

1. Why is it difficult to formulate a simple definition of life? *CI*
2. Name the molecule of inheritance in cells. *1.1*
3. Write out simple definitions of the following terms: *1.1*
   a. cell c. metabolism
   b. energy d. ATP
4. How do organisms sense changes in their surroundings? *1.1*
5. Study Figure 1.5. Then, on your own, arrange and define the levels of biological organization. *1.2*
6. Study Figure 1.6. Then, on your own, make a sketch of the one-way flow of energy and the cycling of materials through the biosphere. To the side of the sketch, write out definitions of producers, consumers, and decomposers. *1.2*
7. List the shared characteristics of life. *CI, 1.3*
8. What are the two parts of the scientific name for each kind of organism? *1.3*
9. List the six kingdoms of species as outlined in this chapter, and name some of their general characteristics. *1.3*
10. Define mutation and adaptive trait. Explain the connection between mutations and the immense diversity of life. *1.4*
11. Write brief definitions of evolution, artificial selection, and natural selection. *1.4*
12. Define and distinguish between: *1.5*
    a. hypothesis and prediction
    b. observational test and experimental test
    c. inductive and deductive logic
    d. speculation and scientific theory
13. With respect to experimental tests, define variable, control group, and experimental group. *1.5*
14. What does sampling error mean? *1.5*

Self-Quiz (*Answers in Appendix III*)

1. ______ is the capacity of cells to extract energy from sources in their environment, and to transform and use energy to grow, maintain themselves, and reproduce.
2. ______ is a state in which the internal environment is being maintained within tolerable limits.
3. The ______ is the smallest unit of life.
4. If a form of a trait improves chances for surviving and reproducing in a given environment, it is a(n) ______ trait.
5. The capacity to evolve is based on variations in heritable traits, which originally arise through ______ .
6. You have some number of traits that also were present in your great-great-great-great-grandmothers and -grandfathers. This is an example of ______ .
   a. metabolism c. a control group
   b. homeostasis d. inheritance
7. DNA molecules ______ .
   a. contain instructions for traits
   b. undergo mutation
   c. are transmitted from parents to offspring
   d. all of the above
8. For many years in a row, a dairy farmer allowed his best milk-producing cows but not the poor producers to mate. Over many generations, milk production increased. This outcome is an example of ______ .
   a. natural selection c. evolution
   b. artificial selection d. both b and c
9. A control group is ______ .
   a. a standard against which experimental groups are compared
   b. identical to experimental groups except for one variable
   c. a standard with several variables against which an experimental group is compared
   d. both a and b are correct
10. A specific aspect of an object or event that may change over time or change among individuals is a ______ .
    a. control group c. variable
    b. experimental group d. sampling error
11. The fewer the individuals from a population that are chosen at random for an experimental group, ______ .
    a. the greater the chance of sampling error
    b. the smaller the chance of sampling error
    c. the less likely differences among them will distort the test results
12. Match the terms with the most suitable descriptions.
    ______ adaptive trait a. statement of what you should find in nature if you were to go looking for it
    ______ natural selection b. educated guess
    ______ theory c. improves chance of surviving and reproducing in environment
    ______ hypothesis d. related set of hypotheses that form a broad, applicable, testable explanation
    ______ prediction e. outcome of differences in survival and reproduction among individuals that differ in details of one or more traits

Figure 1.13 A spider (*Dolomedes*) and its prey: a tiny minnow. The spider delivered paralyzing venom as well as digestive enzymes into its captive and is now sucking predigested juices from it. Such spiders can move about below the water's surface. Hairs on their body trap oxygen (for aerobic respiration) during their hunting expeditions.

Critical Thinking

1. Some spiders (*Dolomedes*) that feed on insects around ponds occasionally capture tadpoles and small fishes, as in Figure 1.13, and isn't that fun to think about? While they are immature, the female spiders confine themselves to a small patch of vegetation next to the pond. When sexually mature, they mate and store sperm that fertilize the eggs. Only then do they move out and occupy larger areas around the pond. Develop hypotheses to explain what might cause the spiders to live in different places at different times. Design an experiment to test each hypothesis.

2. A scientific theory about some aspect of nature rests upon inductive logic. The assumption is that, because an outcome of some event has been observed to happen with great regularity, it will happen again. However, we cannot know this for certain, because there is no way to account for all possible variables that may affect the outcome. To illustrate this point, Garvin McCain and Erwin Segal offer a parable:

> Once there was a highly intelligent turkey. It lived in a pen, attended by a kind, thoughtful master, and had nothing to do but reflect on the world's wonders and regularities. It observed some major regularities. Morning always began with the sky getting light, followed by the clop, clop, clop of its master's friendly footsteps, then by the appearance of delicious food. Other things varied—sometimes the morning was warm and sometimes cold—but food always followed footsteps. The sequence of events was so predictable, it became the basis of the turkey's theory about the goodness of the world. One morning, after more than one hundred confirmations of the theory, the turkey listened for the clop, clop, clop, heard it, and had its head chopped off.

The turkey learned the hard way that explanations about the world only have a high or low probability of not being wrong. Today, some people take this uncertainty to mean that "facts are irrelevant—facts change." If that is so, should we just stop doing scientific research? Why or why not?

3. Witnesses in a court of law are asked to "swear to tell the truth, the whole truth, and nothing but the truth." What are some of the problems inherent in the question? Can you think of a better alternative?

4. Many popular magazines publish an astounding number of articles on diet, exercise, and other health-related topics. Some authors recommend a specific diet or dietary supplement. What kinds of evidence do you think the articles should include so that you can decide whether to accept their recommendations?

5. Although scientific information often is used when making a decision, it cannot tell an individual what is "right" or "wrong." Give an example of this from your own experience.

6. As the old saying goes, Everybody complains about the weather, but nobody does anything about it. Maybe you can start thinking about how we humans might change at least one aspect of local weather conditions. Two researchers at Arizona State University found that, at least for their investigation of the northeast coast of North America, it really does rain more on weekends, when we'd rather be out having fun!

As R. Cerveny and R. Balling, Jr., reasoned, if the amount of rain falling each day of the week is a random event, then rules of probability should apply. (*Probability* means the chance that each possible outcome of an event will occur is proportional to the number of ways it can be reached.) Thus, each day of the week should get one-seventh (14.3 percent) of the total rainfall for the week. But it turns out Monday is driest (13.1 percent). Days of the week are wetter, and Saturday is the wettest of all (with 16 percent of the total). Sunday is a bit above average.

What causes this effect? As the researchers hypothesized, if *air pollution* is greater during weekdays than on weekends, then cyclic human activities may influence regional patterns of rainfall. Monday through Friday, coal-burning factories and gas-powered vehicles release quantities of tiny particles into the air. The particles can promote updrafts and act as "platforms" for the formation of water droplets. Air pollution must build up during the week and carry over into Saturday. Over the weekend, fewer factories operate and fewer commuters are on the road, so by Monday, the air must be cleaner—and drier.

What kind of evidence do you suppose Cerveny and Balling gathered to test the hypothesis? Jot down a few ideas and check them against the researchers' article in the 6 August 1998 issue of *Nature*. Whether or not you continue in biology, this exercise will be good practice for searching through scientific literature for actual data that you can evaluate on topics of interest to you.

7. Scientists devised experiments to shed light on whether different fish species of the same genus compete in their natural habitat. They constructed twelve ponds, identical in chemical composition and physical characteristics. Then they released the following individuals in each pond:

Ponds 1, 2, 3:	Species A	(300 individuals each pond)
Ponds 4, 5, 6:	Species B	(300 individuals each pond)
Ponds 7, 8, 9:	Species C	(300 individuals each pond)
Ponds 10, 11, 12:	Species A, B, C	(300 individuals of each species in each pond)

Does this experimental design take into consideration all factors that can affect the outcome? If not, how would you modify it?

Selected Key Terms

For this chapter and subsequent chapters, these are the **boldface** terms that appear in the text, in the sections indicated here by *italic* numbers (or *CI*, short for Chapter Introduction). As a study aid, make a list of the terms, write a definition for each, and check it against the one in the text. You will use these terms later on. Becoming familiar with each one will help give you a foundation for understanding the material in later chapters.

adaptive trait *1.4*
Animalia *1.3*
Archaebacteria *1.3*
artificial selection *1.4*
biology *CI*
biosphere *1.2*
cell *1.2*
community *1.2*
consumer *1.2*
control group *1.5*
decomposer *1.2*
deductive logic *1.5*
development *1.1*
DNA *1.1*
ecosystem *1.2*
energy *1.1*
Eubacteria *1.3*
evolution *1.4*
experiment *1.5*
Fungi *1.3*
genus *1.3*
homeostasis *1.1*
hypothesis *1.5*
inductive logic *1.5*
inheritance *1.1*
metabolism *1.1*
model *1.5*
multicelled organism *1.2*
mutation *1.4*
natural selection *1.4*
Plantae *1.3*
population *1.2*
prediction *1.5*
producer *1.2*
Protista *1.3*
receptor *1.1*
reproduction *1.1*
species *1.3*
stimulus *1.1*
test, scientific *1.5*
theory, scientific *1.5*
variable *1.5*

Readings *See also www.infotrac-college.com*

Carey, S. 1994. *A Beginner's Guide to the Scientific Method*. Belmont, California: Wadsworth. Paperback.

Committee on the Conduct of Science. 1989. *On Being a Scientist*. Washington, D.C.: National Academy of Sciences. Paperback.

McCain, G., and E. Segal. 1988. *The Game of Science*. Fifth edition. Pacific Grove, California: Brooks/Cole. Paperback.

Moore, J. 1993. *Science as a Way of Knowing—The Foundations of Modern Biology*. Cambridge, Massachusetts: Harvard University Press.

FACING PAGE: *Living cells of a plant (Elodea), observed with the aid of a microscope. Each rectangular cell contains efficient chemical factories called chloroplasts (the round, bright green parts inside).*

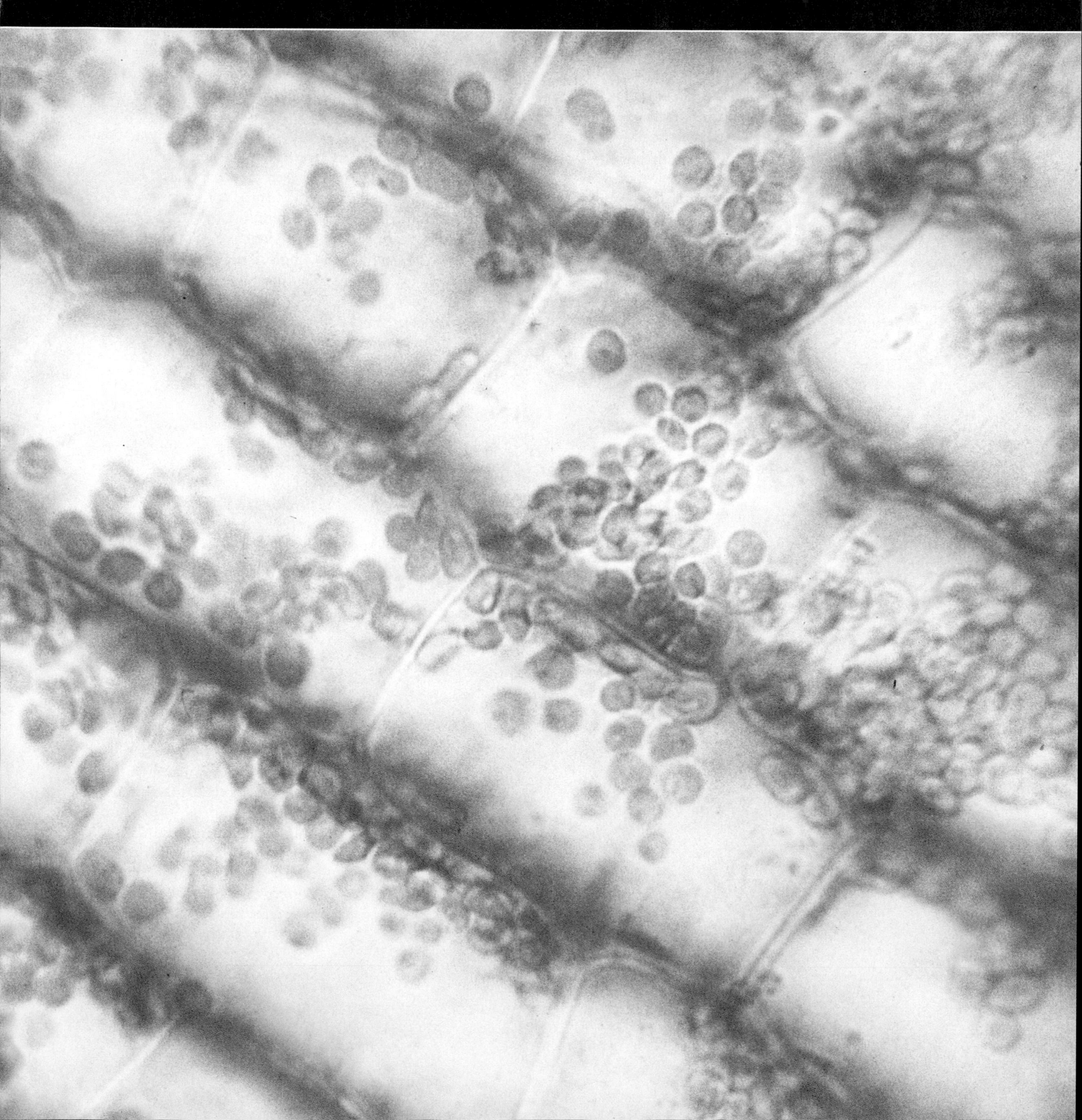

2

CHEMICAL FOUNDATIONS FOR CELLS

Checking Out Leafy Clean-Up Crews

Right now you are breathing in oxygen. You would die without it. Two centuries ago, no one had a clue to what oxygen is, where it comes from, and how it helps keep people alive. Then researchers started unlocking the secrets of this chemical substance and others. As their understanding of chemistry deepened, they began to conjure up such amazing things as nuclear power, synthetic fertilizers, nylons, lipsticks, fabric cleaners, aspirin, antibiotics, and plastic parts of refrigerators, computers, television sets, jet planes, and cars.

Today, our chemical "magic" brings us benefits *and* problems. For example, the scale of agriculture required to sustain the human population, which now surpasses 6 billion, is astounding. Synthetic fertilizers boost crop yields by providing plants with nitrogen, phosphorus, and other growth-enhancing nutrients. The upside is that fertilizer-fed crops help keep more people than you might imagine from starving to death. The downside is that plants do not take up every last bit of fertilizer. Nutrients in runoff from the fields enter lakes, rivers, and seas—where they "feed" such organisms as the protistans that cause huge fish kills (Section 23.10).

What about diverting the runoff to holding ponds, where water can evaporate? Farmers commonly do this. However, evaporation ponds are like magnets that pull selenium from the soil. High concentrations of selenium are toxic to grazing animals and waterfowl.

In 1996, Norman Terry and his coworkers grew cattails and other grasses in ten experimental plots that had become contaminated by runoff from agricultural fields (Figure 2.1). As they knew, plants can incorporate selenium into their tissues and convert some of it into dimethyl selenide—a gas about 600 times less toxic than selenium. They measured how much selenium settled in the mud and how much became incorporated into plant tissues or escaped into the air. As they predicted, before runoff trickled away from the plots, the plants significantly reduced selenium levels.

Terry's research is an example of **bioremediation**—the use of living organisms to withdraw harmful substances from the environment. Another example is the use of sunflowers to clean a pond at Chernobyl, in the Ukraine. Following a total meltdown at a nuclear power plant, strontium 90, cesium 137, and other extremely nasty radioactive elements contaminated the surroundings, including the pond (Section 51.8).

Fertilizers, wetlands, people, pumpkins, the air you breathe—*everything in and around you is "chemistry."* Every solid, liquid, or gaseous substance you care to think about is a collection of one or more elements. Think of the **elements** as fundamental forms of matter that have mass and take up space. Here on Earth, we can't break an element apart into something else except in physics laboratories. Break a chunk of copper into smaller and smaller bits, and the smallest bit you end up with is still copper.

Figure 2.1 At experimental wetlands in Corcoran, California, cattails, bulrushes, *Spartina*, and other grasses help clean up the water from agricultural runoff when they take up selenium.

Ninety-two elements occur naturally on Earth. Four of these—oxygen, hydrogen, carbon, and nitrogen—are the most abundant elements in your body, as they are in all other living organisms (Figure 2.2). Your body also contains some phosphorus, potassium, sulfur, calcium, sodium, and chlorine, plus a number of trace elements such as iodine. A *trace* element simply is one that represents less than 0.01 percent of body weight.

The structure and function of each organism depend on the availability of certain kinds and amounts of elements. For example, without daily intakes of magnesium, your muscles will become sore and weak, and your brain won't work properly. Without magnesium, older leaves of plants droop, turn yellow, and die. The herbicide 2,4-D stimulates weeds to grow faster than vital elements can be absorbed, so the weeds literally grow themselves to death.

As you might deduce from this story, safeguarding the environment, our food supplies, and our health depends on knowledge of chemistry. So do efforts to minimize side effects of its applications on biological systems. You owe it to yourself and others to gain insight into the structure and behavior of chemical substances. By demystifying chemistry's "magic," you will be better equipped to assess its benefits and risks.

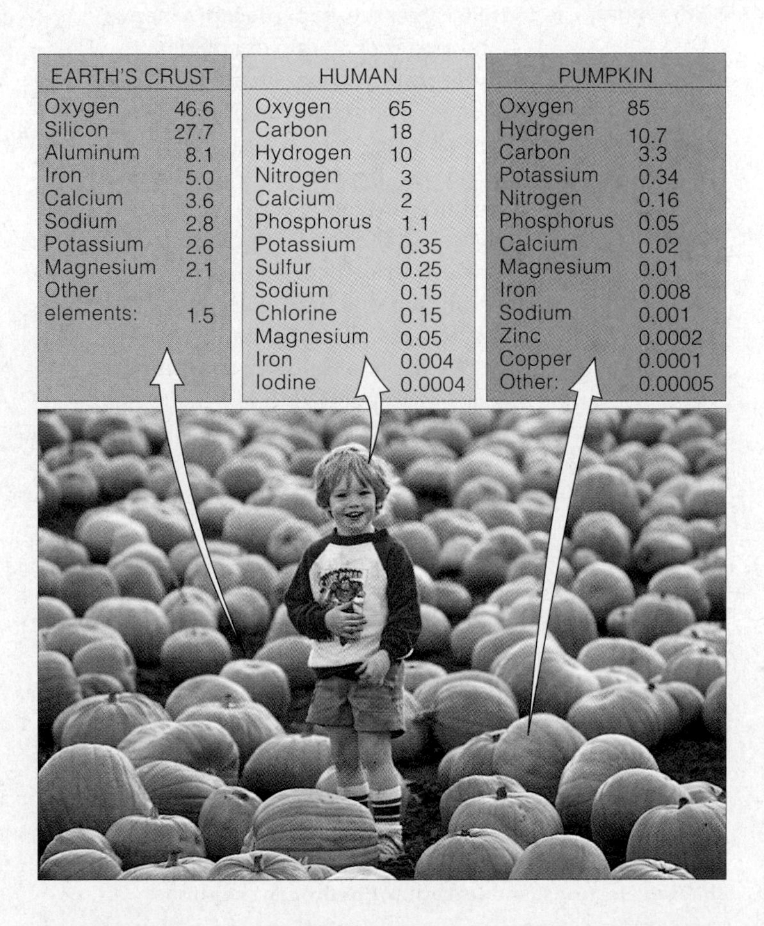

Figure 2.2 Proportions of elements in a human body and the fruit of pumpkin plants, compared to proportions of elements in the materials of the Earth's crust. How are the proportions similar? How do they differ?

KEY CONCEPTS

1. All substances consist of one or more elements, such as hydrogen, oxygen, and carbon. Each element is composed of atoms, which are the smallest units of matter that still display the element's properties. The atoms of elements are composed of protons, electrons, and (except for the hydrogen atom) neutrons.

2. The atoms that make up each kind of element have the same number of protons and electrons, but they may vary slightly from one another in their number of neutrons. Variant forms of an element's atoms are called isotopes.

3. Atoms have no overall electric charge unless they become ionized—that is, unless they lose electrons or acquire more of them. An ion is an atom or molecule that has lost or gained one or more electrons and thereby has acquired an overall positive or negative charge.

4. Whether a given atom will interact with other atoms depends on how many electrons it has and how they are structurally arranged within the atom. When energetic interactions unite two or more atoms, this is a chemical bond.

5. The molecular organization and activities of living things arise largely from ionic, covalent, and hydrogen bonds between atoms.

6. Life probably originated in water and is exquisitely adapted to its properties. Foremost among those properties are its temperature-stabilizing effects, cohesiveness, and capacity to dissolve or repel a variety of substances.

7. All organisms depend on the controlled formation, use, and disposal of hydrogen ions (H^+). The pH scale is a measure of the concentration of H^+ ions in solutions.

2.1 REGARDING THE ATOMS

Structure of Atoms

What are the smallest particles that retain the properties of an element? **Atoms**. A line of about a million of them would fit in the period ending this sentence. Small as atoms are, physicists have split them into more than a hundred kinds of smaller particles. The only subatomic particles you will need to consider in this book are the ones called **protons**, **electrons**, and **neutrons**.

All atoms have one or more protons, which carry a positive electric charge (p^+). Except for hydrogen, atoms also have one or more neutrons, which are uncharged. Protons and neutrons make up the atom's core region, or atomic nucleus (Figure 2.3). Zipping about the nucleus and occupying most of the atom's volume are one or more electrons, which carry a negative charge (e^-). Each atom has just as many electrons as protons. This means that an atom carries no net charge, overall.

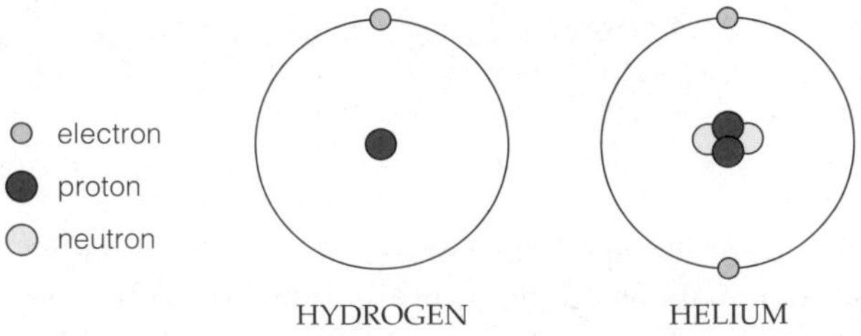

Each element has a unique *atomic* number, which refers to the number of protons in its atoms. To give examples, the atomic number is 1 for the hydrogen atom, which has a single proton. And it is 6 for the carbon atom, which has six protons (Table 2.1).

Also, each element has a *mass* number, which is the combined number of protons and neutrons in the atomic nucleus. A carbon atom, with six protons and six neutrons, has a mass number of 12.

Why bother with atomic numbers and mass numbers? They can give you an idea of whether and how substances will interact. *That knowledge may help you predict how substances might behave in individual cells, in multicelled organisms, and in the environment, under many conditions.*

Table 2.1 Atomic Number and Mass Number of Elements Common in Living Things

Element	Symbol	Atomic Number	Most Common Mass Number
Hydrogen	H	1	1
Carbon	C	6	12
Nitrogen	N	7	14
Oxygen	O	8	16
Sodium	Na	11	23
Magnesium	Mg	12	24
Phosphorus	P	15	31
Sulfur	S	16	32
Chlorine	Cl	17	35
Potassium	K	19	39
Calcium	Ca	20	40
Iron	Fe	26	56
Iodine	I	53	127

electron
proton
neutron
HYDROGEN
HELIUM

Figure 2.3 The hydrogen atom and helium atom according to one model of atomic structure. The model is highly simplified; the nucleus of these two representative atoms really would be an invisible speck at the scale employed here.

Isotopes—Variant Forms of Atoms

Samples of most naturally occurring elements contain two or more **isotopes**. The word indicates that, even though all atoms of an element have the same number of protons, they don't all have the same number of neutrons. Carbon has three isotopes, nitrogen has two, and so on. If the element's symbol has a superscript number to the left of it, this signifies which isotope is being discussed. For example, carbon's three isotopes are abbreviated ^{12}C (six protons and six neutrons, this being the most common form), ^{13}C (six protons, seven neutrons), and ^{14}C (six protons, eight neutrons). We also write these out as carbon 12, carbon 13, and carbon 14.

All isotopes of an element interact with other atoms in the same way. As you will see, this means cells are able to use any isotope of an element for metabolic activities.

Have you heard of radioactive isotopes (radioisotopes)? A physicist, Henri Becquerel, discovered them in 1896, after he had placed a heavily wrapped rock on top of an unexposed photographic plate located in a desk drawer. The rock contained isotopes of uranium, which emit energy. A few days after the plate was exposed to those energetic emissions, a faint image of that rock appeared on it. Marie Curie, Becquerel's coworker, gave the name "radioactivity" to the substance's chemical behavior.

As we now know, a **radioisotope** is an isotope that has an unstable nucleus and that stabilizes itself by spontaneously emitting energy and particles. (These particles are much smaller than protons, electrons, and neutrons.) This process, radioactive decay, transforms a radioisotope into an atom of a different element at a known rate. For instance, over a predictable time span, carbon 14 becomes nitrogen 14. You will look at some uses of this radioisotope and others in Unit III.

Elements are forms of matter that occupy space, have mass, and cannot be degraded to something else by ordinary means.

Atoms, the smallest particles that are unique to each element, have one or more positively charged protons, negatively charged electrons, and (except for hydrogen) neutrons.

Most elements have two or more isotopes (atoms that differ in the number of neutrons). A radioisotope has an unstable nucleus and stabilizes itself by spontaneously emitting particles and energy at a known rate.

Further reading: Student Guide to InfoTrac on web site

2.2 USING RADIOISOTOPES TO TRACK CHEMICALS AND SAVE LIVES

Radioisotopes make splendid tracers. A **tracer** is any substance with a radioisotope attached to it, rather like a shipping label, that researchers can track after they deliver it into a cell, a body, an ecosystem, or some other system. Laboratory devices can detect emissions from the tracer and precisely follow its movement through a pathway or pinpoint its final destination.

Melvin Calvin and some other botanists gave us a classic example. They used tracers to figure out the steps by which plants synthesize carbohydrates during photosynthesis. As they knew, all isotopes of an element have the same number of electrons, so all the isotopes must interact with other atoms the same way. Plant cells, they hypothesized, should be able to use any isotope of carbon when they build carbon compounds. By putting plant cells in a medium enriched with a tracer (^{14}C instead of ^{12}C), they were able to track the uptake of carbon through each reaction step leading to the formation of sugars and starches.

As another example, botanists use a radioisotope of phosphorus (^{32}P) as a tracer to identify how plants take up and use soil nutrients and synthetic fertilizers. Such findings are used to improve crop yields.

Besides being research tools, radioisotopes also are diagnostic tools in medicine. For safety considerations, clinicians use only the kinds that rapidly decay into harmless elements. Consider how clinicians analyze a human thyroid. This gland of ours, located in front of the windpipe, is the only one that takes up iodine. Iodine is a building block for thyroid hormones, which have great influence over growth and metabolism. If a patient's symptoms point to abnormal outputs of the thyroid hormones, clinicians may inject a trace amount of an iodine radioisotope (^{123}I) into the patient's blood. Then they use a photographic imaging device to scan the gland. Figure 2.4 shows examples of the images.

Radioisotopes also have uses in *PET* (for *P*ositron-*E*mission *T*omography). With this device, clinicians obtain images of particular body tissues. Suppose clinicians attach a tracer to glucose (or some other molecule). They inject the labeled glucose into the patient, who is then moved into a PET scanner (Figure 2.5*a*). Because all cells require glucose, cells throughout the patient take up the labeled glucose. Uptake is greater in some tissues than in others, depending on the tissue's metabolic activity at the time of examination. Laboratory devices can detect the radioactive emissions and use them to form an image, such as the one in Figure 2.5*b,c*. Such images can reveal variations or abnormalities in metabolic activity.

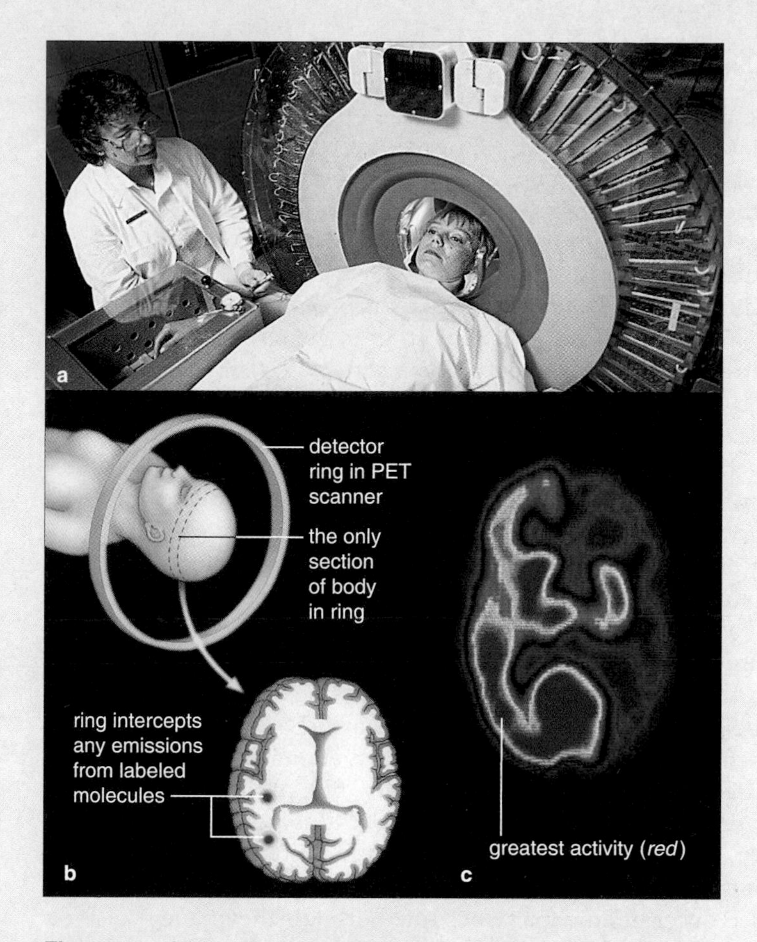

Figure 2.5 (**a**) Patient being moved into a PET scanner. (**b**) Inside, a ring of detectors intercepts the radioactive emissions from labeled molecules that were injected into the patient. Computers analyze and color-code the number of emissions from each location in the scanned body region.

(**c**) Brain scan of a child who has a neurological disorder. Different colors in a brain scan signify differences in metabolic activity. The right half of this brain shows very little activity. By comparison, cells of the left half absorbed and used the labeled molecule at expected rates.

Figure 2.4 Scans of the thyroid gland from three patients.

Radioisotopes that are energetic enough to destroy cells have uses in some therapies. For example, emissions from plutonium 238 drive artificial pacemakers, which smooth out irregular heartbeats. The radioisotope is sealed in a case before the pacemaker is inserted into a patient so its dangerous emissions won't damage body cells.

By contrast, with *radiation therapy*, the idea is to allow radioisotopes to destroy or impair the activity of targeted living cells that are not functioning properly. For example, bombardment with emissions from a source of radium 226 or cobalt 60 may destroy small, localized cancers.

2.3 WHAT HAPPENS WHEN ATOM BONDS WITH ATOM?

Electrons and Energy Levels

Figure 2.6 How even a lone electron can make things happen. Physicists confined electrons inside a bubble of liquid helium, then used fluctuating sound waves to pop the bubble. When an electron escaped, it made the white-centered red flash shown here.

Cells stay alive because energy inherent in all electrons makes things happen (Figure 2.6). Countless atoms in cells acquire electrons, share them, and donate them to other atoms. Atoms of certain elements do this easily, but others do not. What determines whether one atom will interact with another in such ways? *The outcome depends on the number and arrangement of their electrons.*

Tinker with magnets and you can get a sense of the attractive force between unlike charges (+ –) as well as the repulsive force between like charges (++ or – –). Electrons carry a negative charge. In an atom, they repel each other but are attracted to the positive charge of protons. They spend as much time as possible near the protons and far away from each other by moving about in different orbitals (Figure 2.7). Orbitals are volumes of space around the atomic nucleus in which electrons are likely to be at any instant. As a rough analogy, picture three preschoolers circling a cookie jar, not yet expert in the art of sharing. Each is drawn inexorably to the cookies but dreads being shoved away by the others. Two of them might bob and weave about on opposite sides of the jar to avoid a direct hit. But all three never, ever will occupy the same space at the same time.

An orbital can house one or at most two electrons. Because atoms differ in their number of electrons, they also differ in how many occupied orbitals they have.

Hydrogen is the simplest atom. It has a lone electron in a spherical orbital, closest to the nucleus. The orbital corresponds to the *lowest available energy level*. In every other atom, two electrons fill that first orbital. And two more electrons are occupying a second spherical orbital around the first one. Larger atoms have even more electrons. These are occupying orbitals farther from the nucleus, at *higher energy levels*.

The **shell model** is a simple way to think about how electrons are distributed in atoms. By this model, a series of "shells" enclose all orbitals available to electrons (Figure 2.8). The first orbital, which is spherical, is inside the first shell. A second shell (at a higher energy level) encloses the first shell. The next four available orbitals fit in it. More orbitals fit in a third shell, a fourth shell, and so on through the large, complex atoms of heavier elements (Appendix VI).

How can you predict whether atoms will interact? Check for electron vacancies in their outermost shell. Vacancies mean an atom might give up, gain, or share electrons under suitable conditions, which of course will change the distribution or number of its electrons. You can use the shell model to visualize what goes on here.

In Figure 2.8, which shows the electron distribution for several atoms, electrons are assigned to an energy level (a circle, or shell). Count the electron vacancies—that is, one or more unfilled orbitals in the outermost shell of each atom. As the figure shows, helium is one of the atoms with no vacancies. It is an *inert* atom, which shows little tendency to enter chemical reactions.

Now look again at Figure 2.2, which lists the most abundant of the elements making up a typical organism (a human). These elements include hydrogen, oxygen, carbon, and nitrogen. Atoms of all four elements have electron vacancies. Because of this, *they tend to fill the vacancies by forming bonds with other atoms.*

From Atoms to Molecules

Each **chemical bond** is a union between the electron structures of atoms. Take a moment to study Figure 2.9.

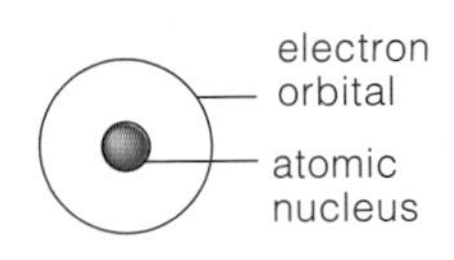

a Hydrogen atom

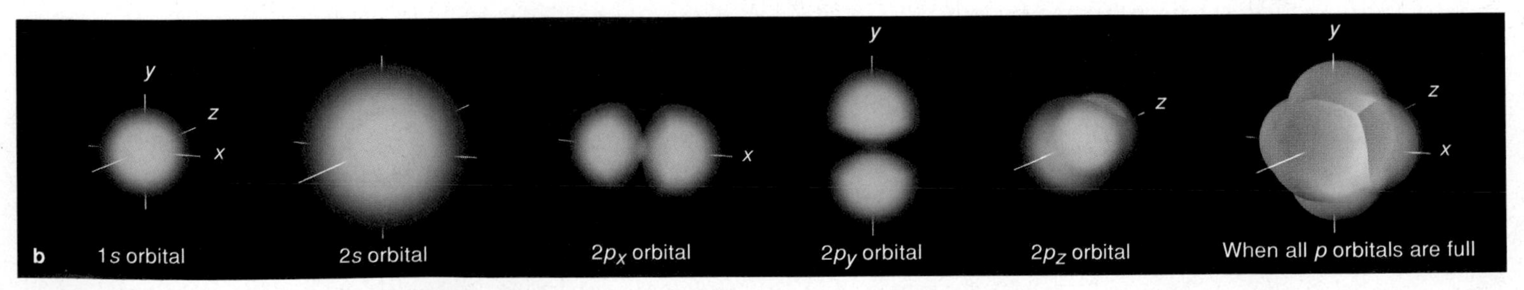

Figure 2.7 Electron arrangements in atoms, which are viewed in terms of three axes. Each axis (*x*, *y*, or *z*) is perpendicular to the other two. (**a**) A hydrogen atom's electron occupies a spherical 1*s* orbital, at the lowest energy level. (**b**) Every atom has a 1*s* orbital, occupied by one or two electrons. The 2*s* orbital and the *p* orbitals at the second energy level can each hold two electrons.

Further reading: Student Guide to InfoTrac on web site →

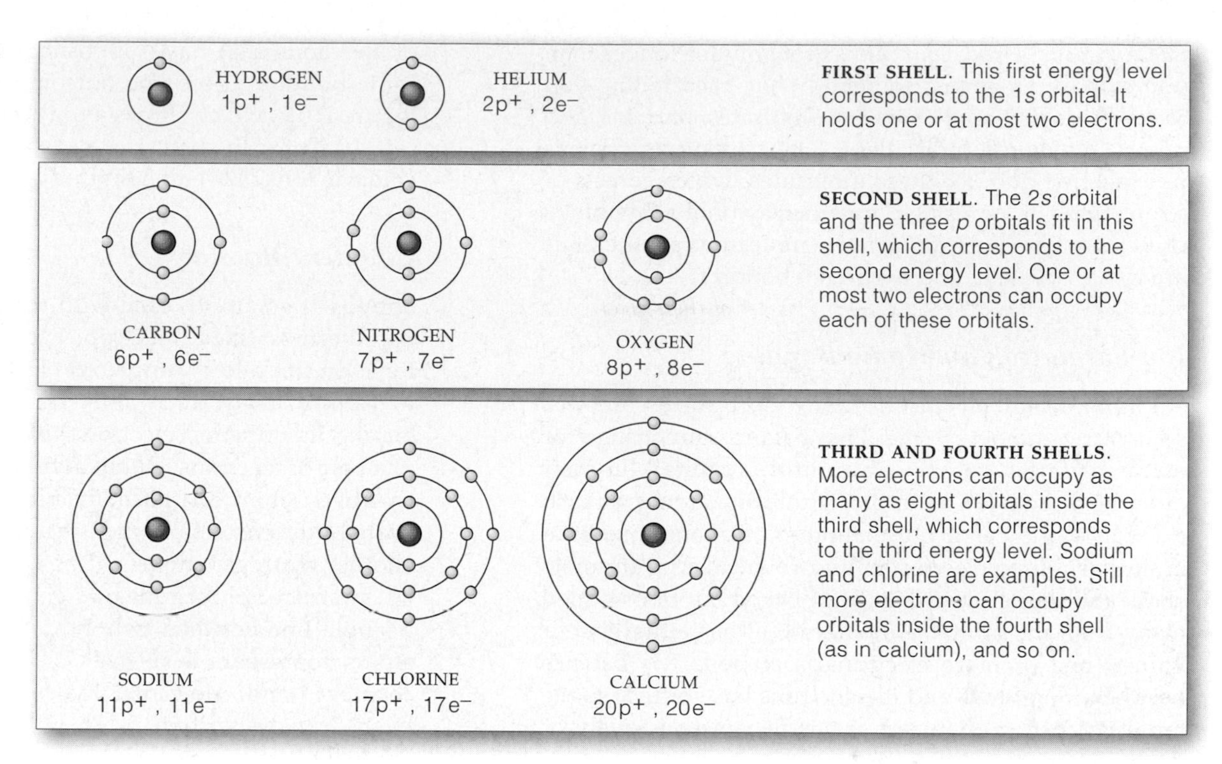

Figure 2.8 Examples of the shell model of how electrons are distributed in atoms. Successive shells correspond to higher energy levels from the nucleus. Hydrogen, carbon, and other atoms having electron vacancies (unfilled orbitals) inside their outermost shell tend to give up, accept, or share electrons. Helium and other atoms that have no electron vacancies in their outermost shell are inert; they show little if any tendency to interact with other atoms.

Figure 2.9 Chemical bookkeeping. We use symbols for elements when writing *formulas*, which identify the composition of compounds. For example, water has the formula H_2O. The subscript indicates two hydrogen (H) atoms are present for every oxygen (O) atom. We use such symbols and formulas when writing *chemical equations*, which are representations of the reactions among atoms and molecules. The substances entering a reaction (reactants) are to the left of the reaction arrow, and the products are to the right, as shown by the following chemical equation for photosynthesis:

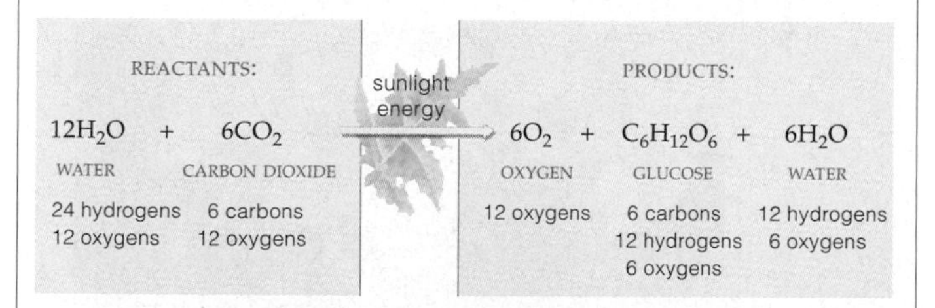

You may read about reactions for which reactants and products are expressed in moles. A *mole* is a certain number of atoms or molecules of any substance, just as "a dozen" can refer to any twelve cats, roses, and so forth. Molar weight, in grams, equals the total atomic weight of all atoms making up that substance.

For example, the atomic weight of carbon is 12, so one mole of carbon weighs 12 grams. A mole of oxygen (atomic weight 16) weighs 16 grams. Can you state why a mole of water (H_2O) weighs 18 grams, and why a mole of glucose ($C_6H_{12}O_6$) weighs 180 grams?

It summarizes some of the conventions used to describe metabolic reactions. When two or more atoms bond, a **molecule** results. Some molecules have one element only. Molecular nitrogen (N_2), with its two nitrogen atoms, is like this. So is molecular oxygen (O_2).

The molecules of **compounds** consist of two or more different elements in proportions that never vary. Water is an example. In each molecule of water you find one oxygen atom bonded to two hydrogen atoms. Water molecules in rainclouds, the ocean, a Siberian lake, your bathtub, the petals of a leaf or flower, or anywhere else always have twice as many hydrogen as oxygen atoms.

By contrast, in a **mixture**, two or more elements are simply intermingling in proportions that can vary (and usually do). Swirl together some water and the sugar sucrose, which is a compound of carbon, hydrogen, and oxygen, and you get a mixture.

In an atom, electrons occupy orbitals, which are volumes of space around the nucleus. By a simplified model, orbitals are arranged as a series of shells that surround the nucleus. The successive shells correspond to levels of energy, which become greater with distance from the nucleus.

One or two electrons at most occupy any orbital. Atoms having unfilled orbitals in their outermost shell tend to interact with other atoms; those with no vacancies do not.

In molecules of an element, all of the atoms are of the same kind. In molecules of a compound, atoms of two or more elements are bonded together, in unvarying proportions.

2.4 IMPORTANT BONDS IN BIOLOGICAL MOLECULES

Eat your peas! Drink your milk! Probably for longer than you care to remember, somebody has been telling you to eat foods that are rich in carbohydrates, proteins, and other "biological molecules." Only living organisms put together and use these molecules, which consist of a few kinds of atoms held together by only a few kinds of bonds. Foremost among the molecular interactions are ionic, covalent, and hydrogen bonds.

Ion Formation and Ionic Bonding

An atom, recall, has just as many electrons as protons, so it carries no net charge. That balance can change for atoms having a vacancy—an unfilled orbital—in their outermost shell. For example, a chlorine atom has such a vacancy and can acquire another electron. A sodium atom has a lone electron in an orbital in its outermost shell, and that electron can be knocked out of or pulled away from the orbital. Any atom that has either lost or gained one or more electrons is an **ion**. The balance between its protons and its electrons has shifted, so the atom has become ionized; it has become positively or negatively charged (Figure 2.10*a*).

In living cells, neighboring atoms commonly accept or donate electrons among one another. When one atom loses an electron and one gains, both become ionized. Depending on cellular conditions, the two ions may not separate; they may remain together as a result of the mutual attraction of opposite charges. An association of two ions that have opposing charges is called an **ionic bond**. You see one outcome of ionic bonding in Figure 2.10*b*, which shows a portion of a crystal of table salt, or NaCl. In such crystals, sodium ions (Na^+) and chloride ions (Cl^-) interact through ionic bonds.

Covalent Bonding

Suppose two atoms, each with an unpaired electron in its outermost shell, meet up. Each exerts an attractive force on the other's unpaired electron but not enough to yank it away. Each atom becomes more stable by *sharing* its unpaired electron with the other. A sharing of a pair of electrons is a **covalent bond**. For example, a hydrogen atom can partially fill the electron vacancy in its outermost shell when it is covalently bonded to another hydrogen atom.

MOLECULAR HYDROGEN (H_2)

In structural formulas, a single line between two atoms represents a *single* covalent bond. Molecular hydrogen has such a bond, which can be written as H—H. In a *double* covalent bond, two atoms share two pairs of electrons. Molecular oxygen (O═O) is like this. In a *triple* covalent bond, two atoms share three pairs of electrons. Molecular nitrogen (N≡N) is like this. All three examples happen to be gaseous molecules. Each time you breathe in some air, you draw a stupendous number of H_2, O_2, and N_2 molecules into your nose.

Covalent bonds are nonpolar or polar. In a *nonpolar* covalent bond, participating atoms exert the same pull on the electrons and both share them equally. "Nonpolar" implies that there is no difference in charge

SODIUM ATOM 11 p^+ 11 e^-

electron transfer

CHLORINE ATOM 17 p^+ 17 e^-

SODIUM ION 11 p^+ 10 e^-

CHLORIDE ION 17 p^+ 18 e^-

a Formation of sodium and chloride ions

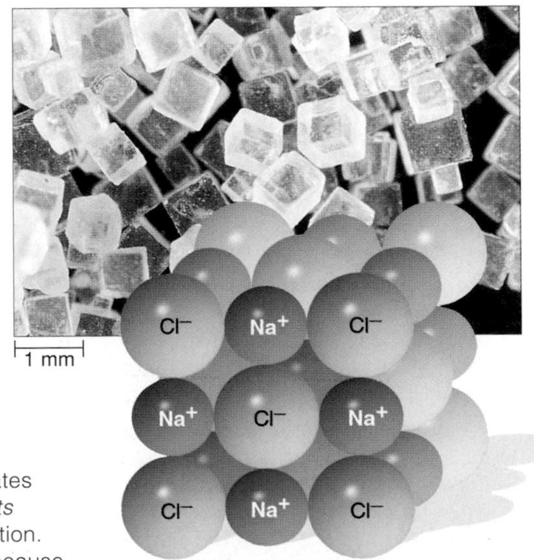

b Crystals of sodium chloride (NaCl)

Figure 2.10 (**a**) Ionization by way of an electron transfer. In this case, a sodium atom donates the single electron in its outermost shell to a chlorine atom, which has an unfilled orbital in *its* outermost shell. A sodium ion (Na^+) and a chloride ion (Cl^-) are the outcome of this interaction. (**b**) In each crystal of table salt, or NaCl, many sodium and chloride ions remain together because of the mutual attraction of opposite charges. Their interaction is a case of ionic bonding.

Further reading: Student Guide to InfoTrac on web site

between two ends of the bond (that is, at its two poles). Molecular hydrogen is a simple example of a nonpolar bond. Its two H atoms, each with one proton, attract the shared electrons equally.

In a *polar* covalent bond, atoms of different elements (which have different numbers of protons) do not exert the same pull on shared electrons. The more attractive atom ends up with a slight negative charge; the atom is "electronegative." Its effect is balanced out by the other atom, which ends up with a slight positive charge. In other words, taken together, the atoms interacting in a polar covalent bond have no *net* charge, but the charge is distributed unevenly between the bond's two ends.

As an example, a water molecule has two polar covalent bonds: H—O—H. In this molecule, electrons are less attracted to the hydrogens than to the oxygen, which has more protons. A water molecule carries no *net* charge, but you will see shortly that its polarity can weakly attract neighboring polar molecules and ions.

Hydrogen Bonding

The patterns of electron sharing in covalent bonds hold atoms together in specific arrangements in molecules. Some of the patterns also give rise to weak attractions and repulsions between charged functional groups of molecules, as well as between molecules and ions. Like interacting skydivers, such interactions break and form easily (Figure 2.11). Yet they have important roles in the structure and functioning of biological molecules.

For example, a **hydrogen bond** is a weak attraction between an electronegative atom (such as an oxygen or nitrogen atom taking part in a polar covalent bond) and a hydrogen atom taking part in a second polar covalent bond. Hydrogen's slight positive charge weakly attracts the atom with the slight negative charge (Figure 2.12).

Figure 2.11 Like skydivers who briefly clasp hands to form an orderly pattern, weak attractions within and between molecules (and between ions and molecules) can form and break easily.

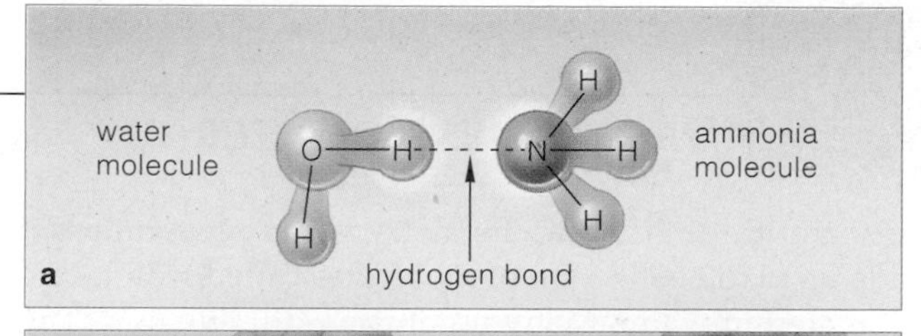

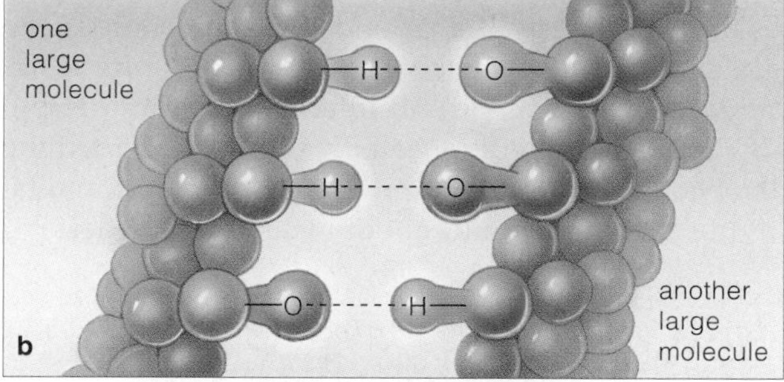

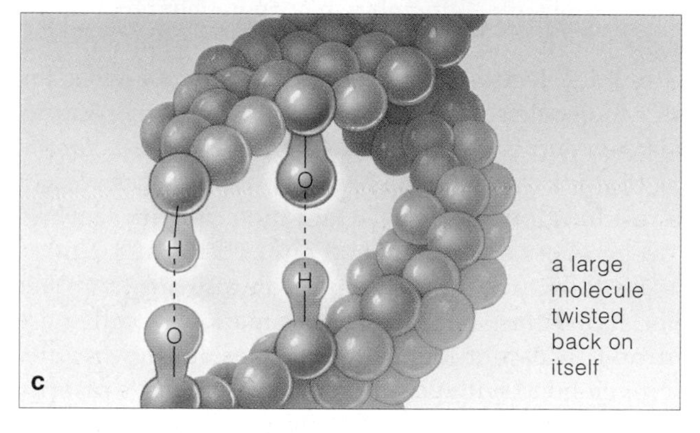

Figure 2.12 Three examples of hydrogen bonds. Compared to covalent bonds, a hydrogen bond is easier to break. Collectively, however, extensive hydrogen bonding has a major role in water, DNA, proteins, and many other substances.

Hydrogen bonds may form between two or more molecules. They also may form in different parts of the same molecule where it twists and folds back on itself. For example, many such bonds form between the two strands of a DNA molecule. Individually, the hydrogen bonds break easily. Collectively, they stabilize DNA's structure. Similarly, hydrogen bonds form between the molecules that make up water. As you will read next, they contribute to water's life-sustaining properties.

In an ionic bond, two ions of opposite charge attract each other and stay together. Ions form when atoms gain or lose electrons and so acquire a net positive or negative charge.

In a covalent bond, atoms share a pair of electrons. When atoms share the electrons equally, the bond is nonpolar. When sharing is not equal, the bond itself is polar—slightly positive at one end, slightly negative at the other.

In a hydrogen bond, a covalently bound atom showing a slight negative charge weakly interacts with a covalently bound hydrogen atom showing a slight positive charge.

2.5 PROPERTIES OF WATER

No sprint through basic chemistry is complete unless it leads us to the collection of molecules called water. Life originated in water. Many organisms still live in it. The ones that don't cart water around with them, in cells and tissue spaces. Many metabolic reactions require water as a reactant. Cell shape and internal structure depend on it. These topics will repeatedly occupy our attention in the book, so you may find it useful to become familiar with the following points about water's properties.

Polarity of the Water Molecule

A water molecule, remember, has no net charge, but the charge that it does carry is unevenly distributed. As a result of its electron arrangements and bond angles, the water molecule's oxygen "end" is a bit negative and its hydrogen end is a bit positive. Figure 2.13*a* is a simple way to think about this charge distribution. Because of the resulting polarity, one water molecule attracts and hydrogen-bonds with others (Figures 2.13*b* and 2.14).

The polarity of the water molecule also attracts other polar molecules, including the sugars. We call these **hydrophilic** (water-loving) **substances**, for they readily hydrogen-bond with water. By contrast, water's polarity repels nonpolar molecules, including oils. We call these **hydrophobic** (water-dreading) **substances**. Observe this for yourself by shaking a bottle that contains water and salad oil, then setting it on a table. Not long afterward, new hydrogen bonds replace the ones that broke apart when you shook the bottle. As the water molecules reunite, they push molecules of oil aside, forcing them to cluster as droplets or as a film at the water's surface.

Life depends on hydrophobic interactions. For example, a thin, oily membrane separates the cell's watery surroundings and watery interior. Membrane organization starts with countless hydrophobic interactions (Section 5.1).

Figure 2.14 Hydrogen bonding pattern of ice, which in vast quantities blankets the habitat of choice for polar bears. Below 0°C, each water molecule is hydrogen-bonded to four others in a three-dimensional lattice. The molecules are spaced farther apart than they would be in liquid water at room temperature, when molecular motion is greater and not as many bonds form. That is why ice floats on water; it has fewer molecules than the same volume of liquid water (the lattice is less dense).

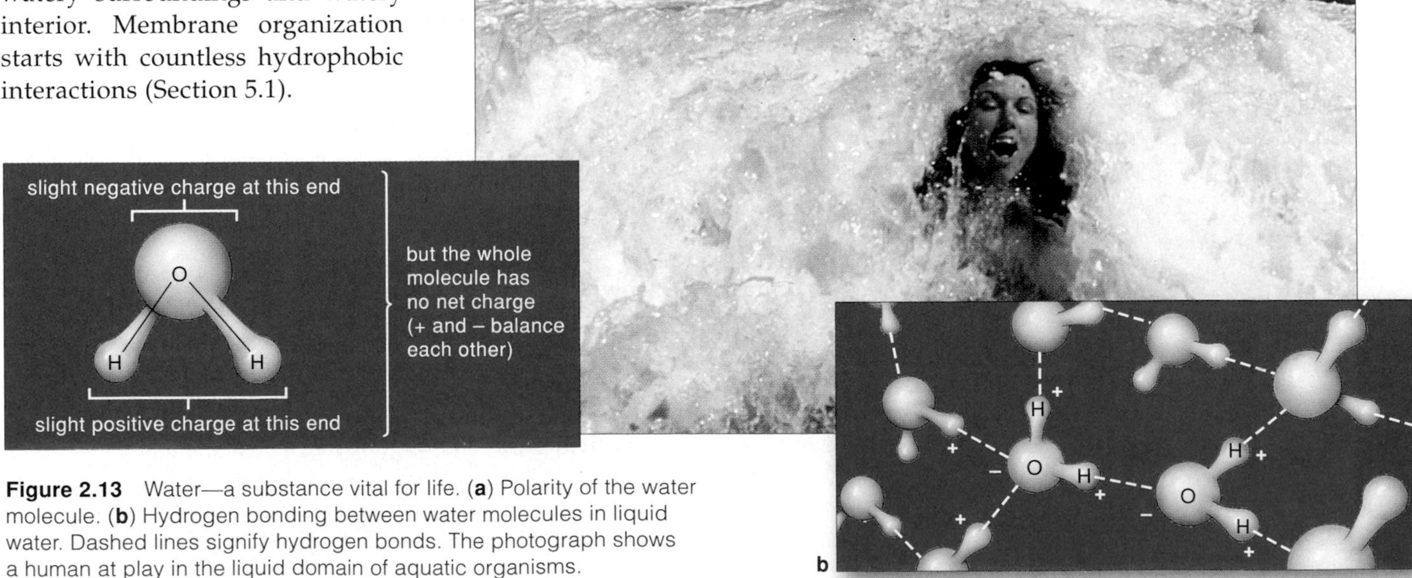

Figure 2.13 Water—a substance vital for life. (**a**) Polarity of the water molecule. (**b**) Hydrogen bonding between water molecules in liquid water. Dashed lines signify hydrogen bonds. The photograph shows a human at play in the liquid domain of aquatic organisms.

Further reading: Student Guide to InfoTrac on web site →

Figure 2.15 Water's cohesion. (**a**) Water strider (*Gerris*). This long-legged bug feeds on insects that land or fall on water. Because of its fine, water-resistant leg hairs and water's high surface tension, the bug scoots across the surface. (**b**) Water's cohesion, combined with its evaporation from leaves, pulls water up to the tops of trees.

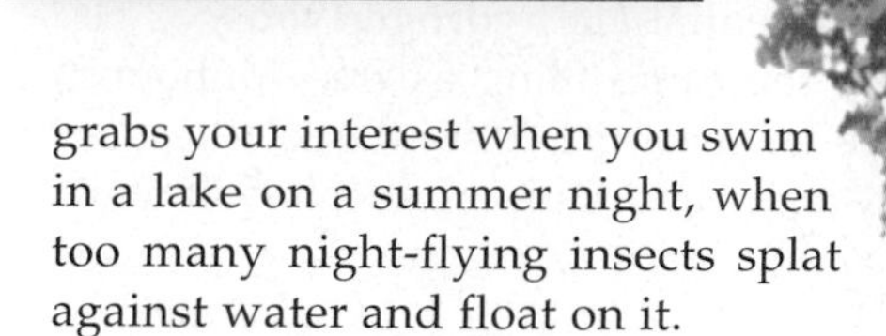

Water's Temperature-Stabilizing Effects

Cells consist mostly of water, and they release a great deal of heat energy during metabolism. If it were not for hydrogen bonds in liquid water, cells might cook in their own juices. To see why this is so, start with these observations: Every molecule vibrates incessantly, and its motion increases when it absorbs heat. **Temperature** is simply a measure of the molecular motion of a given substance. Compared to most other fluids, water can absorb more heat energy before its temperature rises measurably. Why? Much of the added energy disrupts hydrogen bonding between neighboring molecules of water rather than causing an increase in the motion of individual molecules. In liquid water, the stupendous number of hydrogen bonds can buffer large swings in temperature. Such bonds help stabilize the temperature of multicelled organisms and of aquatic habitats.

Even when the temperature of liquid water is not shifting much, hydrogen bonds are constantly breaking. But they also are forming again just as fast. By contrast, a large energy input can increase molecular motion so much that hydrogen bonds stay broken, and individual molecules at the water's surface escape into the air. By this process, called **evaporation**, heat energy converts liquid water to the gaseous state. As large numbers of molecules break free and depart, they carry away some energy and lower the water's surface temperature.

Evaporative water loss can help cool you and some other mammals when you work up a sweat on hot, dry days. Under such conditions, sweat—which is about 99 percent water—evaporates from your skin.

Below 0°C, hydrogen bonds resist breaking and lock water molecules in the latticelike bonding pattern of ice (Figure 2.14). Ice is less dense than water. During winter freezes, ice sheets may form near the surface of ponds, lakes, and streams. Like a blanket, ice "insulates" the liquid water beneath it and helps protect many fishes, frogs, and other aquatic organisms against freezing.

Water's Cohesion

Life also depends on water's cohesion. **Cohesion** means something has a capacity to resist rupturing when placed under tension—that is, stretched—as by the weight of a bug's legs (Figure 2.15*a*). Think of a lake, pool, or some other body of liquid water. Uncountable episodes of hydrogen bonding exert a continual inward pulling on water molecules at or near the surface. The hydrogen bonding results in a high surface tension. That tension grabs your interest when you swim in a lake on a summer night, when too many night-flying insects splat against water and float on it.

Cohesion works inside organisms as well as on the outside. For example, trees and other plants require nutrient-laden water for growth and metabolism. Largely because of the cohesion, narrow columns of liquid water move through pipelines of vascular tissues, from roots to leaves and all other parts of the plant. On sunny days, water evaporates from leaves; individual molecules break free (Figure 2.15*b*). Hydrogen bonding "pulls up" more water molecules into leaf cells as replacements, in ways that you will read about in Section 30.3.

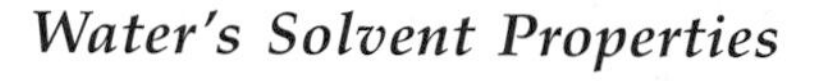

Water's Solvent Properties

Finally, water is a fine solvent; ions and polar molecules easily dissolve in it. Any dissolved substance is called a **solute**. In general, we say a substance is *dissolved* after water molecules cluster around its ions or molecules and keep them dispersed in fluid. Such clusters are "spheres of hydration." This is what happens to solutes in cellular fluid, in the sap of maple trees, in blood, in fluid traveling through your gut, and in every other fluid associated with life.

Watch this happen when you pour some table salt (NaCl) into a cup of water. After a while, the salt crystals separate into Na^+ and Cl^-. Each Na^+ attracts the negative end of some of the water molecules at the same time Cl^- attracts the positive end of others (Figure 2.16). The spheres of hydration keep the ions dispersed in the fluid.

Figure 2.16 Two spheres of hydration.

A water molecule has no net charge, but it shows polarity. Polarity allows water molecules to hydrogen-bond to each other and to other polar (hydrophilic) substances. Water molecules tend to repel nonpolar (hydrophobic) substances.

Water has temperature-stabilizing effects, internal cohesion, and a capacity to dissolve many substances. These properties influence the structure and functioning of organisms.

2.6 ACIDS, BASES, AND BUFFERS

A great variety of ions dissolved in the fluids inside and outside cells influence cell structure and functioning. Among the most influential are **hydrogen ions**, or H^+. Hydrogen ions are the same thing as free (unbound) protons. They have far-reaching effects largely because they are chemically active and there are so many of them.

The pH Scale

At any instant in liquid water, some water molecules break apart into hydrogen ions and **hydroxide ions** (OH^-). This ionization of water is the basis of the **pH scale**, as in Figure 2.17. Biologists use this scale when measuring the H^+ concentration of seawater, tree sap, blood, and other fluids. Pure water (not rainwater or tapwater) always contains just as many H^+ as OH^- ions. This condition also may occur in other fluids, and it signifies neutrality. We assign neutrality a value of 7 at the midpoint of the pH scale, which ranges from 0 (the highest H^+ concentration) to 14 (the lowest). *The greater the H^+ concentration, the lower the pH.*

Starting at neutrality, each change by one unit of the pH scale corresponds to a tenfold increase or decrease in H^+ concentration. An easy way to sense the differences is to dissolve a bit of baking soda (pH 9) on your tongue, then water (7), then lemon juice (2.3).

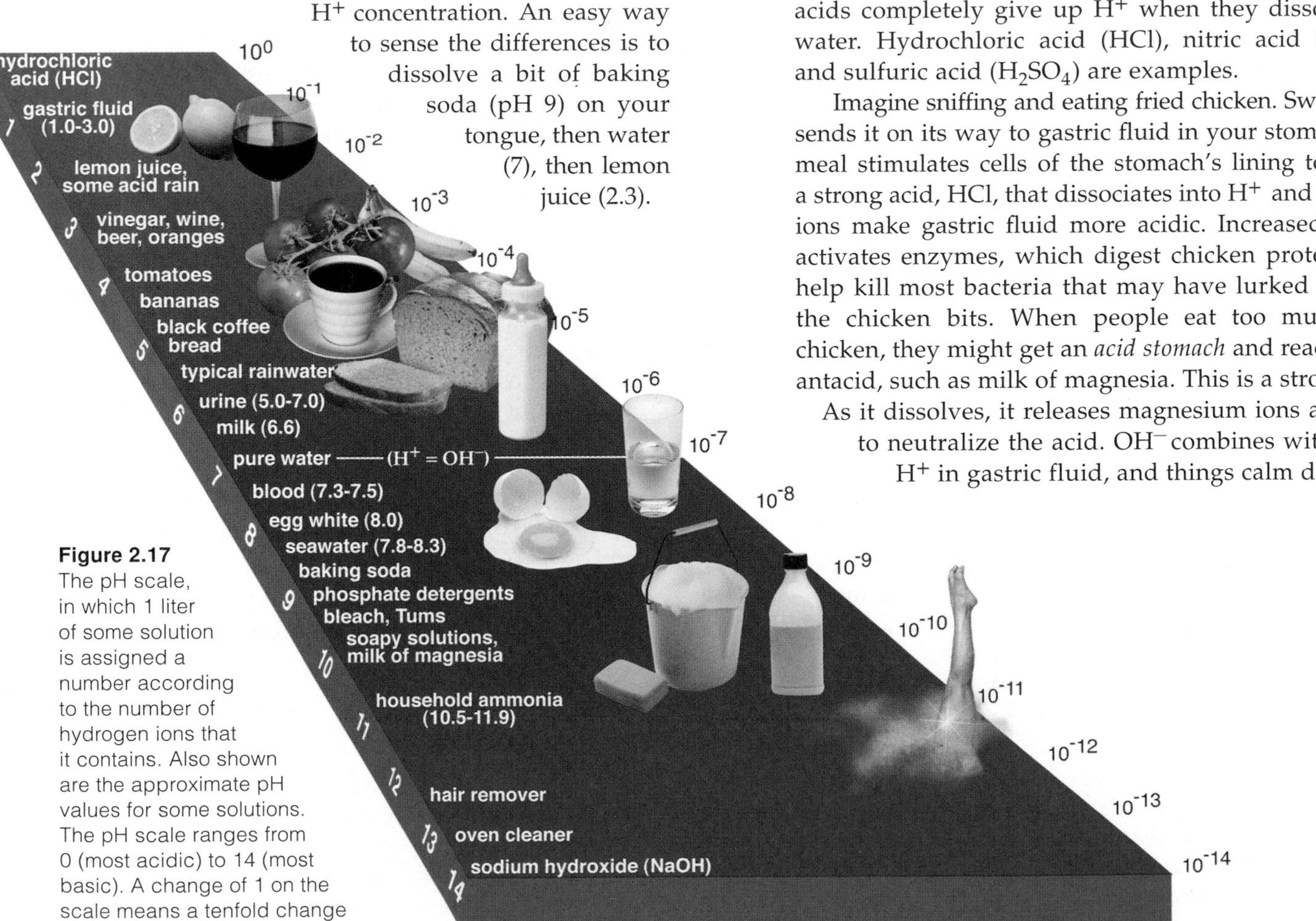

Figure 2.17 The pH scale, in which 1 liter of some solution is assigned a number according to the number of hydrogen ions that it contains. Also shown are the approximate pH values for some solutions. The pH scale ranges from 0 (most acidic) to 14 (most basic). A change of 1 on the scale means a tenfold change in H^+ concentration.

How Do Acids Differ From Bases?

When they dissolve in water, substances categorized as **acids** *donate* protons (H^+) to other solutes or to water molecules. By contrast, the substances we categorize as **bases** *accept* H^+ when dissolved in water, and OH^- forms directly or indirectly after they do this. *Acidic* solutions, such as lemon juice, gastric fluid, and coffee, release more H^+ than OH^-; their pH is below 7. *Basic* solutions, such as seawater, baking soda, and egg white, release more OH^- than H^+. Such solutions are also called "alkaline" fluids; they have a pH above 7.

The fluid inside most human cells is about 7 on the pH scale. The pH values of most fluids bathing these cells are slightly higher; they range between 7.3 and 7.5. This also is the case for the fluid portion of human blood. By contrast, seawater is more alkaline than the body fluids of organisms that live in it.

Think of most acids as being either weak or strong. Weak ones such as carbonic acid (H_2CO_3) are reluctant H^+ donors. Depending on the pH, they just as easily accept H^+ after giving it up, so they alternate between acting as an acid and acting as a base. By contrast, strong acids completely give up H^+ when they dissociate in water. Hydrochloric acid (HCl), nitric acid (HNO_3), and sulfuric acid (H_2SO_4) are examples.

Imagine sniffing and eating fried chicken. Swallowing sends it on its way to gastric fluid in your stomach. The meal stimulates cells of the stomach's lining to secrete a strong acid, HCl, that dissociates into H^+ and Cl^-. The ions make gastric fluid more acidic. Increased acidity activates enzymes, which digest chicken proteins and help kill most bacteria that may have lurked in or on the chicken bits. When people eat too much fried chicken, they might get an *acid stomach* and reach for an antacid, such as milk of magnesia. This is a strong base. As it dissolves, it releases magnesium ions and OH^- to neutralize the acid. OH^- combines with excess H^+ in gastric fluid, and things calm down.

Further reading: Student Guide to InfoTrac on web site

Figure 2.18 Sulfur dioxide emissions from a coal-burning power plant. Camera lens filters revealed the otherwise invisible emissions. Sulfur dioxide and other airborne pollutants dissolve in water vapor to form acidic solutions. They are a major component of acid rain.

High concentrations of strong acids or bases also can disrupt the external environment and pose dangers to life. Read the labels on bottles of ammonia, drain cleaner, and other products that are often stored in households. Many can cause severe *chemical burns*. So can sulfuric acid in car batteries. Smoke from fossil fuels, exhaust from motor vehicles, and nitrogen fertilizers release strong acids, which alter the pH of rain (Figure 2.18). Some regions are sensitive to the pH of this *acid rain*, owing to their soil type and vegetation cover. The altered chemistry of habitats in such regions drastically affects the functioning of organisms. We will return to this topic in Section 51.1.

Buffers Against Shifts in pH

Metabolic reactions are sensitive to even slight shifts in pH, for H^+ and OH^- can combine with many different molecules and alter their functions. Normally, control mechanisms minimize unsuitable shifts in pH, as they do when HCl enters the stomach in response to a meal. Many of the controls involve buffer systems.

A **buffer system** is a partnership between a weak acid and the base that forms when the acid dissolves in water. The two work as a pair to counter slight shifts in pH. Remember, when a strong base enters a fluid, the OH^- level rises. But a weak acid neutralizes part of the added OH^- by combining with it. By this interaction, the weak acid's partner forms. Later, if a strong acid floods in, the base will accept H^+ and thereby become its partner in the system.

Bear in mind, the action of a buffer system cannot make new hydrogen ions or eliminate ones that already are present. It can only bind or release them.

In all complex, multicelled organisms, diverse buffer systems operate in the internal environment—in blood and tissue fluids. For example, metabolic reactions in the vertebrate lungs and kidneys help control the acid–base balance of this environment, at levels suitable for life (Sections 41.4, 43.4, and 43.6). For now, simply think of what happens when the blood level of H^+ decreases and the blood is not as acidic as it should be. At such times, carbonic acid that is dissolved in blood releases H^+ and so becomes the partner base, bicarbonate:

$$\underset{\text{CARBONIC ACID}}{H_2CO_3} \longrightarrow \underset{\text{BICARBONATE}}{HCO_3^-} + H^+$$

When blood becomes more acidic, more H^+ becomes bound to the base, thus forming the partner acid:

$$\underset{\text{BICARBONATE}}{HCO_3^-} + H^+ \longrightarrow \underset{\text{CARBONIC ACID}}{H_2CO_3}$$

Uncontrolled shifts in pH have drastic outcomes. If blood's pH (7.3–7.5) declines even to 7, an individual will enter into a *coma*, a sometimes irreversible state of unconsciousness. An increase to 7.8 can lead to *tetany*, a potentially lethal condition in which skeletal muscles enter a state of uncontrollable contraction. In *acidosis*, carbon dioxide builds up in blood, too much carbonic acid forms, and blood pH severely decreases. *Alkalosis* is an uncorrected increase in blood pH. Both conditions weaken the body and can be lethal.

Salts

Salts are compounds that release ions *other than* H^+ and OH^- in solutions. Salts and water often form when a strong acid and strong base interact. Depending on a solution's pH value, salts can form and dissolve easily. Consider how sodium chloride forms, then dissolves:

$$\underset{\text{HYDROCHLORIC ACID}}{\text{HCl (acid)}} + \underset{\text{SODIUM HYDROXIDE}}{\text{NaOH (base)}} \longrightarrow \underset{\text{SODIUM CHLORIDE}}{\text{NaCl (salt)}} + H_2O$$

$$\text{NaCl} \rightarrow Na^+ \quad Cl^- \text{ (ionization)}$$

Many salts dissolve into ions that serve key functions in cells. For example, nerve cell activity depends on ions of sodium, potassium, and calcium. Muscles contract with the help of calcium ions. And water absorption by plant cells depends largely on potassium ions.

Hydrogen ions (H^+) and other ions dissolved in the fluids inside and outside cells affect cell structure and function.

When dissolved in water, acidic substances release H^+, and basic (alkaline) substances accept them. Certain acid–base interactions, as in buffer systems, help maintain the pH value of a fluid—that is, its H^+ concentration.

A buffer system counters slight shifts in pH by releasing hydrogen ions when their concentration is too low or by combining with them when the concentration is too high.

Salts are compounds that release ions other than H^+ and OH^-, and many of those ions have key roles in cell functions.

2.7 SUMMARY

1. Chemistry helps us understand the nature of all the substances that make up cells, organisms, and the Earth, its waters, and the atmosphere. Each substance consists of one or more elements. Of the ninety-two naturally occurring elements, the most common in organisms are oxygen, carbon, hydrogen, and nitrogen. Organisms also have lesser amounts of many other elements, such as calcium, phosphorus, potassium, and sulfur.

2. Table 2.2 summarizes some key chemical terms that you will encounter throughout this book.

3. Elements consist of atoms. An atom has one or more positively charged protons, an equal number of negatively charged electrons, and (except for hydrogen atoms) one or more uncharged neutrons. The protons and neutrons occupy the core region, the atomic nucleus. Most elements have isotopes: two or more forms of atoms with the same number of protons but different numbers of neutrons.

4. An atom has no net charge. But it may lose or gain one or more electrons and so become an ion, which has an overall positive or negative charge.

5. Whether an atom interacts with others depends on the number and arrangement of its electrons, which occupy orbitals (volumes of space) inside a series of shells around the atomic nucleus. Atoms with unfilled orbitals in the outermost shell tend to bond with other elements.

6. Generally, a chemical bond is a union between the electron structures of atoms.

a. In an ionic bond, a positive ion and negative ion stay together because of a mutual attraction of their opposite charges.

b. Various atoms often share one or more pairs of electrons in single, double, or triple covalent bonds. Such electron sharing is equal in nonpolar covalent bonds, and it is unequal in polar covalent bonds. The interacting atoms have no net charge, but the bond is slightly negative at one end and slightly positive at the other.

c. In a hydrogen bond, one covalently bonded atom (such as oxygen or hydrogen) that displays a slight negative charge is weakly attracted to the slight positive charge of a hydrogen atom that is taking part in a different polar covalent bond.

7. Polar covalent bonds join together three atoms in water molecules (two hydrogens, one oxygen). The polarity of a water molecule invites hydrogen bonding between water molecules. Such hydrogen bonding is the basis of liquid water's ability to resist temperature changes more than other fluids do, display internal cohesion, and easily dissolve polar or ionic substances. These properties greatly influence the metabolic activity, structure, shape, and internal organization of cells.

8. By the pH scale, a solution is assigned a number that reflects its H^+ concentration. This ranges from 0 (highest concentration) to 14 (lowest). At pH 7, the H^+ and OH^- concentrations are equal. Acids release H^+ in water, and bases combine with them. Buffer systems help maintain pH values of blood, tissue fluids, and the fluid inside cells.

Table 2.2 Summary of Key Players in the Chemical Basis of Life

ELEMENT	Fundamental form of matter that occupies space, has mass, and cannot be broken apart into a different form of matter by ordinary physical or chemical means.
ATOM	Smallest unit of an element that still retains the characteristic properties of that element.
Proton (p^+)	Positively charged particle of the atomic nucleus. All atoms of an element have the same number of protons, which is the atomic number. A proton without an electron zipping around it is a hydrogen ion (H^+).
Electron (e^-)	Negatively charged particle that can occupy a volume of space (orbital) around an atomic nucleus. All atoms of an element have the same number of electrons. Electrons can be shared or transferred among atoms.
Neutron	Uncharged particle of the nucleus of all atoms except hydrogen. For a given element, the mass number is the number of protons and neutrons in the nucleus.
MOLECULE	Unit of matter in which two or more atoms of the same element, or different ones, are bonded together.
Compound	Molecule composed of two or more different elements in unvarying proportions. Water is an example.
Mixture	Intermingling of two or more elements in proportions that can and usually do vary.
ISOTOPE	One of two or more forms of atoms of an element that differ in their number of neutrons.
Radioisotope	Unstable isotope, having an unbalanced number of protons and neutrons, that emits particles and energy.
Tracer	Molecule of a substance to which a radioisotope is attached. In conjunction with tracking devices, it can be used to follow the movement or destination of that substance in a metabolic pathway, the body, etc.
ION	Atom that has gained or lost one or more electrons, thus becoming positively or negatively charged.
SOLUTE	Any molecule or ion dissolved in some solvent.
Hydrophilic substance	Polar molecule or molecular region that can readily dissolve in water.
Hydrophobic substance	Nonpolar molecule or molecular region that strongly resists dissolving in water.
ACID	Substance that donates H^+ when dissolved in water.
BASE	Substance that accepts H^+ when dissolved in water; OH^- forms directly or indirectly afterward.
SALT	Compound that releases ions other than H^+ or OH^- when dissolved in water.

Review Questions

1. What is an element? Name four elements (and their symbols) that make up more than 95 percent of the body weight of all living organisms. *CI*

2. Define atom, isotope, and radioisotope. *2.1*

3. How many electrons can occupy each orbital around an atomic nucleus? Using the shell model, explain how the orbitals available to electrons are distributed in an atom. *2.3*

4. Define molecule, compound, and mixture. *2.3*

5. Distinguish between:
   a. ionic and hydrogen bonds *2.4*
   b. polar and nonpolar covalent bonds *2.4*
   c. hydrophilic and hydrophobic interactions *2.5*

6. If a water molecule has no net charge, then why does it attract polar molecules and repel nonpolar ones? *2.5*

7. Label the atoms in each water molecule in the sketch below. Indicate which parts of each molecule carry a slight positive charge (+) and which carry a slight negative charge (−). *2.5*

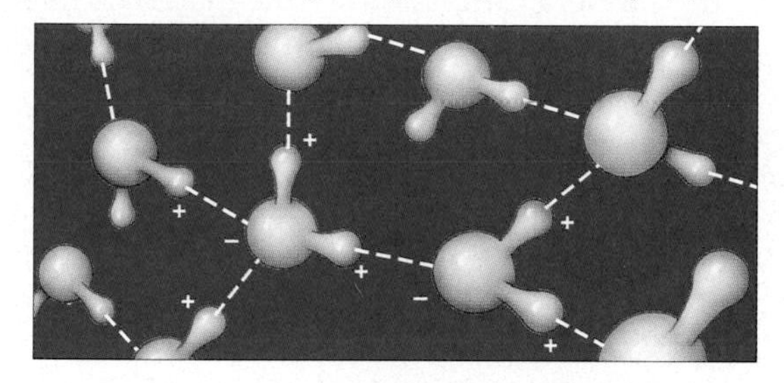

8. Define acid and base. Then describe the behavior of a weak acid in solutions having a high or low pH value. *2.6*

Self-Quiz (*Answers in Appendix III*)

1. Electrons carry (a) ______ charge.
   a. positive b. negative c. zero

2. Atoms share electrons unequally in a(n) ______ bond.
   a. ionic c. polar covalent
   b. nonpolar covalent d. hydrogen

3. A water molecule shows ______.
   a. polarity d. solvency
   b. hydrogen-bonding capacity e. a and b
   c. heat resistance f. all of the above

4. In liquid water, spheres of hydration form around ______.
   a. nonpolar molecules d. solvents
   b. polar molecules e. b and c
   c. ions f. all of the above

5. Hydrogen ions (H^+) are ______.
   a. the basis of pH values d. dissolved in blood
   b. unbound protons e. both a and b
   c. targets of certain buffers f. all of the above

6. When dissolved in water, a(n) ______ donates H^+; however, a(n) ______ accepts H^+.

7. Match the terms with the most suitable descriptions.
   ______ trace element
   ______ buffer system
   ______ chemical bond
   ______ temperature
   a. weak acid and its partner base work as a pair to counter pH shifts
   b. union between electron structures of two atoms
   c. less than 0.01% of body weight
   d. measure of molecular motion in some defined region

Critical Thinking

1. An ionic compound forms when calcium combines with chlorine. Referring to Figure 2.8, give the compound's formula. (Hint: Be sure the outermost shell of each atom is filled.)

2. David, an inquisitive three-year-old, poked his fingers in the water inside a metal pan on the stove and discovered that it was warm. Then he touched the pan itself and got a nasty burn. Devise a hypothesis to explain why water in a metal pan heats up far more slowly than the pan itself.

3. When molecules absorb microwaves, which are a form of electromagnetic radiation, they move more rapidly. Explain why a microwave oven can heat foods.

4. From what you know about cohesion, devise a hypothesis to explain why water forms droplets.

5. Edward is trying to study a chemical reaction that an enzyme catalyzes (speeds up). H^+ forms in the reaction, but the enzyme is destroyed at low pH. What can he include in his reaction mix to protect the enzyme while he studies the reaction? Explain how your suggestion might solve the problem.

6. Many reactions occur on molecular parts of enzymes and other proteins. Cells must have access to those parts. Through interactions with water and ions, a soluble protein can become dispersed in cellular fluid rather than settling against some cell structure. An electrically charged cushion around it makes this happen. Using the sketch below as a guide, explain the chemical interactions by which such a cushion forms, starting with major bonds in the protein itself.

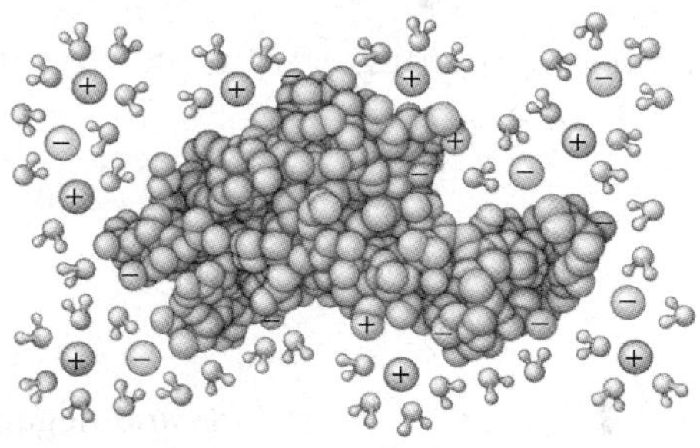

Selected Key Terms

acid *2.6*
atom *2.1*
base *2.6*
bioremediation *CI*
buffer system *2.6*
chemical bond *2.3*
cohesion *2.5*
compound *2.3*
covalent bond *2.4*
electron *2.1*
element *CI*
evaporation *2.5*
hydrogen bond *2.4*
hydrogen ion (H^+) *2.6*
hydrophilic substance *2.5*
hydrophobic substance *2.5*
hydroxide ion (OH^-) *2.6*
ion *2.4*
ionic bond *2.4*
isotope *2.1*
mixture *2.3*
molecule *2.3*
neutron *2.1*
pH scale *2.6*
proton *2.1*
radioisotope *2.1*
salt *2.6*
shell model *2.3*
solute *2.5*
temperature *2.5*
tracer *2.2*

Readings *See also www.infotrac-college.com*

Ritter, P. 1996. *Biochemistry: A Foundation*. Pacific Grove, California: Brooks/Cole. Good, easy-to-read introduction to biochemistry, with plenty of human-interest applications.

3

CARBON COMPOUNDS IN CELLS

Carbon, Carbon, in the Sky—Are You Swinging Low and High?

High in the mountains of the Pacific Northwest, vast forests of conifers have endured another murderously cold winter (Figure 3.1). Like all other living organisms, these evergreen, cone-bearing trees cannot grow or reproduce in the absence of liquid water. Yet through the winter months, water in their habitat is locked away from them, in the form of snow and ice.

The trees get by anyway. Their needle-shaped leaves have an epidermis—a layer of interconnected cells with a thick, waxy wall that restricts water loss. About the only way precious water can escape is through small gaps across the epidermis that can close or open in response to changing conditions. During the cool, dry days of autumn, the trees enter dormancy. Metabolic activities idle and growth ceases, but water conserved inside them is enough to keep their leaf cells alive.

With the arrival of spring, rising temperatures and water from melting snow stimulate renewed growth. Tree roots soak up mineral-laden water. And carbon dioxide, a gaseous molecule of one carbon atom and two oxygen atoms, moves in from the air, through gaps in the leaf epidermis. With their photosynthetic magic, the conifers turn those simple materials into sugars, starches, and other carbon-based compounds. They are premier producers of the northern forests.

Producers, recall, are organisms that use self-made organic compounds as their structural materials and as packets of energy. One way or another, compounds made by producers of forests and other ecosystems all over the world nourish every consumer and decomposer.

Plants of the great prevailing forests and plains at northern latitudes of Canada, the United States, and other parts of the Northern Hemisphere are now breaking dormancy sooner than they did just two decades ago. Why? No one knows for sure, but this change in their life cycles may be one outcome of long-term change in the global climate.

Figure 3.1 Conifers beneath the first snows of winter on Silver Star Mountain, Washington. As is the case for all other organisms, the structure, activities, and very survival of these trees start with the carbon atom and its diverse molecular partners in organic compounds.

Researchers have looked at atmospheric concentrations of carbon dioxide since the early 1950s. Among other things, they found that the concentration shifts with the seasons. It declines during spring and summer, when photosynthesizers take up huge amounts of carbon dioxide from their surroundings. It rises at other times of the year, when huge populations of decomposers that release the gas as a metabolic by-product undergo rapid increases in their population size.

According to the researchers' measurements, the spring decline starts a full week earlier than it did in the mid-1970s. Besides this, the seasonal swings are becoming more pronounced—by as much as 20 percent in Hawaii and a whopping 40 percent in Alaska.

Swings in the atmospheric concentration of carbon dioxide were greatest in 1981 and 1990. Intriguingly, temperatures of the lower atmosphere are rising also, and in 1981 and 1990, they were uncommonly high. Are we in the midst of a long-term, worldwide rise in atmospheric temperature? And is this *global warming* promoting a longer growing season, hence the wider seasonal swings?

The picture gets more intricate. We humans burn great quantities of coal, gasoline, and other fossil fuels for energy. Fossil fuels are rich in carbon, which is released (as carbon dioxide) by the burning processes. The released carbon may be contributing to the global warming, in ways you will read about in later chapters.

For now, the point to keep in mind is this: *Carbon permeates the world of life—from the energy-requiring activities and structural organization of individual cells, to physical and chemical conditions that span the globe and that influence life everywhere.*

With this chapter, we turn to life-giving properties that emerge out of the molecular structure of carbon-rich compounds. Study the chapter well, especially the summary in Table 3.1. It will serve as your foundation for understanding how different organisms put such compounds together, how they use them, and how the effects of those uses ripple through the biosphere.

KEY CONCEPTS

1. Organic compounds have a backbone of one or more carbon atoms to which hydrogen, oxygen, nitrogen, and other atoms are attached. We define cells partly by their capacity to assemble the organic compounds known as carbohydrates, lipids, proteins, and nucleic acids.

2. Cells put together large biological molecules from their pools of smaller organic compounds, which include simple sugars, fatty acids, amino acids, and nucleotides.

3. Glucose and other simple sugars are carbohydrates. So are organic compounds composed of two or more sugar units, of one or more types, that are covalently bonded together. The most complex carbohydrates that cells assemble are polysaccharides, many of which consist of hundreds or thousands of sugar units.

4. Lipids are greasy or oily compounds that show little tendency to dissolve in water but dissolve in nonpolar compounds, including other lipids. They include the neutral fats (triglycerides), phospholipids, waxes, and sterols.

5. Cells use carbohydrates and lipids as building blocks and as their major sources of energy.

6. Proteins have truly diverse roles. Many are structural materials. Many are enzymes, a type of molecule that enormously increases the rate of specific metabolic reactions. Other kinds transport cell substances, contribute to cell movements, trigger changes in cell activities, and defend the body against injury and disease.

7. For living organisms, ATP and other nucleotides are crucial players in metabolism. DNA and RNA, strandlike nucleic acids assembled from nucleotide subunits, are the basis of inheritance and reproduction.

3.1 PROPERTIES OF ORGANIC COMPOUNDS

The Molecules of Life

Under present-day conditions in nature, *only living cells synthesize carbohydrates, lipids, proteins, and nucleic acids.* These are the molecules characteristic of life. Different classes of biological molecules act as the cells' packets of instantly available energy, energy stores, structural materials, metabolic workers, libraries of hereditary information, and cell-to-cell signals.

The molecules of life are **organic compounds**, with hydrogen and often other elements covalently bonded to carbon atoms. The term is a holdover from a time when chemists thought "organic" substances were the ones they got from animals and vegetables, as opposed to "inorganic" substances they got from minerals. The term persists, even though researchers now synthesize organic compounds in laboratories. And it persists even though there are reasons to believe organic compounds were present on Earth *before* organisms were.

Carbon's Bonding Behavior

By far, organisms consist mainly of oxygen, hydrogen, and carbon (Figure 2.2). Much of the oxygen and the hydrogen is in the form of water. Remove the water, and carbon makes up more than half of what's left.

Carbon's importance in life arises from its versatile bonding behavior. As shown in the sketch below, *each carbon atom can share pairs of electrons with as many as four other atoms.* Each covalent bond formed in this way is relatively stable. Such bonds join carbon atoms together in chains. These form a backbone to which many other elements, including hydrogen, oxygen, and nitrogen, become attached.

single covalent bond

carbon atom

The bonding arrangement is the start of wonderful three-dimensional shapes of organic compounds. A chain of carbon atoms, bonded covalently one after another, forms a backbone from which other atoms can project:

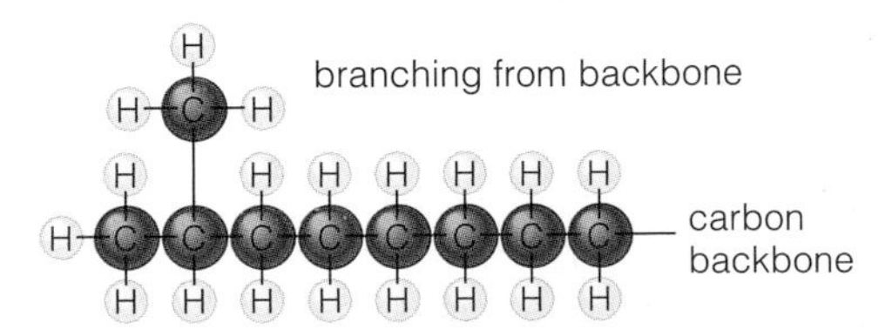

The backbone can coil back on itself in a ring structure, which we can diagram in such ways as:

or

carbon rings

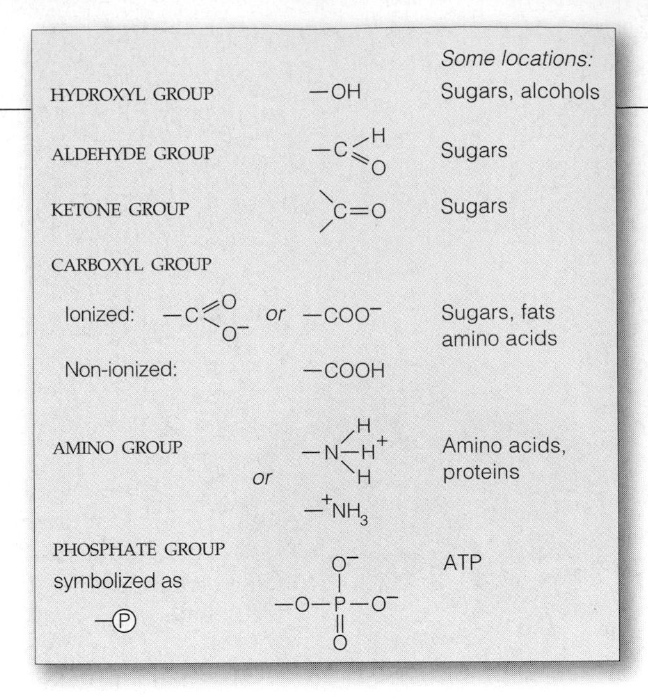

Figure 3.2 Examples of functional groups. 

Functional Groups

A carbon backbone with only hydrogen atoms bonded covalently to it is a hydrocarbon, which is a very stable structure. Besides hydrogen atoms, biological molecules also have **functional groups**: various kinds of atoms or clusters of them covalently bonded to the backbone.

To get a sense of their importance, consider a few of the functional groups shown in Figure 3.2. Sugars and other organic compounds classified as **alcohols** have one or more hydroxyl groups (—OH). Alcohols dissolve quickly in water, because water molecules easily form hydrogen bonds with —OH groups. The backbone of a protein forms by reactions between amino groups and carboxyl groups. As you will see, the backbone is the start of bonding patterns that produce the protein's three-dimensional structure. As other examples, amino groups can combine with H^+ and act as buffers against decreases in pH. And the functional groups shown in Figure 3.3 are a molecular starting point for differences between the males and females of many species.

How Do Cells Build Organic Compounds?

Using carbon (from carbon dioxide), water, and sunlight (as an energy source), the photosynthetic cells you read about earlier put together simple sugar molecules. Like all living cells, they also use such molecules as a starting point for assembling other small molecules, especially the fatty acids, amino acids, and nucleotides. As you will read shortly, cells use some assortment of these four classes of small organic compounds as subunits for building all the organic compounds they require for their structure and functioning.

How do they do it? It will take more than one chapter to sketch out answers (and best guesses) to the question.

Further reading: Student Guide to InfoTrac on web site

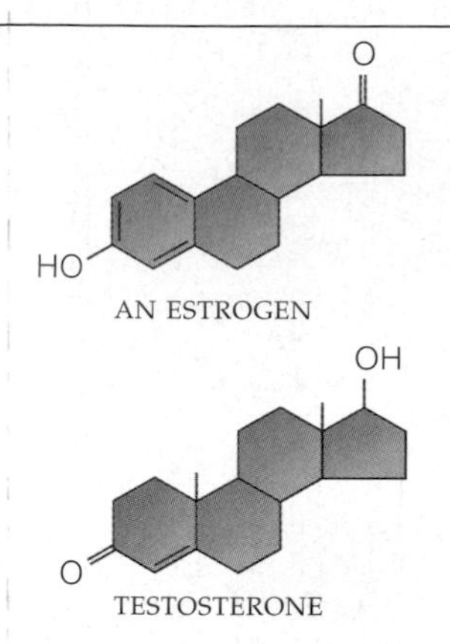

Figure 3.3 Notable differences in traits between male and female wood ducks (*Aix sponsa*). Two different sex hormones have key roles in the development of feather color and other traits that help males and females recognize each other (and therefore influence reproductive success). Both of the hormones, testosterone and one of the estrogens, have the same carbon ring structure. As you can see, however, the ring structures have different functional groups attached to them.

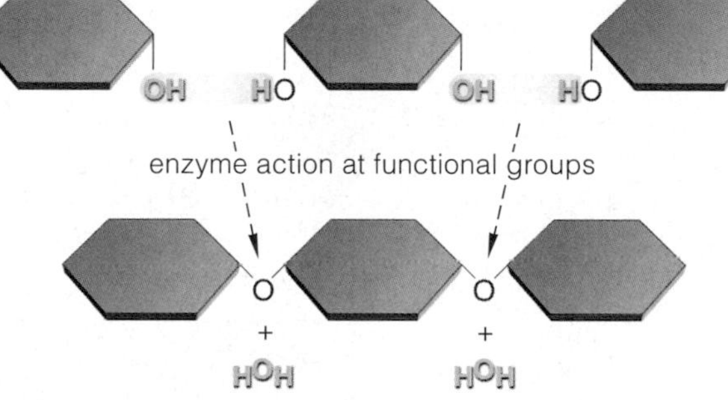

a Two condensation reactions. Enzymes remove an —OH group and H atom from two molecules, which covalently bond as a larger molecule. Two water molecules form.

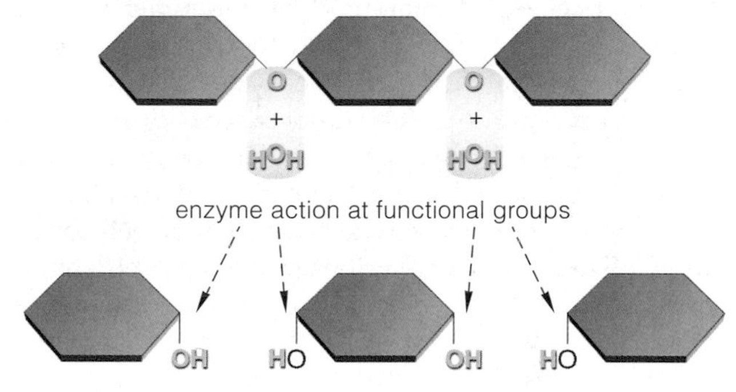

b Hydrolysis, a water-requiring cleavage reaction. Enzyme action splits a molecule into three parts, then attaches an H atom and an —OH group derived from a water molecule to each exposed site.

Figure 3.4 Examples of metabolic reactions by which most biological molecules are put together, rearranged, and broken apart.

At this point in your reading, simply become aware that the reactions by which cells build organic compounds, and even rearrange them and break them apart, require more than an energy input. They also require the class of proteins called **enzymes**, which make specific metabolic reactions proceed faster than they would on their own. Different enzymes mediate different kinds of reactions. Most of the reactions fall into five categories, which you will encounter in chapters to come:

1. *Functional-group transfer*. One molecule gives up a functional group, which another molecule accepts.

2. *Electron transfer*. One or more electrons stripped from one molecule are donated to another molecule.

3. *Rearrangement*. A juggling of its internal bonds converts one type of organic compound into another.

4. *Condensation*. Through covalent bonding, two molecules combine to form a larger molecule.

5. *Cleavage*. A molecule splits into two smaller ones.

To get a sense of what goes on, consider two examples of these events. First, in many **condensation reactions**, enzymes remove a hydroxyl group from one molecule and an H atom from another, then speed the formation of a covalent bond between the two molecules at their exposed sites (Figure 3.4*a*). As a typical but incidental outcome of the reaction, the discarded atoms join to form a molecule of water. A series of condensation reactions can produce starches and other polymers. A polymer is a large molecule having three to millions of subunits, which may or may not be identical. Biologists call the subunits monomers, as in the sugar monomers of starch.

As the second example, a cleavage reaction called **hydrolysis** is like condensation in reverse (Figure 3.4*b*). Enzymes that recognize specific functional groups split molecules into two or more parts, then attach an —OH group and a hydrogen atom derived from a molecule of water to the exposed sites. With hydrolysis, cells can cleave large polymers such as starch into smaller units when these are required for building blocks or energy.

Organic compounds have diverse, three-dimensional shapes and functions. The diversity starts at their carbon backbones and with bonding arrangements that arise from it.

Functional groups that are covalently bonded to the carbon backbone add enormously to the structural and functional diversity of organic compounds.

Carbohydrates, lipids, proteins, and nucleic acids are the main biological molecules, the organic compounds that only living cells can assemble under conditions that now occur in nature.

Cells assemble, rearrange, and degrade organic compounds mainly through enzyme-mediated reactions involving the transfer of functional groups or electrons, rearrangement of internal bonds, and a combining or splitting of molecules.

3.2 CARBOHYDRATES

Consider first the carbohydrates—the most abundant of all biological molecules, which cells use as structural materials and transportable or storage forms of energy. Most **carbohydrates** consist of carbon, hydrogen, and oxygen in a 1:2:1 ratio, which also may be written as $(CH_2O)_n$. Three classes are called the **monosaccharides**, **oligosaccharides**, and **polysaccharides**.

The Simple Sugars

"Saccharide" comes from a Greek word meaning sugar. A *mono*saccharide, meaning "one monomer of sugar," is the simplest carbohydrate. It has at least two —OH groups joined to the carbon backbone plus an aldehyde or a ketone group. Most monosaccharides are sweet tasting and readily dissolve in water. The most common have a backbone of five or six carbon atoms that tends to form a ring structure when dissolved in cells or body fluids. Ribose and deoxyribose, the sugar components of RNA and DNA, respectively, have five carbon atoms. Glucose has six (Figure 3.5*a*). Besides being the main energy source for most organisms, glucose is a precursor (parent molecule) of many compounds and a building block for larger carbohydrates. Vitamin C (a sugar acid) and glycerol (an alcohol with three —OH groups) are other examples of compounds having sugar monomers.

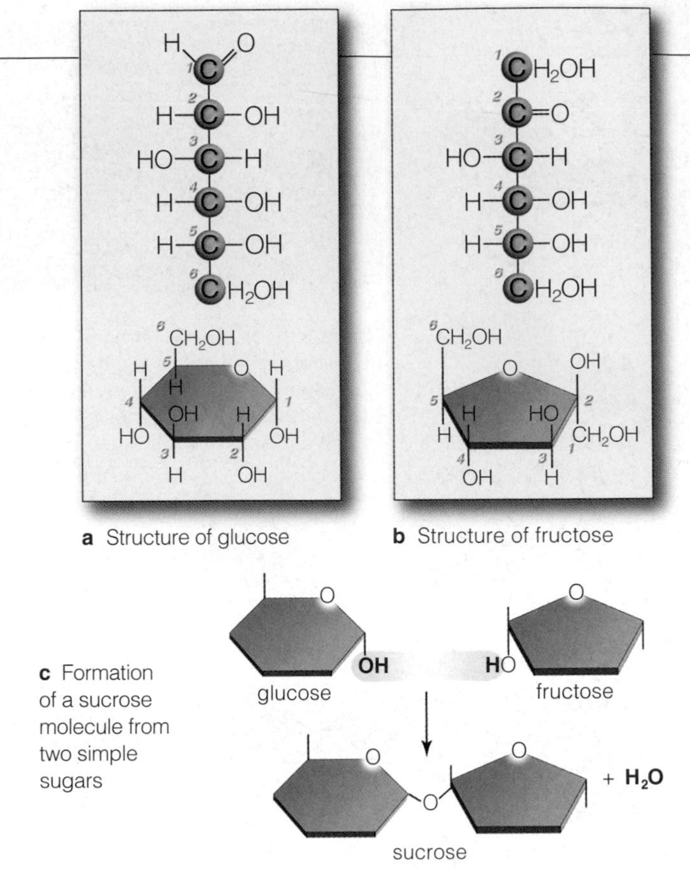

Figure 3.5 Straight-chain and ring forms of (**a**) glucose and (**b**) fructose. For reference purposes, carbon atoms of simple sugars are commonly numbered in sequence, starting at the end closest to the molecule's aldehyde or ketone group. (**c**) Condensation of two monosaccharides into a disaccharide.

Short-Chain Carbohydrates

Unlike the simple sugars, an *oligo*saccharide is a short chain of two or more sugar monomers that are bonded covalently. (*Oligo-* means "a few.") The type known as *di*saccharides consists of just two sugar units. Lactose, sucrose, and maltose are examples. Lactose (a glucose and a galactose unit) is a milk sugar. Sucrose, the most plentiful sugar in nature, consists of one glucose and one fructose unit, as in Figure 3.5*c*. Plants convert their stores of carbohydrates to sucrose, which can be easily transported through their leaves, stems, and roots. Table sugar is sucrose crystallized from sugarcane and sugar beets. Proteins and other large molecules often have oligosaccharides attached as side chains to the carbon backbone. Some chains take part in the body's defenses against disease, others in cell membrane functions.

Complex Carbohydrates

The "complex" carbohydrates, or *poly*saccharides, are straight or branched chains of many sugar monomers (often hundreds or thousands) of the same or different types. Cellulose, starch, and glycogen, the most common polysaccharides, consist only of glucose. In cell walls of plants, cellulose is a structural material (Figure 3.6*a*). Like steel rods in reinforced concrete, fibers of cellulose are tough and insoluble; they withstand considerable weight and stress. Plants store their photosynthetically produced sugars as large starch molecules (Figure 3.6*b*). Enzymes can readily hydrolyze starch to glucose units.

If both cellulose and starch consist of glucose, why are their properties so different? The answer starts with differences in covalent bonding patterns between their monomers, which in both are bonded together in chains.

In cellulose, many glucose chains stretch out side by side and hydrogen-bond to one another at —OH groups (Figure 3.7*a*). This bonding arrangement stabilizes the chains into a tightly bundled pattern that resists being digested, at least by most enzymes.

In starch, the pattern of covalent bonding positions each sugar monomer at an angle relative to the next monomer in line. The chain ends up coiling like a spiral staircase (Figure 3.7*b*). The coils are not particularly stable. In the kinds of starches with branched chains, they are even less stable. A great many —OH groups project outward from the coiled chains, and this makes them readily accessible to enzymes.

In animals, glycogen is the sugar-storage equivalent of starch in plants. Muscle and liver cells house large stores of it. When the level of sugar in blood decreases, liver cells degrade glycogen, so glucose is released and

Further reading: Student Guide to InfoTrac on web site →

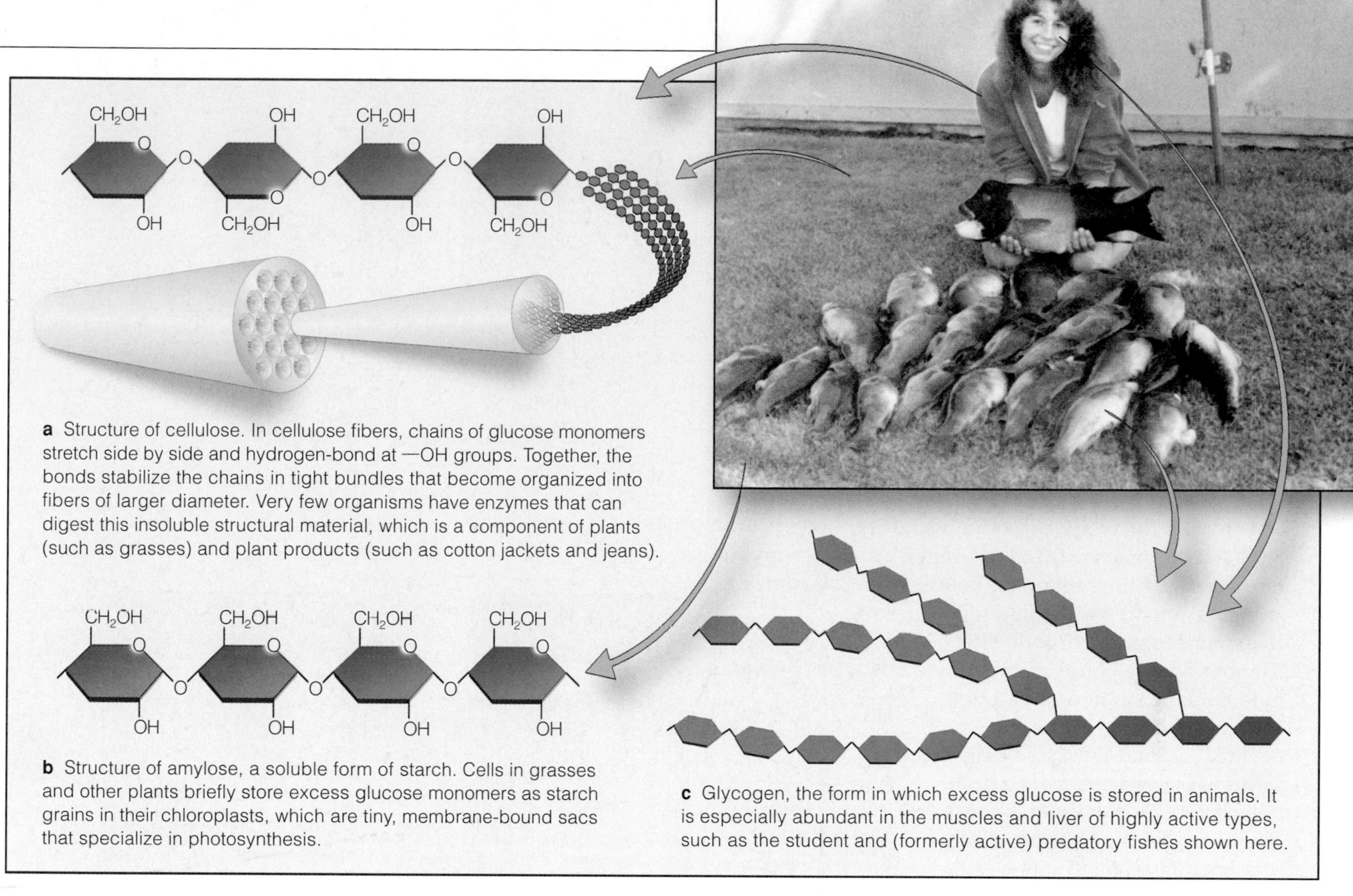

a Structure of cellulose. In cellulose fibers, chains of glucose monomers stretch side by side and hydrogen-bond at —OH groups. Together, the bonds stabilize the chains in tight bundles that become organized into fibers of larger diameter. Very few organisms have enzymes that can digest this insoluble structural material, which is a component of plants (such as grasses) and plant products (such as cotton jackets and jeans).

b Structure of amylose, a soluble form of starch. Cells in grasses and other plants briefly store excess glucose monomers as starch grains in their chloroplasts, which are tiny, membrane-bound sacs that specialize in photosynthesis.

c Glycogen, the form in which excess glucose is stored in animals. It is especially abundant in the muscles and liver of highly active types, such as the student and (formerly active) predatory fishes shown here.

Figure 3.6 Molecular structure of cellulose, starch, and glycogen, and their typical locations in a few organisms. All three carbohydrates consist only of glucose monomers.

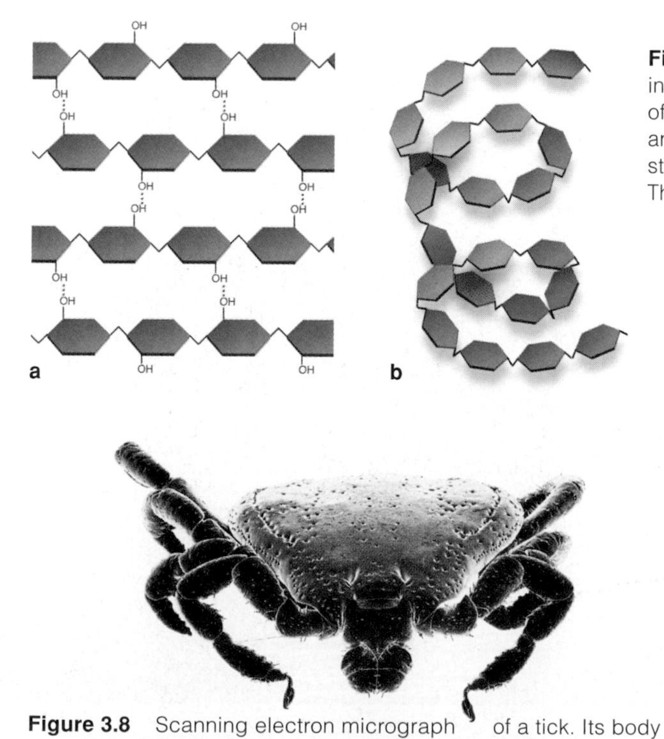

Figure 3.7 Comparison of bonding patterns between glucose monomers in cellulose and in starch. (**a**) In cellulose, bonds form between monomers of neighboring glucose chains. This bonding pattern stabilizes the chains and allows them to become tightly bundled. (**b**) In amylose, a form of starch, the bonding pattern between monomers causes the chains to coil. The coiling orients the bonds in a way that is accessible to enzymes.

Figure 3.8 Scanning electron micrograph of a tick. Its body covering is a protective cuticle reinforced with chitin.

enters the bloodstream. When you exercise strenuously but briefly, your muscle cells tap their glycogen stores for a rapid burst of energy. Figure 3.6*c* represents a few of glycogen's many branchings.

The polysaccharide chitin has nitrogen-containing groups attached to its glucose monomers. It is the main structural material for the external skeletons and other hard body parts of crabs, earthworms, insects, ticks, and many other animals (Figure 3.8). Chitin also is the main structural material in the cell walls of many fungi.

The simple sugars (such as glucose), oligosaccharides, and polysaccharides (such as starch) are carbohydrates. Every cell requires carbohydrates as structural materials, stored forms of energy, and transportable packets of energy.

3.3 LIPIDS

Being mostly hydrocarbon, **lipids** show little tendency to dissolve in water, although they readily dissolve in nonpolar substances. All are greasy or oily to the touch. Cells use different kinds as their main energy reservoirs, as structural materials (for example, in cell membranes and surface coatings), and as signaling molecules. Let's look first at the kinds with fatty acid components—the fats, phospholipids, and waxes. We also will consider the sterols, each with a backbone of four carbon rings.

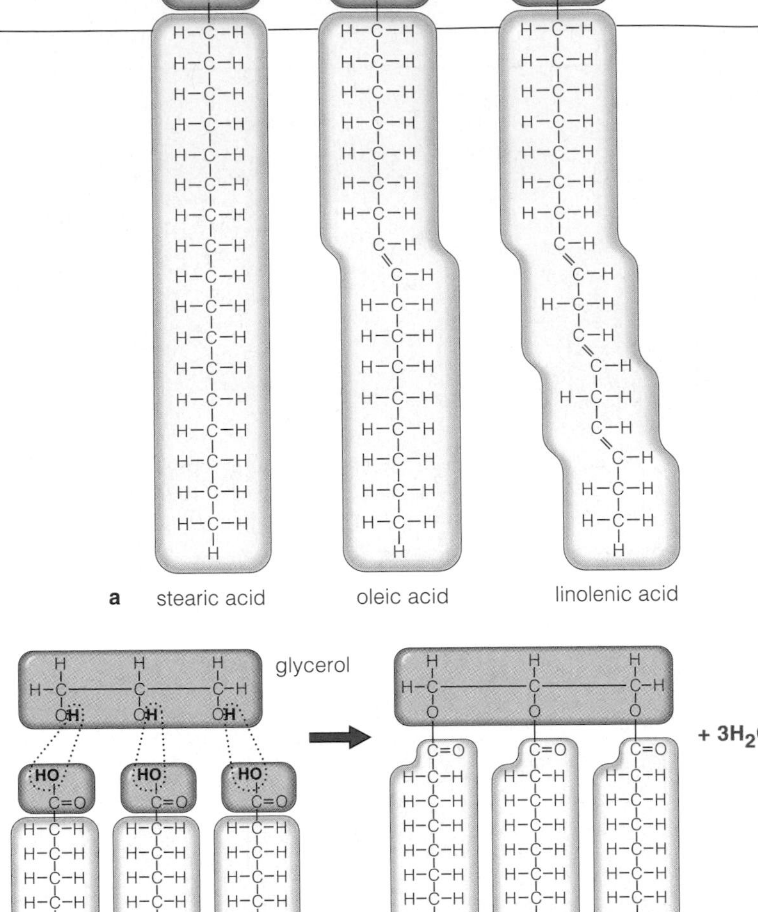

Fats and Fatty Acids

Lipids called **fats** have one, two, or three fatty acids attached to glycerol. Each **fatty acid** has a backbone of as many as thirty-six carbon atoms, a carboxyl group (—COOH) at one end, and hydrogen atoms occupying most or all of the remaining bonding sites. It typically stretches out like a flexible tail. *Unsaturated* tails incorporate one or more double bonds. *Saturated* tails contain single bonds only. Figure 3.9*a* shows examples.

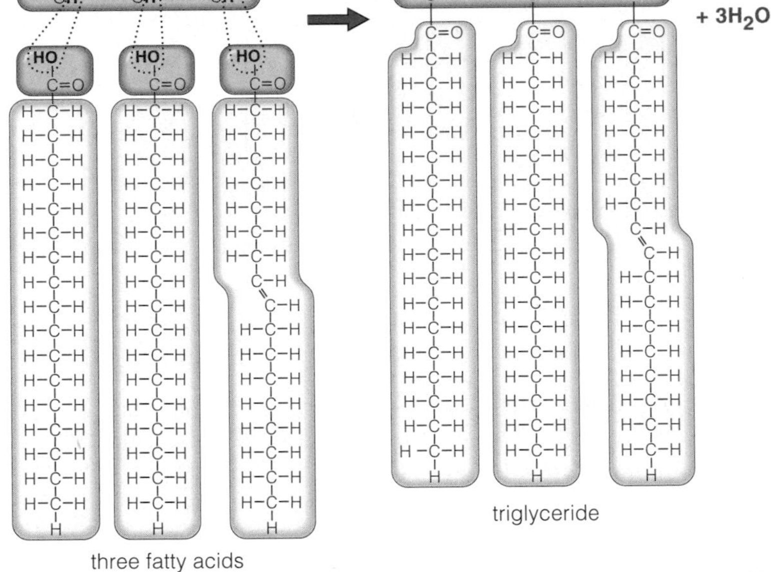

Figure 3.9 (**a**) Structural formulas for three fatty acids. In stearic acid, the carbon backbone is fully saturated with hydrogen atoms. Oleic acid, with a double bond in its backbone, is an unsaturated fatty acid. Linolenic acid, with its three double bonds, is a "polyunsaturated" fatty acid. (**b**) Condensation of fatty acids and glycerol into a triglyceride.

Most animal fats have many saturated fatty acids, which pack together by weak interactions. They stay solid at temperatures that keep most plant fats liquid, as "vegetable oils." The packing interactions in plant fats are not as stable because of rigid kinks in their fatty acid tails. That is why vegetable oils flow freely.

Butter, lard, vegetable oils, and other natural fats consist mostly of **triglycerides**: neutral fats having three fatty acid tails attached to glycerol (Figure 3.9*b*). Triglycerides are the body's most abundant lipids and its richest energy source. Gram for gram, they yield more than twice as much energy when broken down, compared to starches and other complex carbohydrates. Notable quantities of triglycerides are stored as droplets in the cells of body fat (that is, adipose tissue) in vertebrates. A thick layer of triglycerides under the skin has survival value for some animals, including the penguins shown in Figure 3.10. It helps insulate their body against near-freezing temperatures of their habitats.

Figure 3.10 Triglyceride-protected penguins taking the plunge.

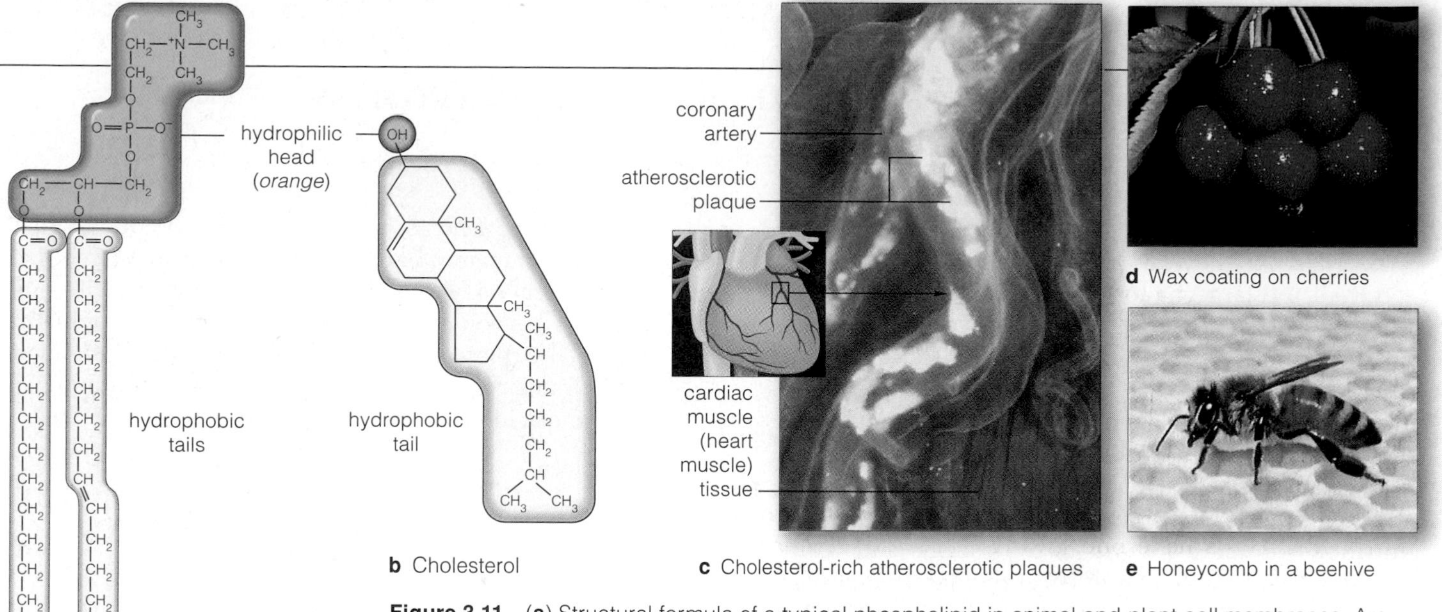

Figure 3.11 (**a**) Structural formula of a typical phospholipid in animal and plant cell membranes. Are its hydrophobic tails saturated or unsaturated? (**b**) Structural formula of cholesterol, the major sterol of animal tissues. (**c**) Your liver synthesizes enough cholesterol for the body. A fat-rich diet may result in excessively high cholesterol levels in blood and the formation of abnormal masses of material in certain blood vessels (arteries). Such *atherosclerotic plaques* may clog the arteries that deliver blood to the heart. (**d**) Demonstration of the water-repelling attribute of a cherry cuticle. (**e**) Honeycomb, a structural material constructed of a firm, water-repellent, waxy secretion called beeswax.

One fatty acid is a precursor of eicosanoids, a class of local signaling molecules that include prostaglandins. Different eicosanoids bind to receptors on cells and help regulate many physiological processes, such as muscle contraction, message transmission through the nervous system, inflammation, and immune responses.

Phospholipids

A **phospholipid** has a glycerol backbone, two fatty acid tails, and a hydrophilic "head" with a phosphate group and another polar group (Figure 3.11*a*). Phospholipids are the main materials of cell membranes, which have two layers of lipids. Heads of one layer are dissolved in the cell's fluid interior, and heads of the other layer are dissolved in the surroundings. Sandwiched between the two are all the fatty acid tails, which are hydrophobic.

Sterols and Their Derivatives

Sterols are among the many lipids that have no fatty acids. Sterols differ in the number, position, and type of their functional groups, but all have a rigid backbone of four fused-together carbon rings:

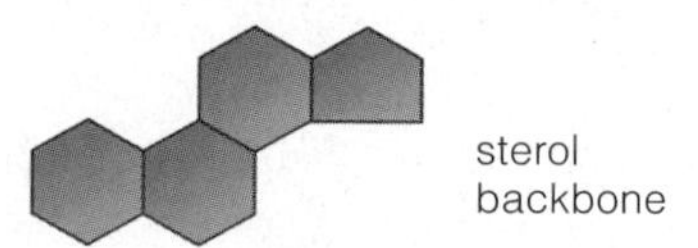

Sterols occur in eukaryotic cell membranes. Cholesterol is the most common type in tissues of animals (Figure 3.11*b*,*c*). Cells also remodel cholesterol into compounds such as vitamin D (required for good bones and teeth), steroids, and bile salts. Steroids include sex hormones, such as estrogens and testosterone, that govern sexual traits and gamete formation. Bile salts have roles in the digestion of fats in the small intestine.

Waxes

The lipids called **waxes** have long-chain fatty acids tightly packed and linked to long-chain alcohols or to carbon rings. They have a firm consistency and repel water. Waxes and another lipid, cutin, make up most of the cuticle that covers the aboveground plant parts. The covering helps plants conserve water and fend off some parasites. A waxy cherry cuticle is an example (Figure 3.11*d*). In many animals, waxy secretions from cells are incorporated in coatings that protect, lubricate, and impart pliability to skin or hair. Among waterfowl and other birds, wax secretions help keep feathers dry. As another example, beeswax is the material of choice when bees construct their honeycombs (Figure 3.11*e*).

Being largely hydrocarbon, lipids can dissolve in nonpolar substances but resist dissolving in water.

Triglycerides (neutral fats), which have a glycerol head and three fatty acid tails, are the body's main energy reservoirs. Phospholipids are the main components of cell membranes.

Sterols such as cholesterol are membrane components as well as precursors of steroid hormones and other compounds. Waxes are firm yet pliable components of water-repelling and lubricating substances.

3.4 AMINO ACIDS AND THE PRIMARY STRUCTURE OF PROTEINS

Have you ever wondered how permanent waves work (Figure 3.12)? The characteristics of many mammalian body parts—including hair—start with the structure of proteins, just as they do for all other organisms.

Of all the large biological molecules, **proteins** are the most diverse. Proteins of the class called enzymes make metabolic events proceed much faster than they otherwise would. Structural proteins are the stuff of spider webs, butterfly wings, feathers, cartilage and bone, and a dizzying array of other body parts and products. Transport proteins move molecules and ions across cell membranes and cart them about through body fluids. Nutritious proteins abound in milk, eggs, and many seeds. Protein hormones and other regulatory types are signals for change in cell activities. Many proteins act as weapons against disease-causing bacteria and other invaders. Amazingly, cells build diverse proteins from their pools of only twenty kinds of amino acids!

Figure 3.12 Appearance of the hair of actress Nicole Kidman (**a**) before and (**b**) after a permanent wave. The difference, as you will see from this section and the next, starts with strings of amino acids in polypeptide chains.

Amino Acid Structure

Every **amino acid** is a small organic compound that consists of an amino group, a carboxyl group (an acid), a hydrogen atom, and one or more atoms known as its R group. As you can see from the structural formula in Figure 3.13, these parts generally are covalently bonded to the same carbon atom. Figure 3.14 shows a number of amino acids that we will consider later in the book.

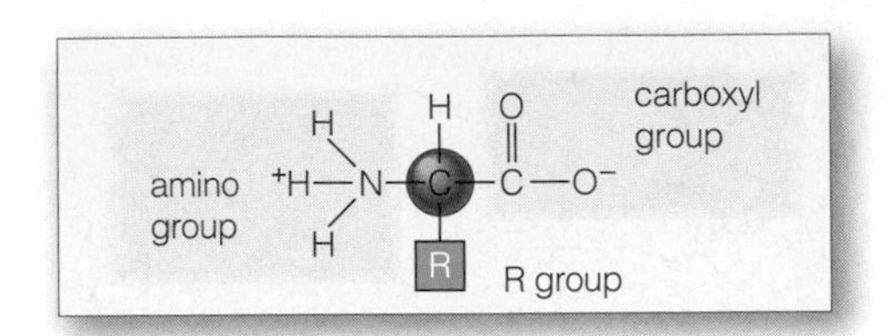

Figure 3.13 Generalized structural formula for amino acids.

What Is A Protein's Primary Structure?

When a cell synthesizes a protein, amino acids become linked, one after the other, by peptide bonds. As Figure 3.15 shows, this is the type of covalent bond that forms between one amino acid's amino group (NH_3^+) and the carboxyl group ($-COO^-$) of the next amino acid.

When peptide bonds join two amino acids together, we have a dipeptide. When they join three or more, we have a **polypeptide chain**. In such chains, the carbon

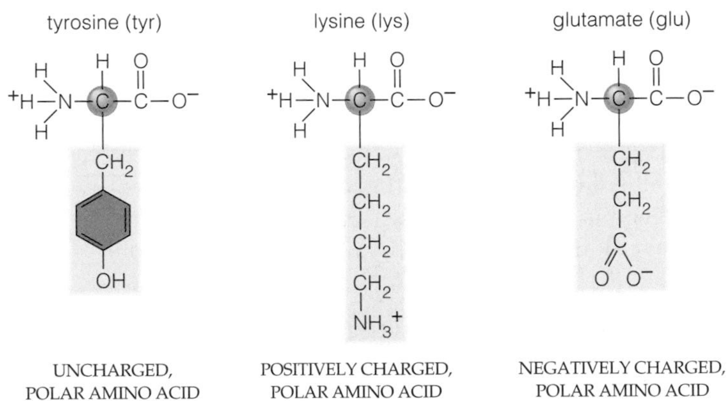

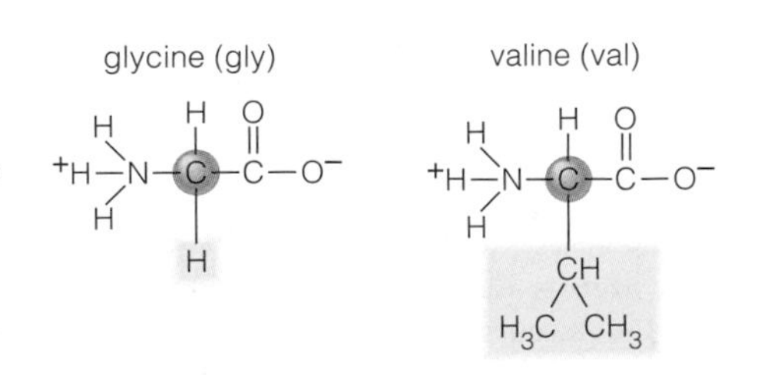

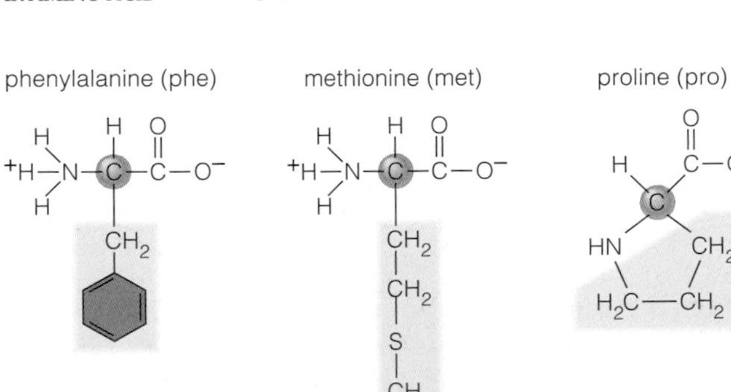

Figure 3.14 Structural formulas for eight of the twenty common amino acids. *Green* boxes highlight the R groups, which are side chains with functional groups. Each type of side chain contributes in a major way to the distinctive properties of each amino acid.

Further reading: Student Guide to InfoTrac on web site

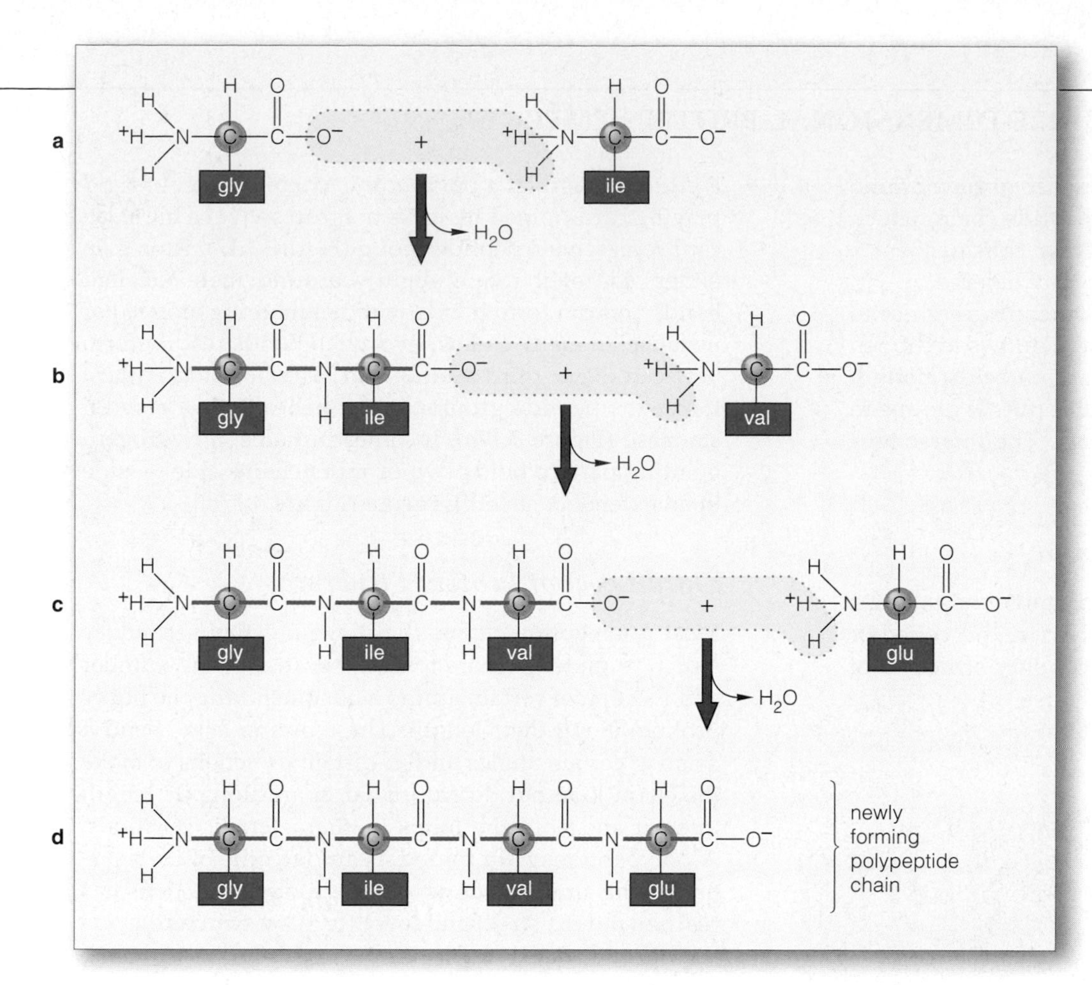

Figure 3.15 Peptide bond formation during protein synthesis.

(**a**) The first two amino acids shown are glycine (gly) and isoleucine (ile). They are at the start of the sequence for one of two polypeptide chains that make up the protein insulin in cattle.

(**b**) Through a condensation reaction, the isoleucine becomes joined to the glycine by a peptide bond. A water molecule forms as a by-product of the reaction.

(**c**) A peptide bond forms between isoleucine and valine (val), another amino acid, and water again forms.

(**d**) Remember, DNA specifies the order in which the different kinds of amino acids follow one another in a growing polypeptide chain. In this case, glutamate (glu) is the fourth amino acid specified. Ultimately, the result is one of the two polypeptide chains of an insulin molecule, as in Figure 3.16. Chapter 14 has more details on the steps that lead from DNA instructions to proteins.

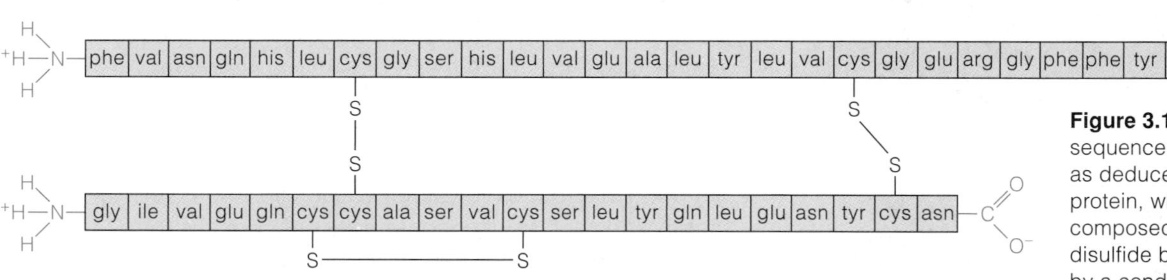

Figure 3.16 Diagram of the amino acid sequence for an insulin molecule (in cattle), as deduced by Frederick Sanger. This protein, which functions as a hormone, is composed of two polypeptide chains. Two disulfide bridges (—S—S—), each formed by a condensation reaction at two R groups (sulfhydryl groups), link the chains together.

backbone has nitrogen atoms positioned in this regular pattern: —N—C—C—N—C—C—.

For each particular kind of protein, different amino acid units are selected one at a time from the twenty kinds available. Their orderly progression is prescribed by the cell's DNA. Overall, the resulting sequence of amino acids is unique for each kind of protein, and it is the protein's *primary* structure (Figures 3.15 and 3.16).

Now consider this: Different cells make thousands of different proteins. Many of the proteins are fibrous, with polypeptide chains organized as strands or sheets. Collectively, many such molecules contribute to the shape and internal organization of cells. Other kinds of proteins are globular, with one or more polypeptide chains folded in compact, rather rounded shapes. Most enzymes are globular proteins. So are actin and other proteins that contribute to cell movement.

Regardless of the type of protein, its shape and its function arise from the primary structure—that is, from information built into its amino acid sequence. As you will see next, that information dictates which parts of a polypeptide chain will coil, bend, or interact with other chains nearby. And the type and arrangement of atoms in coiled, stretched-out, or folded regions determine whether that protein will function as, say, an enzyme, a transporter, a receptor, or even an inadvertent target for a bacterium or virus.

A protein consists of one or more polypeptide chains in which amino acids are strung together. The amino acid sequence (which kind of amino acid follows another in the chain) is unique for each kind of protein and gives rise to its unique structure, chemical behavior, and function.

3.5 HOW DOES A THREE-DIMENSIONAL PROTEIN EMERGE?

The preceding section gave you a sense of how amino acids are strung together in a polypeptide chain, which is a protein's primary structure. Now consider just a few examples of the protein shapes that emerge.

For the most part, the primary structure gives rise to a protein's shape in two ways. First, it allows hydrogen bonds to form between different amino acids along a polypeptide chain's length. Second, it puts R groups in positions that allow them to interact. The interactions force the chain to bend and twist.

Second Level of Protein Structure

Hydrogen bonds form at regular, short intervals along a new polypeptide chain, and they give rise to a coiled or extended pattern known as the secondary structure of a protein. Think of a polypeptide chain as a set of rigid playing cards joined by links that can swivel a bit. Each card represents a peptide group (Figure 3.17). Atoms on either side of it rotate slightly around their covalent bonds and can form bonds with neighboring atoms. For instance, in many chains, hydrogen bonds readily form between every third amino acid. The bonding pattern forces the peptide groups to coil helically, like a spiral staircase (Figure 3.17*a*). In other proteins, a hydrogen-bonding pattern holds two or more chains side by side in an extended, sheetlike array (Figure 3.17*b*).

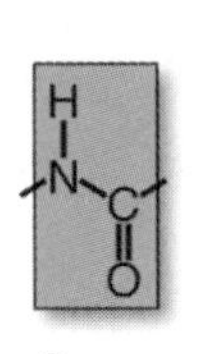

a

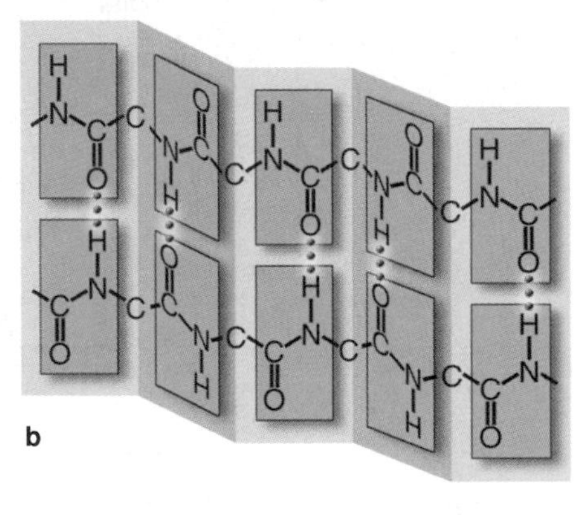

Figure 3.17 Bonding patterns between peptide groups of polypeptide chains. Extensive hydrogen bonding (*dotted* lines) can bring about (**a**) coiling of one chain or (**b**) sheetlike arrays of two or more chains.

Third Level of Protein Structure

Most polypeptide chains that have a coiled secondary structure undergo more folding because of the number and location of certain amino acids (including the bulky proline) along their length. These amino acids bend a chain at certain angles and in certain directions to make the chain loop out. R groups far apart along its length interact and hold the loops in characteristic positions. A polypeptide chain folded as an outcome of its bend-producing amino acids and its R-group interactions has reached a third structural level (tertiary structure).

Figure 3.18*a* shows how a polypeptide chain became folded into the compact, tertiary structure of the protein globin. Hydrogen bonds between particular functional groups along the chain's length caused the folding.

Fourth Level of Protein Structure

Imagine that bonds form between *four* molecules of globin and that an iron-containing functional group, a heme group, is positioned near the center of each. The outcome is **hemoglobin**, an oxygen-transporting protein

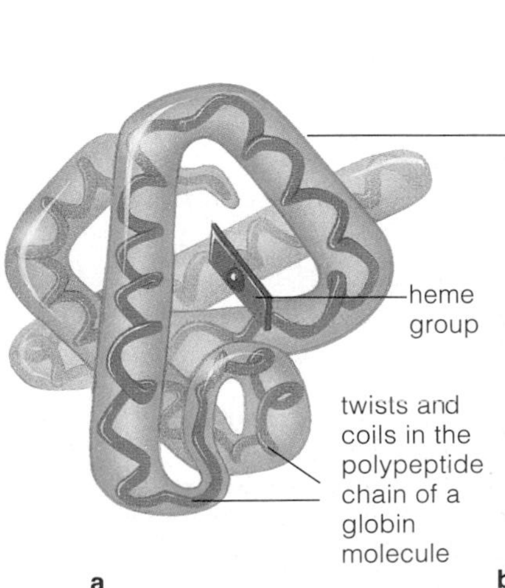

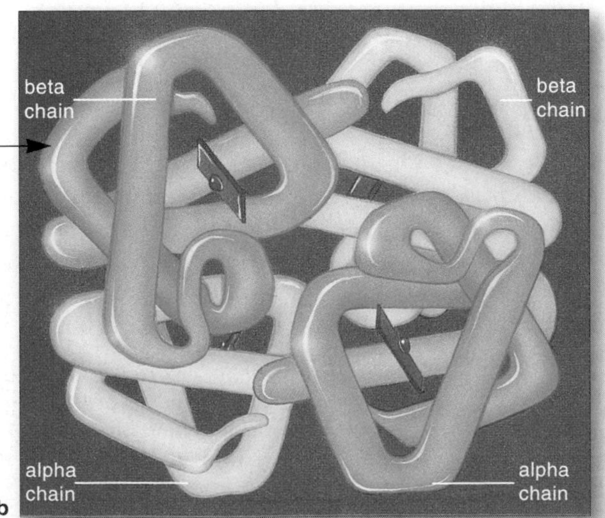

Figure 3.18 (**a**) Globin molecule. This coiled polypeptide chain associates with a heme group, an iron-containing group that strongly attracts oxygen. There is a whole family of globin molecules, identical in most parts of their amino acid sequences but unique in other parts. In this diagram, the artist drew a transparent "green noodle" around each chain to help you visualize how it folds in three dimensions.

(**b**) Hemoglobin, a protein that transports oxygen in blood, consists of four polypeptide chains (globin molecules) and four heme groups. Two chains, designated *alpha*, differ slightly from the other two (the *beta* chains) in their amino acid sequence. To keep this sketch from looking like a tangle of noodles, the two chains in the background are tinted differently from those in the foreground.

Further reading: Student Guide to InfoTrac on web site →

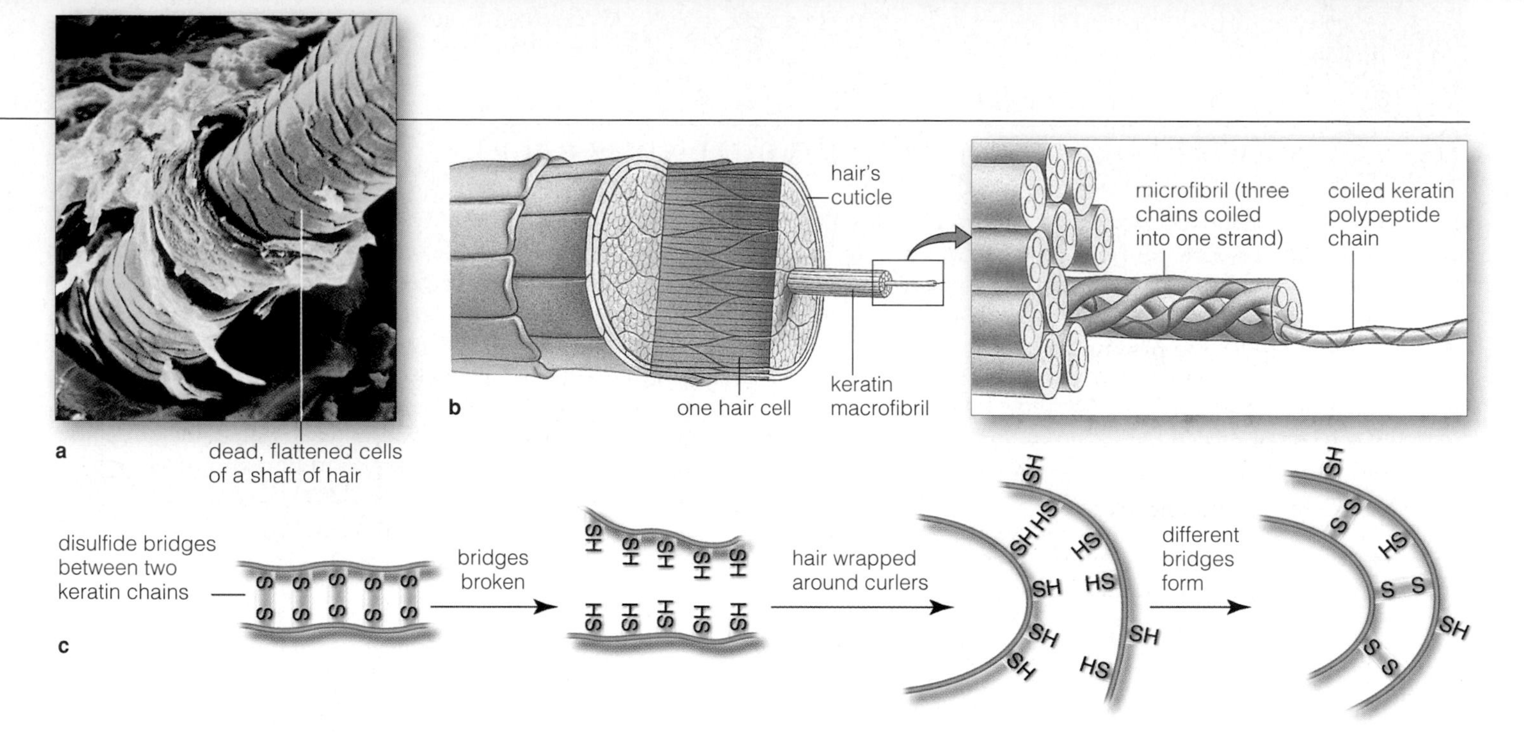

Figure 3.19 Structure of hair. (**a**,**b**) Hair cells develop from modified skin cells. They synthesize polypeptide chains of the protein keratin. Disulfide bridges link three chains together as fine fibers, which get bundled into larger, cable-like fibers. The larger fibers almost fill the cells, which eventually die off. Dead, flattened cells form a tubelike cuticle around the developing hair shaft.

(**c**) For a permanent wave, hair is exposed to chemicals that break the disulfide bridges. When hairs wrap around curlers, their polypeptide chains are held in new positions. Now exposure to a different chemical causes new disulfide bridges to form. But they form between different sulfur-bearing amino acids than before. The displaced bonding locks the hair in curled positions (compare Figure 3.12).

(Figure 3.18*b*). As you read this, each of your mature red blood cells is transporting a billion molecules of oxygen, bound to 250 million hemoglobin molecules.

Hemoglobin is at the fourth (quaternary) level of protein structure. In every protein at this level, two or more polypeptide chains are joined by numerous weak interactions (such as hydrogen bonds) and sometimes by covalent bonds between sulfur atoms of R groups.

Proteins with quaternary structure are globular or fibrous. Hemoglobin is one of the **globular proteins**, which have one or more polypeptide chains folded in a rounded shape. Most enzymes are globular proteins, also. The **fibrous proteins** are long strands or sheets of polypeptide chains. Examples are keratin (Figure 3.19) and collagen, the most common animal protein. Skin, bone, corneas, arteries, and many other animal body parts depend on the strength inherent in collagen.

Glycoproteins and Lipoproteins

Some proteins have other organic compounds attached to their polypeptide chains. For example, lipoproteins form when certain proteins circulating in blood combine with cholesterol, triglycerides, and phospholipids that were absorbed from the gut after a meal. Similarly, most glycoproteins have linear or branched oligosaccharides bonded to them. Nearly all the proteins at the surface of animal cells are glycoproteins. So are most protein secretions from cells and many proteins in blood.

Structural Changes by Denaturation

Breaking weak bonds of a protein or any other large molecule disrupts its three-dimensional shape, an event called **denaturation**. For example, weak hydrogen bonds are sensitive to increases or decreases in temperature and pH. If the temperature or pH exceeds a protein's range of tolerance, its polypeptide chains will unwind or change shape, and the protein will lose its function. Consider the protein albumin, concentrated in the "egg white" of uncooked chicken eggs. When you cook eggs, the heat does not disrupt the strong covalent bonds of albumin's primary structure. But it destroys weaker bonds contributing to the three-dimensional shape. For some proteins, denaturation might be reversed when normal conditions are restored—but albumin isn't one of them. There is no way to uncook a cooked egg.

Proteins have a primary structure, which is the sequence of different kinds of amino acids along a polypeptide chain. An individual's DNA specifies that sequence.

Proteins have a secondary structure: a coiled pattern or an extended, sheetlike pattern that arises by hydrogen bonding at short, regular intervals along a polypeptide chain.

At the third level of protein structure, bonding at certain amino acids makes a coiled chain bend and loop at certain angles, in certain directions. Interactions among R groups in the chain hold the loops in characteristic positions.

At the fourth level of protein structure, numerous hydrogen bonds and other interactions join two or more polypeptide chains. Many proteins are this structurally complex.

3.6 FOOD PRODUCTION AND A CHEMICAL ARMS RACE

The next time you shop for groceries, reflect on what it takes to provide you with your daily supply of organic compounds. For example, those heads of lettuce typically grew in fertilized cropland. Possibly they competed with weeds, which didn't know the nutrients in the fertilizer were meant for the lettuce. Leaf-chewing insects didn't know the lettuce wasn't meant to be their salad bar. Each year, those food pirates and others ruin or gobble up nearly half of what people all over the world try to grow.

Most plants aren't entirely defenseless. They evolved under the intense selection pressures of attacks by insects, fungi, and other organisms, and in natural settings they often can repel the attackers with toxins. A **toxin** is an organic compound, a normal metabolic product of one species, but its chemical effects can harm or kill individuals of a different species that come in contact with it. Humans, too, encounter traces of natural toxins in most of what they eat, even in such familiar edibles as hot peppers, potatoes, figs, celery, rhubarb, and alfalfa sprouts. Still, we do not die in droves from these natural toxins, so apparently our bodies have chemical defenses against them.

Well over 2,500 years ago, farmers realized that sulfur kills insects. About a thousand years later, desperate but misguided farmers were dispensing arsenic, lead, and mercury through croplands. Toxic metal applications continued until the late 1920s, when someone made the connection between the toxins and human poisonings and deaths. These toxins resist breakdown, and we still find traces of them in vegetables and other plants grown on contaminated lands.

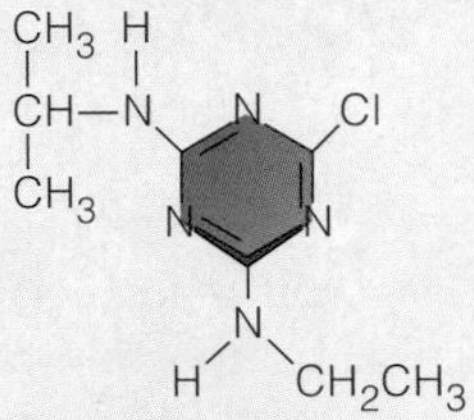

b Atrazine. Farmers spray this herbicide directly on foliage of corn and other crop plants. It kills weeds within a few days but does not affect the plants. Other herbicides with low persistence include glyphosate (Roundup), daminozide (Alar), and alachlor (Lasso).

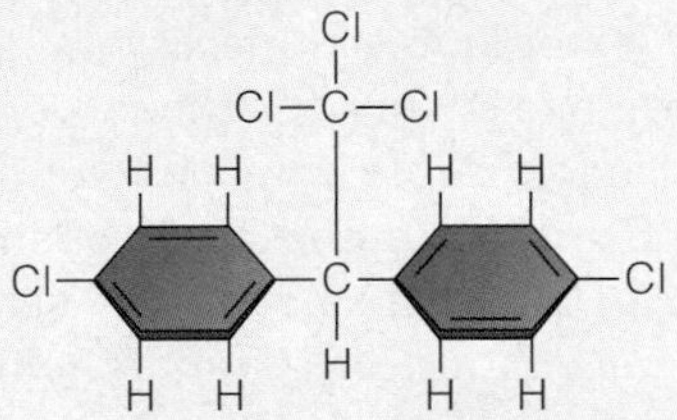

c DDT (dichlorodiphenyltrichloroethane). This nerve cell poison does not break down until two to fifteen years after its application. Other high-persistence insecticides include chlordane.

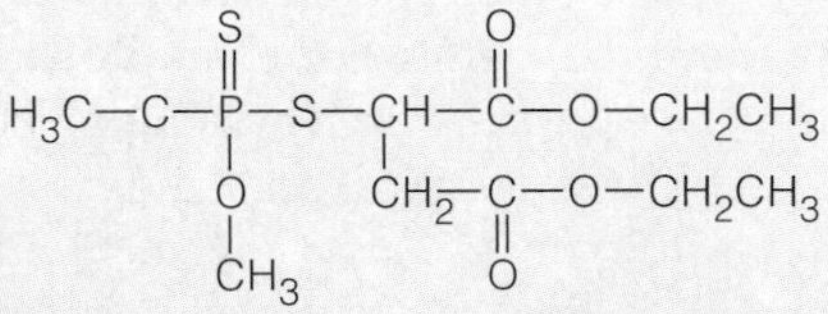

d Malathion. Like other organophosphates, it is a nerve cell poison that persists for one to twelve weeks only. Although organophosphates break down more rapidly than chlorinated hydrocarbons, they are much more toxic. About four dozen organophosphates are cheap and effective. In the United States they account for half of all insecticide uses. The Environmental Protection Agency wants to minimize pesticide residues on food. At this writing methyl parathion has been banned for food crops; azenphos methyl applications must end at least three weeks before harvest time. Farmers are contesting these actions and want the EPA to consider economic and trade issues as well as human health.

a

Figure 3.20 (**a**) Lettuce, a common crop plant, and a low-flying crop duster with its rain of pesticides. (**b–d**) A few of the major pesticides, some more toxic than others. (**e**) And think about *that* the next time you sidle up to a salad bar.

Further reading: Student Guide to InfoTrac on web site →

During the 1600s, farmers took a cue from plants and started using an *organic* compound extracted from tobacco leaves as a natural pesticide. By the mid-1800s, toxins from chrysanthemum flowers and from the roots of some tropical forest plants were added to the arsenal. Then, in 1945, scientists started developing synthetic toxins to protect crops, food stores, freshwater supplies, pets, and ornamental plants. And they started to understand the mechanisms by which toxins exert their effects.

The *herbicides* kill weeds by disrupting metabolism and growth. Most of the *insecticides* clog a target insect's airways, disrupt functioning of its nerves and muscles, or block its reproduction. *Fungicides* work against harmful fungi, including a mold that makes aflatoxin, one of the deadliest poisons. By 1995, people in the United States were spraying or spreading more than 1.25 billion pounds of toxins through fields, gardens, homes, and commercial and industrial sites (Figure 3.20).

e

There are downsides to pesticide applications. Right along with the pests, some toxic organic compounds kill birds and other predators which, in natural settings, help control pest population sizes. Besides, targeted pests have been developing resistance to the chemical arsenal for reasons that you read about earlier, in Chapter 1.

Also, pesticides cannot be released haphazardly into the environment because they can be inhaled, ingested with food, or absorbed through the skin. Different types stay active for weeks or years. Some trigger rashes, hives, sickening headaches, asthma, and joint pain in millions of people. And some trigger life-threatening allergic reactions in abnormally sensitive individuals.

Presently, DDT and other long-lived pesticides are banned in the United States. Even rapidly degradable ones, such as malathion and other organophosphates, are subjected to rigorous application standards and ongoing safety tests. With so many pests around us, however, many farmers resist limits on applications (Figure 3.20*d*).

3.7 NUCLEOTIDES

Nucleotides are small organic compounds with major roles in metabolism. Each consists of a sugar, at least one phosphate group, and a base. The sugar is either ribose or deoxyribose. Both sugars have a five-carbon ring structure. The only difference is that ribose has an oxygen atom attached to carbon 2 in its ring structure and deoxyribose does not. The bases have a single or double carbon ring structure that incorporates nitrogen.

One nucleotide, **ATP** (adenosine triphosphate), has a string of three phosphate groups attached to its sugar component (Figure 3.21*a*). ATP can readily transfer a phosphate group to many other molecules inside cells, and the acceptor molecules become energized enough to enter a reaction. Although an ATP molecule acquires its phosphate groups at certain reaction sites, it is able to deliver them to nearly all other reaction sites in a cell. In this respect ATP is central to metabolism.

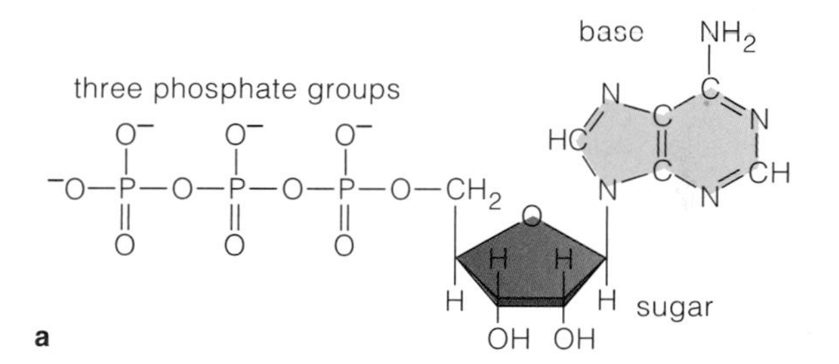

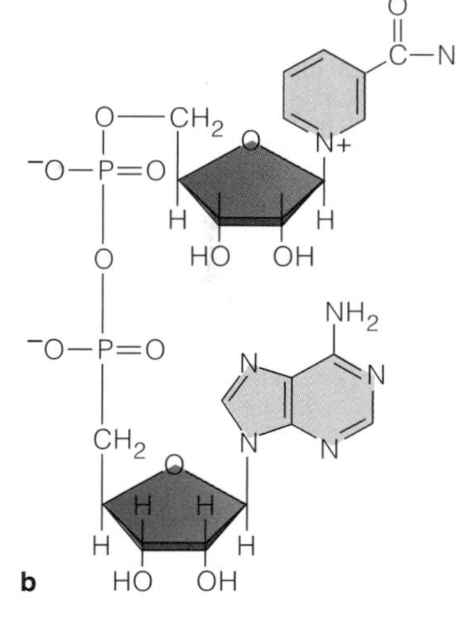

Figure 3.21 Structural formulas for (**a**) ATP and (**b**) NAD^+. Both of these nucleotides have central roles in cell metabolism.

Other nucleotides are subunits of **coenzymes**, enzyme helpers that can accept hydrogen atoms plus electrons stripped away from molecules at a reaction site and then transfer them elsewhere. One coenzyme is NAD^+ (short for nicotinamide adenine dinucleotide; see Figure 3.21*b*). FAD (flavin adenine dinucleotide) is another kind. Still other nucleotides function as chemical messengers between cells and within the cell interior. You will read about one (cyclic adenosine monophosphate, or cAMP) later in the book. As you will see, nucleotides also are building blocks for the large, single- or double-stranded molecules known as nucleic acids.

Nucleotides serve as energy carriers, chemical messengers, and subunits for coenzymes and for nucleic acids.

3.8 NUCLEIC ACIDS

As you read in the preceding section, some nucleotides carry out specific tasks for metabolic reactions. A few kinds have a different—and absolutely vital—function in all cells. These particular nucleotides serve as the monomers for single- or double-stranded molecules classified as **nucleic acids**. In such strands, a covalent bond connects the sugar component of one nucleotide with the phosphate group of the next nucleotide in the sequence (Figure 3.22*a*).

All cells start out life and then maintain themselves by way of instructions they inherited in some number of double-stranded molecules of deoxyribonucleic acid, or **DNA**. This nucleic acid consists of four different kinds of nucleotides, the structural formulas for which are shown in Figure 3.22*b*. The four differ only in their component base, which is adenine, guanine, thymine, or cytosine. The sequence of bases encodes heritable information about how to synthesize all of the diverse proteins that give each new cell the potential to grow, maintain itself, and reproduce. The particular bases that occur in at least some portions of the sequence are unique to each species.

Figure 3.22*c* shows how hydrogen bonds between bases join the two strands together along the length of a DNA molecule. Think of each "base pair" as one rung of a ladder, and the two sugar–phosphate backbones as the ladder's two posts. The ladder twists spirally, in the shape of a double helix.

Like DNA, the **RNAs** (ribonucleic acids) consist of four kinds of nucleotide monomers. Unlike DNA, the bases are adenine, guanine, cytosine, and uracil. Also

Figure 3.22 (**a**) Simplified example of how a series of nucleotides is covalently bonded in a single strand of DNA or RNA. (**b**) The four kinds of nucleotides that can serve as monomers for DNA molecules. Two nucleotide bases, adenine and guanine, have a double ring structure. The other two nucleotide bases, thymine and cytosine, have a single ring structure. (**c**) DNA double helix.

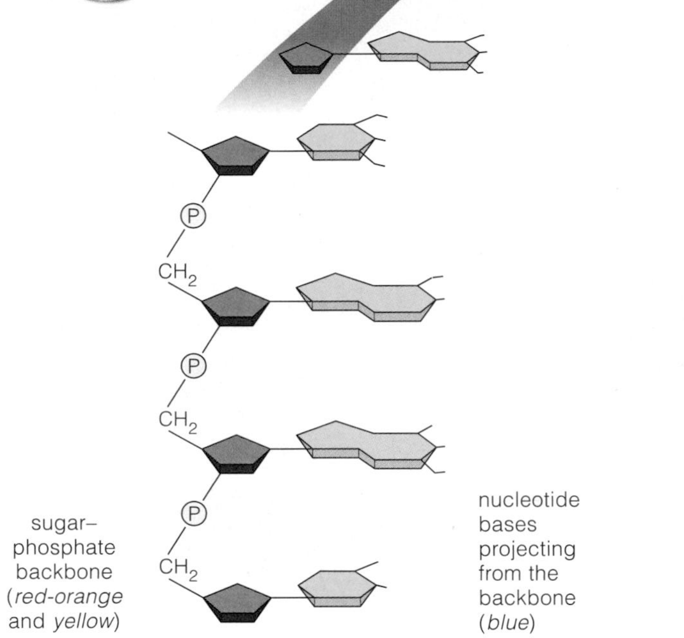

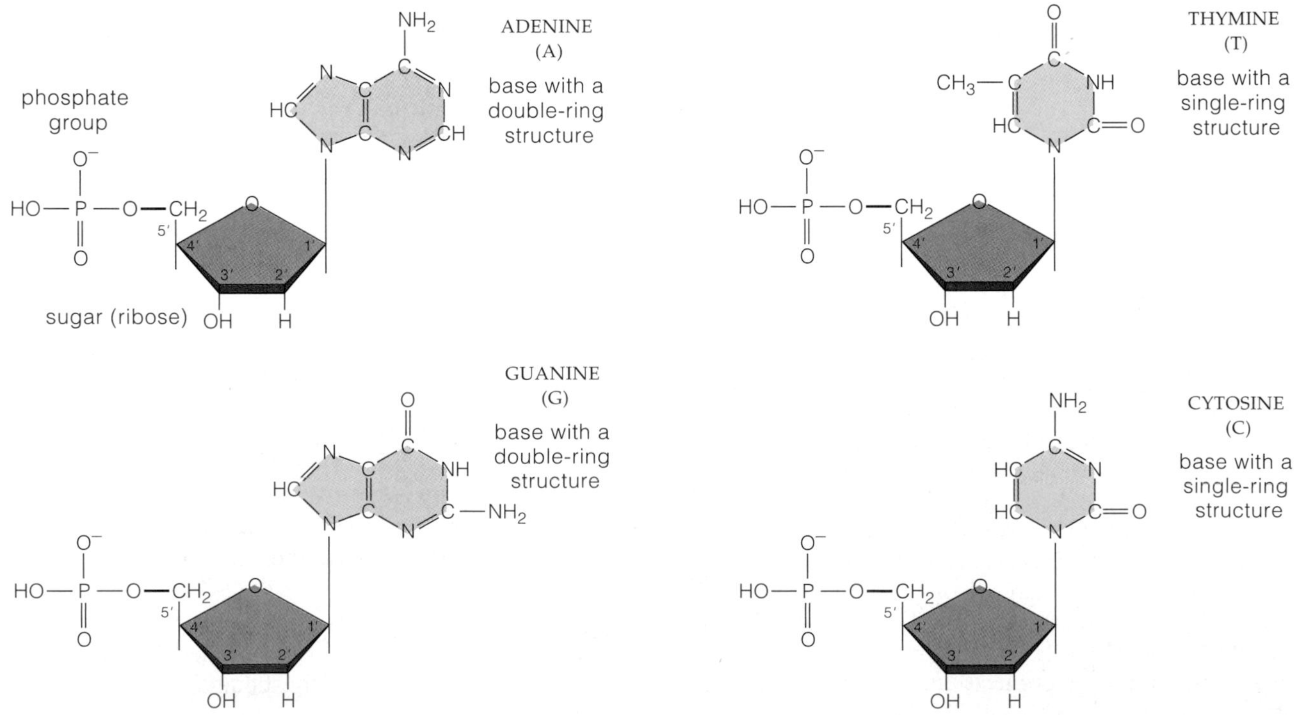

Further reading: Student Guide to InfoTrac on web site

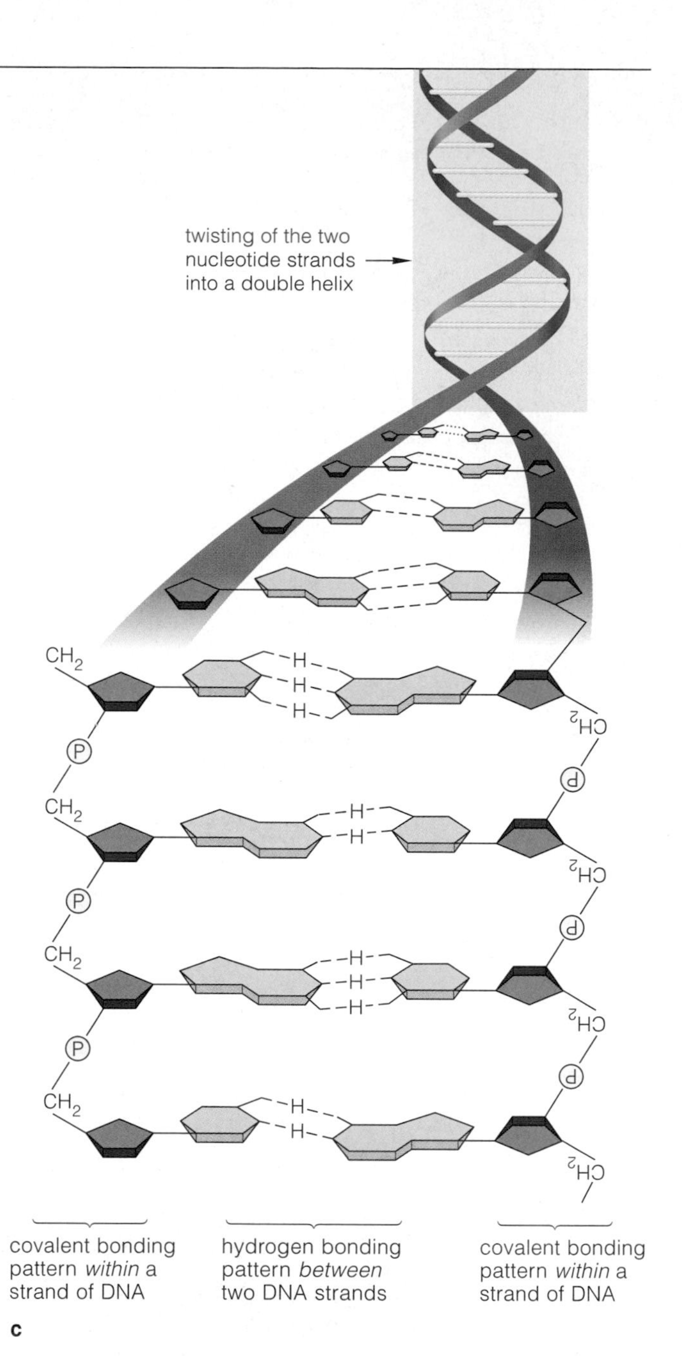

unlike DNA, RNA molecules are usually single strands of nucleotides. RNAs do not encode protein-building instructions, but different classes are the key players in carrying out those instructions in cells. Ever since cells first appeared on Earth, RNAs have served in processes by which genetic information is used to build proteins.

Nucleic acids consist of one or two strands of nucleotides joined in strandlike array by covalent bonds.

DNA is a double-stranded nucleic acid. Encoded in its nucleotide sequence is heritable information about all proteins a new cell requires to survive and reproduce.

The RNAs are single-stranded nucleic acids with roles in processes by which a cell uses genetic information in DNA to build proteins.

3.9 SUMMARY

1. Organic compounds consist of one or more elements covalently bonded to carbon atoms that commonly form linear and ring-shaped backbones. The properties that are characteristic of life emerge out of the molecular structure of carbon-rich compounds.

2. The central role of carbon in the world of life arises from its versatile bonding behavior. Each carbon atom can form relatively stable covalent bonds with as many as four other atoms. The bonding arrangement is the basis of carbon backbones, which can extend as straight chains or coil back on themselves as ring structures.

3. A carbon backbone that has only hydrogen atoms bonded to it is a hydrocarbon, a highly stable structure. Besides hydrogen, biological molecules have different functional groups: a variety of atoms or clusters of them covalently bonded to the carbon backbone. Functional groups impart diverse properties to organic compounds. Hydroxyl (—OH) and phosphate groups are examples.

4. In nature, only living cells can assemble the large organic compounds called biological molecules—the complex carbohydrates and lipids, proteins, and nucleic acids. To do so, they draw from small pools of organic compounds, which include simple sugars, fatty acids, amino acids, and nucleotides.

5. Cells assemble, rearrange, and break down organic compounds mainly through enzyme-mediated reactions involving the transfer of functional groups or electrons, the rearrangement of internal bonds, and a combining or splitting of molecules. As an example, in condensation reactions, enzymes remove a hydroxyl group from one molecule and an H atom from another, then speed the formation of a covalent bond between the two molecules at their exposed sites. A molecule of water usually forms also. Hydrolysis is like condensation in reverse.

6. Most carbohydrates consist of carbon, hydrogen, and oxygen in a 1:2:1 ratio. They are simple sugars, short-chain oligosaccharides, and polysaccharides.

a. Cells use the simple sugars (such as glucose) and oligosaccharides (such as the disaccharide sucrose) as building blocks and transportable forms of energy.

b. Complex carbohydrates (e.g., cellulose and starch) are structural materials and forms of energy storage.

7. Lipids are hydrocarbons that show little tendency to dissolve in water, but they readily dissolve in nonpolar substances. Cells use lipids (including triglycerides) as energy storage forms, as structural components of cell membranes, and as precursors of other compounds.

a. Fats are lipids with one, two, or three fatty acids attached to glycerol. Saturated tails have single covalent bonds only. Unsaturated tails have some double bonds. Triglycerides (three fatty acid tails) are the body's most abundant lipids and its richest energy source.

Table 3.1 Summary of the Main Organic Compounds in Living Things

Category	Main Subcategories	Some Examples	Their Functions
CARBOHYDRATES . . . contain an aldehyde or a ketone group, and one or more hydroxyl groups	**Monosaccharides** (simple sugars)	Glucose	Energy source
	Oligosaccharides	Sucrose (a disaccharide)	Most common form of sugar transported through plants
	Polysaccharides (complex carbohydrates)	Starch, glycogen Cellulose	Energy storage Structural roles
LIPIDS . . . are largely hydrocarbon; generally do not dissolve in water but dissolve in nonpolar substances, such as other lipids	**Lipids with fatty acids**		
	Glycerides: one, two, or three fatty acid tails attached to glycerol backbone	Fats (e.g., butter), oils (e.g., corn oil)	Energy storage
	Phospholipids: phosphate group, one other polar group, and (often) two fatty acids attached to glycerol backbone	Phosphatidylcholine	Key component of cell membranes
	Waxes: long-chain fatty acid tails attached to alcohol	Waxes in cutin	Water conservation at aboveground surfaces of plants
	Lipids with no fatty acids		
	Sterols: four carbon rings; the number, position, and type of functional groups differ among various sterols	Cholesterol	Component of animal cell membranes; precursor of many steroids and of vitamin D
PROTEINS . . . are one or more polypeptide chains, each with as many as several thousand covalently linked amino acids	**Fibrous proteins** Long strands or sheets of polypeptide chains; often tough, water-insoluble	Keratin Collagen	Structural element of hair, nails Structural element of bone, cartilage
	Globular proteins One or more polypeptide chains folded and linked into globular shapes; many roles in cell activities	Enzymes Hemoglobin Insulin Antibodies	Great increase in rates of reactions Oxygen transport Control of glucose metabolism Tissue defense
NUCLEIC ACIDS (AND NUCLEOTIDES) . . . are chains of units (or individual units) that each consist of a five-carbon sugar, phosphate, and a nitrogen-containing base	**Adenosine phosphates**	ATP cAMP (Section 37.2)	Energy carrier Messenger in hormone regulation
	Nucleotide coenzymes	NAD^+, $NADP^+$, FAD	Transport of protons (H^+), electrons from one reaction site to another
	Nucleic acids Chains of thousands to millions of nucleotides	DNA, RNAs	Storage, transmission, translation of genetic information

b. Phospholipids are the main components of cell membranes. Sterols such as cholesterol are components of cell membranes and precursors of steroid hormones and other compounds. Waxes are pliable components of water-repelling, protective, and lubricating substances.

8. Proteins are the most diverse biological molecules. They function in enzyme activity, structural support, transport, cell movements, changes in cell shape and activities, and defense against disease.

a. Each protein consists of one or more polypeptide chains in which amino acids are joined one after another by peptide bonds. The amino acid sequence, which is unique for each kind of protein, gives rise to its unique structure, chemical behavior, and function.

b. The amino acid sequence is the protein's primary structure. An individual's DNA specifies the sequence.

c. A protein has a secondary structure (sheetlike or coiled) arising from hydrogen bonding along the chain. A third-level structure arises from interactions among amino acids that hold the chain in bends and loops. Proteins with a fourth-level structure consist of two or more polypeptide chains bonded together.

9. Nucleotides serve as energy carriers (mostly ATP), coenzymes (such as NAD^+), and monomers for strands of DNA and RNA, which are the basis of inheritance and reproduction.

10. Table 3.1 summarizes the main biological molecules.

Review Questions

1. Define organic compound. Name the type of chemical bond that predominates in the backbone of such a compound. *3.1*

2. Define hydrocarbon. Also define functional group and give an example. *3.1*

3. Name the "molecules of life." Do they break apart most easily at their hydrocarbon portion or at functional groups? *3.1*

4. Describe the difference between a condensation reaction and hydrolysis. *3.1*

5. Select one of the carbohydrates, lipids, proteins, or nucleic acids described in this chapter. Speculate on how its functional groups and bonds between the carbon atoms in its backbone contribute to its final shape and function. *3.2 through 3.8*

6. Which item listed includes all of the other items listed? *3.3*
 a. triglyceride c. wax e. lipid
 b. fatty acid d. sterol f. phospholipid

7. Explain how a hemoglobin molecule's three-dimensional shape arises, starting with its primary structure. *3.4, 3.5*

Self-Quiz (*Answers in Appendix III*)

1. Each carbon atom can share pairs of electrons with as many as ______ other atoms.
 a. one b. two c. three d. four

2. Hydrolysis is a ______ reaction.
 a. functional group transfer
 b. electron transfer
 c. rearrangement
 d. condensation
 e. cleavage
 f. both b and d

3. ______ is a simple sugar (monosaccharide).
 a. glucose c. ribose e. both a and b
 b. sucrose d. chitin f. both a and c

4. In unsaturated fats, fatty acid tails have one or more ______
 a. single covalent bonds b. double covalent bonds

5. ______ are to proteins as ______ are to nucleic acids.
 a. sugars; lipids c. amino acids; hydrogen bonds
 b. sugars; proteins d. amino acids; nucleotides

6. A denatured protein or DNA molecule has lost its ______.
 a. hydrogen bonds c. function
 b. shape d. all of the above

7. Nucleotides include ______.
 a. ATP c. RNA subunits
 b. DNA subunits d. all of the above

8. Match each molecule with the most suitable description.
 ____ long sequence of amino acids a. carbohydrate
 ____ the main energy carrier b. phospholipid
 ____ glycerol, fatty acids, phosphate c. protein
 ____ two strands of nucleotides d. DNA
 ____ one or more sugar monomers e. ATP

Critical Thinking

1. Jack decided to celebrate summer by making a crab salad and a peach pie for some friends. He had such a good time, he forgot to put the cap on the bottle of olive oil after making the salad. He also didn't tighten the lid on the shortening can after making the pie crust. A few weeks later, the oil had turned rancid but the shortening still smelled okay. Both substances are fats. Both were stored in a dark cupboard, at room temperature. Yet one spoiled and the other stayed fresh. Speculate on which molecular bonds in these substances were the starting point for the difference.

2. In the following list, identify which is the carbohydrate, fatty acid, amino acid, and polypeptide:
 a. $^{+}NH_3$—CHR—COO^{-} c. $(glycine)_{20}$
 b. $C_6H_{12}O_6$ d. $CH_3(CH_2)_{16}COOH$

3. A clerk in a health-food store tells you that certain "natural" vitamin C tablets extracted from rose hips are better for you than synthetic vitamin C tablets. Given your understanding of the structure of organic compounds, what would be your response? Design an experiment to test whether the vitamins differ.

4. Rabbits that eat green, leafy vegetables containing xanthophyll accumulate this yellow pigment molecule in body fat but not in muscles. What chemical properties of the molecule might cause this selective accumulation?

5. Cows can digest grasses, but people cannot. The cow's four-chambered stomach houses populations of certain microorganisms that are not normal residents of the human stomach. What kind of metabolic reactions do you think the microorganisms carry out? What do you think might happen to a cow undergoing treatment with an antibiotic that killed the microorganisms?

6. A Gary Larsen cartoon states that sheep with steel wool have no natural enemies. You might laugh at the thought of such a preposterous animal without knowing what wool is. Wool is the keratin-rich, soft undercoat of sheep, Angora goats, llamas, and other hairy mammals. Keratin is a strandlike, fibrous protein. Many hydrogen bonds hold it in a helical shape. Visualize three such chains coiled together, with many disulfide bridges in the end regions stabilizing the trio in a tight, ropelike array. Each hair has linear aggregates of the ropelike fibers, which resist stretching (Figure 3.19). Given keratin's structure, speculate why, when you run a wool sweater through the hot cycle of a dryer, it shrinks pathetically and permanently.

7. Reflect on the component atoms of the different classes of organic compounds. Now look at the structural formulas for the pesticides shown in Figure 3.20. Which one is a glycoprotein? An organophosphate? A chlorinated hydrocarbon?

8. Many farmers believe crop losses will be disastrous if pesticides are banned without sufficient research and effective alternatives. Many environmental activists are pushing for rapid bans to protect people and the environment. What do you think?

Selected Key Terms

alcohol *3.1*
amino acid *3.4*
ATP *3.7*
carbohydrate *3.2*
coenzyme *3.7*
condensation reaction *3.1*
denaturation *3.5*
DNA *3.8*
enzyme *3.1*
fat *3.3*
fatty acid *3.3*
fibrous protein *3.5*
functional group *3.1*
globular protein *3.5*
hemoglobin *3.5*
hydrolysis *3.1*
lipid *3.3*
monosaccharide *3.2*
nucleic acid *3.8*
nucleotide *3.7*
oligosaccharide *3.2*
organic compound *3.1*
phospholipid *3.3*
polypeptide chain *3.4*
polysaccharide *3.2*
protein *3.4*
RNA *3.8*
sterol *3.3*
toxin *3.6*
triglyceride *3.3*
wax *3.3*

Readings *See also www.infotrac-college.com*

Atkins, P. 1987. *Molecules*. New York: Scientific American Library. A molecular "glossary."

Ritter, P. 1996. *Biochemistry: An Introduction*. Pacific Grove, California: Brooks/Cole. Great human applications.

Wolfe, S. 1995. *Introduction to Molecular and Cellular Biology*. Belmont, California: Wadsworth.

4

CELL STRUCTURE AND FUNCTION

Animalcules and Cells Fill'd With Juices

Early in the seventeenth century, a scholar by the name of Galileo Galilei arranged two glass lenses within a cylinder. With this instrument he happened to look at an insect, and later he described the stunning geometric patterns of its tiny eyes. Thus Galileo, who was not a biologist, was among the first to record a biological observation made through a microscope. The study of the cellular basis of life was about to begin. First in Italy, then in France and England, scholars set out to explore a world whose existence had not even been suspected.

At midcentury Robert Hooke, Curator of Instruments for the Royal Society of England, was at the forefront of these studies. When Hooke first turned a microscope to thinly sliced cork from a mature tree, he observed tiny compartments (Figure 4.1*c*). He gave them the Latin name *cellulae*, meaning small rooms—hence the origin of the biological term "cell." They actually were the interconnecting walls of dead plant cells, which is what cork is made of, but Hooke did not think of them as being dead because neither he nor anyone else at the time knew that cells could be alive. In still other plant tissues, he observed cells "fill'd with juices" but could not imagine what they represented.

Given the simplicity of their instruments, it is just amazing that the pioneers in microscopy observed as much as they did. Antony van Leeuwenhoek, a Dutch shopkeeper, had exceptional skill in constructing lenses and possibly the keenest vision of all (Figure 4.1*a*). By the late 1600s, he was discovering natural wonders everywhere, including "many very small animalcules, the motions of which were very pleasing to behold," in scrapings of tartar from his own teeth. Elsewhere he observed a variety of protistans, sperm, and even a bacterium—an organism so small that it would not be seen again for another two centuries!

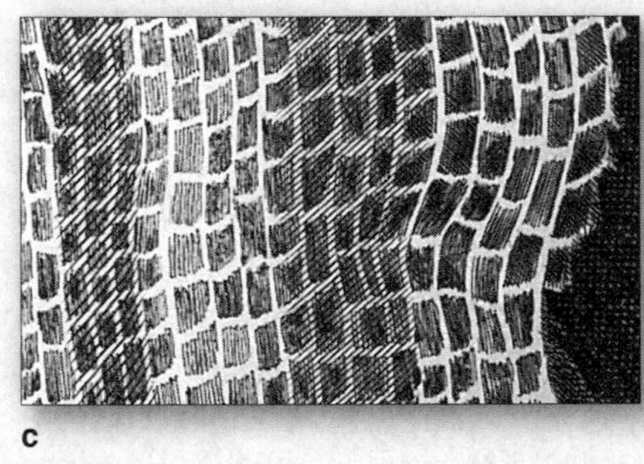

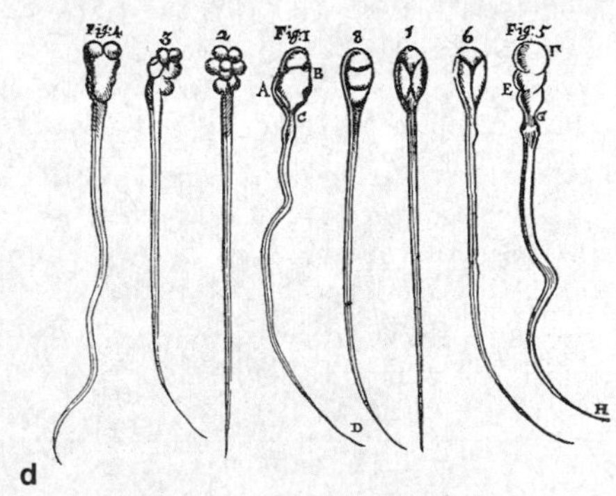

Figure 4.1 Early glimpses into the world of cells. (**a**) Antony van Leeuwenhoek, with microscope in hand. (**b**) Robert Hooke's compound microscope and (**c**) his drawing of cell walls from cork tissue. (**d**) One of van Leeuwenhoek's early sketches of sperm cells. (**e**) Cartoon evidence of the startling impact of microscopic observations on nineteenth-century London.

In the 1820s, improvements in lenses brought cells into sharper focus. Robert Brown, a botanist, was noticing an opaque spot in a variety of cells. He called it a nucleus. In 1838 still another botanist, Matthias Schleiden, wondered if the nucleus had something to do with a cell's development. As he hypothesized, each plant cell must develop as an independent unit even though it is part of the plant.

By 1839, after years of studying animal tissues, the zoologist Theodor Schwann had this to say: Animals as well as plants consist of cells and cell products—and even though the cells are part of a whole organism, to some extent they have an individual life of their own.

A decade later a question remained: Where do cells come from? Rudolf Virchow, a physiologist, completed his own studies of a cell's growth and reproduction—that is, its division into two daughter cells. Every cell, he reasoned, comes from a cell that already exists.

And so, by the middle of the nineteenth century, microscopic analysis had yielded three generalizations, which together constitute the **cell theory**. *First, every organism is composed of one or more cells. Second, the cell is the smallest unit having the properties of life. Third, the continuity of life arises directly from the growth and division of single cells.* All three insights still hold true.

This chapter is not meant for memorization. Read it simply to gain an overview of current understandings of cell structure and function. In later chapters, you can refer back to it as a road map through the details. With its images from microscopy, this chapter and others in the book can transport you into spectacular worlds of juice-fill'd cells and animalcules.

e

KEY CONCEPTS

1. All organisms consist of one or more cells. The cell is the smallest unit that still retains the characteristics of life. Since the origin of life, each new cell has arisen from a preexisting cell. These are the three generalizations of the cell theory.

2. All cells have an outermost, double-layered membrane. This plasma membrane separates the cell interior from the surroundings. In addition, cells contain cytoplasm, an organized internal region where energy conversions, protein synthesis, movements, and other activities necessary for survival proceed. In eukaryotic cells only, DNA is enclosed within a membrane-bound nucleus. In prokaryotic cells (bacteria), DNA is concentrated in part of the cell interior.

3. The plasma membrane and internal cell membranes consist largely of phospholipid and protein molecules. The phospholipids form two adjacent layers that give the membrane its basic structure and prevent water-soluble substances from freely crossing it. Proteins embedded in cell membranes or positioned at their surfaces perform most membrane functions.

4. The nucleus is one of many organelles. Organelles are membrane-bound compartments inside eukaryotic cells. They physically separate different metabolic reactions and allow them to proceed in orderly fashion. Bacteria do not have comparable organelles.

5. When cells grow, they increase faster in volume than in surface area. This physical constraint on increases in size influences cell size and shape.

6. Different microscopes modify light rays or accelerated beams of electrons in ways that allow us to form images of incredibly small specimens. They are the foundation for our current understanding of cell structure and function.

4.1 BASIC ASPECTS OF CELL STRUCTURE AND FUNCTION

Inside your body and at its moist surfaces, trillions of cells live in interdependency. In northern forests, four-cell structures—pollen grains—escape from pine trees. In scummy pondwater, a single-celled amoeba moves freely, thriving on its own. For humans, pines, amoebas, and all other organisms, the **cell** is the smallest unit that retains the properties of life (Figure 4.2). Each living cell can survive on its own or has the potential to do so. Its structure is highly organized for metabolism. It senses and responds to its environment. And, based on inherited instructions in its DNA, it has the potential to reproduce.

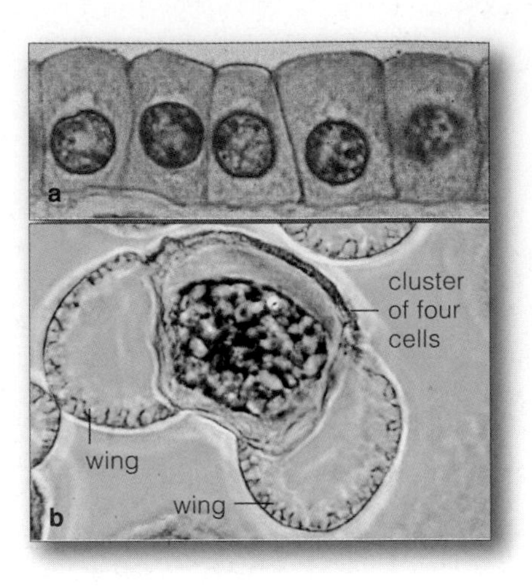

Figure 4.2 (**a**) A row of cells at the surface of a tiny tube inside a kidney. Each has a nucleus (the sphere stained darker red). (**b**) A winged pollen grain. One of its cells will give rise to a sperm that may meet up with an egg and help start a new pine tree.

Structural Organization of Cells

Cells differ enormously in size, shape, and activities, as you might gather by comparing a tiny bacterium with one of your relatively giant liver cells. Yet they are alike in three respects. All cells start out life with a plasma membrane, a region of DNA, and a region of cytoplasm:

1. **Plasma membrane**. This thin, outermost membrane maintains the cell as a distinct entity. By doing so, it allows metabolic events to proceed apart from random events in the environment. A plasma membrane does not isolate the cell interior. Substances and signals continually move across it in highly controlled ways.

2. **Nucleus** or **nucleoid**. Depending on the species, the DNA occupies a membranous sac in the cell interior (nucleus) or simply an interior region (nucleoid).

3. **Cytoplasm**. Cytoplasm is everything between the plasma membrane and region of DNA. It consists of a semifluid matrix and other components, such as **ribosomes** (structures on which proteins are built).

This chapter introduces two fundamentally different kinds of cells. The cytoplasm of **eukaryotic cells** includes organelles, which are tiny sacs and other compartments bounded by membranes. One of these is the nucleus, the defining feature of eukaryotic cells. **Prokaryotic cells** do not have a nucleus. No cell membranes intervene between the nucleoid (where DNA is located) and the cytoplasm surrounding it. The only prokaryotic cells are bacteria. Beyond the bacterial realm, all other organisms —from amoebas to peach trees to puffball mushrooms to zebras—consist of one to trillions of eukaryotic cells.

Lipid Bilayer of Cell Membranes

All cell membranes have the same structural framework of two sheets of lipid molecules. Figure 4.3 shows this **lipid bilayer** arrangement for the plasma membrane, the continuous boundary that bars the free passage of water-soluble substances into and out of the cell. Within eukaryotic cells, membranes subdivide the cytoplasm into specific zones in which substances are synthesized, processed, stockpiled, or degraded.

Diverse proteins embedded in the lipid bilayer or positioned at one of its surfaces carry out most of the

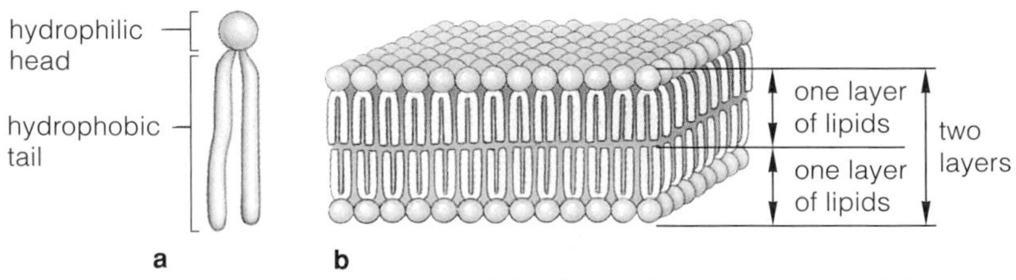

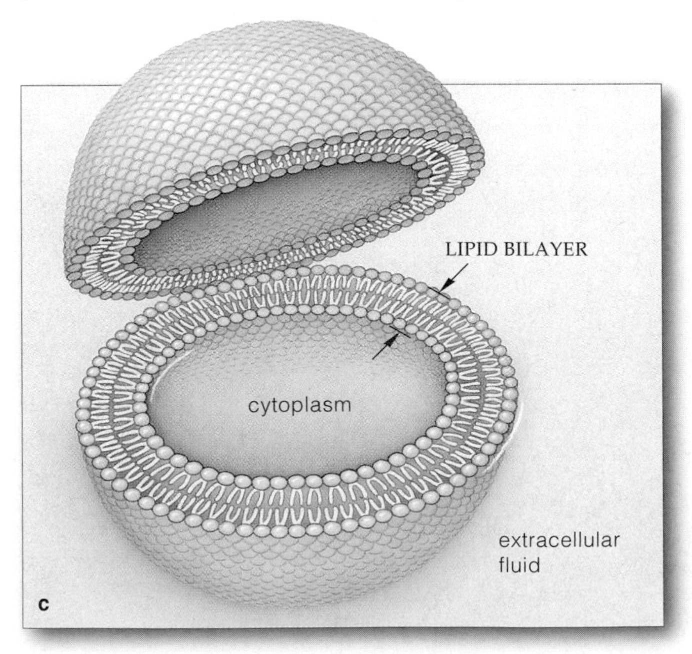

Figure 4.3 Organization of the lipid bilayer of cell membranes. (**a**) Icon for phospholipid molecules, the most abundant components of cell membranes. These and other lipids are arranged as two layers. (**b**) The hydrophobic tails of the molecules are sandwiched between their hydrophilic heads, which are dissolved in cytoplasm on one side of the bilayer and in extracellular fluid on the other (**c**).

Further reading: Student Guide to InfoTrac on web site

Diameter (cm):	0.5	1.0	1.5
Surface area (cm^2):	0.79	3.14	7.07
Volume (cm^3):	0.06	0.52	1.77
Surface-to-volume ratio:	13.17:1	6.04:1	3.99:1

a

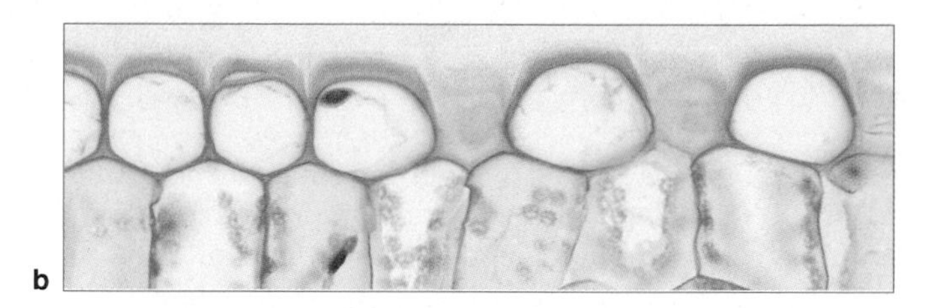

b

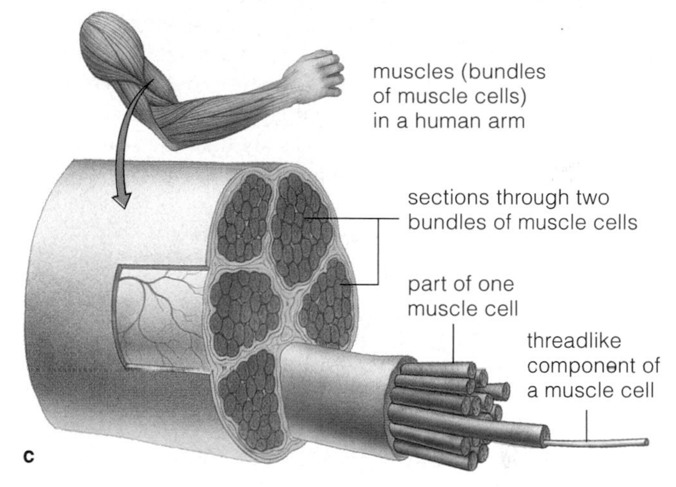

Figure 4.4 (**a**) An example of the surface-to-volume ratio. This physical relationship between increases in volume and in surface area imposes restrictions on the size and shape of cells, including the ones near the surface of leaves (**b**) and in skeletal muscles (**c**).

membrane functions. For example, some proteins act as passive channels for water-soluble substances. Others pump substances across the bilayer. Still others serve as receptors, which can latch onto hormones and other types of signaling molecules that trigger changes in cell activities. You will read more about membrane proteins in the chapter to follow.

Cell Size and Cell Shape

You may be wondering how small cells really are. Can any be observed with the unaided human eye? There are a few, including the "yolks" of bird eggs, cells in the red part of watermelons, and the fish eggs we call caviar. However, most cells cannot be observed without microscopes. To give you a sense of cell sizes, red blood cells are about 8 millionths of a meter across. You could fit a string of 2,000 of them across your thumbnail!

Why are most cells so small? A physical relationship called the **surface-to-volume ratio** constrains increases in a cell's size. By this relationship, an object's volume increases with the cube of the diameter, but its surface area increases only with the square. Figure 4.4*a* gives an example of this. Simply put, *if a cell expands in diameter during growth, its volume will increase more rapidly than its surface area will.*

Suppose you figure out a way to make a round cell grow four times wider than it normally would. Its volume increases 64 times (4^3), but its surface area only increases 16 times (4^2). As a result, each unit of the cell's plasma membrane must now serve four times as much cytoplasm as it did previously! Moreover, past a certain point, the inward flow of nutrients and outward flow of wastes will not be fast enough, and the cell will die.

A large, round cell also would have trouble moving materials *through* its cytoplasm. In small cells, such as those shown in Figure 4.4*b*, the random, tiny motions of molecules can easily distribute materials. If a cell is not small, you usually can expect it to be long and thin or to have outfoldings and infoldings that increase its surface relative to its volume. *The smaller or narrower or more frilly-surfaced the cell, the more efficiently materials cross its surface and become distributed through the interior.*

Multicelled body plans show evidence of surface-to-volume constraints, also. For example, cells attach end to end in strandlike algae, and each one interacts directly with its surroundings. Your skeletal muscle cells are very thin, but each one is as long as a biceps or some other muscle of which it is part (Figure 4.4*c*). Your circulatory system delivers materials to muscle cells and trillions of others, and it also sweeps away their metabolic wastes. Its many "highways" cut through the volume of body tissues, and by doing so they shrink the distance to and from individual cells.

All cells have an outermost plasma membrane, an internal region of cytoplasm, and an internal region of DNA.

Besides the plasma membrane, eukaryotic cells have internal, membrane-bound compartments, including a nucleus.

Each cell membrane has a lipid bilayer structure, consisting largely of phospholipid molecules. The hydrophobic parts of the lipids are sandwiched between the hydrophilic parts, which are dissolved in the fluid surroundings.

A lipid bilayer imparts structure to the cell membrane and serves as a barrier to water-soluble substances.

Proteins embedded in the bilayer or positioned at one of its surfaces carry out most membrane functions.

During their growth, cells increase faster in volume than they do in surface area. This physical constraint on increases in size influences the size and shape of cells. It also influences the body plans of multicelled organisms.

4.2 MICROSCOPES—GATEWAYS TO CELLS

Modern microscopes are gateways to astounding worlds. Some even afford glimpses into the structure of molecules. Different kinds use wavelengths of light or accelerated electrons (Figure 4.5). The micrographs in Figures 4.6 and 4.7 hint at the details now being observed. A **micrograph** simply is a photograph of an image that came into view with the aid of a microscope.

LIGHT MICROSCOPES Picture a series of waves moving across an ocean. Each **wavelength** is the distance from one wave's peak to the peak of the wave behind it. Light also travels as waves from sources such as the sun and illuminated specimens. In a *compound light microscope*, two or more sets of glass lenses bend light emanating from a cell or some other specimen in ways that form an enlarged image of it (Figure 4.5*a*,*b*).

A living cell must be small or thin enough for light to pass through. It would help if cell parts differed in color and density from the surroundings, but most are nearly colorless and appear uniformly dense. To get around this problem, microscopists stain cells (expose them to dyes that react with some parts of a specimen but not others). Staining may alter and kill cells. Dead cells break down fast, so cells typically are pickled or preserved before staining.

Suppose you use the best glass lens system. When you magnify the diameter of a specimen by 2,000 times or more, you discover that cell parts appear larger but are not clearer. To understand why, think about the distance between the two crests of a wavelength of light. To give two examples, that distance is about 750 nanometers for red light and 400 nanometers for violet (wavelengths of all other colors fall in between). If a cell structure is less than one-half of a wavelength long, that structure will not be able to disturb the rays of light streaming past, and it will not be visible.

ELECTRON MICROSCOPES Better resolution of very fine details is possible with the assistance of electrons. Remember, electrons are particles of matter, but they also behave like waves. In electron microscopy, streams of electrons are accelerated to wavelengths of about 0.005 nanometer—about 100,000 times shorter than wavelengths of visible light. The electrons cannot pass through glass lenses, but a magnetic field can bend them from their path and focus them.

In a *transmission electron microscope*, a magnetic field is the "lens." Accelerated electrons are directed through a specimen, focused into an image, and magnified. With *scanning electron microscopes*, a narrow beam of electrons moves back and forth across a specimen to which a thin coat of metal was applied. The metal responds by emitting some of its own electrons. A detector tied into electronic circuitry then transforms the energy of electrons into an image of the specimen's surface on a television screen. Most of the scanning images have fantastic depth (Figure 4.7*d*).

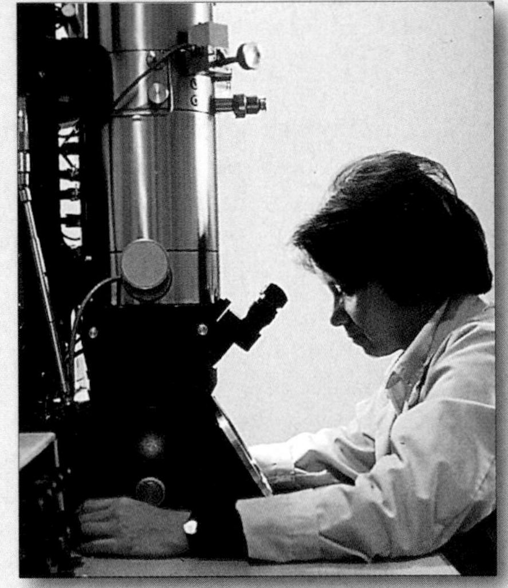

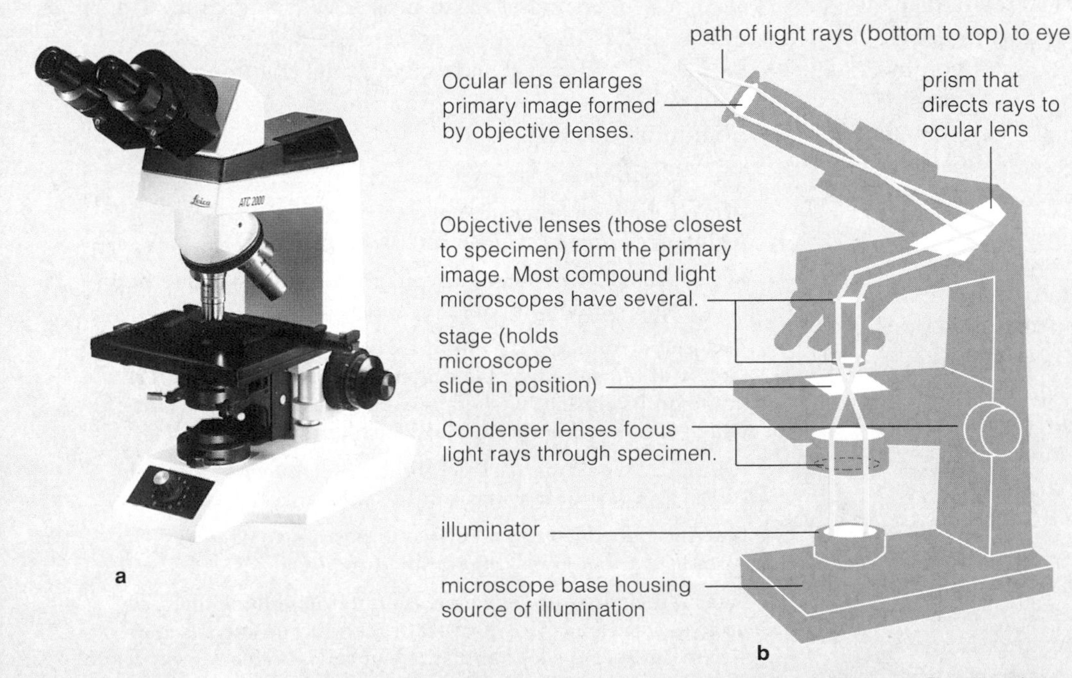

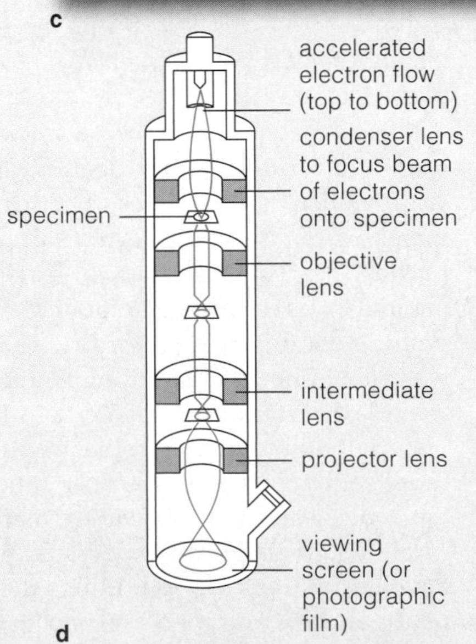

Figure 4.5 (**a**) Exterior view and (**b**) diagram of a compound light microscope. Exterior view (**c**) and diagram (**d**) of an electron microscope.

Further reading: Student Guide to InfoTrac on web site

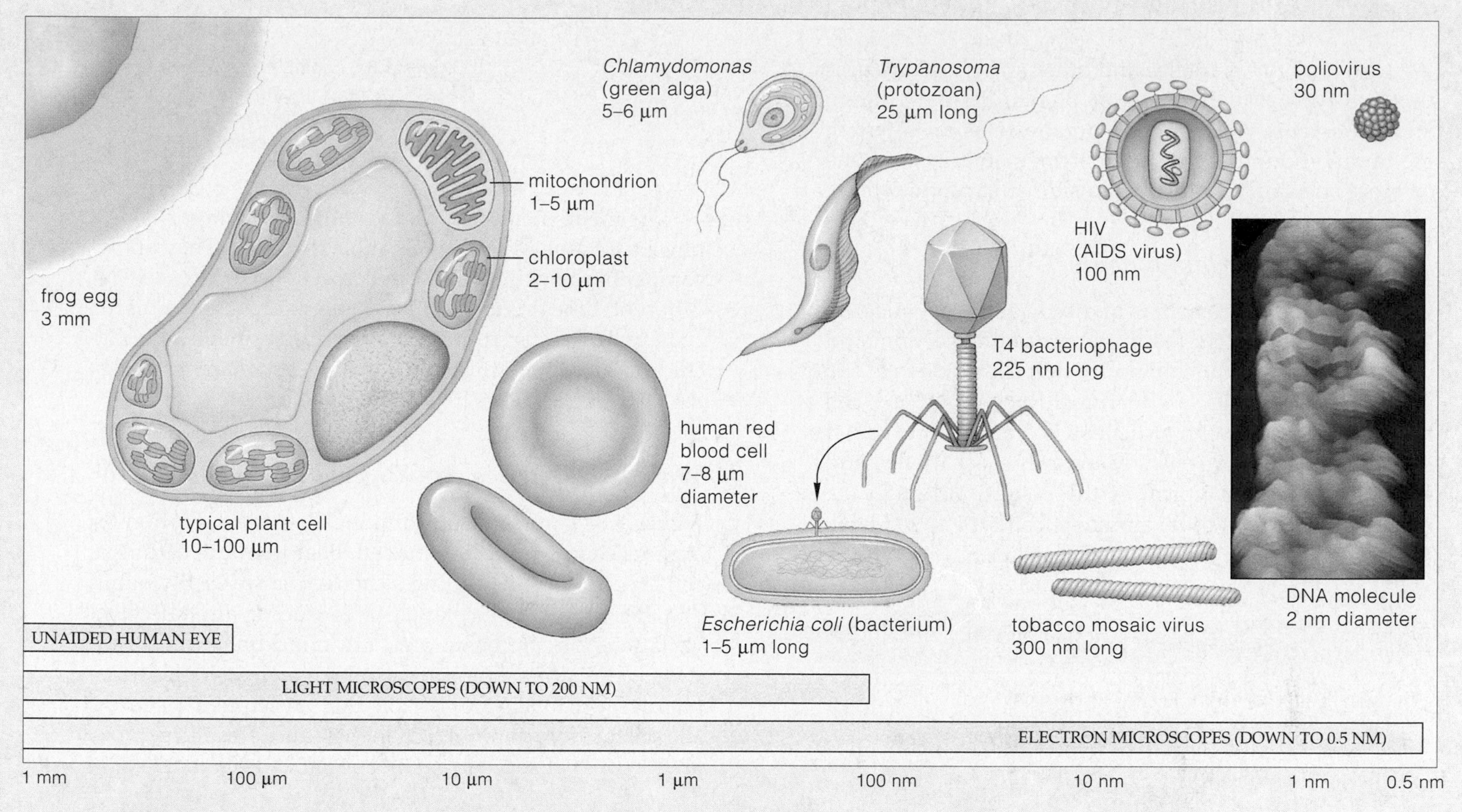

Figure 4.6 Units of measure used in microscopy. A *scanning tunneling microscope*, which can provide magnifications up to 100 million, gave us the photomicrograph of part of a DNA molecule. The scope's needlelike probe has a single atom at its tip. When voltage is applied between the tip and an atom at a specimen's surface, it tunnels into the electron orbitals. As the tip moves over a specimen's contours, a computer analyzes the tunneling motion and creates a three-dimensional view of the surface atoms.

1 centimeter	(cm)	= 1/100 meter, or 0.4 inch
1 millimeter	(mm)	= 1/1,000 meter
1 micrometer	(µm)	= 1/1,000,000 meter
1 nanometer	(nm)	= 1/1,000,000,000 meter

1 meter = 10^2 cm = 10^3 mm = 10^6 µm = 10^9 nm

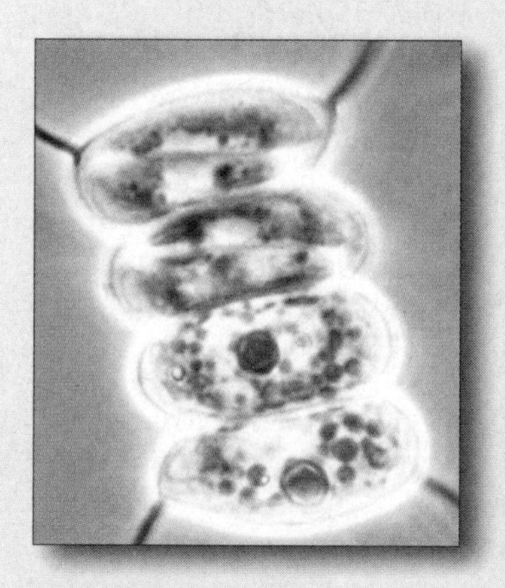

a Light micrograph (phase-contrast process)

b Light micrograph (Nomarski process)

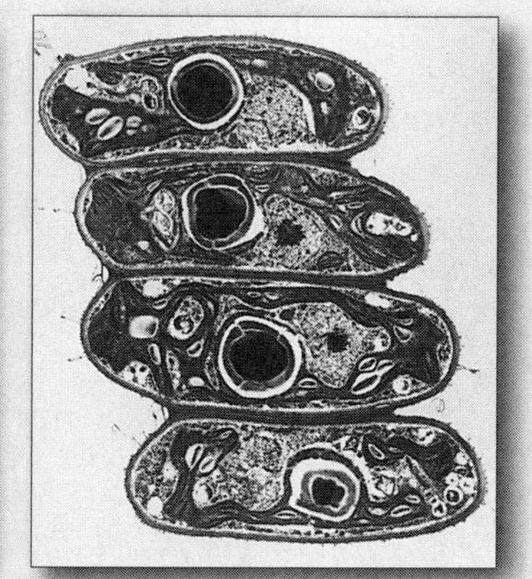

c Transmission electron micrograph, thin section

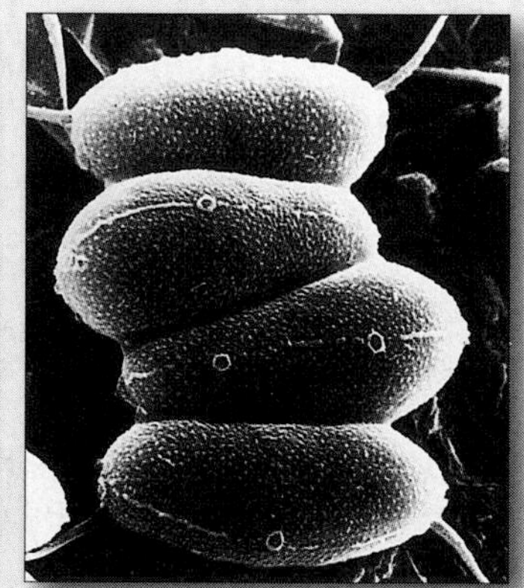

d Scanning electron micrograph

10 µm

Figure 4.7 How different microscopes can reveal different aspects of the same organism—in this case, a green alga (*Scenedesmus*). The images of all four specimens are at the same magnification. The phase-contrast and Nomarski processes mentioned in (**a**) and (**b**) can create optical contrasts without staining the cells. Both processes enhance the usefulness of light micrographs. As for other micrographs in the book, the short horizontal bar below the micrograph in (**d**) provides you with a visual reference for size. A micrometer (µm) is 1/1,000,000 of a meter. Using the scale bar, can you estimate the length and width of *Scenedesmus*?

4.3 DEFINING FEATURES OF EUKARYOTIC CELLS

We turn now to organelles and other structural features that are typical of the cells of plants, animals, fungi, and protistans. We define an **organelle** as an internal, membrane-bound sac or compartment that serves one or more specialized functions inside eukaryotic cells.

Major Cellular Components

Observe some micrographs of a typical eukaryotic cell, such as the ones included in the preceding section, and you probably will quickly notice that the nucleus is one of the most conspicuous features. Remember, any cell that starts out life with a nucleus is a eukaryotic cell; it contains a "true nucleus." Many other organelles and structures also are typical of these cells, although the numbers and kinds differ from one cell type to the next. Table 4.1 lists the most common features.

Table 4.1 Common Features of Eukaryotic Cells

ORGANELLES AND THEIR MAIN FUNCTIONS	
Nucleus	Localizing the cell's DNA
Endoplasmic reticulum	Routing and modifying newly formed polypeptide chains; also, synthesizing lipids
Golgi body	Modifying polypeptide chains into mature proteins; sorting and shipping proteins and lipids for secretion or for use inside cell
Various vesicles	Transporting or storing a variety of substances; digesting substances and structures in the cell; other functions
Mitochondria	Producing many ATP molecules in highly efficient fashion
NON-MEMBRANOUS STRUCTURES AND THEIR FUNCTIONS	
Ribosomes	Assembling polypeptide chains
Cytoskeleton	Imparting overall shape and internal organization to the cell; moving the cell and its internal structures

Think about the list, and you might find yourself asking: What is the advantage of partitioning the cell interior with such organelles? *The compartmentalization allows a large number of activities to occur simultaneously in very limited space.* Consider a photosynthetic cell in a leaf. It can put together starch molecules by one set of reactions and break them apart by another set. Yet the cell would get nothing if the synthesis and breakdown reactions proceeded at the exact same time on the same starch molecule. Without membranes of organelles, the balance of diverse chemical activities that helps keep eukaryotic cells alive would spiral out of control.

Figure 4.8 *Facing page*: Generalized sketches showing some of the features that are typical of (**a**) many plant cells and (**b**) many animal cells.

Organelle membranes have another function besides physically separating incompatible reactions. They allow compatible and interconnected reactions to proceed at different times. For instance, a plant's photosynthetic cells produce starch molecules in an organelle called a chloroplast, then store and later release starch for use in different reactions inside the same organelle.

Which Organelles Are Typical of Plants?

Figure 4.8*a* can start you thinking about the location of organelles in a typical plant cell. Bear in mind, calling a cell "typical" is like calling a cactus or a water lily or an elm tree a "typical" plant. As is true of animal cells, variations on the basic plan are mind-boggling. With this qualification in mind, also take a close look at the micrograph in Figure 4.9, on the subsequent page. It shows the locations of organelles and structures you are likely to observe in many specialized plant cells.

Which Organelles Are Typical of Animals?

Now start thinking about organelles of a typical animal cell, such as the one shown in Figures 4.8*b* and 4.10. Like plant cells, it has a nucleus, many mitochondria, and the other features listed in Table 4.1. *The structural similarities point to basic functions that are necessary for survival, regardless of the cell type.* We will return to this concept throughout the book.

Comparing Figures 4.8 through 4.10 also will give you an initial idea of how plant and animal cells differ in their structure. For example, you will never observe an animal cell surrounded by a cell wall. (You might see many different kinds of fungal and protistan cells with one, however.) What other differences can you identify?

Eukaryotic cells contain a number of organelles, which are internal, membrane-bound sacs and compartments that serve specific metabolic functions.

Organelles physically separate chemical reactions, many of which are incompatible.

Organelles separate different reactions in time, as when certain molecules are put together, stored, then used later in other reaction sequences.

All eukaryotic cells contain certain organelles (such as the nucleus) and structures (such as ribosomes) that perform functions essential for survival. Specialized cells also may incorporate additional kinds of organelles and structures.

Further reading: Student Guide to InfoTrac on web site →

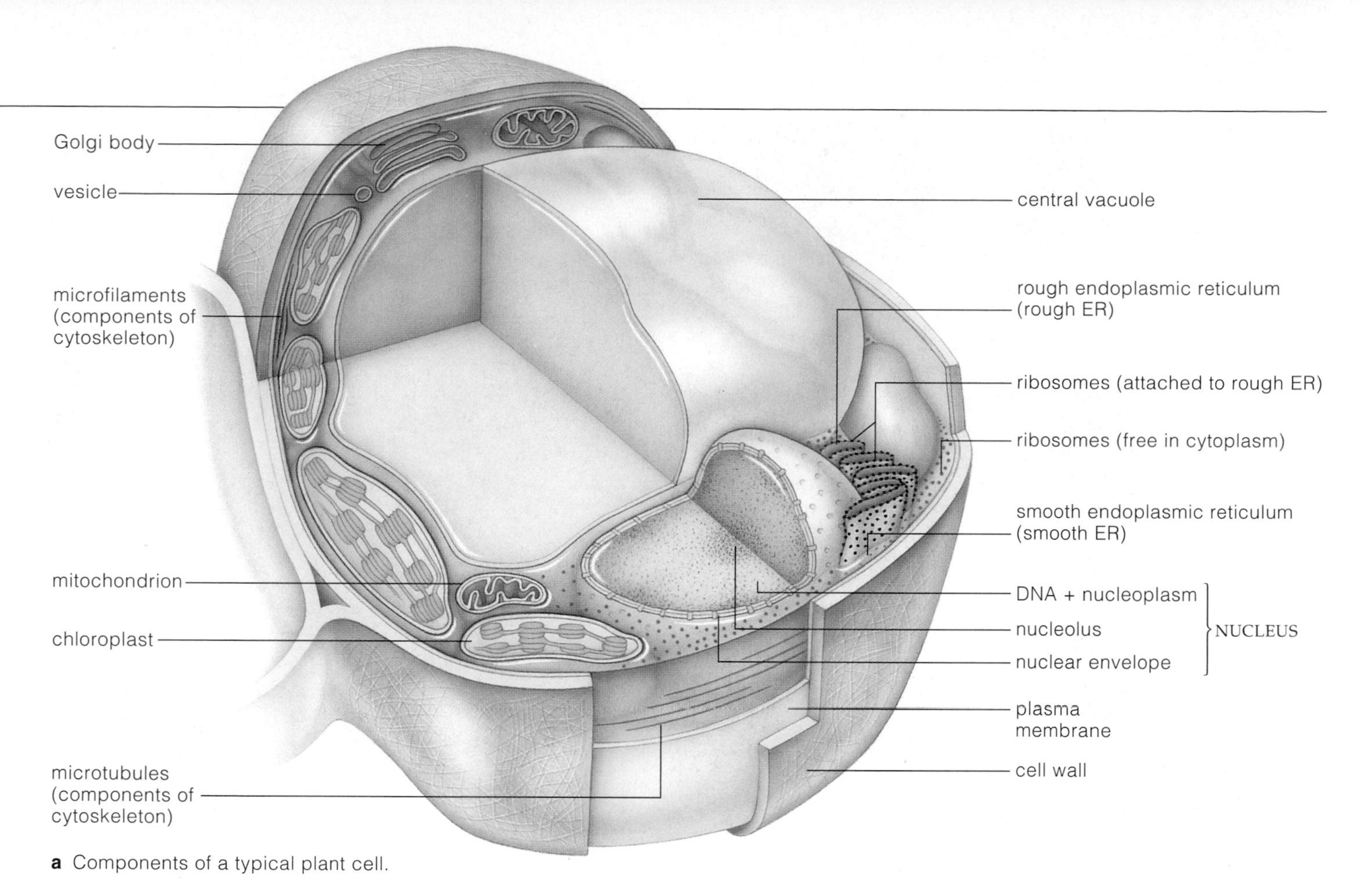

a Components of a typical plant cell.

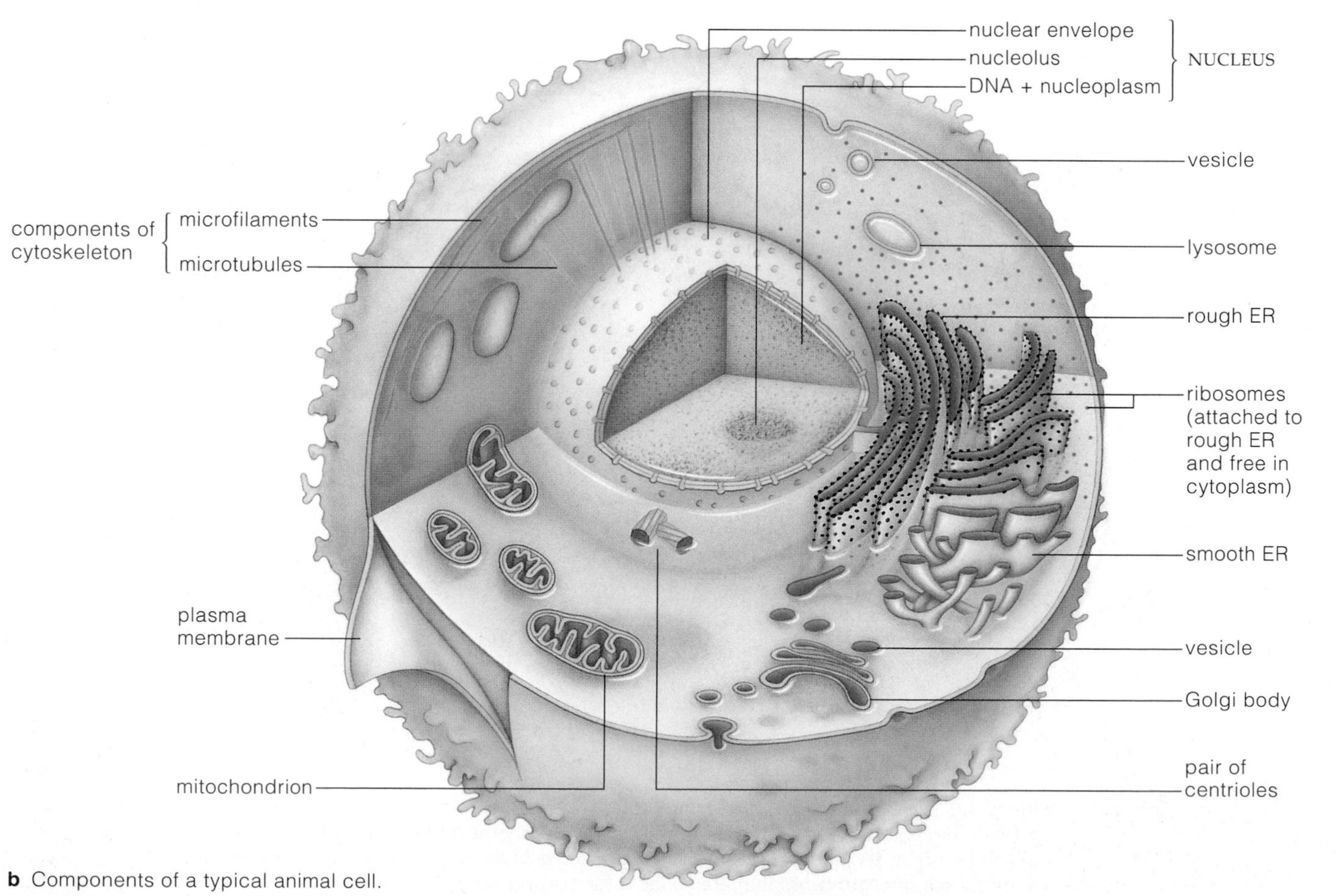

b Components of a typical animal cell.

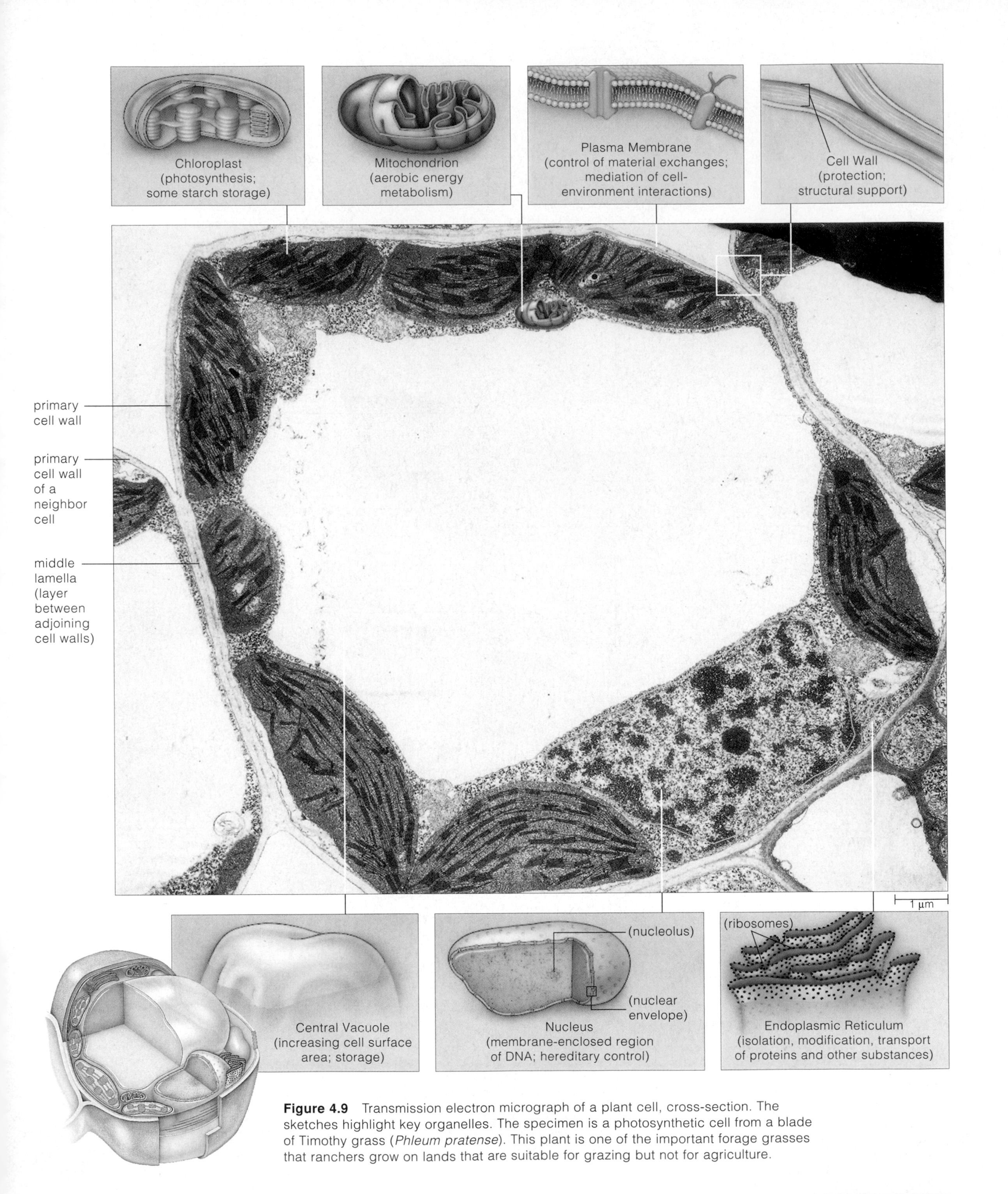

Figure 4.9 Transmission electron micrograph of a plant cell, cross-section. The sketches highlight key organelles. The specimen is a photosynthetic cell from a blade of Timothy grass (*Phleum pratense*). This plant is one of the important forage grasses that ranchers grow on lands that are suitable for grazing but not for agriculture.

Further reading: Student Guide to InfoTrac on web site →

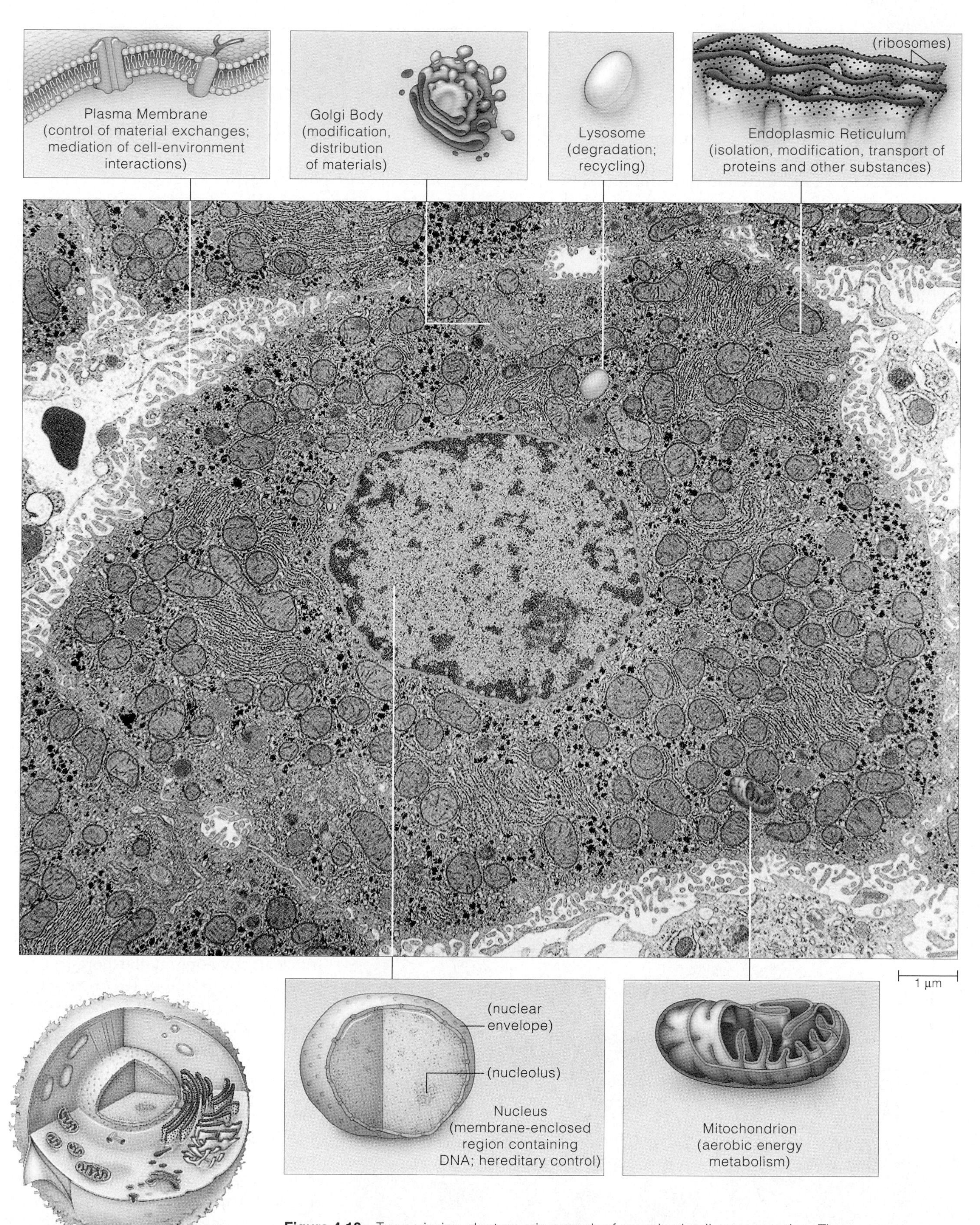

Figure 4.10 Transmission electron micrograph of an animal cell, cross-section. The sketches highlight key organelles. The specimen is a cell from the liver of a rat.

THE NUCLEUS

Constructing, operating, and reproducing cells simply cannot be done without carbohydrates, lipids, proteins, and nucleic acids. It takes a class of proteins—enzymes—to build and use these molecules. Said another way, a cell's structure and function begin with proteins. *And instructions for building proteins are contained in DNA.*

Unlike bacteria, eukaryotic cells have their genetic instructions distributed among several to many DNA molecules of different lengths. For example, each human body cell has forty-six molecules of DNA. Stretch them end to end, and they would be about a meter long. Similarly, stretch the twenty-six DNA molecules in a frog cell end to end and they would extend ten meters. We have no idea what frogs are doing with so much of it, but that's a lot of DNA!

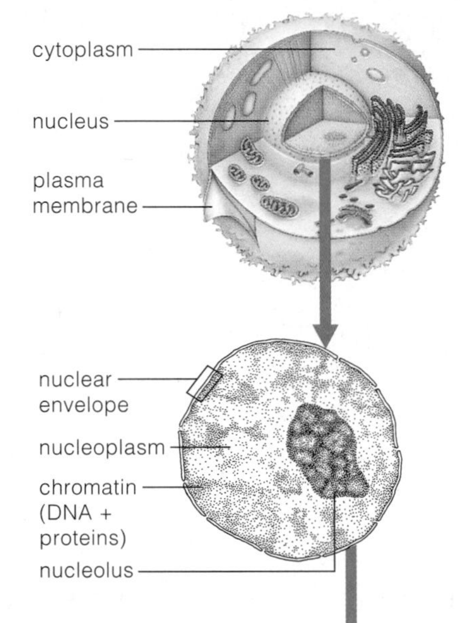

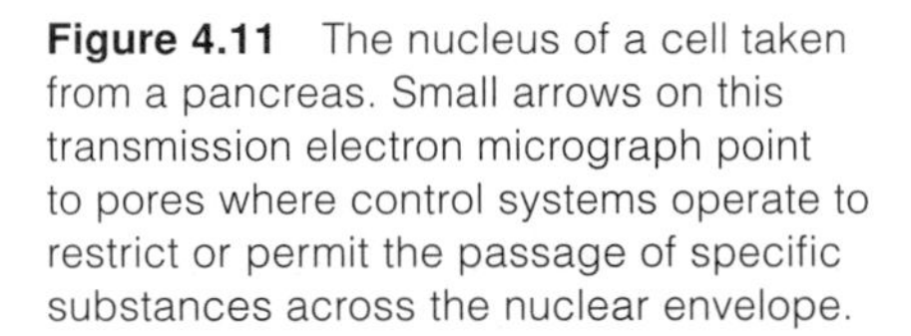

Figure 4.11 The nucleus of a cell taken from a pancreas. Small arrows on this transmission electron micrograph point to pores where control systems operate to restrict or permit the passage of specific substances across the nuclear envelope.

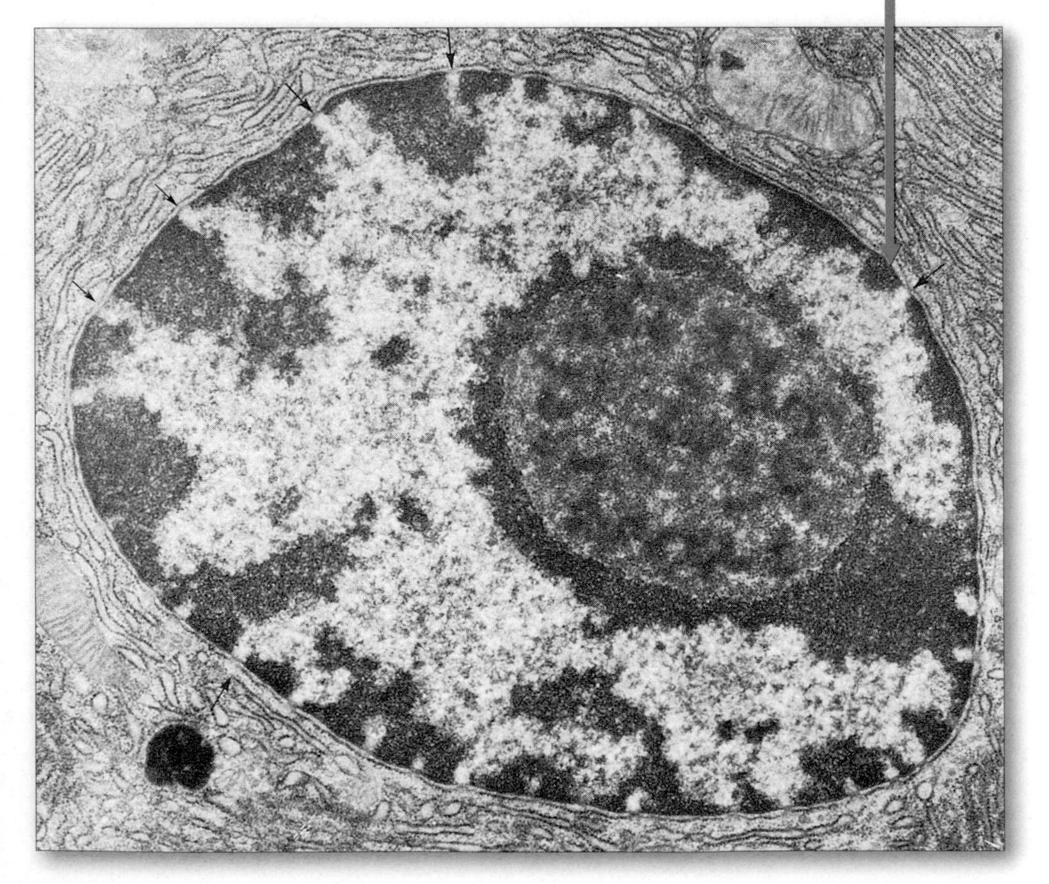

Eukaryotic cells protect their DNA inside a **nucleus**. Figure 4.11 shows the distinctive structure of this type of organelle, which has two functions. *First*, the nucleus physically tucks away all the DNA molecules, apart from the complex metabolic machinery of the cytoplasm. This localization of DNA makes it easier for parent cells to copy their genetic instructions before the time comes for them to divide. DNA molecules can be sorted out into parcels—one for each daughter cell that forms. *Second*, outer membranes of the nucleus form a boundary where cells can control the passage of substances and signals to and from the cytoplasm.

Nuclear Envelope

Unlike the cell itself, a nucleus has two outer membranes, one wrapped around the other. This double-membrane system is called a **nuclear envelope**. It consists of two lipid bilayers in which numerous protein molecules are embedded (Figure 4.12). It surrounds the fluid portion of the nucleus, which is the nucleoplasm.

Stitched on the innermost surface of the nuclear envelope are attachment sites for protein filaments. These anchor the molecules of DNA to the envelope and help keep them organized. On the outer surface of the envelope is a profusion of ribosomes. All proteins are built either on such membrane-bound ribosomes or on ribosomes located in the cytoplasm.

As is the case for all cell membranes, a nuclear envelope's lipid bilayers keep water-soluble substances from moving freely into and out of the nucleus. But many pores, each composed of clusters of proteins, span both bilayers. Ions and small, water-soluble molecules cross the nuclear envelope at the pores. As you will see, large molecules (including the subunits of ribosomes) cross the bilayers at pores in highly controlled ways.

Nucleolus

Take a look at the nucleus in Figure 4.11. Inside is a rather round, dense mass of material. What is it? While eukaryotic cells are growing, one or more of these masses form inside its nucleus. Each is a **nucleolus** (plural, nucleoli). Here, a large number of protein and RNA molecules

Further reading: Student Guide to InfoTrac on web site

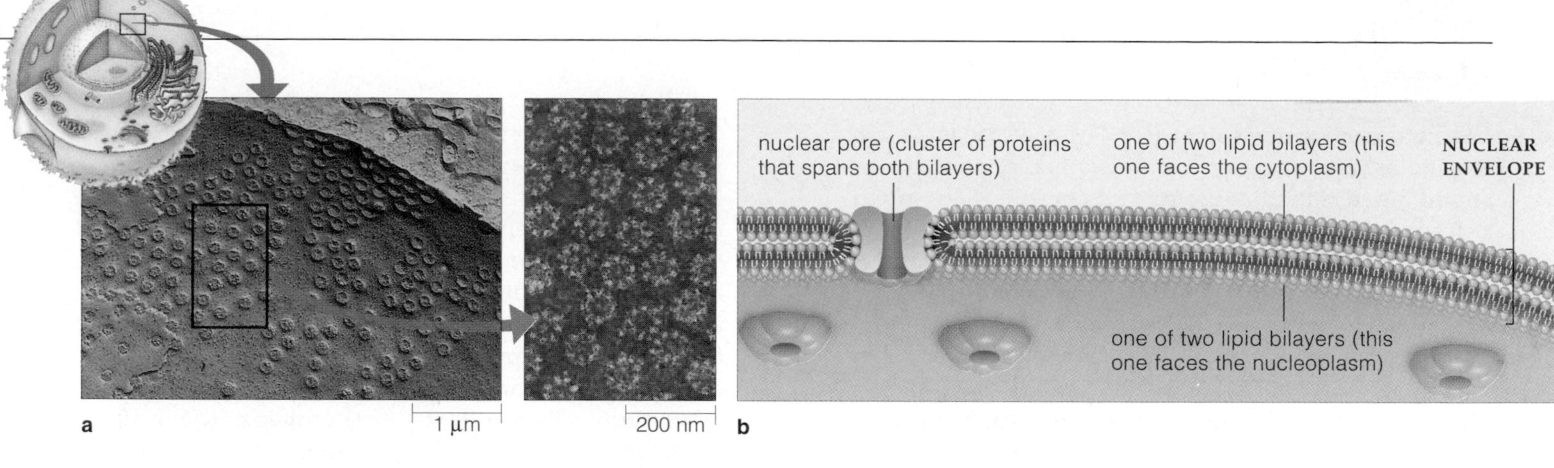

Figure 4.12 (**a**) Part of the outer surface of a nuclear envelope. *Left*: This specimen was fractured to reveal the intimate layering of its two lipid bilayers. *Right*: Nuclear pores. Each pore across the envelope is an organized array of membrane proteins. It permits the selective transport of substances into and out of the nucleus. (**b**) Sketch of the nuclear envelope's structure.

are being constructed. These particular materials are subunits from which ribosomes are built. The subunits pass through nuclear pores and reach the cytoplasm. At times of protein synthesis, intact ribosomes form (each from two subunits) in the cytoplasm.

Chromosomes

When eukaryotic cells are not busy dividing, their DNA looks like thin threads inside the nucleus. Many protein molecules are attached to the threads, a bit like beads on a string. Except at extreme magnification, the beaded threads have a grainy appearance, as in Figure 4.11. However, when the cell is preparing to divide, it duplicates its DNA molecules so each of its daughter cells will get all of the required hereditary instructions. In addition, each DNA molecule becomes folded and twisted into a condensed structure, proteins and all.

Early microscopists bestowed the name *chromatin* on the seemingly grainy substance and *chromosomes* on the condensed structures. We now define **chromatin** as the cell's collection of DNA, together with all proteins associated with it (Table 4.2). Each **chromosome** is one DNA molecule and its associated proteins, regardless of whether it is in threadlike or condensed form:

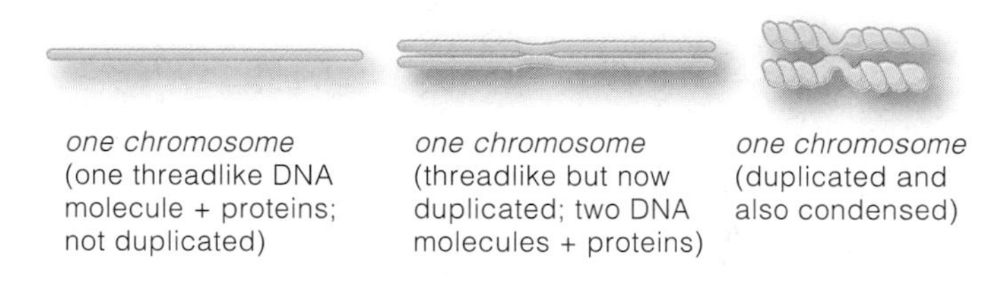

In other words, the appearance of "the chromosome" changes over the life of a eukaryotic cell. In chapters to come, you will look at different aspects of chromosomes, and you may find it useful to remember this point.

Table 4.2 Summary of Components of the Nucleus

Nuclear envelope	Pore-riddled double-membrane system that selectively controls the passage of various substances into and out of the nucleus
Nucleoplasm	Fluid interior portion of the nucleus
Nucleolus	Dense cluster of RNA and proteins that will be assembled into subunits of ribosomes
Chromosome	One DNA molecule and many proteins that are intimately associated with it
Chromatin	Total collection of all DNA molecules and their associated proteins in the nucleus

What Happens to the Proteins Specified by DNA?

Outside the nucleus, polypeptide chains for proteins are assembled on ribosomes. What happens to the new chains? Many become stockpiled in the cytoplasm or get used at once. Many others enter a cytomembrane system. As described in the next section, this system consists of different organelles, including endoplasmic reticulum, Golgi bodies, and vesicles.

Thanks to DNA's instructions, many proteins take on particular, final forms in the cytomembrane system. Lipids also are packaged and assembled in the system by enzymes and other proteins (which also were built according to DNA's instructions). As you will see next, vesicles deliver the proteins and lipids to specific sites within the cell or to the plasma membrane, for export.

The nucleus, an organelle with two outer membranes, keeps the DNA molecules of eukaryotic cells separated from the metabolic machinery of the cytoplasm.

The localization makes it easier to organize the DNA and to copy it before a parent cell divides into daughter cells.

Pores across the nuclear envelope help control the passage of many substances between the nucleus and cytoplasm.

4.5 THE CYTOMEMBRANE SYSTEM

The **cytomembrane system** is a series of organelles in which lipids are assembled and new polypeptide chains are modified into final proteins. Its products are sorted and shipped to different destinations. Figure 4.13 shows how its organelles—the ER, Golgi bodies, and various vesicles—functionally interconnect with one another.

Endoplasmic Reticulum

The functions of the cytomembrane system begin with **endoplasmic reticulum**, or **ER**. In animal cells, the ER is continuous with the nuclear envelope and extends through cytoplasm. Its membrane regions appear rough or smooth, depending mainly on whether ribosomes are attached to the membrane facing the cytoplasm.

We typically observe *rough* ER arranged into stacks of flattened sacs with many ribosomes attached (Figure 4.14*a*). Every new polypeptide chain is synthesized on ribosomes. But only the newly forming chains having a built-in signal can enter the space within rough ER or become incorporated into ER membranes. (The signal is a string of fifteen to twenty specific amino acids.) Once the chains are in rough ER, enzymes may attach oligosaccharides and other side chains to them. Many specialized cells secrete the final proteins. Rough ER is abundant in such cells. For example, in your pancreas, ER-rich gland cells make and secrete enzymes that end up in the small intestine and help digest your meals.

Smooth ER is free of ribosomes and curves through cytoplasm like connecting pipes (Figure 4.14*b*). Many cells assemble most lipids inside the pipes. Smooth ER is well developed in seeds. In liver cells, some drugs and toxic metabolic wastes are inactivated in it. Sarcoplasmic reticulum, a type of smooth ER in skeletal muscle cells, functions in muscle contraction.

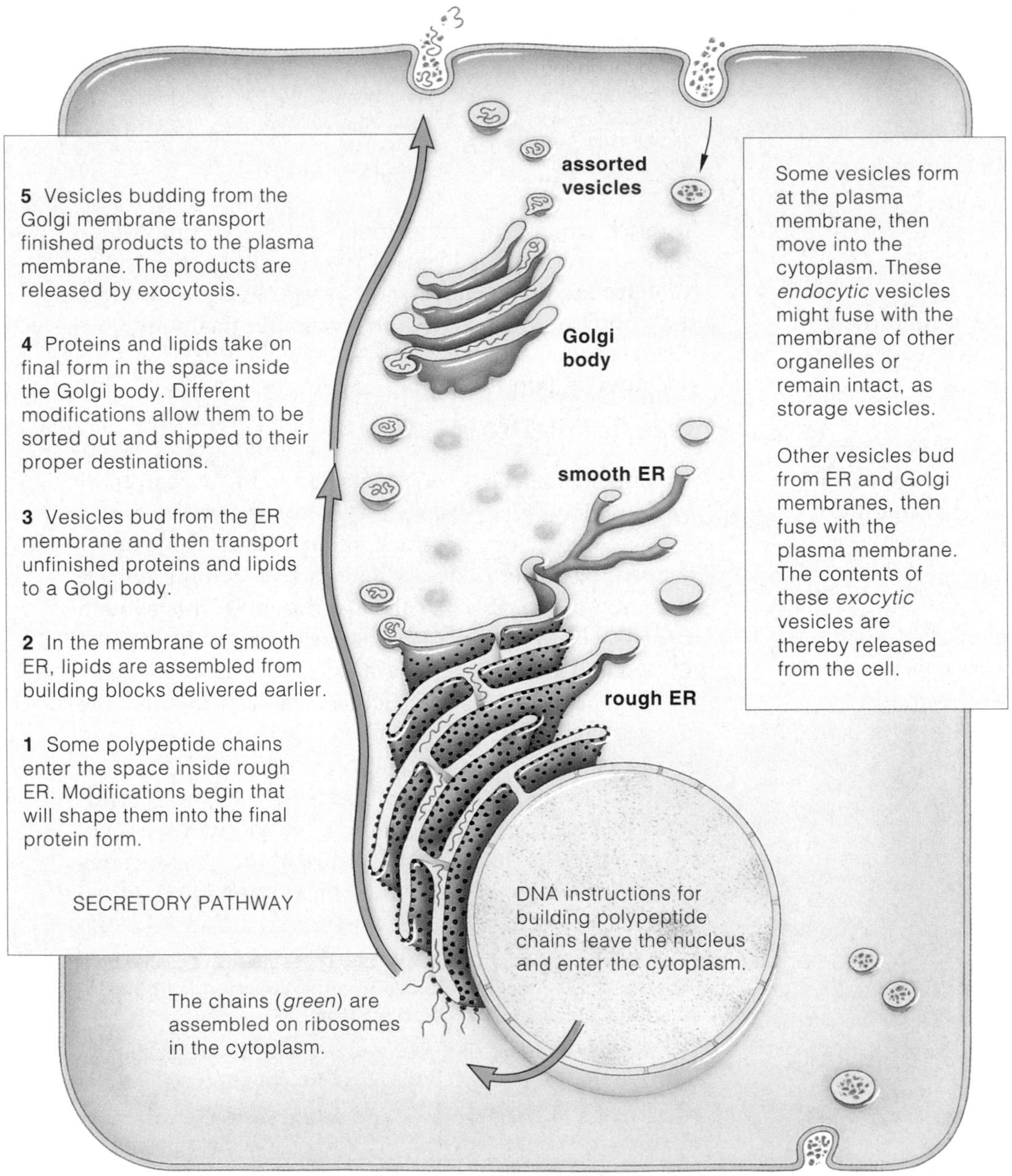

Golgi Bodies

In **Golgi bodies**, enzymes put the finishing touches on proteins and lipids, sort them out, and package them inside vesicles for shipment to specific locations. For example, an enzyme in one Golgi region might attach a phosphate group to a new protein, thereby giving it a mailing tag to its proper destination.

Commonly, a Golgi body looks vaguely like a stack of pancakes; it is composed of a series of flattened membrane-bound sacs (Figure 4.15). In functional terms, the last portion of a Golgi body corresponds to the top pancake. Here, vesicles form as patches of the membrane bulge out, then break away into the cytoplasm.

Figure 4.13 Cytomembrane system, a membrane system in the cytoplasm that assembles, modifies, packages, and ships proteins and lipids. *Green* arrows highlight a secretory pathway by which certain proteins and lipids are packaged and released from many types of cells, including gland cells that secrete mucus, sweat, and digestive enzymes.

Further reading: Student Guide to InfoTrac on web site →

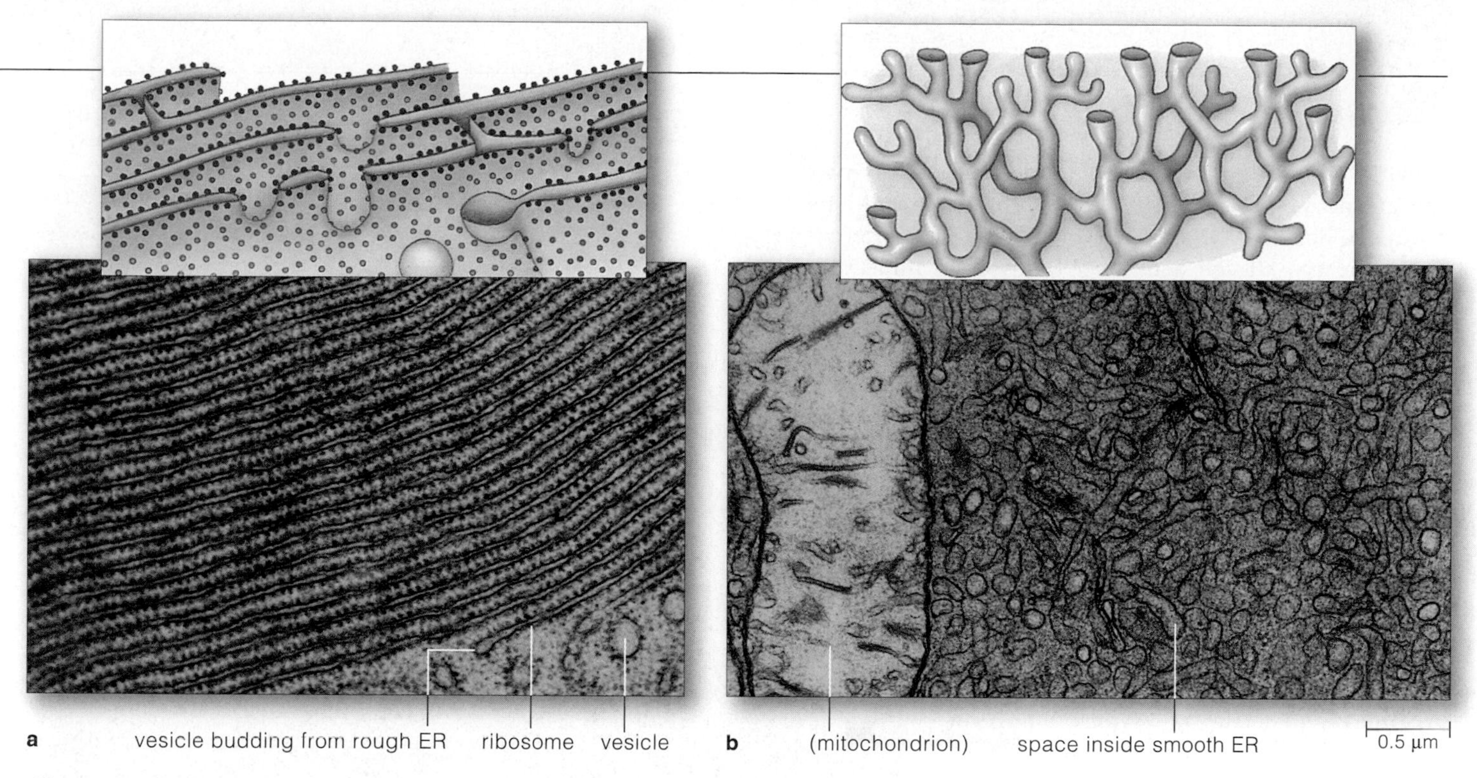

Figure 4.14 Transmission electron micrographs and sketches of endoplasmic reticulum. (**a**) Many ribosomes dot the flattened surfaces of rough ER that face the cytoplasm. (**b**) This section shows the diameters of the many interconnected, pipelike regions of smooth ER.

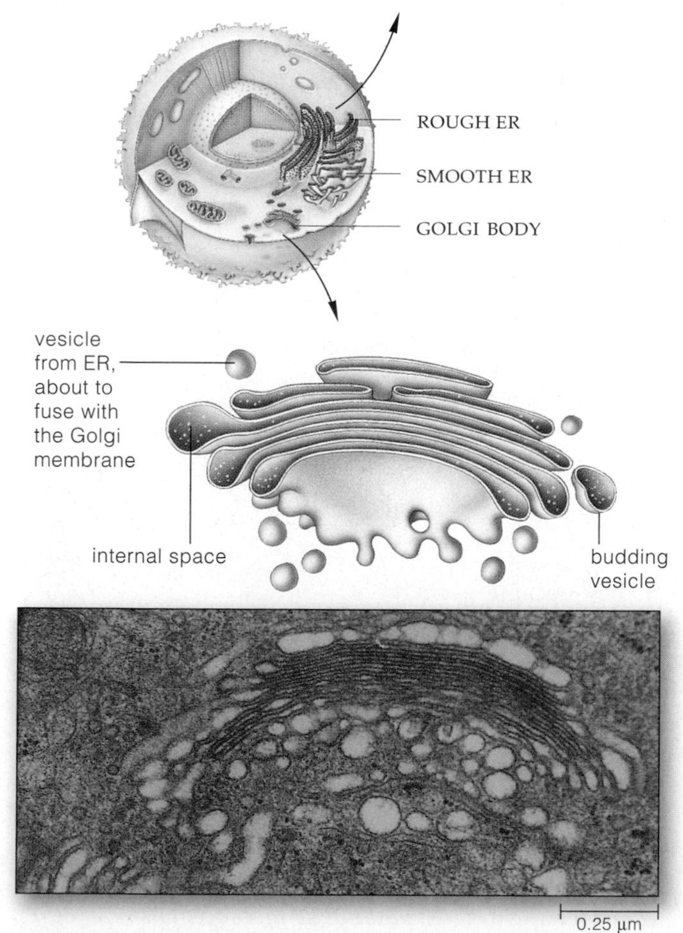

Figure 4.15 Sketch and micrograph of a Golgi body from an animal cell.

A Variety of Vesicles

Vesicles are tiny, membranous sacs that move through the cytoplasm or take up positions in it. A common type, the lysosome, buds from Golgi membranes of animal cells and certain fungal cells. Lysosomes are organelles of intracellular digestion. They contain a potent brew, rich with diverse enzymes that speed the breakdown of proteins, complex carbohydrates, nucleic acids, and some lipids. Often, lysosomes fuse with vesicles that formed at the plasma membrane. The vesicles typically contain molecules, bacteria, or other items that docked at the plasma membrane. Lysosomes even digest whole cells or cell parts. For example, as a tadpole is developing into an adult frog, its tail slowly disappears. Lysosomal enzymes are responding to developmental signals and are helping to destroy cells that make up the tail.

Peroxisomes, another type, are tiny sacs of enzymes that break down fatty acids and amino acids. Hydrogen peroxide, a potentially harmful product, forms during the reactions. Enzyme action converts it to water and oxygen or channels it into reactions that break down alcohol. After someone drinks alcohol, nearly half of it is degraded in peroxisomes of liver and kidney cells.

Many proteins take on final form and lipids are synthesized in the ER and Golgi bodies of the cytomembrane system.

Lipids, proteins (such as enzymes), and other items become packaged in vesicles destined for export, storage, membrane building, intracellular digestion, and other cell activities.

4.6 MITOCHONDRIA

Recall, from Section 3.7, that ATP molecules are premier energy carriers. Energy associated with their phosphate groups is delivered to nearly all reaction sites, where it drives nearly all cell activities. Many ATP molecules form when organic compounds are completely broken down to carbon dioxide and water in a **mitochondrion** (plural, mitochondria).

Only eukaryotic cells contain these organelles. The example in Figure 4.16 will give you an idea of their structure. The kind of ATP-forming reactions that proceed in mitochondria extract far more energy from organic compounds than can be done by any other means. They cannot run to completion without plenty of oxygen. Like all other land-dwelling vertebrates, every time you breathe in, you take in oxygen mainly for mitochondria in cells—in your case, many trillions of individual cells.

Each mitochondrion has a double-membrane system. As you can tell from Figure 4.8, the outermost membrane faces the cytoplasm. Most commonly, the inner membrane repeatedly folds back on itself. Each inner fold is a crista (plural, cristae).

What is the function of the intricate membrane system? It forms two distinct compartments within a mitochondrion. Enzymes and other proteins stockpile hydrogen ions in the outer compartment. Electron transfers drive the stockpiling, and oxygen helps keep the machinery running by binding and thus removing the spent electrons. Hydrogen ions flow out of the compartment in controlled ways. Energy inherent in the flow drives ATP formation, as Chapter 8 describes.

All eukaryotic cells contain one or more mitochondria. You may find only one in a single-celled yeast. You might find a thousand or more in energy-demanding cells, such as those of muscles. Take a look at the profusion of mitochondria in Figure 4.10, which is a micrograph of merely one thin slice from a liver cell. It alone tells you that the liver is an exceptionally active, energy-demanding organ.

In terms of size and biochemistry, mitochondria resemble bacteria. They even have their own DNA and some ribosomes, and they divide on their own. Perhaps they evolved from ancient bacteria that were engulfed by a predatory, amoebalike cell yet managed to escape digestion. Perhaps they were able to reproduce inside the cell and continued doing so in the descendant cells. If they became permanent, protected residents, they might have lost structures and functions required for independent life while they were becoming mitochondria. We return to this topic in Section 21.3.

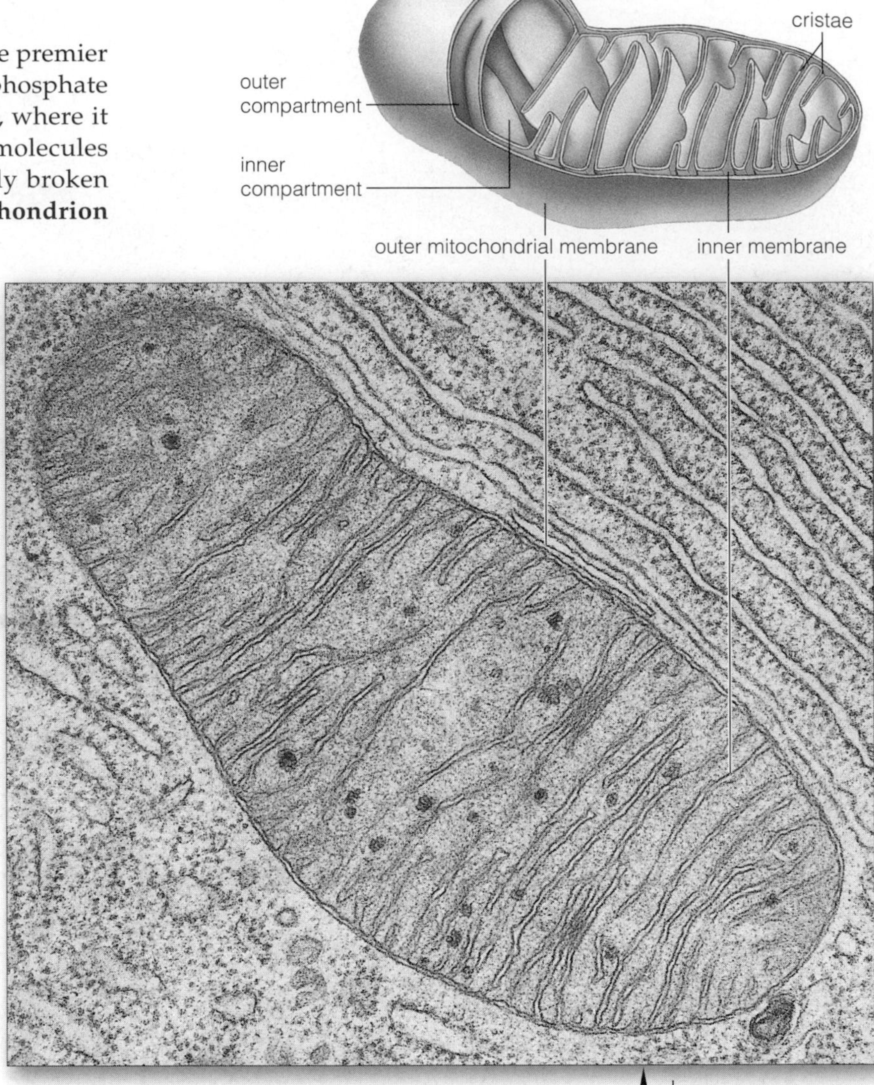

Figure 4.16 Sketch and transmission electron micrograph, thin section, of a typical mitochondrion. This organelle specializes in the production of large quantities of ATP, the main energy-carrying molecule between different reaction sites in cells. The process that produces ATP in mitochondria cannot proceed without free oxygen.

The organelles called mitochondria are the ATP-producing powerhouses of all eukaryotic cells.

Energy-releasing reactions proceed at the compartmented, internal membrane system of mitochondria. The reactions, which require oxygen, produce far more ATP than can be made by any other cellular reactions.

Further reading: Student Guide to InfoTrac on web site

SPECIALIZED PLANT ORGANELLES

Chloroplasts and Other Plastids

Many plant cells contain plastids, a general category of organelles that specialize in photosynthesis or function in storage. Three types are common in different parts of plants. They are the chloroplasts, chromoplasts, and amyloplasts.

Of all eukaryotic cells, only the photosynthetic ones have **chloroplasts**. These organelles convert sunlight energy into the chemical energy of ATP, which is used to make sugars and other organic compounds. Chloroplasts commonly are oval or disk-shaped. Their semifluid interior, the stroma, is enclosed by two outer membrane layers. In the stroma is a third membrane called the thylakoid membrane. It is folded into a system of interconnecting, disk-shaped compartments. In many chloroplasts, these compartments stack, one atop the other, as in Figure 4.17. Each stack is a granum (plural, grana).

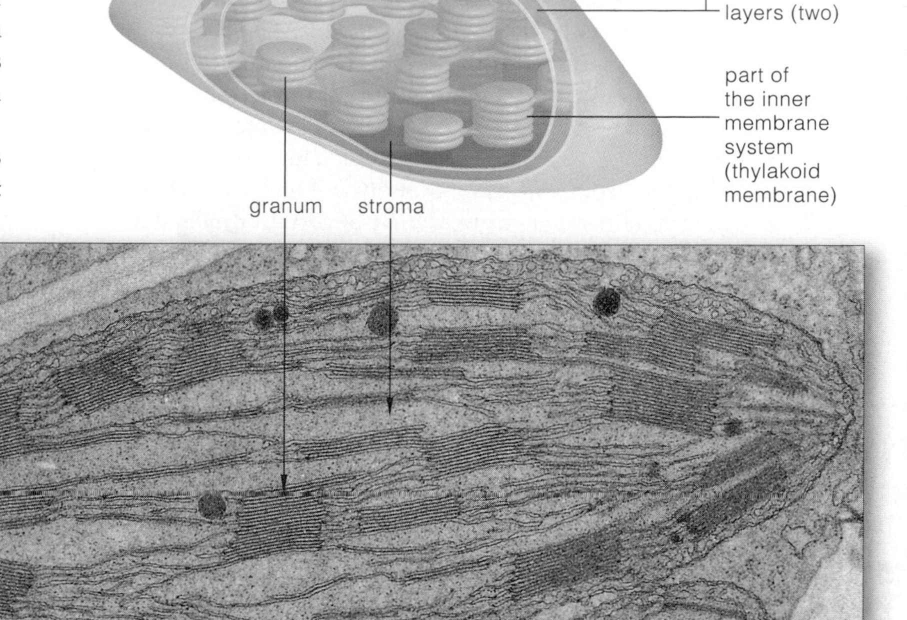

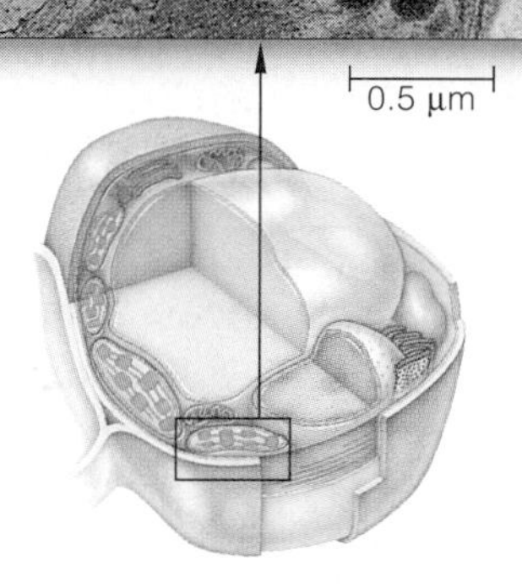

Figure 4.17 Generalized sketch of a chloroplast, the key defining feature of every photosynthetic eukaryotic cell. The transmission electron micrograph shows a thin section of a chloroplast from a photosynthetic corn cell.

The first stage of photosynthesis starts and ends at a thylakoid membrane. Many light-trapping pigments, enzymes, and other proteins carry out the reactions. They work together to absorb light energy and "store" it in the form of ATP. Inside the stroma, ATP energy is used to make sugars, and then starch and other organic compounds, from carbon dioxide and water. Clusters of the new starch molecules (starch grains) may briefly accumulate inside the stroma.

The most abundant photosynthetic pigments are chlorophylls, which reflect or transmit green light. Others include carotenoids, which reflect or transmit yellow, orange, and red light. The relative abundances of the different pigments influence the colors of plant parts.

In many ways, chloroplasts resemble photosynthetic bacteria. Like mitochondria, they might have evolved from bacteria that were engulfed by predatory cells but escaped digestion and became permanent residents in them. We return to this idea in Section 21.3.

Unlike chloroplasts, chromoplasts lack chlorophylls but have an abundance of carotenoids. They are the source of red-to-yellow colors of many flowers, autumn leaves, ripening fruits, and carrots and other roots. The pigment colors commonly attract animals that pollinate plants or disperse seeds. Amyloplasts lack pigments. Often they store starch grains and are abundant in cells of stems, potato tubers (underground stems), and seeds.

Central Vacuole

Many mature, living plant cells have a **central vacuole** (Figure 4.8). This fluid-filled organelle stores amino acids, sugars, ions, and toxic wastes. As it enlarges, it causes fluid pressure to build up inside the cell and so forces the cell's still-pliable cell wall to enlarge. Hence the cell itself enlarges. As the cell surface area increases, so does the rate at which water and other substances can be absorbed across the plasma membrane.

In most cases, the central vacuole increases so much in volume that it takes up 50 to 90 percent of the cell's interior. The cytoplasm ends up as a very narrow zone between the central vacuole and plasma membrane.

Photosynthetic eukaryotic cells contain chloroplasts and other plastids that function in food production and storage.

Many plant cells have a central vacuole. When this storage vacuole enlarges during growth, cells are forced to enlarge, and this increases the surface area available for absorption.

4.8 COMPONENTS OF THE CYTOSKELETON

An interconnected system of fibers, threads, and lattices extends between the nucleus and plasma membrane of eukaryotic cells. This system, the **cytoskeleton**, gives cells their internal organization, shape, and capacity to move. Some elements reinforce the plasma membrane and nuclear envelope. Others form scaffolds that hold protein clusters in membranes or cytoplasmic regions. Others are like railroad tracks upon which organelles are shipped from one site to another! Many elements are permanent; others appear only at certain times in a cell's life. Before a cell divides, for instance, many new microtubules form a "spindle" structure that moves its chromosomes, then disassemble when the task is done.

Figure 4.18 shows the cytoskeleton of an animal cell, isolated from its home tissue, as it was stretching out from left to right across a glass slide. Researchers were studying its **microtubules** and **microfilaments**. These two classes of cytoskeletal elements, acting singly or collectively, are responsible for nearly all movements of eukaryotic cells. In addition, some animal cells have **intermediate filaments**, which are ropelike cytoskeletal elements that impart mechanical strength to cells and tissues. All three classes grow through polymerization, by which many monomers become joined together.

Microtubules

A microtubule is a long, hollow cylinder, twenty-five nanometers wide, made of tubulin monomers (Figure 4.19*a*). It is the largest cytoskeletal element. Tubulin is a protein made of two chemically distinct polypeptide chains, each folded into a globular shape. In a growing microtubule, all tubulin subunits become oriented in the same direction, which puts slightly different chemical and electrical properties at opposite ends of the cylinder. Thus a microtubule shows polarity. One end (the *minus* end) tends to lose monomers. Usually it is stabilized by becoming anchored in a centrosome, a type of MTOC. (MTOCs, short for *M*icrotubule *O*rganizing *C*enters, are sites of dense material that give rise to microtubules.) The cylinder's other end (the *plus* end) is free to grow rapidly through the cytoplasm.

Each microtubule shows dynamic instability: it can suddenly shorten (depolymerize) as well as lengthen. By one model, GTP becomes bound to free tubulin. (GTP is an organic compound made of guanine, ribose, and three phosphate groups.) Binding puts tubulin in a shape that fits right in at the plus end, like a brick being added to a wall. Then hydrolysis removes one phosphate group from GTP to form GDP. Regions of the cylinder that contain GDP are not stable. However, when "bricks" stack up at the growing end faster than hydrolysis is occurring, they act as a cap that stabilizes the cylinder. When conditions trigger loss of the GTP cap, the cylinder curves, weakens, and disassembles.

Microtubules govern the division of cells and some aspects of their shape as well as many cell movements. Cells cannot do much without them. As you might well imagine, microtubules have become prime targets in the chemical warfare between vulnerable species and their attackers. For example, autumn crocus and other plants of the genus *Colchicum* synthesize colchicine, a poison that inhibits the assembly and promotes disassembly of microtubules, with dire effects on browsing animals. The plant cells themselves have an evolved insensitivity to colchicine, which cannot bind well to the type of tubulin monomers they produce.

Taxol, a poison offered up by the western yew (*Taxus brevifolia*), can stabilize assembled microtubules and so prevent formation of new ones that a cell might require for other tasks. For example, like colchicine and some other microtubule poisons, taxol is able to block cell division by interfering with the formation of

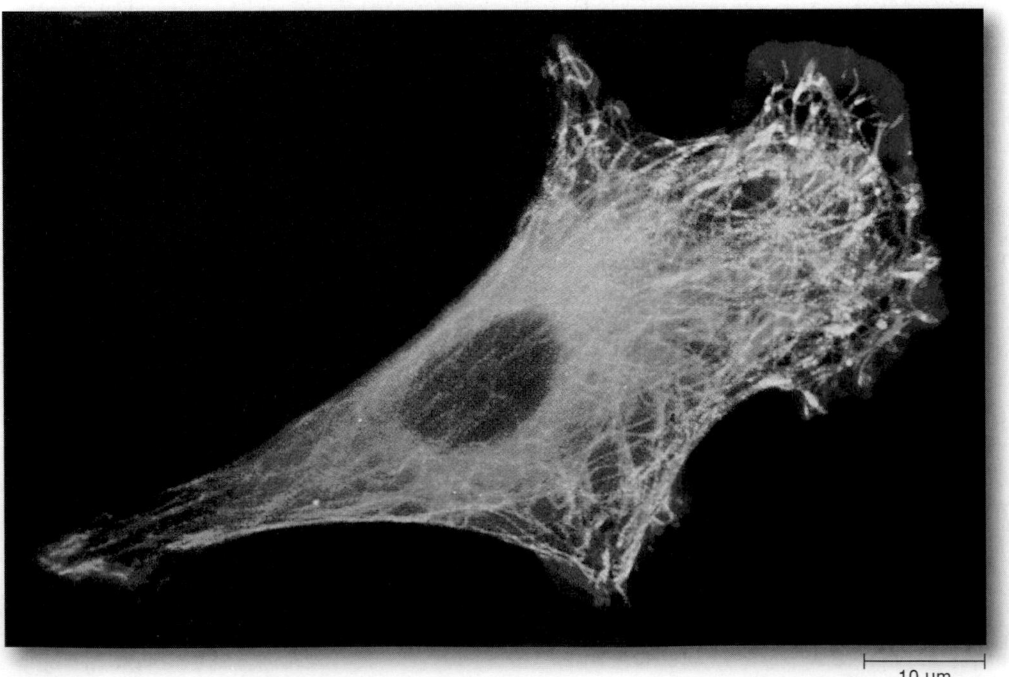

Figure 4.18 Locations of three types of structural elements in the cytoskeleton of a fibroblast, a type of animal cell that gives rise to certain animal tissues. This composite of three micrographic images reveals the presence of many microtubules (tinted *green*). Two kinds of microfilaments are tinted *blue* and *red*.

Further reading: Student Guide to InfoTrac on web site

microtubular spindles. Doctors have used it to decrease the uncontrolled cell divisions that underlie the growth of some benign and malignant tumors.

Microfilaments

Microfilaments, the thinnest of the cytoskeletal elements, are five to seven nanometers wide (Figure 4.19*b*). Each is made of two polypeptide chains that are helically twisted together. The chains are assembled from monomers of the protein actin.

Like microtubules, the actin filaments are polar and have a dynamic structure. Monomers tend to be added to one end and removed from the other end. Unlike microtubules, however, they generally are organized in bundles or networks. As explained in several chapters in the book, microfilaments take part in a great variety of movements, especially the kinds that affect the cell surface. They also contribute to the development and maintenance of animal cell shapes.

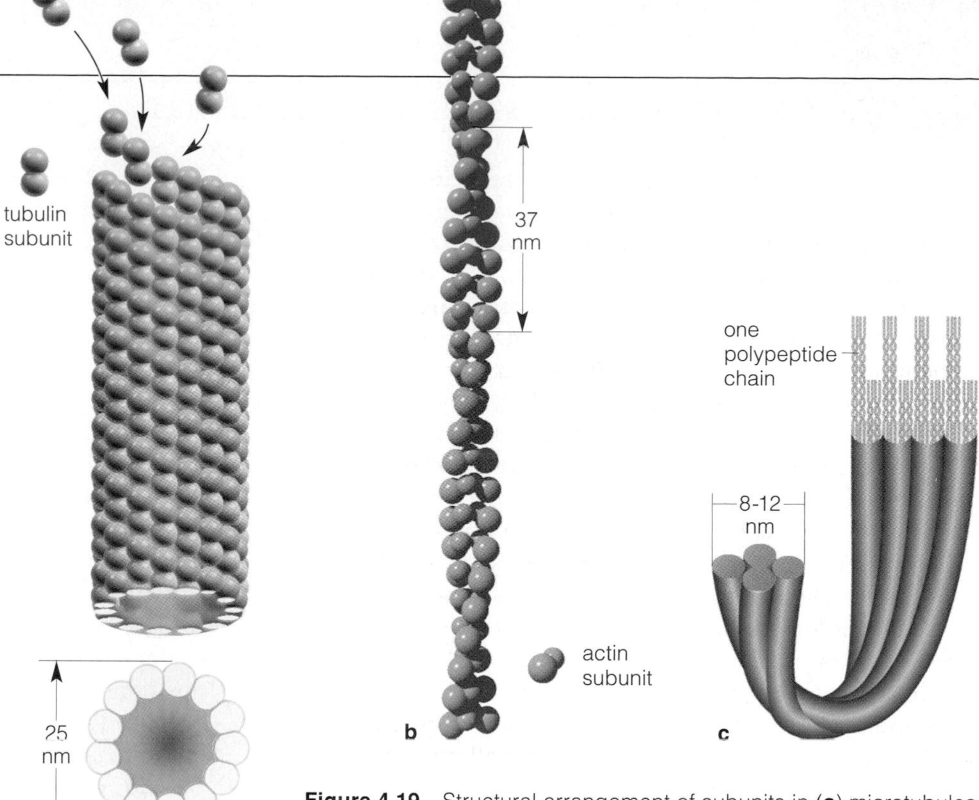

Figure 4.19 Structural arrangement of subunits in (**a**) microtubules, (**b**) microfilaments, and (**c**) one of the intermediate filaments.

Myosin and Other Accessory Proteins

Both tubulin and actin have been highly conserved over evolutionary time; the microtubules and microfilaments of all eukaryotic species are assembled from them. In spite of the uniformity, monomers of a variety of other proteins become attached to them. Thus embellished, microtubules and microfilaments that were assembled according to the same basic pattern can serve different functions in different regions of the cell.

For example, monomers of myosin, dynein, or some other **motor protein** typically attach to the surface of microtubules and microfilaments in ways that cause cell movement. Myosin is abundant in muscle cells, and it is part of the machinery by which they contract. "Crosslinking proteins" splice adjacent microfilaments together. Some kinds take part in the formation of the **cell cortex**—an extensive, three-dimensional network of microfilaments and other proteins just beneath the plasma membrane. The cortex reinforces the cell surface and facilitates movements as well as changes in shape. Other crosslinking proteins splice microfilaments in a stable network having the properties of a gel. Spectrin, another accessory protein, attaches microfilaments to the plasma membrane. As one more example, the integrins span this outermost membrane, attach to microfilaments inside the cell, and connect them with proteins outside.

Intermediate Filaments

Intermediate filaments, the most stable elements of the cytoskeleton, are between eight and twelve nanometers wide (Figure 4.19*c*). The six known groups mechanically strengthen cells or cell parts and help maintain their shape. For example, desmins and vimentins help hold the contractile units of muscle cells in organized arrays. Lamins help form a scaffold that reinforces the nucleus. Diverse cytokeratins structurally reinforce the cells that give rise to nails, claws, horns, and hairs.

Unlike the other two classes of cytoskeletal elements, intermediate filaments are present only in animal cells of specific tissues. Moreover, because each cell usually has only one or sometimes two kinds, researchers can use intermediate filaments to identify cell type. Such typing has proved to be a useful tool in diagnosing the tissue origin of different forms of cancer.

Every eukaryotic cell has a cytoskeleton, the diverse elements of which are the basis of its shape, internal structure, and capacity for movement.

Microtubules are key organizers of the cytoskeleton and help move certain cell structures. Microfilaments take part in diverse movements and in the formation and maintenance of cell shape. Intermediate filaments structurally reinforce certain animal cells and internal cell structures.

4.9 THE STRUCTURAL BASIS OF CELL MOTILITY

If a cell lives, it moves. This is true of free-living single cells, such as the sperm and ciliated cells in Figure 4.20. It is true also of cells with fixed positions in the tissues of multicelled organisms. Eukaryotic cells rearrange or shunt organelles and chromosomes, contract (shorten), thrust out long or wide lobes of cytoplasm, stir fluids, or bend a tail and propel the cell body forward. These and all other movements of cell structures, the cell itself, and entire multicelled organisms are forms of motility.

Section 4.8 introduced the major players in cellular motility—the microtubules, microfilaments, and other proteins that interact with them. Energy inputs, as from ATP, trigger motions at the molecular level. Many such motions combine to produce movement at the cellular level, as when a muscle cell contracts. The coordinated movements of many cells cause movements of tissues or organs, as when bundles of skeletal muscle cells interact with bones to make a thumb and finger turn a page.

Mechanisms of Cell Movements

Microfilaments, microtubules, or both take part in most aspects of motility. They do so by three mechanisms.

First, *the length of a microtubule or microfilament can grow or diminish by the controlled assembly or disassembly of its subunits.* When either lengthens or shortens at one end, a chromosome or some other structure attached to the other end is pushed or dragged through cytoplasm.

Some cells crawl about on protrusions of the body surface that form by microfilament assembly. *Amoeba proteus*, a soft-bodied protistan, has **pseudopods** ("false feet"): temporary, lobelike protrusions from the body. Inside each lobe, microfilaments grew rapidly in length, and the attached plasma membrane was dragged along with them. Similarly, the animal cell in Figure 4.18 was migrating on sheetlike extensions of microfilaments.

Second, *parallel rows of microfilaments or microtubules actively slide in specific directions.* For instance, a muscle cell has a series of contractile units. Microfilaments of actin are attached to one end or the other of each unit, and many myosin strands lie parallel with them. ATP energizes the myosin, which has oarlike projections that repeatedly bind and release microfilament neighbors. The short, repeated strokes make the microfilaments slide over the myosin strands, toward the center of the contractile unit—which shortens (contracts) as a result.

A similar sliding mechanism might be operating as cells crawl about. If the microfilament network beneath the plasma membrane at a cell's trailing end contracts, then some cytoplasmic gel would be squeezed into the leading end, which would bulge forward in response.

Third, *microtubules or microfilaments shunt organelles or parts of the cytoplasm from one location to another.* For example, chloroplasts move to new light-intercepting positions in response to the changing angle of the sun's overhead position. How? They are attached to myosin monomers that are "walking" over microfilaments and carrying chloroplasts with them. This shunt mechanism also causes cytoplasmic streaming: a dynamic and often rapid flowing of certain components in the cytoplasmic gel. Observe a living plant cell with a light microscope, and you may see pronounced streaming.

Flagella and Cilia

To biologists, the **flagellum** (plural, flagella) and **cilium** (plural, cilia) are classic examples of structures for cell motility. Both motile structures have a ring of nine pairs of microtubules and a central pair. A system of spokes and links stabilizes this "9 + 2 array." The array arises from a **centriole**, a barrel-shaped structure that is one type of microtubule-producing center. A centriole remains at the base of the completed array, as in Figure 4.21. In that location it is known as a **basal body**.

How do flagella and cilia differ? Flagella typically are longer and less profuse than cilia. Sperm and many other free-living cells use flagella as whiplike tails when swimming (Figure 4.20*a*). In many multicelled species, ciliated epithelial cells stir air or fluid. Many thousands line the airways in your chest (Figure 4.20*b*). When they beat, they direct mucus-trapped particles away from the lungs.

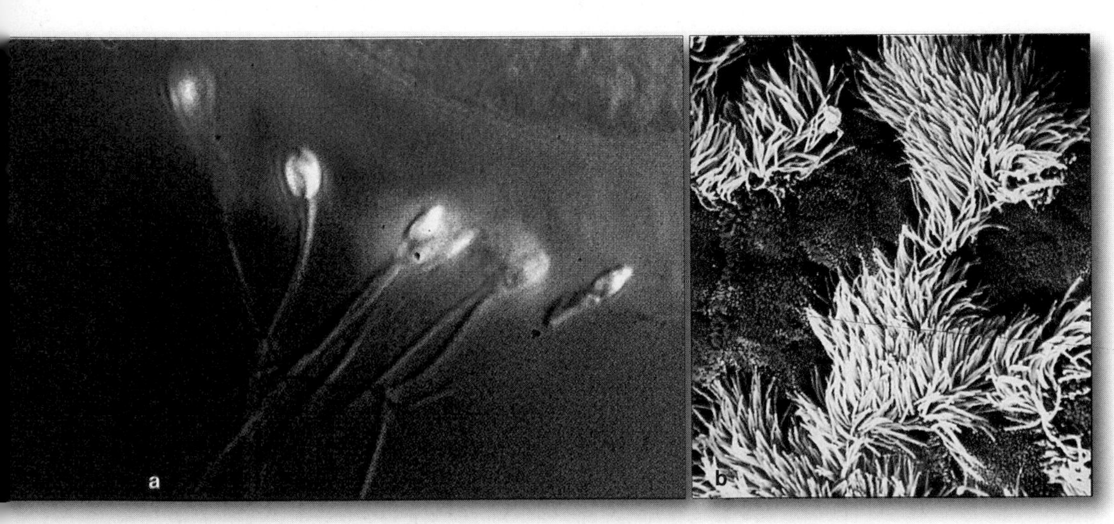

Figure 4.20 Flagella and cilia. (**a**) Human sperm consisting of a DNA-packed head, mitochondria, and a flagellum. (**b**) From the lining of an airway to the lungs, a view of the free surface of mucus-secreting cells (*brown*) and ciliated cells (*yellow*).

Further reading: Student Guide to InfoTrac on web site

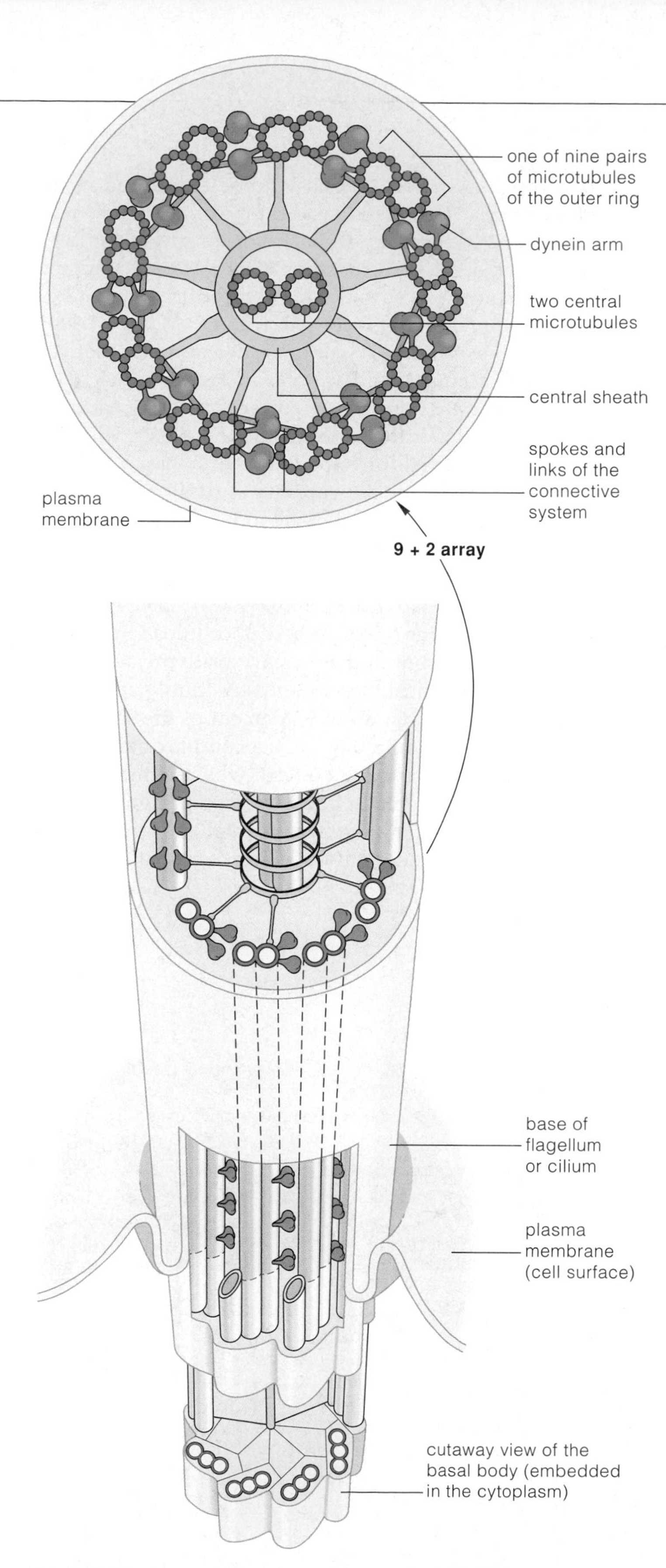

Figure 4.21 Internal organization of flagella and cilia. Both motile structures have a system of microtubules and a connective system of spokes and linking elements. Nine pairs (doublets) of the microtubules are arranged as an outer ring around two central microtubules. This is a 9 + 2 array.

Figure 4.22 Model of a sliding mechanism responsible for the beating of flagella and cilia, as proposed by cell biologists K. Summers and R. Gibbons. Inside an unbent flagellum (or cilium), all microtubule doublets extend the same distance into the tip. When the flagellum bends, doublets on the side that is bending the most are being displaced farthest from the tip, as shown here:

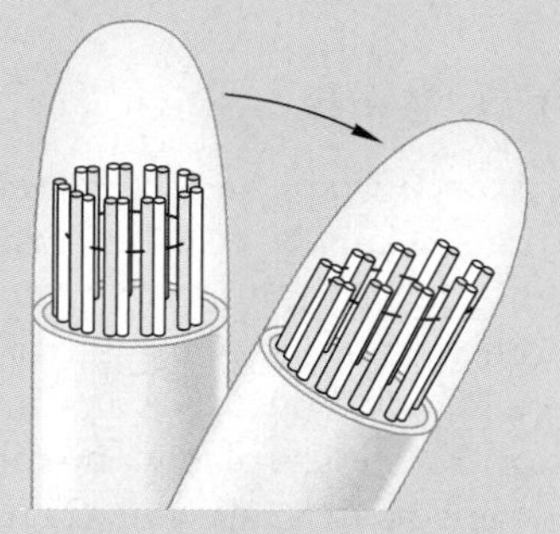

A sliding mechanism operates between the microtubule doublets of the outer ring. A series of motor proteins (dynein) form short arms that extend from each doublet toward the next doublet in the ring. Inputs of ATP energy cause the arms of one doublet to attach to the doublet in front of them, tilt in a short power stroke that pulls on the attached doublet, then release their hold. Repeated power strokes and cross-bridgings force the doublet to slide in a direction toward the base of the flagellum. As the attached doublet moves, *its* arms force the *next* doublet in line to slide down a bit, which forces the *next* doublet to slide down a bit also, and so on in sequence.

The system of interconnecting spokes and links extends through the length of the flagellum and prevents doublets from sliding totally out of the 9 + 2 array. This restriction forces the flagellum to bend and thus accommodate the internal displacement of the sliding doublets.

In short, in the 9 + 2 array of flagella and cilia, energized motor proteins make the nine doublets slide in sequence, and restrictions imposed by a system of spokes and links convert the doublet sliding into a bending motion.

Flagella and cilia beat by a sliding mechanism, as outlined in Figure 4.22. Extending from each pair of microtubules in the outer ring are short arms of motor proteins (dynein). When energized by ATP, the arms attach to the pair in front of them, tilt in a downward-directed short stroke, then release their hold. Repeated strokes make the pairs slide down in sequence. Spokes and links connect all the pairs. The collective downward displacement makes the flagellum or cilium bend.

Cell contractions and migrations, chromosome movements, and all other forms of cell motility arise at organized arrays of microtubules, microfilaments, and accessory proteins.

Different mechanisms cause these cytoskeletal elements to assemble or disassemble, slide past one another, and shunt structures to new locations.

4.10 CELL SURFACE SPECIALIZATIONS

This survey of eukaryotic cells concludes with a look at cell walls and some other specialized surface structures. Many of these architectural marvels are constructed of various secretions from the cells themselves. Others are cytoplasmic bridges or sets of membrane proteins that connect neighboring cells and allow them to interact.

Eukaryotic Cell Walls

Single-celled eukaryotic species are directly exposed to their surroundings. Many have a **cell wall**, a structural component that wraps continuously around the plasma membrane. A cell wall protects and physically supports its owner. The wall is porous, so water and solutes can easily move to and from the plasma membrane. A great variety of protistans have a cell wall. (Figure 4.23 shows one.) So do plant cells and many types of fungal cells.

For instance, young plant cells in actively growing regions secrete gluelike polysaccharides (such as pectin), glycoproteins, and cellulose. The cellulose molecules join together into ropelike strands, which become embedded in the gluey matrix. The secretions combine as a **primary wall** (Figure 4.24*a*). Primary walls are quite sticky, and they cement adjacent cells together. They also are thin and pliable, so the cell surface area can continue to enlarge under the pressure of incoming water.

Figure 4.23 From a freshwater habitat, one of the single-celled, walled protistans (*Ceratium*, a dinoflagellate).

At cell surfaces exposed to air, waxes and other cell secretions accumulate. The deposits form a cuticle. This semitransparent, protective surface covering restricts evaporative water loss from plants (Figure 4.25*a*).

Many plant cells develop only a thin wall. These are the cells that retain the capacity to divide or change shape during growth and development. As other plant cells mature, they stop enlarging and start secreting material on the primary wall's inner surface. The deposits combine to form a rigid, **secondary wall** that reinforces cell shape (Figure 4.24*e*). Whereas cellulose makes up less than 25 percent of the primary wall, the additional deposits now contribute more to structural support.

In woody plants, up to 25 percent of the secondary wall consists of lignin. Lignin is a complex molecule; it has a six-carbon ring structure to which a three-carbon chain and an oxygen atom are attached. Lignin makes plant parts stronger, more waterproof, and less inviting to insects and other plant-attacking organisms.

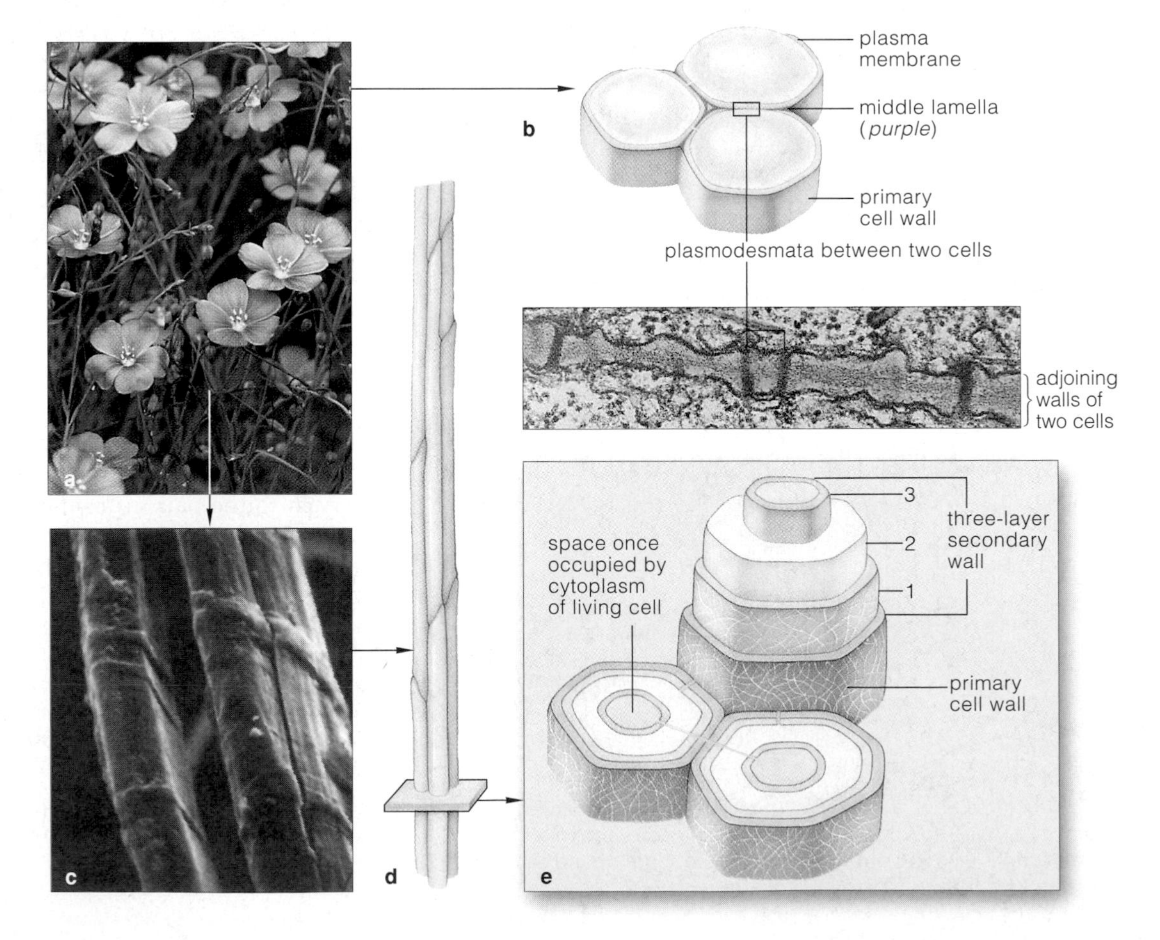

Figure 4.24 Examples of cell walls in flax plants (**a**). (**b**) Primary cell wall of young cells in flower petals. Cell secretions form the middle lamella, a layer between the walls of adjoining cells. The layer is thickest in adjoining corners. Plasmodesmata, membrane-lined channels across the adjacent walls, connect the cytoplasm of neighboring cells. (**c**,**d**) Part of the lustrous fibers in a cell from a flax stem. We make linen from such fibers, which are three times stronger than cotton fibers. (**e**) In flax fibers, as in many other types of plant cells, more layers become deposited inside the primary wall. The layers stiffen the wall and help maintain its shape. Later, the cell dies, leaving the stiffened walls behind. This also happens in water-conducting pipelines that thread through most plant tissues. Interconnected, stiffened walls of dead cells form the tubes.

Further reading: Student Guide to InfoTrac on web site →

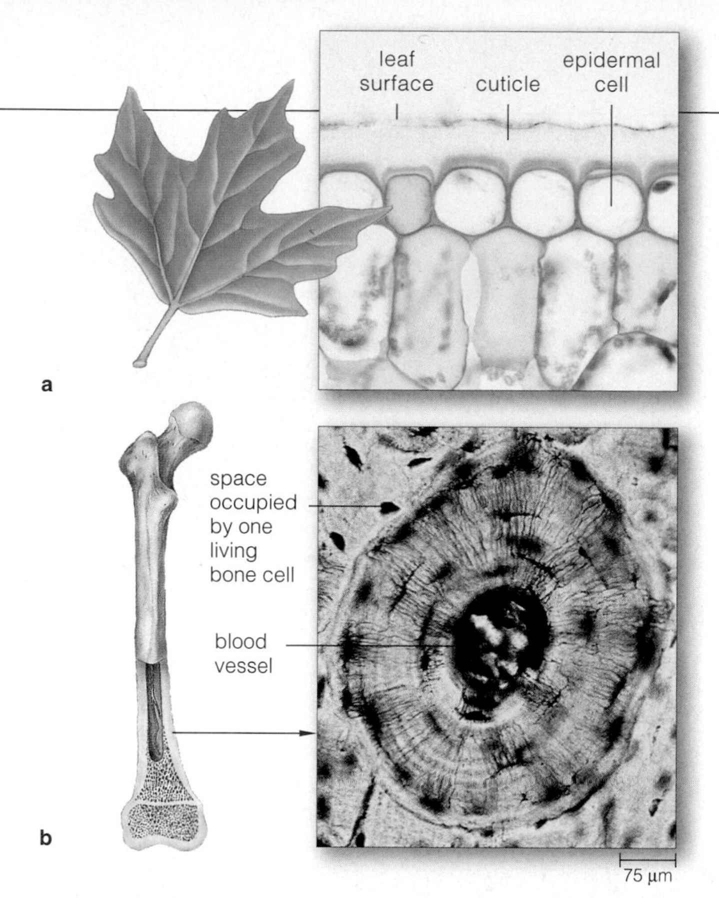

Figure 4.25 (**a**) Section through a plant cuticle, a surface layer composed of cell secretions. (**b**) Section through compact bone tissue, stained for microscopy.

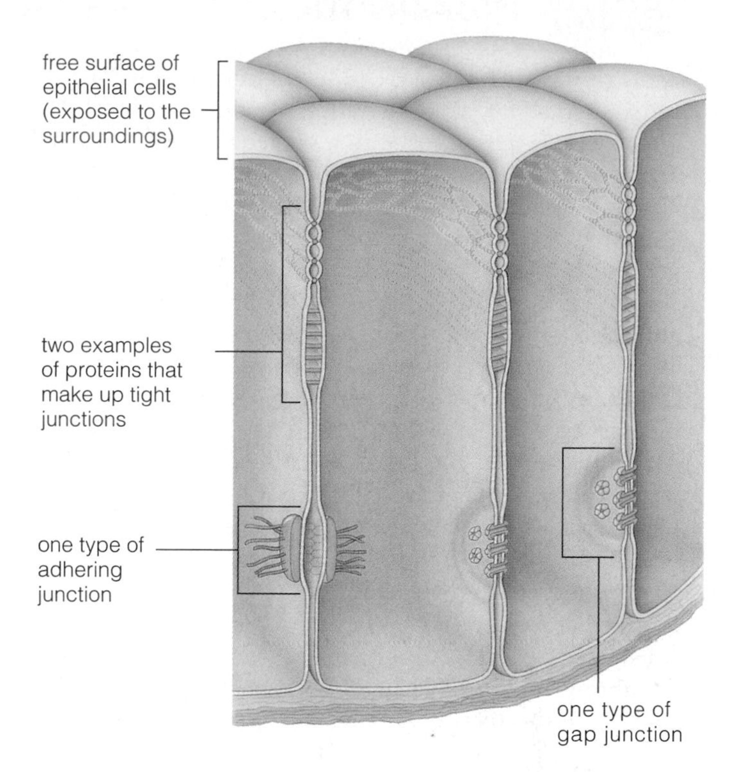

Figure 4.26 Composite drawing of the most common types of cell junctions in animals. These are cells from epithelial tissue.

Matrixes Between Animal Cells

Although animal cells have no walls, diverse matrixes composed of cell secretions and even materials drawn from the surroundings intervene between many of them. Think of cartilage at the knobby ends of your leg bones. Cartilage consists of scattered cells and their secretions, which form collagen or elastin fibers embedded in a "ground substance" of modified polysaccharides. As another example, an extensive matrix widely separates the living bone cells in bone tissue (Figure 4.25*b*).

Cell-to-Cell Junctions

Even when a wall or some other structure imprisons a cell in its own secretions, the only contact that cell has with the outside world is *through* its plasma membrane. In multicelled species, membrane components project into adjacent cells as well as the surrounding medium. Among the components are junctions where the cell sends and receives diverse signals and materials, where it recognizes and cements itself to cells of the same type.

In plants, for instance, many tiny channels cross the adjacent primary walls of living cells and interconnect their cytoplasm. Figure 4.24*b* shows a few. Each channel is a plasmodesma (plural, plasmodesmata). The plasma membranes of adjoining cells have merged and fully line the channels, so there can be an uninterrupted flow of substances between cells. Thus, living cells inside the plant body have the potential to exchange substances.

In most animal tissues, three categories of cell-to-cell junctions are common (Figure 4.26). *Tight* junctions link the cells of epithelial tissues, which line the body's outer surface and internal cavities and organs. They seal adjoining cells together; water-soluble substances can't leak between them. Thus gastric fluid cannot leak across the stomach's lining and damage surrounding tissues. *Adhering* junctions join cells in tissues of the skin, heart, and other organs subjected to stretching. *Gap* junctions link the cytoplasm of neighboring cells. They are open channels for a rapid flow of signals and substances.

We will be returning to the cell walls, intercellular substances, and cell-to-cell interactions in later chapters. For now, these are the points to remember:

A variety of protistan, plant, and fungal cells have a porous but protective wall that surrounds the plasma membrane. The cells themselves secrete wall-forming materials.

Cell secretions form a cuticle at the surfaces of plants and some animals, extracellular matrixes in tissues, and other specialized structures.

In multicelled organisms, coordinated cell activities depend on cell-to-cell junctions, which are protein complexes or cytoplasmic bridges that serve as physical links and sites of communication between cells.

4.11 PROKARYOTIC CELLS—THE BACTERIA

We turn now to bacteria. Unlike the cells you have considered so far, all bacterial cells are prokaryotic; their DNA is *not* enclosed in a nucleus. *Prokaryotic* means "before the nucleus." Biologists selected the word as a reminder that bacterial cells already had appeared on the Earth before the nucleus evolved in the forerunners of eukaryotic cells.

As a group, the bacteria are the smallest cells, although a rare exception was recently discovered. Most are not much more than a micrometer wide; even the rod-shaped species are only a few micrometers in length. In terms of their structure, bacteria are the simplest kinds of cells to think about. Most species have a semirigid or rigid cell wall that wraps around the plasma membrane, structurally supports the cell, and imparts shape to it (Figure 4.27*a*). Dissolved substances can move freely to and from the plasma membrane because the wall is permeable. Often, sticky polysaccharides surround the cell wall. They help the bacterium attach to interesting surfaces, such as river rocks, teeth, and the vagina. In many of the disease-causing (pathogenic) bacteria, the polysaccharides form a thick, jellylike capsule that surrounds and helps protect the wall.

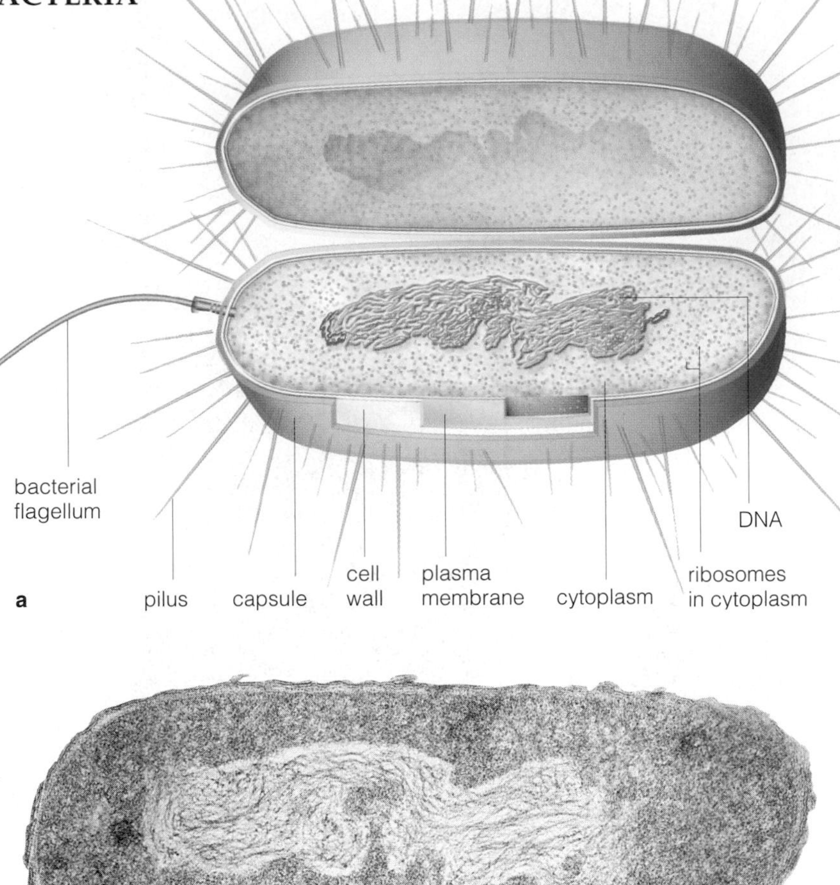

Figure 4.27 (**a**) Generalized sketch of a typical prokaryotic body plan. (**b**) Micrograph of the bacterium *Escherichia coli.*

Your own gut is home to a large population of a normally harmless strain of *E. coli.* A harmful strain has repeatedly contaminated large quantities of meat sold commercially. The same strain also has contaminated hard apple cider sold at a few roadside stands. Cooking the meat thoroughly or boiling the cider would have killed the bacterial cells. Where this was not done, people who ate the meat or drank the cider became quite sick. Some died.

Facing page: (**c**) Researchers manipulated this *E. coli* cell to release its single, circular molecule of DNA. (**d**) Cells of different bacterial species are shaped like balls, rods, or corkscrews. The ball-shaped cells of *Nostoc*, a photosynthetic bacterium, stick together inside a thick, gelatin-like sheath of their own secretions. Chapter 22 gives other fine examples. (**e**) Like this *Pseudomonas marginalis* cell, many species have one or more bacterial flagella, motile structures that propel cells through fluid environments.

Like eukaryotic cells, bacteria have a plasma membrane that helps to control the movement of substances to and from the cytoplasm. A bacterial plasma membrane, too, has proteins that serve as channels, transporters, and receptors for signals and substances. It incorporates built-in machinery for metabolic reactions, such as the breakdown of energy-rich compounds. In photosynthetic types, clusters of some membrane proteins harness light energy and convert it to chemical energy in ATP.

Bacterial cells are too small to contain more than a small volume of cytoplasm, but they have many ribosomes upon which polypeptide chains are assembled. Apparently, these cells are small enough and so internally simple that they do not require a cytoskeleton.

The cytoplasm of a bacterium is distinct from that of eukaryotic cells. It is continuous with an irregularly shaped region of DNA. Membranes do not surround the region, which is named a nucleoid (Figure 4.27*c*). A circular molecule of DNA, also called the bacterial chromosome, occupies this region.

Extending from the surface of many bacterial cells are one or more threadlike motile structures known as bacterial flagella (singular, flagellum). These are not the same as eukaryotic flagella, because the 9 + 2 array of microtubules is absent. Bacterial flagella help a cell move rapidly in its fluid surroundings. Other surface projections include pili (singular, pilus). These are the main protein filaments that help many kinds of bacteria attach to various surfaces, even to one another.

There are two kingdoms of prokaryotic cells: the Archaebacteria and Eubacteria (Section 1.3). Together, they contain the most metabolically diverse organisms. Different kinds have managed to exploit energy and raw materials in just about every kind of environment.

Further reading: Student Guide to InfoTrac on web site

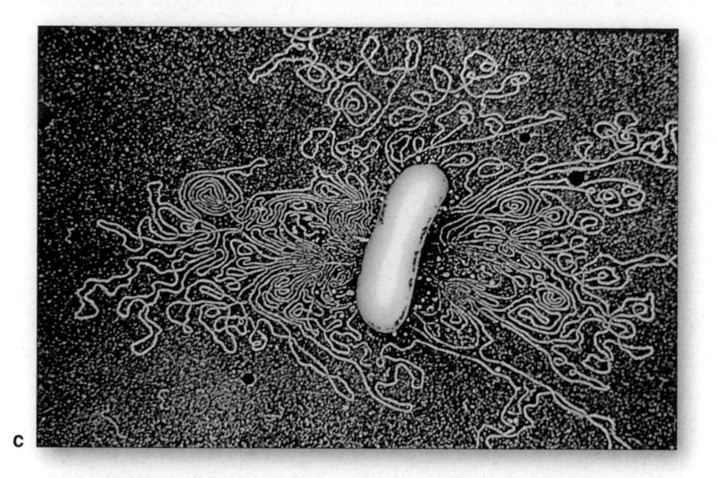

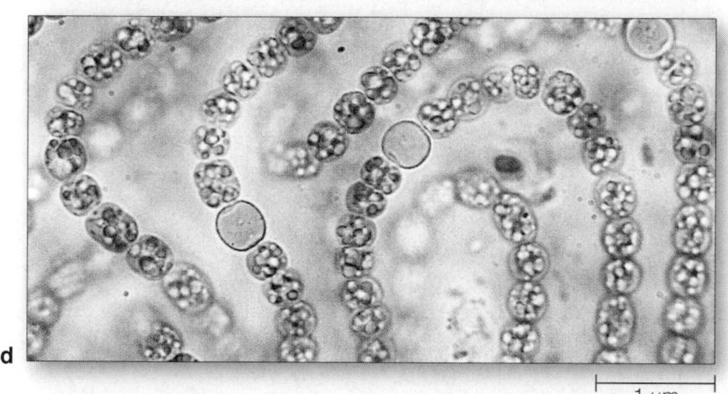

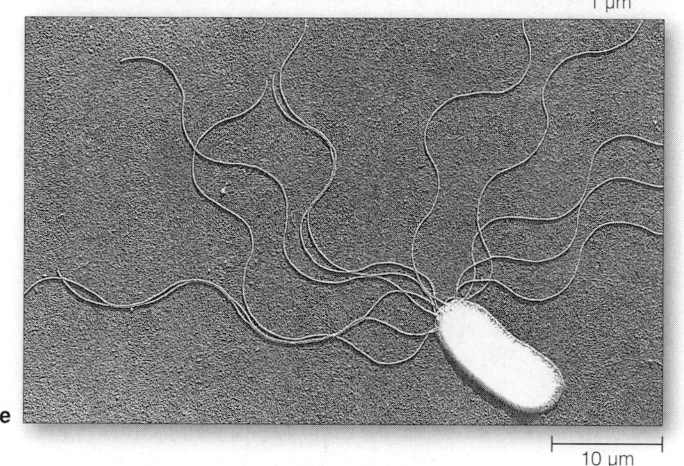

In addition, ancient prokaryotic cells gave rise to all the protistans, plants, fungi, and animals ever to appear on Earth. The evolution, structure, and functions of these remarkable cells are topics of later chapters.

Bacteria alone are prokaryotic cells; their DNA is not housed inside a nucleus. Most species have a cell wall around the plasma membrane. Generally, their cytoplasm does not have any organelles comparable to those of eukaryotic cells.

Bacteria are the simplest cells, but as a group they show the most metabolic diversity. Their metabolic activities proceed at the plasma membrane and within the cytoplasm.

4.12 SUMMARY

1. Three generalizations constitute the cell theory:
   a. All living things are composed of one or more cells.
   b. The cell is the smallest entity that still retains the properties of life. That is, it either lives independently or has a built-in, genetic capacity to do so.
   c. New cells arise only from cells that already exist.

2. At the minimum, a newly formed cell has a plasma membrane, a region of cytoplasm, and a region of DNA.
   a. The plasma membrane (a thin, outer membrane) maintains the cell as a distinct, separate entity. It allows metabolic events to proceed apart from random events in the environment. Many substances and signals are continually moving across it in highly controlled ways.
   b. Cytoplasm is all the fluids, ribosomes, structural elements, and (in eukaryotic cells) organelles between the plasma membrane and the region of DNA.

3. Membranes are vital to cell structure and function. They consist of lipids (phospholipids, for the most part) and proteins. The lipids are arrayed as two layers, with all the hydrophobic tails of both layers sandwiched in between all the hydrophilic heads. This lipid bilayer imparts structure to the membrane and bars passage of water-soluble substances across it. Diverse proteins are embedded in the bilayer or attached to its surfaces.

4. Proteins carry out most cell membrane functions. For example, many serve as channels or pumps that allow or promote passage of water-soluble substances across the lipid bilayer. Others are receptors for extracellular substances that trigger changes in cell activities.

5. Cell membranes divide the cytoplasm of eukaryotic cells into functional compartments called organelles. Prokaryotic cells do not have comparable organelles.

6. Organelle membranes separate metabolic reactions in the space of the cytoplasm and allow different kinds to proceed in orderly fashion. (In bacteria, many similar reactions proceed at the plasma membrane.)
   a. The nuclear envelope functionally separates the DNA from the metabolic machinery of the cytoplasm.
   b. The cytomembrane system includes the ER, Golgi bodies, and vesicles. Many new proteins are modified into final form and lipids are assembled in this system. Finished products are packaged and then shipped off to destinations inside or outside the cell.
   c. Mitochondria are specialists in oxygen-requiring reactions that produce many ATP molecules.
   d. Chloroplasts trap sunlight energy and produce organic compounds in photosynthetic eukaryotic cells.

7. The cytoskeleton of eukaryotic cells functions in cell shape, internal organization, and movements.

8. Table 4.3 on the next page summarizes the defining features of both prokaryotic and eukaryotic cells.

Table 4.3 Summary of Typical Components of Prokaryotic and Eukaryotic Cells

		PROKARYOTIC	EUKARYOTIC			
Cell Component	Function	Archaebacteria, Eubacteria	Protistans	Fungi	Plants	Animals
Cell wall	Protection, structural support	✓*	✓*	✓	✓	*None*
Plasma membrane	Control of substances moving into and out of cell	✓	✓	✓	✓	✓
Nucleus	Physical separation and organization of DNA	*None*	✓	✓	✓	✓
DNA	Encoding of hereditary information	✓	✓	✓	✓	✓
RNA	Transcription, translation of DNA messages into polypeptide chains of specific proteins	✓	✓	✓	✓	✓
Nucleolus	Assembly of subunits of ribosomes	*None*	✓	✓	✓	✓
Ribosome	Protein synthesis	✓	✓	✓	✓	✓
Endoplasmic reticulum (ER)	Initial modification of many of the newly forming polypeptide chains of proteins; lipid synthesis	*None*	✓	✓	✓	✓
Golgi body	Final modification of proteins, lipids; sorting and packaging them for use inside cell or for export	*None*	✓	✓	✓	✓
Lysosome	Intracellular digestion	*None*	✓	✓*	✓*	✓
Mitochondrion	ATP formation	**	✓	✓	✓	✓
Photosynthetic pigments	Light–energy conversion	✓*	✓*	*None*	✓	*None*
Chloroplast	Photosynthesis; some starch storage	*None*	✓*	*None*	✓	*None*
Central vacuole	Increasing cell surface area; storage	*None*	*None*	✓*	✓	*None*
Bacterial flagellum	Locomotion through fluid surroundings	✓*	*None*	*None*	*None*	*None*
Flagellum or cilium with 9 + 2 microtubular array	Locomotion through or motion within fluid surroundings	*None*	✓*	✓*	✓*	✓
Cytoskeleton	Cell shape; internal organization; basis of cell movement and, in many cells, locomotion	*None*	✓*	✓*	✓*	✓

* Known to be present in cells of at least some groups.
** Many groups use oxygen-requiring (aerobic) pathways of ATP formation, but mitochondria are not involved.

Review Questions

1. State the three key points of the cell theory. *CI*

2. Suppose you wish to observe the three-dimensional surface of an insect's eye. Would you see more details with the aid of a compound light microscope, transmission electron microscope, or scanning electron microscope? *4.2*

3. Label the organelles in the two diagrams below of a plant cell and an animal cell. What are the main differences between the two kinds of cells? *4.3*

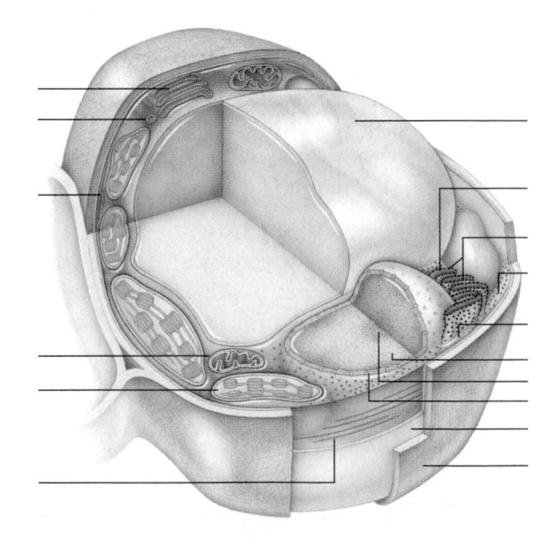

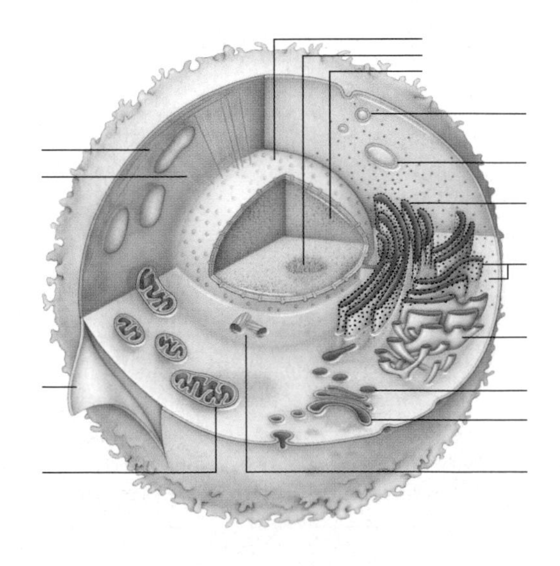

4. Label the parts of this bacterial cell. *4.11*

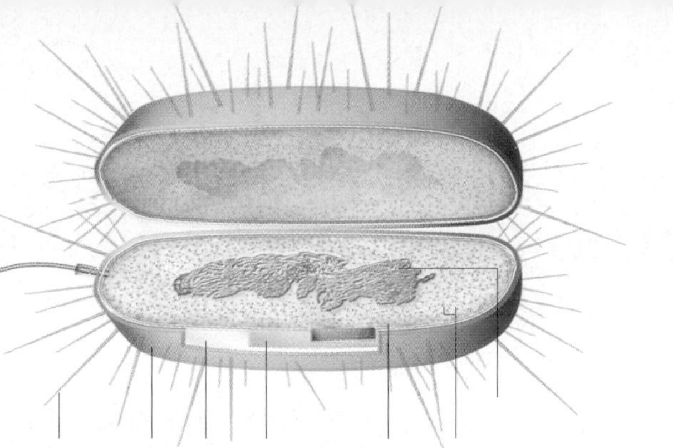

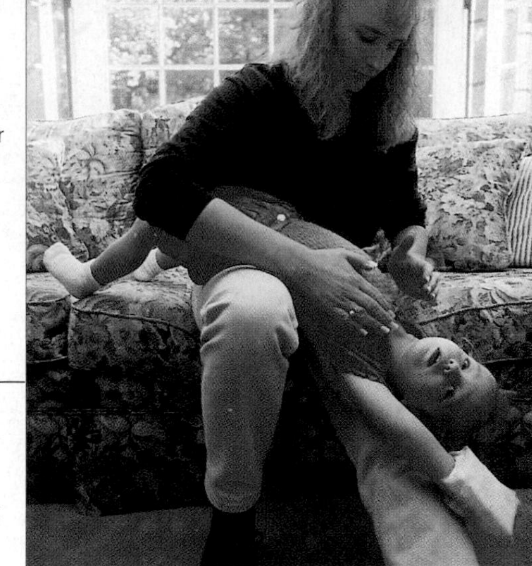

Figure 4.28 Daily chest thumping for a child who is affected by cystic fibrosis.

5. Describe three features that all cells have in common. After reviewing Table 4.3, write a paragraph on the key differences between prokaryotic and eukaryotic cells. *4.1, 4.3, 4.11*

6. Briefly characterize the structure and function of the cell nucleus, nuclear envelope, and nucleolus. *4.4*

7. Define chromosome and chromatin. Do chromosomes always have the same appearance during a cell's life? *4.4*

8. Which organelles are part of the cytomembrane system? *4.5*

9. Is this statement true or false: Plant cells have chloroplasts, but not mitochondria. Explain your answer. *4.6, 4.7*

10. What are the functions of the central vacuole? *4.7*

11. Define cytoskeleton. How does it aid in cell functioning? *4.8*

12. What gives rise to the 9 + 2 array of cilia and flagella? *4.9*

13. Cell walls are typical of which organisms: bacteria, protistans, fungi, plants, animals? Are the walls impermeable? *4.10*

14. In certain plant cells, is a secondary wall deposited inside or outside the surface of the primary wall? *4.10*

15. In multicelled organisms, coordinated interactions depend on linkages and communications between cells. What types of junctions occur between adjacent animal cells? Plant cells? *4.10*

Self-Quiz (*Answers in Appendix III*)

1. Cell membranes consist mainly of a ______ .
   a. carbohydrate bilayer and proteins
   b. protein bilayer and phospholipids
   c. lipid bilayer and proteins
   d. none of the above

2. Organelles ______ .
   a. are membrane-bound compartments
   b. are typical of eukaryotic cells, not prokaryotic cells
   c. separate chemical reactions in time and space
   d. all of the above are features of the organelles

3. Cells of many protistans, plants, and fungi, but not animals, commonly have ______ .
   a. mitochondria
   b. a plasma membrane
   c. ribosomes
   d. a cell wall

4. Is this statement true or false: The plasma membrane is the outermost component of all cells. Explain your answer.

5. Unlike eukaryotic cells, prokaryotic cells ______ .
   a. lack a plasma membrane
   b. have RNA, not DNA
   c. do not have a nucleus
   d. all of the above

6. Match each cell component with its function.
   ____ mitochondrion
   ____ chloroplast
   ____ ribosome
   ____ rough ER
   ____ Golgi body

   a. synthesis of polypeptide chains
   b. initial modification of new polypeptide chains
   c. final modification of proteins; sorting, shipping tasks
   d. photosynthesis
   e. formation of many ATP

Critical Thinking

1. Why is it likely that you will never encounter a predatory two-ton living cell on the sidewalk?

2. In compound light microscopes having blue filters, the lens transmits only blue light. Think about the spectrum of visible light (as in Figure 7.4). Then speculate on why blue light is efficient for viewing objects at high magnification.

3. Your professor shows you an electron micrograph of a cell with large numbers of mitochondria and Golgi bodies. You notice that this particular cell also contains a great deal of rough endoplasmic reticulum. What kinds of cellular activities would require such an abundance of the three kinds of organelles?

4. *Cystic fibrosis* is a fatal genetic disorder. Affected glands secrete far more than they should, with far-reaching effects. In time, digestive enzymes clog a duct between the pancreas and small intestine, food can't be digested properly, and even if food intake increases, malnutrition results. Cysts form in the pancreas, which degenerates and becomes fibrous (hence the disorder's name). Thick mucus builds up in the respiratory tract; affected people have trouble expelling airborne bacteria and particles that enter lungs (Figure 4.28). The disorder may arise from a defective protein in the plasma membrane of gland cells that secrete mucus, digestive enzymes, and sweat. Review Section 4.5, then name the organelles involved in the secretory pathway in those cells.

Selected Key Terms

basal body *4.9*
cell *4.1*
cell cortex *4.8*
cell theory *CI*
cell wall *4.10*
central vacuole *4.7*
centriole *4.9*
chloroplast *4.7*
chromatin *4.4*
chromosome *4.4*
cilium *4.9*
cytomembrane system *4.5*
cytoplasm *4.1*
cytoskeleton *4.8*
ER (endoplasmic reticulum) *4.5*
eukaryotic cell *4.1*
flagellum *4.9*
Golgi body *4.5*
intermediate filament *4.8*
lipid bilayer *4.1*
microfilament *4.8*
micrograph *4.2*
microtubule *4.8*
mitochondrion *4.6*
motor protein *4.8*
nuclear envelope *4.4*
nucleoid *4.1*
nucleolus *4.4*
nucleus *4.1, 4.4*
organelle *4.3*
plasma membrane *4.1*
primary wall *4.10*
prokaryotic cell *4.1*
pseudopod *4.9*
ribosome *4.1*
secondary wall *4.10*
surface-to-volume ratio *4.1*
vesicle *4.5*
wavelength *4.2*

Readings *See also www.infotrac-college.com*

Alberts, B. et al. 1994. *Molecular Biology of the Cell.* Third edition. New York: Garland. See Chapter 16: The Cytoskeleton.

deDuve, C. 1985. *A Guided Tour of the Living Cell.* New York: Freeman. Beautiful introduction to the cell; two short volumes.

5

A CLOSER LOOK AT CELL MEMBRANES

It Isn't Easy Being Single

As small as it may be, a cell is a living thing engaged in the risky business of survival. Think of a single-celled amoeba and how something as ordinary as water can challenge its very existence. Water bathes the amoeba inside and out, donates its individual molecules to many metabolic reactions, and dissolves the ions that are necessary for cell functioning. If all goes well, the amoeba holds onto enough water and dissolved ions—not too little, not too much—to survive. But who is to say that life consistently goes well?

Next, think of the cells making up a goose barnacle. Not long ago, this marine organism attached itself to the submerged side of a log near a coastline. Now the log is drifting offshore, at the mercy of ocean currents. At feeding time, the barnacle opens its hinged shell and extends many featherlike appendages, which trap bacteria and other bits of food suspended in the water (Figure 5.1*a*).

The fluids bathing each cell of the barnacle body are salty, rather like the salt composition of seawater. And seawater normally is in balance with the salty fluid inside the barnacle's cells.

However, suppose the log drifts into a part of the ocean where meltwater from a glacier has diluted the water (Figure 5.1*b*). For reasons you will explore in this chapter, salts inevitably move out of the barnacle's body—and out of its cells. The previously exquisite salt–water balance gradually spirals out of control, and cells die. So, in time, does the barnacle.

The same thing happens to burrowing worms and other soft-bodied organisms that live between the high and low tide marks along a rocky shore. For instance, after an unpredictably fierce storm along the coast, seawater becomes highly dilute with runoff from the land. Dilution upsets the balance between body fluids and the water moving in with the tides. At such times, the functions of cells, tissues, and organs are disrupted, and the resulting death toll in the intertidal zone can be catastrophic.

With these examples we begin to see the cell for what it is: a tiny, organized bit of life in a world that is, by comparison, unorganized and sometimes harsh. How finely adapted a living cell must be to its environment! It must be built in such a way that it can bring in certain substances, release or keep out other substances, and conduct its internal activities with great precision.

For this bit of life, precision begins at the plasma membrane—a flimsy bilayer of lipids, dotted with diverse proteins, that surrounds the cytoplasm. Across the membrane, the cell exchanges substances with its surroundings in highly selective ways.

For eukaryotic cells, precision continues at the membranes of internal compartments called organelles. Aerobic respiration, photosynthesis, and many other metabolic processes depend on the selective movement of substances across those internal boundary layers. Gain insight into the structure and function of cell membranes, and you will gain insight as well into survival at life's most fundamental level.

Figure 5.1 (**a**) Goose barnacles, a type of marine animal that glues itself to logs and other floating objects in the sea. As is the case for other organisms, drastic changes in salt concentrations can threaten their body cells. Such changes would occur if the barnacles were to accidentally end up in glacial meltwaters.

(**b**) The dark-blue seawater in this photograph is quite salty. The lighter blue water has been diluted by meltwater from glaciers, including the one in the background. It is much, much lower in salts than seawater.

KEY CONCEPTS

1. Cell membranes consist mainly of lipids, organized as a double layer. The lipid "bilayer" gives the membrane its basic structure and prevents the haphazard movement of water-soluble substances across it. Proteins embedded in the bilayer or associated with one of its surfaces carry out most membrane functions.

2. The plasma membrane has receptor proteins (which receive chemical signals that can trigger changes in cell activities) and recognition proteins (which identify a cell as being of a certain type). Spanning all cell membranes are transport proteins through which specific water-soluble substances cross the bilayer.

3. A concentration gradient is a difference in the number of molecules (or ions) of a substance between two regions. The molecules tend to show a net movement down the gradient, to the region where they are less concentrated. This behavior is called diffusion.

4. Metabolism depends on concentration gradients that drive the directional movements of substances. Cells have mechanisms for increasing or decreasing water and solute concentrations across the plasma membrane and internal cell membranes.

5. With passive transport, a solute crosses a membrane simply by diffusing through the interior of transport proteins. With active transport, membrane proteins pump a solute across a membrane, against its concentration gradient. Active transport requires an energy input.

6. By a molecular behavior called osmosis, water diffuses across any selectively permeable membrane to a region where its concentration is lower.

7. Larger packets of material move across the plasma membrane by processes of endocytosis and exocytosis.

5.1 MEMBRANE STRUCTURE AND FUNCTION

Earlier chapters provided you with a brief look at the structure of cell membranes and the general functions of their component parts. Here, we incorporate some of the background information in a more detailed picture.

The Lipid Bilayer of Cell Membranes

Fluid bathes the two surfaces of a cell membrane and is vital for its functioning. The membrane, too, has a fluid quality; it is not a solid, static wall between cytoplasmic and extracellular fluids. For instance, puncture a cell with a fine needle, and its cytoplasm will not ooze out. The membrane will flow over the puncture site and seal it!

How does a fluid membrane remain distinct from its fluid surroundings? To arrive at the answer, start by reviewing what we have already learned about its most abundant components, the phospholipids. Recall that a **phospholipid** has a phosphate-containing head and two fatty acid tails attached to a glycerol backbone (Figure 5.2*a*). The head is hydrophilic; it easily dissolves in water. Its tails are hydrophobic; water repels them. Immerse a number of phospholipid molecules in water, and they interact with water molecules and with one another until they spontaneously cluster in a sheet or film at the water's surface. Their jostlings may even force them to become organized in two layers, with all fatty acid tails sandwiched between all hydrophilic heads. This **lipid bilayer** arrangement, remember, is the structural basis of cell membranes (Section 4.1 and Figure 5.2*c*).

The organization of each lipid bilayer minimizes the total number of hydrophobic groups exposed to water, so the fatty acid tails do not have to spend a lot of energy fighting water molecules, so to speak. A "punctured" membrane exhibits sealing behavior precisely because a puncture is energetically unfavorable. It leaves far too many hydrophobic groups exposed to the surrounding fluid.

Ordinarily, few cells get jabbed by fine needles. But the self-sealing behavior of membrane phospholipids is good for more than damage control. Among other things, it functions in vesicle formation. For example, as vesicles bud away from ER or Golgi membranes, phospholipids interact hydrophobically with cytoplasmic water. They get pushed together, and the rupture seals. You will read more about vesicle formation later in the chapter.

Fluid Mosaic Model of Membrane Structure

Figure 5.3 shows a bit of membrane that corresponds to the **fluid mosaic model**. By this model, cell membranes are a mixed composition—a *mosaic*—of phospholipids, glycolipids, sterols, and proteins. The phospholipid heads as well as the length and saturation of the tails are not all the same. (Recall that unsaturated fatty acids have one or more double bonds in their backbone and fully saturated ones have none.) The glycolipids are structurally similar to phospholipids, but their head incorporates one or more sugars. In animal cell membranes, cholesterol is the most abundant sterol (Figure 5.2*b*). Phytosterols are their equivalent in plant cell membranes.

Also by this model, the membrane is *fluid* as a result of the motions and interactions of its component parts.

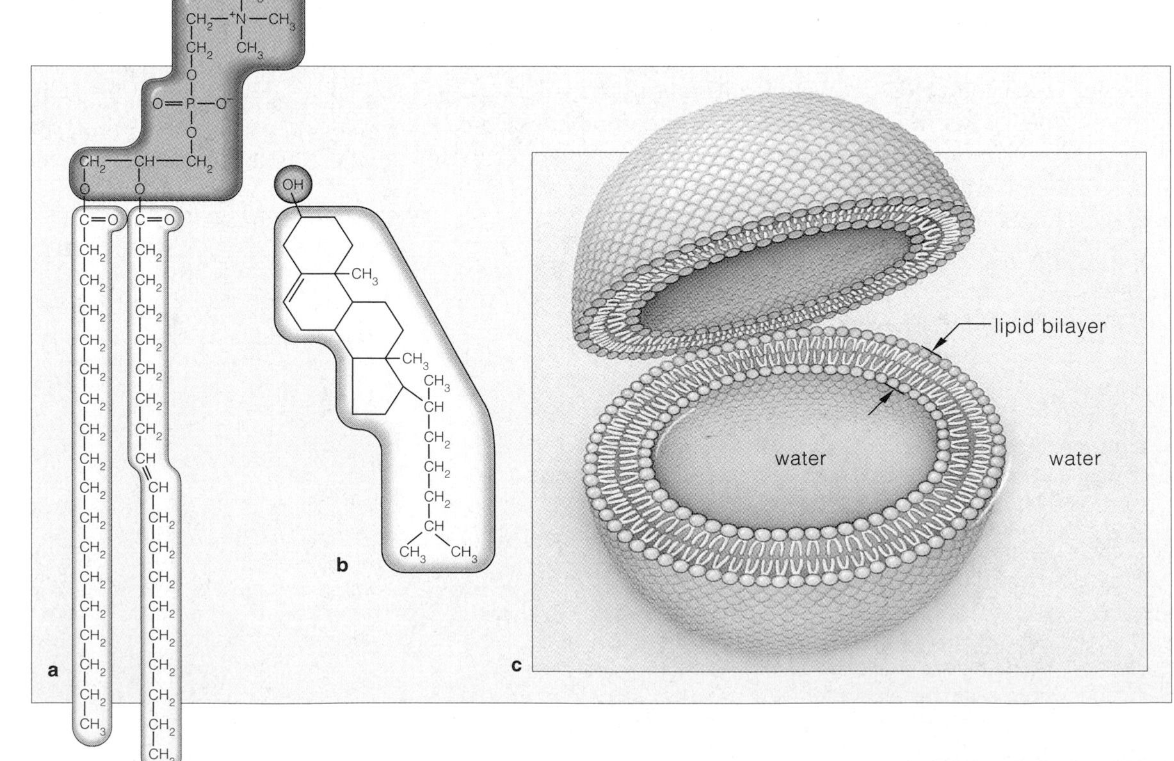

Figure 5.2 (**a**) The structural formula for phosphatidylcholine, a phospholipid that is one of the most common components of the membranes of animal cells. *Orange* indicates its hydrophilic head; *yellow* indicates its hydrophobic tails.

(**b**) Structural formula for cholesterol, the major sterol in animal tissues.

(**c**) Diagram of how lipids spontaneously organize themselves into a bilayer structure when placed in liquid water.

Further reading: Student Guide to InfoTrac on web site

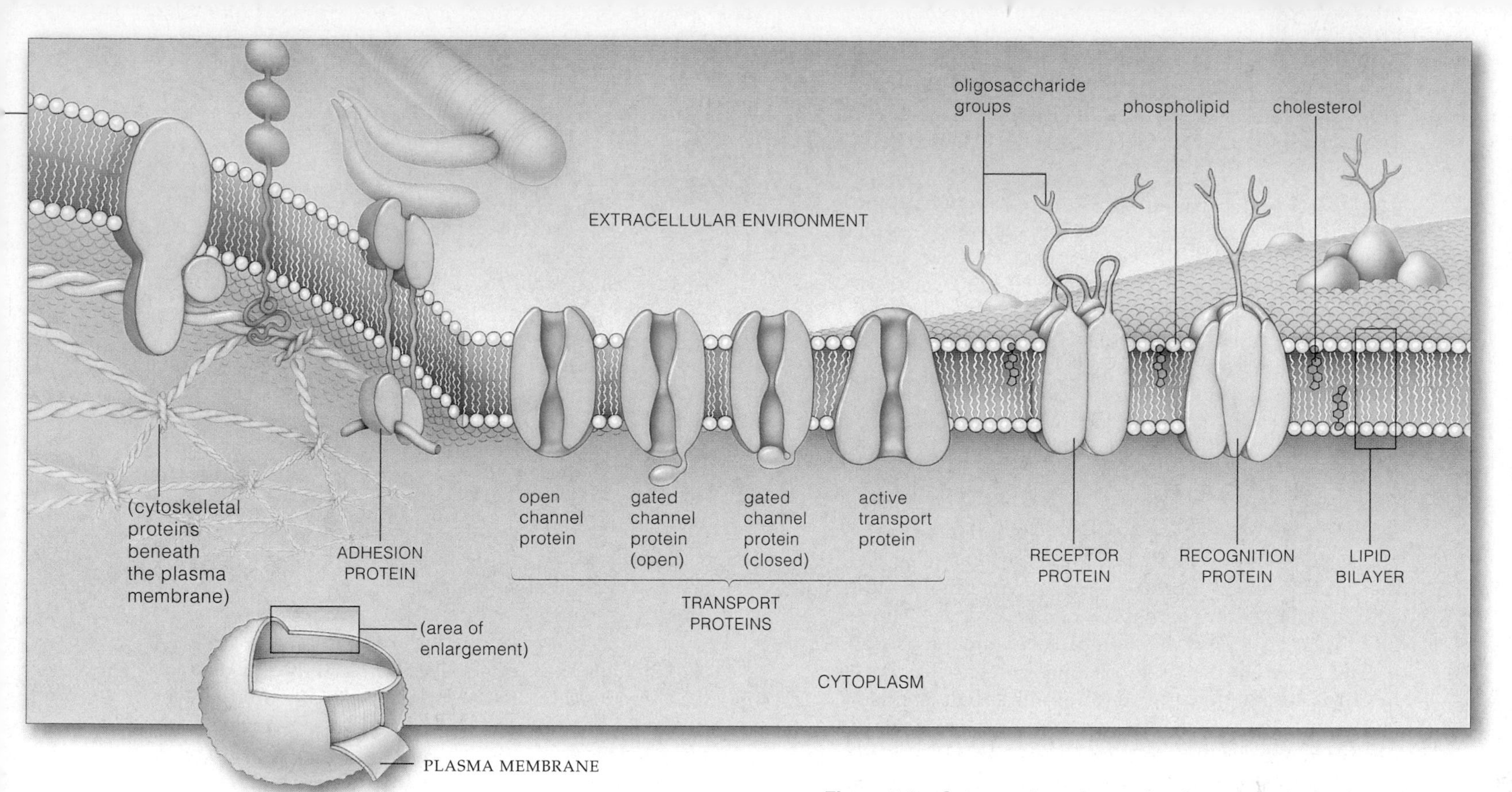

Figure 5.3 Cutaway view of part of a plasma membrane, based on the fluid mosaic model. In addition to the specialized proteins shown, enzymes also are associated with cell membranes.

The hydrophobic interactions that give rise to most of a membrane's structure are weaker than covalent bonds. This means most phospholipids and some proteins are free to drift sideways. Also, the phospholipids can spin about their long axis and flex their tails, which keeps neighboring molecules from packing together in a solid layer. Short or kinked (unsaturated) hydrophobic tails contribute to the membrane fluidity.

The fluid mosaic model is a good starting point for exploring membranes. But bear in mind, cell membranes differ in composition and molecular arrangements. They are not even the same on both surfaces of their bilayer. For example, oligosaccharides and other carbohydrates are covalently bonded to protein and lipid components of a plasma membrane, but only on its outward-facing surface (Figure 5.3). Also, these carbohydrates differ in number and kind from one species to the next, even among different cells of the same individual.

Overview of Membrane Proteins

The proteins embedded in a lipid bilayer or attached to one of its surfaces carry out most membrane functions. Many are enzyme components of metabolic machinery. Others are **transport proteins** that span the bilayer and allow water-soluble substances to move through their interior. They bind molecules or ions on one side of the membrane, then release them on the other side.

The **receptor proteins** bind extracellular substances, such as hormones, that trigger changes in cell activities. For example, certain enzymes that crank up machinery for cell growth and division become switched on when somatotropin, a hormone, binds with receptors for it. Different cells have different combinations of receptors.

Diverse **recognition proteins** at the cell surface are like molecular fingerprints; their oligosaccharide chains identify a cell as being of a specific type. For example, the plasma membrane of your cells bristles with "self" proteins. Certain white blood cells chemically recognize self proteins and leave your cells alone, but they attack invading bacterial cells bearing "nonself" proteins at their surface. Finally, **adhesion proteins** of multicelled organisms help cells of the same type locate and stick to one another and stay positioned in the proper tissues. They are glycoproteins, with oligosaccharides attached. After tissues form, the sites of adhesion may become a type of cell junction, as described in Section 4.10.

A cell membrane has two layers composed mainly of lipids, phospholipids especially. This lipid bilayer is the structural foundation for the membrane and also serves as a barrier to water-soluble substances.

The hydrophobic parts of membrane lipids are sandwiched between the hydrophilic parts, which are dissolved either in cytoplasmic fluid or extracellular fluid.

Diverse proteins associated with the bilayer carry out most membrane functions. Many of the membrane proteins are enzymes, transporters of substances across the bilayer, or receptors for extracellular substances. Other types function in cell-to-cell recognition or adhesion.

TESTING IDEAS ABOUT CELL MEMBRANES

TESTING MEMBRANE MODELS Recall, from Chapter 1, that scientific methods are used to reveal nature's secrets. To gain insight into how researchers discovered some structural details of cell membranes, imagine yourself duplicating some of their test methods. You can start by attempting to identify the molecular components of the plasma membrane.

Your first challenge is to secure a membrane sample that is large enough to study and not contaminated with organelle membranes. Using red blood cells will simplify the task. These cells are abundant, easy to collect, and structurally simple. As they are maturing, their nucleus disintegrates. Ribosomes, hemoglobin, and just a few other components remain, but these are enough to keep the cells functional for their brief, four-month life span.

By placing a sample of red blood cells in a test tube filled with distilled water, you can separate the plasma membranes from the other cell components. Cells have more solutes and fewer water molecules than distilled water does. For reasons that you will read about shortly, the difference in solute concentrations between the two regions causes water to move across a plasma membrane, into the cells. These particular cells have no mechanisms for actively expelling the excess water, so they swell. In time they burst, and their contents spill out.

Now you have a mixture of membranes, hemoglobin, and other components in water. How do you separate all the bits of membrane? You place a tube that holds a solution of the cell parts in a **centrifuge**, a motor-driven rotary device that can spin test tubes at very high speed (Figure 5.4). Structures and molecules move in response to the centrifugal force. How far they move depends on their mass, density, and shape—as well as on the mass, density, and fluidity of the solution in the tubes. If the bits of membrane have greater mass and density than the solution, they will move downward in the tube. If they have less mass or density, they will remain near the top of the tube.

Each cell component has a molecular composition that gives it a characteristic density. As the centrifuge spins at an appropriate speed, the components having the greatest density move toward the bottom of the tube. Other cell components take up positions in layers above them, according to their relative densities.

You perform centrifugation properly, so one layer in the solution consists only of membrane shreds. You draw off this layer carefully and examine it with a microscope to check whether the membrane sample is contaminated.

Afterward, by using standard procedures of chemical analysis, you discover that the plasma membrane of red blood cells consists of lipids and proteins.

In the past, two competing structural models guided research into membranes. According to the "protein coat" model, a cell membrane consists of a lipid bilayer that has a layer of proteins coating both of its outward-facing surfaces. According to the "fluid mosaic" model, proteins are largely embedded within the bilayer.

You decide to test the protein coat model. First you calculate how much protein it would take to coat the inner and outer surfaces of a known number of red blood cells. Then you separate a membrane sample into its lipid and protein fractions and measure the amount of each. In this way, you can compare the *observed* ratio of proteins to lipids against the ratio *predicted* on the basis of the model.

Such calculations were performed with membrane samples. There are indeed enough lipids for a bilayer arrangement. And there *might* have been enough proteins to cover both of its surfaces *if* all the membrane proteins had a fibrous (not globular) structure. That possibility was discarded for two reasons. First, it became clear that proteins stretched out in a thin layer over lipids would be an unfavorable arrangement, in terms of the energy cost required to maintain it. Second, biochemical analysis showed that the proteins are globular. Such evidence does not favor the protein coat model.

Now you perform an observational test of the fluid mosaic model. You prepare cell samples for microscopy by two research methods, called *freeze–fracturing* and *freeze–etching*. First you immerse the cell sample in liquid nitrogen, an extremely cold fluid. The cells freeze instantly. Then you strike them with an exceedingly fine blade, as in Figure 5.5. A suitably directed blow can fracture cells in such a way that one lipid layer of the plasma membrane will separate from the other. Such preparations can be observed under extremely high magnifications.

If the protein coat model were correct, then you would observe a perfectly smooth, pure lipid layer. But the freeze–fractured cell membranes you observe reveal many bumps and other irregularities in the

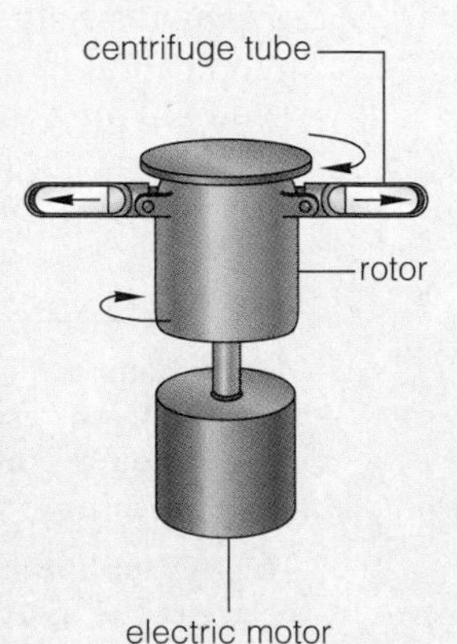

Figure 5.4 One example of a centrifuge. The sketch shows the arrangement of the tubes and the rotor of this high-speed spinning device.

Further reading: Student Guide to InfoTrac on web site

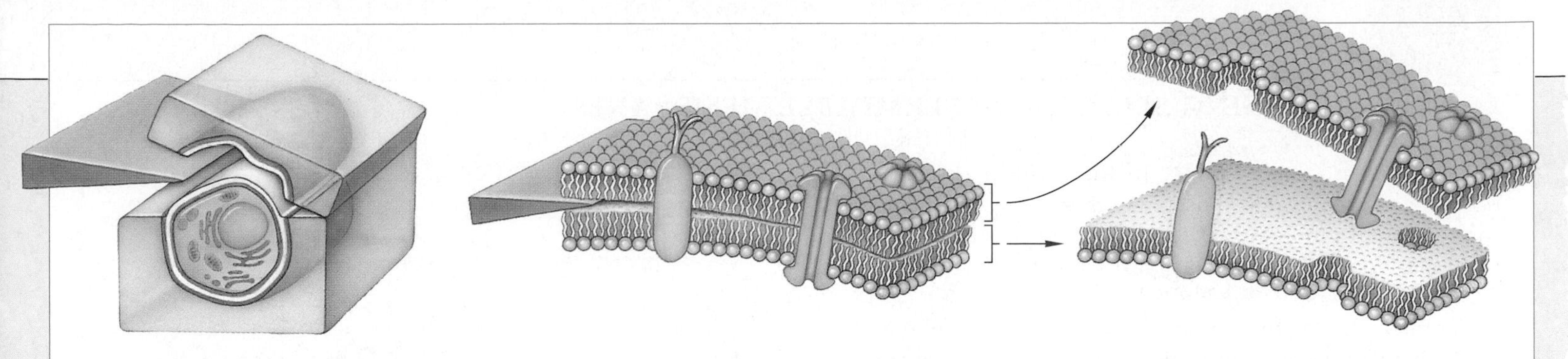

a With freeze–fracturing, specimens being prepared for electron microscopy are rapidly frozen, then fractured by a sharp blow from the edge of an ultrafine blade.

b A fractured cell membrane often splits down the middle of its lipid bilayer. Typically, many particles and depressions dot the inner surface of one exposed layer, and a complementary pattern of particles and depressions dots the other. The particles are membrane proteins. Depressions are sites where they projected into the facing layer.

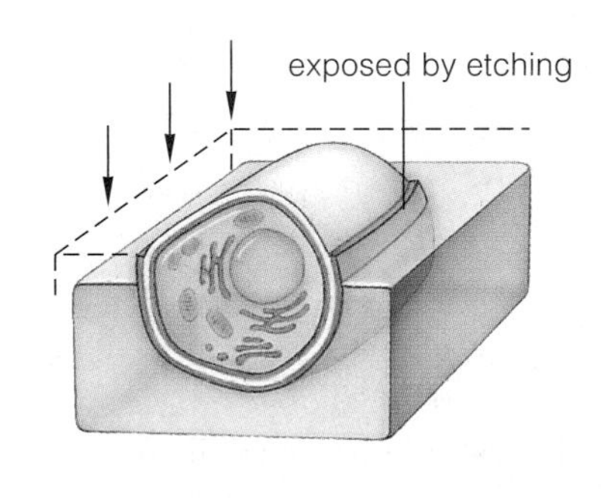

c Commonly, the specimens are also freeze–etched. With this method, more ice is forced to evaporate away from the fracture face, so that a portion of the outward-facing surface of the cell membrane is exposed.

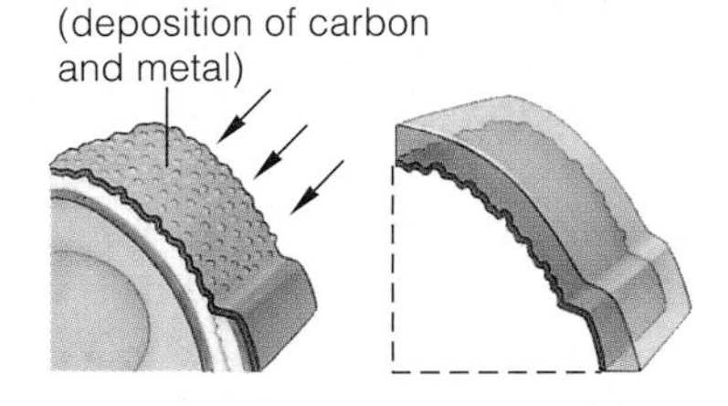

d By metal-shadowing methods, the fractured surface is coated with a layer of carbon and heavy metal, such as platinum. The coat is thin enough to replicate details of the exposed specimen surface. The metal replica, not the specimen, is used to make micrographs of the details.

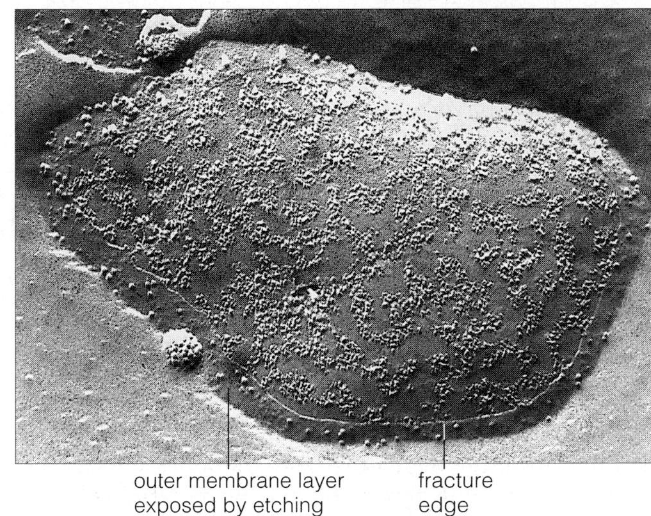

Figure 5.5 Freeze–fracturing and freeze–etching methods. The micrograph shows part of a replica of a red blood cell that was prepared by the techniques described in the text.

separated layers of lipids. The "bumps" are proteins, incorporated directly in the bilayer—just as the fluid mosaic model predicts.

OBSERVING MEMBRANE FLUIDITY Consider now an experiment that yielded evidence of the fluid motion of membrane proteins. Researchers induced an isolated human cell and an isolated mouse cell to fuse. The plasma membranes of the cells from the two species merged to form a single, continuous membrane in the new, hybrid cell. Analysis showed that, in less than one hour, most of the membrane proteins were mixed together. They had been free to drift laterally through the hybrid membrane (Figure 5.6).

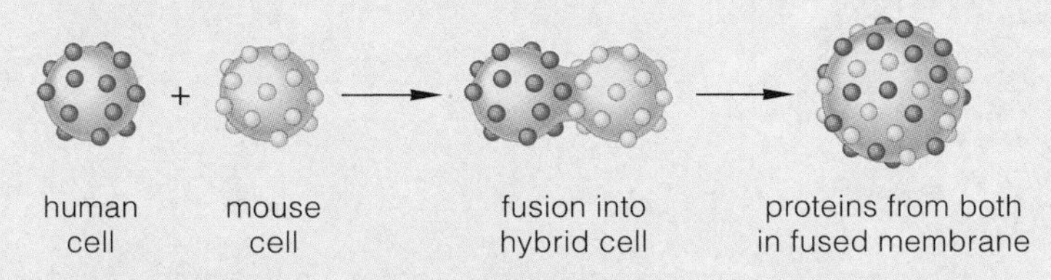

Figure 5.6 Result of an experiment in which the plasma membranes from cells of different species were induced to fuse. Membrane proteins drifted laterally and became mixed.

Through observational tests, we also know that the fluidity of a membrane is subject to change, depending on environmental temperatures—and that at least some cells can respond to the danger.

For example, after some researchers decreased the temperature of a solution that contained live bacterial or yeast cells, the membranes of those cells started to stiffen. Such stiffening can disrupt the functioning of membrane proteins. However, the cells responded by rapidly synthesizing unsaturated fatty acids, and the infusion of those kinked lipids into the cell membranes countered the stiffening effect. However, past a certain low temperature, the membranes did solidify. The temperature at which this occurs varies, depending on the lipid composition of a given membrane.

5.3 CROSSING SELECTIVELY PERMEABLE MEMBRANES

All molecules contain energy in the form of chemical bonds. Cell membranes help make that energy available for cell functioning. How? They selectively concentrate reactants and products in amounts most favorable for specific reactions.

Think of the water bathing both sides of the bilayer of a cell membrane. Plenty of substances are dissolved in it, but the kinds and amounts are not the same near the two surfaces of the bilayer. The membrane itself is establishing and maintaining the differences, which are vital for metabolism. How? Like all cell membranes, it shows **selective permeability**: Owing to its molecular structure, the membrane permits some substances but not others to cross it in certain ways, at certain times.

For instance, being largely nonpolar, the hydrocarbon chains of the lipid bilayer passively let carbon dioxide, molecular oxygen, and other small, nonpolar molecules cross the membrane. Some water molecules cross, also. Even though they are polar, they slip through gaps that open up when the chains flex and bend (Figure 5.7*a*).

By contrast, large, polar molecules such as glucose almost never move freely across the bilayer. Neither do ions (Figure 5.7*b*). Such water-soluble substances cross the bilayer passively or actively, through the interior of transport proteins. The water molecules in which they are dissolved cross with them.

Membrane structure also allows cells to import and export substances in bulk across the plasma membrane. The processes are called endocytosis and exocytosis.

At any time in a cell's life, all of these mechanisms are operating simultaneously. They help cells increase, decrease, and maintain concentrations of the molecules and ions that are crucial for metabolism.

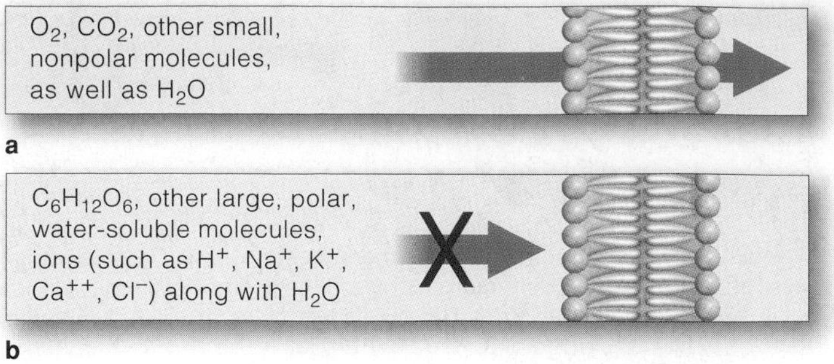

Concentration Gradients and Diffusion

Concentration refers to the number of molecules or ions of a substance in a specified region, as in a volume of fluid or air. *Gradient* means the number in one region is not the same as it is in another. Thus, a **concentration gradient** is a difference in the number of molecules or ions of a given substance between adjoining regions.

In the absence of other forces, a substance moves from a region where it is more concentrated to a region where it is less concentrated. *The energy inherent in its individual molecules, which keeps them in constant motion, drives the directional movement.* Although the molecules collide randomly and careen back and forth millions of times a second, the *net* movement is away from the place of greater concentration (and the most collisions).

Diffusion is the name for the net movement of like molecules or ions down a concentration gradient. It is a key factor in the movement of substances across cell membranes and through cytoplasmic fluid. In multicelled organisms, diffusion also moves substances from one body region to another, and between the body and its environment. For example, when oxygen builds up in photosynthetic leaf cells, it diffuses across their plasma membrane, into air inside the leaf, then into air outside the leaf—where the oxygen concentration is lower.

a: O_2, CO_2, other small, nonpolar molecules, as well as H_2O

b: $C_6H_{12}O_6$, other large, polar, water-soluble molecules, ions (such as H^+, Na^+, K^+, Ca^{++}, Cl^-) along with H_2O

Figure 5.7 Selective permeability of cell membranes. (**a**) Small, nonpolar molecules and some water molecules cross the lipid bilayer. (**b**) Ions and large, polar, water-soluble molecules and the water dissolving them cannot cross the bilayer; transport proteins spanning the bilayer must help them across.

Like other substances, oxygen tends to diffuse in a direction established by its *own* concentration gradient, not those of any other substances dissolved in the same fluid. You see the outcome of this tendency when you squeeze a drop of dye into water. Molecules of the dye diffuse to the region where they are less concentrated. And the water molecules move to the region where *they* are not as concentrated (Figure 5.8).

Several factors affect the rate at which molecules or ions move down a concentration gradient. A gradient's steepness is a factor. So are molecular size, temperature, and electric or pressure gradients that may be present.

Diffusion is faster when a gradient is steep. Then, far more molecules are moving outward from the region of greatest concentration compared to the number that are moving in. The difference in how many molecules are moving in either direction declines as the gradient decreases. If the gradient vanishes, individual molecules will still be in motion. But

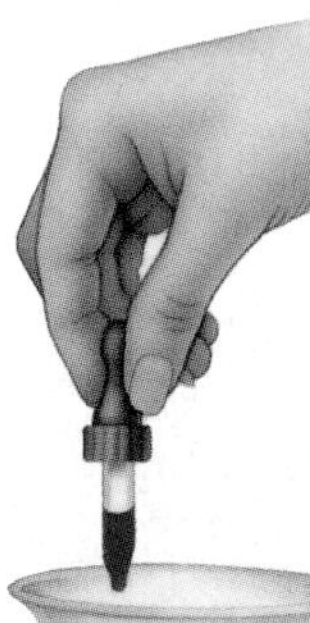

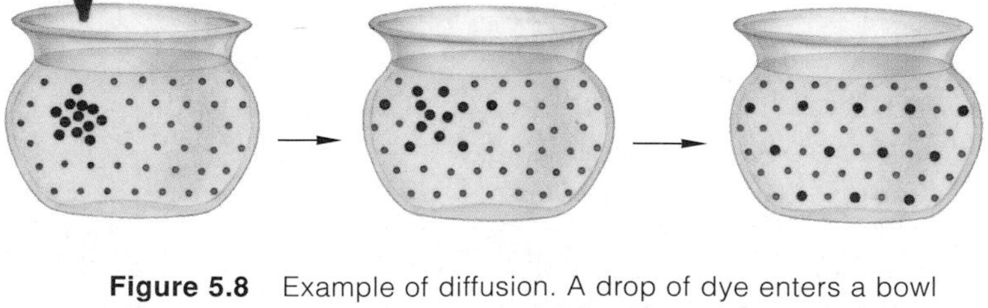

Figure 5.8 Example of diffusion. A drop of dye enters a bowl of water. The dye molecules *and* the water molecules slowly become evenly dispersed, because each substance shows a net movement down its own concentration gradient.

Further reading: Student Guide to InfoTrac on web site

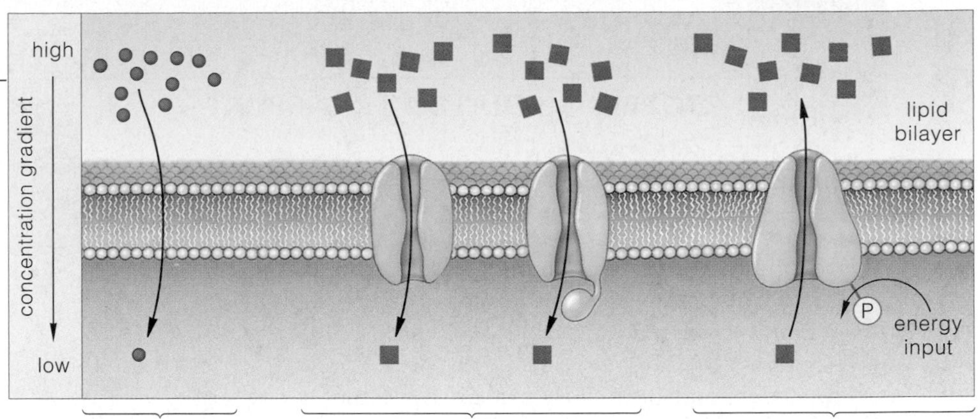

DIFFUSION ACROSS LIPID BILAYER
Lipid-soluble substances as well as water diffuse across.

PASSIVE TRANSPORT
Water-soluble substances, and water, diffuse through interior of transport proteins. No energy boost required. Also called facilitated diffusion.

ACTIVE TRANSPORT
Specific solutes are pumped through interior of transport proteins. Requires energy boost.

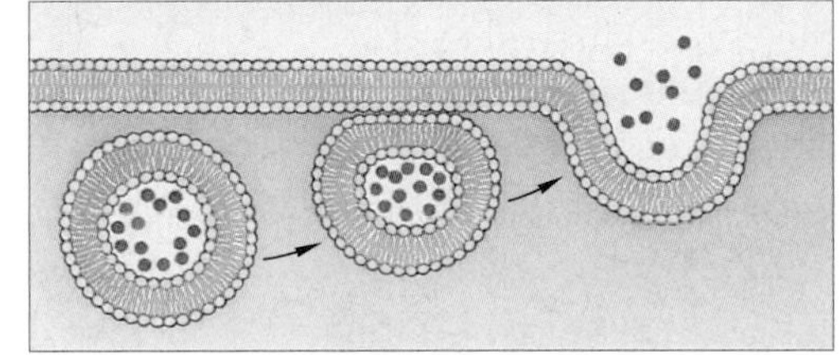

EXOCYTOSIS
Vesicle in cytoplasm moves to plasma membrane, fuses with it; contents released to the outside.

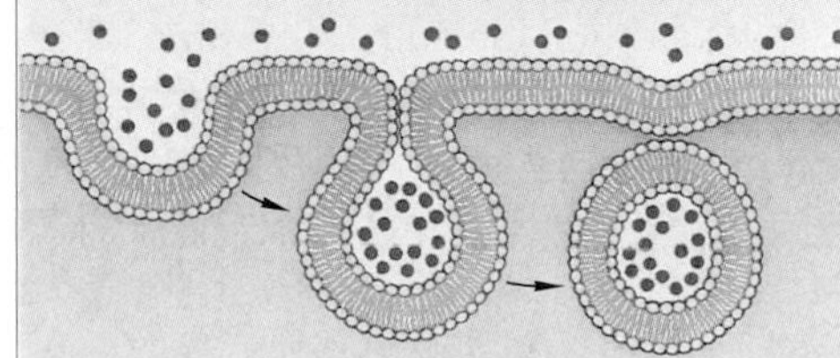

ENDOCYTOSIS
Vesicle forms from a patch of inward-sinking plasma membrane, enters cytoplasm.

Figure 5.9 Overview of major mechanisms by which solutes can cross cell membranes. Exocytosis and endocytosis proceed only at the plasma membrane.

then the total number moving one way or the other during a given interval will be about the same. A net distribution of molecules that is nearly uniform through two adjoining regions is called "dynamic equilibrium."

In addition, the rates of diffusion are faster at higher temperatures. More heat energy causes molecules to move faster and therefore to collide more frequently. Hence, diffusion is more rapid than in a cooler adjoining region.

Molecular size also affects diffusion rates. Generally, small molecules tend to move down their concentration gradient more rapidly than larger ones do.

Besides this, the rate and direction of diffusion might be under the influence of an electric gradient. An **electric gradient** simply is a difference in electric charges of adjoining regions. For example, in the fluid bathing the surface of each cell membrane are many kinds of dissolved ions. Each ion is contributing to the fluid's overall electric charge. Opposite charges attract. So the fluid having the more negative charge, overall, tends to exert the greatest pull on positively charged substances, such as sodium ions. Many biological events, including information flow through nervous systems, involve the combined force of electric and concentration gradients across membranes.

Finally, as you will see shortly, the presence of a pressure gradient also may affect the rate and direction of diffusion. A **pressure gradient** is a difference in the pressure being exerted in adjoining regions.

Overview of Membrane Crossing Mechanisms

Before taking a closer look at how substances cross membranes, study Figure 5.9, which is an overview of the key mechanisms. Together, these mechanisms help supply cells and organelles with raw materials and rid them of wastes, at controlled rates. They contribute to maintaining the volume and pH of a cell or organelle within functional ranges.

Again, small nonpolar molecules (such as oxygen) and water can simply diffuse across the lipid bilayer. Polar molecules cross by diffusing through the interior of transport proteins that span the bilayer. Some kinds of proteins help the substance follow its concentration gradient simply by functioning as a channel across the membrane. They are the basis of *passive transport*, or "facilitated" diffusion. Other proteins engage in *active transport*. They, too, allow polar molecules to cross the membrane through their interior, but the net direction of movement is against the concentration gradient. Active transport only operates when transport proteins become energized, as by a phosphate-group transfer from ATP.

Certain mechanisms also move substances in bulk across a plasma membrane. *Exocytosis* involves fusion between the plasma membrane and a membrane-bound vesicle that formed inside the cytoplasm. *Endocytosis* involves an inward sinking of a small patch of plasma membrane, which seals back on itself to form a vesicle in the cytoplasm.

Molecules or ions of a substance constantly collide because of their inherent energy of motion. The collisions result in diffusion, a net outward movement of a substance from one region into an adjoining region where it is less concentrated.

A concentration gradient is a form of energy. It can drive the directional movement of a substance across a cell membrane. The steepness of the gradient, temperature, molecular size, electric gradients, and pressure gradients influence the rates of diffusion.

Metabolic reactions depend on the chemical energy inherent in concentrated amounts of molecules and ions. Cells have mechanisms for increasing or decreasing those concentrations across the plasma membrane and internal cell membranes.

5.4 PROTEIN-MEDIATED TRANSPORT

Let's now take a look at how water-soluble substances diffuse into and out of cells or organelles. Whether by passive or active transport mechanisms, a great variety of proteins have roles in moving molecules and ions across cell membranes. These proteins span the lipid bilayer, and their interior is able to open on both sides of it (Figure 5.10).

To understand how such protein molecules work, you must know they are not rigid blobs of atoms. When such a protein interacts with a solute, it changes from one shape to another, then back again. The changes start when a solute weakly binds to a site in the protein's interior. Part of the protein closes in behind the bound solute, and part of the interior opens up on the other side of the cell membrane. Then the solute dissociates (separates) from the binding site. Think of it as hopping onto the transport protein on one side of the membrane and hopping off on the other side.

Passive Transport

Many transport proteins permit ions and other solutes to diffuse freely across a membrane. **Passive transport** is the name for a flow of solutes through the interior of transport proteins, down their concentration gradients. Energetically, the flow costs only what the cell already spent to produce and maintain the gradients. Passive transport itself adds nothing more to the energy cost.

Transport proteins allow solutes to move both ways across a cell membrane. In cases of passive transport, the *net* direction of movement during a given interval depends on how many molecules or ions of the solute are making random contact with vacant binding sites in the interior of the proteins (Figure 5.11). The binding and transport simply proceed more often on the side of the membrane where the solute is more concentrated. Because there are more molecules around, the random encounters with binding sites are more frequent than they are on the other side of the membrane.

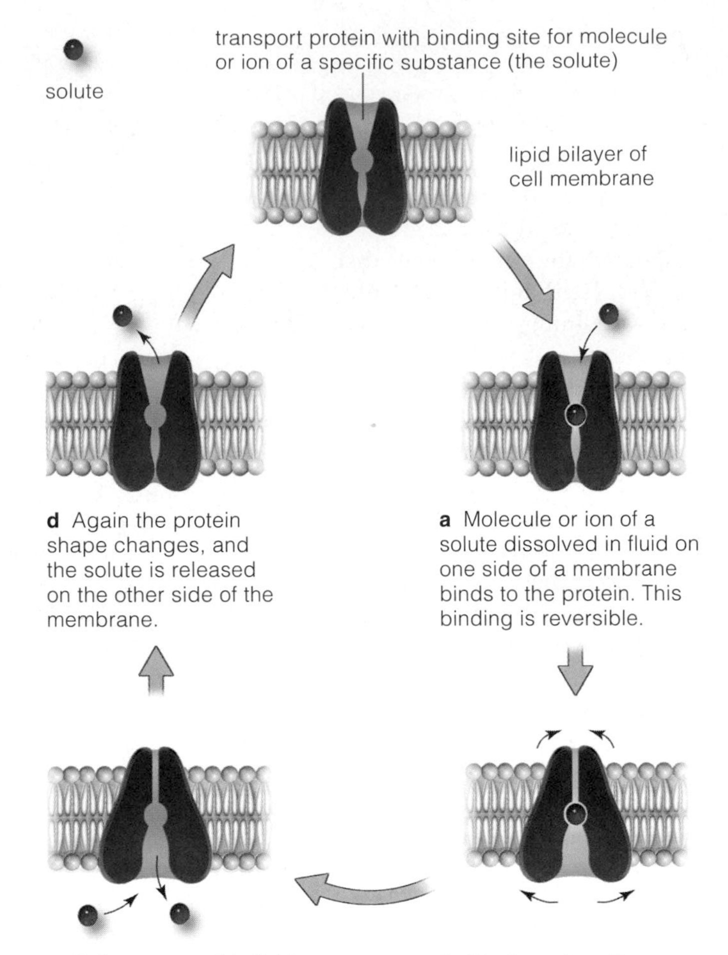

Figure 5.11 Passive transport across a cell membrane. Solutes are able to move in both directions through transport proteins. In passive transport, *net* movement will be down the concentration gradient (from higher to lower) until concentrations are the same on both sides of the membrane.

Figure 5.10 Sketch of transport proteins that span the lipid bilayer of a plasma membrane, cutaway view. Such proteins passively allow or actively assist ions and large polar molecules such as glucose to move through their interior, from one side of the cell membrane to the other.

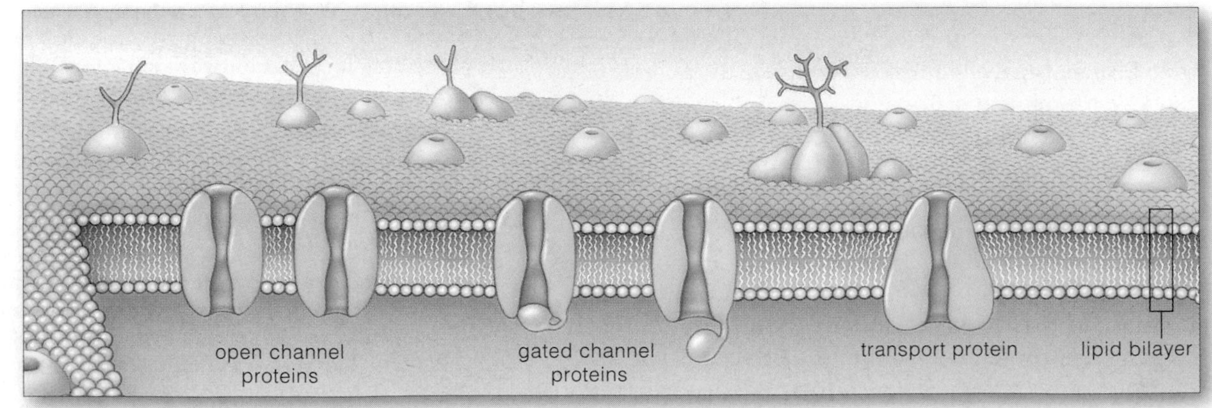

Further reading: Student Guide to InfoTrac on web site

Figure 5.12 Active transport across a cell membrane. ATP transfers a phosphate group to a transport protein. The transfer sets in motion reversible changes in the protein's shape that result in a greater net movement of solute particles *against* the concentration gradient.

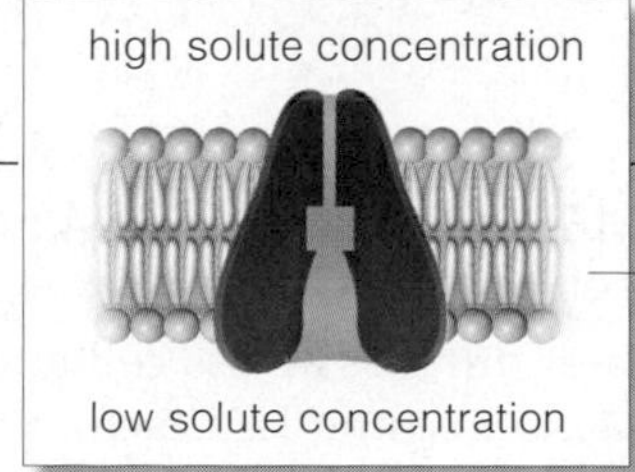

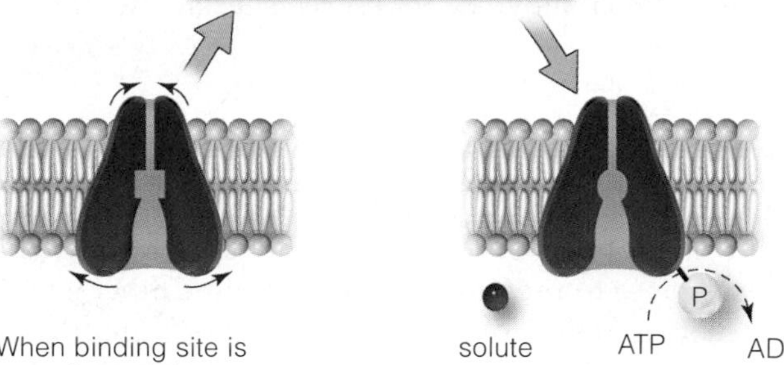

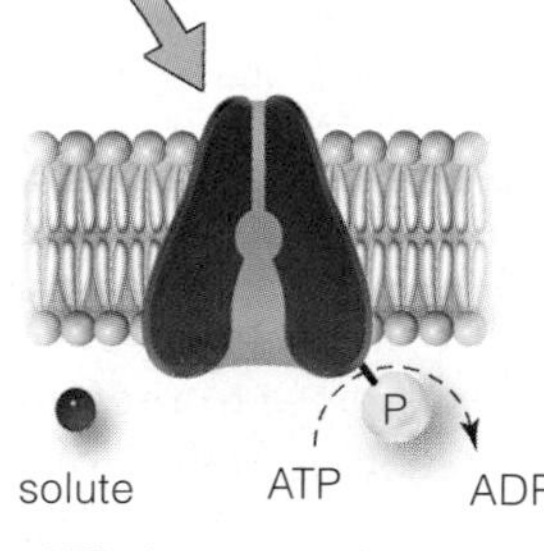

a ATP donates a phosphate group to a transport protein.

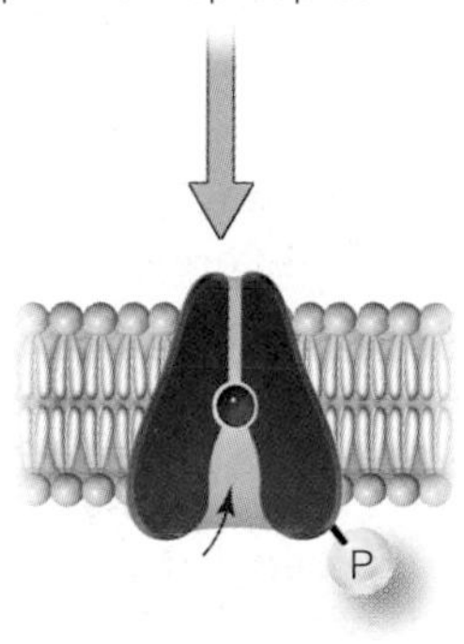

b The chemical fit between protein and solute improves on one side of the membrane when the protein is phosphorylated (energized). The solute binds to the protein.

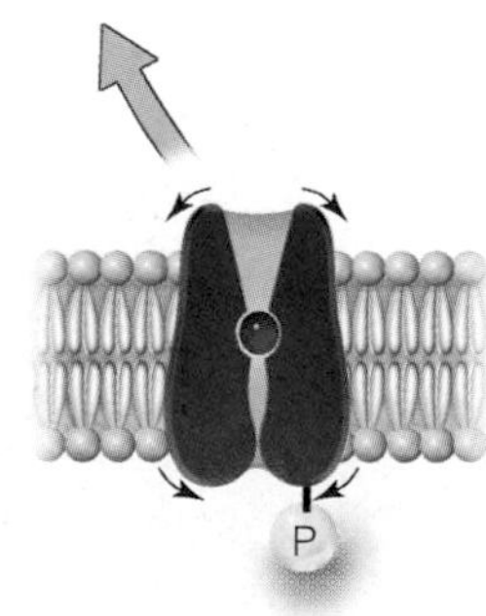

c When the solute is bound, the protein shape changes and the bound solute ends up exposed to fluid on the other side of the membrane.

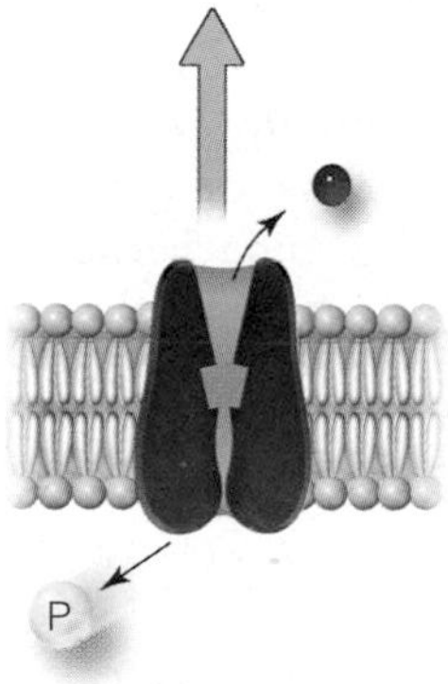

d The binding site reverts to its original shape. The solute and phosphate group are released.

e When binding site is vacant, the protein can revert to its original shape.

If nothing else were going on, the passive two-way transport would proceed until a solute's concentrations were equal across the membrane. But other processes usually affect the outcome. For example, blood delivers glucose to all body tissues, where nearly all cells use it as an energy source and as a building block. Cells can rapidly take up glucose when its blood concentration is high, and still the gradient is maintained. How? As fast as some glucose molecules diffuse into a cell, others are entering metabolic reactions. Thus, when cells quickly use glucose, they help maintain a concentration gradient that favors the uptake of *more* glucose.

Active Transport

Only in dead cells do solute concentrations become equal on both sides of membranes. Living cells never stop expending energy to pump potassium and other solutes to and from their interior. In **active transport**, energy-driven mechanisms called "membrane pumps" make solutes cross membranes *against* concentration gradients. ATP provides most of the energy to do this.

When ATP gives up one of its phosphate groups to a transport protein, the chemical fit between the solute and the protein binding site improves on one side of the membrane (Figure 5.12*a*,*b*). After a solute particle binds at the site, the protein's folded shape changes in such a way that the bound solute becomes exposed to fluid bathing the other side of the membrane (Figure 5.12*c*). Now the binding site reverts to its less attractive state, and the solute is released. A less attractive site means fewer molecules or ions of the solute can make the return trip. The *net* movement is to the side of the membrane where the solute is more concentrated.

One type of active transport system, the **calcium pump**, helps keep the calcium concentration in a cell at least a thousand times lower than outside. Another, the **sodium–potassium pump**, moves potassium ions (K^+) across the plasma membrane. Activation of this protein facilitates binding of a sodium ion (Na^+) on one side of a cell membrane. After the Na^+ makes the crossing, it is released. Its release facilitates the binding of K^+ at a different binding site on the protein, which reverts to its original shape after K^+ is delivered to the other side.

Through operation of such active transport systems, concentration and electric gradients can be maintained across membranes. The gradients are vital to many cell activities and physiological processes, including muscle contraction and nerve cell (neuron) function. We return to the mechanisms of active transport in later chapters. For now, keep these points in mind:

Transport proteins span the lipid bilayer of cell membranes. When they bind a solute on one side of a membrane, they undergo a reversible change in shape that shunts the solute through their interior, to the other side.

With passive transport, a solute simply diffuses through the protein; its net movement is down its concentration gradient.

With active transport, the net diffusion of a solute is uphill, against its concentration gradient. The transporting protein must be activated, as by ATP energy, to counter the energy inherent in the gradient.

5.5 MOVEMENT OF WATER ACROSS MEMBRANES

By far, more water diffuses across cell membranes than any other substance, so the key factors that influence its directional movement deserve special attention.

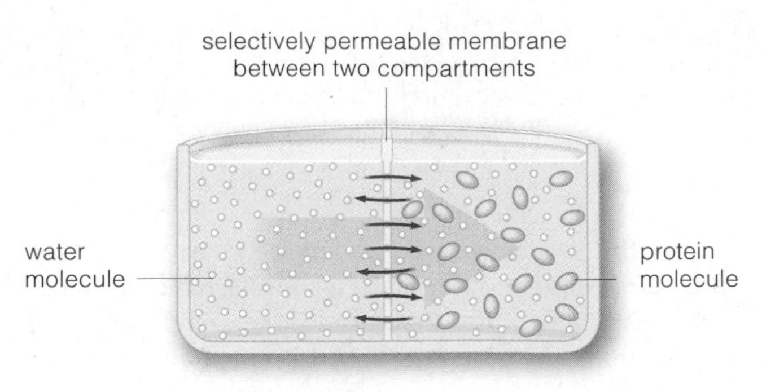

Figure 5.13 Demonstration of how a solute concentration gradient influences osmotic movement. Start with a container divided by a membrane that water but not proteins can cross. Pour water into the left compartment. Pour the same volume of a protein-rich solution into the right compartment. There, proteins occupy some of the space. The net diffusion of water in this example is from left to right (large *gray* arrow).

Osmosis

Turn on a faucet or watch a waterfall, and the moving water provides a demonstration of bulk flow. **Bulk flow** is the mass movement of one or more substances in response to pressure, gravity, or some other external force. It accounts for some movement of water through complex plants and animals. With each beat, your heart creates fluid pressure that drives a volume of blood, which is mainly water, through interconnected blood vessels. Sap runs inside conducting tissues that thread through maple trees, and this, too, is bulk flow.

What about the movement of water into and out of cells or organelles? A membrane intervening between two regions allows the small, polar water molecules to cross but restricts the passage of ions and large polar molecules. **Osmosis** is the name for the diffusion of water molecules in response to a water concentration gradient between two regions separated by a selectively permeable membrane.

Concentrations of solutes in water on both sides of a membrane influence osmosis, as you can see from Figure 5.13. *The side that has more solute particles has a lower concentration of water.* To gain insight into why this is so, imagine dissolving a small amount of glucose in water. Compared to an equivalent volume of water, the glucose solution has fewer water molecules—because each molecule of glucose occupies some of the space formerly occupied by molecules of water.

It is mainly the *total number* of molecules or ions, not the type of solute, that dictates the concentration of water. Dissolve one mole of an amino acid or urea in 1 liter of water, and the water concentration decreases about as much as it did in the glucose solution. Add one mole of sodium chloride (NaCl) to 1 liter of water, and it dissociates into equal numbers of sodium ions and chloride ions. There are now two moles of solute particles—twice as many as in the glucose solution—so the water concentration has decreased proportionately.

Effects of Tonicity

Given that water molecules tend to move osmotically to a region where water is less concentrated, the direction of their movement tends to be toward a region where solutes are more concentrated. Figure 5.14 illustrates this tendency. Suppose you decide to make a simple observational test of this statement. You construct three sacs out of a membrane that water but not sucrose can

2M sucrose solution

a

1 liter of distilled water

10M sucrose solution

2M sucrose solution

b

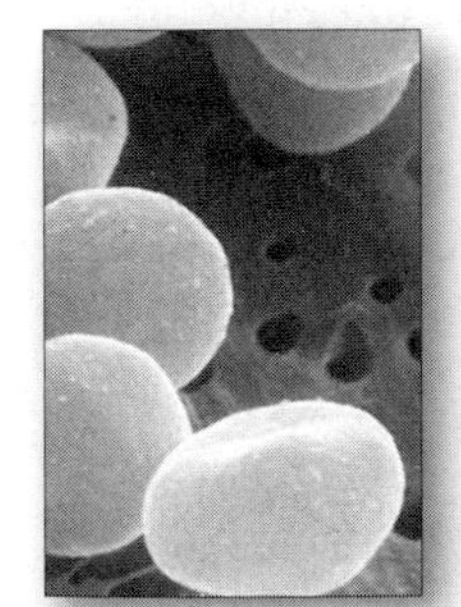

HYPOTONIC CONDITIONS

Water diffuses into red blood cells, which swell up

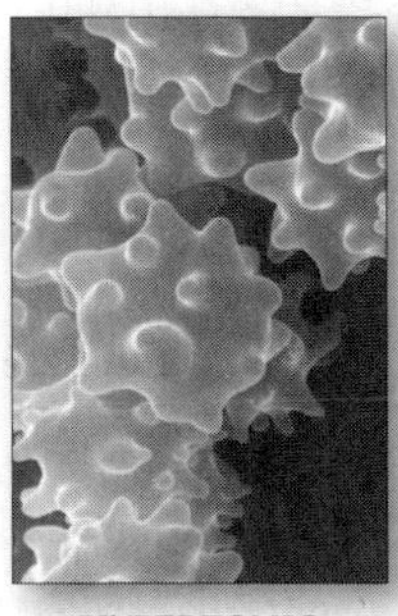

HYPERTONIC CONDITIONS

Water diffuses out of the cells, which shrink

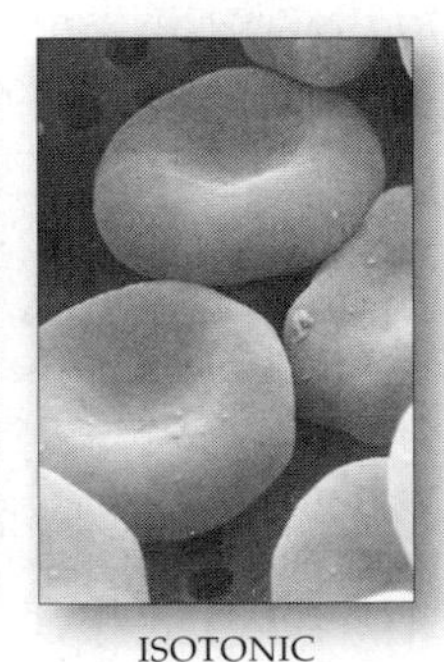

ISOTONIC CONDITIONS

No net movement of water, no change in cell size or shape

Figure 5.14 Effect of tonicity on water movement. (**a**) The arrow widths show the direction and relative amounts of water movement. (**b**) The micrographs correspond to the sketches. They show shapes that human red blood cells assume when you place them in fluids of higher, lower, and equal solute concentrations. Normally, solutions inside and outside of red blood cells are in balance. This type of cell does not have any built-in mechanisms that would help it adjust to drastic changes in solute levels in its fluid surroundings.

Further reading: Student Guide to InfoTrac on web site

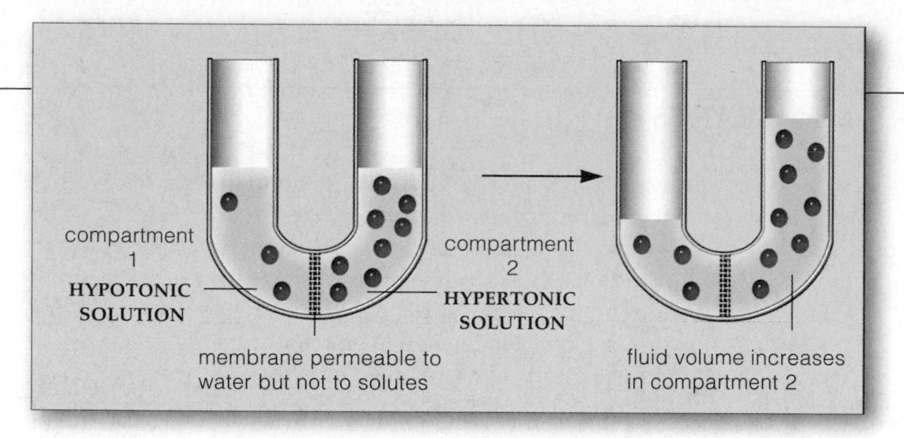

Figure 5.15 Increase in fluid volume owing to osmosis. In time, the net diffusion across a membrane separating two compartments is equal. Then, the fluid volume in compartment 2 is greater because the membrane is impermeable to solutes.

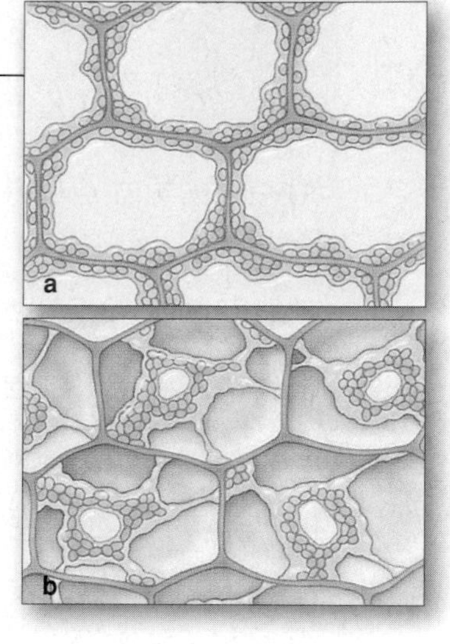

Figure 5.16 An osmotically induced loss of internal fluid pressure (called plasmolysis) from young plant cells such as those in (**a**). The cytoplasm and central vacuole shrink; the plasma membrane moves away from the wall, as in (**b**).

cross, and you fill each with a 2M sucrose solution. (*M* stands for *M*olarity, the number of moles of a solute in 1 liter of fluid.) Next you immerse one of the sacs in 1 liter of distilled water (which has no solutes), one in a 10M sucrose solution, and the third in a 2M sucrose solution. In each case, the extent and direction of water movement are dictated by tonicity (Figure 5.14).

Tonicity refers to the relative solute concentrations of two fluids. When two fluids on opposing sides of a membrane differ in solute concentration, the one having fewer solutes is called the **hypotonic solution**, and the one with more is the **hypertonic solution**. Water tends to diffuse from hypotonic to hypertonic fluids. **Isotonic solutions** have the same solute concentrations, so water shows no net osmotic movement from one to the other.

Normally, the fluid inside your cells and the tissue fluid bathing them are isotonic. If a tissue fluid became drastically hypotonic, so much water would diffuse into the cells that they would burst. If the fluid became too hypertonic, an outward diffusion of water would shrivel the cells. Most cells have built-in mechanisms that adjust to shifts in tonicity. Red blood cells do not; Figure 5.14 shows what happened to them during a demonstration of the effects of tonicity differences. That is why patients who are severely dehydrated are given infusions of a solution isotonic with blood. Such solutions move by bulk flow from a bottle positioned above the patient, through a tube, and directly into an incised vein.

Effects of Fluid Pressure

Animal cells generally can avoid bursting by engaging in the ongoing selective transport of solutes across the plasma membrane. Cells of plants and many protistans, fungi, and bacteria also avoid that unpleasant prospect with the help of pressure exerted on their cell walls.

Pressure differences as well as solute concentrations influence the osmotic movement of water. Take a look at Figure 5.15. It shows how water continues to diffuse across a membrane between a hypotonic solution and a hypertonic solution until the water concentration is the same on both sides. As you can see, the *volume* of the formerly hypertonic solution has increased (because its solutes cannot diffuse out). Any volume of fluid exerts **hydrostatic pressure**, or a force directed against a wall, membrane, or some other structure that encloses the fluid. The greater the solute concentration of the fluid, the greater will be the hydrostatic pressure it exerts.

Living cells cannot increase in volume indefinitely (Section 4.1). At some point, hydrostatic pressure that develops in a cell counters the inward diffusion of water. That point is the **osmotic pressure**, the amount of force that prevents further increase in a solution's volume.

Think of a young plant cell, with its pliable primary wall. As it grows, water diffuses into it and hydrostatic pressure increases against the wall. The wall expands, and the cell volume increases. The thin walls are strong enough for the cell's internal fluid pressure to develop to the point where it counterbalances water uptake.

Plant cells are vulnerable to water loss, which can happen when soil dries or becomes too salty. Water stops diffusing in, the cells lose water, and internal fluid pressure drops. Such osmotically induced shrinkage of cytoplasm is called plasmolysis (Figure 5.16). Plants can adjust somewhat to the loss of pressure, as when they actively take up potassium ions against a concentration gradient by mechanisms outlined in the next section.

As you will read in Chapters 39 and 41, hydrostatic and osmotic pressure also influence the distribution of water in the blood, tissue fluid, and cells of animals.

Osmosis is the net diffusion of water between two solutions that differ in water concentration and that are separated by a selectively permeable membrane. The greater the number of molecules and ions dissolved in a solution, the lower its water concentration will be.

Water tends to move osmotically to regions of greater solute concentration (from hypotonic to hypertonic solutions). There is no net diffusion between isotonic solutions.

The fluid pressure that a solution exerts against a membrane or wall also influences the osmotic movement of water.

BULK TRANSPORT ACROSS MEMBRANES

Exocytosis and Endocytosis

Transport proteins can move only small molecules and ions into or out of cells. When it comes to taking in or expelling large molecules or particles, cells use vesicles that form through exocytosis and endocytosis.

By **exocytosis**, a vesicle moves to the cell surface, and the protein-studded lipid bilayer of its membrane fuses with the plasma membrane. While this exocytic vesicle is losing its identity, its contents are released to the surroundings (Figure 5.17*a*).

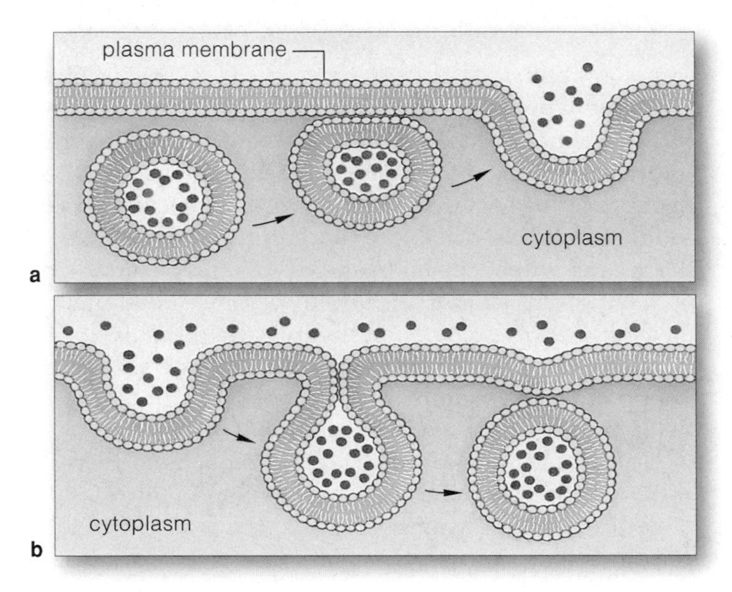

Figure 5.17 (**a**) Exocytosis. Cells release substances when an exocytic vesicle's membrane fuses with the plasma membrane. (**b**) Endocytosis. A bit of plasma membrane balloons inward beneath water and solutes outside, then it pinches off as an endocytic vesicle that moves into the cytoplasm.

By three pathways of **endocytosis**, a cell takes in substances next to its surface. In all three cases, a small indentation forms at the plasma membrane, balloons inward, and pinches off. The resulting endocytic vesicle transports its contents or stores them in the cytoplasm (Figure 5.17*b*). By *receptor-mediated* endocytosis, the first pathway, membrane receptors chemically recognize and bind specific substances, such as lipoproteins, vitamins, iron, peptide hormones, growth factors, and antibodies. The receptors become concentrated in tiny indentations in the plasma membrane (Figure 5.18). Each of these pits looks like a woven basket on its cytoplasmic side. The basket consists of protein filaments (clathrin) that are interlocked into stable, geometric patterns. When the pit sinks in the cytoplasm, the basket closes back on itself and becomes the vesicle's structural framework.

The second pathway, *bulk-phase* endocytosis, is less selective. An endocytic vesicle forms around a small volume of extracellular fluid regardless of what kinds of substances happen to be dissolved in it. Bulk-phase endocytosis operates at a fairly constant rate in nearly all eukaryotic cells. By continually pulling patches of plasma membrane into the cytoplasm, this pathway compensates for membrane that steadily departs from the cytoplasm in the form of exocytic vesicles.

The third pathway, **phagocytosis**, is an active form of endocytosis by which a cell engulfs microorganisms, large edible particles, and cellular debris. (Phagocytosis literally means "cell eating.") Amoebas and some other protistans get food this way. In multicelled organisms, macrophages and some other white blood cells engage in phagocytosis when they defend the body against

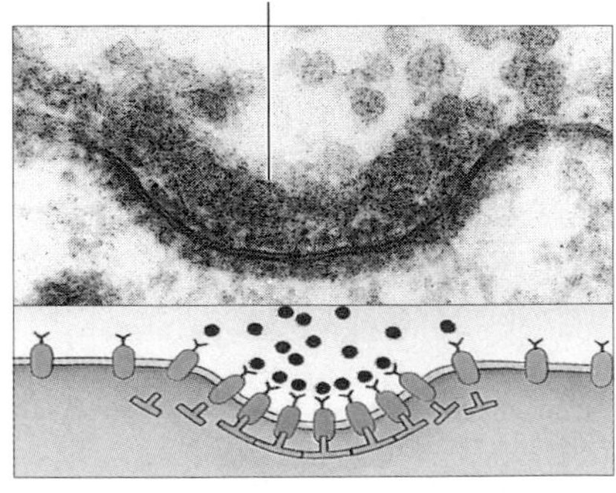

b clathrin filaments of coated pit

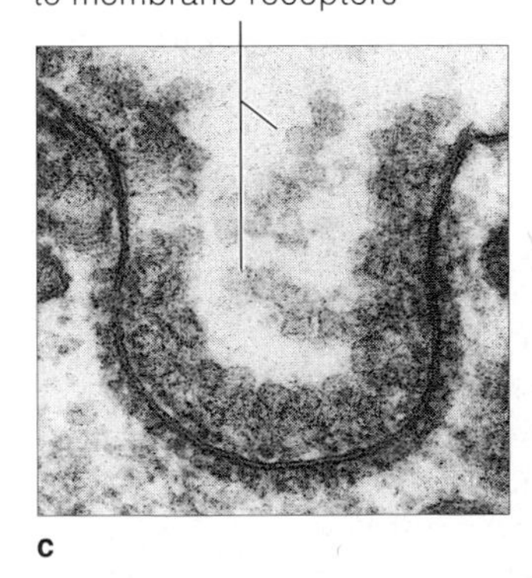

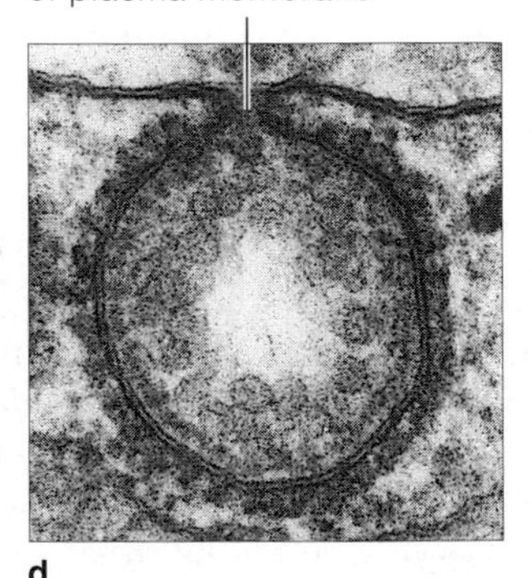

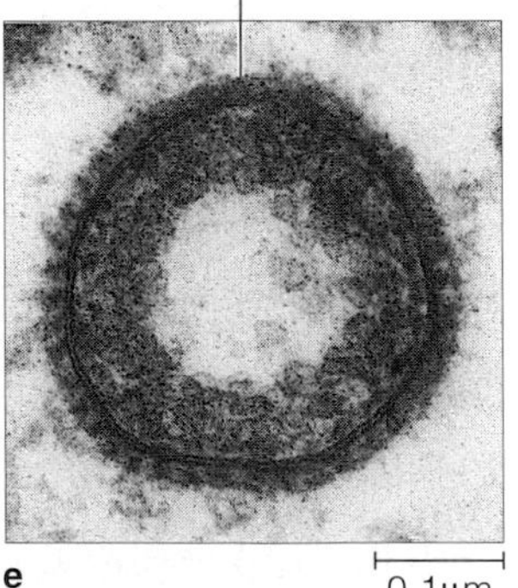

Figure 5.18 Example of receptor-mediated endocytosis, at the plasma membrane of an immature chicken egg. (**a**) This shallow indentation is a coated pit. (**b**) The cytoplasmic surface of each pit has a basketlike array of clathrin filaments. (**c**) Receptor proteins at the pit's outer surface preferentially bind lipoprotein particles. (**d**) The pit deepens and rounds out. (**e**) The formed endocytic vesicle contains lipoproteins that the cell will use or store.

Further reading: Student Guide to InfoTrac on web site

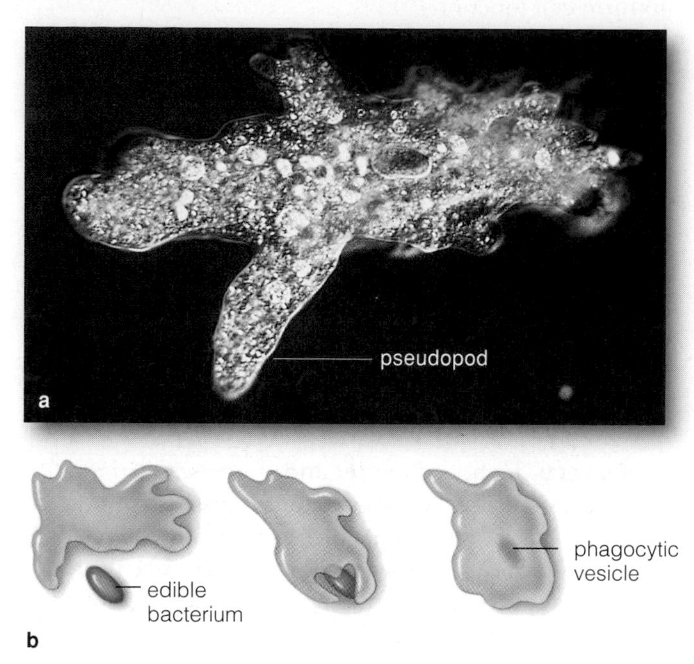

Figure 5.19 Phagocytosis in amoebas (including the *Amoeba proteus* cell in **a**), certain white blood cells such as macrophages, and some other cells. (**b**) Lobes of the amoeba's cytoplasm extend outward and surround the target. The plasma membrane of the extensions fuses together, thereby forming a phagocytic vesicle. Such vesicles move deeper into the cytoplasm and then fuse with lysosomes. Their contents are digested and, along with the vesicle's membrane components, are recycled elsewhere.

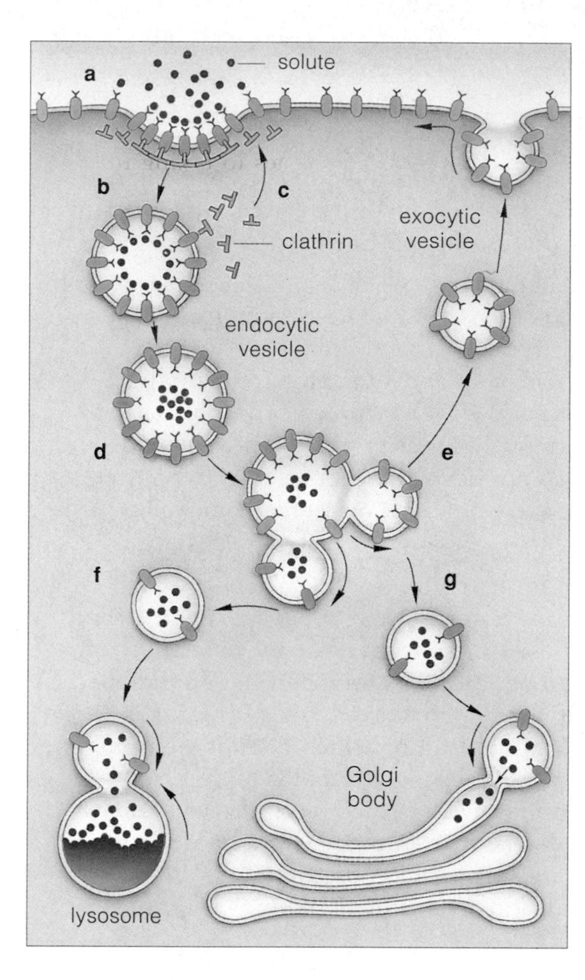

Figure 5.20 Cycling of membrane lipids and proteins. This example starts with receptor-mediated endocytosis. The plasma membrane gives up small patches of itself to endocytic vesicles that form from coated pits. It gets membrane back from exocytic vesicles that budded from ER membranes and Golgi bodies. The membrane initially appropriated by endocytic vesicles will cycle receptor proteins and lipids back to the plasma membrane.

invasions of harmful viruses, bacteria, cancerous body cells, and other threats to health.

A phagocytic cell gets busy after a target binds with certain receptors that bristle from its plasma membrane. Binding sends signals into the cell. The signals trigger a directional assembly and crosslinking of microfilaments into a dynamic, ATP-requiring network just beneath the plasma membrane. The network contracts in ways that squeeze some cytoplasm toward the cell margins, thus forming lobes called pseudopods (Figure 5.19). The pseudopods flow over the target and fuse at their tips. The result is a phagocytic vesicle, which sinks into the cytoplasm. There it fuses with lysosomes, the organelles of intracellular digestion in which trapped items are digested to fragments and smaller, reusable molecules.

Membrane Cycling

As long as a cell stays alive, exocytosis and endocytosis continually replace and withdraw patches of its plasma membrane. And they apparently do so at rates that can maintain the plasma membrane's total surface area.

As an example, neurons release neurotransmitters in bursts of exocytosis. Each neurotransmitter is a type of signaling molecule released from one cell that acts on neighboring cells. Cell biologist John Heuser and T. S. Reese documented an intense burst of endocytosis that immediately followed an intense episode of exocytosis in neurons—and that counterbalanced it. Figure 5.20 gives more examples of the ways in which cells cycle their membrane lipids and proteins.

Whereas transport proteins in a cell membrane deal only with ions and small molecules, exocytosis and endocytosis move larger packets of material across the plasma membrane.

By exocytosis, a cytoplasmic vesicle fuses with the plasma membrane, so that its contents are released outside the cell. By endocytosis, a small patch of the plasma membrane sinks inward and seals back on itself, forming a vesicle inside the cytoplasm. Membrane receptors often mediate this process.

5.7 SUMMARY

Membrane Structure and Function

1. The plasma membrane is a structural and functional boundary between the cytoplasm and the surroundings of all cells. In eukaryotic cells, organelle membranes subdivide the fluid portion of the cytoplasm into many functionally diverse compartments.

2. A cell membrane consists of two water-impermeable layers of lipids (phospholipids especially) and proteins associated with the layers, as shown in Figure 5.21.

a. Fatty acid tails and other hydrophobic parts of the lipids are sandwiched between hydrophilic heads.

b. Many different kinds of proteins are embedded in the lipid bilayer or positioned at one of its two surfaces. The proteins carry out most membrane functions.

3. These are the key features of the fluid mosaic model of membrane structure:

a. A cell membrane shows fluid behavior, mainly because its lipid components twist, move laterally, and flex hydrocarbon tails. Also, some of these lipids have ring structures and many have kinked (unsaturated) or short fatty acid tails, all of which disrupt what might otherwise be tight packing within the bilayer.

b. A membrane is a mosaic, or composite, of diverse lipids and proteins. The proteins are embedded in the bilayer and are at its surface. Its two layers differ in the number, kind, and arrangement of lipids and proteins.

4. All cell membranes include transport proteins and proteins that structurally reinforce the membrane. The plasma membrane also has proteins that serve in signal reception, cell recognition, and adhesion. Differences in the number and types of proteins among cells affect metabolism, cell volume, pH, and responsiveness to substances that make contact with the membrane.

a. Transport proteins help water-soluble substances cross the membrane by passing through their interior, which opens to both sides of the membrane.

b. Receptor proteins bind extracellular substances, and binding triggers alterations in metabolic activities. Recognition proteins are like molecular fingerprints; they identify cells as being of a given type. Adhesion proteins help cells of tissues adhere to one another and to form cell junctions.

Movement of Substances Into and Out of Cells

1. Molecules or ions of a substance tend to move from a region of higher to lower concentration. Movement in response to a concentration gradient is called diffusion.

a. Diffusion rates are influenced by the steepness of the concentration gradient, temperature, and molecular size, as well as by gradients in electrical charge and pressure that may occur between two regions.

b. Cells have built-in mechanisms that work with or against gradients to move solutes across membranes.

c. Metabolism requires chemical energy inherent in concentration and electric gradients across cell membranes.

2. Oxygen, carbon dioxide, and other small nonpolar molecules diffuse across a membrane's lipid bilayer. Ions and large, polar molecules such as glucose cross it with the passive or active help of transport proteins. Water moves through proteins and through the bilayer.

3. Transport proteins bind specific solutes on one side of a cell membrane, and they shunt solutes to the other side as an outcome of reversible changes in their shape.

a. Passive transport does not require energy inputs; the protein allows a solute simply to diffuse through its interior in the direction of the concentration gradient.

b. Active transport requires energy boosts, as from ATP. The protein pumps a solute across the membrane against its concentration gradient.

4. Osmosis is the diffusion of water across a selectively permeable membrane in response to a concentration gradient.

5. By exocytosis, vesicles in the cytoplasm move to the plasma membrane. When their membrane fuses with it, their contents are automatically released to the outside.

6. By endocytosis, a small patch of plasma membrane sinks into the cytoplasm and seals back on itself to form a vesicle. Three pathways are called receptor-mediated endocytosis (requires recognition of specific solutes), bulk-phase endocytosis (indiscriminate uptake of some extracellular fluid), and phagocytosis (active uptake of large particles, cell parts, or whole cells).

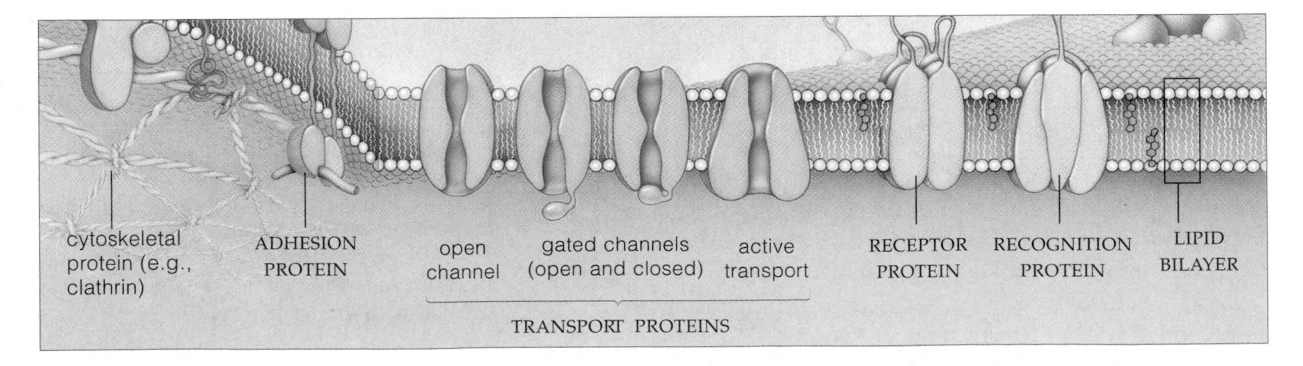

Figure 5.21 Summary of the main types of proteins associated with plasma membranes.

Review Questions

1. Describe the fluid mosaic model of cell membranes. What imparts fluidity to the membrane? What makes it a mosaic? *5.1*

2. State the functions of transport proteins, receptor proteins, recognition proteins, and adhesion proteins. *5.1*

3. Define diffusion. Does diffusion occur in response to a solute concentration gradient, an electric gradient, a pressure gradient, or some combination of these? *5.3*

4. If all transport proteins work by changing shape, then how do passive transporters differ from active transporters? *5.4*

5. Define osmosis. Explain how solute concentrations of two solutions on either side of a membrane influence the osmotic movement of water down the water concentration gradient. *5.5*

6. Define hypertonic, hypotonic, and isotonic solutions. Does each term refer to a property inherent in a given solution? Or are the terms used only when comparing solutions? *5.5*

7. Define exocytosis and endocytosis. Describe the main features of the three pathways of endocytosis. *5.6*

Self-Quiz (*Answers in Appendix III*)

1. Cell membranes consist mainly of a ______.
   a. carbohydrate bilayer and proteins
   b. protein bilayer and phospholipids
   c. lipid bilayer and proteins

2. In a lipid bilayer, ______ of lipid molecules are sandwiched between ______.
   a. hydrophilic tails; hydrophobic heads
   b. hydrophilic heads; hydrophilic tails
   c. hydrophobic tails; hydrophilic heads
   d. hydrophobic heads; hydrophilic tails

3. Most membrane functions are carried out by ______.
   a. proteins
   b. phospholipids
   c. nucleic acids
   d. hormones

4. All cell membranes incorporate ______.
   a. transport proteins
   b. adhesion proteins
   c. recognition proteins
   d. all of the above

5. Immerse a living cell in a hypotonic solution, and water will tend to ______.
   a. move into the cell
   b. move out of the cell
   c. show no net movement
   d. move in by endocytosis

6. A ______ is a device that spins test tubes and separates cell components according to their relative densities.

7. ______ can readily diffuse across a lipid bilayer.
   a. Glucose
   b. Oxygen
   c. Carbon dioxide
   d. b and c

8. Sodium ions cross a membrane through transport proteins that receive an energy boost. This is an example of ______.
   a. passive transport
   b. active transport
   c. facilitated diffusion
   d. a and c

Critical Thinking

1. The bacterium *Vibrio cholerae* causes a disease, *cholera*. Infected people have severe diarrhea and may lose up to twenty liters of fluid a day. *V. cholerae* enters the body when someone drinks contaminated water, then it adheres to the intestinal lining. It secretes a metabolic product that is toxic to cells of the lining, and they in turn start secreting chloride ions (Cl^-). Sodium ions (Na^+) follow the chloride ions into the intestinal fluid. Explain how the series of ion movements causes the massive fluid loss.

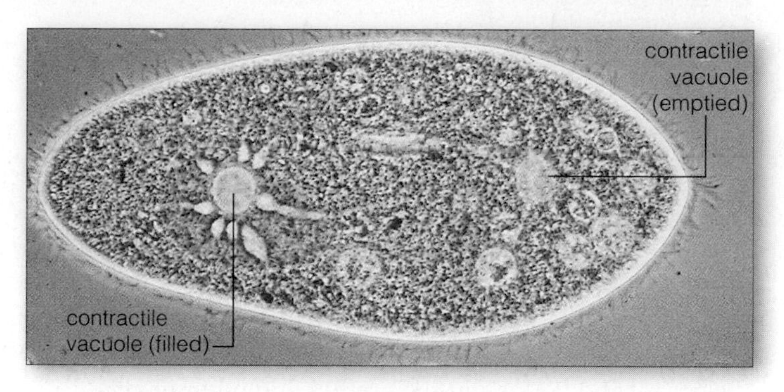

Figure 5.22 Contractile vacuoles of *Paramecium*, a protistan.

2. Many cultivated fields in California require heavy irrigation. Over the years, most of the water has evaporated from the soil, leaving behind all of the irrigation water's solutes. What kinds of problems might the altered soil conditions cause for plants?

3. Imagine that you are a juvenile shrimp living in an *estuary*, where freshwater draining from the land mixes with saltwater from the sea. Many people who own homes around a large lake want boat access to the sea, so they ask their city for permission to build a canal between the lake and estuary. If they succeed, what might happen to you and other estuary inhabitants?

4. Water moves osmotically into *Paramecium*, a single-celled protistan of aquatic habitats. If unchecked, the influx would bloat the cell and rupture its plasma membrane. An energy-requiring mechanism involving contractile vacuoles expels the excess (Figure 5.22). Water enters tubelike extensions of this organelle and collects in a central space in the vacuole. When full, the vacuole contracts and squirts excess water out of a pore that opens to the outside. Are the fluid surroundings hypotonic, hypertonic, or isotonic relative to *Paramecium*'s cytoplasm?

5. Certain species of bacteria thrive in environments where the temperatures approach the boiling point of water—for example, in the steam-venting fissures of volcanoes and in hot springs of Yellowstone National Park. Assume that the lipid bilayer of the bacterial cell membranes consists mainly of phospholipids. What features might the fatty acid tails of the phospholipids have that help stabilize the membranes at such extreme temperatures?

Selected Key Terms

active transport *5.4*
adhesion protein *5.1*
bulk flow *5.5*
calcium pump *5.4*
centrifuge *5.2*
concentration gradient *5.3*
diffusion *5.3*
electric gradient *5.3*
endocytosis *5.6*
exocytosis *5.6*
fluid mosaic model *5.1*
hydrostatic pressure *5.5*
hypertonic solution *5.5*
hypotonic solution *5.5*
isotonic solution *5.5*
lipid bilayer *5.1*
osmosis *5.5*
osmotic pressure *5.5*
passive transport *5.4*
phagocytosis *5.6*
phospholipid *5.1*
pressure gradient *5.3*
receptor protein *5.1*
recognition protein *5.1*
selective permeability *5.3*
sodium–potassium pump *5.4*
transport protein *5.1*

Readings *See also www.infotrac-college.com*

Alberts, B. et al. 1994. *Molecular Biology of the Cell*. Third edition. New York: Garland.

6 GROUND RULES OF METABOLISM

You Light Up My Life

Find yourself out and about at night or snorkeling in the ocean, and you might see fireflies and other insects, squids, or certain other organisms emitting light. Figure 6.1*a* shows fireflies (actually a type of beetle) engaging in social displays in a tropical forest. Different varieties, known locally as kittyboos, light up the night with green, yellow-green, yellow, or orange flashes.

Kittyboos emit light when enzymes—luciferases—convert chemical energy to light energy. The reactions get under way when a phosphate group from ATP is transferred to luciferin, a highly fluorescent substance. Luciferases, with a little help from oxygen, convert the activated luciferin to a different molecule. Their action boosts electrons of the molecule to a higher energy level. As the excited electrons quickly return to a lower energy level, they release energy as *fluorescent* light. Fluorescent light is emitted when a destabilized molecule reverts to a stable configuration. When organisms flash with it, this is called **bioluminescence**.

Imaginative biologists learned how to borrow light from such flashers to make *bioluminescent gene transfers*. They now insert copies of the genes for bioluminescence into bacteria, plants, and other organisms (Figure 6.1*b*)! Besides being fun to think about, these transfers have practical applications.

Each year, for example, 3 million people die from a lung disease caused by *Mycobacterium tuberculosis*. No one antibiotic is effective against all the different strains of this bacterium, so an infected person cannot receive effective treatment until the strain causing the infection is identified. A fast way to do this is to expose bacterial cells in samples taken from a patient to luciferase genes. The genes typically slip into the bacterial DNA of some cells. Clinicians isolate those cells, then expose colonies of the descendants to different antibiotics. If a particular antibiotic doesn't work, the colonies glow. Their cells have churned out gene products—including luciferase. If the colonies don't glow, the antibiotic works.

Christopher and Pamela Contag, two postdoctoral students at Stanford University, wanted to light up bacteria that cause *Salmonella* infections in laboratory mice. Why? Researchers of viral or bacterial diseases typically infect dozens to hundreds of laboratory mice for experiments. The only option has been to kill infected mice and examine tissues to find out whether infection occurred—a costly, tediously painstaking practice that also requires dispatching the experimental animals.

The Contags approached David Benaron, a medical imaging researcher at Stanford, with this hypothesis: If live, infectious bacteria were made bioluminescent, then flashes would shine through the tissues of infected animals. In a preliminary test of this novel idea, the researchers put glowing *Salmonella* cells into a thawed chicken breast from a market. A glow showed through.

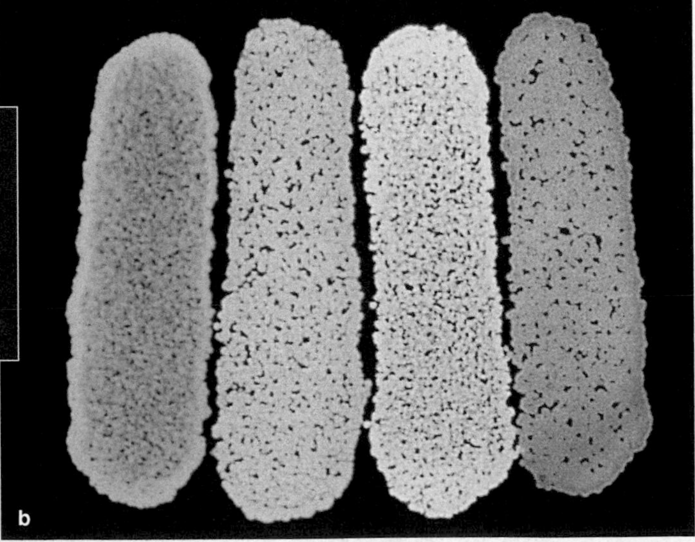

Figure 6.1 (**a**) Jamaican fireflies, known locally as kittyboos (*Pyrophorus noctilucus*). This type of beetle lights up the night with brief bioluminescent flashes, which help potential mates find each other in the dark. (**b**) Micrograph of four colonies of bacterial cells. Each colony started with a parent bacterial cell that had taken up a kittyboo gene for a glowing color.

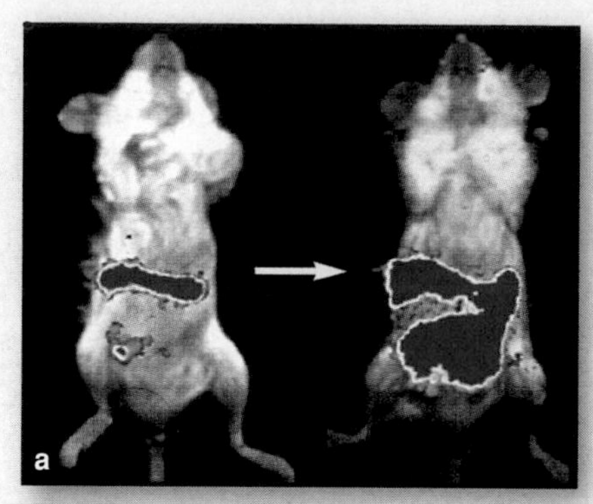

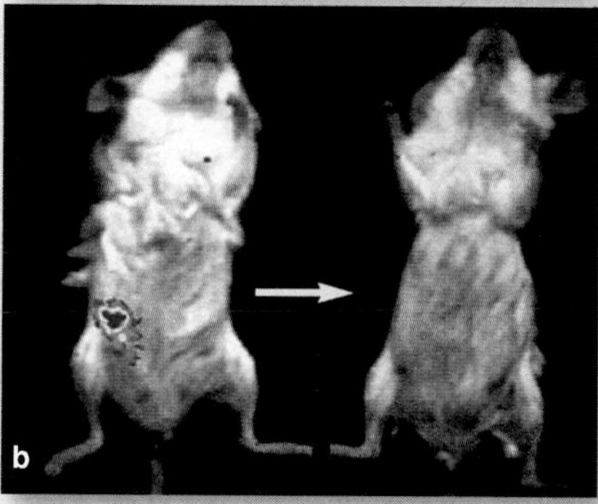

Figure 6.2 Using bioluminescent bacterial cells to chart the location of infectious bacteria inside living laboratory mice and their spread through body tissues. (**a**) False-color images in this pair of photographs show how the infection spread in a control group that had not been given a dose of antibiotics. (**b**) This pair shows how antibiotics had killed most of the infectious bacterial cells.

Next the Contags transferred bioluminescence genes into three strains of *Salmonella*. Then they injected the strains into mice of three experimental groups and used a digital imaging camera to track the infection in each group. The first strain was weak; the mice were able to fight off the infection in less than six days and did not glow. The second strain was not as weak but could not spread through the mouse body; it remained localized. The third strain was dangerous. It spread very rapidly through the mouse gut—and the entire gut glowed.

Thus bioluminescent gene transfer, combined with imaging of enzyme activity, can be used to track the course of infection and to evaluate the effectiveness of drugs in living organisms (Figure 6.2). It may also have uses in gene therapy, whereby copies of functional genes replace defective or cancer-causing genes in patients.

Why use bioluminescent organisms to introduce a chapter? They give us visible signs of **metabolism**—of the cell's capacity to acquire energy and use it to build, break apart, store, and release substances in controlled ways. Each flash reminds us that living cells are taking in energy-rich solutes, building membranes, storing things, replenishing enzymes, and checking out their DNA. A constant supply of energy drives all of these activities. The story of metabolism starts with ways in which cells get energy and channel it into the reactions by which they stay alive, grow, and reproduce.

KEY CONCEPTS

1. Cells engage in metabolism, or chemical work. That is, they use energy to stockpile, build, rearrange, and break apart substances. Cells also use energy for mechanical work, as when they move cell structures such as flagella. They also channel energy into electrochemical work, as when they move charged substances into or out of the cytoplasm or an organelle compartment.

2. All organisms are adapted to secure energy from their environment, such as energy from the sun and from inorganic or organic molecules. Their cells couple the energy inputs to thousands of energy-requiring reactions.

3. ATP is the main carrier of energy in cells. It couples reactions that release energy from the sun's ray's or from molecules with reactions that require energy.

4. ATP forms when energy inputs lead to the transfer of a phosphate group or inorganic phosphate to a molecule called ADP. Later, ATP transfers a phosphate group to enzymes, glucose, and other molecules, which primes them to enter a reaction (activates them).

5. Many metabolic reactions involve electron transfers from one substance to another. Such transfers, or oxidation–reduction reactions, occur singly or in a series of small steps through electron transport systems.

6. Thousands of reactions proceed in cells in orderly, enzyme-mediated sequences called metabolic pathways. Operation of the pathways is coordinated in ways that maintain, increase, or decrease the relative amounts of various substances in cells.

7. Chemical reactions proceed far too slowly on their own to sustain life. In living organisms, the action of specific enzymes greatly increases the rate of specific reactions. Control of enzyme activity is central to metabolism.

6.1 ENERGY AND THE UNDERLYING ORGANIZATION OF LIFE

Defining Energy

If you have ever watched a house cat stalking a mouse, you know it can "freeze" its position to avoid detection before springing at its unsuspecting prey. Like anything else in the universe that is stationary, the cat has a store of **potential energy**—a capacity to do work, simply owing to its position in space and the arrangement of its parts. As a cat springs, some of its potential energy is transformed into **kinetic energy**, the energy of motion.

Energy on the move does work when it imparts motion to other things. In skeletal muscle cells inside the cat, ATP gave up some of its potential energy to molecules of contractile units and set them in motion. The combined motions in many muscle cells resulted in the movement of whole muscles. The transfer of energy from ATP also resulted in the release of another form of energy called **heat**, or *thermal* energy.

The potential energy of molecules has its own name: **chemical energy**. It is measurable, as in kilocalories. A kilocalorie is the same thing as 1,000 calories, which is the amount of energy it takes to heat 1,000 grams of water from 14.5°C to 15.5°C at standard pressure.

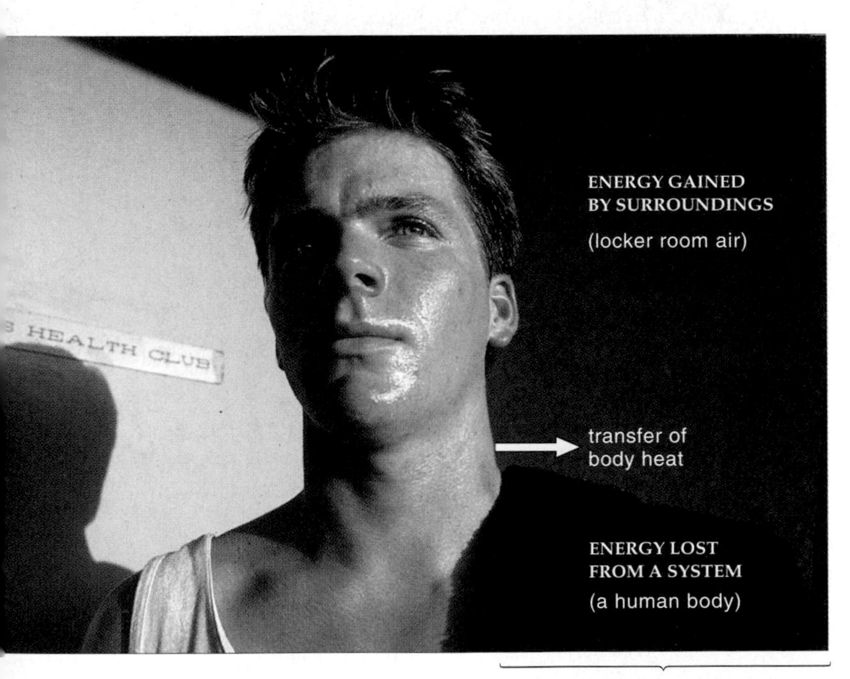

Figure 6.3 Example of how the total energy content of any system *together with its surroundings* remains constant.

"System" means all matter in a specific region, such as a human body, a plant, a DNA molecule, or a galaxy. "Surroundings" can be a small region in contact with the system or as vast as the entire universe. The system shown (a human male) is giving off heat to the surroundings (a locker room) by evaporative water loss from sweat. What one region loses, the other region gains, so the total energy content of both does not change.

What Can Cells Do With Energy?

All organisms have specific adaptations for securing energy from their environment. Some harness energy from the sun, and others extract energy from inorganic or organic substances in the environment. Regardless of the source, energy inputs become *coupled* to thousands of energy-requiring processes in cells. Cells use energy for *chemical* work—to stockpile, build, rearrange, and break apart substances. They channel it into *mechanical* work—to move flagella and other cell structures and (in multicelled species) the whole body or portions of it. They channel it into *electrochemical* work—to move charged substances into or out of the cytoplasm or an organelle compartment.

How Much Energy Is Available?

Like single cells, we cannot create energy from scratch; we must get it from someplace else. Why? According to the **first law of thermodynamics**, the total amount of energy in the universe remains constant. More energy cannot be created; existing energy cannot vanish. It can only be converted from one form to some other form.

Think about what the law means. The universe has only so much energy, distributed in a variety of forms. One form can be converted to another, as when corn plants absorb energy from the sun and convert it to the chemical energy of starch. After you eat and digest corn, your cells extract energy from starch and convert it to other forms, such as kinetic energy for moving about.

With each metabolic conversion, some of the energy escapes to the surroundings, as heat. Even when you "do nothing," your body gives off about as much heat as a 100-watt lightbulb because of conversions in your cells. The energy being released is transferred to atoms and molecules making up the air, and the conversion of thermal to kinetic energy "heats up" the surroundings (Figure 6.3). The kinetic energy increases the number of ongoing, random collisions among molecules in the air. And with each collision, a bit more energy is released as heat. However, none of the energy ever vanishes.

The One-Way Flow of Energy

Energy available for conversions in cells resides mainly in covalent bonds. Glucose, glycogen, starches, fatty acids, and other organic compounds have organized arrangements of many of these bonds and are said to have a high energy content. When the compounds enter metabolic reactions, specific bonds break or become rearranged. During that molecular commotion, some amount of heat energy is lost to the surroundings. In general, cells cannot recapture energy lost as heat.

Further reading: Student Guide to InfoTrac on web site →

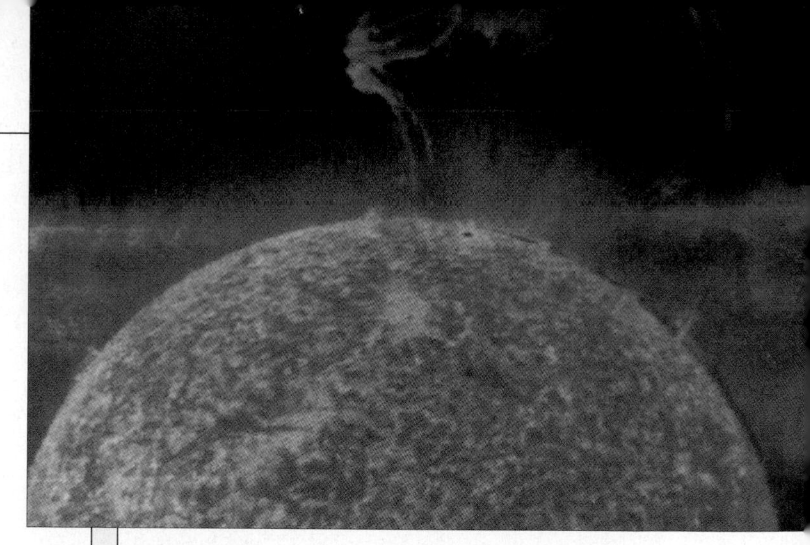

Figure 6.4 An example of the one-way flow of energy into the world of life that compensates for the one-way flow of energy out of it. The sun continuously loses energy, much of it in the form of wavelengths of light (Section 7.2). Living cells intercept some of the energy and convert it to useful forms of energy, stored in bonds of organic compounds. Each time a metabolic reaction proceeds in cells, stored energy is released—and some inevitably is lost to the surroundings, mostly as heat.

The lower photograph shows green, water-dwelling, photosynthetic cells (*Volvox*). They live in tiny, spherical colonies. The orange cells function in reproduction. They form new colonies inside the parent sphere.

ENERGY LOST
one-way flow of energy from sun to Earth's environment

ENERGY GAINED
one-way flow of energy from environment to organisms

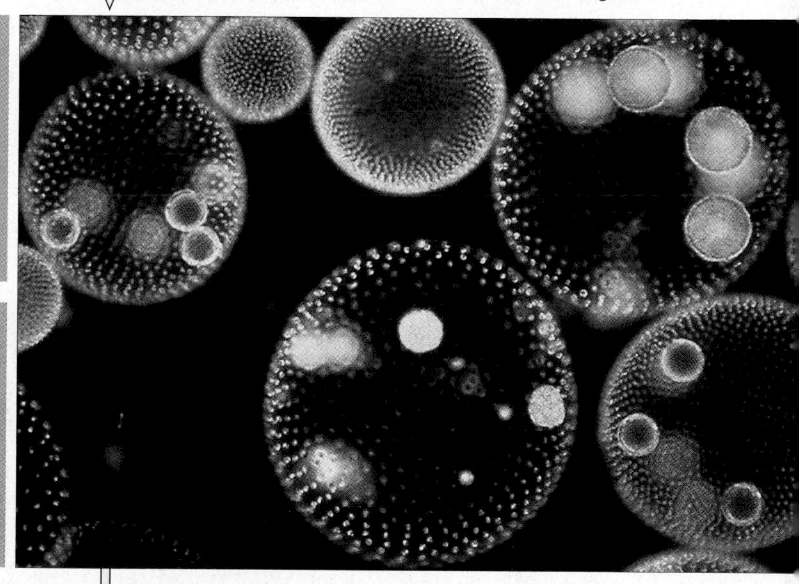

Producer organisms harness sun's energy, use it to build organic compounds from simple raw materials available in their environment.

All organisms tap potential energy stored in organic compounds to drive energy conversions that keep them alive. Some energy is lost with each conversion.

ENERGY LOST
one-way flow of energy from organisms back to the environment

For example, your cells release usable energy from glucose by breaking all of its covalent bonds. After many steps, six molecules of carbon dioxide and six of water remain. Compared with glucose, those leftovers have more stable arrangements of atoms, but chemical energy in their bonds is much less than the total chemical energy of glucose. Why? *Some energy was lost at each step leading to their formation.* Said another way, glucose is a better source of usable energy.

What about the heat that was transferred from cells to their surroundings when carbon dioxide formed? It is not at all useful. Cells cannot convert it to other forms, so it cannot be used to do work.

Bad news for cells of the remote future: The amount of "low-quality" energy in the universe is increasing. No energy conversion can ever be 100 percent efficient—even highly efficient ones lose heat—so the total amount of energy in the universe is spontaneously flowing from forms rich in energy to forms having less of it. Billions of years from now, all of the energy available for conversions will be dissipated.

Without energy inputs to maintain it, any organized system tends to become more and more disorganized over time. **Entropy** is a measure of the degree of a system's disorder. Think of the Egyptian pyramids—originally organized, presently crumbling, and many thousands of years from now, dust. It seems that the ultimate destination of those pyramids and everything else in the universe is a state of maximum entropy. That, basically, is the point to remember about the **second law of thermodynamics**.

Can life be one glorious pocket of resistance to the depressing flow toward maximum entropy? After all, in each new organism, new bonds form and hold atoms together in precise arrays. So molecules become more organized and have a richer store of energy, not poorer!

Yet a simple example will show that the second law does indeed apply to life on Earth. The primary energy source for life is the sun, which has been releasing energy since it first formed. Plants can capture sunlight energy, convert it to other forms, then lose energy to other organisms that feed, directly or indirectly, on the plants. At each energy transfer, some energy is lost as heat that joins the universal pool. *Overall, energy still flows in one direction.* The world of life maintains its amazing degree of organization only because it is being resupplied with energy that is being lost from someplace else (Figure 6.4).

The amount of energy in the universe remains constant. Energy can undergo conversions from one form to another, but it cannot be created out of nothing or destroyed.

The total amount of energy in the universe is spontaneously flowing from usable to nonusable forms.

A steady flow of sunlight energy into the interconnected web of life compensates for the steady flow of energy leaving it.

6.2

ENERGY CHANGES AND CELLULAR WORK

Energy Inputs, Energy Outputs

When cells convert one form of energy to another, there is a change in the amount of potential energy available to them. The greater the initial amount of potential energy a cell taps into, the larger the energy change can be—and the more work can be done.

Imagine a Martian who is not happy that the NASA Rover is inching around her planet. She decides to push it to the top of a rocky hill (Figure 6.5*a*). To do this, she converts some potential energy stored in her muscles to kinetic energy. Once the Rover is precariously perched on the hill, it has potential energy (owing to its position) and it tends to roll down on its own, spontaneously. The higher up the Rover has been pushed relative to its final position at the base of the hill, the greater the energy change and the more work done—in this case, a bigger impact and more broken parts (Figure 6.5*b*).

The same ground rule applies to the cellular world, in which temperature and pressure remain constant. *Energy changes in cells tend to proceed spontaneously in the direction that results in a decrease in usable energy.*

For instance, glucose ($C_6H_{12}O_6$) is built from carbon dioxide ($6CO_2$) and water ($12H_2O$). Each substance has potential energy in chemical bonds. But the bond energy of glucose exceeds those of the other two substances combined. Thus, the assembly of glucose from carbon dioxide and water is not something that will proceed spontaneously. Visualize carbon dioxide and water at the base of an energy hill. On their own, they simply don't have enough energy for an uphill run to make glucose.

In photosynthetic cells, *energy inputs* from the sun drive reactions that synthesize glucose. The outcome is a net increase in usable energy for the cell (Figure 6.5*c*). Said another way, the reaction sequence by which glucose forms is **endergonic** (meaning "energy in").

Now picture the reactions running in reverse, from glucose (at the top of the energy hill) to carbon dioxide and water (at the base). Energetically, a downhill run is favorable. It proceeds spontaneously and ends with a net loss in energy. So the reaction sequence that breaks down glucose is **exergonic**, which means "energy out" (Figure 6.5*d*). All cells, including photosynthetic ones, release energy from glucose and other large molecules.

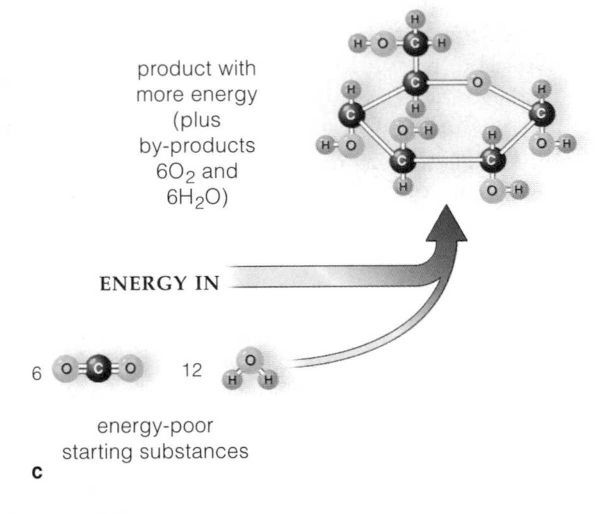

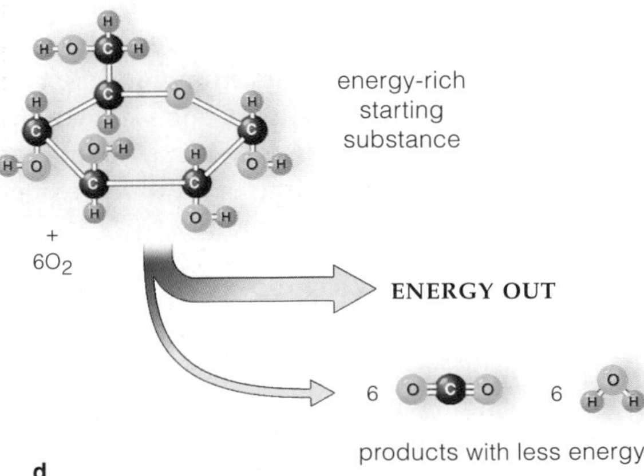

Figure 6.5 Examples of energy changes involved in (**a**,**b**) mechanical work and (**c**,**d**) chemical work.

Further reading: Student Guide to InfoTrac on web site

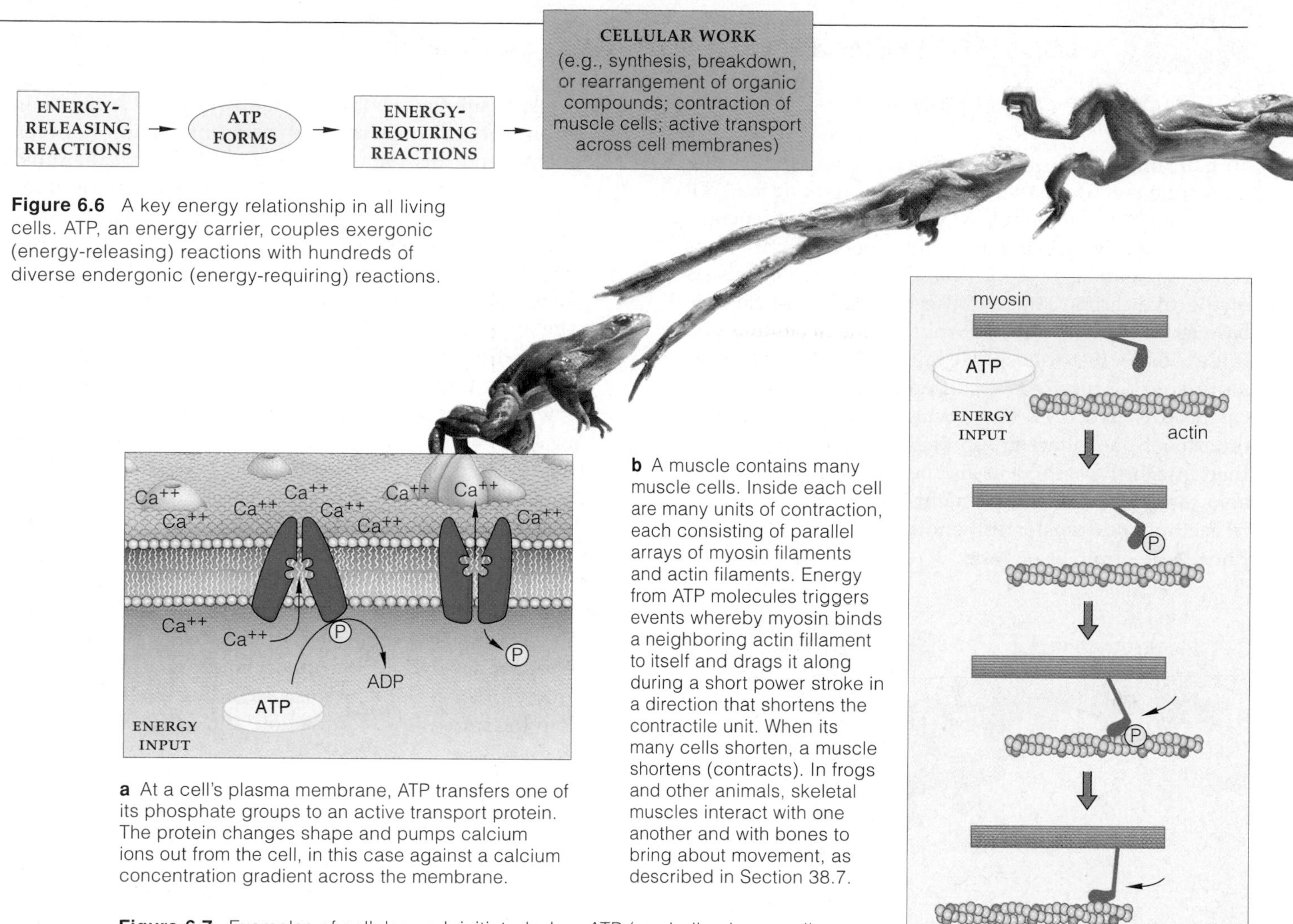

Figure 6.6 A key energy relationship in all living cells. ATP, an energy carrier, couples exergonic (energy-releasing) reactions with hundreds of diverse endergonic (energy-requiring) reactions.

a At a cell's plasma membrane, ATP transfers one of its phosphate groups to an active transport protein. The protein changes shape and pumps calcium ions out from the cell, in this case against a calcium concentration gradient across the membrane.

b A muscle contains many muscle cells. Inside each cell are many units of contraction, each consisting of parallel arrays of myosin filaments and actin filaments. Energy from ATP molecules triggers events whereby myosin binds a neighboring actin fillament to itself and drags it along during a short power stroke in a direction that shortens the contractile unit. When its many cells shorten, a muscle shortens (contracts). In frogs and other animals, skeletal muscles interact with one another and with bones to bring about movement, as described in Section 38.7.

Figure 6.7 Examples of cellular work initiated when ATP (symbolized as a yellow coin) gives up energy to molecules that participate in specific metabolic reactions.

The Role of ATP

Rays of sunlight are a form of kinetic energy, released during stupendous exergonic reactions at the surface of the sun. Here on Earth, photoautotrophs can convert some of that energy into chemical bond energy of ATP. Also, all organisms can make ATP through exergonic reactions during aerobic respiration and other glucose-degrading processes. Regardless of how it forms, ATP can activate many different molecules by transferring one of its phosphate groups to them. Section 6.3 explains this process, which is called phosphorylation.

ATP's role is like currency in an economy: cells earn it by exergonic reactions and can spend it in endergonic reactions that drive hundreds of cellular activities, such as those shown in Figures 6.6 and 6.7. That is why we often use a cartoon coin to symbolize ATP.

The Role of Electron Transfers

Think back on earlier chapters that describe the making and breaking of chemical bonds. Remember how atoms, ions, and molecules can be made to accept or give up electrons? Electrons are transferred in virtually every reaction that harnesses energy for use in the formation of ATP. For example, in plant cells, energy from the sun drives electrons from water molecules. Being attracted to the oppositely charged electrons, hydrogen ions go along for the ride—and both join up with carbon and oxygen to form new organic compounds. In all cells, ATP jump-starts the release of electrons from glucose to form more ATP, then oxygen or some other substance picks up the "spent" electrons. Section 6.4 takes a closer look at the nature of electron transfers in cells.

On their own, energy changes in cells spontaneously run in the direction that results in a decrease in usable energy.

ATP, an energy carrier in all living cells, couples energy-releasing reactions with energy-requiring reactions.

Electron transfers are involved in all of the reactions that harness the energy necessary to make ATP.

A LOOK AT TYPICAL PHOSPHATE-GROUP TRANSFERS

How Does ATP Give Up Energy?

To gain insight into how energy is transferred during metabolic reactions, let's take a closer look at the ATP molecule. As you know, **ATP** is short for adenosine triphosphate, which is one of the organic compounds called nucleotides (Section 3.7). Each molecule consists of the five-carbon sugar ribose to which a nucleotide base (adenine) and three phosphate groups are attached (Figure 6.8*a*). Its triphosphate tail is where the action is, so to speak. Hundreds of different kinds of enzymes can readily hydrolyze the covalent bond between the outermost phosphate group of the molecule's tail and then attach that group to another substance. Enzymes also can break the next covalent bond in the tail and transfer still another phosphate group elsewhere.

a

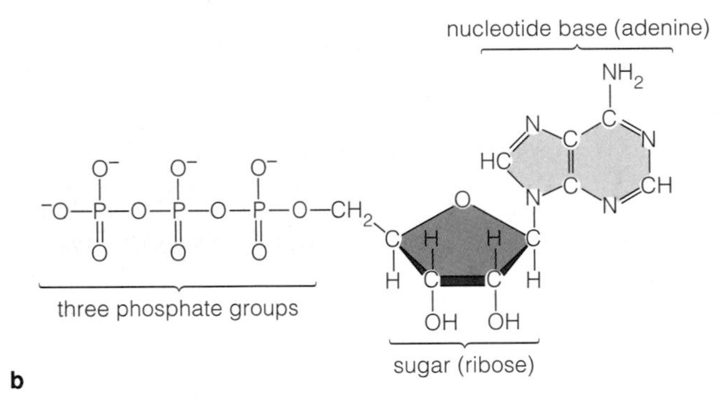

b

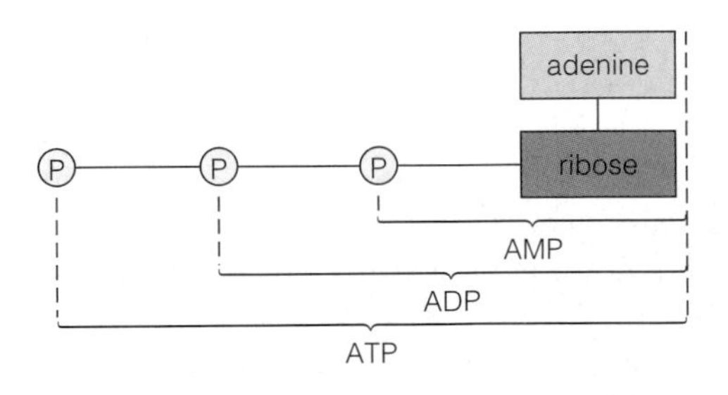

c

Figure 6.8 (**a**) Three-dimensional model of an ATP molecule showing its component atoms. (**b**) Structural formula for ATP. (**c**) Molecular relationship between ATP and two structurally similar nucleotides. When ATP gives up one of its phosphate groups, adenosine diphosphate (ADP) forms. Two phosphate-group transfers leave adenosine monophosphate (AMP).

Why are those bonds so readily broken? After all, in itself, a covalent bond is a stable interaction between atoms, and this same type of bond holds all the other component parts of ATP together. However, the three phosphate groups are clustered together. By studying Figure 6.8, you see that the clustering puts a number of negative charges in proximity. Remember, like charges repel each other. In this case, the repulsions destabilize the molecule's tail.

Getting rid of a phosphate group helps stabilize the molecule and results in more stable products of much lower energy. The products are adenosine diphosphate, or **ADP**, and a free inorganic phosphate atom, which is abbreviated $\mathbf{P_i}$ (Figure 6.8*c*). In other words, hydrolysis of ATP is an energetically favorable reaction, one that releases a large quantity of usable energy. This is the energy that can drive endergonic reactions. It can put glucose, enzymes, and other molecules in an activated state, meaning they become primed to react.

Any transfer of a phosphate group to a molecule is a **phosphorylation**. Phosphorylation is at the heart of many diverse cell activities, including photosynthesis, aerobic respiration, contraction, and active transport.

Renewing Supplies of ATP

ATP is the main energy carrier for so many metabolic reactions, you might speculate that cells must have a way to renew it, and you would be right. After ATP gives up a phosphate group to become ADP, enzymes can use available energy to attach phosphate to ADP, as shown in Figure 6.9.

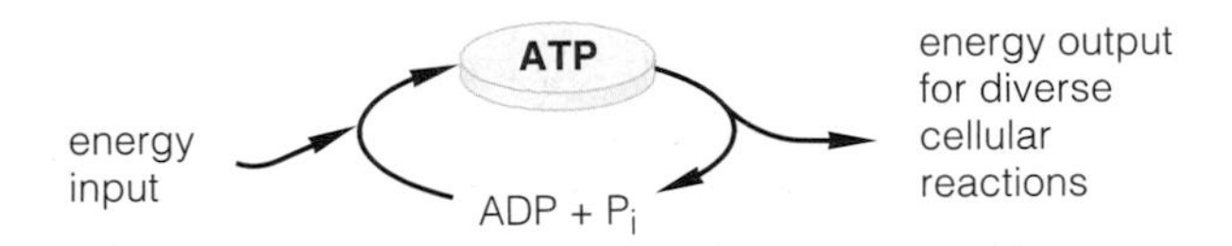

Figure 6.9 The ATP/ADP cycle, by which an ATP molecule that gives up a phosphate group is regenerated.

Regenerating ATP this way is called the **ATP/ADP cycle**. It is an important aspect of metabolism.

Phosphorylation is the transfer of a phosphate group from one molecule, such as ATP, to another molecule.

The transfer releases enough usable energy to activate an acceptor molecule, meaning the molecule becomes primed to enter a reaction.

Because the covalent bonds between its phosphate groups lend themselves to hydrolysis by so many enzymes, ATP is the main renewable energy carrier in cells.

Further reading: Student Guide to InfoTrac on web site

6.4 A LOOK AT TYPICAL ELECTRON TRANSFERS

Recall, from Section 6.2, that many cellular activities involve electron transfers from one substance to another. An electron transfer is also called an **oxidation–reduction reaction**. "Oxidation" simply refers to the removal of electrons from substances. "Reduction" refers to the addition of electrons to substances.

As a molecule is being reduced, it often acquires a hydrogen ion (H^+) at the same time. The ions, which are abundant in fluids bathing cell membranes, are attracted to the opposite charge of the electrons and go along for the ride. Such reductions are known as hydrogenations. Every saturated fat is fully hydrogenated. When its fatty acid tails were being synthesized, electron transfers caused two hydrogen ions to become attached to every carbon atom in their backbone.

Often electron transfers occur singly, as when ATP phosphorylates a protein dealing with active transport and thereby primes it for reaction. Electron transfers also proceed one after another, in a series of tiny steps.

For example, this happens when cells are running low on energy and oxidize glucose, fatty acids, and other organic compounds. Remember, a molecule such as glucose is energy-rich yet unstable, compared with the most stable products (CO_2 and H_2O) of its complete breakdown. Atmospheric oxygen is so abundant that CO_2 is the most stable form of carbon. H_2O the most stable form of hydrogen. Cells extract the most energy from glucose when its carbon and hydrogen atoms combine with oxygen in air.

The more stable forms cannot be reached all at once in cells. Imagine throwing some glucose into a wood-burning fire. The carbon atoms and hydrogen atoms of the glucose molecules would very quickly let go of one another and combine with oxygen in air—but all of the released energy would be lost as heat. Compared to the wood-burning fire, cells are able to release energy from glucose far more efficiently. How? They strip electrons from it and send them on through transport systems built into cell membranes.

An **electron transport system** consists of enzymes, coenzymes, and other molecules organized for electron transfers at a cell membrane. As one molecule becomes oxidized, the next in line becomes reduced. Electrons entering the system are at a higher energy level than ones leaving. Think of the electrons as dropping down a staircase and releasing energy at each step (Figure 6.10). At certain steps, energy is used to do work, as when it leads to the formation of H^+ concentration and electric gradients. Such gradients are used to produce ATP, in ways described in later chapters.

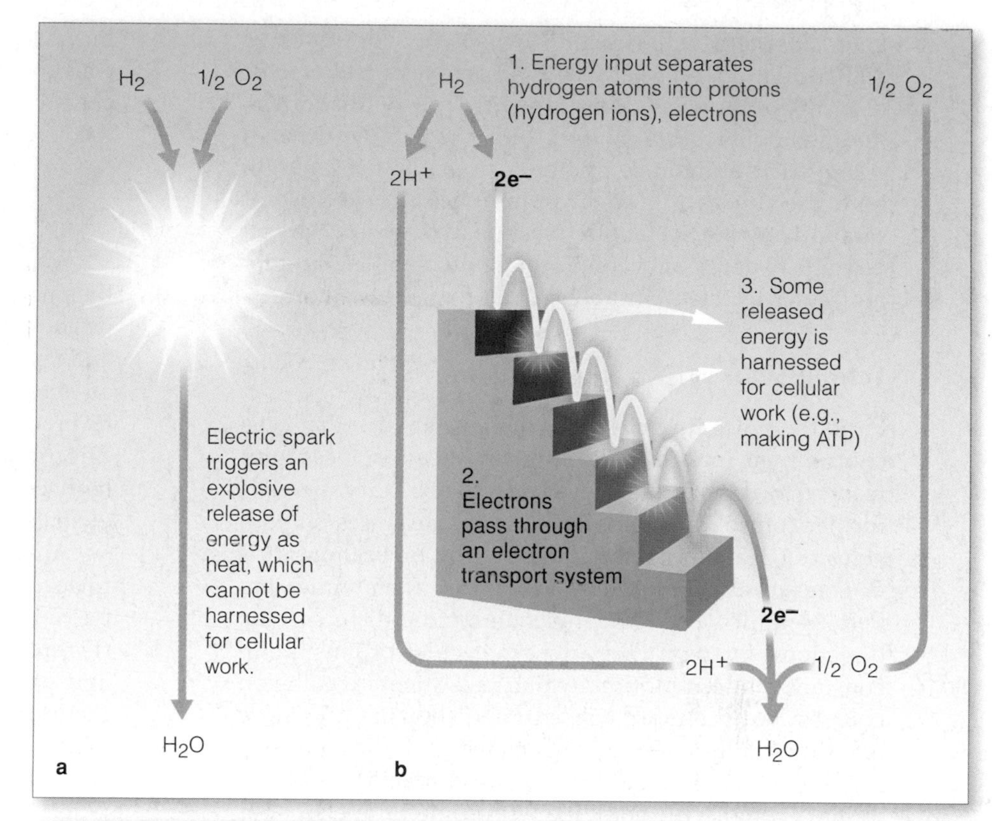

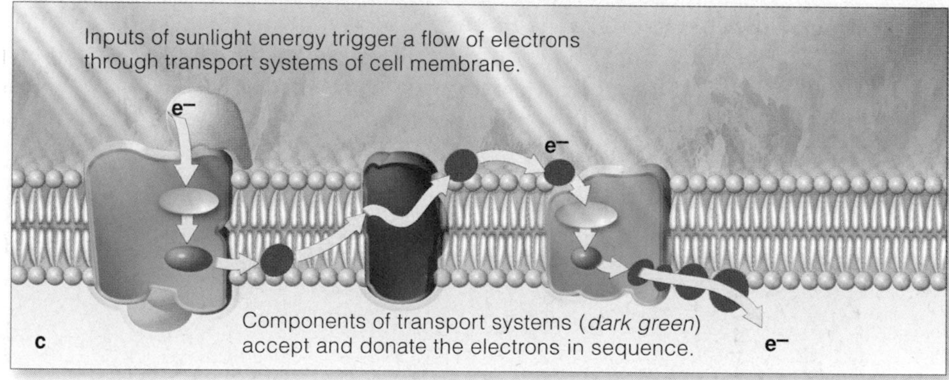

Figure 6.10 Examples of uncontrolled and controlled energy release. (**a**) Hydrogen and oxygen exposed to an electric spark react and release energy all at once. (**b**) Electron transport systems allow the same reaction to occur in small steps that let cells harness some of the released energy. (**c**) Electron-donating complexes and electron transport systems of a chloroplast's thylakoid membrane. The transport systems accept electrons that were excited to a higher energy level by sunlight.

Many electron transfers, or oxidation–reduction reactions, occur singly. Also, when cells degrade organic compounds as fuel, electron transfers proceed as a series of small steps to make more efficient use of the energy being released.

6.5 THE DIRECTIONAL NATURE OF METABOLISM

Later chapters will explain how energy inputs drive ATP formation, especially in photosynthesis and aerobic respiration. Photosynthesis is an example of **anabolism**, the metabolic reactions by which the cell synthesizes energy-rich organic compounds from smaller, energy-poor precursors. Aerobic respiration is an example of **catabolism**, the metabolic reactions by which the cell degrades energy-rich compounds into smaller, simpler molecules to release usable energy for cellular work.

Which Way Will a Reaction Run?

From the preceding chapter, you have an idea that cells control their internal concentrations of substances with respect to the surroundings, and that eukaryotic cells further control the concentrations of substances on both sides of organelle membranes. At any time, thousands of concentration gradients across cell membranes are helping to drive specific molecules and ions in specific directions. Even on their own, molecules or ions are in constant random motion, which puts them on collision courses. But the more concentrated they are, the more often they collide. Energy associated with the collisions might be enough to cause a **chemical reaction**—that is, to make a molecule combine with something else, split into smaller parts, or change its shape.

Nearly all chemical reactions in cells are reversible. In other words, they might start out in the "forward" direction, from starting substances to products. But they also can run in "reverse," with the products being converted back to starting substances. When you see a chemical equation with opposing arrows, this signifies the reaction is reversible. Each arrow means *yields*:

$$A + B \rightleftharpoons C$$

STARTING SUBSTANCES — PRODUCT

Which way such a reaction runs depends partly on the energy content of the participants. It also depends on the reactant-to-product ratio. When the energy level and concentration of reactant molecules is high, this is an energetically favorable state, and the reaction tends to proceed spontaneously and strongly in the forward direction. However, when the product concentration is high enough, more molecules or ions of product are available to revert spontaneously to reactants.

Any reversible reaction tends to run spontaneously toward **chemical equilibrium**, the time at which it will be running at about the same pace in both directions (Figure 6.11). Almost always, the *amounts* of the reactant and product molecules are not the same at that time. Picture a party with just as many people drifting in as drifting out of two rooms. The number in each room stays the same overall—say, thirty in one and ten in the other—even as the mix of people in each room changes.

Each reaction has a characteristic ratio of reactant to product molecules at equilibrium. For instance, glucose–1–phosphate can form from glucose–6–phosphate, which is nineteen times higher in energy. (As Figure 6.12 shows, both are "glucose," but their phosphate group is attached to a different carbon atom.) Depending on the reactant-to-product ratio at the outset, this reaction runs in the forward or the reverse direction. In time it will run both ways at the same rate, but it will do so only when there are nineteen glucose–6–phosphate molecules for each glucose–1–phosphate molecule. For this reaction, the equilibrium ratio is 19:1.

RELATIVE CONCENTRATION OF REACTANT

RELATIVE CONCENTRATION OF PRODUCT

HIGHLY SPONTANEOUS

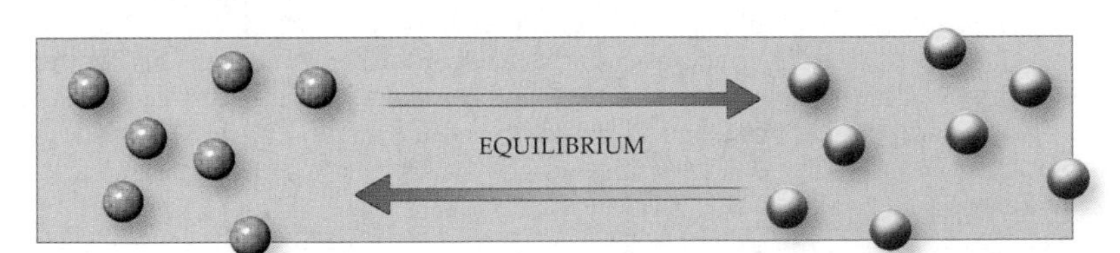

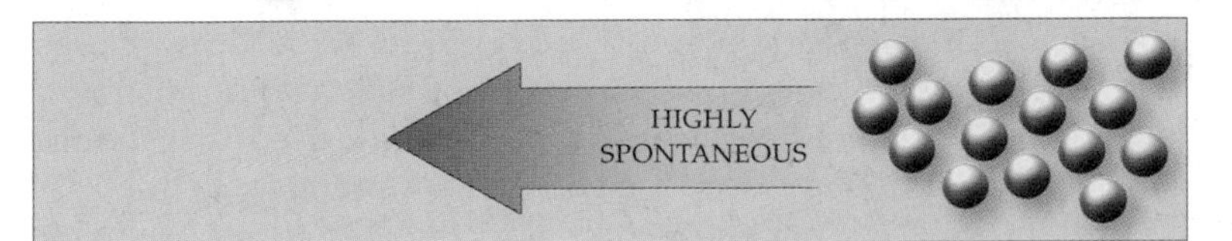

Figure 6.11 Chemical equilibrium. When the concentration of reactant molecules is high, a reaction runs most strongly in the forward direction (to products). When the concentration of product molecules is high, it runs most strongly in reverse. At equilibrium, the rates of the forward and reverse reactions are the same.

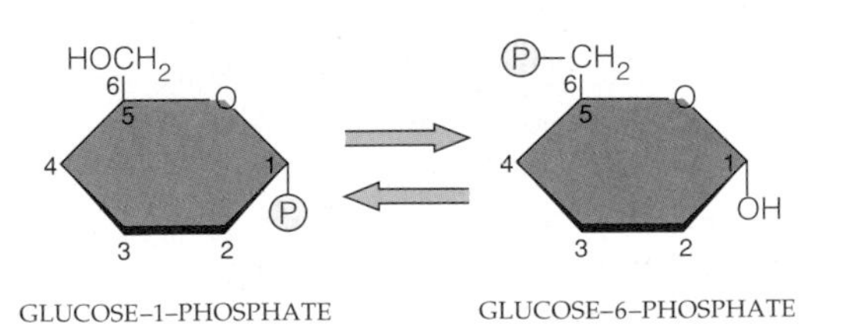

Figure 6.12 A reversible reaction. Glucose is primed to enter reactions when a phosphate group becomes attached to it. When there is a high concentration of glucose–1–phosphate, the reaction tends to run in the forward direction. With a high glucose–6–phosphate concentration, it runs in reverse. (The 1 and 6 of these names simply identify which particular carbon atom of the glucose ring has a phosphate group attached to it.)

Further reading: Student Guide to InfoTrac on web site →

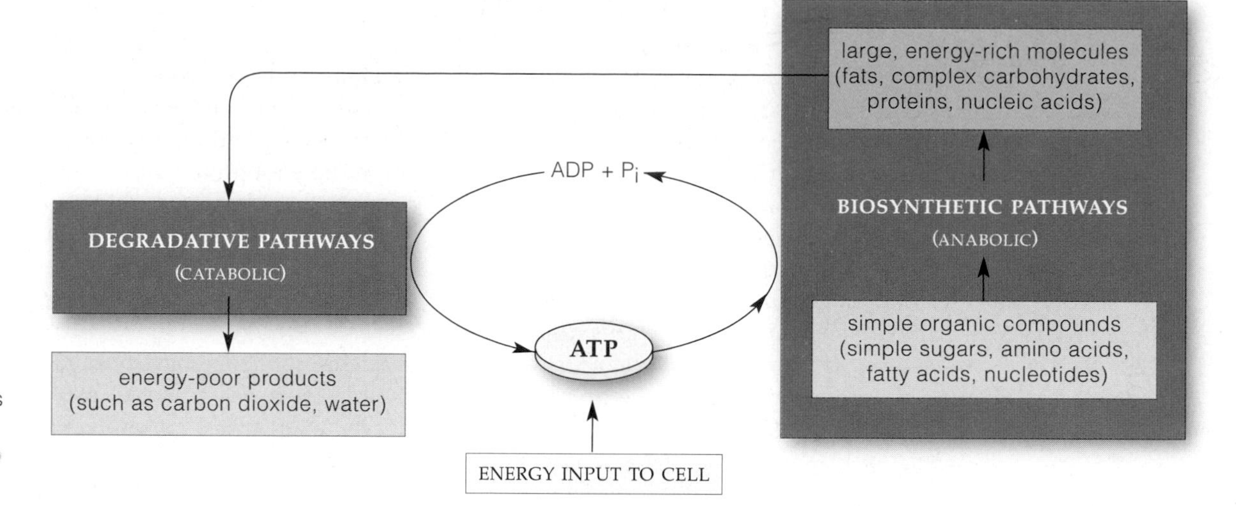

Figure 6.13 Energy relationship between the main biosynthetic and degradative pathways in cells. ATP forms by inputs into the cell and by catabolic pathways. ATP drives synthesis of large, energy-rich biological molecules from simple precursors by anabolic pathways.

No Vanishing Atoms at the End of the Run

When metabolic reactions run in either the forward or reverse direction, they rearrange atoms, but they never destroy them. By the **law of conservation of mass**, the total mass of all substances entering a reaction equals the total mass of all the products. When you study any chemical equation, count up the individual atoms of each reactant and product molecule. There should be as many atoms of each element to the right of the arrow as there are to the left, even though they are combined in different forms. When you write out equations for metabolic reactions, they must balance this way.

Metabolic Pathways

Energy inputs drive reactions that involve thousands of substances within the confines of a cell. Most of the reactions occur in orderly, enzyme-mediated sequences called **metabolic pathways**.

In the *biosynthetic* pathways, small molecules are assembled into molecules having higher bond energies, such as complex carbohydrates, lipids, and proteins. Such pathways require energy inputs (Figure 6.13). In *degradative* pathways, large molecules are broken down to products with lower bond energies. Such pathways yield energy. Photosynthesis is the main biosynthetic pathway in the world of life. Aerobic respiration is the main degradative pathway.

The participants of metabolic pathways go by these names: **Substrates** are substances that enter a reaction. They are also called reactants or precursors. Between the start and conclusion of a pathway, any substance that forms is an **intermediate**. Those remaining at the end of a reaction or a pathway are **end products**. ATP and a few other compounds are **energy carriers**; they activate enzymes and other molecules with phosphate-group transfers. ATP is the coupling agent between the main biosynthetic and degradative pathways. Most **enzymes** are proteins that speed specific reactions. (A few RNAs also display enzyme activity.) **Cofactors** are coenzymes (NAD^+ and some other organic compounds) and metal ions. They either assist enzymes or pick up electrons, atoms, or functional groups from one reaction site and taxi them to a different site. **Transport proteins**, recall, help substances cross membranes in controlled ways. Controls over these proteins help adjust concentrations on both sides of a cell membrane and thereby influence metabolic reactions.

Many pathways advance step by step in a straight line from substrates to end products. Many others are cyclic; the steps proceed in a circle, with end products becoming reactants (Figure 6.14). As you will see soon enough, intermediates or end products of one pathway also can enter different metabolic pathways.

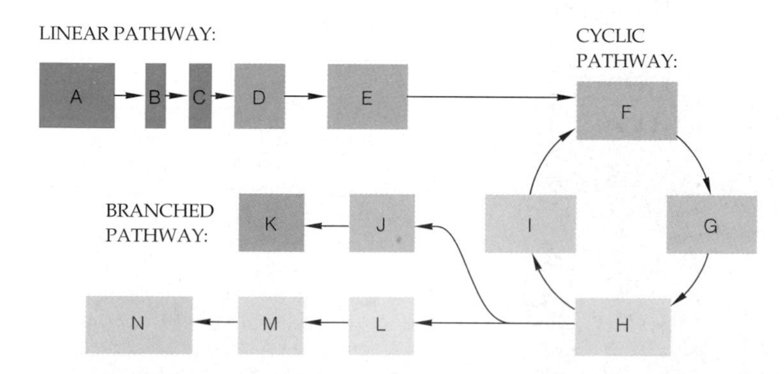

Figure 6.14 Types of reaction sequences of metabolic pathways.

A cell can simultaneously increase, decrease, and maintain the concentrations of thousands of different substances by coordinating thousands of metabolic reactions.

The coupling of energy-releasing reactions with energy-requiring reactions, as by ATP, is central to metabolism. Metabolic pathways are enzyme-mediated reaction sequences from substrates and intermediates to end products.

6.6 ENZYME STRUCTURE AND FUNCTION

Without enzymes, the dynamically steady state called "you" would quickly cease to exist. Reactions simply wouldn't proceed fast enough for your body to process food, build and tear down hemoglobin and other vital molecules, send signals to and from brain cells, make muscles contract, and do everything else to stay alive.

To see how enzymes work, start with this concept: *Cells control their internal concentrations of substances with respect to the surroundings, and eukaryotic cells also control the concentrations across membranes of organelles.* Remember, molecules or ions of any substance are in constant, random motion that puts them on collision courses. The more concentrated they are, the more often they collide. And energy associated with the collisions might be enough to cause a metabolic reaction—that is, to make a molecule combine with something else, split into smaller parts, or change its shape.

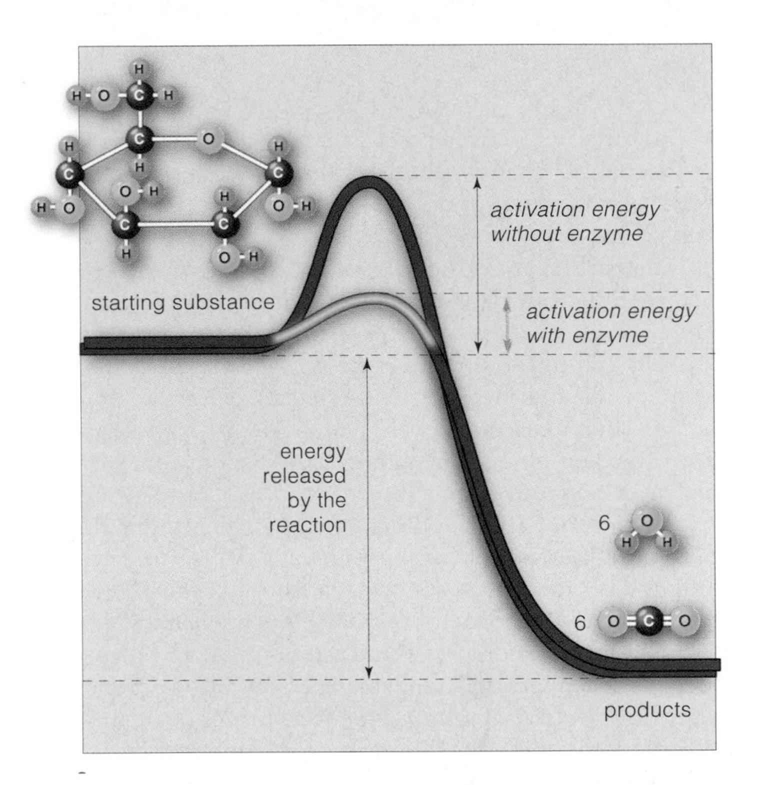

Four Features of Enzymes

By definition, enzymes are catalytic molecules; *they speed the rate at which reactions approach equilibrium.* Again, nearly all enzymes are proteins. All share four features. First, enzymes do not make anything happen that could not happen on its own, but they usually make it happen hundreds to millions of times faster. Second, reactions do not permanently alter or use up enzyme molecules; the same enzyme may act repeatedly. Third, the same type of enzyme usually works for the forward and the reverse directions of a reaction. Fourth, each type of enzyme is very picky about its substrates. Its substrates are specific substances that it can chemically recognize, bind, and modify in certain ways. For instance, thrombin is one of the enzymes necessary to clot blood. It only recognizes a side-by-side arrangement of arginine and glycine (two amino acids) in a protein molecule. When it does so, it cleaves the peptide bond between them.

Enzyme–Substrate Interactions

Take a look at Figure 6.15. A metabolic reaction occurs when participating molecules collide—provided they collide with some minimum amount of energy called the **activation energy**. Collision can be spontaneous or enzymes can promote it; this makes no difference. The activation energy is like a hill, an energy barrier, that must be surmounted *one way or another* before a reaction will proceed.

Enzymes make the energy barrier smaller, so to speak. How? Every enzyme has one

Figure 6.15 Activation energy. (**a**) Before reactants can enter a metabolic reaction, an energy input must activate them. Only then can they spontaneously proceed to end products. (**b**) An enzyme enhances the rate of a specific reaction by lowering the amount of activation energy required to boost reactants to the transition state. Going back to the analogy in Figure 6.5, they lower the hill (energy barrier).

Further reading: Student Guide to InfoTrac on web site →

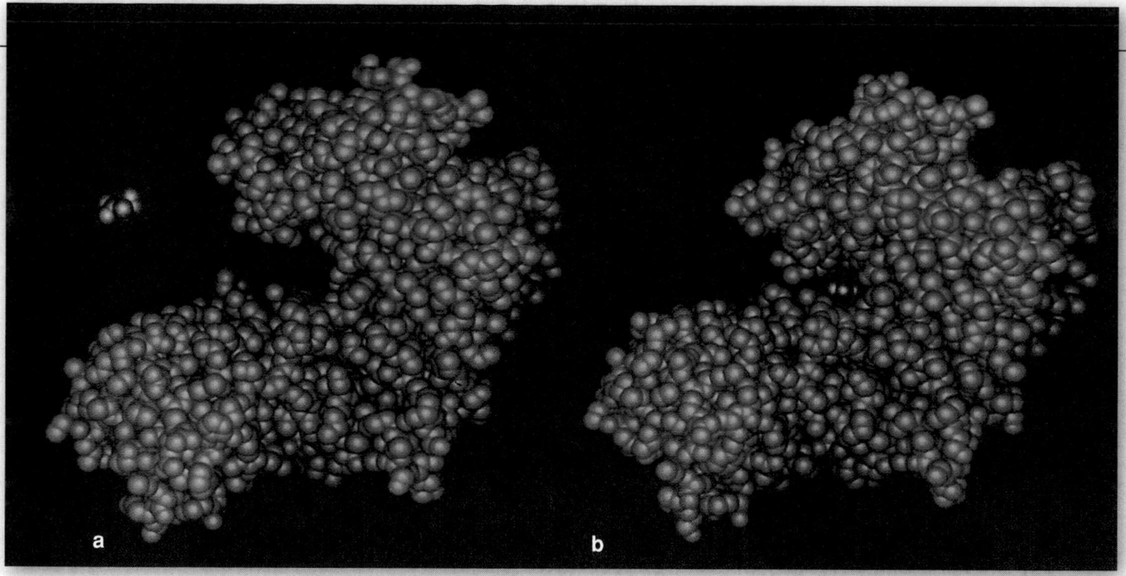

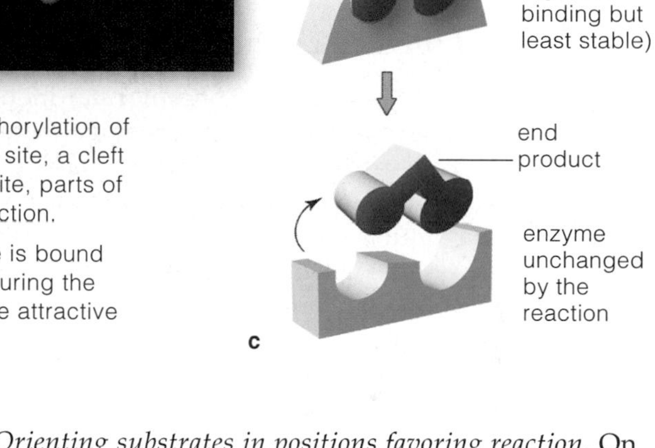

Figure 6.16 Model of the enzyme hexokinase. Hexokinase catalyzes the phosphorylation of glucose. (**a**) A glucose molecule (color-coded *red*) is heading toward the active site, a cleft in the enzyme (*green*). (**b**) When the glucose molecule makes contact with the site, parts of the enzyme briefly close in around it and prod the molecule to enter into the reaction.

(**c**) Induced-fit model of enzyme–substrate interactions. Only when the substrate is bound in place is an enzyme's active site complementary to it. The fit is most precise during the transition state of a reaction. An enzyme–substrate complex is short-lived, for the attractive forces holding it together are usually weak.

or more **active sites**. At these crevices in the molecule's surface, the enzyme interacts with its substrates and catalyzes a reaction. Figure 6.16 shows the active site of one enzyme, which activates glucose by catalyzing the attachment of a phosphate group to it.

According to Daniel Koshland's **induced-fit model**, a surface region of each substrate has chemical groups that are almost but not quite complementary to chemical groups in an active site. When substrates first settle into the site, the contact strains some of their bonds. Strained bonds are easier to break, so they promote the formation of new bonds (in the products). Also in the active site, interactions among charged or polar groups often shift the electric charge in substrates. And that redistribution primes substrates for conversion to an activated state.

When substrates fit most precisely in the active site of an enzyme, they are in an activated, *transition* state and will now react (Figure 6.16*c*). And so the reaction must proceed, just as NASA's Rover must roll down the Martian hill if something pushes it over the crest (Figure 6.5).

What induces the transition state or gets substrates over the energy barrier once that state is reached? The following are among the mechanisms involved:

1. *Helping substrates get together*. Substrate molecules rarely collide if their concentrations are low. Binding at an active site is like a localized boost in concentration. The boost increases the rate by 10,000 to 10,000,000 times, depending on the particular reaction.

2. *Orienting substrates in positions favoring reaction*. On their own, substrates collide from random directions. By contrast, weak but extensive bonding at an active site puts reactive chemical groups on precise collision courses much more frequently.

3. *Promoting acid–base reactions.* In many active sites, acidic or basic side groups of amino acids are poised to donate or accept hydrogen atoms from substrates. The loss or addition destabilizes covalent bonds in a substrate and makes them easier to break. Hydrolysis works this way.

4. *Shutting out water.* Some active sites bind substrates so tightly that some or all of the water molecules that bathed the site are shut out. A nonpolar environment lowers the activation energy for certain reactions, such as the attachment of a carboxyl group (—COO^-) to a molecule, by as much as 500,000 times.

Depending on the enzyme, such mechanisms work alone or in combination with one another to bring about the transition state.

Enzymes catalyze (speed) the rate at which specific reactions reach equilibrium. They do so by lowering the amount of activation energy necessary to make substrates react.

Enzymes change the rate, not the outcome, of a reaction. They only act on specific substrates. And they may catalyze the same reaction repeatedly, as long as substrates are available.

6.7 FACTORS INFLUENCING ENZYME ACTIVITY

You probably don't get much done when you feel too hot or cold or out of sorts because you ate too many sour plums or salty potato chips. When the cupboard is bare, you focus on food. Maybe you call a friend to go shopping with you, and if you drive too fast to the grocery store, police tend to slow you down. In such respects, you have a lot in common with the enzymes. They, too, respond to shifts in temperature, pH, and salinity, and to the relative abundances of particular substances. Many even engage helpers for specific tasks. And all normal enzymes respond to metabolic police.

Enzymes and the Environment

Temperature, recall, is a measure of molecular motion. You may think increases in temperature must increase the rate of enzyme-mediated reactions by making the substrates collide more frequently with active sites. That is so, but only until some point on the temperature scale. Past that point—which differs among enzymes—the increased molecular motion disrupts weak bonds that hold the enzyme in its three-dimensional shape. Substrates can no longer bind to the active site, so the reaction rate declines sharply, as in Figure 6.17*a*.

Expose an organism to temperatures that are far higher than it normally encounters. Its enzymes will change in shape and thereby throw metabolic activities into turmoil. This is what happens when sick people develop dangerously high fevers. They usually die when their internal temperature reaches 44°C (112°F).

Similarly, pH values that rise or sink beyond each enzyme's range of tolerance disrupt enzyme structure and function (Figure 6.18). Most enzymes work best when the pH is between 6 and 8. For example, trypsin is active in a mammal's small intestine, where the pH is 8 or so. Pepsin, a protein-digesting enzyme, is one of the exceptions. It functions in gastric fluid, a highly acidic fluid (pH of about 1–2) that denatures most enzymes.

Enzyme activity also will suffer if the environment gets far saltier than is normally encountered. Extremely high ion concentrations disrupt interactions that help hold most enzymes in their three-dimensional shapes.

How Is Enzyme Action Controlled?

Each cell controls its enzyme activity. By coordinating control mechanisms, it maintains, lowers, or raises the concentrations of substances. Controls that adjust how fast enzyme molecules are synthesized affect how many are available for a metabolic pathway. Other controls boost or slow the action of enzyme molecules that were synthesized earlier. For example, enzymes are activated or inhibited through **allosteric control** when a specific substance combines with them at a binding site *other than* the active site. (*Allo-* means different, *steric* means structure, or state.) Figure 6.19 shows two models of the binding, which is reversible.

Picture a bacterium, busily synthesizing tryptophan (and other amino acids) used to construct its proteins. After a bit, protein synthesis slows, so tryptophan is no longer required. But the tryptophan pathway is in full swing. The concentration of its end product (tryptophan) continues to rise. Now **feedback inhibition** kicks in: a cellular change, caused by a specific activity, *shuts down*

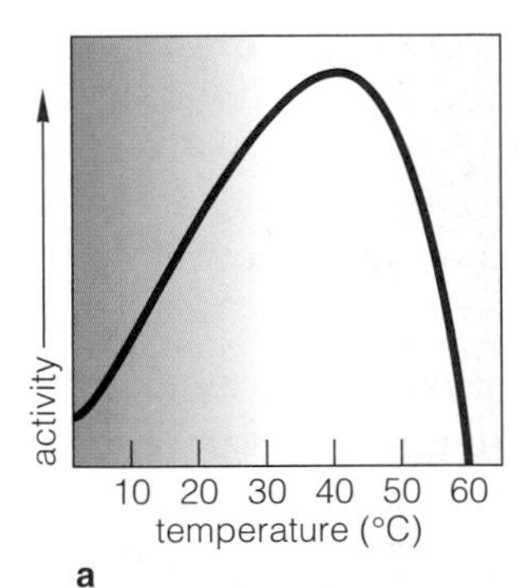

Figure 6.17 (**a**) Example of how temperatures outside the range of tolerance for one enzyme influence its activity. (**b**) Siamese cats show observable effects of such changes. Fur on the ears and paws has more of a dark brown pigment, melanin, than the rest of the body does. A heat-sensitive enzyme controlling melanin production is less active in warmer parts of the body, which end up with lighter fur.

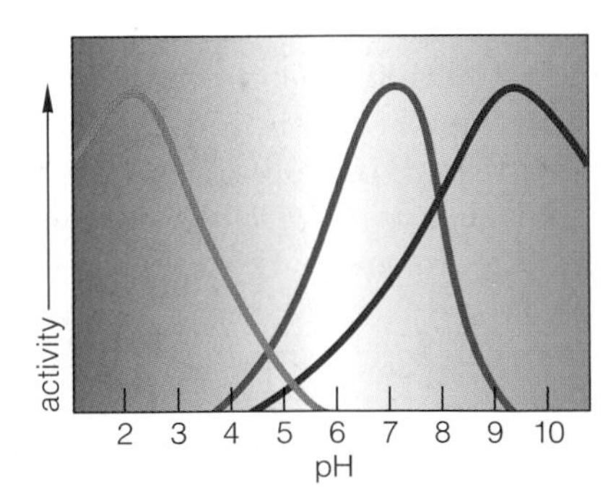

Figure 6.18 Diagram comparing changes in activity of three kinds of enzymes at different pH values. One of the enzymes functions best in neutral solutions, as indicated by the *brown* line. (Remember, 7 is neutrality on the pH scale.) Another enzyme (*red* line) functions best in basic solutions, and another (*purple* line) in acidic solutions.

Further reading: Student Guide to InfoTrac on web site

the activity that brought it about. In this case, a feedback loop starts and ends with a specific allosteric enzyme in the pathway. When tryptophan molecules accumulate, the unused ones bind with the allosteric site. This shuts down the enzyme and blocks the pathway.

By contrast, if tryptophan molecules are scarce when the demand for them increases, then the enzyme will remain free of inhibition, so the production of tryptophan will increase. Such feedback loops quickly adjust the concentrations of many substances (Figure 6.20).

In humans and other multicelled organisms, control of enzyme activity is just amazing. Cells not only work to keep themselves alive, they work with other cells in ways that can benefit the whole body! As an example, this vast enterprise relies on hormones, a type of signaling molecule. Specialized cells release hormones into the surrounding tissue. Any cell having receptors for a given hormone responds to it, then its program for building a protein or some other activity changes. The hormone trips cell controls into action—and the activities of specific enzymes change.

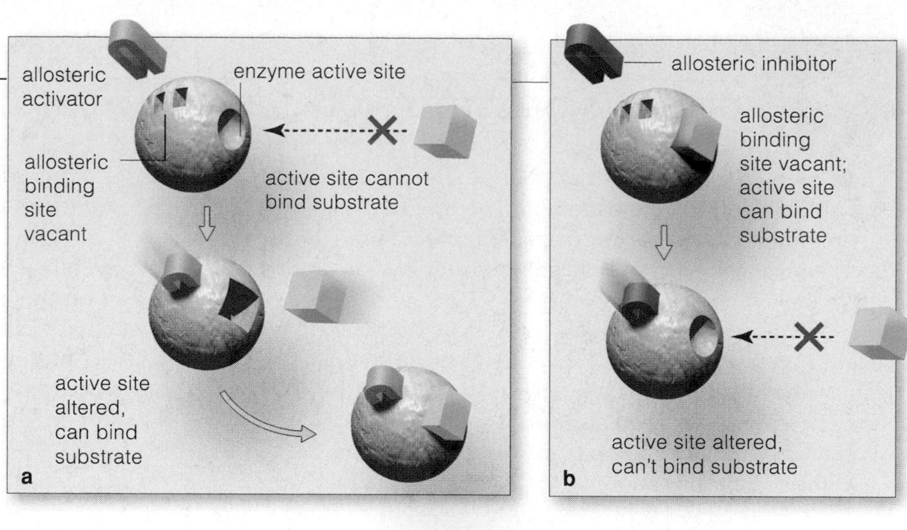

Figure 6.19 Example of allosteric control of enzymes. (**a**) Binding of an activator protein to a vacant allosteric site causes a change in the active site's shape that allows it to bind substrate. (**b**) Binding of an inhibitor protein to a vacant allosteric site causes a change in the active site's shape that shuts out substrate.

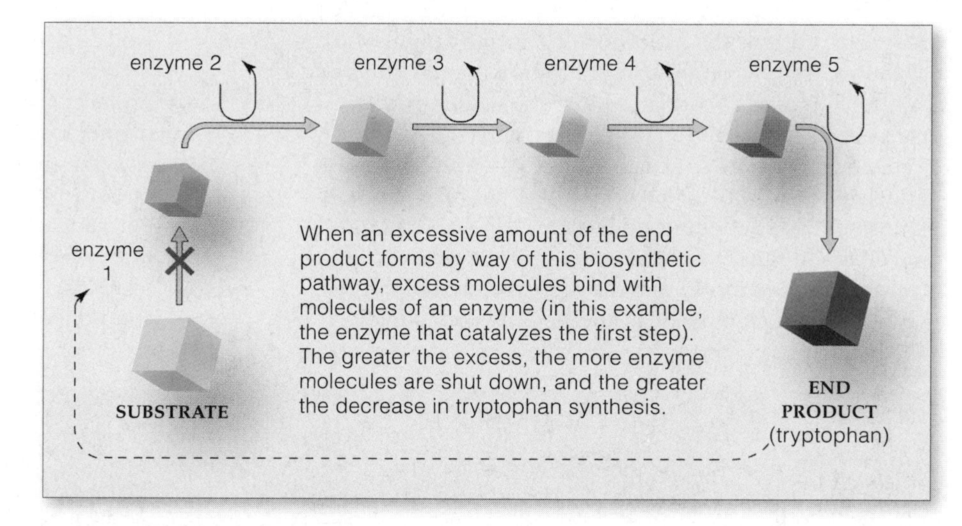

Figure 6.20 Example of feedback inhibition of a metabolic pathway. Five kinds of enzymes act in sequence to convert a substrate to the end product, tryptophan. When end product accumulates, some of the excess molecules bind to molecules of the first enzyme and block the entire pathway.

Enzyme Helpers

In many metabolic reactions, enzymes enormously hasten the transfer of one or more electrons, atoms, or functional groups from one substrate to another. Cofactors either assist the reactions or briefly act as the transferring agents. Cofactors include complex organic compounds called **coenzymes** as well as metal ions that associate with the enzyme molecule.

NAD^+ (for nicotinamide adenine dinucleotide) and FAD (for flavin adenine dinucleotide) are examples of coenzymes that are derived from vitamins. Both accept electrons and hydrogen atoms released during many degradative reactions, such as glucose breakdown, and then transfer the electrons to other reaction sites. Being attracted to the electrons, the unbound protons (H^+) go along for the ride. When reduced—that is, after they've accepted electrons (and hydrogen)—NAD^+ and FAD are abbreviated NADH and $FADH_2$, respectively. Another coenzyme, $NADP^+$ (nicotinamide adenine dinucleotide phosphate), is a key player in photosynthesis and other synthesis reactions. Like NAD^+, it is derived from the vitamin niacin. When reduced, it is abbreviated NADPH.

Ferrous iron (Fe^{++}), one of the metal ions that serve as cofactors, is a component of cytochrome molecules. Cytochromes are proteins that show enzyme activity. They are embedded in cell membranes, including the membranes of chloroplasts and mitochondria.

Enzymes function best when the cellular environment stays within limited ranges of temperature, pH, and salinity. The actual ranges depend on the type of enzyme.

Control mechanisms govern the synthesis of new enzymes and stimulate or inhibit the activity of existing enzymes. By controlling enzymes, cells control the concentrations and kinds of substances available to them.

Some enzymes require cofactors (coenzymes and metal ions), which help catalyze reactions or transfer electrons, atoms, and functional groups from one substrate to another.

GROWING OLD WITH MOLECULAR MAYHEM

Somewhere in those slender strands of DNA in your cells are snippets of instructions for constructing two enzymes: superoxide dismutase and catalase (Figure 6.21*a,b*). Both kinds of enzymes may help keep you from growing old before your time.

The two enzymes help your body clean house, so to speak. Together, they use oxygen (O_2) to strip hydrogen from substrate molecules, thereby helping to neutralize many potentially toxic wastes. The same reactions occur in other organisms, ranging from bacteria to plants.

At the end of the reactions, O_2 is supposed to pick up electrons. Sometimes it picks up only one—which is not enough to complete the reaction but is enough to give the oxygen a negative charge (O_2^-).

Like other unbound molecular fragments that have the wrong number of electrons, O_2^- is a **free radical**. Free radicals can slip away from a variety of enzyme-catalyzed reactions, such as the oxidation of fats and amino acids. They can form when ionizing radiation (gamma rays and x-rays) bombards water and other molecules. And they can escape from electron transport chains.

Free radicals are so reactive, they can even become attached to molecules that usually do not take part in just any reaction. These molecules include DNA, the lipids of cell membranes, and membrane receptors that trigger cell suicide (Section 15.6). Free radicals can disrupt the structure of such molecules and destroy their function.

Enter superoxide dismutase. Under its biochemical prodding, two rogue oxygen molecules will combine with hydrogen ions. Hydrogen peroxide (H_2O_2) and O_2 are the outcome. Hydrogen peroxide is a normal by-product of certain aerobic reactions, and its accumulation can be lethal to cells.

Enter catalase. Under *its* prodding, two molecules of hydrogen peroxide react and split into ordinary water and ordinary oxygen:

$$2H_2O_2 \longrightarrow 2H_2O + O_2$$

This crucial reaction is supposed to occur before H_2O_2 can do major damage.

Thus both superoxide dismutase and catalase belong to a class of molecules called **antioxidants**, which protect the body by scavenging for free radicals. Vitamins E and C, as well as beta-carotene, serve similar protective functions.

When people age, their capacity to make functional proteins starts to falter. Among the proteins are superoxide dismutase and catalase. Cells synthesize copies of both enzymes in ever diminishing numbers, in crippled form, or both. When that happens, free radicals and hydrogen peroxide can accumulate. Like loose cannons, they careen through cells as tiny blasts at the structural integrity of proteins, DNA, membrane lipids, and other essential components.

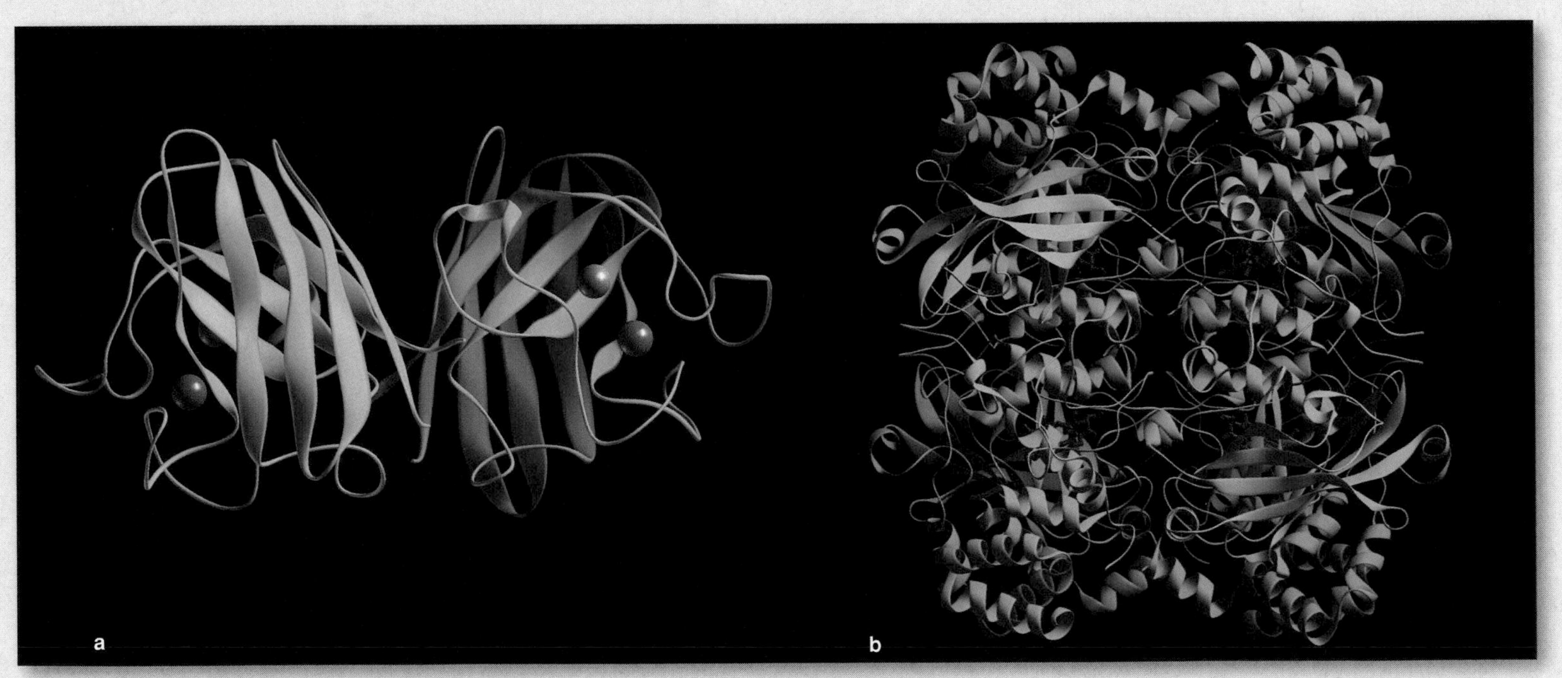

Figure 6.21 Computer models of the molecular structure of (**a**) superoxide dismutase and (**b**) catalase. (**c**) Adult owner of skin with a spattering of age spots, visible evidence of free radicals on the loose. At one time he, like the boy shown in (**d**), had of a good supply of smoothly functioning molecules of superoxide dismutase and catalase. Both enzymes help keep free radicals inside the body in check.

Further reading: Student Guide to InfoTrac on web site →

Cells under attack suffer or die outright. Those brown "age spots" you may have noticed on an older person's skin are evidence of assaults by free radicals. The irregular dark spots on the skin of the adult male in Figure 6.21*c* are examples. Each age spot is a mass of brownish-black pigment molecules that build up in cells whenever free radicals take over—all for the want of specific enzymes.

And so with this chapter, we have considered the kinds of activities that help keep all cells alive and functioning smoothly. At times they might have seemed remote from your interests. But these activities help define who *you* are and who you will become, age spots and all.

6.9 SUMMARY

1. Cells store, break down, and dispose of substances by acquiring and using energy and raw materials from outside sources. Metabolism, the sum of these energy-driven activities, underlies the survival of organisms.

2. Two laws of thermodynamics affect life. *First*, energy undergoes conversion from one form to another, but its total amount never increases or decreases as a result of conversions. Thus, the total amount of energy in the universe holds constant. *Second*, energy spontaneously flows in one direction, from usable forms to forms that are less and less usable.

a. All matter has some amount of potential energy (as measured by the capacity to do work) by virtue of its position in space and the arrangement of its parts.

b. Potential energy may be transformed into kinetic energy, the energy of motion. Mechanical movements and heat, which correspond to the degree of molecular motion, are two common forms of kinetic energy.

c. Chemical energy (the potential energy inherent in molecular bonds) is often measured in kilocalories.

3. Like all organized systems, the cell tends to become disorganized without energy. It inevitably loses some of its chemical potential energy during every metabolic reaction, mainly in the form of heat. It stays organized and alive as long as it counters its energy expenditures (outputs) with energy replacements (inputs).

4. The sun is life's primary energy source. In plants and other photosynthetic organisms, cells trap energy from the sun and convert it to chemical bond energy of organic compounds. Plants, then organisms that feed on plants and one another, use energy that was stored in these organic compounds to do cellular work.

5. Exergonic (energy out) reactions end with a net loss in energy. Endergonic (energy in) reactions end with a net gain in energy. Cells conserve energy by coupling energy-releasing reactions with energy-requiring ones.

6. ATP is the main energy carrier in every cell. It forms when a phosphate group or inorganic phosphate is attached to ADP. ATP gives up energy at many reaction sites when it phosphorylates reactants or intermediates, which often become primed to enter specific reactions.

7. ATP couples energy-releasing and energy-requiring reactions in cells.

8. Much of the cellular work leading to ATP formation involves electron transfers between substances. We call such transfers oxidation–reduction reactions.

a. Molecules that lose electrons are oxidized; those that gain electrons are reduced.

b. Hydrogen ions often are attracted to free electrons and are simultaneously transferred to molecules along with them. These are called hydrogenation reactions.

c. Electron transfers may occur singly or in a series of small steps through electron transport systems.

9. Metabolic reactions may release energy when they run toward equilibrium, or they may require energy inputs to drive them away from equilibrium.

10. Most metabolic reactions proceed strongly in the forward direction (to products) when concentrations of starting substances are high. They also run strongly in reverse (to starting substances) when concentrations of products are high.

a. When left to themselves, reversible reactions run toward chemical equilibrium, at which time they are proceeding at about the same rate in both directions.

b. Cellular mechanisms work with and against this tendency. During a given interval, reversible reactions help increase, decrease, and maintain concentrations of the thousands of different substances that are required for specific reactions.

c. Regardless of the direction of a metabolic reaction or the extent of the molecular cleavages, combinations, and rearrangements, the total number of atoms that end up in all the products will equal the total number of atoms that were present in the starting substances.

11. Metabolic pathways are orderly, stepwise sequences of enzyme-mediated reactions. Table 6.1 summarizes the main participants.

a. A substrate (or reactant) is a substance that enters a reaction or pathway. Intermediates are the substances that form between the reactants and end products.

b. By the end of a biosynthetic pathway, energy-rich organic compounds have been assembled from smaller molecules of lower energy content.

c. By the end of a degradative pathway, energy-rich molecules have been broken down to smaller ones of lower energy content.

12. Overall, the main metabolic pathways are anabolic (biosynthetic) and catabolic (degradative).

a. Cells assemble energy-rich organic compounds from smaller molecules of lower energy content in the biosynthetic pathways, such as photosynthesis.

b. Cells break down energy-rich molecules to small ones of lower energy content in degradative pathways, such as aerobic respiration.

c. ATP is the main coupling agent between energy-releasing and energy-requiring pathways.

13. Enzymes are catalysts; they greatly enhance the rate of a reaction that involves specific substrates but do not change the outcome. Nearly all enzymes are proteins, although some RNAs also show catalytic activity.

a. Enzymes lower the activation energy required to start a reaction. They bind substrates at an active site, strain its bonds, and make the bonds easier to break.

b. Each kind of enzyme functions best within limited ranges of temperature, pH, and salinity.

c. Cofactors assist enzymes in speeding a reaction, or they carry electrons, hydrogen, or functional groups stripped from substrates to other reaction sites. They include coenzymes (such as NAD^+) and metal ions.

d. Controls stimulate or inhibit enzyme activity at key steps in metabolic pathways. They help coordinate the kinds and amounts of substances available at any given time in the cell.

Table 6.1 Main Participants in Metabolic Pathways

SUBSTRATE	Substance that enters a metabolic reaction or pathway; also called a reactant
INTERMEDIATE	Substance formed between reactants and end products of a pathway
END PRODUCT	Substance remaining at end of reaction or pathway
ENZYME	Usually a protein that enhances reaction rates
COFACTOR	Coenzyme (such as NAD^+) or metal ion; assists enzymes or taxis electrons, hydrogen, or functional groups between reaction sites
ATP	Main energy carrier in cells; couples energy-releasing reactions with energy-requiring ones
TRANSPORT PROTEIN	Protein that passively assists substances across a cell membrane or actively pumps them across

Review Questions

1. State the first and second laws of thermodynamics. Does life violate the second law? *6.1*
2. Define and give a few examples of potential energy and kinetic energy. What is the name for the potential energy of molecules? *6.1*
3. Define mechanical work, chemical work, and electrical work as accomplished by cells. *6.1*
4. Give examples of a change in potential energy involved in (a) mechanical work and (b) chemical work. *6.2*
5. Make a simple diagram of the ATP molecule. Highlight which parts of an ATP molecule can be transferred to another molecule and then later replaced. *6.3*
6. What is an oxidation–reduction reaction? *6.4*
7. Define and describe four key features of enzymes. *6.6*
8. Define activation energy, then state four ways in which enzymes may lower it. *6.6*
9. Briefly describe the induced-fit model of enzyme–substrate interactions. *6.6*
10. Define feedback inhibition as it relates to the activity of an allosteric enzyme. *6.7*
11. Define free radical. How does it relate to aging? *6.8*
12. Fill in the blanks in the sketch on the facing page. *6.5*

Self-Quiz (*Answers in Appendix III*)

1. ______ is life's primary source of energy.
   a. Food
   b. Water
   c. Sunlight
   d. ATP

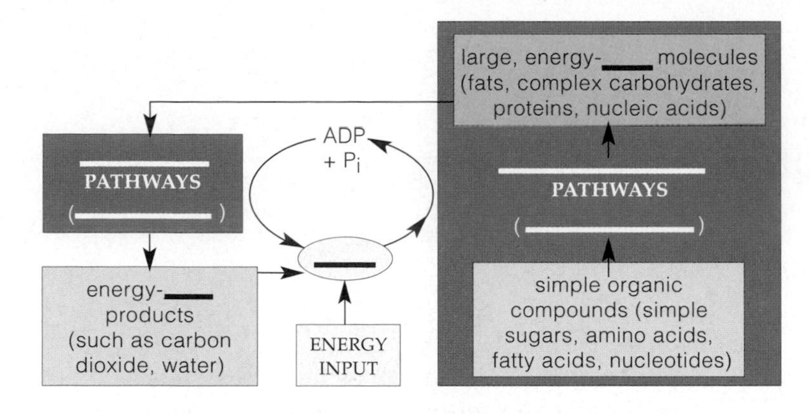

2. If we liken chemical equilibrium to the bottom of an energy hill, then an ______ reaction is an uphill run.
   a. endergonic   c. ATP-assisted
   b. exergonic   d. both a and c

3. Phosphate-group transfers from ATP to another molecule are a ______ mechanism for delivering energy.
   a. rapid   c. near-universal
   b. renewable   d. all of the above

4. Electron transport systems involve ______ .
   a. enzymes and cofactors   c. cell membranes
   b. electron transfers   d. all of the above

5. Which statement is *not* correct? A metabolic pathway ______ .
   a. has an orderly sequence of reaction steps
   b. is mediated by one enzyme, which initiates the reactions
   c. may be biosynthetic or degradative, overall
   d. all of the above

6. Which is *not* true of chemical equilibrium?
   a. Product and reactant concentrations are always equal.
   b. The rates of the forward and reverse reactions are the same.
   c. There is no further net change in product and reactant concentrations.

7. Enzymes are ______ .
   a. enhancers of reaction rates   d. not influenced by salinity
   b. influenced by temperature   e. a through c
   c. influenced by pH   f. all of the above

8. All enzymes incorporate a(n) ______ .
   a. active site   c. metal ion
   b. coenzyme   d. all of the above

9. The coenzymes NAD^+, FAD, and $NADP^+$ are ______ .
   a. cofactors   c. metal ions
   b. allosteric enzymes   d. both a and b

10. Match each substance with the most suitable description.

___ coenzyme or metal ion	a. reactant or substrate
___ adjusts gradients at membrane	b. enzyme
___ substance entering a reaction	c. cofactor
___ substance formed while a reaction is proceeding	d. intermediate
___ substance at end of reaction	e. end product
___ enhances reaction rate	f. energy carrier
___ mainly ATP	g. transport protein

Critical Thinking

1. When cyanide, a toxic organic compound, binds to an enzyme that is a component of electron transport systems, the outcome is *cyanide poisoning*. Binding prevents the enzyme from donating electrons to a nearby acceptor molecule in the system. What effect will this have on ATP production? From what you know of ATP function, what effect will this have on a person's health?

2. AZT, or azidothymidine, is a drug used to alleviate symptoms of *AIDS* (acquired immunodeficiency syndrome). AZT is very similar in molecular structure to thymidine, one of the DNA nucleotides. AZT also can fit into the active site of an enzyme produced by the virus that causes AIDS. Infection puts the viral enzyme and viral genetic material (RNA) into a cell. There, the RNA is used as a template (structural pattern) for joining nucleotides to form a strand of DNA. The viral enzyme takes part in the assembly reactions. Propose a model to explain how AZT might inhibit replication of the virus inside cells.

3. The bacterium *Clostridium botulinum* is an obligate anaerobe, meaning it dies quickly in the presence of oxygen. It lives in oxygen-free pockets in soil, and it can enter a metabolically inactive, resting state by forming spores. The spores may end up on the surfaces of garden vegetables. When the picked vegetables are being canned, the spores must be destroyed. If they are not destroyed, *C. botulinum* may grow and produce botulinum, a toxin. When tainted canned foods are not cooked properly, the toxin remains active and causes a type of food poisoning called *botulism*. This bacterium is not able to produce either superoxide dismutase or catalase. Develop a hypothesis to explain how the absence of these enzymes is related to its anaerobic life-style.

4. *Pyrococcus furiosus* thrives at 100°C, the boiling point of water. This species of bacterium was discovered growing in a volcanic vent in Italy. Enzymes isolated from *P. furiosus* cells do not function well below 100°C. What is it about the structure of these enzymes that allows them to remain stable and active at such high temperatures? (Hint: Review Section 3.5, which summarizes the interactions that maintain protein structure.)

5. Hydrogen peroxide is a free radical that attacks cell structure and function by reacting with and disrupting nearly all molecules it contacts. Some research indicates its cumulative effects over time may contribute to the gradual deterioration associated with aging. For example, decreases in fruit fly longevity are directly proportional to a decrease in a specific type of enzyme in a specific organelle. Reflect on the organelles introduced earlier, in Chapter 4, then identify the enzyme and the organelle.

Selected Key Terms

activation energy *6.6*
active site *6.6*
ADP *6.3*
allosteric control *6.5*
anabolism *6.5*
antioxidant *6.3*
ATP *6.8*
ATP/ADP cycle *6.3*
bioluminescence *CI*
catabolism *6.5*
chemical energy *6.1*
chemical equilibrium *6.5*
chemical reaction *6.5*
coenzyme *6.7*
cofactor *6.5*
electron transport system *6.4*
end product *6.5*
endergonic reaction *6.2*
energy carrier *6.5*
entropy *6.1*
enzyme *6.5*
exergonic reaction *6.2*
feedback inhibition *6.7*
first law of thermodynamics *6.1*
free radical *6.8*
heat (thermal energy) *6.1*
induced-fit model *6.6*
intermediate *6.5*
kinetic energy *6.1*
law of conservation of mass *6.5*
metabolic pathway *6.5*
metabolism *CI*
oxidation–reduction reaction *6.4*
phosphorylation *6.3*
P_i *6.3*
potential energy *6.1*
second law of thermodynamics *6.1*
substrate (reactant) *6.5*
transport protein *6.5*

Readings *See also www.infotrac-college.com*

Fenn, J. *Engines, Energy, and Entropy*. 1982. New York: Freeman. Deceptively simple paperback.

7 HOW CELLS ACQUIRE ENERGY

Sunlight and Survival

Think about the last time you were hungry and craved a bit of apple, maybe, or lettuce, chicken, pizza, bread, or any other kind of food. Where did it come from? For the answer, look past the refrigerator, the market or restaurant, or even the farm. Look to individual plants—the starting point for nearly all of the food you put into your mouth. Plants use environmental sources of energy and raw materials to build glucose and all other organic compounds necessary for survival. Organic compounds, recall, are built on a framework of carbon atoms. So the questions become these:

1. *Where does the carbon come from in the first place?*
2. *Where does the energy come from to drive the synthesis of carbon-based compounds?*

The answers vary according to an organism's mode of nutrition.

Plants generally are "self-nourishing" organisms, or **autotrophs**. As their carbon source, they use carbon dioxide (CO_2), a gaseous compound present in the air and dissolved in aquatic habitats. Plants, some bacteria, and many protistans are *photo*autotrophs, which means they capture sunlight energy to drive a metabolic process called **photosynthesis**. By this process, energy from the sun is converted to chemical bond energy in ATP, then ATP gives up energy at sites where glucose and other organic compounds are synthesized. An enzyme helper, the coenzyme $NADP^+$, typically picks up electrons and hydrogen, then delivers them to those same sites.

Many other organisms are **heterotrophs**, meaning they must feed on autotrophs, one another, and organic wastes. (*Hetero-* means other, as in "being nourished by other organisms.") That is how most bacteria, many protistans, and all fungi and animals stay alive. Unlike plants, heterotrophs cannot nourish themselves with sunlight and raw materials from their environment.

How do we know such things? We didn't have a clue until observational and experimental tests began in the mid-seventeenth century. Before then, most people assumed that plants got the raw materials they needed to make food from the soil they grew in. By 1882 a few chemists had an inkling that plants use sunlight, water, and something in the air to make food. T. Englemann, a botanist, was curious: What parts of sunlight do plants favor? As he already knew, when plants and algae are photosynthesizing, they release oxygen. He also knew that, like many other organisms, certain free-living bacterial cells require oxygen for aerobic respiration. Those cells move toward places where conditions favor their activities and away from unfavorable conditions.

Englemann hypothesized: If bacterial cells require oxygen, then they will move toward places where photosynthesis is most effective. He put a strand of a green alga, *Spirogyra,* in a water droplet that contained such bacteria. He mounted the strand on a microscope slide, then used a crystal prism to break up a beam of sunlight and to cast a spectrum of colors across it. As Figure 7.1 shows, bacteria congregated mainly where violet and red light fell on the algal strand. Englemann concluded that algal cells were releasing more oxygen in the area illuminated by light of those colors—which is the most effective light for photosynthesis.

Such observations yielded insight into the process of photosynthesis. Ultimately, they also helped reveal a great pattern in nature, because nearly all organisms depend on photosynthesis—*the main pathway by which carbon and energy enter the web of life.* Once a cell makes or takes up organic compounds, it uses or stores them. *All* autotrophs and heterotrophs store energy in organic compounds, and that energy can be released by other processes. Aerobic respiration, the most common energy-releasing process, requires oxygen to run to completion. Use Figure 7.2 as your preview of the chemical links between photosynthesis and aerobic respiration—the focus of this chapter and the next.

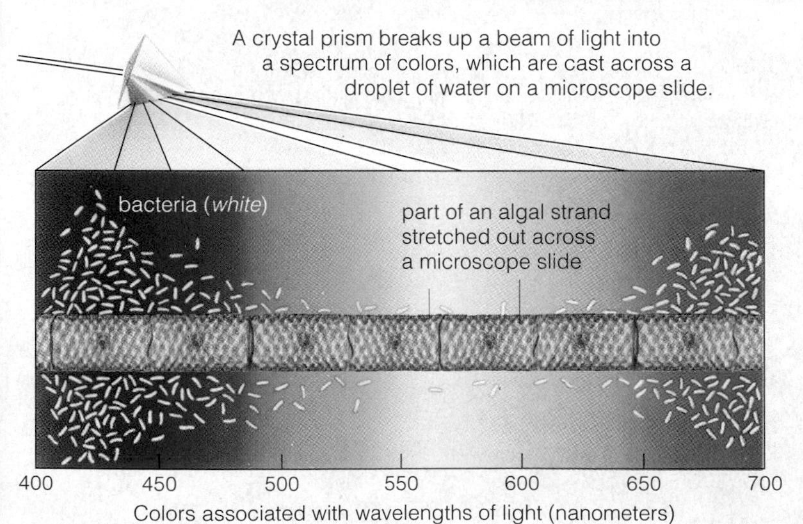

Figure 7.1 Results from T. Englemann's observational test that correlated portions of visible light with photosynthesis in *Spirogyra*, a strandlike green alga. Many oxygen-requiring bacterial cells moved to the colors where algal cells released the most oxygen, which is a by-product of their photosynthetic activity.

Figure 7.2 Links between photosynthesis—the main energy-requiring process in the world of life—and aerobic respiration, the main energy-releasing process.

At times, such processes might seem far removed from your daily interests. But remember this: The food that nourishes you and most other organisms cannot be produced or used without them. You will be returning to this point in later chapters. It provides perspective on many important issues, including nutrition and dieting, agriculture and human population growth, genetic engineering, as well as the impact of pollution on our sources of food—hence on our survival.

KEY CONCEPTS

1. Plants, some bacteria, and many protistans use sunlight energy, carbon dioxide, and water to produce glucose and other organic compounds, which have a backbone of carbon atoms. The metabolic process by which they accomplish this is called photosynthesis.

2. Photosynthesis is the main route by which carbon and energy enter the web of life.

3. In the first stage of photosynthesis, sunlight energy is trapped and converted to chemical bond energy in ATP molecules. Typically, water molecules are split, their electrons and hydrogen atoms are picked up by $NADP^+$ to form NADPH, and their oxygen atoms are released as a by-product.

4. In the second stage of photosynthesis, ATP delivers energy to reaction sites where glucose is synthesized. Carbon dioxide provides carbon and oxygen for the reactions, and NADPH provides the electrons and hydrogen.

5. In the photosynthetic cells of plants and in many protistans, both stages of the reactions proceed inside organelles called chloroplasts.

6. Often we summarize photosynthesis this way:

$$\underset{\text{WATER}}{12H_2O} + \underset{\text{CARBON DIOXIDE}}{6CO_2} \xrightarrow{\text{LIGHT ENERGY}} \underset{\text{OXYGEN}}{6O_2} + \underset{\text{GLUCOSE}}{C_6H_{12}O_6} + \underset{\text{WATER}}{6H_2O}$$

PHOTOSYNTHESIS—AN OVERVIEW

Where the Reactions Take Place

Let's start with a look at **chloroplasts**, the organelles of photosynthesis in plants and in the protistans called algae. Chloroplasts, recall, specialize in producing and briefly storing food (Section 4.7). Figure 7.3 shows the structure and functional zones of one type. All chloroplasts have two outermost membranes that surround a largely fluid interior, the **stroma**. Still another membrane weaves through the stroma; we call it a thylakoid membrane system. In many species, parts of this system are often folded repeatedly into disk-shaped sacs, or **thylakoids**, stacked one on top of the other as in Figure 7.3*d*, *e*. Each stack of disks is a granum (plural, grana). Membranous channels interconnect the stacks.

Two stages of photosynthesis—the *light-dependent* and *light-independent* reactions—proceed in the interior of a chloroplast. The first stage occurs at the thylakoid membrane system (Figure 7.3*e*,*f*). The interconnecting spaces within all thylakoids and channels form a single compartment in which hydrogen ions (H^+) accumulate. ATP forms when these ions flow across the membrane. The second stage of photosynthesis, the set of reactions by which sugars are assembled, occurs in the stroma.

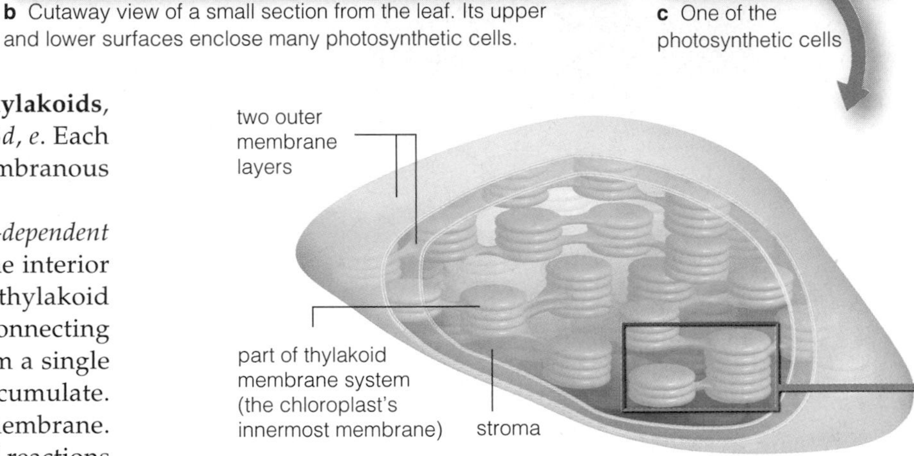

a Part of a leaf

b Cutaway view of a small section from the leaf. Its upper and lower surfaces enclose many photosynthetic cells.

c One of the photosynthetic cells

d Cutaway view of one of the chloroplasts inside the photosynthetic cell shown in (**c**).

Figure 7.3 Zooming in on sites of photosynthesis inside one of the leaves of a typical plant.

Energy and Materials for the Reactions

In the light-dependent reactions, absorption of energy from sunlight drives the formation of ATP from the chloroplast's pool of ADP and inorganic phosphate (P_i). In this way, energy from the sun becomes temporarily stored as chemical bond energy of ATP. Typically, water molecules are split and a coenzyme, $NADP^+$, picks up their electrons and hydrogen atoms to form NADPH.

In the light-independent reactions, the ATP donates energy to sites where carbon, hydrogen, and oxygen are assembled into glucose ($C_6H_{12}O_6$). NADPH delivers the electrons and hydrogen atoms. And carbon dioxide (CO_2) provides the carbon and oxygen atoms.

Overall, the reactions of photosynthesis are often summarized as a simple equation:

$$12H_2O + 6CO_2 \xrightarrow{\text{LIGHT ENERGY}} 6O_2 + C_6H_{12}O_6 + 6H_2O$$

Remember the description of tracers in Section 2.2? By attaching radioisotopes to atoms of the two kinds of reactants given in the summary equation, researchers showed where each atom ends up in the three products:

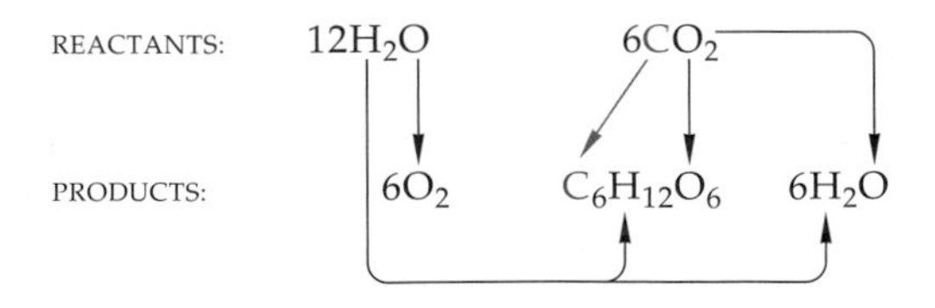

This summary equation shows *glucose* as the carbon-rich end product of the reactions, to keep the chemical bookkeeping simple. But you will find very little glucose in a chloroplast. Each newly formed glucose molecule has an attached phosphate group; it is primed to react

Further reading: Student Guide to InfoTrac on web site

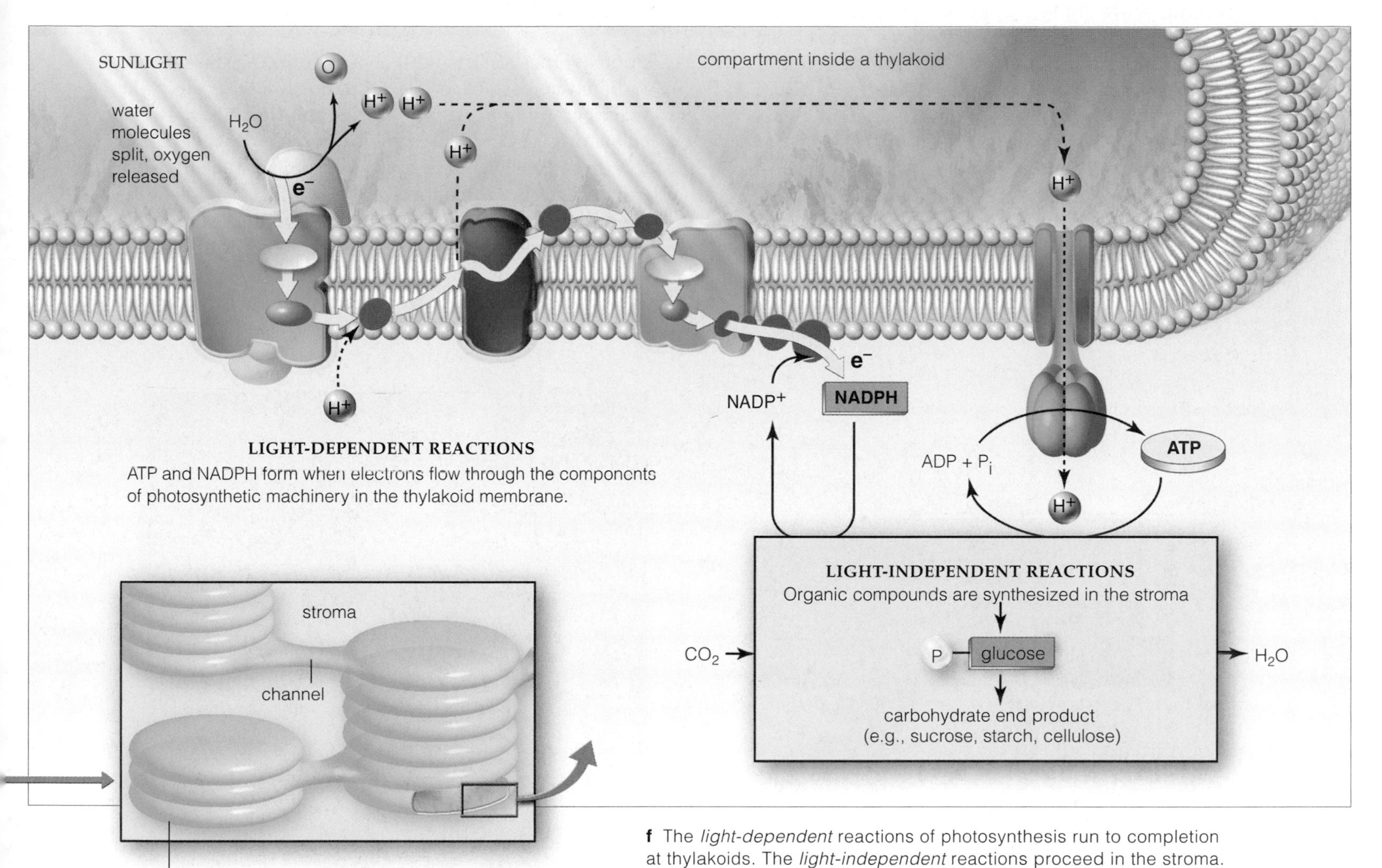

e One granum (a stack of disks, or thylakoids, that are part of the thylakoid membrane system)

f The *light-dependent* reactions of photosynthesis run to completion at thylakoids. The *light-independent* reactions proceed in the stroma.

with something else. Almost always, it is used at once in reactions that form sucrose, starch, and cellulose. Just keep in mind that these are the main end products of photosynthesis even though we "stop" with glucose.

The diagram below is a simplified version of Figure 7.3. Where you see it repeated in sections to follow, use it as a reminder to refer back to Figure 7.3 to reinforce your grasp of how the details fit into the big picture:

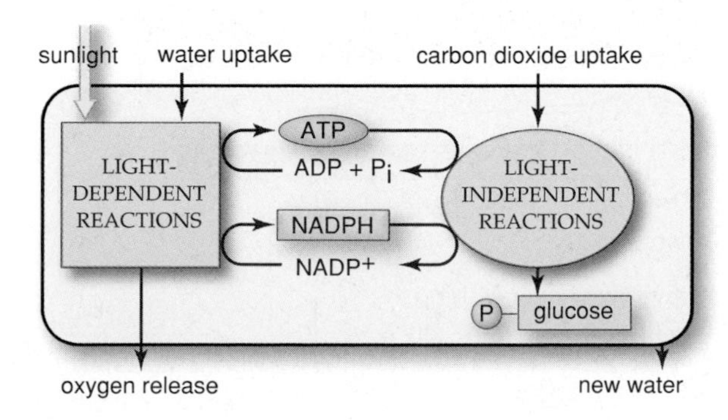

We turn next to the details of photosynthesis. Before we do, think about this: Two thousand chloroplasts, lined up single file, would be no wider than a dime. Imagine all of the chloroplasts in just one corn or rice plant—each a small factory for producing sugars and starch—and you may get an inkling of the magnitude of the metabolic events required to feed you and every other organism on this planet.

Chloroplasts, the organelles of photosynthesis in plants and many protistans, specialize in food production.

In the first stage of photosynthesis, sunlight energy drives the formation of ATP and NADPH, and oxygen is released. In chloroplasts, this stage occurs at thylakoids.

The second stage occurs in the stroma, where energy from ATP drives glucose formation. For these synthesis reactions, carbon dioxide provides the carbon and oxygen atoms, and NADPH provides the electrons and hydrogen atoms.

7.2 SUNLIGHT AS AN ENERGY SOURCE

Each second, more than 2 million metric tons of the sun's mass enter thermonuclear reactions that release stupendous amounts of energy. It takes a mere eight minutes for a fraction of the released energy to reach the Earth's atmosphere, 160 million kilometers away. Each day, that fraction averages 7,000 kilocalories per square meter of the Earth as a whole. Almost a third is reflected back into space. Of the amount of sunlight energy that does reach the Earth's surface, only about 1 percent is intercepted by photoautotrophs—which are the entry point for a one-way flow of energy through nearly all webs of life (Figure 7.4*a*).

Properties of Light

Photosynthesis starts with energy from the sun that has radiated across space in undulating motion, a bit like waves crossing a sea. The horizontal distance between crests of every two successive waves is a **wavelength** (Figure 7.4*b*). There are many different wavelengths of radiant energy. The entire range of all the wavelengths represents the **electromagnetic spectrum** (Figure 7.4*c*).

Collectively, different kinds of photoautotrophs can absorb wavelengths between 380 and 750 nanometers. That is the range of *visible* light, which we humans and

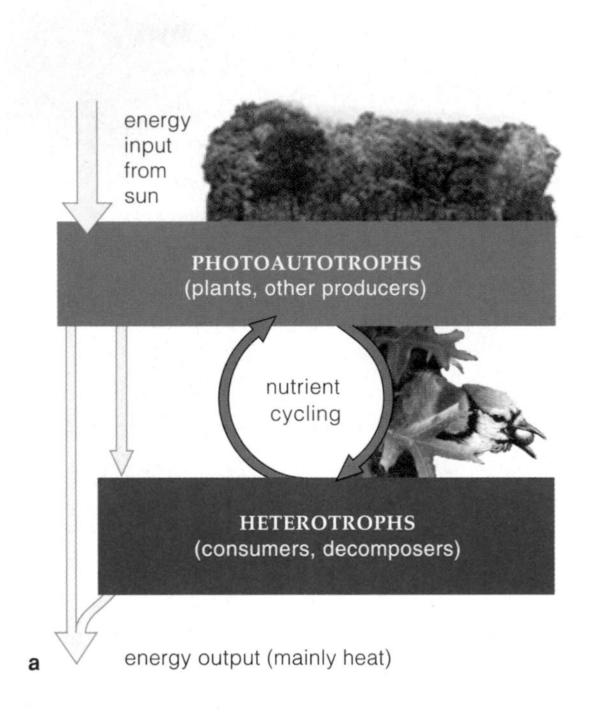

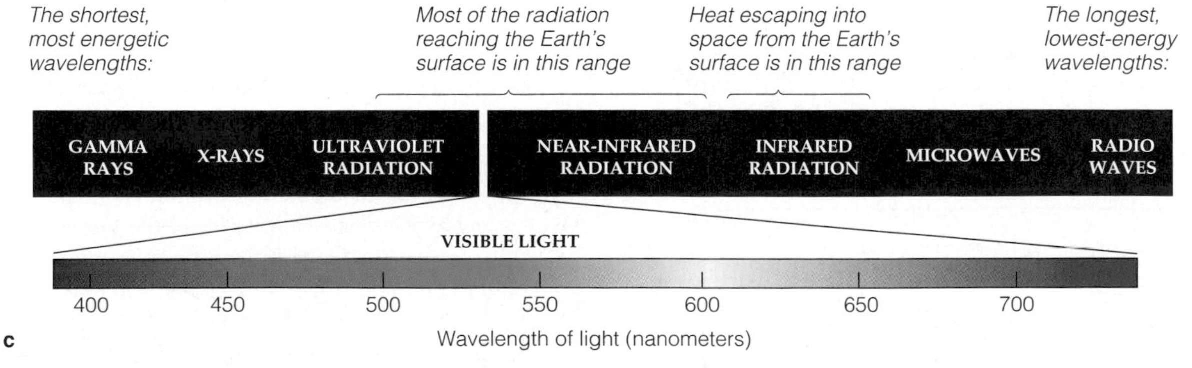

Figure 7.4 (**a**) Model of how photoautotrophs are the basis of a one-way flow of energy through nearly all webs of life, starting with energy inputs from the sun. (**b**) Two examples of wavelengths, the horizontal distance between crests of successive waves. (**c**) The electromagnetic spectrum. Visible light is one of many forms of electromagnetic radiation in the spectrum.

Further reading: Student Guide to InfoTrac on web site →

Figure 7.5 (**a**) Two absorption spectra that indicate the efficiency with which chlorophylls *a* and *b* respond to different wavelengths. (**b**) Absorption spectra for a beta-carotene (a carotenoid) and for a phycobilin.

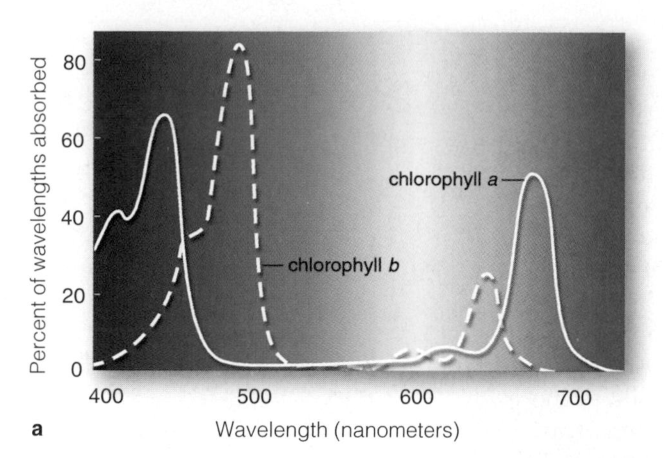

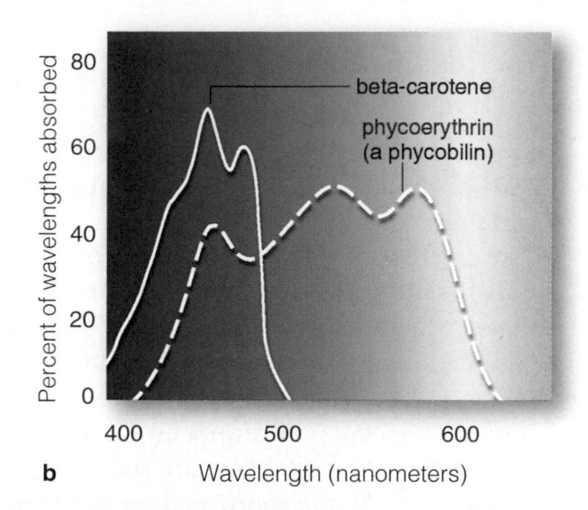

many other organisms perceive as different colors. Wavelengths shorter than this, such as ultraviolet (UV) radiation, are energetic enough to break the bonds of organic compounds—hence to kill cells. Hundreds of millions of years ago, before a layer of ozone (O_3) gradually accumulated in the upper atmosphere, only water shielded the surface of the Earth from UV radiation. The lethal bombardment kept photoautotrophs below the surface of the seas.

When absorbed by matter, the energy of visible light can be measured as if it were organized in packets, which we call **photons**. Each type of photon has a fixed amount of energy. Those having the most energy travel as the shortest wavelengths, which correspond to blue-violet light. Photons having the least energy travel as long wavelengths that correspond to red light (Figure 7.4*b*).

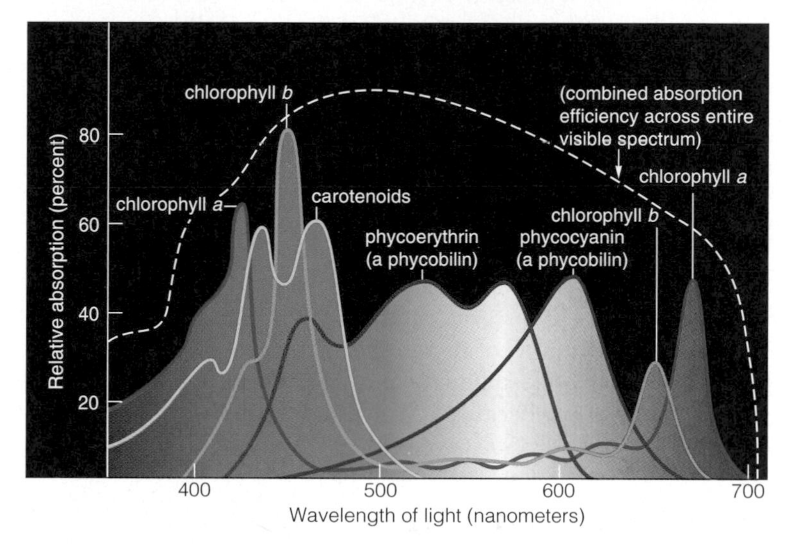

Figure 7.6 Combined energy absorption efficiency, indicated by the dashed line, of certain chlorophylls (*green* lines), carotenoids (*yellow-orange* lines), and phycobilins (*purple and blue* lines).

Pigments—Molecular Bridge From Sunlight to Photosynthesis

Collectively, wavelengths of visible light look white to us. A crystal prism intercepting white light can sort it out into its component colors by bending the different wavelengths by different degrees. In the 1800s, recall, T. Englemann used a prism as part of an experiment to identify which wavelengths drive photosynthesis in a green alga (Figure 7.1). But molecular biology was far in the future, and Englemann was not able to identify the actual bridge between sunlight and photosynthetic activity.

Pigments are the molecular bridge. **Pigments** are different molecules that absorb wavelengths of light, and organisms put them to many uses. Most pigments absorb only some wavelengths and transmit the rest. A few, such as the melanins in animals, absorb so many wavelengths they appear dark or black.

An **absorption spectrum** is a diagram that shows how effectively a pigment molecule absorbs different wavelengths in the spectrum of visible light. Consider chlorophylls, the main pigments of photosynthesis. As Figure 7.5*a* shows, chlorophylls absorb all wavelengths except very little of the green and yellow-green ones, which they mainly transmit. (That is why plant parts with an abundance of chlorophylls look green to us.) Other, *accessory* pigments that impart different colors also occur in photoautotrophs. Figure 7.5*a* and 7.6 show how their assistance extends the range of wavelengths that drive photosynthesis. As you can see, the combined absorption efficiency is high indeed.

Radiation from the sun travels in waves that differ in length and energy content. Short wavelengths pack the most energy.

We perceive wavelengths of visible light as different colors and measure their energy content in packets called photons.

Chlorophylls and certain other pigments absorb specific wavelengths of visible light. They are the molecular bridge between the sun's energy and photosynthetic activity.

7.3 THE RAINBOW CATCHERS

The Chemical Basis of Color

How is it that pigment molecules can impart color to plants and other organisms? Each has a light-catching array of atoms, which often are joined by alternating single and double bonds. Figure 7.7 shows examples. Electrons that are distributed around the atomic nuclei of such arrays can absorb photons of specific energies, which correspond to specific colors of light.

Remember, when an atom's electrons absorb energy, they move to a higher energy level (Section 2.3). In a pigment molecule, an input of energy destabilizes the distribution of electrons inside the light-catching array. Within 10^{-15} of a second, excited electrons return to a lower energy level, the electron distribution stabilizes, and energy is emitted in the form of light. When any destabilized molecule emits light as it reverts to its more stable configuration, this is called a **fluorescence**.

Excitation occurs only when the quantity of energy of an incoming photon matches the amount of energy required to boost an electron to a higher energy level. Suppose a sunbeam strikes a pigment. If photons of its red or orange wavelengths match the amount of energy necessary for the boost, the pigment will absorb them. However, the photons of its blue or violet wavelengths are a mismatch. The pigment molecule cannot absorb these wavelengths. Because it transmits (reflects) them, it takes on a blue or violet color.

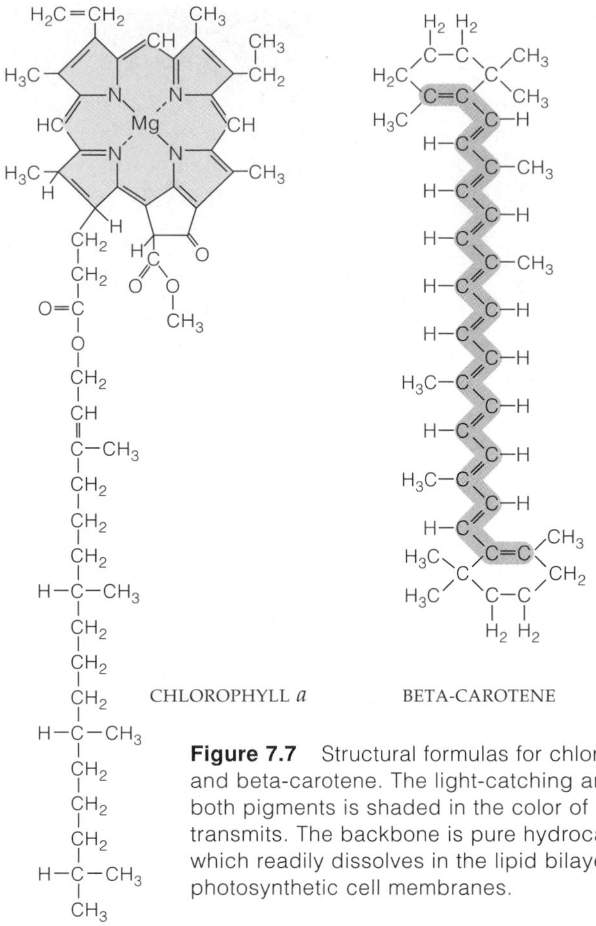

CHLOROPHYLL *a*

BETA-CAROTENE

Figure 7.7 Structural formulas for chlorophyll *a* and beta-carotene. The light-catching array of both pigments is shaded in the color of light it transmits. The backbone is pure hydrocarbon, which readily dissolves in the lipid bilayer of photosynthetic cell membranes.

On the Variety of Photosynthetic Pigments

Most pigments respond to only part of the rainbow of visible light. If acquiring energy is so vital for life, why doesn't each photosynthetic pigment go after the whole rainbow? In other words, why isn't each one black?

If the first photoautotrophs evolved in the seas, then so did their pigments. Ultraviolet and red wavelengths do not penetrate water as deeply as green and blue wavelengths do. Possibly natural selection favored the evolution of diverse pigments at different depths. Many red algae live in deep water, and indeed they are nearly black. Green algae live in shallow water, where their chlorophylls absorb red wavelengths. Their accessory pigments harvest more wavelengths, and some even function as shields against ultraviolet radiation.

Today, **chlorophylls** are the main pigments in all but one marginal group of photoautotrophs. Remember, photosynthetic pigments respond best to red and blue-to-violet light. Chlorophyll *a*, a grass-green pigment that absorbs blue-violet and red wavelengths, is a key player in the light-dependent reactions (Figure 7.7). Chlorophyll *b* absorbs blue and red-orange wavelengths. In chloroplasts, it is one of several accessory pigments, busily harvesting wavelengths that chlorophyll *a* misses. Chlorophyll *b*, a bluish-green pigment, occurs in plants, green algae, and a few photoautotrophic bacteria.

All photoautotrophs also contain **carotenoids**. These accessory pigments absorb blue-violet and blue-green wavelengths that chlorophylls miss. They reflect red, orange, and yellow wavelengths. You can expect flowers, fruits, and vegetables in this color range to have an abundance of carotenoids.

Usually the backbone of a carotenoid molecule has a carbon ring at each end. Many types, including beta-carotene, are pure hydrocarbons (Figure 7.7). The type called xanthophylls also contain oxygen. Some of the carotenoids have proteins attached, in which case they appear purple, violet, blue, green, brown, or black.

Carotenoids are less abundant than the chlorophylls in green leaves, but in many plants they become visible in autumn (Figure 7.8). Each year, tourists spend about a billion dollars just to watch the three-week demise of chlorophyll in the deciduous trees of New England.

Other accessory pigments include the red and blue **anthocyanins** and **phycobilins**. The anthocyanins are pigments in many flowers, and the phycobilins are the signature pigments of red algae and cyanobacteria.

Further reading: Student Guide to InfoTrac on web site →

Figure 7.8 Leaf color. Inside intensely green leaves, photosynthetic cells are continually synthesizing chlorophyll molecules. The chlorophylls mask the presence of carotenoids and other accessory pigments. In autumn, however, chlorophyll synthesis lags behind chlorophyll breakdown in many species. When that happens, the other pigments in the leaf are unmasked, and more colors show through.

Also in autumn, anthocyanins accumulate in leaf cells. These water-soluble pigments appear red if fluids moving through plants are slightly acidic, blue if the fluids are basic (alkaline), or colors in between if the fluids are of intermediate pH. Soil conditions contribute to the pH values.

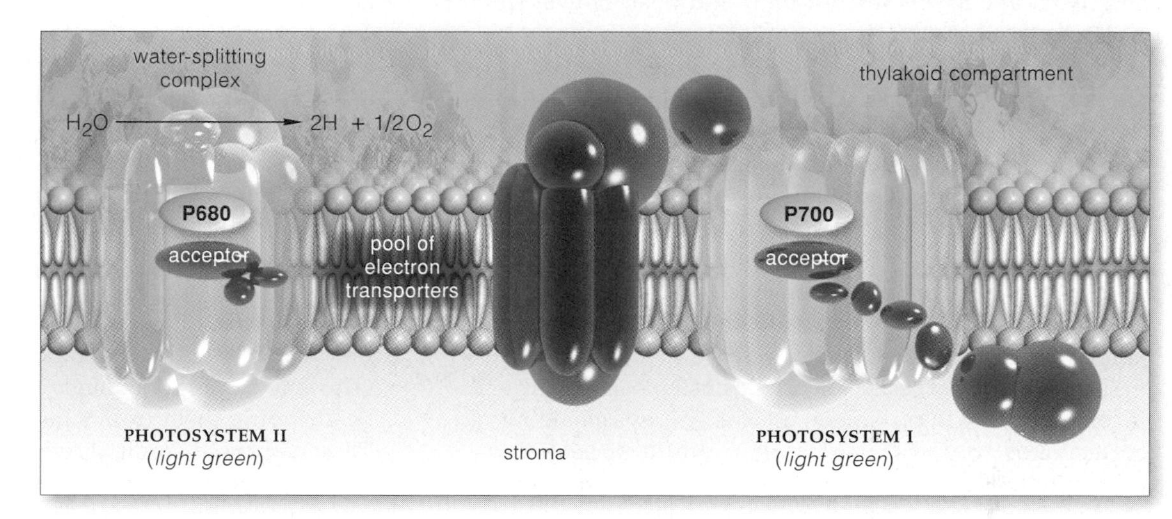

Figure 7.9 How two kinds of photosystems, designated I and II, are arranged in the thylakoid membrane system of chloroplasts. Components of neighboring electron transport systems are coded *dark green*. One transport system extends from photosystem II to P700 of photosystem I. A second transport system that extends from P700 gives up electrons and hydrogen to a coenzyme ($NADP^+$).

Where Are Photosynthetic Pigments Located?

We find photosynthetic pigments among some bacteria of ancient lineages. For example, the archaebacterium *Halobacterium halobium* has bacterorhodopsin (a purple pigment) in its plasma membrane. Cyanobacteria have chlorophyll *a* and other pigments embedded in internal foldings of the plasma membrane.

The thylakoid membrane system of chloroplasts has pigments organized in clusters called **photosystems**. In some plant species, the membrane has many thousands of these clusters. Each photosystem consists of proteins and 200 to 300 pigment molecules. As Figure 7.9 shows, it is located next to its functional partner—an electron transport system. How the two systems work together is the topic of the next two sections.

About Those Roving Pigments

There are only about eight classes of pigments, but this limited group gets around in the world. For example, animals synthesize some pigments (such as melanin) but not carotenoids. These originate with photoautotrophs and move up through food webs, as when tiny aquatic snails feed on green algae and flamingos subsequently eat the snails. Flamingos modify ingested carotenoids in diverse ways. For instance, their cells can split beta-carotene molecules to form two molecules of vitamin A. Vitamin A is the precursor of retinol, a visual pigment that transduces light to electric signals in the flamingo's eyes. Beta-carotene molecules that become dissolved in fat reservoirs under the skin are taken up by skin cells that differentiate and give rise to bright pink feathers.

The chlorophylls are the main photosynthetic pigments in photoautotrophs. Carotenoids and other accessory pigments enhance their light-harvesting function.

Photosystems (pigment clusters) are the functional partners of electron transport systems in the thylakoid membrane.

7.4 THE LIGHT-DEPENDENT REACTIONS

We are now ready to look more closely at the first stage of photosynthesis, the **light-dependent reactions**, by tracking three events that proceed in the chloroplast. *First*, pigments absorb light energy and give up excited electrons, which enter electron transport systems. *Second*, water molecules are split, ATP and NADPH form, and oxygen is released. *Third*, the pigments that gave up electrons in the first place get electron replacements.

What Happens to the Absorbed Energy?

Photon absorption, recall, can boost electrons of atoms of a photosynthetic pigment to a higher energy level, but the electrons quickly emit the excitation energy as they return to a lower energy level. If nothing else were around to intercept it, all of the emitted energy would simply escape as fluorescent light and heat. However, photosynthetic pigments are not isolated.

As you read earlier, hundreds of pigment molecules are clustered together in each of the photosystems that are embedded in the thylakoid membrane. Most of the pigments of each cluster only harvest the energy from photons. Instead of releasing the excitation energy as a fluorescent afterglow, a harvester directly transfers it to another pigment molecule, which randomly passes it on to a neighboring pigment molecule, and so on. Figure 7.10 illustrates the "random walk" of excitation energy among the pigments of a photosystem.

Each time a harvester makes a transfer, some energy is lost (as heat). In no time at all, the energy remaining corresponds to a wavelength that only a specialized chlorophyll *a* can trap. That chlorophyll is the **reaction center** for the photosystem. It accepts excitation energy,

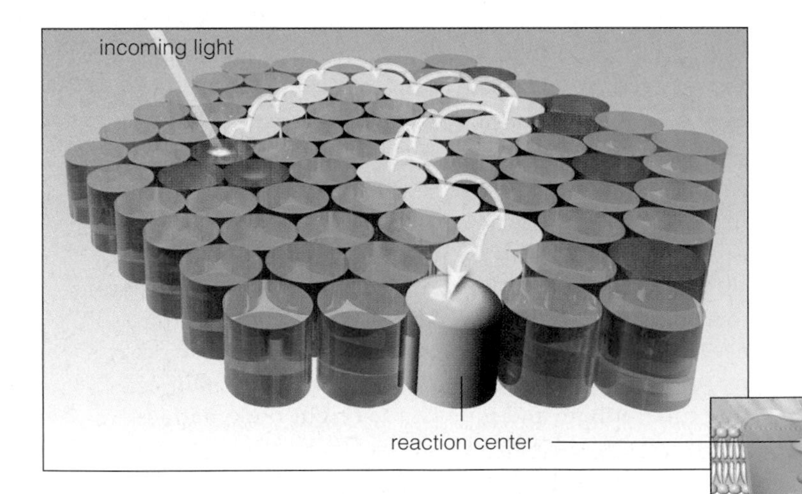

Figure 7.10 A small sampling of harvester pigments of a photosystem. Notice the random flow of energy from photons among them. The energy of excitation quickly reaches a reaction center which, when suitably activated, releases electrons for photosynthesis.

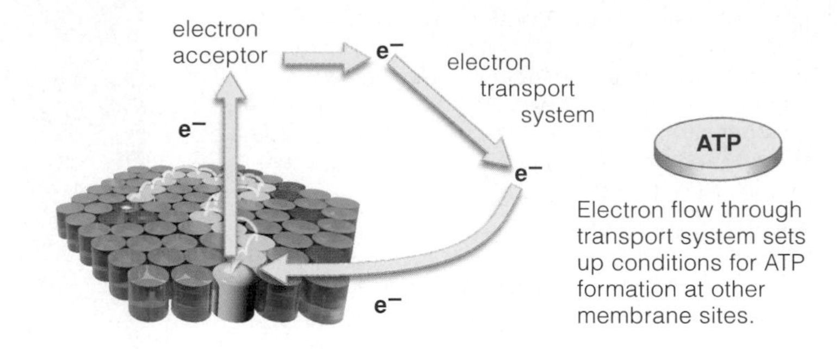

Figure 7.11 Cyclic pathway of ATP formation.

but it does not pass energy on to another pigment. A suitably activated reaction center donates electrons to an acceptor molecule near an electron transport system.

Electron transport systems are organized arrays of enzymes, coenzymes, and other proteins associated with a cell membrane (Section 6.4). Electrons are transferred step-by-step through the array, and some energy escapes at each step. In chloroplasts, much of the energy drives the machinery that produces ATP and NADPH.

Cyclic and Noncyclic Electron Flow

In ancient bacteria, photosynthesis evolved as a way to produce ATP, not to synthesize organic compounds. (Those bacteria got enough electrons to make NADPH by stripping them from simple inorganic compounds.) They harnessed sunlight energy to cycle electrons from a photosystem, to a transport system, then back to the photosystem (Figure 7.11). This *cyclic* pathway still runs in all photoautotrophs. It requires a *type I* photosystem, which has a reaction center designated P700.

The electron flow from P700 does not pack enough punch to make NADPH, too. But more than 2 billion years ago, it seems that the energy for electron transfers increased as if two batteries had been hooked together. Another photosystem, *type II,* entered the picture. And the two photosystems operating together could harness enough energy to strip electrons from water molecules.

Take a look at Figure 7.12*a.* As you see, the newer pathway of electron flow is *noncyclic*. There is a linear flow of electrons from water to photosystem II, through a transport system to photosystem I, then on through a transport system that delivers the electrons to $NADP^+$.

The machinery works when photosystem II absorbs enough photon energy to make its designated reaction center—P680—give up electrons. The energy input also triggers **photolysis**. By this reaction sequence, water molecules split into oxygen, hydrogen ions (H^+), and electrons. When P680 releases the excited electrons, new electrons released from water replace them (Figure 7.12*a*).

Further reading: Student Guide to InfoTrac on web site

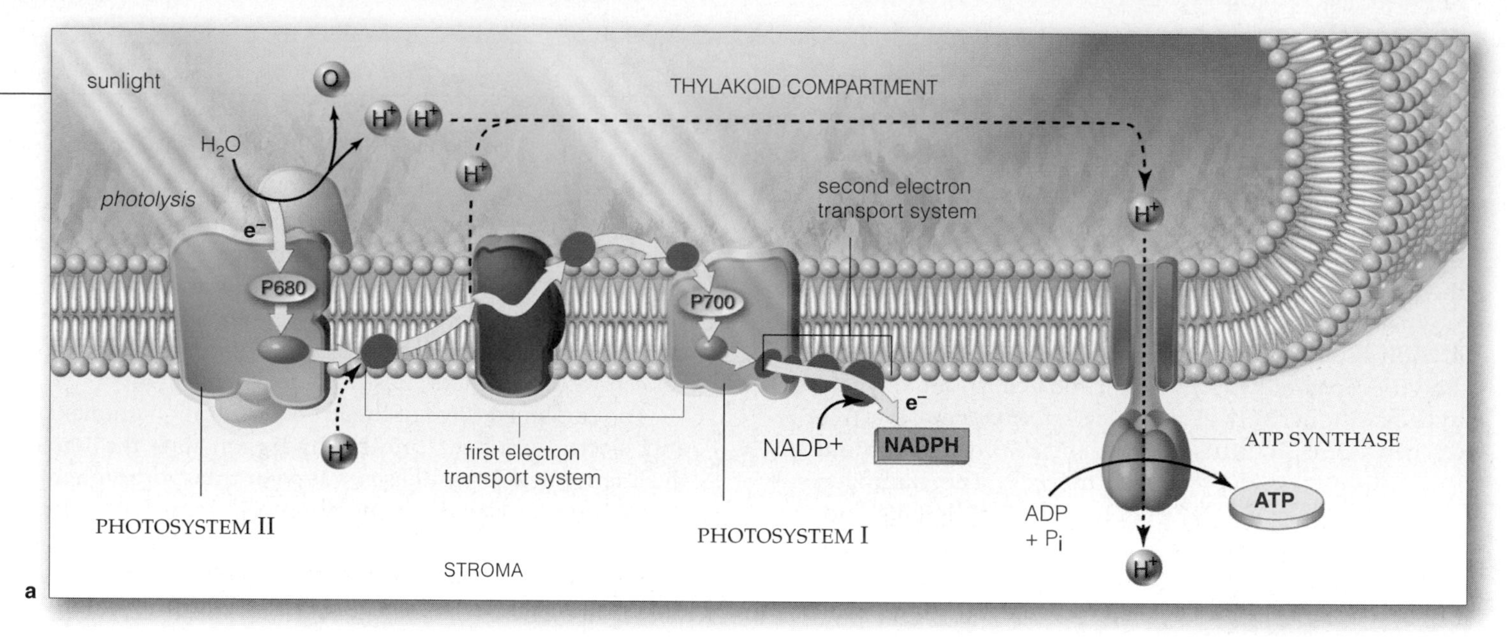

Figure 7.12 (**a**) Filling in the details for Figure 7.3—the noncyclic pathway of photosynthesis, by which ATP and NADPH form. *Yellow* arrows show electron flow. Electrons released from water molecules split by photolysis move through two photosystems (*light green*) and two transport systems (*dark green*). The joint operation of the two photosystems boosts electrons to an energy level high enough to drive NADPH formation. Also, as you will read in Section 7.5, hydrogen ions move across the thylakoid membrane in ways that lead to ATP formation at proteins called ATP synthases.

(**b**) Diagram the energy changes associated with the noncyclic pathway, and you will end up with this "Z scheme."

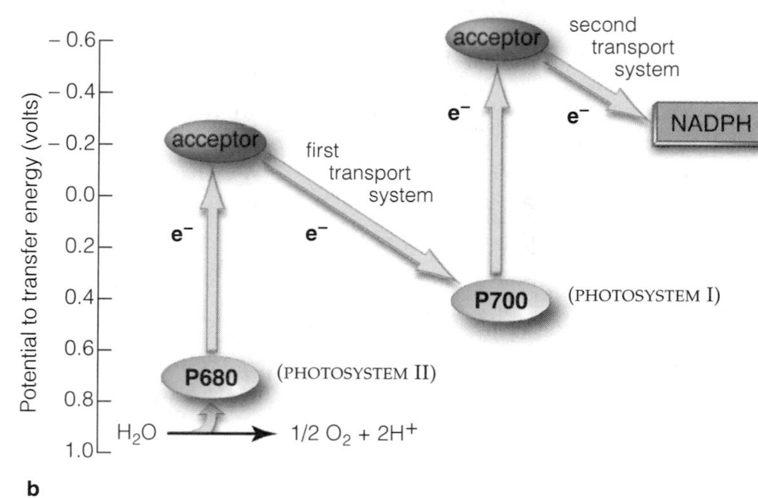

The excited electrons move on to a transport system, then to P700 of photosystem I. They still have a bit of extra energy. And they get more upon their arrival, for photons are also bombarding photosystem I. The energy boost puts electrons at a higher energy level that allows them to enter a transport system. There, $NADP^+$ accepts two electrons and a hydrogen ion to become NADPH.

Today, both the cyclic and noncyclic pathways of electron flow operate in all photosynthetic organisms. Which one dominates at a given time depends on the organism's metabolic demands for ATP and NADPH.

The Legacy—A New Atmosphere

On sunny days, on the surfaces of aquatic plants, you see bubbles of oxygen, a by-product of the noncyclic pathway (Figure 7.13). When the pathway first evolved, the oxygen being released dissolved in seawater, wet mud, and similar bacterial habitats. By 1.5 billion years ago, enormous quantities of dissolved oxygen were escaping to what had been an oxygen-free atmosphere. The accumulation of oxygen changed the atmosphere forever. That vast, global change made possible aerobic respiration, which became the most efficient pathway for releasing usable energy from organic compounds. Ultimately, the emergence of the noncyclic pathway allowed you and every other animal to be around today, breathing the oxygen that helps keep your cells alive.

Figure 7.13 Visible evidence of photosynthesis—bubbles of oxygen escaping from the leaves of *Elodea*, a plant of freshwater habitats.

In the light-dependent reactions, sunlight energy drives the release of electrons from photosystems in thylakoid membranes. The electron flow through nearby transport systems results in the formation of ATP, NADPH, or both.

ATP can form when electrons cycle from and back to a type I photosystem. Both ATP and NADPH can form by a noncyclic pathway in which electrons flow from water, through two photosystems (types II and I), and finally to $NADP^+$.

All photosynthetic species employ one or both pathways in response to the cell's changing needs for ATP and NADPH.

Oxygen, a by-product of the noncyclic pathway, changed the early atmosphere and made aerobic respiration possible.

7.5 A CLOSER LOOK AT ATP FORMATION IN CHLOROPLASTS

As you probably have noticed, we saved the trickiest question for last. Exactly how does ATP form during the noncyclic pathway of photosynthesis? To arrive at the answer, let's walk through Figure 7.14.

Inside the chloroplast, photon absorption triggers photolysis, the pathway's first step. At this step, enzyme activity repeatedly splits water molecules into oxygen, hydrogen ions, and electrons. O_2 forms from the free oxygen atoms. It diffuses out of the chloroplast and out of the cell. The hydrogen ions (H^+) are left behind. And they accumulate in the fluid inside the thylakoid compartment (Figure 7.14*a*).

Therefore, *by a combination of photolysis and electron transport,* the concentration of hydrogen ions becomes greater inside the thylakoid compartment, compared to the stroma. The unequal distribution of these positively charged ions also creates a difference in electric charge across the membrane. An electric gradient, as well as a concentration gradient, has become established.

The combined force of the H^+ concentration gradient and electric gradient propels hydrogen ions through the interior of ATP synthases, a type of transport protein that spans the thylakoid membrane (Figure 7.14*c*). In other words, the ions flow out from the compartment, through ATP synthases, into the stroma. ATP synthases have built-in enzymatic machinery. And the flow of ions through them drives the machinery, which catalyzes the attachment of unbound phosphate to a molecule of ADP. In this way, ATP forms.

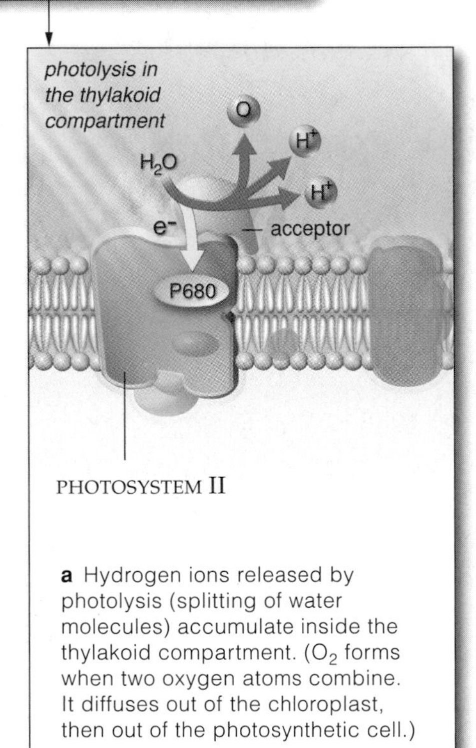

a Hydrogen ions released by photolysis (splitting of water molecules) accumulate inside the thylakoid compartment. (O_2 forms when two oxygen atoms combine. It diffuses out of the chloroplast, then out of the photosynthetic cell.)

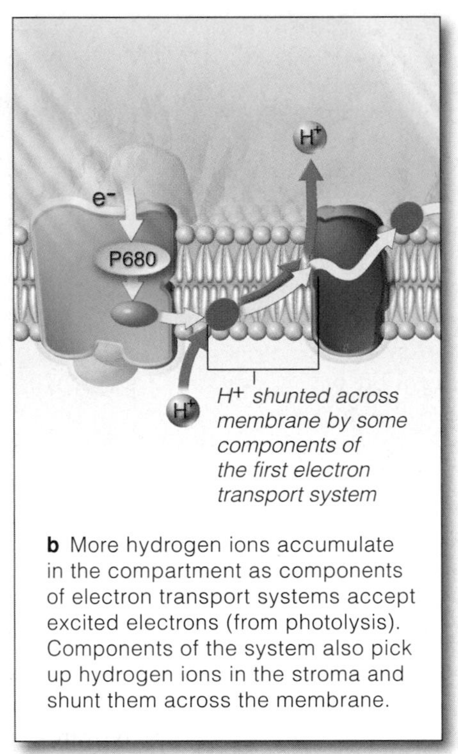

b More hydrogen ions accumulate in the compartment as components of electron transport systems accept excited electrons (from photolysis). Components of the system also pick up hydrogen ions in the stroma and shunt them across the membrane.

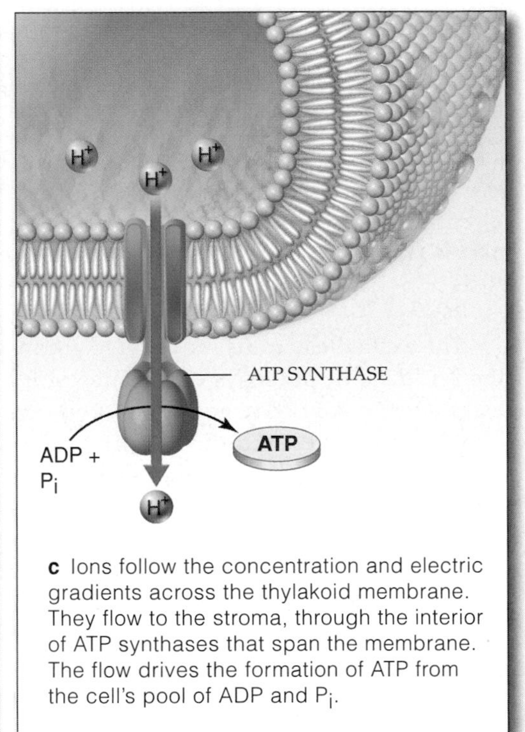

c Ions follow the concentration and electric gradients across the thylakoid membrane. They flow to the stroma, through the interior of ATP synthases that span the membrane. The flow drives the formation of ATP from the cell's pool of ADP and P_i.

Figure 7.14 How ATP forms in chloroplasts during the noncyclic pathway of photosynthesis. 

The released electrons are picked up by a primary acceptor molecule, which transfers them to the electron transport system positioned next to photosystem I in the thylakoid membrane. More hydrogen ions flow into the thylakoid compartment while the electron transport system is operating. This is the case for both the cyclic and noncyclic pathways. The thylakoid membrane has many photosystems and many transport systems. At the same time that certain components of the transport systems accept electrons, they also pick up hydrogen ions from the stroma. They immediately shunt those ions across the membrane, as shown in Figure 7.14*b*.

The sequence of events just described is called the chemiosmotic model of ATP formation in chloroplasts. As you will see in the next chapter, the same model applies to ATP formation in mitochondria.

In chloroplasts, H^+ concentration and electric gradients form across the thylakoid membrane. The flow of ions from the thylakoid compartment to the stroma drives ATP formation.

Further reading: Student Guide to InfoTrac on web site →

7.6 THE LIGHT-INDEPENDENT REACTIONS

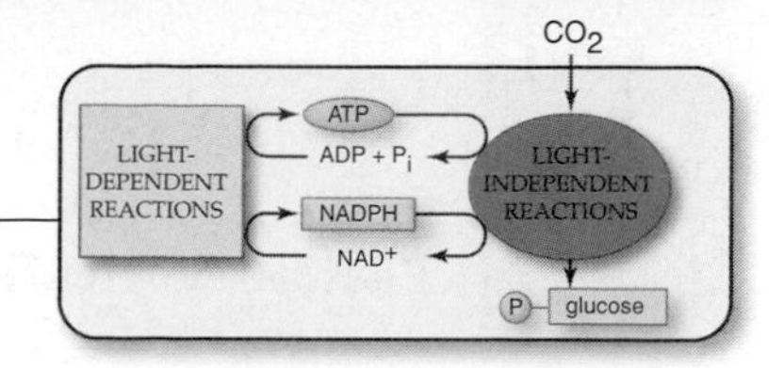

The **light-independent reactions** are the "synthesis" part of photosynthesis. They occur as a cyclic pathway called the **Calvin–Benson cycle**. The reactions require energy from ATP, hydrogen and electrons from NADPH, and carbon and oxygen from carbon dioxide (CO_2), which is present in the air (or water) that surrounds photosynthetic cells. We say these reactions are light-independent because they do not depend directly on sunlight. They can proceed just as well in the dark, as long as ATP and NADPH are available.

How Do Plants Capture Carbon?

Let's track a CO_2 molecule that diffuses into air spaces inside a leaf and ends up next to a photosynthetic cell. It diffuses into the cell, then into a chloroplast's stroma. An enzyme attaches the carbon atom of CO_2 to **RuBP** (ribulose bisphosphate), a compound with a backbone of five carbon atoms. **Rubisco** (for RuBP carboxylase) is the enzyme's name. Its action produces an unstable six-carbon intermediate that splits into two molecules of **PGA** (phosphoglycerate). PGA is a stable molecule with a three-carbon backbone. Incorporating a carbon atom from CO_2 into a stable organic compound is called **carbon fixation**. Quite simply, food can't be produced without this first step of the Calvin–Benson cycle.

The cycle yields phosphorylated glucose, and it also regenerates the RuBP. For our purposes, we can focus on the carbon atoms of the substrates, intermediates, and end products, as in Figure 7.15.

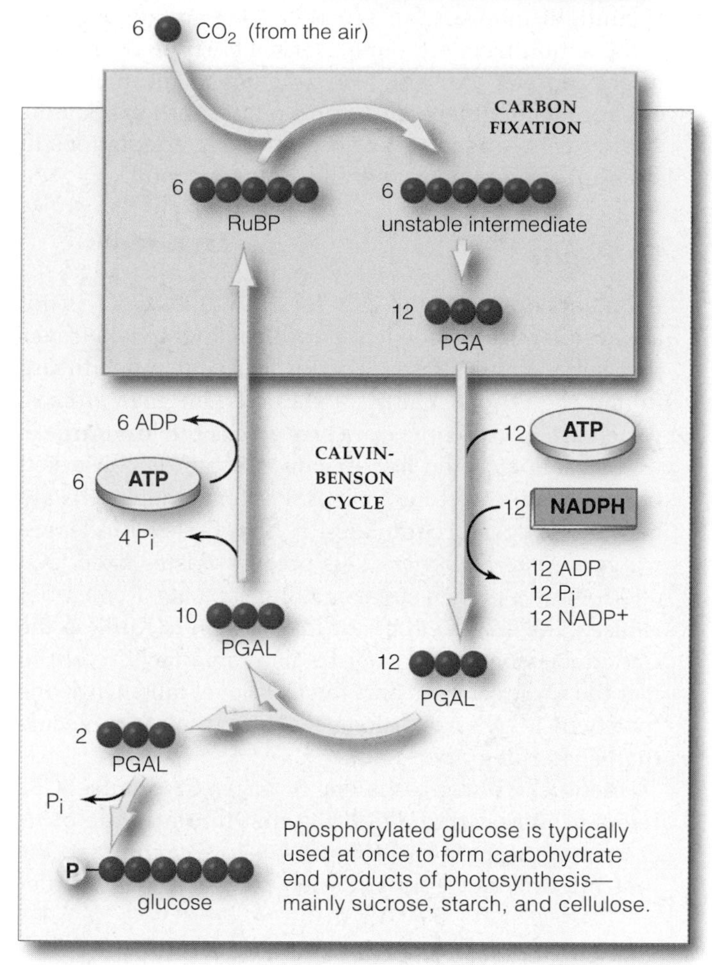

Figure 7.15 Summary of the light-independent reactions of photosynthesis. *Red* circles are carbon atoms of key molecules. All intermediates have one or two phosphate groups attached, but to keep things simple, we show only the phosphate group on the resulting glucose. Also, many water molecules that formed in the light-dependent reactions enter this pathway, and six remain at its conclusion. Appendix V (Figure C) shows details.

How Do Plants Build Glucose?

Each PGA accepts a phosphate group from ATP, plus hydrogen and electrons from NADPH. The resulting intermediate is called **PGAL** (phosphoglyceraldehyde). To build *one* six-carbon sugar phosphate, carbon atoms from six CO_2 must be fixed and twelve PGAL must form. Most of the PGAL becomes rearranged to form new RuBP, which can be used to fix more carbon. But two of the PGAL combine, thus forming glucose with a phosphate group attached to its six-carbon backbone.

When phosphorylated this way, glucose is primed to enter other reactions. Plants use it as a building block for their main carbohydrates—sucrose, cellulose, and starch. *The synthesis of these organic compounds by other pathways marks the end of the light-independent reactions.*

With six turns of the cycle, enough RuBP molecules form to replace the ones used in carbon fixation. The ADP, $NADP^+$, and phosphate remaining diffuse through the stroma, to sites of the light-dependent reactions. There they are converted back to NADPH and ATP.

Photosynthetic cells convert phosphorylated glucose to sucrose or starch during daylight hours. Of all plant carbohydrates, sucrose is the most easily transported. Starch is the most common storage form. The cells also convert excess PGAL to starch. They briefly store starch (as starch grains) inside the stroma. After the sun goes down, they convert starch to sucrose for export to other living cells in leaves, stems, and roots. Ultimately, the products and intermediates of photosynthesis end up as energy sources and as building blocks for all of the lipids, amino acids, and other organic compounds that are required for growth, survival, and reproduction.

In the light-independent reactions (the Calvin–Benson cycle), carbon is "captured" from carbon dioxide, glucose forms during reactions that require ATP and NADPH, and RuBP (necessary to capture the carbon) is regenerated.

7.7 FIXING CARBON—SO NEAR, YET SO FAR

If sunlight intensity, air temperature, rainfall, and soil composition were the same everywhere, photosynthesis might proceed the same way in every plant. However, environments differ—and so do photosynthetic details. A brief comparison of two carbon-fixing adaptations to stressful environments will illustrate this point.

C4 Plants

All plants must take up CO_2 for growth. But CO_2 is not always plentiful inside leaves, which have a waxy cover that helps plants conserve water. The CO_2 must diffuse in and the O_2 out mainly at **stomata** (singular, stoma), which are microscopic openings across the leaf surface.

Stomata close on hot, dry days. Water is conserved but CO_2 cannot get into leaves. Photosynthetic cells are busy, so oxygen accumulates. A high O_2 level in leaves triggers *photorespiration*. This process wastes fixed CO_2 and lowers a plant's sugar-making capacity. Remember rubisco, the enzyme that attaches carbon to RuBP in the Calvin–Benson cycle? Rubisco also can attach *oxygen* to it if the O_2 level rises and the CO_2 level falls. Only one (not two) PGA forms, along with a glycolate molecule that is later degraded to CO_2.

Kentucky bluegrass is one of many **C3 plants**. "C3" refers to *three*-carbon PGA, the first intermediate of its carbon-fixing pathway (Figure 7.16*a*). By contrast, the first intermediate formed when corn and many other plants fix carbon is the *four*-carbon oxaloacetate. Hence *their* name, **C4 plants**.

When C4 plants close stomata on hot, dry days, they still maintain enough CO_2 in leaves by *fixing carbon twice*, in two types of photosynthetic cells. Mesophyll cells, the first type, use CO_2 to form oxaloacetate—which is transferred to bundle-sheath cells at leaf veins (Figure 7.16*b*). There, CO_2 is released and its local concentration rises, so rubisco cannot use oxygen. In the mesophyll cells, CO_2 is fixed again during the Calvin–Benson cycle. With this pathway, C4 plants get by with tinier

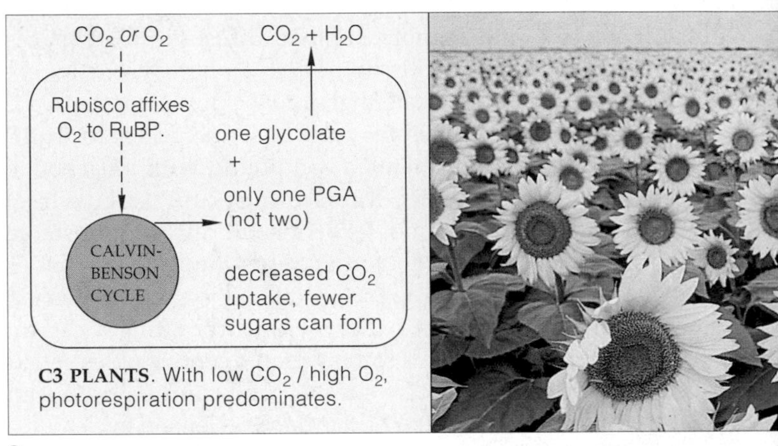

C3 PLANTS. With low CO_2 / high O_2, photorespiration predominates.

a

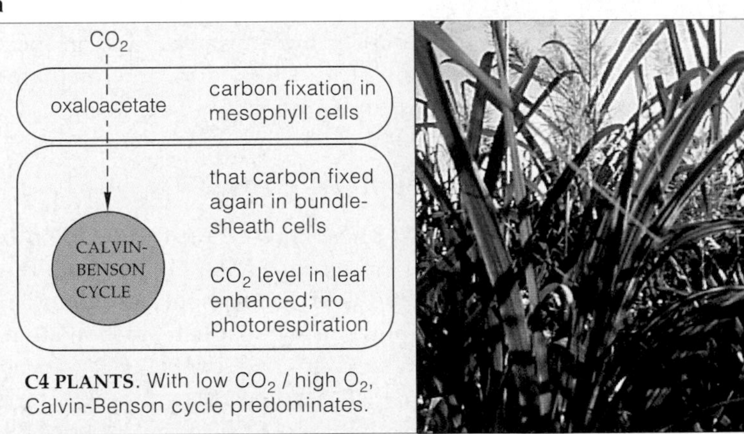

C4 PLANTS. With low CO_2 / high O_2, Calvin-Benson cycle predominates.

b

CAM PLANTS. With low CO_2 / high O_2, Calvin-Benson cycle predominates.

c

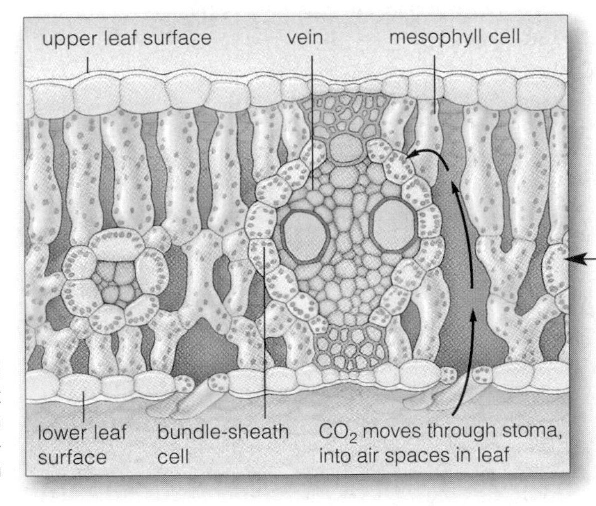

Typical C4 plant leaf, in cross-section

Figure 7.16 Three ways of fixing carbon in hot, dry weather, when the CO_2 level is low and the O_2 level is high in leaves. (**a**) The C3 pathway is common among evergreen trees and shrubs as well as many nonwoody plants of temperate zones, such as sunflowers. (**b**) The C4 pathway is common among grasses and other plants that evolved in the tropics and fix CO_2 twice. Corn, crabgrass, and the sugarcane shown here are examples. (**c**) CAM plants open stomata and fix carbon at night. They include pineapple, cacti and many other succulents (plants with a low surface-to-volume ratio), and orchids.

Further reading: Student Guide to InfoTrac on web site

stomata, lose less water, and make more glucose than C3 plants can when conditions are hot, bright, and dry.

Experiments show that photorespiration lowers the photosynthetic efficiency of many C3 crop plants. For example, when hothouse tomatoes were grown at CO_2 levels high enough to block photorespiration, growth rates increased by as much as five times. If the process is so wasteful, why hasn't natural selection eliminated it? The answer may lie with rubisco, the enzyme that sets the whole process in motion. That switch-hitting enzyme evolved long ago, when atmospheric levels of oxygen were still low and carbon dioxide levels high. Maybe a gene coding for its structure cannot mutate without adversely affecting the carbon-fixing activity. Or maybe the pathway that degrades glycolate to carbon dioxide has proved so adaptive that it cannot be eliminated. Glycolate can be toxic at high concentrations.

The C4 pathway evolved separately in many lineages of flowering plants over the past 50 million to 60 million years. Before then, atmospheric CO_2 levels were higher, and they gave C3 plants a selective advantage.

Which pathway will be most adaptive in years to come? Atmospheric CO_2 levels have been increasing for decades. Some ecologists predict that they will double over the next fifty years. If that happens, photorespiration will decline again—and many vital crops might benefit.

CAM Plants

We see a carbon-fixing adaptation to desert conditions in cacti. A cactus is one of the succulents; it has juicy, water-storing tissues and thick surface layers that restrict water loss. It cannot open stomata on hot days without losing precious water. It opens them and fixes CO_2 *at night*. Its cells store the resulting intermediate in their central vacuoles, then use it for photosynthesis the next day when stomata are closed. Many plants are adapted this way. They are **CAM plants** (short for Crassulacean Acid Metabolism). Unlike C4 species, CAM plants do not fix carbon twice in different types of cells. They fix it in the same cells but at different times (Figure 7.16*c*).

Many plants die during prolonged droughts, but some CAM plants survive by keeping stomata closed even at night. They repeatedly fix the CO_2 that forms by aerobic respiration. Not much forms, but it is enough to allow these plants to maintain very low rates of metabolism. CAM plants grow slowly. Try growing a cactus plant in Seattle or some other place with a mild climate, and it will compete poorly with C3 and C4 plants.

C4 plants and CAM plants both have modified ways of fixing carbon for photosynthesis. The modifications counter the stress imposed by hot, dry conditions in their environments.

Focus on Science

7.8 LIGHT IN THE DEEP DARK SEA?

During the late 1980s a graduate student, Cindy Lee Van Dover, was puzzling over an eyeless shrimp. Divers had discovered it at a hydrothermal vent deep in the Atlantic Ocean. **Hydrothermal vents** are fissures in the seafloor where molten rock rises and mixes with cold seawater. Mineral-rich water, superheated by 650 degrees, spews into the perpetually dark surroundings. The sides of the vents teem with life—including shrimps and bacteria.

When Van Dover saw videotapes of shrimps clinging to the sides of the vent, she noticed a pair of bright strips on their back. By examining specimens in the lab, she saw that the strips connected to a nerve. The eyeless shrimps, it seemed, were equipped with a novel sensory organ.

EYELESS SHRIMP

Steven Chamberlain, a neuroscientist, confirmed the strips are a sensory organ with photoreceptors, or light-absorbing pigmented cells. Later, Ete Szuts, an expert on pigments, isolated the pigment. Its absorption spectrum is the same as that for rhodopsin, a visual pigment in eyes as complex as yours.

Van Dover had a hunch: If the strips serve as an eye, then we should find *light* on the seafloor. On later dives, special cameras confirmed her hypothesis. Like coils in a toaster oven, vents release heat (infrared radiation). They also release faint radiation at the low end of the visible spectrum—light up to nineteen times brighter than expected for infrared. A billion billion photons per square inch per second strike a typical sunlit tree. Only a trillion photons per square inch per second reach photoautotrophic bacteria living 72 meters (240 feet) below the surface of the Black Sea. *But just as many photons are available to organisms at hydrothermal vents.*

About this time Van Dover casually asked a colleague, "Hey, what if there's enough light for photosynthesis?" The response was, "What a stupid idea."

Euan Nisbet, who studies ancient environments, thought about this. The first cells arose about 3.8 billion years ago. As Nisbet and Van Dover hypothesized, What if they arose at hydrothermal vents? They could have used inorganic compounds such as hydrogen sulfide (for their hydrogen and electrons) and carbon dioxide (for carbon). Survival probably depended on being able to move away from light at vents and thereby avoid being boiled alive. Millions of years later, some bacterial descendants were evolving in hot springs near the ocean's surface. In time they used their light-detecting machinery to absorb light from the sun. They were adapted to a new energy source; they had become photosynthetic.

Some observations support the intriguing hypothesis that light-sensing machinery of deep-sea bacteria became modified for shallow-water photosynthesis. For example, absorption spectra for the chlorophyll in evolutionarily ancient photosynthetic bacteria correspond to the light measured at hydrothermal vents. Also, photosynthetic machinery contains iron, sulfur, manganese, and other minerals—which are abundant at hydrothermal vents.

7.9 AUTOTROPHS, HUMANS, AND THE BIOSPHERE

We conclude this chapter with a story that reinforces how photosynthesizers and other autotrophs fit in the world of living things. It is a story of mind-boggling numbers of single-celled and multicelled species on land and in the sunlit waters of the Earth.

Each spring, you sense renewed growth of autotrophs on land as leaves unfurl and lawns and fields turn green. You might not be aware that uncountable numbers of single-celled autotrophs also drift through the surface waters of the world ocean. You can't see them without a microscope; a row of 7 million cells of one aquatic species would be less than a quarter-inch long. In some regions, a cupful of seawater might hold 24 million cells of one species, and that wouldn't include cells of all the other aquatic species.

Nearly all of the drifters are photoautotrophic bacteria and protistans. Together they are the "pastures of the seas," the producers that ultimately feed all other marine organisms. The pastures "bloom" in the spring, when nutrient inputs sustain rapid reproduction. At that time seawater becomes warmer and enriched with nutrients churned up from the deep by winter currents.

Until NASA gathered data from satellites in space, we had no idea of their numbers and distribution. Figure 7.17*a* shows visual evidence of their activities one winter in the surface waters of the North Atlantic Ocean. Figure 7.17*b* shows a springtime bloom stretching from North Carolina all the way past Spain!

Collectively, the cells help shape the global climate, for they deal in staggering numbers of reactant and product molecules. For instance, they sponge up nearly half the carbon dioxide that we humans release each year, as when we burn fossil fuels or burn vast tracts of forests to clear land for farming. Without aquatic photoautotrophs, atmospheric carbon dioxide would accumulate more rapidly and possibly contribute to global warming, as described in Section 49.9. If the atmosphere warms by only a few degrees, then sea levels will rise and all lowlands near the coasts of continents and islands will be submerged.

Even though such global change is a real possibility, tons of industrial wastes, raw sewage, and fertilizers in runoff from croplands drain into the ocean each day. The pollutants seriously alter the chemical composition of seawater. How long will marine photoautotrophs be able to function in this new chemical brew? The answer will have impact on your own life in more ways than one.

a Photosynthetic activity in winter.

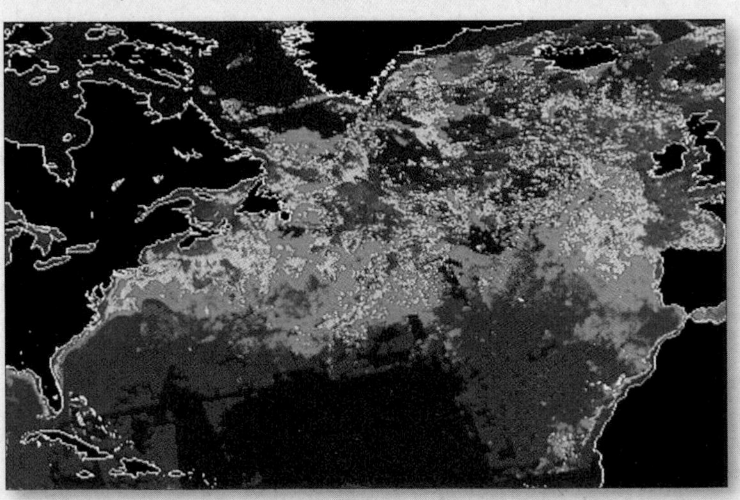

b Photosynthetic activity in spring.

Figure 7.17 Two satellite images that help convey the magnitude of photosynthetic activity during springtime in the North Atlantic portion of the world ocean. In these color-enhanced images, *red-orange* shows where chlorophyll is most concentrated.

Other autotrophs also influence your life in ways you might not expect. At hydrothermal vents, in hot springs, even in waste heaps of coal mines are diverse bacteria classified as *chemo*autotrophs. They, too, use carbon dioxide as a carbon source. But they use inorganic compounds in their environment as energy sources.

For example, some of those bacteria living on the sides of hydrothermal vents strip hydrogen and electrons from hydrogen sulfide (Section 7.8). Other chemoautotrophs living in soil obtain energy from nitrogen-containing wastes and remains of animals and other organisms.

Although they, too, are microscopically small, the chemoautotrophs also exist in monumental numbers. They influence the cycling of nitrogen, phosphorus, and other elements through the biosphere. We will be returning to their impact on the environment. In this unit, we turn next to pathways by which cells release the chemical bond energy of glucose and other biological molecules—the chemical legacy of autotrophs everywhere.

Further reading: Student Guide to InfoTrac on web site

7.10 SUMMARY

1. Cell structure and function are based on organic compounds, the synthesis of which depends on sources of carbon and energy. Plants and other autotrophs use carbon dioxide as their source of carbon. Some types use sunlight as the energy source. Others use inorganic compounds. Animals and other heterotrophs are not self-nourishing; they must get carbon and energy from organic compounds already synthesized by autotrophs.

2. Photosynthesis is the main process by which carbon and energy enter the web of life. It has two stages: the light-dependent and light-independent reactions. Key reactants, intermediates, and products are summarized in Figure 7.18 and in this simplified equation:

$$\underset{\text{WATER}}{12H_2O} + \underset{\text{CARBON DIOXIDE}}{6CO_2} \xrightarrow{\text{LIGHT ENERGY}} \underset{\text{OXYGEN}}{6O_2} + \underset{\text{GLUCOSE}}{C_6H_{12}O_6} + \underset{\text{WATER}}{6H_2O}$$

3. In plants and algae, photosynthesis proceeds inside the organelles called chloroplasts.

a. Two outermost membranes enclose a chloroplast's semifluid interior. A third membrane extends through the stroma as a system of interconnected channels and disks (which are often arranged in stacks).

b. Light-dependent reactions proceed at thylakoids: disk-shaped sacs of the inner membrane system. ATP and NADPH form. Oxygen is released as a by-product.

c. Light-independent reactions occur in the stroma. Glucose molecules form. Each has a phosphate group attached and typically is used at once for the assembly of sucrose, starch, and cellulose. Those three organic compounds are the main end products of photosynthesis.

4. The light-dependent reactions start after pigments absorb wavelengths of visible light from the sun.

a. Chlorophyll *a*, the main photosynthetic pigment, absorbs wavelengths that are most efficient for driving photosynthesis. Like other chlorophylls, it absorbs nearly all wavelengths of visible light except for some yellow-green and green ones, which it reflects or transmits.

b. Accessory pigments can absorb wavelengths that chlorophyll *a* cannot absorb. They include chlorophyll *b* and the carotenoids (such as beta-carotene).

5. In chloroplasts, photosynthetic pigments are part of photosystems. Each photosystem has a cluster of 200 to 300 pigments, most of which are harvesters of energy. Light absorption raises some electrons of the harvester pigments to a higher energy level. Then the energy of excitation passes randomly among the pigments until it reaches a chlorophyll *a* that acts as a reaction center. When suitably activated, that chlorophyll alone releases excited electrons to a nearby acceptor molecule, which is poised at the start of an electron transport system.

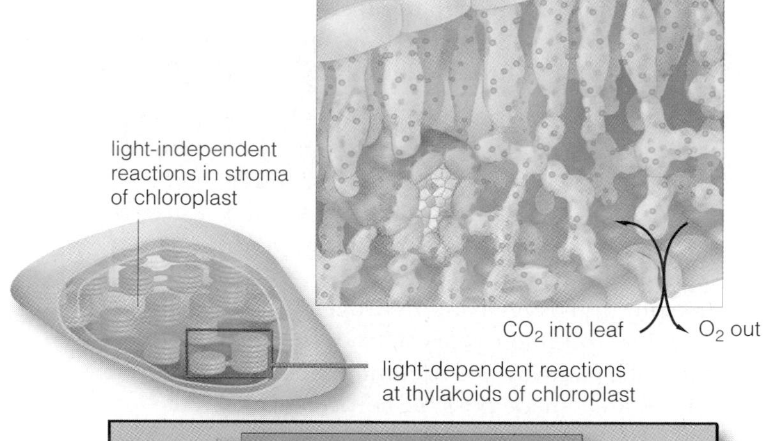

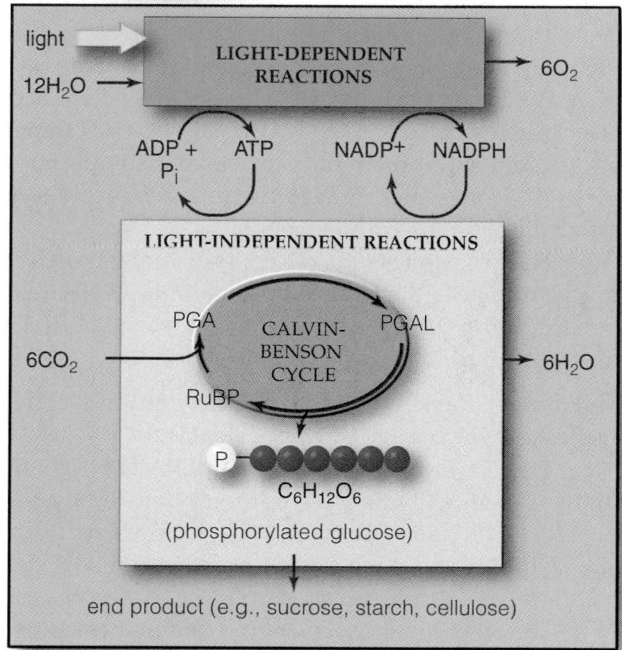

Figure 7.18 Summary of photosynthesis. In the light-dependent reactions, sunlight energy is converted to chemical bond energy of ATP. With a noncyclic pathway, water molecules are split by photolysis. $NADP^+$, a coenzyme, accepts released electrons and hydrogen to form NADPH. Oxygen is released as a by-product.

With the light-independent reactions, a phosphorylated glucose molecule forms in the Calvin–Benson cycle, the RuBP on which the cycle turns is regenerated, and new water molecules form. *Each turn* of the cycle requires one CO_2, three ATP, and two NADPH. Because each glucose molecule has a backbone of six carbon atoms, its formation requires six turns of the cycle.

6. The thylakoid membrane incorporates two types of photosystems: I and II. An ancient pathway of electron flow leading to ATP formation uses only photosystem I. A chlorophyll *a* molecule designated P700 is at this photosystem's reaction center. Photosystems I and II both operate in a noncyclic pathway by which ATP and NADPH form. The reaction center of photosystem II has a chlorophyll *a* designated P680.

7. In the cyclic pathway, sunlight energy is converted to chemical bond energy of ATP when excited electrons

cycle from P700, through a transport system, then back to photosystem I. New electrons are not required; the electrons released from P700 simply return to P700.

8. Noncyclic electron flow has these features:

a. There is a *linear* flow of excited electrons from water to P680 (photosystem II), into a transport system to P700 (photosystem I), then on through a transport system that delivers electrons to $NADP^+$.

b. Water molecules give up the electrons when they are split into oxygen and hydrogen (photolysis).

c. By 1.5 billion years ago, oxygen released by the noncyclic pathway had accumulated in the atmosphere. It ultimately made possible aerobic respiration.

9. With the light-dependent reactions, ATP forms from the cell's pools of ADP and inorganic phosphate.

a. While components of the transport systems accept electrons, they also pick up hydrogen ions (H^+) from the stroma and shunt them into a thylakoid compartment.

b. H^+ from water molecules split during photolysis also collects in the compartment.

c. The activity sets up H^+ concentration and electric gradients across the thylakoid membrane. So H^+ flows out through ATP synthases (into the stroma), and the energy behind this flow drives the formation of ATP.

10. Here are the key points about the light-independent reactions, which proceed in the stroma:

a. The reactions require ATP and NADPH. The ATP delivers energy (by phosphate-group transfers), and the NADPH delivers hydrogen and electrons to reaction sites. Glucose forms and is used in the assembly of starch, cellulose, and other end products of photosynthesis.

b. The reactions are steps of the Calvin–Benson cycle. They start when an enzyme, rubisco, affixes carbon from CO_2 to five-carbon RuBP. The intermediate formed splits into two PGA. ATP phosphorylates each one. NADPH gives electrons and H^+ to the formation of two PGAL.

c. For every six carbon atoms that enter the cycle by way of carbon fixation, twelve PGAL form. Two PGAL are used to produce a six-carbon sugar phosphate. The remainder are used to regenerate the RuBP.

11. Rubisco, a carbon-fixing enzyme, evolved when the atmosphere had far more CO_2 and far less oxygen. Today, when the O_2 level is higher than the CO_2 level in leaves, it attaches oxygen rather than carbon to RuBP. This results in the formation of only one PGA (not two) and glycolate, a compound that cannot be used to form sugars but instead is degraded to carbon dioxide and water. This wasteful process is called photorespiration.

12. Photorespiration predominates in C3 plants such as sunflowers during hot, dry conditions. Then, stomata close and O_2 from photosynthesis builds up inside the leaf to levels higher than the CO_2 levels. Sugarcane and other C4 plants can raise the CO_2 level by fixing carbon twice, in two cell types. CAM plants such as cacti raise it by fixing carbon at night, when stomata are open.

Review Questions

1. A cat eats a bird, which earlier ate a caterpillar that chewed on a weed. Which organisms are autotrophs? Heterotrophs? *CI*

2. Summarize the photosynthesis reactions as an equation. Name where each stage takes place inside a chloroplast. *7.1*

3. Which of the following pigments are most visible in a maple leaf in summer? Which become the most visible in autumn? *7.3*
   a. chlorophylls
   b. phycobilins
   c. anthocyanins
   d. carotenoids

4. How does chlorophyll *a* differ in function from accessory pigments during the light-dependent reactions? *7.3*

5. Fill in Figure 7.19 for the light-dependent reactions. *7.4*

6. With respect to the light-dependent reactions, how do the cyclic and noncyclic pathways of electron flow differ? *7.4*

7. What substance does *not* take part in the Calvin–Benson cycle: ATP, NADPH, RuBP, carotenoids, O_2, CO_2, or enzymes? *7.6*

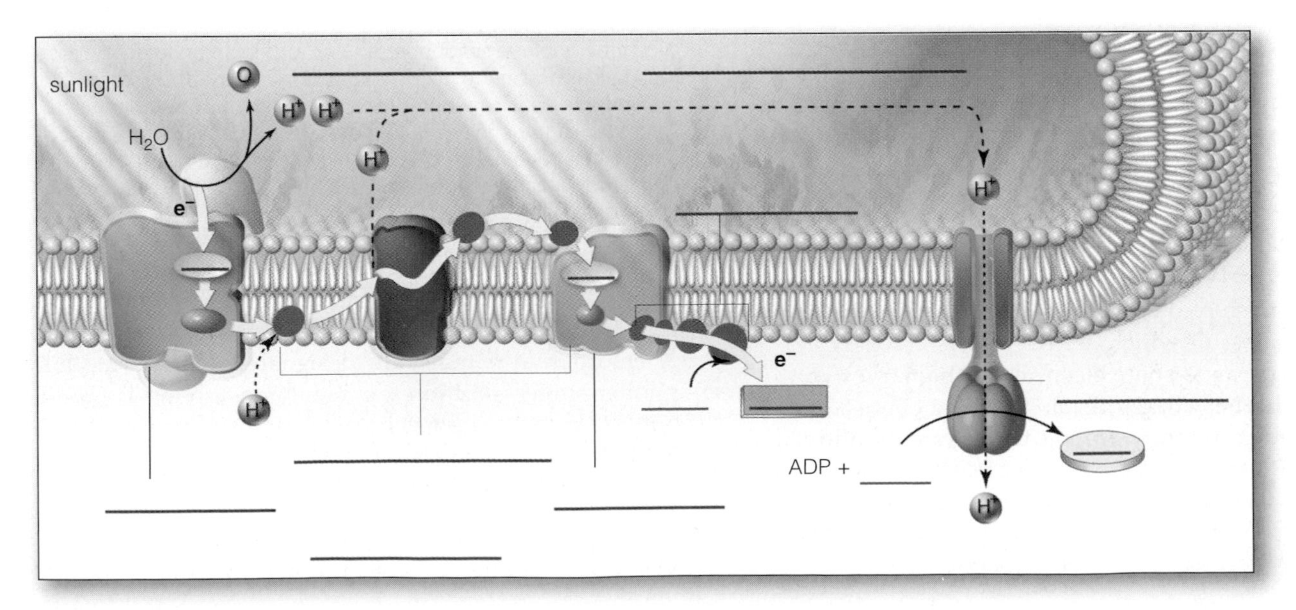

Figure 7.19 Review Figure 7.12, which shows the light-dependent reactions of photosynthesis. Then, on your own, fill in the blanks (*red* lines) with the names of key components and activities.

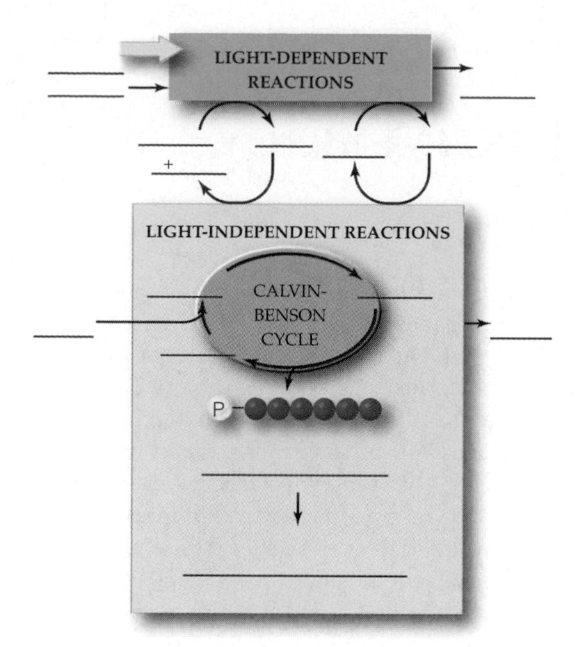

Figure 7.20 Review Figure 7.18. Then, on your own, fill in the blanks (*red* lines) with the names of the key reactants, intermediates, and products of photosynthesis.

8. Fill in the blanks for Figure 7.20. Which substances are the original sources of carbon atoms and hydrogen atoms used in the synthesis of glucose in the Calvin–Benson cycle? *7.1, 7.6*

9. How many carbon atoms from CO_2 must enter the Calvin–Benson cycle to produce one glucose molecule? Why? *7.6*

Self-Quiz (*Answers in Appendix III*)

1. Photosynthetic autotrophs use ______ from the air as a carbon source and ______ as their energy source.

2. In plants, light-dependent reactions proceed at the ______.
   a. cytoplasm c. stroma
   b. plasma membrane d. thylakoid membrane

3. In the light-*dependent* reactions, ______.
   a. carbon dioxide is fixed c. CO_2 accepts electrons
   b. ATP and NADPH form d. sugar phosphates form

4. Identify which of the following substances accumulates inside the thylakoid compartment of chloroplasts during the light-dependent reactions:
   a. glucose c. chlorophyll e. hydrogen ions
   b. carotenoids d. fatty acids

5. When a photosystem absorbs light, ______.
   a. sugar phosphates are produced
   b. electrons are transferred to ATP
   c. RuBP accepts electrons
   d. light-dependent reactions begin

6. The light-*independent* reactions proceed in the ______.
   a. cytoplasm b. plasma membrane c. stroma d. grana

7. The Calvin–Benson cycle starts when ______.
   a. light is available
   b. light is not available
   c. carbon dioxide is attached to RuBP
   d. electrons leave a photosystem

8. ATP phosphorylates ______ in the light-independent reactions.
   a. RuBP b. $NADP^+$ c. PGA d. PGAL

9. Match each event with its most suitable description.
   ____ RuBP used; PGA forms a. photon absorption
   ____ ATP and NADPH used b. noncyclic pathway
   ____ NADPH forms c. CO_2 fixation
   ____ ATP and NADPH form d. PGAL forms
   ____ only ATP forms e. H^+ and e^- to $NADP^+$
   ____ energy matches amount needed to excite electrons f. cyclic pathway

Critical Thinking

1. About 200 years ago, Jan Baptista van Helmont performed an experiment on the nature of photosynthesis. He wanted to know where growing plants acquire the raw materials necessary to increase in size. For his experiment, he planted a tree seedling weighing 5 pounds in a barrel filled with 200 pounds of soil. He watered the tree regularly. Five years passed. Van Helmont again weighed the tree and the soil. The tree weighed 169 pounds, 3 ounces. The soil weighed 199 pounds, 14 ounces. Because tree weight had increased so much and soil weight had decreased so little, he concluded the tree had gained weight after absorbing the water he had added to the barrel.

   Given what you know about the composition of biological molecules, why was van Helmont's conclusion misguided? Reflect on the current model of photosynthesis and then give a more plausible explanation of his results.

2. Like other accessory pigments, carotenoids extend the effective range of light absorption beyond chlorophyll *a*, the main pigment of photosynthesis. They also protect plants from *photo-oxidation*. This destructive process begins when the excitation energy of chlorophylls drives the conversion of oxygen into free radicals, which affect carotenoid synthesis. In fact, plants damaged by photo-oxidation cannot make cartenoids. Grow them in light and they will bleach white and die. Given this result, which molecules in the plant cells are among the first to go?

3. Suppose a garden in your neighborhood is filled with red, white, and blue petunias. Explain the floral colors in terms of which wavelengths of light they are absorbing and reflecting.

4. A busily photosynthesizing plant takes up molecules of CO_2 that have incorporated radioactively labeled carbon atoms ($^{14}CO_2$). Identify the compound in which the labeled carbon will appear first: NADPH, PGAL, pyruvate, or PGA.

Selected Key Terms

absorption spectrum *7.2*
anthocyanin *7.3*
autotroph *CI*
C3 plant *7.7*
C4 plant *7.7*
Calvin–Benson cycle (light-independent reactions) *7.6*
CAM plant *7.7*
carbon fixation *7.6*
carotenoid *7.3*
chlorophyll *7.3*
chloroplast *7.1*
electromagnetic spectrum *7.2*
electron transport system *7.4*
fluorescence *7.3*
heterotroph *CI*
hydrothermal vent *7.8*
light-dependent reactions *7.4*
PGA *7.6*
PGAL *7.6*
photolysis *7.4*
photon *7.2*
photosynthesis *CI*
photosystem *7.3*
phycobilin *7.3*
pigment *7.2*
reaction center *7.4*
rubisco *7.6*
RuBP *7.6*
stoma (stomata) *7.7*
stroma *7.1*
thylakoid *7.1*
wavelength *7.2*

Readings *See also www.infotrac-college.com*

Bazzaz, F., and E. Jajer. January 1992. "Plant Life in a CO_2-Rich World." *Scientific American*, 266:68–74.

Zimmer, C. November 1996. "The Light at the Bottom of the Sea." *Discover*, 63–73.

8 HOW CELLS RELEASE STORED ENERGY

The Killers Are Coming! The Killers Are Coming!

In 1990, thanks to selective breeding experiments gone wrong, descendants of "killer" bees that flew out of South America a few decades earlier buzzed across the border between Mexico and Texas. By 1995, they had invaded 13,287 square kilometers of southern California and were busily setting up colonies. By 1998, when nectar-rich desert flowers bloomed profusely after heavy El Niño storms, the invasion extended even farther west and north than scientists had predicted.

When provoked, the bees behave in a terrifying way. For example, thousands flew into action simply because a construction worker started up a tractor a few hundred yards away from their hive. Agitated bees entered a nearby subway station and started stinging passengers on the platform and in trains. They killed one person and injured a hundred others.

Where did these bees come from? Some queen bees were shipped from Africa to Brazil for breeding experiments in the 1950s. Why? As it happens, honeybees are big business. Besides being a source of nutritious honey, bees are rented to commercial orchards, where their collective pollinating activities may significantly enhance fruit production. For example, enclose a blossoming orchard tree in a pollinator-excluding cage, and less than 1 percent of the tree's flowers will set fruit. But put a hive of honeybees inside the cage with the tree and 40 percent of the flowers will set fruit. Compared to their relatives in Africa, bees in Brazil are rather sluggish pollinators and honey producers. By cross-breeding the two, researchers thought they might come up with a strain of mild-mannered but zippier bees. So they put local bees and imported ones together in netted enclosures, complete with artificial hives. Then they let nature take its course.

Figure 8.1 One of the mild-mannered honeybees buzzing in for a landing on a flower, wings beating with energy provided by ATP. If this were one of its Africanized relatives protecting a hive, possibly you would not stay around to watch the landing. Both kinds of bees look alike. How can we tell them apart? From our own biased perspective, Africanized bees are the ones with an attitude problem.

Twenty-six African queen bees escaped. That was bad enough. Then beekeepers got wind of preliminary experimental results. After learning that the first few generations of offspring were more energetic but not overly aggressive, they imported hundreds of African queens and encouraged them to mate with the locals. And they set off a genetic time bomb.

Before long, African bees became established in commercial hives—and in wild bee populations. Their traits became dominant. The "Africanized" bees do everything other bees do, but they do more of it faster. Their eggs develop into adults more quickly. Adults fly more rapidly, outcompete other bees for nectar, and even die sooner.

When something disturbs their hives or swarms, Africanized bees become extremely agitated. They can remain that way for as long as eight hours. Whereas a mild-mannered honeybee might chase an intruding animal fifty yards or so, a squadron of Africanized bees will chase it a quarter of a mile. If they catch up to it, they collectively can sting it to death.

Doing things faster means having a continuous supply of energy and efficient ways of using it. An Africanized bee's stomach can hold thirty milligrams of sugar-rich nectar—which is enough fuel to fly sixty kilometers. That's more than thirty-five miles! Besides this, compared to other kinds of bees, an Africanized bee's flight muscle cells have larger mitochondria. These organelles specialize in releasing a great deal of energy from sugars and other organic compounds, then converting it to the energy of ATP.

Whenever they tap into the stored energy of organic compounds, Africanized bees reveal their biochemical connection with other organisms. Study a primrose or puppy, a mold growing on stale bread, an amoeba in pondwater, or a bacterium living on your skin, and you will discover that their energy-releasing pathways differ in some details. But all of the pathways require characteristic starting materials. They yield predictable products and by-products. And they yield the universal energy currency of life—ATP.

In fact, throughout the biosphere, organisms put energy and raw materials to use in amazingly similar ways. *At the biochemical level, we find undeniable unity among all forms of life.* We will return to this idea in the concluding section of the chapter.

KEY CONCEPTS

1. The cells of all organisms can release energy stored in glucose and other organic compounds, then use it in ATP production. The energy-releasing pathways differ from one another. But the main types all start with the breakdown of glucose to pyruvate.

2. The initial breakdown reactions, known as glycolysis, can proceed in the presence of oxygen or in its absence. Said another way, these reactions can be the first stage of either aerobic or anaerobic pathways.

3. Two kinds of energy-releasing pathways are entirely anaerobic, from start to finish. We call them fermentation and anaerobic electron transport. They proceed only in the cytoplasm, and none yields more than a small amount of ATP for each glucose molecule metabolized.

4. Another pathway, aerobic respiration, also starts in the cytoplasm. But this one runs to completion in organelles called mitochondria. Compared with the other pathways, aerobic respiration releases far more energy from glucose.

5. Aerobic respiration has three stages. First, pyruvate forms from glucose (during glycolysis). Second, different reactions break down the pyruvate to carbon dioxide. These reactions liberate electrons and hydrogen, which coenzymes deliver to an electron transport system. Third, stepwise electron transfers through the system help set up the conditions that favor ATP formation. Free oxygen accepts the electrons at the end of the line and combines with hydrogen, thereby forming water.

6. Over evolutionary time, photosynthesis and aerobic respiration became linked on a global scale. The oxygen-rich atmosphere, a long-term outcome of photosynthetic activity, sustains aerobic respiration, which has become the dominant energy-releasing pathway. And most kinds of photosynthesizers use carbon dioxide and water from aerobic respiration as raw materials when they synthesize organic compounds:

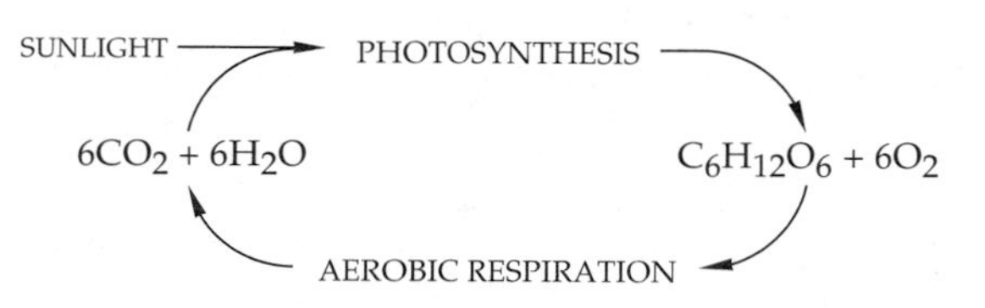

8.1 HOW DO CELLS MAKE ATP?

Organisms stay alive by taking in energy. Plants and all other organisms that engage in photosynthesis get energy from the sun. Animals get energy secondhand, thirdhand, and so on, by eating plants and one another. Regardless of its source, energy must be in a form that can drive thousands of life-sustaining reactions. Energy that becomes converted into the chemical bond energy of adenosine triphosphate—ATP—serves that function.

Plants make ATP during photosynthesis, which they then use to produce glucose and other carbohydrates. Plants and all other organisms also can make ATP by breaking down carbohydrates (glucose especially), fats, and proteins. During the breakdown reactions, electrons are stripped from intermediates, then energy associated with the liberated electrons drives the formation of ATP. Electron transfers of the sort outlined in Section 6.4 are central to these energy-releasing pathways.

Comparison of the Main Types of Energy-Releasing Pathways

The first energy-releasing pathways evolved about 3.8 billion years ago, when conditions were very different on Earth. Because the atmosphere had little free oxygen, the pathways must have been *anaerobic*, which means they could run to completion without utilizing oxygen. Many bacteria and protistans still live in places where oxygen is absent or not always available. They make ATP by anaerobic routes, mainly fermentation pathways and anaerobic electron transport. Some cells in your own body use an anaerobic route for short periods, but only when they are not receiving enough oxygen. Your cells, like most others, rely primarily on **aerobic respiration**, an oxygen-dependent pathway of ATP formation. With each breath you take, you are providing your actively respiring cells with a fresh supply of oxygen.

Make note of this point: *The main energy-releasing pathways all start with the same reactions in the cytoplasm.* During this initial stage of reactions, called **glycolysis**, enzymes cleave and rearrange a glucose molecule into two molecules of **pyruvate**, which has a backbone of three carbon atoms. Once glycolysis is completed, the energy-releasing pathways differ. Most importantly, only the aerobic pathway continues inside a mitochondrion (Figure 8.2). There, oxygen serves as the final acceptor of electrons used during the reactions. The anaerobic pathways start and end in the cytoplasm. A substance other than oxygen is the final electron acceptor.

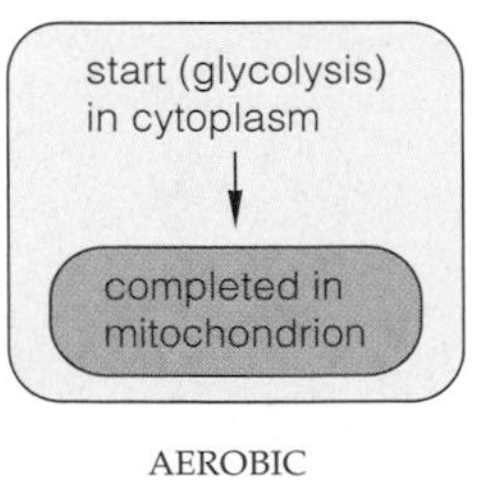

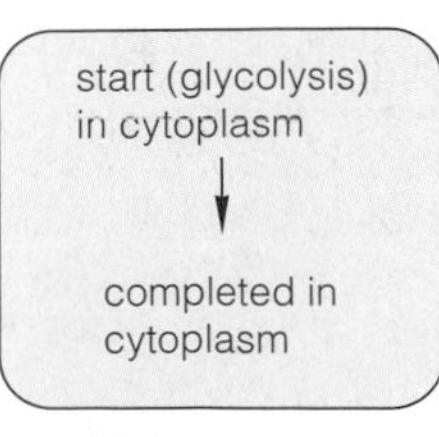

Figure 8.2 Where the aerobic and anaerobic pathways of ATP formation start and finish.

As you examine the energy-releasing pathways in sections to follow, keep in mind that the reaction steps do not proceed by themselves. Enzymes catalyze each step, and the intermediate molecules formed at one step serve as substrates for the next enzyme in the pathway.

Overview of Aerobic Respiration

Of all energy-releasing pathways, aerobic respiration gets the most ATP for each glucose molecule. Whereas anaerobic routes typically have a net yield of two ATP molecules, the aerobic route commonly yields thirty-six or more. If you were a bacterium, you wouldn't require much ATP. Being far larger, more complex, and highly active, you rely absolutely on the aerobic route's high yield. When a glucose molecule is the starting material, aerobic respiration can be summarized this way:

$$\underset{\text{GLUCOSE}}{C_6H_{12}O_6} + \underset{\text{OXYGEN}}{6O_2} \longrightarrow \underset{\text{CARBON DIOXIDE}}{6CO_2} + \underset{\text{WATER}}{6H_2O}$$

However, as you can see, the summary equation only tells us what the substances are at the start and finish of the pathway. In between are three reaction stages.

Further reading: Student Guide to InfoTrac on web site →

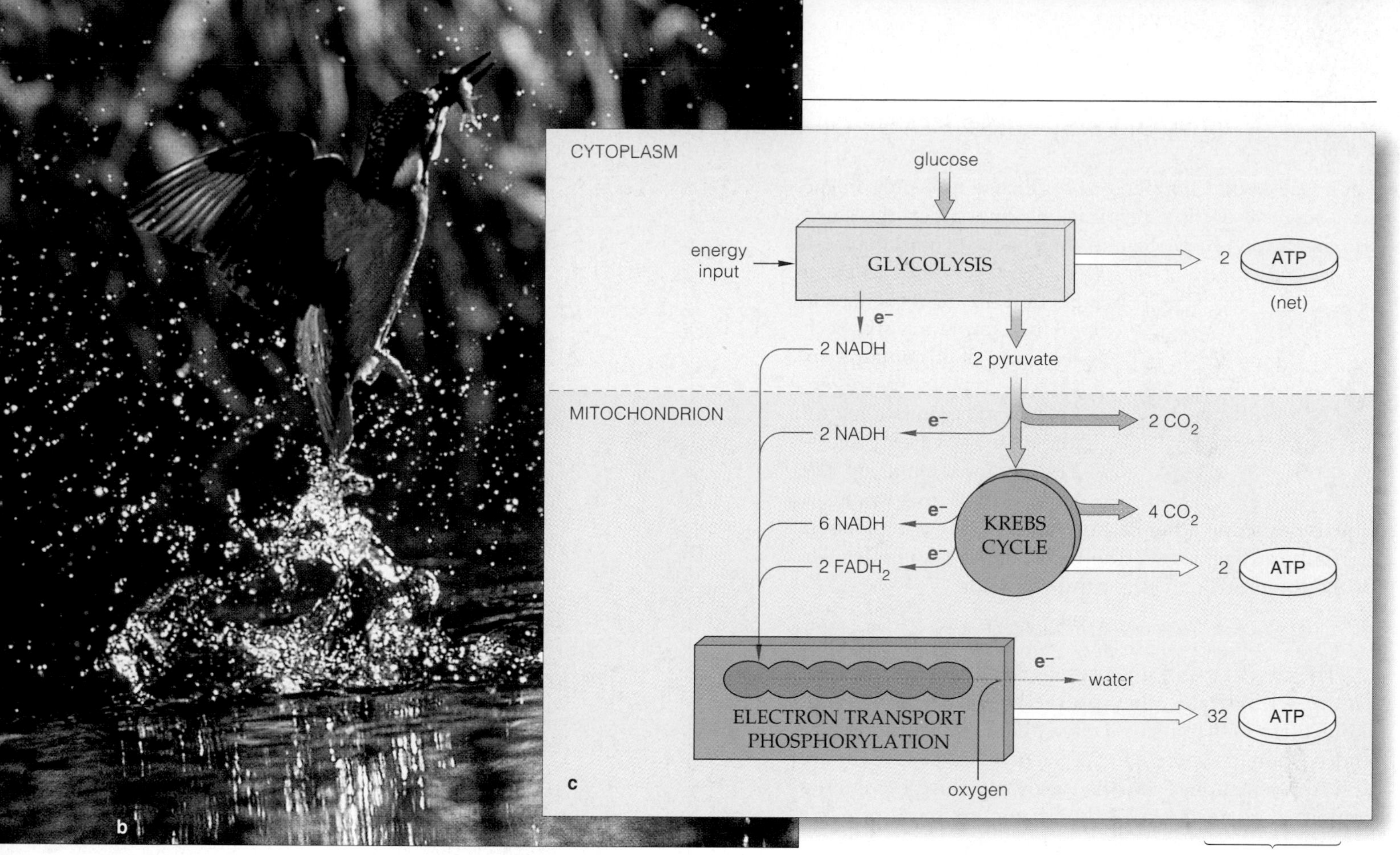

Figure 8.3 Overview of aerobic respiration, which has three stages. From start to finish, the typical net energy yield from each glucose molecule is thirty-six ATP. Only this pathway delivers enough ATP to construct and maintain giant redwoods and all other large, multicelled organisms. It alone delivers enough ATP for highly active animals, such as bees, humans (**a**), and kingfishers (**b**) and other birds.

(**c**) In the first stage (glycolysis), enzymes partially break down glucose to pyruvate. In the second stage, which is mainly the Krebs cycle, enzymes completely break down pyruvate to carbon dioxide. NAD^+ and FAD pick up electrons and hydrogen stripped from intermediates at both stages. In the final stage (electron transport phosphorylation), those reduced coenzymes (NADH and $FADH_2$) give up electrons and hydrogen to a transport system. Energy released during the flow of electrons through the system drives ATP formation. Oxygen accepts the electrons at the end of the third stage.

Let's use Figure 8.3 and the following descriptions as a brief overview of these reactions. The initial stage, again, is glycolysis. Most of the second stage consists of a cyclic pathway, the **Krebs cycle**, when enzymes break down pyruvate to carbon dioxide and water.

NAD^+ (nicotinamide adenine dinucleotide) and **FAD** (flavin adenine dinucleotide) take part in glycolysis and the Krebs cycle. These organic compounds, derived from vitamins, are *coenzymes* (Section 6.7). They help enzymes by accepting electrons (e^-) and hydrogen removed from intermediates, then transferring the electrons elsewhere. Unbound hydrogen atoms, recall, are naked protons, or hydrogen ions (H^+). So they tag along with the oppositely charged electrons. In their reduced form, the coenzymes are abbreviated NADH and $FADH_2$.

Not much ATP forms during glycolysis or the Krebs cycle. The large energy harvest comes *after* coenzymes deliver their cargo to an electron transport system.

In the third stage, the transport system functions as machinery for **electron transport phosphorylation**. It sets up H^+ concentration and electric gradients, which drive ATP formation at nearby membrane proteins. It is during this final stage that so many ATP molecules are produced. As it ends, oxygen inside the mitochondrion accepts the "spent" electrons from the last component of the transport system. Oxygen picks up hydrogen at the same time and thereby forms water.

Cells drive nearly all metabolic activities by releasing energy from glucose and other organic compounds and converting it to chemical bond energy of ATP.

All of the main energy-releasing pathways start inside the cytoplasm with glycolysis, a stage of reactions that break down glucose to pyruvate.

The most common anaerobic pathways, which include the fermentation routes, end in the cytoplasm. Each has a net energy yield of two ATP.

Aerobic respiration, an oxygen-dependent pathway, runs to completion in the mitochondrion. From start (glycolysis) to finish, it commonly has a net energy yield of thirty-six ATP.

8.2 GLYCOLYSIS: FIRST STAGE OF ENERGY-RELEASING PATHWAYS

Let's track what happens to a glucose molecule in the first stage of aerobic respiration. Remember, the same things happen to glucose in the anaerobic pathways.

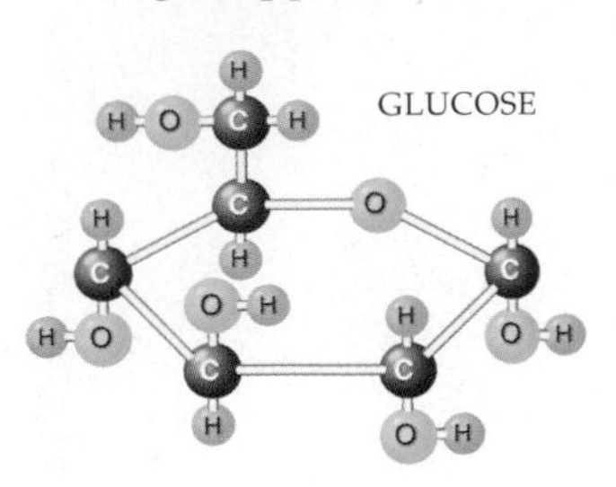

As described earlier, in Section 3.2, glucose is one of the simple sugars. Each molecule of it contains six carbon, twelve hydrogen, and six oxygen atoms, all joined by covalent bonds. The carbons make up the backbone. With glycolysis, glucose or some other carbohydrate in the cytoplasm is partially broken down, the result being two molecules of the three-carbon compound pyruvate:

glucose → glucose–6–phosphate → 2 pyruvate

The first steps of glycolysis are *energy-requiring*. As Figure 8.4 indicates, they proceed only when two ATP molecules each transfer a phosphate group to glucose and so donate energy to it. Such transfers, recall, are "phosphorylations." In this case, they raise the energy content of glucose to a level high enough to allow entry into the *energy-releasing* steps of glycolysis.

The first energy-releasing step cleaves the activated glucose into two molecules. We can call each of these PGAL (phosphoglyceraldehyde). Each PGAL becomes converted to an unstable intermediate that allows ATP to form by giving up a phosphate group to ADP. The next intermediate in the sequence does the same thing.

Thus, a total of four ATP form by **substrate-level phosphorylation**. We define this metabolic event as the direct transfer of a phosphate group from a substrate of a reaction to some other molecule—in this case, ADP. Remember, though, two ATP were invested to jump-start the reactions. So the *net* energy yield is only two ATP.

Meanwhile, the coenzyme NAD^+ picks up electrons and hydrogen atoms liberated from each PGAL, thus becoming NADH. When the NADH gives up its cargo at a different reaction site, it reverts to NAD^+. Said another way, like other coenzymes, NAD^+ is reusable.

In sum, glycolysis converts energy stored in glucose to a transportable form of energy, in ATP. NAD^+ picks up electrons and hydrogen stripped from glucose. These have roles in the next stage of reactions. So do the end products of glycolysis—the two molecules of pyruvate.

Glycolysis is an energy-releasing stage of reactions in which glucose or some other carbohydrate is partially broken down to two molecules of pyruvate.

Two NADH and four ATP form. However, when we subtract the two ATP required to start the reactions, the *net* energy yield of glycolysis is two ATP per glucose molecule.

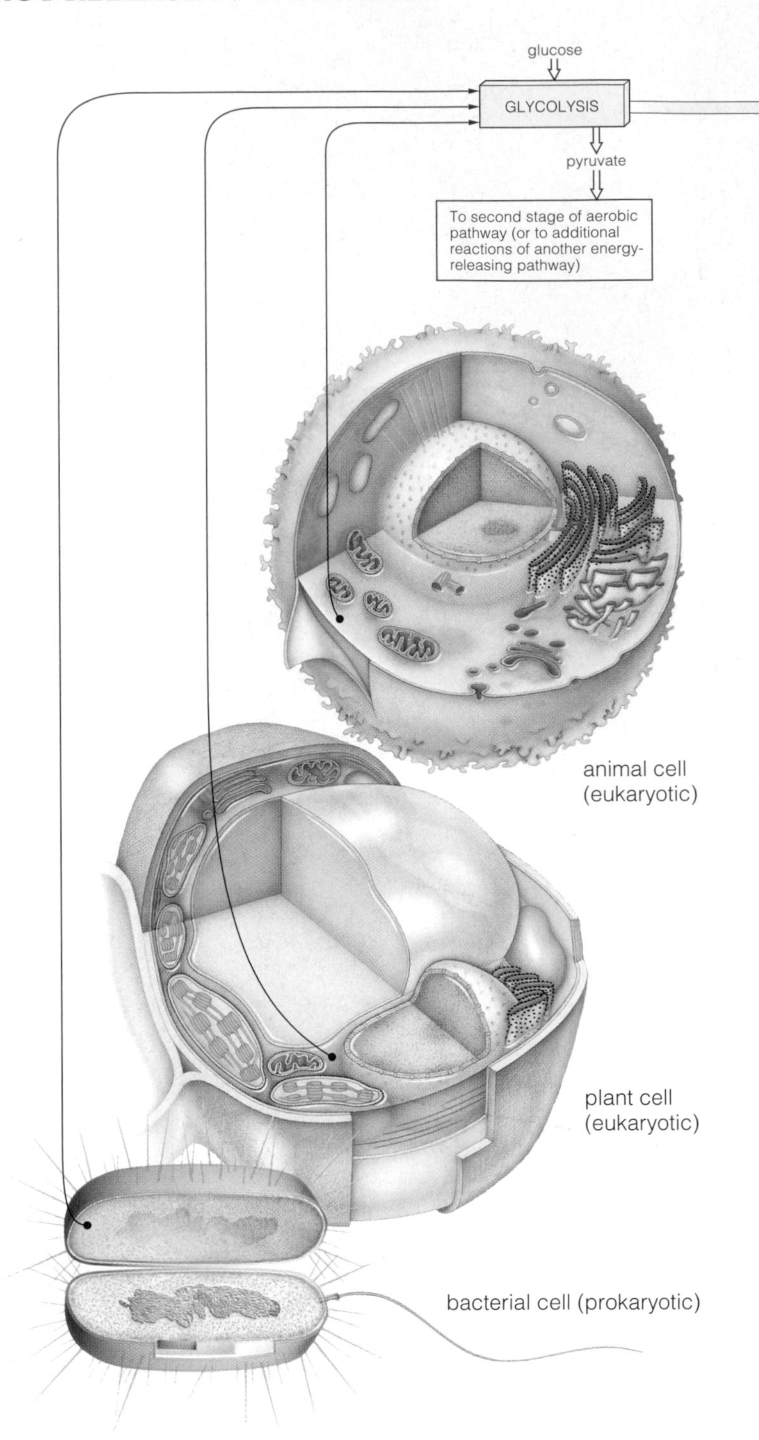

Figure 8.4 Glycolysis, first stage of the main energy-releasing pathways. The reaction steps proceed inside the cytoplasm of every living prokaryotic and eukaryotic cell. In this example, glucose is the starting material. By the time the reactions end, two pyruvate, two NADH, and four ATP have been produced. Cells invest two ATP to start glycolysis, however, so the *net* energy yield of glycolysis is two ATP. For an expanded picture of the reactions, refer to Appendix V (Figure A).

Depending on the type of cell and on environmental conditions, the pyruvate may enter the second set of reactions of the aerobic pathway, which includes the Krebs cycle. Or it may be used in other reactions, such as those of fermentation pathways.

Further reading: Student Guide to InfoTrac on web site

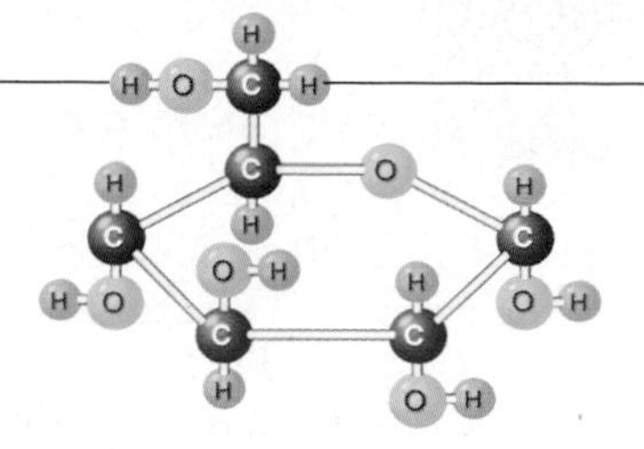

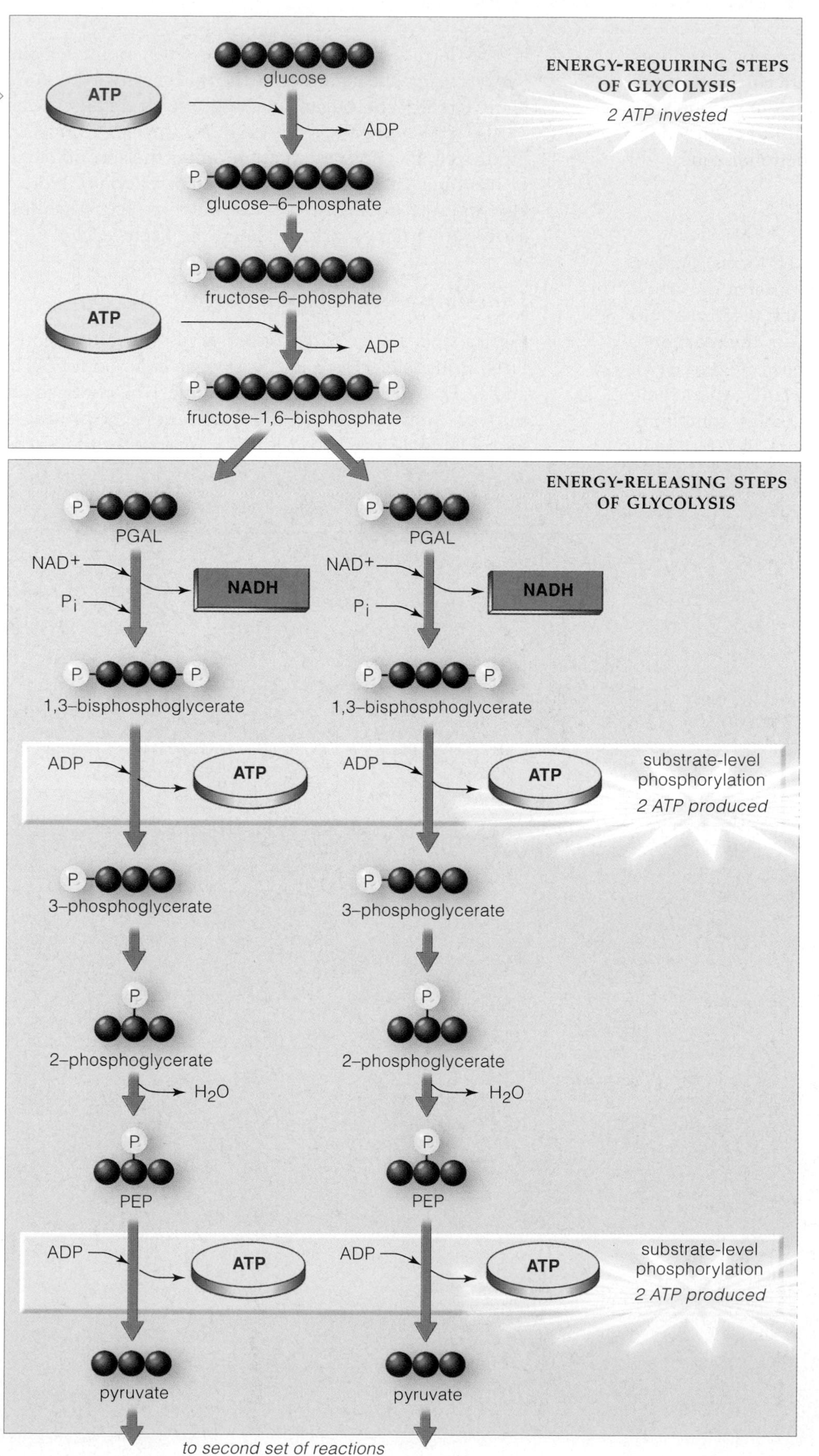

a This diagram tracks the fate of the six carbon atoms (*red* spheres) of a glucose molecule, shown above. Glycolysis starts with an energy investment of two ATP.

b A phosphate-group transfer from one of the ATP molecules to glucose causes atoms in glucose to undergo rearrangements.

c A phosphate-group transfer from another ATP causes rearrangements that form the intermediate fructose–1,6–bisphosphate.

d The intermediate splits at once into two molecules, each with a three-carbon backbone. We can call these two PGAL.

e Each PGAL gives up two electrons and a hydrogen atom to NAD^+, thus forming two NADH.

f Each PGAL also combines with inorganic phosphate (P_i) in the cytoplasm and then makes a phosphate-group transfer to ADP.

g *Thus two ATP have formed by the direct transfer of phosphate from two intermediate molecules that served as substrates for the reactions.* Formation of two ATP means the original energy investment of two ATP has been paid off.

h During the next two enzyme-mediated reactions, the two intermediates each release a hydrogen atom and an —OH group, which combine to form water.

i Two 3–phosphoenolpyruvate (PEP) molecules result. Each PEP makes a phosphate-group transfer to ADP.

j *Once again, two ATP have formed by substrate-level phosphorylation.*

k In sum, the net energy yield from glycolysis is two ATP for each glucose molecule entering the reactions. Two molecules of pyruvate, the end product, may enter the next set of reactions in an energy-releasing pathway.

8.3

SECOND STAGE OF THE AEROBIC PATHWAY

Suppose two pyruvate molecules formed by glycolysis leave the cytoplasm and enter a **mitochondrion** (plural, mitochondria). In this organelle, both the second and third stages of the aerobic pathway run to completion. Figure 8.5 shows its structure and functional zones.

Preparatory Steps and the Krebs Cycle

During the second stage, a bit more ATP forms. Carbon atoms depart from the pyruvate, in the form of carbon dioxide. And coenzymes latch onto the electrons and hydrogen stripped from intermediates of the reactions.

In a few preparatory reactions, an enzyme removes a carbon atom from each pyruvate molecule. An enzyme helper, coenzyme A, becomes **acetyl–CoA** by combining with the remaining two-carbon fragment. The fragment is transferred to **oxaloacetate**, the entry point for the Krebs cycle. The name of this cyclic pathway honors Hans Krebs, who began working out its details in the 1930s. It also is called the citric acid cycle. Notice, in Figure 8.6, that *six* carbon atoms enter the second stage of reactions (three in each pyruvate backbone). Notice also that *six* depart, in six carbon dioxide molecules, during the preparatory steps and the Krebs cycle.

Functions of the Second Stage

The second stage of reactions serves three functions. First, it loads electrons and hydrogen onto both NAD^+ and FAD, which results in NADH and $FADH_2$. Second, through substrate-level phosphorylations, it produces two ATP molecules. And third, it rearranges the Krebs

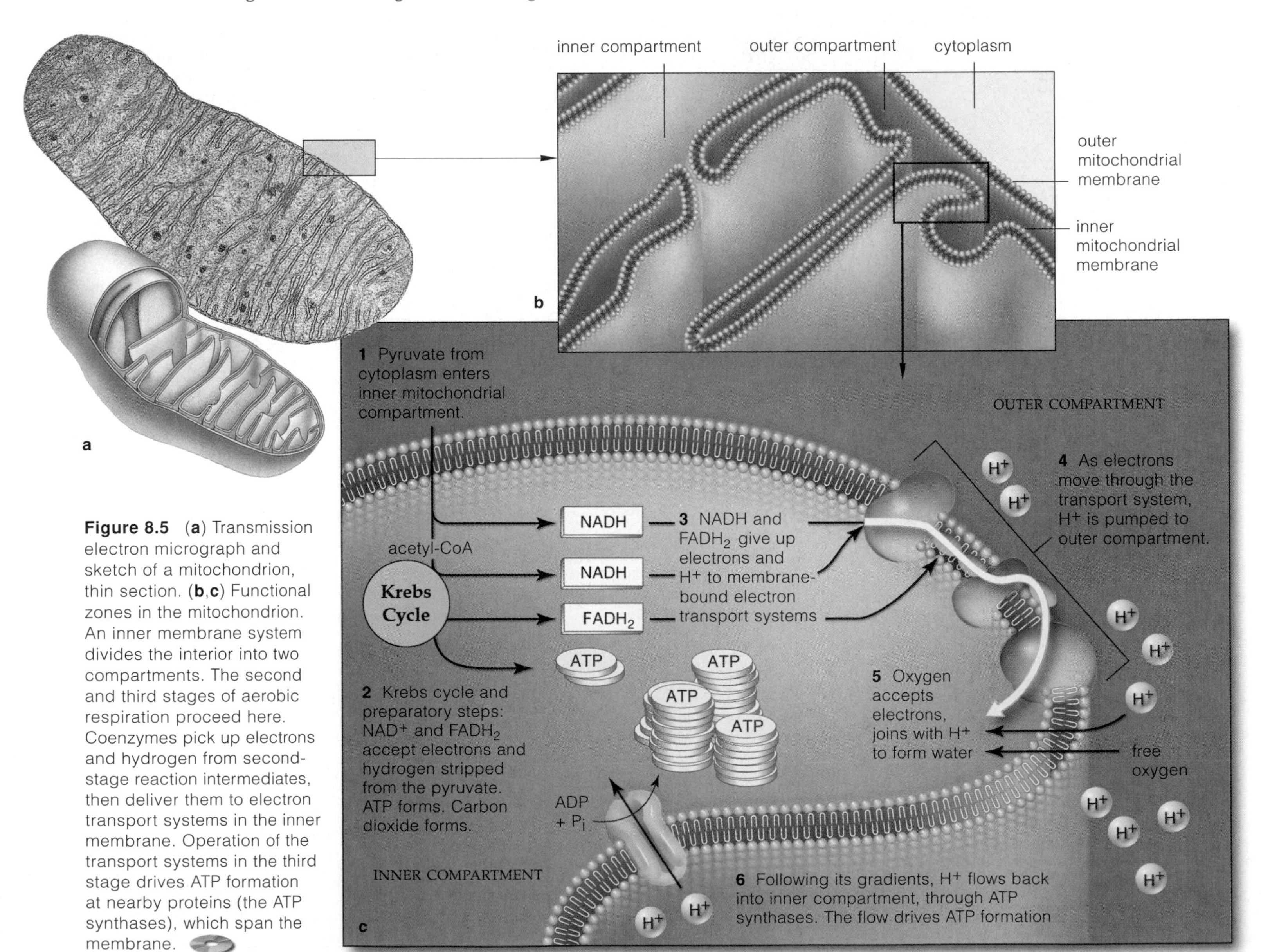

Figure 8.5 (**a**) Transmission electron micrograph and sketch of a mitochondrion, thin section. (**b**,**c**) Functional zones in the mitochondrion. An inner membrane system divides the interior into two compartments. The second and third stages of aerobic respiration proceed here. Coenzymes pick up electrons and hydrogen from second-stage reaction intermediates, then deliver them to electron transport systems in the inner membrane. Operation of the transport systems in the third stage drives ATP formation at nearby proteins (the ATP synthases), which span the membrane.

Further reading: Student Guide to InfoTrac on web site →

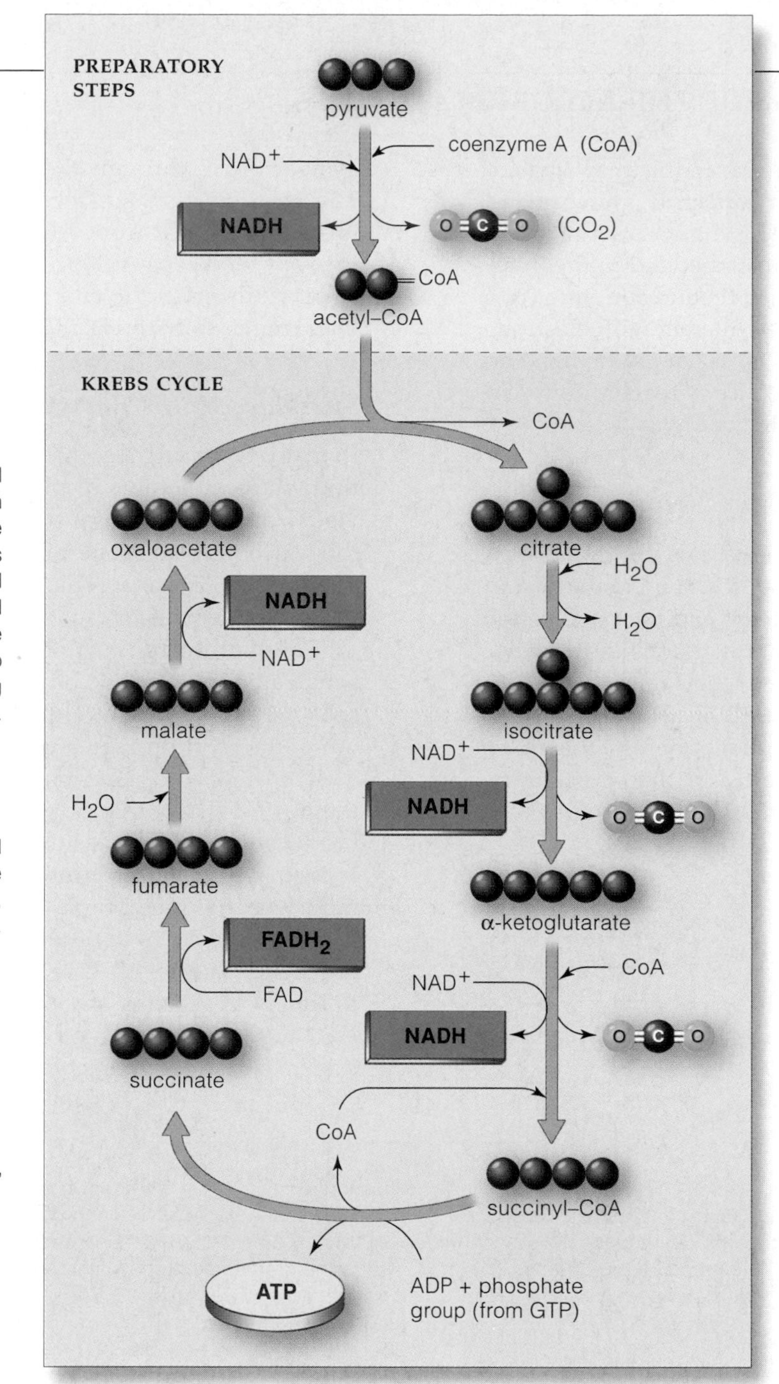

a Pyruvate from glucose enters a mitochondrion. It undergoes initial conversions before entering cyclic reactions (the Krebs cycle).

b Pyruvate is stripped of a carboxyl group (COO^-), which departs as CO_2. It also gives up hydrogen and electrons to NAD^+, forming NADH. Then a coenzyme joins with the remaining two-carbon fragment, forming acetyl–CoA.

c The acetyl–CoA transfers its two-carbon group to oxaloacetate, a four-carbon compound that is the entry point into the Krebs cycle. The result is citrate, with a six-carbon backbone. Addition and then removal of H_2O changes citrate to isocitrate.

d Isocitrate enters conversion reactions, and a COO^- group departs (as CO_2). Hydrogen and electrons are transferred to NAD^+, forming NADH.

e Another COO^- group departs (as CO_2) and another NADH forms. The resulting intermediate attaches to a coenzyme A molecule, forming succinyl–CoA. *At this point, three carbon atoms have been released, balancing out the three that entered the mitochondrion (in pyruvate).*

f Now the attached coenzyme is replaced by a phosphate group (donated by the substrate GTP). That phosphate group becomes attached to ADP. Thus, for each turn of the cycle, one ATP forms by substrate-level phosphorylation.

g Electrons and hydrogen are transferred to FAD, forming $FADH_2$.

h In the final conversion reactions, the oxaloacetate is regenerated, and hydrogen and electrons are transferred to NAD^+, forming NADH.

Figure 8.6 Second stage of aerobic respiration: the Krebs cycle and reaction steps that precede it. For each three-carbon pyruvate molecule entering the cycle, three CO_2, one ATP, four NADH, and one $FADH_2$ molecules form. The steps shown proceed *twice*. Why? Glucose, remember, was broken down earlier to *two* pyruvate molecules.

cycle intermediates into oxaloacetate. Cells have only so much oxaloacetate, which must be regenerated to keep the cyclic reactions going.

The two ATP that form in the second stage do not add much to the small yield from glycolysis. However, for each glucose molecule, *many* coenzymes pick up electrons and hydrogen for transport to the sites of the third and final stage of the aerobic pathway:

Glycolysis:	2 NADH
Pyruvate conversion before Krebs cycle:	2 NADH
Krebs cycle:	2 $FADH_2$ + 6 NADH
Coenzymes sent to third stage:	2 $FADH_2$ + 10 NADH

Overall, these are the key points to remember about the second stage of aerobic respiration:

During the second stage of aerobic respiration, two pyruvate molecules from glycolysis enter a mitochondrion.

Each pyruvate molecule gives up a carbon atom, then its two-carbon remnant enters the Krebs cycle. All of the carbon atoms of pyruvate eventually end up in carbon dioxide.

The second stage yields only two ATP. But the reactions regenerate oxaloacetate, the entry point for the Krebs cycle. And many coenzymes pick up electrons and hydrogen that were stripped from substrates, for delivery to the final stage of the pathway.

THIRD STAGE OF THE AEROBIC PATHWAY

ATP production goes into high gear in the third stage of the aerobic pathway. Electron transport systems and neighboring proteins called ATP synthases serve as the production machinery. They are embedded in the inner membrane that divides the mitochondrion into two compartments (Figure 8.7). They interact with electrons and unbound hydrogen—that is, H^+ ions. Remember, coenzymes deliver this bounty from reaction sites of the first two stages of the aerobic pathway.

Electron Transport Phosphorylation

Briefly, during the final stage, electrons get transferred from one molecule of each transport system to the next in line. When certain molecules accept and then donate the electrons, they also pick up hydrogen ions in the inner compartment. Shortly afterward, they release them to the outer compartment. Their shuttling action sets up H^+ concentration and electric gradients across the inner mitochondrial membrane. Nearby in that membrane, the ions follow the gradients and flow back to the inner compartment, through the interior of ATP synthases. The H^+ flow through these transport proteins drives formation of ATP from ADP and unbound phosphate. Free oxygen keeps ATP production going. It withdraws spent electrons at the end of the transport systems and then combines with H^+. Water is the result.

Summary of the Energy Harvest

In many types of cells, thirty-two ATP form during the third stage of aerobic respiration. Add these to the net yield from the preceding stages, and the total harvest is thirty-six ATP from one glucose molecule (Figure 8.8). That's a lot! An anaerobic pathway may use eighteen glucose molecules to produce the same amount of ATP.

Think of thirty-six ATP as a typical yield only. The actual amount depends on cellular conditions, as when cells require a given intermediate elsewhere and pull it out of the reaction sequence.

The yield also depends on how particular cells use the NADH that formed during glycolysis. Any NADH produced in the cytoplasm can't enter a mitochondrion. It can only deliver electrons and hydrogen *to* the outer mitochondrial membrane, where proteins shuttle them across to NAD^+ or FAD molecules already inside this organelle. Then both coenzymes deliver the electrons to transport systems of the inner membrane. However, FAD puts them at a *lower* entry point in the transport system, so *its* deliveries produce less ATP (Figure 8.8).

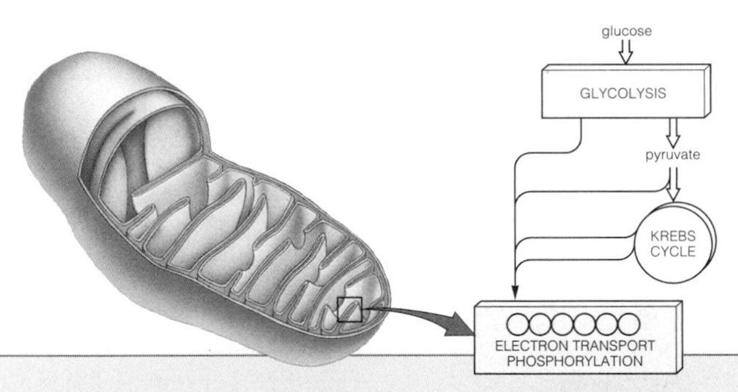

Figure 8.7 Electron transport phosphorylation, the third and final stage of aerobic respiration. The reactions proceed at electron transport systems and at ATP synthases, a type of transport protein, in the inner mitochondrial membrane. Each electron transport system consists of specific enzymes, cytochromes, and other proteins that act in sequence.

The inner membrane functionally divides the mitochondrion into two compartments. The third-stage reactions start in the inner compartment, when NADH and $FADH_2$ give up electrons and hydrogen to transport systems. Electrons are transferred *through* the system, but electron transport proteins drive unbound hydrogen (H^+) *to* the outer compartment:

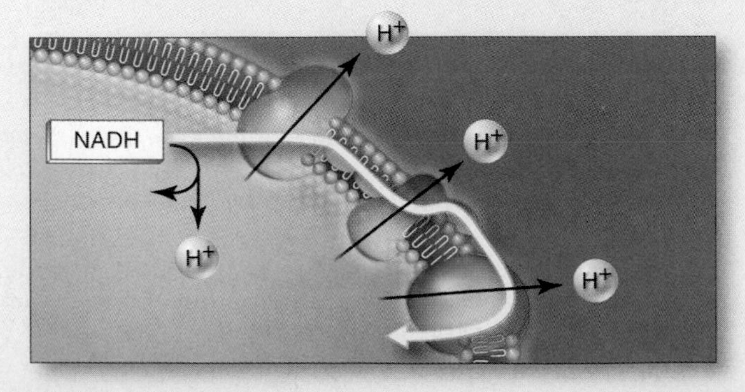

In short order, there is a higher concentration of H^+ in the outer compartment compared to the inner one. Concentration and electric gradients now exist across the membrane. The ions follow the gradients and flow across the membrane, through the interior of the ATP synthases. Energy associated with the flow drives the formation of ATP from ADP and unbound phosphate (P_i). Hence the name, electron transport *phosphorylation*:

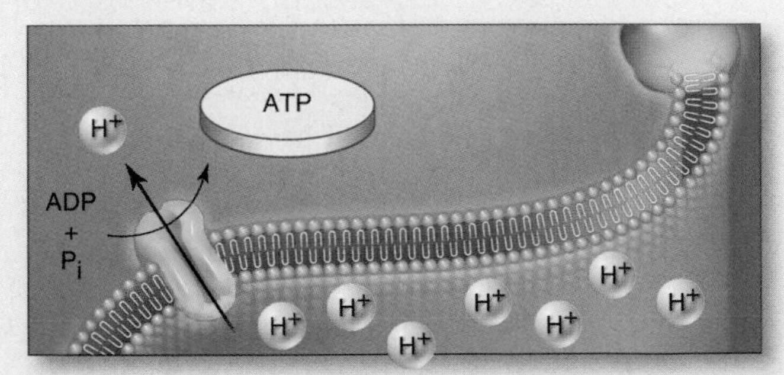

Do these metabolic events sound familiar? They should. Recall, from Section 7.5, that ATP forms in much the same way inside chloroplasts. By the *chemiosmotic* model, H^+ concentration and electric gradients across a cell membrane drive ATP formation. The model applies also to mitochondria, although the ions flow in the opposite direction compared to chloroplasts.

Further reading: Student Guide to InfoTrac on web site →

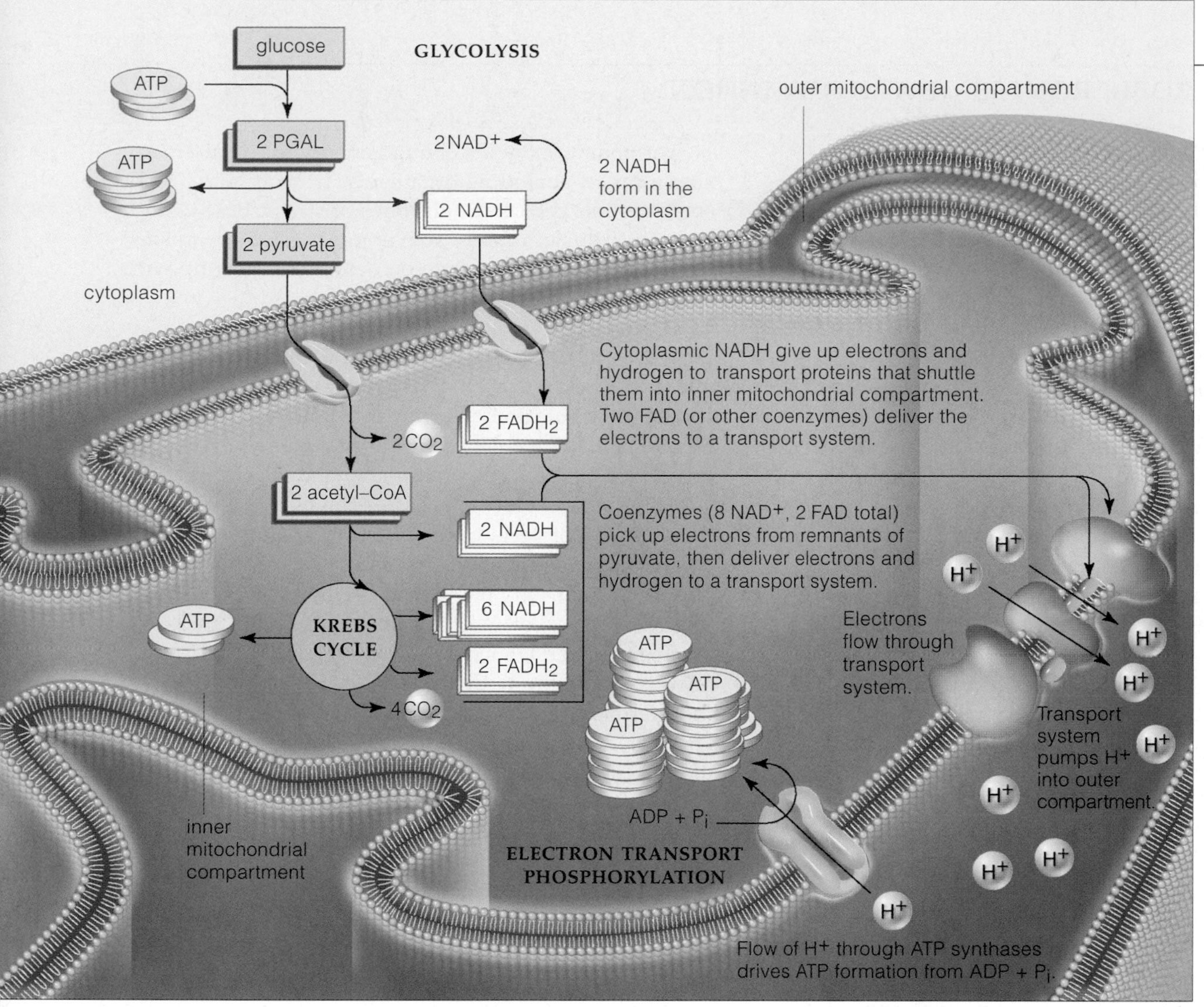

a Two ATP formed in first stage in cytoplasm (during glycolysis, by *substrate-level* phosphorylations).

b NADH that formed in cytoplasm during first stage delivers electrons and hydrogen that help drive the formation of four ATP during third stage at the inner mitochondrial membrane (by *electron transport* phosphorylations).

c Two ATP form at second stage in mitochondrion (by *substrate-level* phosphorylations of Krebs cycle).

d Coenzymes from Krebs cycle and its preparatory steps deliver electrons and hydrogen that drive formation of twenty-eight ATP at third stage (by *electron transport* phosphorylations at the inner mitochondrial membrane).

36 ATP

TYPICAL NET ENERGY YIELD

Figure 8.8 Summary of the harvest from the energy-releasing pathway of aerobic respiration. Commonly, thirty-six ATP form for each glucose molecule that enters the pathway. But the net yield varies according to shifting concentrations of reactants, intermediates, and end products of the reactions. It also varies among different types of cells.

As you read earlier, cells differ in how they use the NADH from glycolysis, which cannot enter mitochondria. At the outer mitochondrial membrane, these NADH give up electrons and hydrogen to transport proteins, which shuttle the electrons and hydrogen across the membrane. NAD^+ or FAD already inside the mitochondrion accept them, thus forming NADH or $FADH_2$.

Any NADH inside the mitochondrion delivers electrons to the highest possible entry point into a transport system. When it does, enough H^+ is pumped across the inner membrane to make *three* ATP. By contrast, any $FADH_2$ delivers them to a lower entry point. Fewer hydrogen ions can be pumped, so only *two* ATP can form.

In liver, heart, and kidney cells, for example, electrons and hydrogen from glycolysis enter the highest entry point of transport systems, so the energy harvest is thirty-eight ATP. More commonly, as in skeletal muscle and brain cells, they are transferred to FAD—so the harvest is thirty-six ATP.

One final point. Glucose, recall, has more energy (stored in more covalent bonds) than carbon dioxide or water. About 686 kilocalories of energy are released when glucose is broken down to those more stable end products. Much of this energy escapes (as heat), but about 7.5 kilocalories are conserved in every mole of ATP. So when 36 ATP form through breakdown of a glucose molecule, the energy-conserving efficiency of aerobic respiration is $36 \times 7.5 / 686 \times 100$, or 39 percent.

In the final stage of the aerobic pathway, coenzymes deliver electrons to transport systems of the inner mitochondrial membrane. As electrons move through the system, they set up H^+ gradients that drive ATP formation at nearby proteins in the membrane. Oxygen is the final electron acceptor.

Again, from start (glycolysis in the cytoplasm) to finish (in mitochondria), the pathway commonly has a net yield of thirty-six ATP for every glucose molecule metabolized.

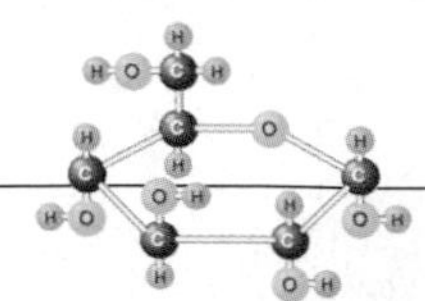

ANAEROBIC ROUTES OF ATP FORMATION

So far, we have tracked the fate of a glucose molecule through the pathway of aerobic respiration. We turn now to its use as a substrate for fermentation pathways. Remember, these are anaerobic pathways; they do *not* use oxygen as the final acceptor of the electrons that ultimately drive the ATP-forming machinery.

Fermentation Pathways

Diverse kinds of organisms use fermentation pathways. Many are bacteria and protistans that make their homes in marshes, bogs, mud, deep-sea sediments, the animal gut, canned foods, sewage treatment ponds, and other oxygen-free settings. Some fermenters actually will die if they are exposed to oxygen. Bacteria responsible for many diseases, including botulism and tetanus, are like this. Other kinds of fermenters, including the bacterial "employees" of yogurt manufacturers, are indifferent to the presence of oxygen. Still others can use oxygen, but they also might use a fermentation pathway when oxygen becomes scarce. Even your muscle cells do this.

As is true of aerobic respiration, glycolysis serves as the first stage of the fermentation pathways. Here also, enzymes split glucose and rearrange the fragments into two pyruvate molecules. Here again, two NADH form, and the net energy yield is two ATP.

However, fermentation reactions do not completely break down glucose to carbon dioxide and water, and they produce no more ATP beyond the tiny yield from glycolysis. *The final steps of fermentation serve only to regenerate NAD^+, the coenzyme with a central role in the breakdown reactions.*

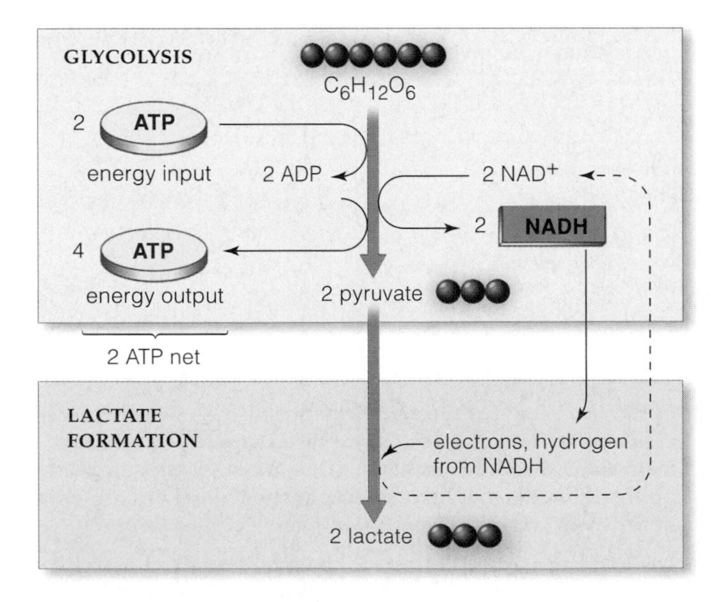

Figure 8.9 Lactate fermentation. In this anaerobic pathway, electrons end up in lactate, the reaction product.

Fermentation yields enough energy to sustain many single-celled anaerobic organisms. It even helps carry some aerobic cells through times of stress. But it is not enough to sustain large, active, multicelled organisms, this being one reason why you never will meet up with an anaerobic elephant.

LACTATE FERMENTATION With these points in mind, take a look at Figure 8.9, which tracks the main steps of **lactate fermentation**. During this anaerobic pathway, the *pyruvate* molecules from the first stage of reactions (glycolysis) accept the hydrogen and electrons from NADH. The transfer regenerates the NAD^+ and, at the same time, converts each pyruvate to a three-carbon compound called lactate. You may hear people refer to this compound as "lactic acid." However, its ionized form (lactate) is far more common in cellular fluids.

Some bacteria, such as *Lactobacillus*, rely exclusively on this anaerobic pathway. Left to their own devices, their fermentation activities often spoil food. Yet certain fermenters have commercial uses, as when they break down glucose in huge vats where cheeses, yogurt, and sauerkraut are produced.

In humans, rabbits, and many other animals, some types of cells also may switch to lactate fermentation for a quick fix of ATP. When your own demands for energy are intense but brief—say, during a short race —muscle cells use this pathway. They cannot do so for long; they would throw away too much of glucose's stored energy for too little ATP. When their stores of glucose become depleted, muscles fatigue and lose their ability to contract.

ALCOHOLIC FERMENTATION In the anaerobic route called **alcoholic fermentation**, enzymes convert each pyruvate molecule that formed during glycolysis to an intermediate form: acetaldehyde. The NADH transfers electrons and hydrogen to this form and so converts it to an alcoholic end product—ethanol (Figure 8.10).

Certain species of single-celled fungi called yeasts are renowned for their use of this pathway. One type, *Saccharomyces cerevisiae,* makes bread dough rise. Bakers mix the yeast with sugar, then blend both into dough. When yeast cells degrade the sugar, they release carbon dioxide. Bubbles of the gas expand the dough (make it rise). Oven heat forces the gas out of the dough, and a porous product remains.

Beer and wine producers use yeasts on a large scale. Vintners use wild yeasts living on grapes and cultivated strains of *S. ellipsoideus,* which remain active until the alcohol concentration in wine vats exceeds 14 percent. (Wild yeast cells die when the concentration exceeds 4 percent.) That is why some birds get drunk on naturally fermenting berries. That is why landscapers don't plant

Further reading: Student Guide to InfoTrac on web site →

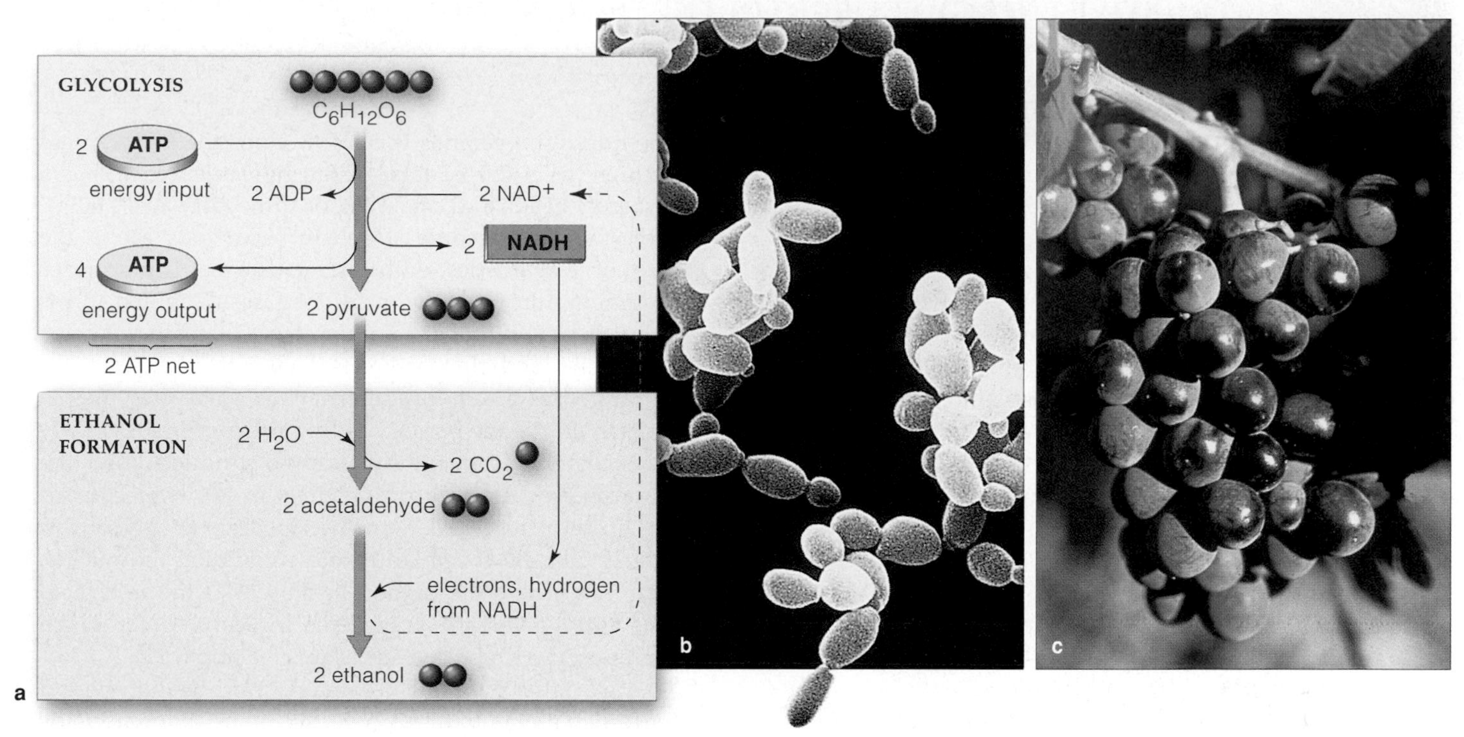

Figure 8.10 (**a**) Alcoholic fermentation, one of the anaerobic pathways. Acetaldehyde, an intermediate of the reactions, is the final acceptor of electrons. Ethanol is the end product. Yeasts, single-celled organisms, use this pathway. (**b**) Activity of one species of *Saccharomyces* makes bread dough rise. (**c**) Another species lives on sugar-rich tissues of ripened grapes.

prodigious berry-producing shrubbery near highways; drunk birds doodle into windshields (Figure 8.11). Wild turkeys similarly have been known to get tipsy when they gobble fermenting apples in untended orchards.

Anaerobic Electron Transport

Especially among the bacteria, we find less common energy-releasing pathways, some of which are topics of later chapters in the book. For example, many bacterial species have key roles in the global cycling of sulfur, nitrogen, and other crucial elements. Collectively, their metabolic activities influence nutrient availability for organisms everywhere.

Figure 8.11 Robin feasting on the fermented berries of a pyracantha bush.

For example, certain bacteria use **anaerobic electron transport**. Electrons stripped from some type of organic compound move on through transport systems of their plasma membrane. Commonly, an inorganic compound in the environment serves as the final electron acceptor. The net energy yield varies, but it is always small.

Even as you read this, some anaerobic bacteria that live in waterlogged soil are stripping electrons from a variety of compounds. They dump electrons on sulfate. Hydrogen sulfide, a putrid-smelling gas, is the result. The sulfate-reducing bacteria also live in many aquatic habitats that are enriched with decomposed organic material. They even live on the deep ocean floor, near hydrothermal vents. As described in Section 50.11, they form the food production base for unique communities.

In fermentation pathways, an organic substance that forms during the reactions serves as the final acceptor of electrons from glycolysis. The reactions regenerate NAD^+, which is required to keep the pathway operational.

In anaerobic electron transport, an inorganic substance (but not oxygen) usually serves as the final electron acceptor.

For each glucose molecule metabolized, anaerobic pathways typically have a net energy yield of two ATP molecules, which only form during glycolysis.

8.6 ALTERNATIVE ENERGY SOURCES IN THE HUMAN BODY

So far, you have looked at what happens after a lone glucose molecule enters an energy-releasing pathway. Now you can start thinking about what cells do when they have too many or too few molecules of glucose.

Carbohydrate Breakdown in Perspective

THE FATE OF GLUCOSE AT MEALTIME Consider what happens to you or any other mammal during a meal. Glucose and certain other small organic molecules are being absorbed across the gut lining, then the blood transports them through the body. A rise in the glucose level in blood prompts the pancreas to release insulin, a hormone that stimulates cells to take up glucose at a faster rate. The cells convert the windfall of glucose to glucose–6–phosphate and so "trap" it in the cytoplasm. (When phosphorylated, glucose cannot be transported back out, across the plasma membrane.) Look again at Figure 8.4, and you see that glucose–6–phosphate is the first activated intermediate of glycolysis.

If your glucose intake exceeds cellular demands for energy, ATP-producing machinery goes into high gear. Unless a cell is rapidly using ATP, its concentration of ATP can rise to high levels. Then, glucose–6–phosphate is diverted into a biosynthesis pathway that assembles glucose units into glycogen, a storage polysaccharide (Section 3.3). This is especially the case for muscle and liver cells, which maintain the largest glycogen stores.

THE FATE OF GLUCOSE BETWEEN MEALS When you are not eating, glucose is not entering your bloodstream, and its level in the blood declines. If the decline were not countered, that would be bad news for the brain, your body's glucose hog. The brain constantly takes up more than two-thirds of the freely circulating glucose because its many hundreds of millions of cells simply use this sugar alone as their preferred energy source.

The pancreas responds to the decline by secreting glucagon, a hormone that prompts liver cells to convert glycogen back to glucose and send it back to the blood. Only liver cells do this; muscle cells won't give it up. The blood glucose level rises, and brain cells keep on functioning. Thus, *hormones control whether the body's cells use free glucose as an energy source or tuck it away.*

A word of caution: Don't let the preceding examples lead you to believe cells squirrel away large amounts of glycogen. In adult humans, glycogen makes up merely 1 percent or so of the body's total energy reserves, the energy equivalent of two cups of cooked pasta. Unless you eat on a regular basis, you will deplete the liver's small glycogen stores in less than twelve hours. Of the total energy reserves in, say, a typical adult American, 78 percent (about 10,000 kilocalories) is concentrated in body fat and 21 percent in proteins.

Energy From Fats

The question becomes this: How does the body access its huge reservoir of fats? A fat molecule, recall, has a glycerol "head" and one, two, or three fatty acid "tails." Most fats that become stored in your body are in the form of triglycerides, with three tails each. Triglycerides accumulate in fat cells of adipose tissue, which forms at buttocks and other strategic places beneath the skin.

When blood glucose levels decline, triglycerides can be tapped as an energy alternative. Then, enzymes in fat cells cleave the bonds holding the glycerol and fatty acids together, and the breakdown products enter the bloodstream. Afterward, enzymes in the liver convert the glycerol to PGAL—an intermediate of glycolysis. Nearly all cells can take up the circulating fatty acids. Enzymes cleave the carbon backbone of the fatty acid tails and convert the fragments to acetyl–CoA, which can enter the Krebs cycle (Figures 8.6 and 8.12).

Each fatty acid tail has many more carbon-bound hydrogen atoms than glucose, so its breakdown yields much more ATP. In between meals or during sustained exercise, fatty acid conversions supply about half of the ATP that muscle, liver, and kidney cells require.

What happens if you eat too many carbohydrates? Exceed the glycogen-storing capacity of your liver and muscle cells, and the excess gets converted to fats. *Too much glucose ends up as excess fat.* Worse yet, 25 percent of the people in the United States are blessed with a combination of genes that allows them to eat as much as they like without gaining weight—but a diet far too rich in carbohydrates keeps the other 75 percent fat. For them, insulin levels remain elevated, which "tells" the body to store fat rather than use it for energy. We will return to this topic in Sections 37.7 and 42.10.

Energy From Proteins

Eat more proteins than your body requires to grow and maintain itself, and its cells won't store them. Enzymes split these proteins into amino acid units. Then they remove the amino group ($-NH_3^+$) from each unit, and ammonia (NH_3) forms. What happens to the leftover carbon backbones? Depending on conditions in the cell, the outcome varies. The backbones can be converted to carbohydrates or fats. Or they may enter the Krebs cycle, as shown in Figure 8.12, where coenzymes can pick up hydrogen and electrons stripped away from the carbon atoms. The ammonia that forms undergoes conversions to become urea. This nitrogen-containing waste product would be toxic if it accumulated to high concentrations. Normally your body excretes urea, in urine.

As this brief discussion makes clear, maintaining and accessing the body's energy reserves is complicated

Further reading: Student Guide to InfoTrac on web site →

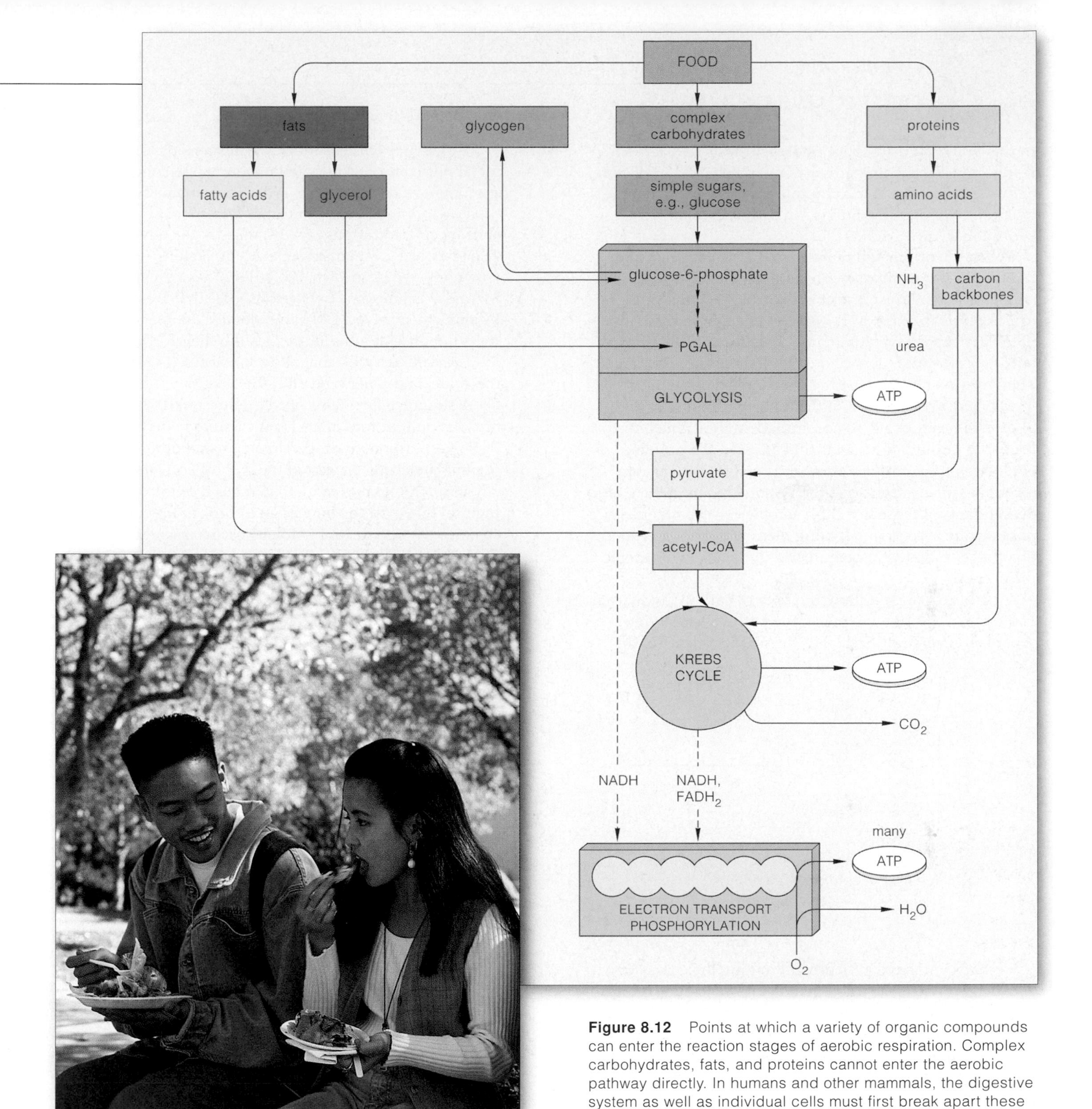

Figure 8.12 Points at which a variety of organic compounds can enter the reaction stages of aerobic respiration. Complex carbohydrates, fats, and proteins cannot enter the aerobic pathway directly. In humans and other mammals, the digestive system as well as individual cells must first break apart these large molecules into simpler, degradable subunits.

business. Hormonal controls over the disposition of glucose are special only because glucose is the fuel of choice for the all-important brain. However, as you will see in later chapters, providing all of your cells, organs, and organ systems with energy starts with the kinds and proportions of food you put in your mouth.

This concludes our look at aerobic respiration and other energy-releasing pathways. The section to follow may help you get a sense of how they fit into the larger picture of life's evolution and interconnectedness.

In humans and other mammals, the entrance of glucose or other organic compounds into an energy-releasing pathway depends on the kinds and proportions of carbohydrates, fats, and proteins in the diet as well as on the type of cell.

PERSPECTIVE ON LIFE

In this unit, you read about photosynthesis and aerobic respiration—the main pathways by which cells trap, store, and release energy. What you might not know is that the two pathways became linked, on a grand scale, over evolutionary time.

When life originated more than 3.8 billion years ago, the Earth's atmosphere had little free oxygen. The earliest single-celled organisms probably used reactions similar to glycolysis to make ATP. Without oxygen, fermentation pathways must have dominated. By about 1.5 billion years later, oxygen-producing photosynthetic cells had emerged. They irrevocably changed the course of evolution.

Oxygen, a by-product of the noncyclic pathway of photosynthesis, began to accumulate in the atmosphere. Probably through mutations that affected the proteins of electron transport systems, some cells started using oxygen as an electron acceptor. At some point in the past, descendants of those fledgling aerobic cells abandoned photosynthesis entirely. Among them were the forerunners of animals and all other organisms that engage in aerobic respiration.

With aerobic respiration, the flow of carbon, hydrogen, and oxygen through the metabolic pathways of living organisms came full circle. For the final products of this aerobic pathway—carbon dioxide and water—are the same materials that are necessary to build organic compounds in photosynthesis:

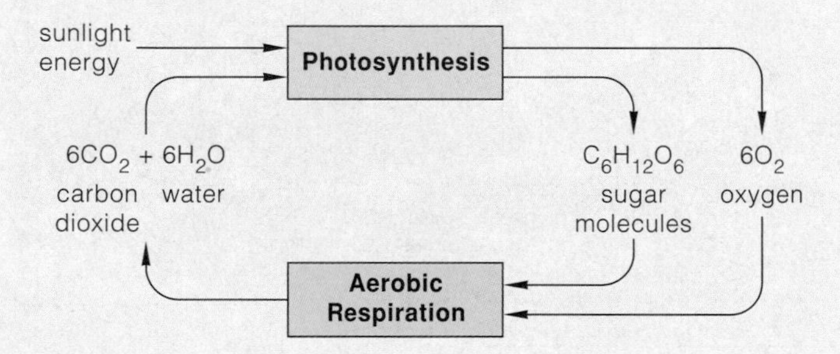

Perhaps you have difficulty fathoming the connection between yourself—an intelligent being—and such remote-sounding events as energy flow and the cycling of carbon, hydrogen, and oxygen. Is this really the stuff of humanity?

Think back, for a moment, on the structure of a water molecule. Two hydrogen atoms sharing electrons with an oxygen atom may not seem close to your daily life. And yet, through that sharing, water molecules show polarity—and they hydrogen-bond with one another. Their chemical behavior is a beginning for the organization of lifeless matter that leads to the organization of all living things.

For now you can imagine other molecules interspersed through water. The nonpolar kinds resist interaction with water; the polar kinds dissolve in it. On their own, the phospholipids among them assemble into a two-layered film. Such lipid bilayers, remember, serve as the very framework of all cell membranes, hence all cells. From the beginning, the cell has been the fundamental *living* unit.

The essence of life is not some mysterious force. It is metabolic control. With a cell membrane to contain them, reactions *can* be controlled. With mechanisms built into their membranes, cells can respond to energy changes and shifting concentrations of substances in the environment. The response mechanisms operate by "telling" proteins—enzymes—when and what to build or tear down.

And it is not some mysterious force that creates the proteins themselves. DNA, the slender double-stranded treasurehouse of inheritance, has the chemical structure—*the chemical message*—that allows molecule to reproduce molecule, one generation after the next. In your own body, DNA strands tell trillions of cells how countless molecules must be built or torn apart for their stored energy.

So yes, carbon, hydrogen, oxygen, and other atoms of organic molecules represent the stuff of you, and us, and all of life. But it takes more than molecules to complete the picture. Life exists as long as an unbroken flow of energy sustains its organization. Molecules are assembled into cells, cells into organisms, organisms into communities, and so on up through the biosphere. It takes energy inputs from the sun to maintain these levels of organization. And energy flows through time in one direction—from organized to less organized forms. Only as long as energy continues to flow into the web of life can life continue in all its rich diversity.

In short, life is no more *and no less* than a marvelously complex system of prolonging order. Sustained by energy transfusions from the sun, life continues onward, through its capacity for self-reproduction. For with the hereditary instructions contained in DNA, energy and materials can be organized, generation after generation. Even with the death of individuals, life elsewhere is prolonged. With each death, molecules are released and may be cycled once more, as raw materials for new generations.

In this flow of energy and cycling of material through time, each birth is affirmation of our ongoing capacity for organization, each death a renewal.

Further reading: Student Guide to InfoTrac on web site →

8.8 SUMMARY

1. Phosphate-group transfers from ATP are central to metabolism. Autotrophic cells alone can tap energy from the environment to make ATP. But *all* cells are able to make ATP by pathways that release chemical energy from organic compounds, such as glucose.

2. After glucose enters these pathways, enzymes strip electrons and hydrogen from intermediates that form along the way. Coenzymes pick these up and deliver them to other reaction sites, where the pathway ends. NAD^+ is the main coenzyme; the aerobic route also uses FAD. The reduced forms of these coenzymes are designated NADH and $FADH_2$.

3. All of the main energy-releasing pathways start with glycolysis, a stage of reactions that begin and end in the cytoplasm. Glycolytic reactions can be completed either in the presence of oxygen or in its absence.

a. During glycolysis, enzymes break down a glucose molecule to two pyruvate molecules. Two NADH and four ATP form during the reactions.

b. The net energy yield is two ATP (because two ATP had to be invested up front to get the reactions going).

4. Aerobic respiration continues on through two more stages: (1) the Krebs cycle and a few steps preceding it, and (2) electron transport phosphorylation. The stages occur only in organelles called mitochondria, which are present only in eukaryotic cells.

5. The second stage of the aerobic pathway starts when an enzyme strips a carbon atom from each pyruvate. Coenzyme A binds the remaining two-carbon fragment (to form acetyl–CoA), then transfers it to oxaloacetate, the entry point of the Krebs cycle. The cyclic reactions, along with the steps immediately preceding them, load up ten coenzymes with electrons and hydrogen (eight NADH and two $FADH_2$). Two ATP form. Three carbon dioxide molecules are released for each pyruvate that entered this second stage.

6. The third stage of the aerobic pathway proceeds at a membrane that divides the interior of a mitochondrion into two compartments. Electron transport systems and ATP synthases are embedded in this inner membrane.

a. Coenzymes deliver electrons from the first two stages to transport systems. In the outer compartment, hydrogen ions accumulate; therefore, concentration and electric gradients form across the membrane.

b. Hydrogen ions follow the gradients and flow from the outer to the inner compartment, through the interior of ATP synthases. Energy released during the ion flow drives the formation of ATP from ADP and unbound phosphate.

c. Oxygen withdraws electrons from the transport system and, at the same time, combines with hydrogen ions to form water molecules. The oxygen is the final acceptor of electrons that initially resided in glucose.

7. Aerobic respiration has a typical net energy yield of thirty-six ATP for each glucose molecule metabolized. Yields vary, according to cell type and cell conditions.

8. The fermentation pathways and anaerobic electron transport also start with glycolysis, but they do not use oxygen. They are anaerobic, start to finish.

a. Lactate fermentation has a net energy yield of two ATP, which form in glycolysis. The remaining reaction regenerates NAD^+. The two NADH from glycolysis transfer electrons and hydrogen to two pyruvate from glycolysis. Two lactate molecules are the end products.

b. Alcoholic fermentation has a net energy yield of two ATP from glycolysis, and its remaining reactions serve to regenerate NAD^+. Enzymes convert pyruvate from glycolysis to acetaldehyde, and carbon dioxide is released. The NADH from glycolysis transfer electrons and hydrogen to the two acetaldehyde molecules, thus forming two ethanol molecules, the end products.

c. Certain bacteria use anaerobic electron transport. Electrons are stripped from various organic compounds and travel through transport systems in the bacterial cell's plasma membrane. An inorganic compound in the environment often serves as the final electron acceptor.

9. In humans and other mammals, simple sugars such as glucose from carbohydrates, glycerol and fatty acids from fats, and carbon backbones of amino acids from proteins can enter ATP-producing pathways.

Review Questions

1. Is this true or false: Aerobic respiration occurs in animals but not plants, which make ATP only by photosynthesis. *8.1*

2. For this diagram of the aerobic pathway, fill in all blanks and write in the number of molecules of pyruvate, coenzymes, and end products. Also write in the net ATP formed in each stage, and the net ATP formed from start (glycolysis) to finish. *8.1*

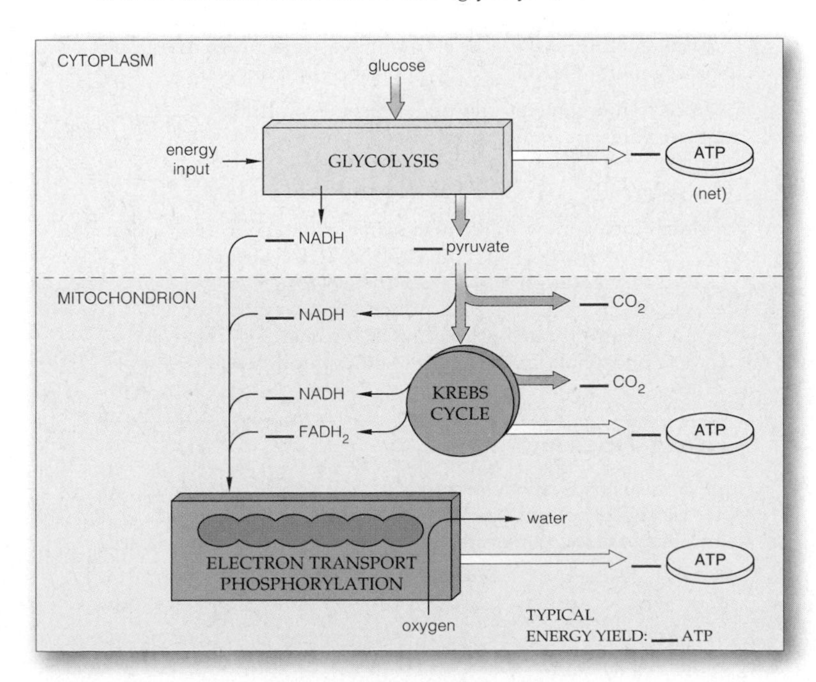

3. Is glycolysis energy-*requiring* or energy-*releasing*? Or do both kinds of reactions occur during glycolysis? *8.2*

4. In what respect does *electron transport* phosphorylation differ from *substrate-level* phosphorylation? *8.2, Figure 8.7*

5. Sketch the double-membrane system of the mitochondrion and show where transport systems and ATP synthases are located. *8.3*

6. Name the compound that is the entry point for the Krebs cycle, and state whether it directly accepts the pyruvate from glycolysis. For each glucose molecule, how many carbon atoms enter the Krebs cycle? How many depart from it, and in what form? *8.3*

7. Is this statement true or false: Muscle cells cannot contract at all when deprived of oxygen. If true, explain why. If false, name the alternative(s) available to them. *8.5*

Self-Quiz (*Answers in Appendix III*)

1. Glycolysis starts and ends in the ______.
   a. nucleus c. plasma membrane
   b. mitochondrion d. cytoplasm

2. Which of the following does *not* form during glycolysis?
   a. NADH c. $FADH_2$
   b. pyruvate d. ATP

3. Aerobic respiration is completed in the ______.
   a. nucleus c. plasma membrane
   b. mitochondrion d. cytoplasm

4. In the last stage of aerobic respiration, ______ is the final acceptor of electrons that originally resided in glucose.
   a. water c. oxygen
   b. hydrogen d. NADH

5. ______ engage in lactate fermentation.
   a. *Lactobacillus* cells c. Sulfate-reducing bacteria
   b. Muscle cells d. a and b

6. In alcoholic fermentation, ______ is the final acceptor of the electrons stripped from glucose.
   a. oxygen c. acetaldehyde
   b. pyruvate d. sulfate

7. The fermentation pathways produce no more ATP beyond the small yield from glycolysis, but the remaining reactions ______.
   a. regenerate ADP c. dump electrons on an inorganic substance (not oxygen)
   b. regenerate NAD^+

8. In certain organisms and under certain conditions, ______ can be used as an energy alternative to glucose.
   a. fatty acids c. amino acids
   b. glycerol d. all of the above

9. Match the event with its most suitable metabolic description.
   ____ glycolysis a. ATP, NADH, $FADH_2$, CO_2, and water form
   ____ fermentation b. glucose to two pyruvate
   ____ Krebs cycle c. NAD^+ regenerated, two ATP net
   ____ electron transport phosphorylation d. H^+ flows through ATP synthases

Critical Thinking

1. Diana suspects that a visit to her family doctor is in order. After eating carbohydrate-rich food, she always experiences sensations of being intoxicated and becomes nearly incapacitated, as if she had been drinking alcohol. She even wakes up with a hangover the next day. Having completed a course in freshman biology, Diana has an idea that something is affecting the way her body is metabolizing glucose. Explain why.

2. The cells of your body absolutely do not use nucleic acids as alternative energy sources. Suggest why.

3. The body's energy needs and its programs for growth depend on balancing the levels of amino acids in blood with proteins in cells. Cells of the liver, kidneys, and intestinal lining are especially important in this balancing act. When the levels of amino acids in blood decline, lysozymes in cells can rapidly digest some of their proteins (structural and contractile proteins are spared, except in cases of malnutrition). The amino acids released this way enter the blood and thereby help maintain the required levels.

Suppose you embark on a body-building program. You already eat plenty of carbohydrates, but a nutritionist advises a protein-rich diet that includes protein supplements. Speculate on how the extra dietary proteins will be put to use, and in which tissues.

4. Each year, Canada geese lift off in precise formation from their northern breeding grounds. They head south to spend the winter months in warmer climates, and then they make the return trip in spring. As is true of other migratory birds, their flight muscle cells are efficient at using fatty acids as an energy source. (Remember, the carbon backbone of fatty acids can be cleaved into fragments that can be converted to acetyl–CoA for entry into the Krebs cycle.)

Suppose a lesser Canada goose from Alaska's Point Barrow has been steadily flapping along for three thousand kilometers and is nearing Klamath Falls, Oregon. It looks down and notices a rabbit sprinting like the wind from a coyote with a taste for rabbit. With a stunning burst of speed, the rabbit reaches the safety of its burrow.

Which energy-releasing pathway predominated in muscle cells in the rabbit's legs? Why was the Canada goose relying on a different pathway for most of its journey? And why wouldn't the pathway of choice in goose flight muscle cells be much good for a rabbit making a mad dash from its enemy?

5. Reflect on this chapter's introduction, then question 4. Now speculate on which energy-releasing pathway is predominating in agitated Africanized bees that are chasing a farmer through a cornfield.

Selected Key Terms

acetyl–CoA *8.3*
aerobic respiration *8.1*
alcoholic fermentation *8.5*
anaerobic electron transport *8.5*
electron transport phosphorylation *8.1*
FAD *8.1*
glycolysis *8.1*
Krebs cycle *8.1*
lactate fermentation *8.5*
mitochondrion *8.3*
NAD^+ *8.1*
oxaloacetate *8.3*
pyruvate *8.1*
substrate-level phosphorylation *8.2*

Readings *See also www.infotrac-college.com*

Levi, P. October 1984. "Travels with C." *The Sciences.* Journey of a carbon atom through the world of life.

Wolfe, S. 1995. *An Introduction to Molecular and Cellular Biology.* Belmont, California: Wadsworth. Exceptional reference text.

FACING PAGE: *Human sperm, one of which will penetrate this mature egg and so set the stage for the development of a new individual in the image of its parents.*

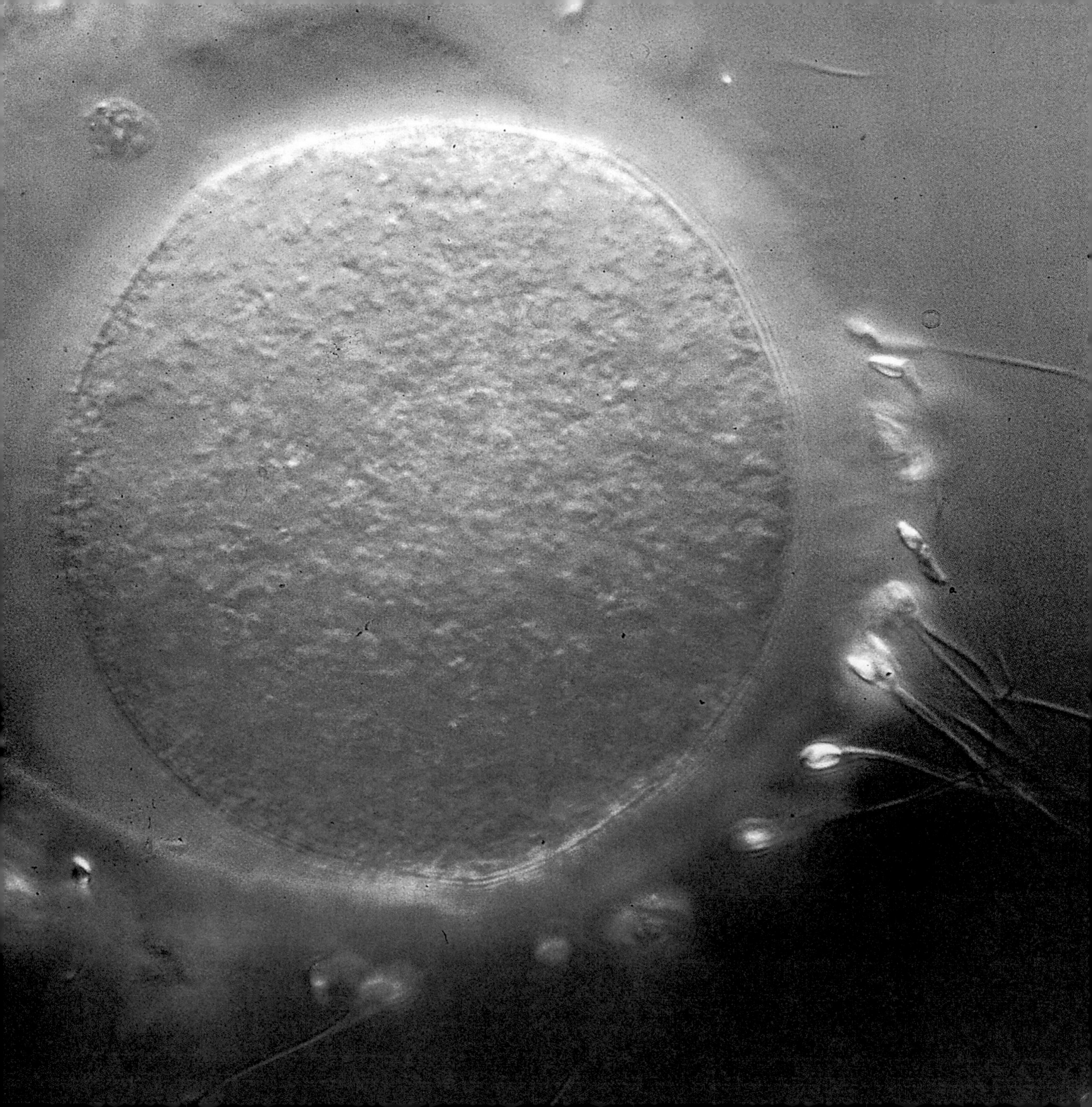

9

CELL DIVISION AND MITOSIS

Silver In the Stream of Time

Five o'clock, and the first rays from the sun dance over the wild Alagnak River of the Alaskan tundra. It is September, and life is ending and beginning in the clear, cold waters. By the thousands, mature silver salmon have returned from the open ocean to spawn in their shallow native home. The females are tinged with red, the color of spawners, and they are dying.

This morning you observe a female salmon releasing translucent pink eggs into a shallow "nest," which her fins hollowed out in the gravel riverbed (Figure 9.1). Within moments a male sheds a cloud of sperm, and fertilization follows. Trout and other predators eat most of the eggs. But if you wait around, you will find that some eggs survive and give rise to a new generation.

Within three years, the pea-sized eggs have become streamlined salmon, fashioned from billions of cells. A few of their cells will develop into eggs or sperm. In time, on some September morning, they will take part in an ongoing story of birth, growth, death, and rebirth.

For you, as for salmon and every other multicelled species, growth as well as reproduction depends on *cell division*. Inside your mother, a fertilized egg divided in two, then the two into four, and so on until billions of cells were growing, developing in specialized ways, and dividing at different times to produce all of your genetically prescribed body parts. Your body now has roughly 65 trillion living cells—and many of them are still dividing. Every five days, for instance, divisions replace the tissue that lines your small intestine.

Understanding cell division—and, ultimately, how new individuals are put together in the image of their parents—begins with answers to three questions. *First*, what instructions are necessary for inheritance? *Second*, how are those instructions duplicated for distribution into daughter cells? *Third*, by what mechanisms are the duplicated instructions parceled out to daughter cells? We will require more than one chapter to consider the nature of cell reproduction and other mechanisms of inheritance. Even so, the points made early in this chapter can help you keep the overall picture in focus.

Begin with the word **reproduction**. In biology, this means that parents produce a new generation of cells or multicelled individuals like themselves. The process starts in cells that are programmed to divide. And the

Figure 9.1 The last of one generation and the first of the next in Alaska's Alagnak River.

ground rule for division is this: *Parent cells must provide their daughter cells with hereditary instructions, encoded in DNA, and enough metabolic machinery to start up their own operation.*

DNA, recall, contains instructions for synthesizing proteins. Some proteins are structural materials. Many are enzymes that speed the assembly of specific organic compounds, such as the lipids that cells use as building blocks and sources of energy. Unless a daughter cell receives the necessary instructions for making proteins, it simply will not be able to grow or function properly.

Also, the cytoplasm of a parent cell already contains enzymes, organelles, and other operating machinery. When a daughter cell inherits what looks like a blob of cytoplasm, it really is getting start-up machinery—which will keep that cell operating until it can use its inherited DNA for growing and developing on its own.

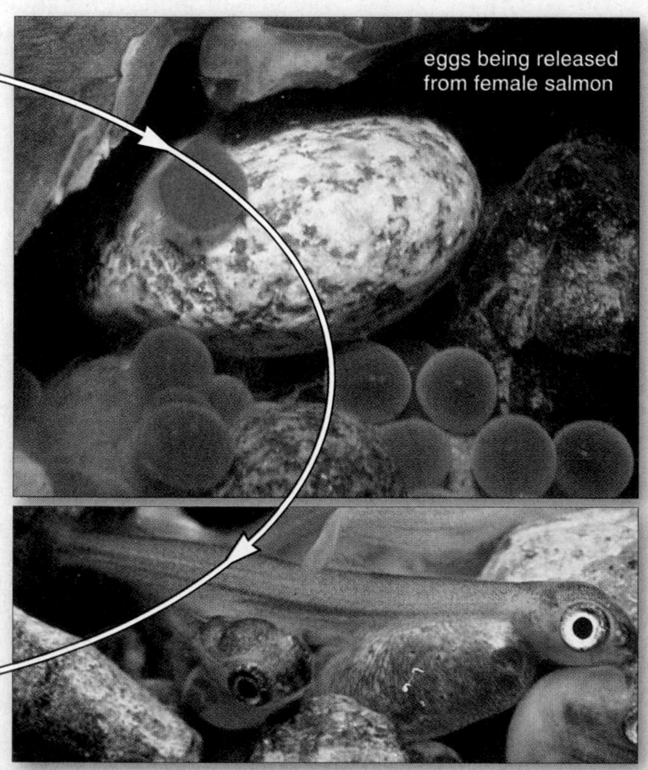

Fingerlings—young fishes growing, from fertilized eggs, by way of mitotic cell divisions

KEY CONCEPTS

1. The continuity of life depends on reproduction, by which parents produce a new generation of cells or multicelled individuals like themselves. Cell division is the bridge between generations.

2. When a cell divides, its two daughter cells must each receive a required number of DNA molecules and some cytoplasm. For eukaryotic cells, a division mechanism called mitosis sorts out the DNA into two new nuclei. A separate mechanism divides the cytoplasm.

3. Mitosis is one part of the cell cycle. The other part is interphase, an interval when each new cell formed by mitosis and cytoplasmic division increases in mass, increases the number of its components, then duplicates its DNA. The cycle ends when that cell divides.

4. In eukaryotic cells, many proteins with structural and functional roles are attached to the DNA. Each DNA molecule, with its attached proteins, is a chromosome.

5. Members of the same species have a characteristic number of chromosomes in their cells. The chromosomes differ from one another in length and shape, and they carry different portions of the hereditary instructions.

6. The body cells of humans and many other organisms have a diploid chromosome number; they contain two of each type of chromosome characteristic of the species.

7. Mitosis keeps the chromosome number constant, from one cell generation to the next. Therefore, if a parent cell is diploid, the daughter cells also will be diploid.

8. Mitotic cell division is the basis of growth and tissue repair in multicelled eukaryotes. It also is the means by which single-celled eukaryotes and many multicelled eukaryotes reproduce asexually.

DIVIDING CELLS: THE BRIDGE BETWEEN GENERATIONS

Overview of Division Mechanisms

In plants, animals, and all other eukaryotic organisms, hereditary instructions are distributed among a number of DNA molecules. Before the cells of such organisms reproduce, they must undergo *nuclear* division. **Mitosis** and **meiosis** are two nuclear division mechanisms. Both sort out and then package a parent cell's DNA into new nuclei for their forthcoming daughter cells. A separate mechanism splits the cytoplasm into daughter cells.

Multicelled organisms grow, replace dead or worn-out cells, and repair tissues by way of mitosis and the cytoplasmic division of body cells, which are called **somatic cells**. (Cut yourself peeling a potato and mitotic cell divisions will replace the cells that the knife sliced away.) Also, many protistans, fungi, plants, and some animals reproduce asexually by mitotic cell division.

By contrast, meiosis occurs only in **germ cells**, a cell lineage set aside for the formation of gametes (such as sperm and eggs) and sexual reproduction. As you will read in the next chapter, meiosis has much in common with mitosis, but the end result is different.

Prokaryotic cells—bacteria—reproduce asexually by a different mechanism, called prokaryotic fission. We will consider the bacteria later, in Section 22.2.

Some Key Points About Chromosomes

In a nondividing cell, the DNA molecules are stretched out like thin threads, with many proteins attached to them. Each DNA molecule, with its attached proteins, is a **chromosome**. When a cell prepares for mitosis, each threadlike chromosome gets duplicated. It now consists of two DNA molecules, which will stay together until late in mitosis. As long as they remain attached, the two are called **sister chromatids** of the chromosome.

Figure 9.2 illustrates a eukaryotic chromosome in the unduplicated and duplicated states. Notice how the duplicated chromosome narrows down in one section along its length. This is the **centromere**, a small region with attachment sites for microtubules that move the chromosome during nuclear division. Bear in mind, the sketch in Figure 9.2 is highly simplified. For example, the centromere's location differs among chromosomes. And the two strands of a DNA molecule do not look like a ladder; they twist rather like a spiral staircase and are much longer than can be shown here.

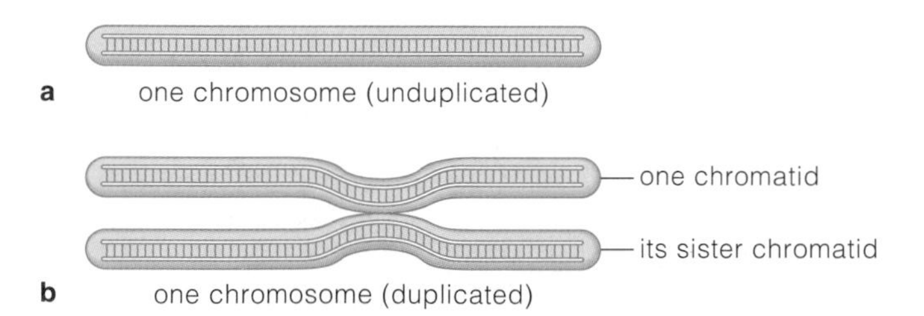

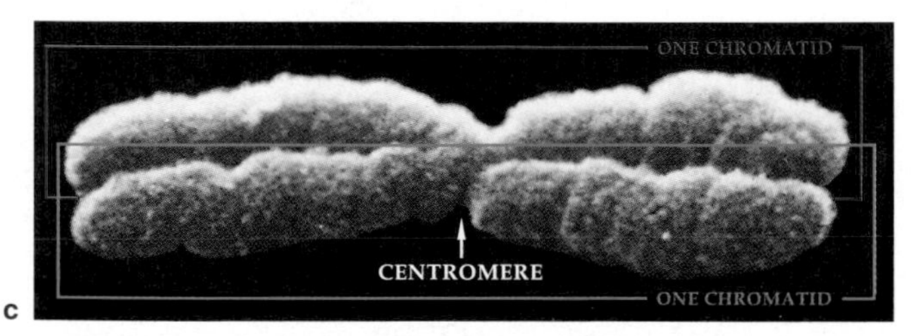

Figure 9.2 (**a**,**b**) Sketches of a chromosome in the unduplicated and duplicated states. Chromosomes are duplicated before cell division. (**c**) This scanning electron micrograph shows a human chromosome in the duplicated state; it consists of two sister chromatids attached at the centromere.

Mitosis and the Chromosome Number

Each species has a characteristic **chromosome number**, which refers to the sum total of chromosomes in cells of a given type. Human somatic cells have 46, those of gorillas have 48, and those of pea plants have 14.

Actually, your 46 chromosomes are like volumes of two sets of books. Each set is numbered, from 1 to 23. For example, you have two "volumes" of chromosome 22—that is, *a pair of them*. Generally, both members of each pair have the same length and shape, and they carry the same portion of hereditary instructions for the same traits. Think of them as two sets of books on how to build a house. Your father gave you one set. Your mother had her own ideas about storage, plumbing, and so on, so she gave you an alternate edition. Her set covers the same topics but says slightly different things about many of them.

We say the chromosome number is **diploid**, or $2n$, if a cell has two of each type of chromosome characteristic of the species. The body cells of humans, gorillas, pea plants, and a great many other organisms are like this. (By contrast, as Chapter 10 describes, eggs and sperm of these organisms have a *haploid* chromosome number, or n. This means they contain only one of each type of chromosome characteristic of the species.)

With mitosis, a diploid parent cell can produce two diploid daughter cells. This doesn't mean each merely gets forty-six or forty-eight or fourteen chromosomes. If only the total mattered, one cell might get, say, two pairs of chromosome 22 and no pairs whatsoever of chromosome 9. Neither would be able to function like the parent cell *without two of each type of chromosome.*

Mitosis keeps the chromosome number constant, division after division, from one cell generation to the next. Thus, if a parent cell is diploid, its daughter cells will be diploid also.

Further reading: Student Guide to InfoTrac on web site

THE CELL CYCLE

Mitosis is only one phase of the **cell cycle**. Such cycles start each time new cells form, and they end when those cells complete their own division. The cycle starts again for each new daughter cell, as in Figure 9.3. Usually, the longest phase of the cell cycle is **interphase**, which has three parts. During interphase, a cell increases its mass, roughly doubles its number of cytoplasmic components, then duplicates its DNA. Biologists abbreviate the four parts of a cell cycle this way:

G1 Of interphase, a "Gap" (interval) of cell growth and functioning before the onset of DNA replication

S Of interphase, a time of "Synthesis" (DNA replication)

G2 Of interphase, a second "Gap" (interval) after DNA replication, when the cell prepares for division

M Mitosis; nuclear division only, usually followed by cytoplasmic division

The cell cycle lasts about the same length of time for cells of the same type. Its duration differs among cells of different types. As examples, all neurons (nerve cells) in your brain are arrested at interphase and usually do not divide again. All red blood cells form (and replace your worn-out ones) at an average rate of 2–3 million every second. Early in the development of a sea urchin embryo, the number of cells doubles every two hours.

Adverse conditions may disrupt a cell cycle. When deprived of a vital nutrient, for instance, the free-living cells called amoebas do not leave interphase. Even so, if any cell proceeds past a certain point in interphase, the cycle normally will continue regardless of outside conditions owing to built-in controls over its duration.

We turn now to mitosis and to how it maintains the chromosome number through turn after turn of the cell cycle. Figure 9.4 only hints at the divisional ballet that begins when a cell leaves interphase.

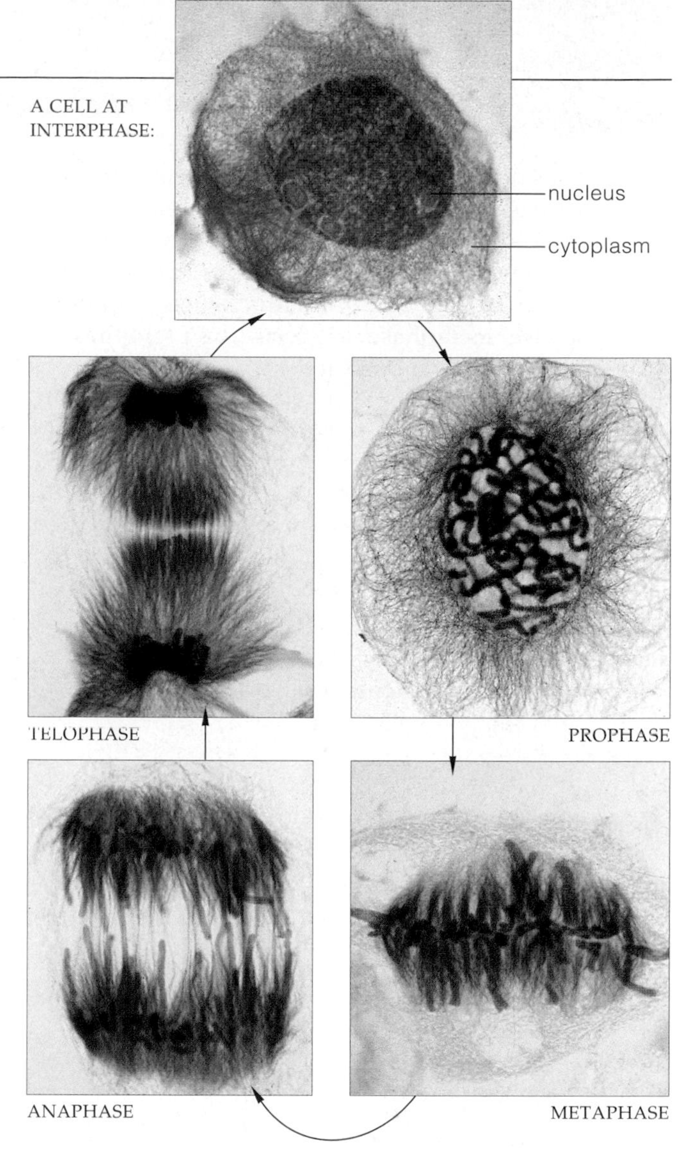

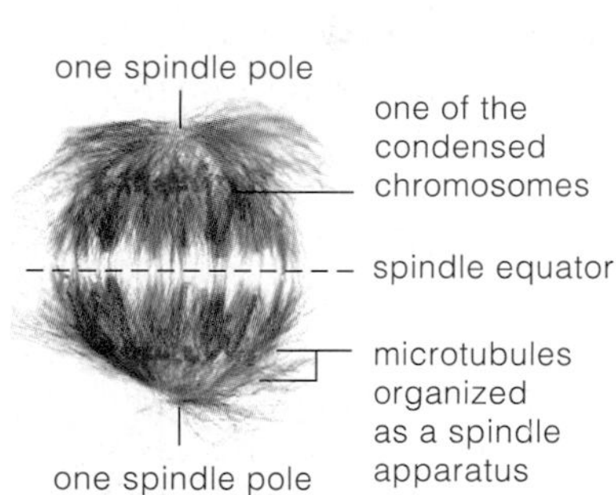

Figure 9.4 Mitosis in a cell from the African blood lily, *Haemanthus*. The chromosomes are stained *blue*, and the many microtubules are stained *red*. Before reading further, take a moment to become familiar with the labels on the micrographs.

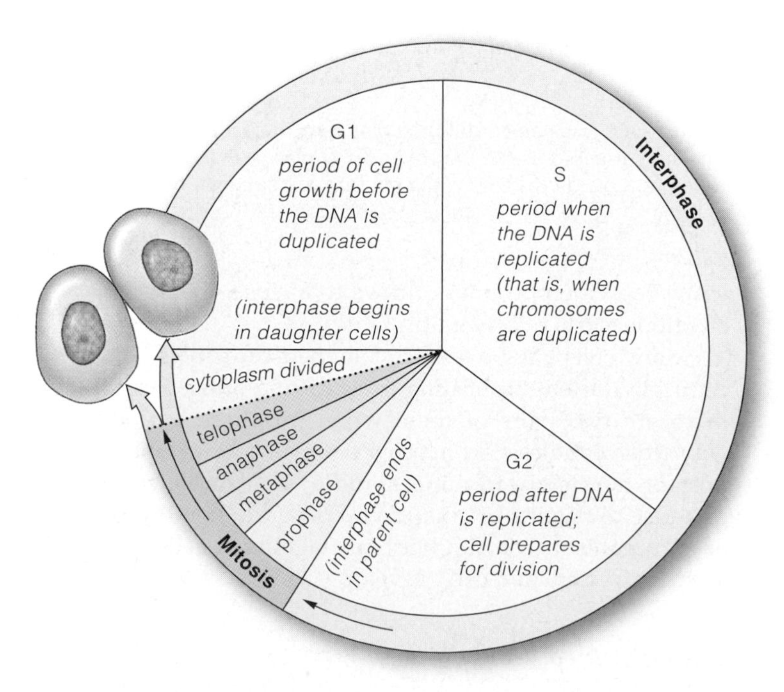

Figure 9.3 Eukaryotic cell cycle, generalized. The length of each part differs among different cell types.

A cell cycle starts at interphase, when a new cell (formed by mitosis and cytoplasmic division) increases its mass and the number of its cytoplasmic components, then duplicates its chromosomes. The cycle ends when the cell divides.

9.3 STAGES OF MITOSIS—AN OVERVIEW

When a cell makes the transition from interphase to mitosis, it has stopped constructing new cell parts, and the DNA has been replicated. Within that cell, major changes will now proceed smoothly, one after the other, through four stages. The sequential stages of mitosis are **prophase**, **metaphase**, **anaphase**, and **telophase**.

Figure 9.5 shows mitosis in an animal cell. By comparing its series of photographs with those of the plant cell in Figure 9.4, you see the chromosomes are changing positions. They are not doing so on their own. A **spindle apparatus** is moving them. Each spindle consists of microtubules organized as two distinct sets. Each set extends from one of the two poles (end points) of the spindle. The two sets overlap each other a bit at the spindle equator, midway between the poles. Formation of this bipolar, microtubular spindle will establish the final destinations of chromosomes during mitosis, as you will see shortly.

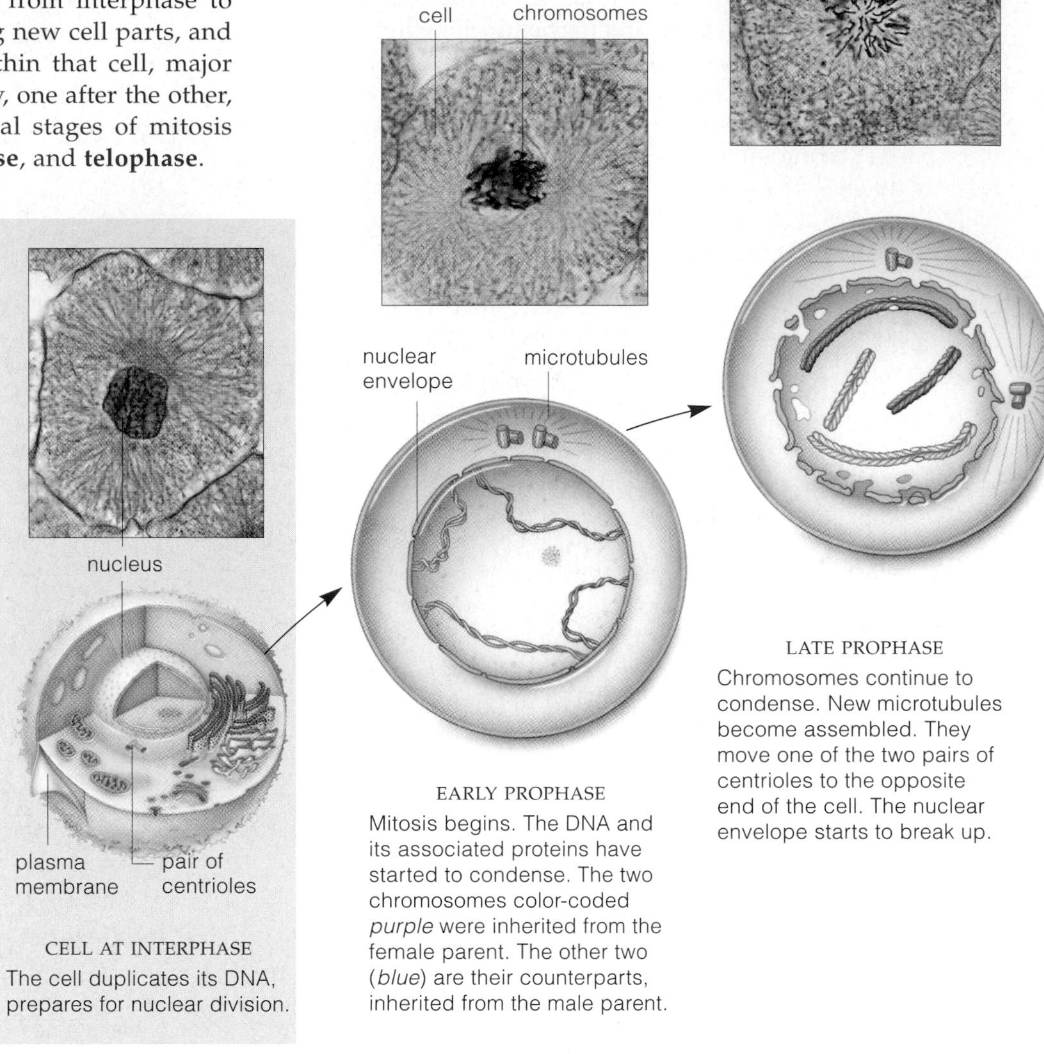

Figure 9.5 Mitosis in a generalized animal cell. This nuclear division mechanism ensures that each daughter cell will have the same chromosome number as the parent cell. For clarity, this diagram shows only two pairs of chromosomes from a diploid ($2n$) cell. With only rare exceptions, the picture is more complicated, as suggested by the micrographs of mitosis in a whitefish cell.

Prophase: Mitosis Begins

We know a cell is in prophase when its chromosomes become visible in the light microscope as threadlike forms. ("Mitosis" comes from the Greek *mitos*, for thread.) Each chromosome was duplicated earlier, during interphase, so each is now two sister chromatids joined at the centromere. During early prophase, sister chromatids of each chromosome twist and fold into a more compact form. By late prophase, all of the chromosomes will be condensed into thicker, rod-shaped forms.

Meanwhile, in the cytoplasm, most microtubules of the cytoskeleton are breaking down to tubulin subunits (Section 4.8). The subunits reassemble near the nucleus as *new* microtubules of the spindle. Many of the new microtubules will extend from one spindle pole or the other to a centromere of a chromosome. The remainder will not interact with chromosomes. Instead, they will extend from the poles and overlap each other.

While new microtubules are assembling, the nuclear envelope is a physical barrier that prevents them from interacting with the chromosomes inside the nucleus. However, when prophase draws to a close, the nuclear envelope starts to break up.

Many cells have two barrel-shaped **centrioles**. Each centriole started duplicating itself during interphase, so there are two pairs of them when prophase is under way. Microtubules start moving one pair to the opposite pole of the newly forming spindle. Centrioles, recall, give rise to flagella or cilia. If you observe them in cells of an organism, you can bet that flagellated cells (such as sperm) or ciliated cells develop during its life cycle.

Transition to Metaphase

So much happens between prophase and metaphase that researchers give this transitional period its own

Further reading: Student Guide to InfoTrac on web site →

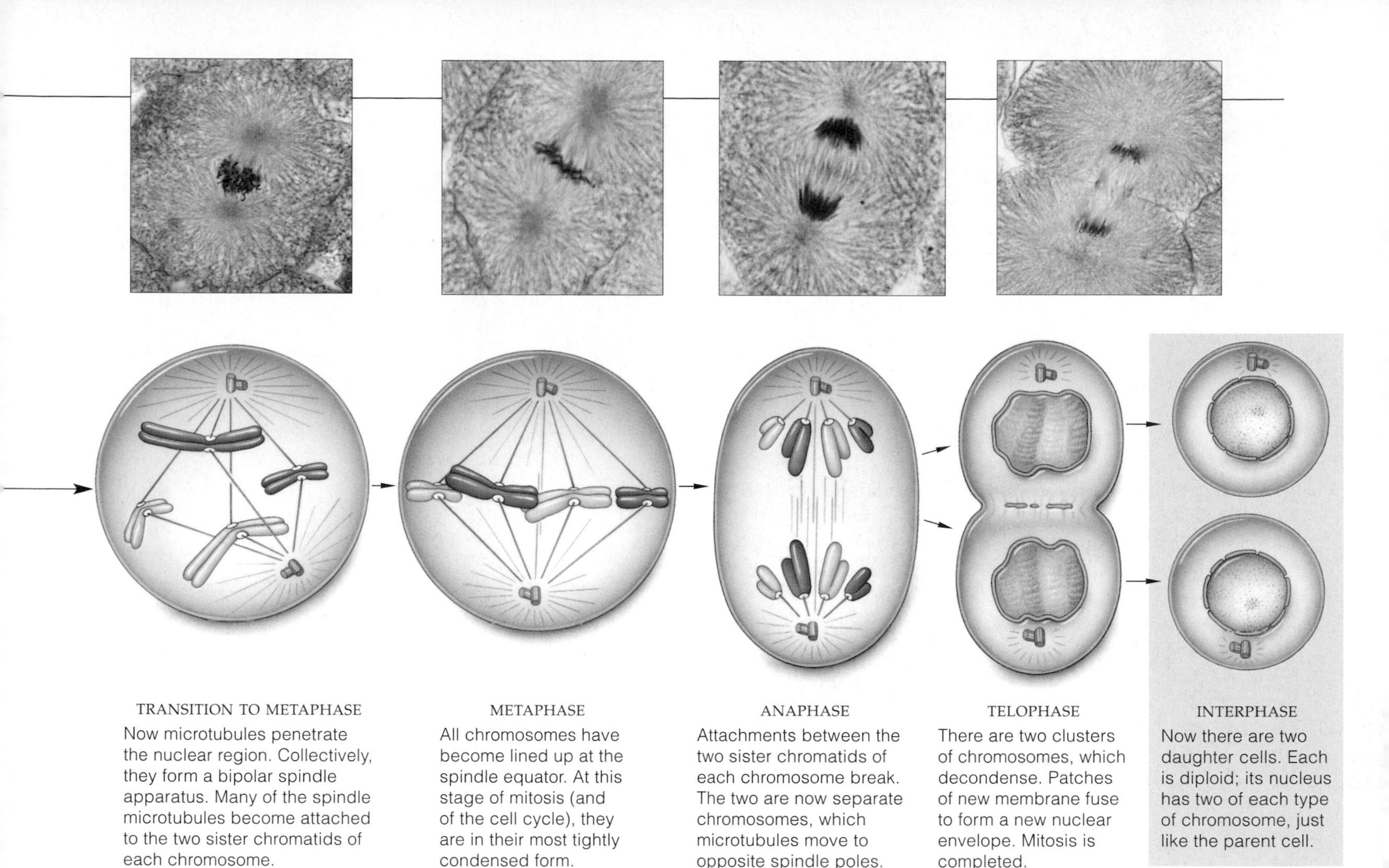

name, "prometaphase." The nuclear envelope breaks up completely into numerous tiny, flattened vesicles. Now the chromosomes are free to interact with microtubules that are extending toward them, from the poles of the forming spindle. Microtubules from both poles harness each chromosome and start pulling on it. The two-way pulling orients the chromosome's two sister chromatids toward opposite poles. Meanwhile, overlapping spindle microtubules ratchet past each other and push the poles of the spindle apart. The push–pull forces are balanced when the chromosomes reach the spindle's midpoint.

When all of the duplicated chromosomes are aligned midway between the poles of a completed spindle, we call this metaphase (*meta-* means "midway between"). The alignment is crucial for the next stage of mitosis.

From Anaphase Through Telophase

At anaphase, the sister chromatids of each chromosome separate from each other and move to opposite poles by two mechanisms. First, microtubules attached to the centromere regions shorten and *pull* the chromosomes to the poles. Second, the spindle elongates as overlapping microtubules continue to ratchet past each other and *push* the two spindle poles even farther apart. Once each chromatid is separated from its sister, it has a new name. It is a separate chromosome in its own right.

Telophase gets under way as soon as each of two clusters of chromosomes arrives at a spindle pole. The chromosomes, no longer harnessed to the microtubules, return to threadlike form. Vesicles derived from the old nuclear envelope fuse and form patches of membrane around the chromosomes. Patch joins with patch, and soon a new nuclear envelope separates each cluster of chromosomes from the cytoplasm. If the parent cell was diploid, each cluster contains two chromosomes of each type. With mitosis, remember, each new nucleus has the same chromosome number as the parent nucleus. Once two nuclei form, telophase is over—and so is mitosis.

Prior to mitosis, each chromosome in a cell's nucleus is duplicated, so that it consists of two sister chromatids.

Mitosis proceeds through four consecutive stages called prophase, metaphase, anaphase, and telophase.

A microtubular spindle moves sister chromatids of every chromosome apart, to opposite spindle poles. Around each of two clusters of chromosomes, new nuclear envelope forms. Both daughter nuclei formed this way have the same chromosome number as the parent cell's nucleus.

9.4 DIVISION OF THE CYTOPLASM

The cytoplasm usually divides at some time between late anaphase and the end of telophase. As you might gather by comparing Figures 9.6 and 9.7, the actual mechanism of **cytoplasmic division** (or cytokinesis, as it is often called) differs among organisms.

Cell Plate Formation in Plants

As described in Section 4.10, most plant cells are walled, which means their cytoplasm cannot be pinched in two. Cytoplasmic division of such cells involves **cell plate formation**, as shown in Figure 9.6. By this mechanism, vesicles packed with wall-building materials fuse with one another and with remnants from the microtubular spindle. Together, they form a disklike structure—a cell plate. At this location, deposits of cellulose accumulate.

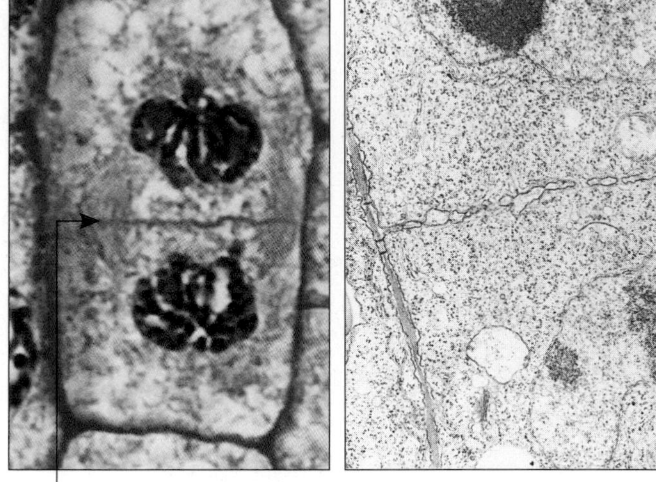

light micrograph and transmission electron micrograph showing cell plate formation in a dividing plant cell

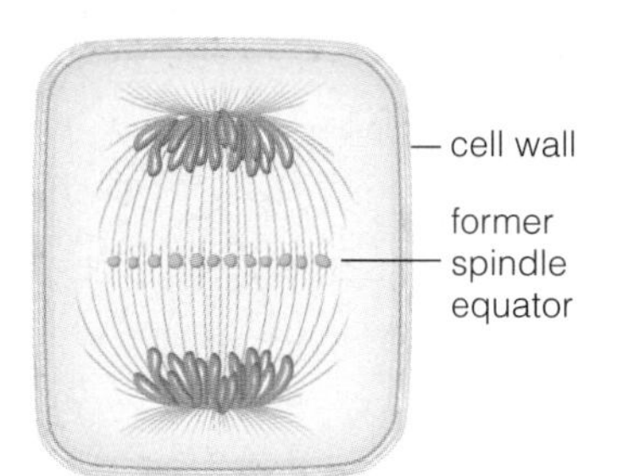

a As mitosis ends, vesicles converge at the spindle equator. They contain cementing materials and structural materials for a new primary cell wall.

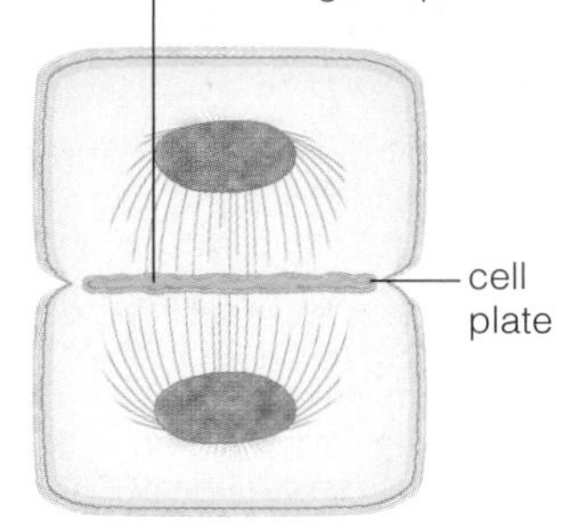

b A cell plate starts forming as membranes of the vesicles fuse. Materials inside the vesicles get sandwiched between two new membranes that elongate along the plane of the cell plate.

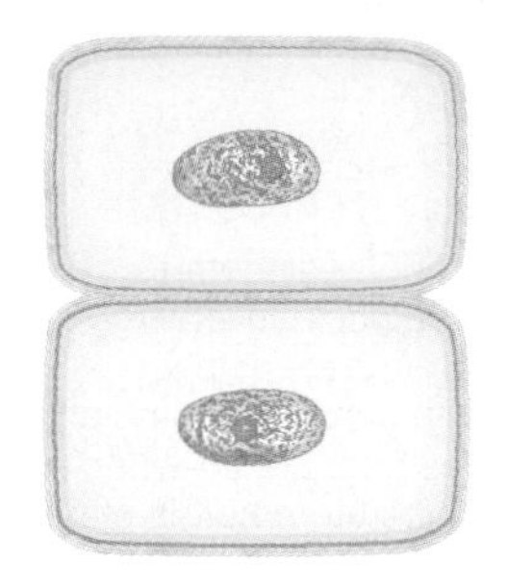

c Cellulose is deposited on the inside of the "sandwich." (In time, deposits will form two cell walls. Other deposits will form a middle lamella and cement the walls together; refer to Section 4.10.)

d A cell plate grows at its margins until it fuses with the parent cell's plasma membrane. During growth, when new plant cells expand and their walls are still thin, new material is deposited on the old primary wall.

Figure 9.6 Cytoplasmic division of a plant cell, as brought about by cell plate formation.

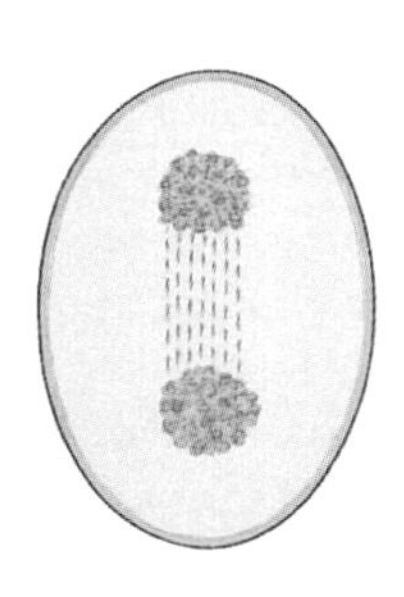

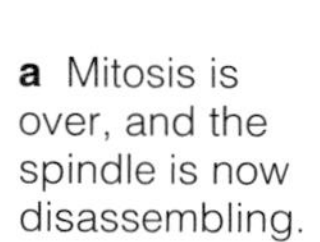

a Mitosis is over, and the spindle is now disassembling.

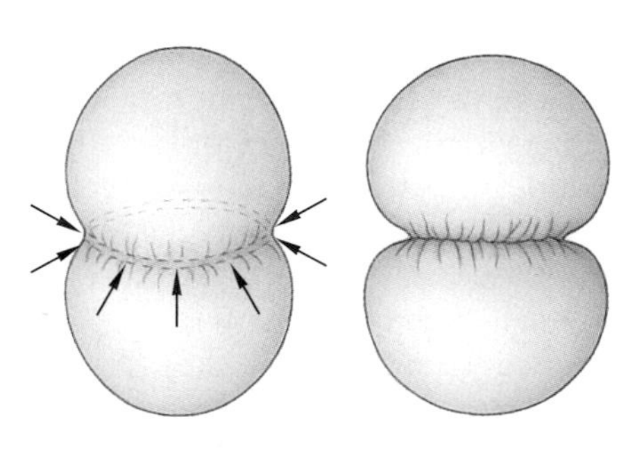

b Just beneath the plasma membrane, a band of microfilaments at the former spindle equator contracts, so that its diameter shrinks all around the cell.

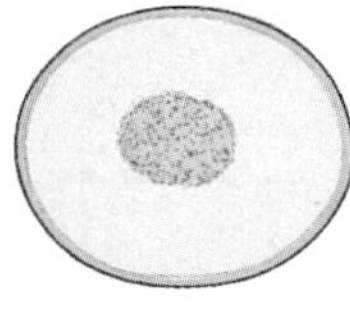

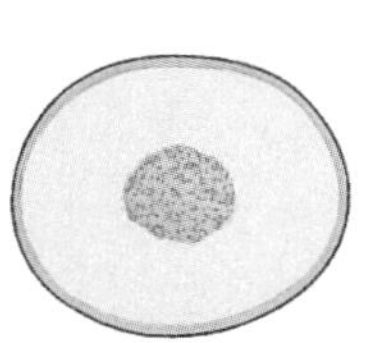

c The contractions continue and cut the cell in two.

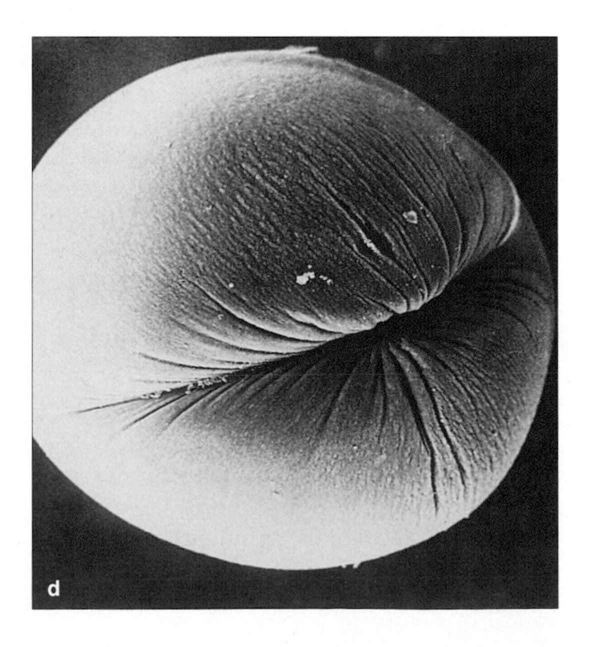

Figure 9.7 (**a–c**) Cytoplasmic division of an animal cell. (**d**) Scanning electron micrograph of the cleavage furrow at the plane of the former spindle's equator. Beneath it, a band of microfilaments attached to the plasma membrane contracts and pulls the surface inward. The furrow deepens until the cell is cut in two.

Further reading: Student Guide to InfoTrac on web site

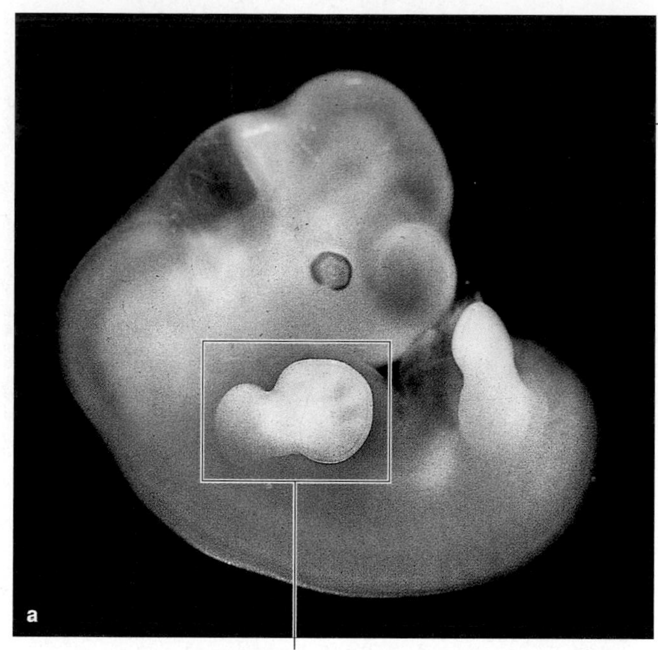

future arm and hand of embryo, five weeks old

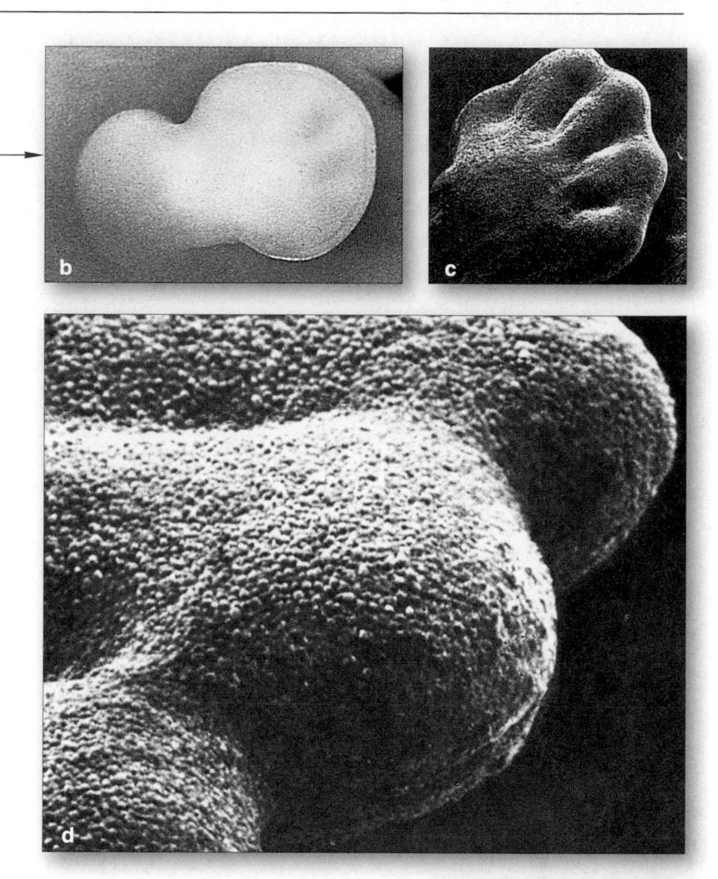

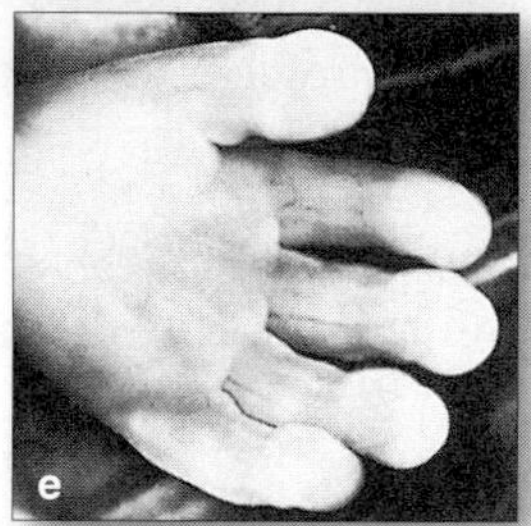

Figure 9.8 A few moments in the emergence of the human hand by way of mitosis, cytoplasmic divisions, and other developmental processes. Many individual cells produced by mitotic cell divisions are visible in (**d**). The photograph in (**e**) shows the digits at a later stage of development.

In time, the cellulose deposits are thick enough to form a crosswall. The new crosswall bridges the cytoplasm and divides the parent cell into two daughter cells.

Cytoplasmic Division of Animal Cells

Unlike plant cells, an animal cell is not confined within a cell wall, and its cytoplasm typically "pinches in two." Look at Figure 9.7, a surface view of a newly fertilized animal egg. An indentation is forming about midway between the egg's two poles ("ends"). The indentation in its plasma membrane is a **cleavage furrow**. Such a furrow is the first visible sign that an animal cell is undergoing cytoplasmic division. It will extend around the cell and continue to deepen along the plane of the former spindle's midpoint until the cell is cut in two.

A band of microfilaments beneath the cell's plasma membrane generates the force for cytoplasmic division. Microfilaments, remember, are threadlike cytoskeletal elements. These particular ones are organized so that they slide past one another (Section 4.9). When they do, they pull the plasma membrane inward until the two daughter nuclei are cut off in separate cells, each with its own cytoplasm and plasma membrane.

This concludes our picture of mitotic cell division. Look now at your two hands and try to visualize all of the cells in your palms, thumbs, and fingers. Imagine the divisions that produced all the generations of cells that preceded them when you were developing early on, inside your mother (Figure 9.8). And be grateful for the astonishing precision of the mechanisms that led to their formation at certain times, in certain numbers, for the alternatives can be terrible indeed. Why? Good health, and survival itself, depends absolutely on the proper timing and the completion of cell cycle events, including mitosis. Some genetic disorders arise from mistakes in the duplication or distribution of even one chromosome. Also, when normal controls that prevent cells from dividing are lost, unchecked cell divisions may destroy surrounding tissues and, ultimately, the organism. Section 9.6 touches on a landmark case of such losses, which we will explore further in Section 15.6.

Following mitosis, a separate mechanism cuts the cytoplasm into two daughter cells, each with a daughter nucleus.

In plants, cytoplasmic division often involves the formation of a cell plate and a crosswall between the adjoining, new plasma membranes of daughter cells.

Cytoplasmic division in animals may involve cleavage. Rings of microfilaments around a parent cell's midsection slide past one another in a way that pinches the cytoplasm in two.

9.5 A CLOSER LOOK AT THE CELL CYCLE

Close this book for a moment. Be sure you have a clear picture of the flow of events in interphase, mitosis, and on through cytoplasmic division. How easily you will get through many later chapters depends on how well you understand the cell division story. If parts of the picture are still not clear, it may be worth your time to read the preceding sections once again before getting into the details presented here.

The Wonder of Interphase

If you could coax the DNA molecules from just one of your somatic cells to stretch in a single line, one after another, that line would extend past the fingertips of your outstretched arms. Salamander DNA is even more amazing. A single line of it would extend ten meters! The wonder is, enzymes and other proteins in the cell selectively scan all the DNA, switch protein-building instructions on and off, and make base-by-base copies of each DNA molecule—all during interphase.

G1, S, and G2 of interphase have distinct patterns of biosynthesis. Most of your cells remain in G1, when they assemble most of the carbohydrates, lipids, and proteins they use or export. The cells destined to divide enter S, when they copy the DNA and the histones and other proteins associated with it. During G2, these cells produce proteins that will drive mitosis to completion.

Once S begins, events normally proceed at about the same rate in all cells of a species and continue through mitosis. Given this observation, you may well assume the cycle has built-in molecular brakes. Apply brakes that operate in G1, and the cycle stalls in G1. Lift the brakes and the cycle runs to completion. Said another way, *control mechanisms govern the rate of cell division.*

Imagine a car losing its brakes just as it starts down a steep mountain road. As you will read later on in the book, that is how cancer starts. Controls over division are lost, and the cell cycle cannot stop turning.

On Chromosomes and Spindles

The precision with which mitosis parcels out DNA for forthcoming daughter cells is impressive. That precision depends on chromosome organization and interactions among microtubules and motor proteins.

ORGANIZATION OF METAPHASE CHROMOSOMES Even during interphase, eukaryotic DNA has many proteins bound tightly to it. **Histones** are among them. Many histones are like spools for winding up small stretches of DNA. Each histone–DNA spool is a single structural unit called a **nucleosome**. Other histones stabilize the spools. During mitosis (and meiosis also), interactions between histones and DNA make the chromosome coil

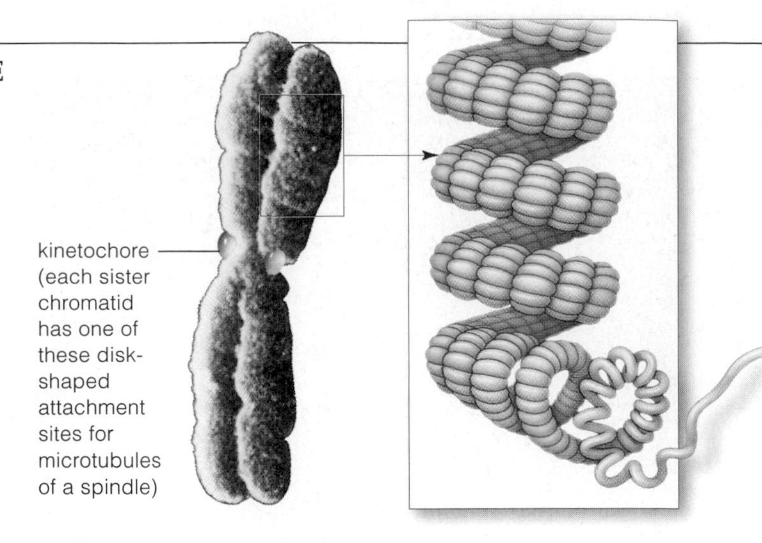

a A duplicated human chromosome at metaphase, at its most condensed. Interactions among some chromosomal proteins may keep loops of DNA tightly packed in a "supercoiled" array.

Figure 9.9 One model of the levels of organization in a human chromosome at metaphase.

back on itself repeatedly. The coiling greatly increases the chromosome's diameter. Other proteins besides the histones form a structural scaffold when the DNA folds even more, possibly into a series of loops (Figure 9.9).

A chromosome acquires its distinct shape and size late in prophase, when condensation is nearly complete. By then, each of its sister chromatids has at least one constricted region, the most prominent of which is the centromere. As you will read next, small, disk-shaped structures at the surface of centromeres serve as the docking sites for spindle microtubules (Figure 9.9*a*). We call them **kinetochores**.

When threadlike molecules of DNA are condensing into such compact chromosome structures, why don't they get tangled up? Actually, it appears that they do, but an enzyme called DNA topoisomerase chemically recognizes such tangling and puts things right. When researchers deliberately interfered with this enzyme's activity, the tangles persisted. Later, sister chromatids failed to separate at anaphase.

SPINDLES COME, SPINDLES GO All through anaphase, the spindle microtubules attached to chromosomes do not change position, yet the distance shrinks between these microtubules and the spindle poles. How? When a kinetochore slides over them, the microtubules shorten by disassembling (Figure 9.10). A kinetochore is like a train chugging along a railroad track—except the track falls apart after the train has passed over it. It contains two motor proteins, dynein and kinesin, that may drive the sliding motion.

Motor proteins, remember, are accessory proteins for cytoskeletal elements. As mentioned in Section 4.8,

Further reading: Student Guide to InfoTrac on web site →

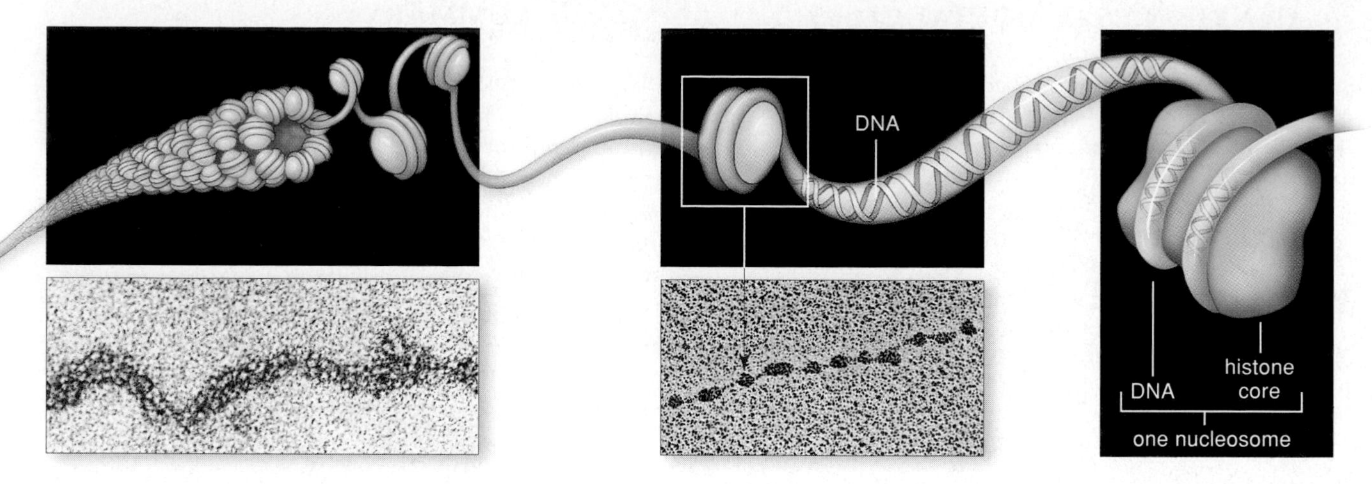

b At a deeper level of structural organization, the chromosomal proteins and the DNA are arranged as a cylindrical fiber (solenoid) thirty nanometers in diameter.

c Immerse a chromosome in saltwater and it loosens up to a beads-on-a-string organization. The "string" is one DNA molecule. Each "bead" is a nucleosome.

d A nucleosome consists of a double loop of DNA around a core of eight histones. Other histones stabilize the structural array.

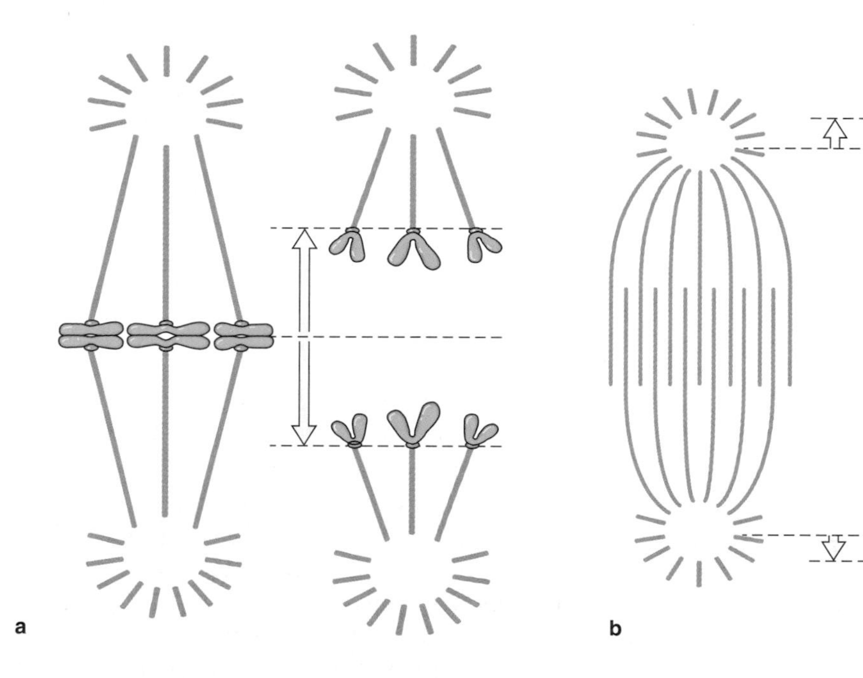

Figure 9.10 Models of two of the mechanisms that separate sister chromatids of a chromosome from each other at anaphase. In (**a**), the microtubules attached to the chromatids *shorten* and thereby decrease the distance between the kinetochores and spindle poles. In (**b**), overlapping microtubules ratchet past each other. As they do this, they move the spindle poles apart and thereby increase the distance between the sister chromatids of each chromosome.

they are attached to and project from microtubules and microfilaments that take part in cell movements.

And what about the spindle microtubules extending from both poles but not attached to kinetochores? They, too, incorporate dynein and kinesin. They actively slide past one another where they overlap in a spindle.

A final point about spindle structure and function: Over evolutionary time, spindle microtubules became targets in the chemical warfare between certain edible plants and animals that browse on them. For instance, plants of the genus *Colchicum* make a poison that blocks assembly and promotes disassembly of microtubules. The poison, colchicine, is a favorite of those who study cancer and other expressions of cell division. A *Critical Thinking* question at the chapter's end mentions another kind. Spindles in cells will disassemble within seconds or minutes after exposure to such microtubule poisons.

Once the S stage of interphase begins, the cell cycle turns at about the same rate in all cells of a given type, all the way through mitosis. Molecular mechanisms control whether a cell enters S and thereby control the rate of cell division.

The condensed form of metaphase chromosomes arises by interactions among DNA and structural proteins, including histones, that associate with DNA throughout the cell cycle.

During mitosis, motor proteins act on microtubules of the spindle to produce chromosomal movements.

Focus on Science

9.6 HENRIETTA'S IMMORTAL CELLS

Each human starts out as a single fertilized egg. By the time of birth, mitotic cell divisions and other processes have resulted in a human body of about a trillion cells. Even in adults, billions of cells still divide. For example, cells of the stomach's lining divide every day. Liver cells usually do not divide, but if part of the liver becomes injured or diseased, repeated cell divisions will yield more new cells until the damaged part is finally replaced.

In 1951, George and Margaret Gey of Johns Hopkins University were trying to develop a way to keep human cells dividing *outside* the body. With such isolated cells, these researchers and others could investigate basic life processes. They also could conduct studies of cancer and other diseases without having to experiment directly on patients and gamble with human lives. The Geys used normal and diseased human cells, which local physicians had sent them. But they couldn't stop the descendants of those precious cells from dying out within a few weeks.

Mary Kubicek, a laboratory assistant, worked with the Geys in their efforts to start a self-perpetuating lineage of cultured human cells. She was about to give up after dozens of failed attempts. Even so, in 1951 she decided to prepare one more sample of cancer cells for culture. She gave the sample the code name **HeLa cells**, for the first two letters of the patient's first and last names.

The HeLa cells began to divide. And divide. And divide again! By the fourth day there were so many cells that Kubicek subdivided them into more culture tubes. Unfortunately, tumor cells in the patient were just as vigorous. Six months after she was diagnosed as having cancer, tumor cells had spread to tissues throughout her body. Only eight months after the diagnosis, Henrietta Lacks, a young woman from Baltimore, was dead.

Although Henrietta passed away, some of her cells lived on in the Geys' laboratory as the first successful human cell culture. In time, HeLa cells were shipped to research laboratories all over the world. Some journeyed into space for experiments on the *Discoverer XVII* satellite. Each year hundreds of scientific papers describe work that was based on HeLa cells.

Henrietta was thirty-one when runaway cell divisions killed her. And now, decades later, her legacy is benefitting humans everywhere—in her cellular descendants that are still alive and dividing, day after day after day.

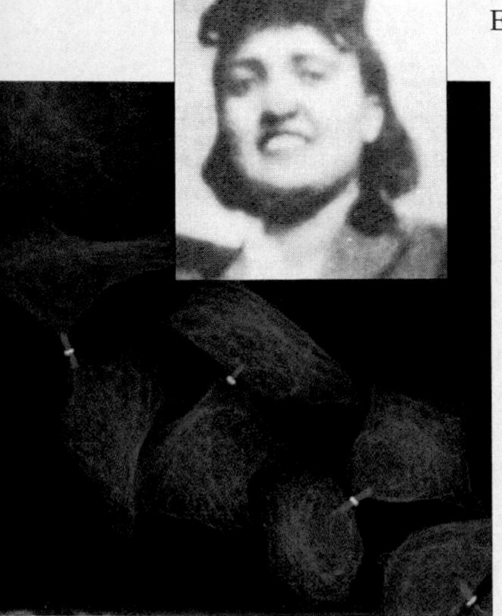

Figure 9.11 *Above:* Henrietta Lacks, a casualty of cancer whose cellular contribution to science is still helping others. *Below:* Dividing HeLa cells.

9.7 SUMMARY

1. Through specific division mechanisms, summarized in Table 9.1, a parent cell provides each daughter cell with the hereditary instructions (DNA) and cytoplasmic machinery necessary to start up its own operation.

a. In eukaryotic cells, the nucleus divides by mitosis or meiosis. Cytoplasmic division typically follows.

b. Prokaryotic cells divide by prokaryotic fission.

2. Each eukaryotic chromosome is one DNA molecule with numerous proteins attached. Chromosomes in a given cell differ in length, shape, and which portion of the hereditary instructions they carry.

a. "Chromosome number" refers to the sum total of chromosomes in cells of a given type. Cells having a diploid chromosome number ($2n$) contain two of each kind of chromosome.

b. Mitosis divides the nucleus into two equivalent nuclei, each with the same chromosome number as the parent cell. It maintains the chromosome number from one cell generation to the next.

c. Mitosis is the basis of growth, tissue repair, and cell replacements among multicelled eukaryotes. It is the basis of asexual reproduction in many single-celled eukaryotes. (Meiosis occurs only in germ cells.)

3. A cell cycle starts when a new cell forms. It proceeds through interphase and ends when the cell reproduces by mitosis and cytoplasmic division. In interphase, a cell carries out its functions. If it is to divide again, the cell increases in its mass and cytoplasmic components and duplicates its chromosomes in preparation for division.

4. A duplicated chromosome has two DNA molecules attached at the centromere. While the two stay attached to each other, they are called sister chromatids.

5. Mitosis proceeds through four continuous stages:

a. Prophase. Duplicated, threadlike chromosomes start to condense. A spindle apparatus starts to form. The nuclear envelope starts to break up and, during the *transition* to metaphase (prometaphase), its remnants form vesicles, and microtubules from opposite poles of

Table 9.1 Summary of Cell Division Mechanisms

Mechanisms	Functions
MITOSIS, CYTOPLASMIC DIVISION	In all multicelled eukaryotes, the basis of increases in body size during growth, tissue repair, and cell replacements. Also the basis of *asexual* reproduction in single-celled and many multicelled eukaryotes.
MEIOSIS, CYTOPLASMIC DIVISION	In single-celled and multicelled eukaryotes, the basis of gamete formation and of *sexual* reproduction.
PROKARYOTIC FISSION	In bacterial cells only, the basis of *asexual* reproduction.

Further reading: Student Guide to InfoTrac on web site →

the developing spindle attach to only one of two sister chromatids of each chromosome. *At* metaphase, all the chromosomes are aligned at the spindle equator.

b. Anaphase. Microtubules pull sister chromatids of each chromosome away from each other, to opposite spindle poles. Now each type of parental chromosome is represented by a daughter chromosome at both poles:

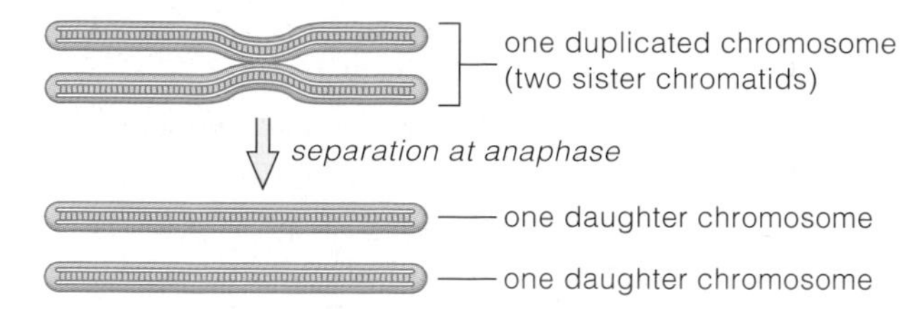

c. Telophase. Chromosomes decondense to threadlike form. A new nuclear envelope forms around them. Each nucleus has the parental chromosome number.

6. Separate mechanisms divide the cytoplasm near the end of nuclear division or afterward.

Review Questions

1. Define mitosis and meiosis, two mechanisms that operate in eukaryotic cells. Does either one divide the cytoplasm? *9.1*

2. Define somatic cell and germ cell. *9.1*

3. What is a chromosome called when it is in the unduplicated state? In the duplicated state (with two sister chromatids)? *9.1*

4. Describe the microtubular spindle and its functions. *9.3*

5. Using Figure 9.5 as a guide, name and describe the key features of the stages of mitosis. *9.3*

6. Briefly explain how cytoplasmic division differs in typical plant and animal cells. *9.4*

Self-Quiz (*Answers in Appendix III*)

1. A somatic cell having two of each type of chromosome has a(n) ______ chromosome number.
   a. diploid b. haploid c. tetraploid d. abnormal

2. A duplicated chromosome has ______ chromatid(s).
   a. one b. two c. three d. four

3. In a chromosome, a ______ is a constricted region with attachment sites for microtubules.
   a. chromatid b. cell plate c. centromere d. cleavage

4. Interphase is the part of the cell cycle when ______.
   a. a cell ceases to function
   b. a germ cell forms its spindle apparatus
   c. a cell grows and duplicates its DNA
   d. mitosis proceeds

5. After mitosis, the chromosome number of a daughter cell is ______ the parent cell's.
   a. the same as c. rearranged compared to
   b. one-half d. doubled compared to

6. Mitosis and cytoplasmic division function in ______.
   a. asexual reproduction of single-celled eukaryotes
   b. growth, tissue repair, and sometimes asexual reproduction in many multicelled eukaryotes
   c. gamete formation in prokaryotes
   d. both a and b

7. Only ______ is not a stage of mitosis.
   a. prophase b. interphase c. metaphase d. anaphase

8. Match each stage with the events listed.
   ______ metaphase a. sister chromatids move apart
   ______ prophase b. chromosomes start to condense
   ______ telophase c. chromosomes decondense and daughter nuclei form
   ______ anaphase d. all duplicated chromosomes are aligned at spindle equator

Critical Thinking

1. Suppose you have a way to measure the amount of DNA in a single cell during the cell cycle. You first measure the amount at the G1 phase. At what points during the rest of the cell cycle would you predict changes in the amount of DNA per cell?

2. The cervix is part of the uterus, in which embryos develop. A screening procedure, the *Pap smear,* detects the earliest stages of *cervical cancer*, when chances for survival are greatest. Freezing precancerous cells or hitting them with a laser beam kills them. A hysterectomy (removing the uterus) rids the uterus of cancer. If the cancer is caught early, treatment is 90+ percent effective. If it spreads, survival chances plummet to less than 9 percent.

Most cervical cancers develop slowly. Unsafe sex increases the risk (Section 45.14). Why? A major risk factor is infection by human papillomaviruses that cause genital warts. In 93 percent of all cases, viral genes coding for tumor-inducing proteins had been inserted into the DNA of previously normal cervical cells.

Not all women request Pap smears. Many wrongly believe the procedure is costly. Many do not recognize the importance of abstinence or "safe" sex. Others simply do not want to know whether they have cancer. Knowing what you have learned so far about the cell cycle and cancer, what would you say to a woman who falls into one or more of these groups?

3. Pacific yews (*Taxus brevifolius*) face extinction. People started stripping its bark and killing the trees when they heard that *taxol,* a chemical extracted from the bark, might be useful for treating breast and ovarian cancer. (Synthesizing taxol in the laboratory may save the species.) Taxol can prevent microtubules from disassembling into tubulin subunits. What does this tell you about its potential as an anticancer drug?

4. X-rays and gamma rays emitted from some radioisotopes chemically damage DNA, especially in cells engaged in DNA replication. High-level exposure can result in *radiation poisoning.* Hair loss and damage to the lining of the stomach and intestines are two early symptoms. Speculate why. Also speculate on why highly focused radiation therapy is used against some cancers.

Selected Key Terms

anaphase *9.3*
cell cycle *9.2*
cell plate formation *9.4*
centriole *9.3*
centromere *9.1*
chromosome *9.1*
chromosome number *9.1*
cleavage furrow *9.4*
cytoplasmic division *9.4*
diploid (chromosome number) *9.1*
germ cell *9.1*
HeLa cell *9.6*
histone *9.5*
interphase *9.2*
kinetochore *9.5*
meiosis *9.1*
metaphase *9.3*
mitosis *9.1*
motor protein *9.5*
nucleosome *9.5*
prophase *9.3*
reproduction *CI*
sister chromatid *9.1*
somatic cell *9.1*
spindle apparatus *9.3*
telophase *9.3*

Readings *See also www.infotrac-college.com*

Murray, A., and M. Kirschner. March 1991. "What Controls the Cell Cycle?" *Scientific American* 264(3): 56–63.

10

MEIOSIS

Octopus Sex and Other Stories

The couple clearly are interested in each other. First he caresses her with one tentacle, then another—and then another and another. She reciprocates with a hug here, a squeeze there. This goes on for hours. Finally the male reaches under his mantle, a fold of tissue that drapes around most of his body. He removes a packet of sperm from a reproductive organ and inserts it into an egg chamber underneath the female's mantle. For every sperm that successfully fertilizes an egg, a new octopus may grow and develop.

Unlike the one-to-one coupling between a male and female octopus, sex for the slipper limpet is a group enterprise. Slipper limpets are marine animals, relatives of the familiar snails on land. Before becoming transformed into a sexually mature adult, a slipper limpet must pass on through a free-living stage of development called a larva. When a limpet larva is about to undergo its programmed transformation, it settles on a rock or pebble or shell. If it settles down all by itself, it will become a female. Then, if another larva settles and develops on the first limpet, the second limpet will function right off as a male. However, if *another* limpet develops into a male on top of it, the second limpet will switch gears and develop into a female. Later, the third limpet will also become a female if still another limpet develops into a male on top of *it*—and so on amongst ten or more limpets!

Slipper limpets typically live in such piles, with the bottom one always being the oldest female and the uppermost one being the youngest male (Figure 10.1*a*).

Figure 10.1 Examples of variations in reproductive modes of eukaryotic organisms. (**a**) Slipper limpets, busily perpetuating the species by group participation in sexual reproduction. The tiny crab in the foreground is merely a passerby. (**b**) Live birth of an aphid, a type of insect that reproduces sexually in autumn but can switch to an asexual mode in summer.

Until they make the gender switch, the male limpets release sperm, which fertilize a female's eggs, which grow up to become males and then, most likely, females—and so it goes, from one sexually flexible generation to the next.

Limpets are not alone in having unusual variations in their mode of reproduction. For example, sexual reproduction is common in many life cycles—and so are asexual episodes that are based on mitotic cell divisions. Orchids, dandelions, and many other plants reproduce very well with or without engaging in sex. Aquatic animals called flatworms can engage in sex or can split their small body into two roughly equivalent parts, which each grow into a new flatworm.

And what about aphids! In summertime, nearly all aphids are females, which produce more females from *unfertilized* egg cells (Figure 10.1*b*). Only when autumn approaches do male aphids finally develop and do their part in the sexual phase of the life cycle. Even then, females that manage to survive through the winter can do without males. Come summer, the females begin another round of producing offspring all by themselves.

These examples only hint at the immense variation in reproductive modes among eukaryotic organisms. And yet, despite the variation, *sexual* reproduction dominates nearly all of the life cycles, and it always involves certain events. Briefly, before the time of cell division, chromosomes are duplicated in cells that are set aside for reproduction. For instance, the immature reproductive cells called **germ cells** develop in male and female animals. Germ cells undergo meiosis and cytoplasmic division. In time, the cellular descendants of germ cells mature and become **gametes**, or sex cells. When gametes manage to get together at fertilization, they form the first cell of a new individual.

With this chapter, we turn to the kinds of cells that serve as the bridge between generations of organisms. Specialized phases of reproduction and development, including asexual episodes, loop out from the basic life cycle of many eukaryotic species. Regardless of the specialized details, all of the life cycles turn on three events: *meiosis, the formation of gametes, and fertilization.* These three interconnected events are the hallmarks of sexual reproduction. As you will see in many chapters throughout the book, they have contributed to the diversity of life.

KEY CONCEPTS

1. Sexual reproduction proceeds through three events: meiosis, gamete formation, and fertilization. Sperm and eggs are familiar gametes.

2. Meiosis, a nuclear division mechanism, occurs only in cells that are set aside for sexual reproduction. The immature germ cells of male and female animals are examples. Meiosis sorts out a germ cell's chromosomes into four new nuclei. After meiosis, gametes form by way of cytoplasmic division and other events.

3. Cells with a diploid chromosome number contain two of each type of chromosome characteristic of the species. The two function as a pair during meiosis. Commonly, one chromosome of the pair is maternal, with hereditary instructions from a female parent. The other is paternal, with the same categories of hereditary instructions from a male parent.

4. Meiosis divides the chromosome number by half for each forthcoming gamete. Thus, if both parents have a diploid chromosome number ($2n$), the gametes that form will be haploid (n). Later, union of two gametes at fertilization restores the diploid number in the new individual ($n + n = 2n$).

5. During meiosis, each pair of chromosomes may swap segments. Each time they do, they exchange hereditary information about certain traits. Also, meiosis assigns one of each pair of chromosomes to a forthcoming gamete—but *which* gamete is its destination is a matter of chance. Hereditary instructions are further shuffled at fertilization. All three reproductive events lead to variations in traits among offspring.

6. In most plants, spore formation and other events intervene between meiosis and gamete formation.

10.1 COMPARING SEXUAL WITH ASEXUAL REPRODUCTION

When an orchid, flatworm, or aphid reproduces all by itself, what sort of offspring does it get? By the process of **asexual reproduction**, one parent alone produces offspring, and each offspring inherits the same number and kinds of genes as its parent. **Genes** are particular stretches of chromosomes—that is, of DNA molecules. Taken together, the genes for every species contain all of the heritable bits of information that are necessary to produce new individuals. Rare mutations aside, this means asexually produced offspring can only be clones: genetically identical copies of the parent.

Inheritance gets much more interesting with **sexual reproduction**. This process involves meiosis, gamete formation, and fertilization (union of the nuclei of two gametes). In humans and other sexually reproducing species, the first cell of a new individual has *pairs of genes* on pairs of homologous chromosomes. Typically, one of each pair is maternal and the other paternal in origin.

If instructions encoded in every pair of genes were identical down to the last detail, sexual reproduction would produce clones, also. Just imagine—you, every single person you know, the entire human population might be a clone, with everybody looking alike. But the two genes of a pair may *not* be identical. Why not? The molecular structure of genes can change; this is what we mean by mutation. Depending on their structure, two genes that happen to be paired in a person's cells may "say" slightly different things about a trait. Each unique molecular form of the same gene is called an **allele**.

Such tiny differences affect thousands of traits. For example, whether your chin has a dimple depends on which pair of alleles you inherited at one chromosome location. One kind of allele at that location says "put a dimple in the chin," another kind says "no dimple." This leads us to a key reason why members of sexually reproducing species don't all look alike. *Through sexual reproduction, offspring inherit new combinations of alleles, which lead to variations in the details of their traits.*

This chapter describes the cellular basis of sexual reproduction. More importantly, it starts you thinking about far-reaching effects of gene shufflings at different stages of the process. The process introduces variations in traits among offspring that are typically acted upon by agents of natural selection. Thus, *variation in traits is a foundation for evolutionary change.*

Asexual reproduction produces genetically identical copies of the parent. Sexual reproduction introduces variations in the details of traits among offspring.

Sexual reproduction dominates the life cycles of eukaryotic species. Meiosis, formation of gametes, and fertilization are the basic events of this process.

10.2 HOW MEIOSIS HALVES THE CHROMOSOME NUMBER

Think "Homologues"

Think back on the preceding chapter and its focus on mitotic cell division. Unlike mitosis, **meiosis** divides chromosomes into separate parcels not once *but twice* prior to cell division. Unlike mitosis, it is the first step leading to the formation of gametes.

Gametes, recall, are sex cells such as sperm or eggs. In nearly all multicelled eukaryotic organisms, gametes develop from cells that arise in specialized reproductive structures or organs. Figure 10.2 shows a few examples of where gametes form.

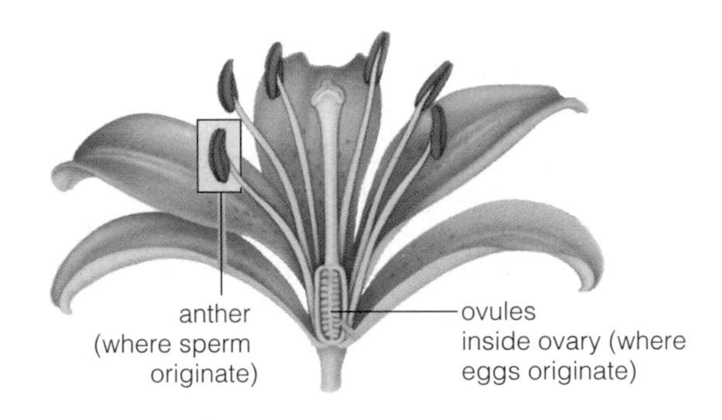

a FLOWERING PLANT

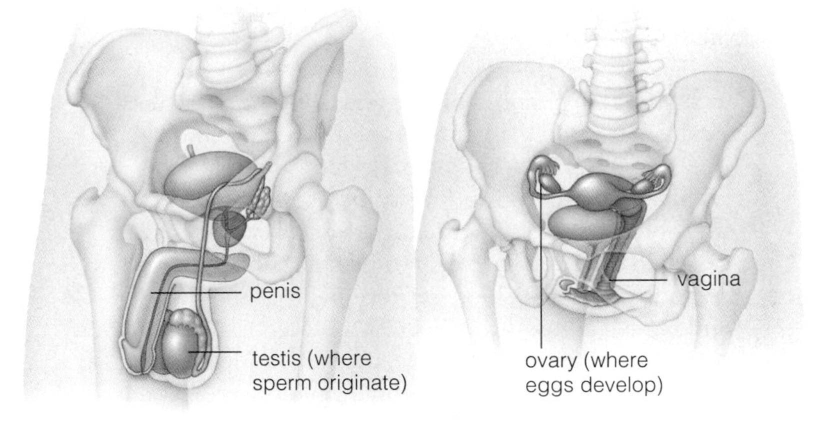

b HUMAN MALE

c HUMAN FEMALE

Figure 10.2 Examples of gamete-producing structures.

As you know, the **chromosome number** is the sum total of chromosomes in cells of a given type (Section 9.1). Germ cells start out with the same chromosome number as somatic cells (the rest of the body's cells). If a cell has a **diploid number** ($2n$), it has a *pair* of each type of chromosome, often from two parents. In general, the chromosomes of each pair have the same length and shape. Their genes deal with the same traits. And they line up with each other during meiosis. We call them **homologous chromosomes** (*hom-* means alike).

Further reading: Student Guide to InfoTrac on web site →

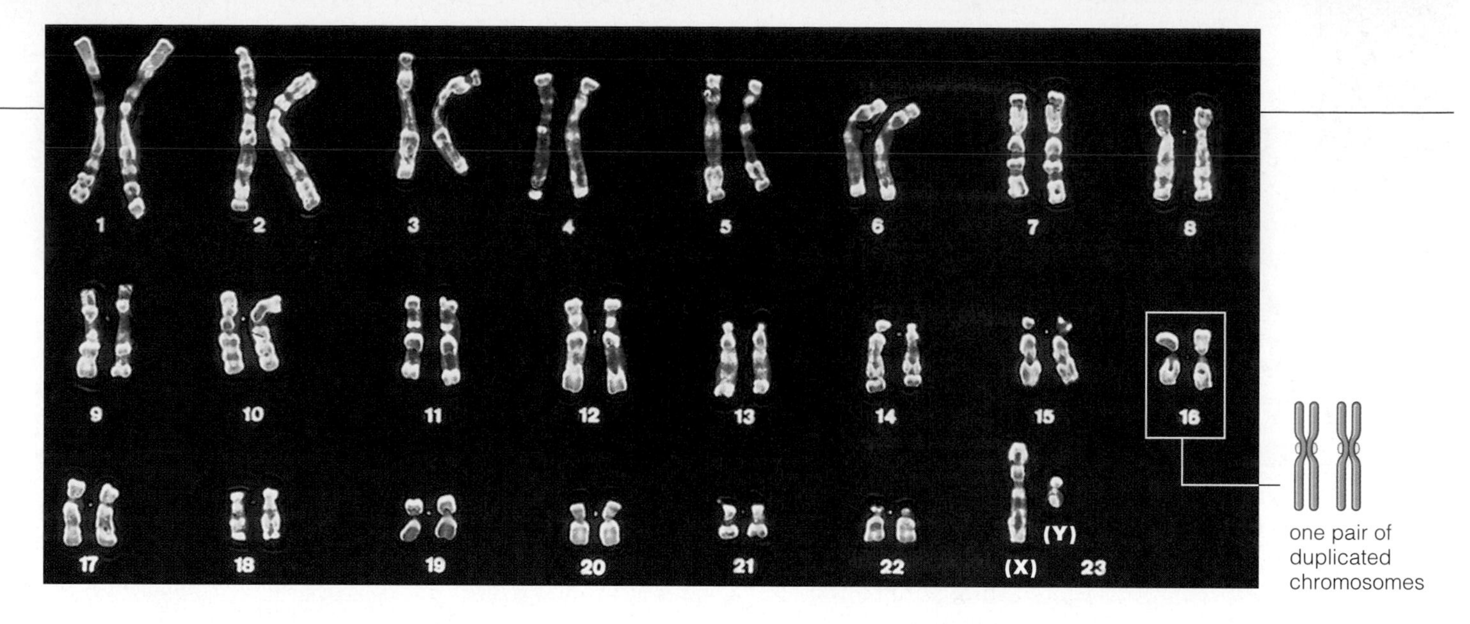

Figure 10.3 From a diploid cell of a human male, twenty-three pairs of homologous chromosomes. The sketch to the right of the photograph corresponds to two chromosomes (in the white box).

As you can probably deduce from Figure 10.3, your own germ cells have 23 + 23 homologous chromosomes. After meiosis, 23 chromosomes—one of each type—end up in gametes. Thus meiosis halves the chromosome number, so that gametes have a **haploid number** (n).

Two Divisions, Not One

Meiosis resembles mitosis in some respects, even though the outcome is different. Before interphase gives way to meiosis, a germ cell duplicates its DNA. Each duplicated chromosome now consists of two DNA molecules. These remain attached at a narrowed-down region called the centromere. For as long as the two stay attached, they are called **sister chromatids** of the chromosome:

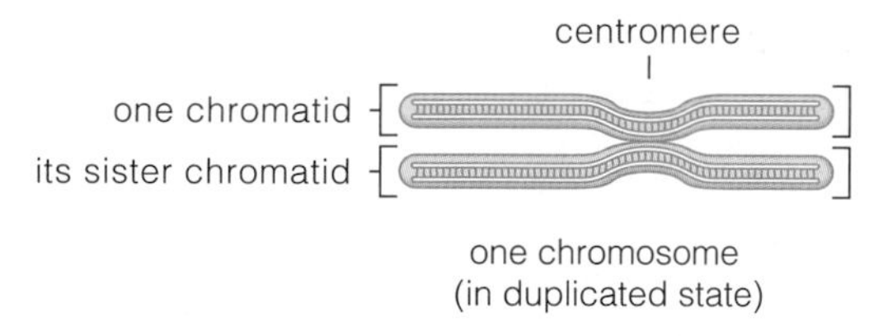

As in mitosis, the microtubules of a spindle apparatus move the chromosomes in prescribed directions.

With meiosis alone, *chromosomes proceed through two consecutive divisions, which conclude with the formation of four haploid nuclei.* We call the two nuclear divisions meiosis I and meiosis II:

DNA is replicated during interphase

MEIOSIS I
PROPHASE I
METAPHASE I
ANAPHASE I
TELOPHASE I

DNA is not replicated between divisions

MEIOSIS II
PROPHASE II
METAPHASE II
ANAPHASE II
TELOPHASE II

During meiosis I, each duplicated chromosome lines up with its partner, *homologue to homologue;* then the two partners are moved apart from each other:

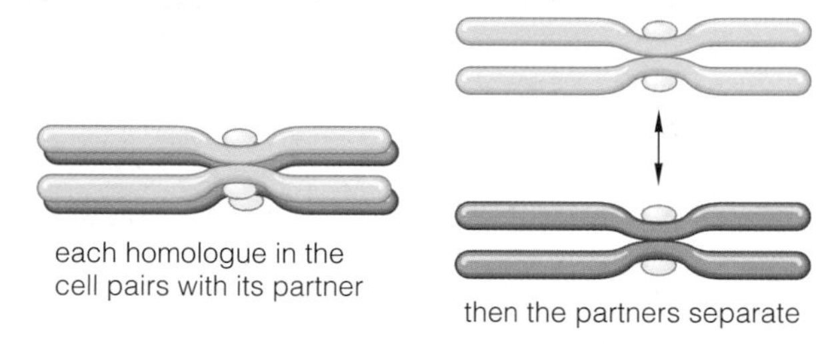

The cytoplasm typically starts to divide at some point after homologues have been separated from each other. The cytoplasmic division results in two daughter cells. Each daughter cell is haploid; it has only one of each type of chromosome. But remember, the chromosomes are still in the duplicated state.

Next, during meiosis II, *the two sister chromatids of each chromosome are separated from each other:*

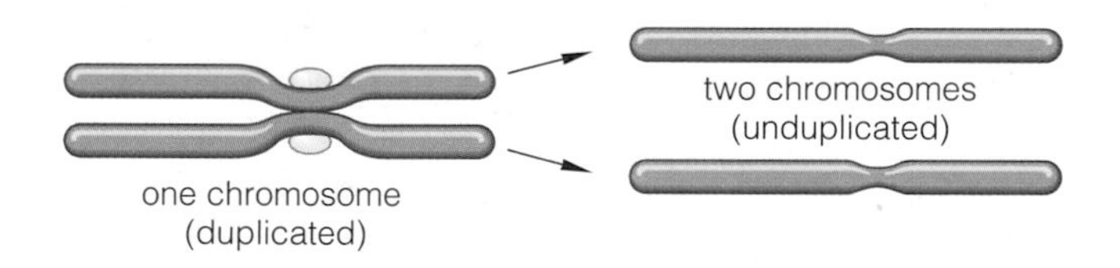

Each sister chromatid is now a chromosome in its own right. Four nuclei now form. Often the cytoplasm divides once more. The final outcome is four haploid cells.

Figure 10.4 on the next two pages illustrates the key events of meiosis I and II.

Meiosis is a type of nuclear division mechanism that reduces the parental chromosome number by half, to the haploid number (n).

Meiosis proceeds only in immature cells, such as germ cells, that are set aside for sexual reproduction. It is the first step leading to the formation of gametes.

10.3 VISUAL TOUR OF THE STAGES OF MEIOSIS

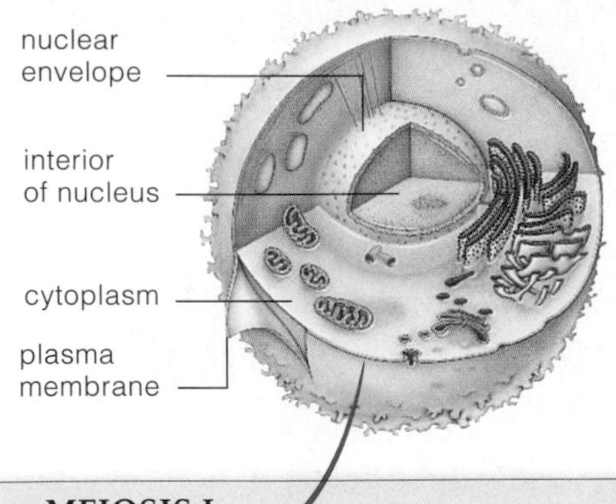

Figure 10.4 Meiosis in a generalized animal cell. This nuclear division mechanism reduces the chromosome number in an immature reproductive cell by half (to the haploid number) for forthcoming gametes. To keep things simple, only two pairs of homologous chromosomes are shown. Maternal chromosomes are shaded *purple* and paternal chromosomes, *blue*.

A GERM CELL AT INTERPHASE

A germ cell with a diploid chromosome number (2*n*) is about to leave interphase and undergo the first division of meiosis. The cell's DNA is already duplicated, so each chromosome is in the duplicated state; it consists of two sister chromatids.

MEIOSIS I

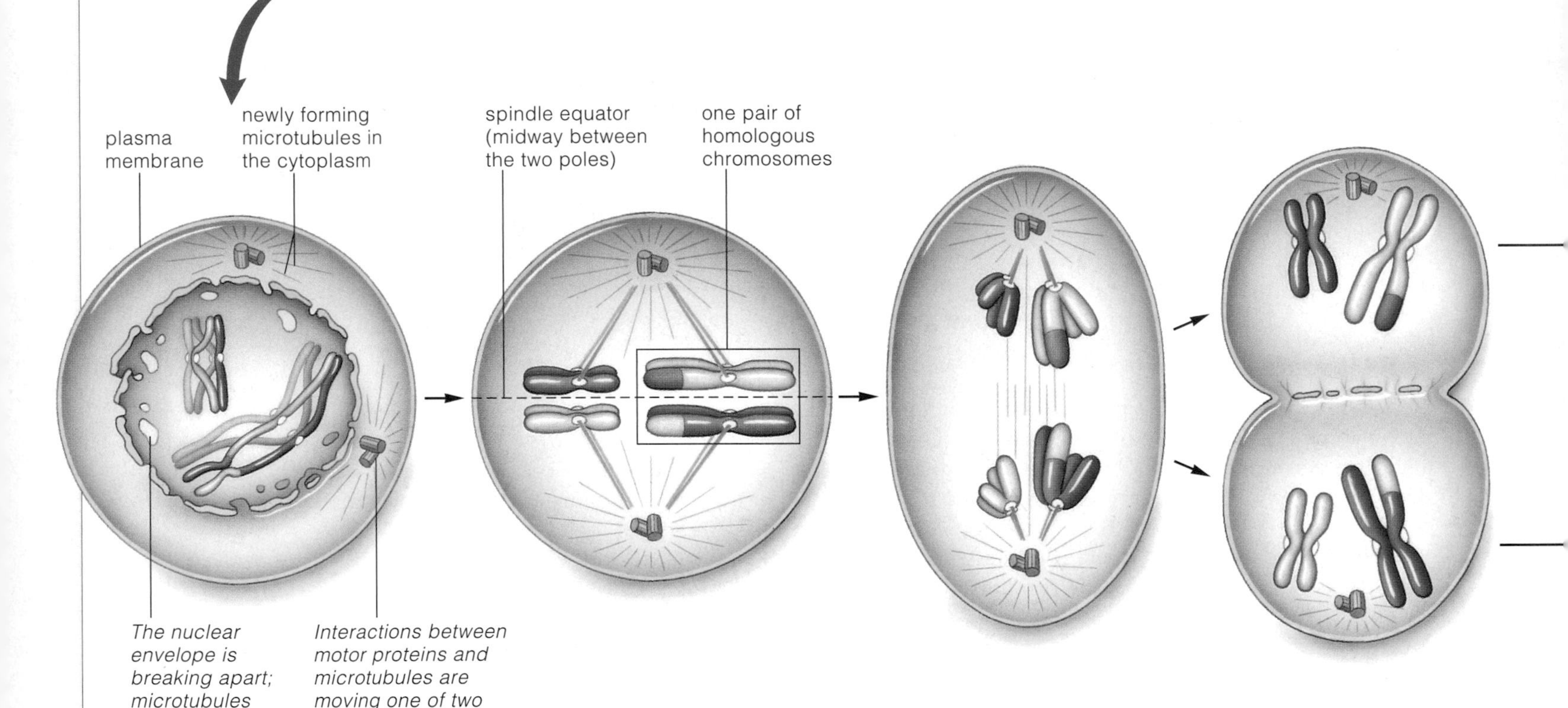

PROPHASE I

Each duplicated chromosome is in threadlike form, but now it starts to twist and fold into more condensed form. It pairs with its homologue, and the two typically swap segments. The swapping, called crossing over, is indicated by the break in color on the pair of larger chromosomes. Each chromosome becomes attached to some microtubules of a newly forming spindle.

METAPHASE I

Motor proteins have been driving the movement of microtubules that became attached to the kinetochores of chromosomes. As a result, the chromosomes have been pushed and pulled into position midway between the spindle poles. Now the spindle is fully formed, owing to dynamic interactions of motor proteins, microtubules, and the chromosomes themselves.

ANAPHASE I

Microtubules extending from the poles and overlapping at the spindle equator *lengthen* and push the poles apart. At the same time, other microtubules extending from the poles to the kinetochores of chromosomes *shorten*, and each chromosome is thereby pulled away from its homologous partner. These motions move the homologous partners to opposite poles.

TELOPHASE I

At some point, the cytoplasm of the germ cell divides. Two cells, each with a haploid chromosome number (*n*), result. That is, the cells have one of each type of chromosome that was present in the parent (2*n*) cell. All chromosomes are still in the duplicated state.

Further reading: Student Guide to InfoTrac on web site →

Of the four haploid cells that form by way of meiosis and cytoplasmic divisions, one or all may proceed to develop into gametes and function in sexual reproduction.

(In plants, the cells that form after meiosis is completed may develop into spores, which take part in a stage of the life cycle that precedes gamete formation.)

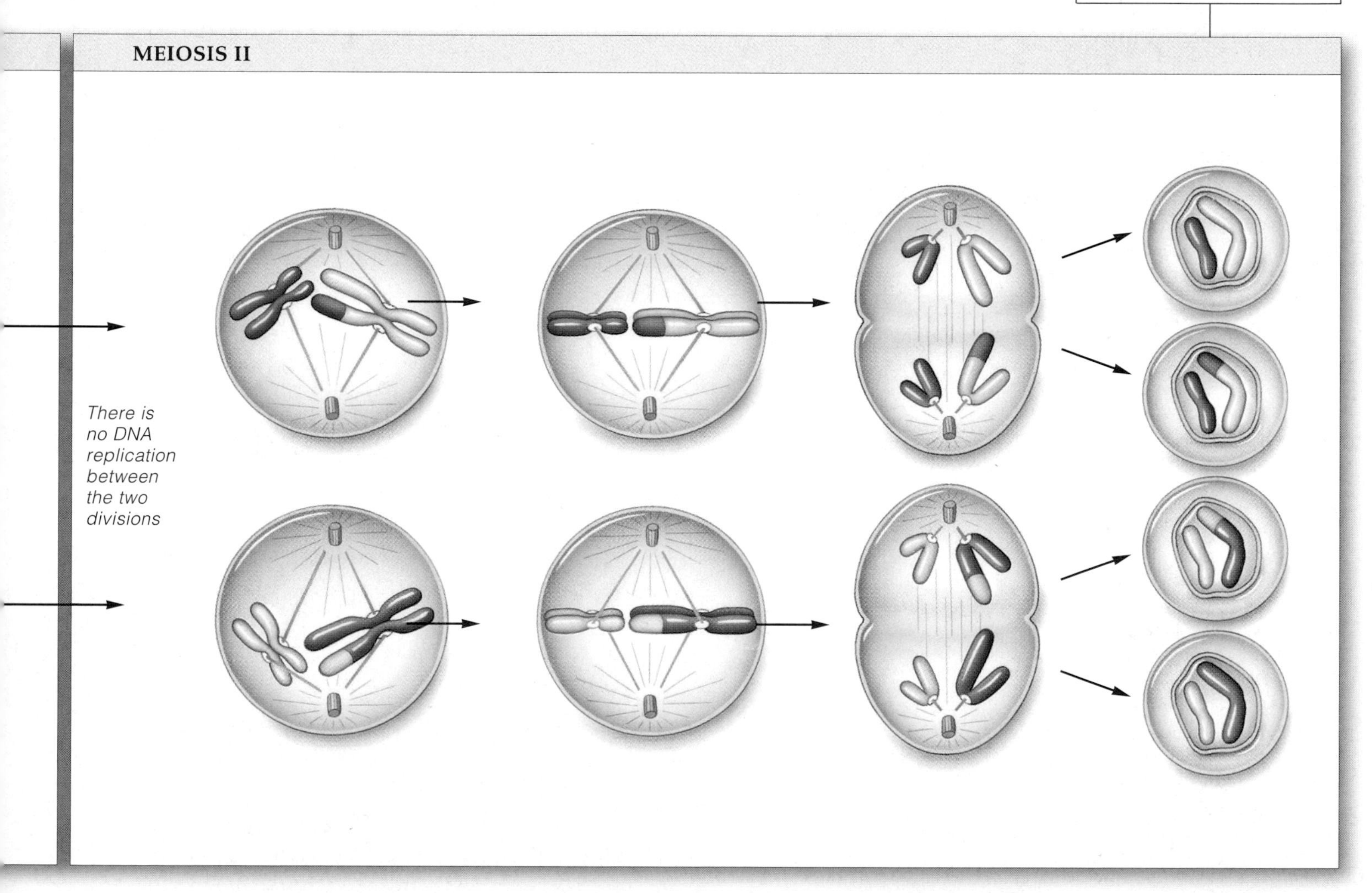

PROPHASE II

In each of the two daughter cells, microtubules have already moved one member of the centriole pair to the opposite pole of the spindle during the transition to prophase II. Now, at prophase II, microtubules attach to the kinetochores of chromosomes, and motor proteins drive the movement of chromosomes toward the spindle's equator.

METAPHASE II

Now, in each daughter cell, interactions among motor proteins, spindle microtubules, and each duplicated chromosome have moved all of the chromosomes so that they are positioned at the spindle equator, midway between the two poles.

ANAPHASE II

The attachment between the two chromatids of each chromosome breaks. Once that happens, each of the former "sister chromatids" is a chromosome in its own right. Motor proteins interact with kinetochore microtubules to move the separated chromosomes to opposite poles of the spindle.

TELOPHASE II

By the time telophase II is completed, there will be four daughter nuclei. Also, at the time when division of the cytoplasm is completed, each new, daughter cell will have a haploid chromosome number (n). All of those chromosomes will now be in the unduplicated state.

10.4 A CLOSER LOOK AT KEY EVENTS OF MEIOSIS I

The preceding overview, in Sections 10.2 and 10.3, is enough to convey the overriding function of meiosis—that is, *the reduction of the chromosome number by half for forthcoming gametes.* However, as you will now read, two other events that take place during prophase and metaphase of meiosis I contribute greatly to the adaptive advantage of sexual reproduction.

That advantage, recall, is production of offspring with new combinations of alleles. During growth and development, those combinations are translated into a new generation of individuals that differ in the details of some number of traits.

What Goes On in Prophase I?

Prophase I of meiosis is a time of major gene shufflings. Reflect on Figure 10.5*a*, which shows two chromosomes condensed to threadlike form. All chromosomes in a germ cell condense this way. As they do, each is drawn close to its homologue. Molecular interactions stitch homologues together point by point along their length, with little space between. The intimate, parallel array favors **crossing over**, a molecular interaction between two of the *non*sister chromatids of a pair of homologous chromosomes. Nonsister chromatids break at the same

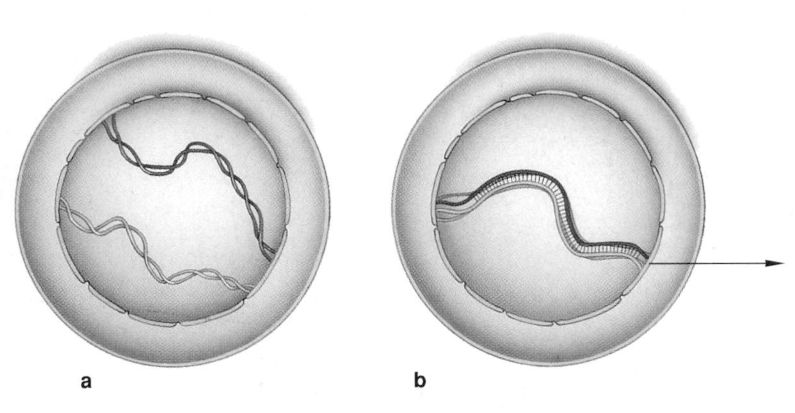

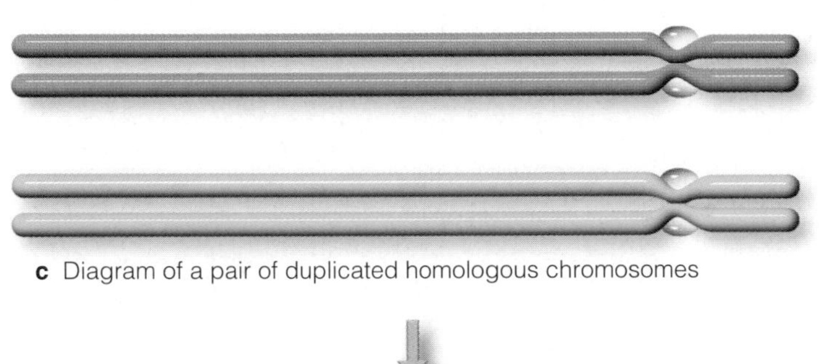

c Diagram of a pair of duplicated homologous chromosomes

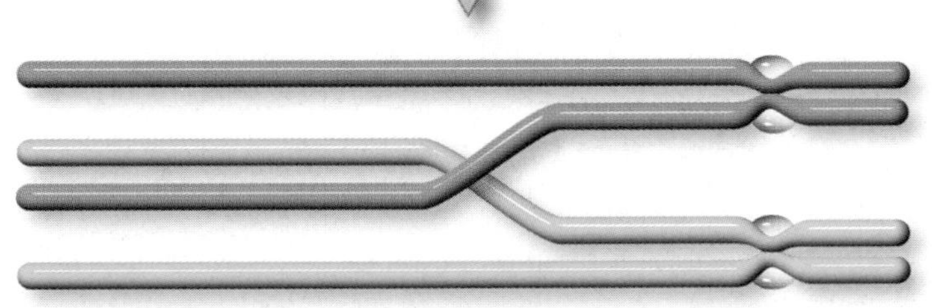

d Crossover between nonsister chromatids of the two chromosomes

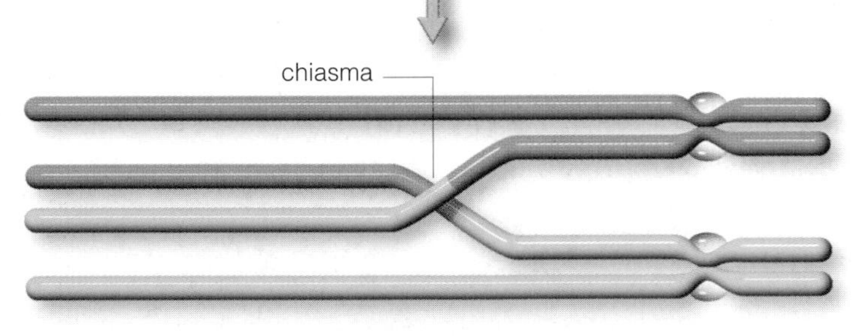

e Nonsister chromatids exchange segments

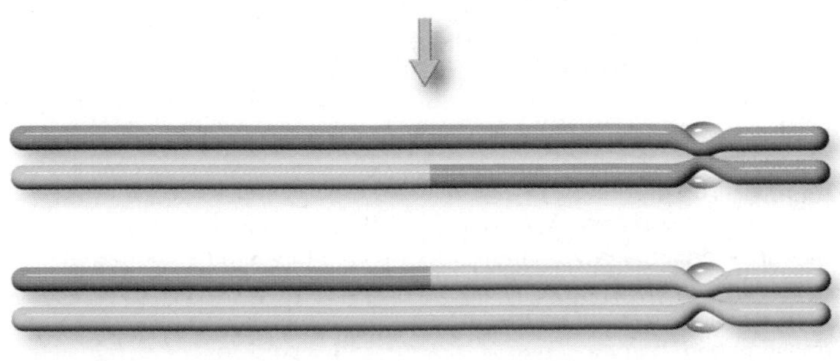

f Homologues have new combinations of alleles

Figure 10.5 Key events during prophase I, the first stage of meiosis. For clarity, this diagram of a cell shows only one pair of homologous chromosomes and one crossover event, although more than one typically occur. *Blue* signifies the paternal chromosome, and *purple* signifies its maternal homologue.

(**a**) Both of the chromosomes were duplicated earlier, during interphase. Early in prophase I, the two sister chromatids of each duplicated chromosome are in thin, threadlike form. They are all positioned so closely together, they look like a single thread.

(**b**) Each chromosome becomes zippered to its homologue, so all four chromatids are intimately aligned. When the two sex chromosomes have different forms (such as X paired with Y) they still get zippered together, although only in a small region along their length.

(**c**,**d**) We show the pair of chromosomes as if they were already condensed, then teased apart so that you can visualize what goes on. Bear in mind, their double-stranded DNA molecules are still tightly aligned at this stage. The intimate contact allows one (and usually more) crossover to occur at intervals along the length of nonsister chromatids.

(**e**) Nonsister chromatids exchange segments. As prophase I ends, the chromosomes continue to condense into thicker, rodlike forms. Then they unzipper from each other except at places where they physically cross each other. Those places are called chiasmata (singular, chiasma, meaning "cross"). Each chiasma does not last long, but it is indirect evidence that a crossover occurred at some place in the chromosomes.

(**f**) What is the function of crossing over? It breaks up old combinations of alleles and puts new ones together in pairs of homologous chromosomes.

Further reading: Student Guide to InfoTrac on web site →

places along their length. At these break points, they exchange corresponding segments—that is, genes.

Gene swapping would be rather pointless if each type of gene never varied from one chromosome to the next. But remember, a gene can have slightly different forms: alleles. You can safely bet that some number of alleles on one chromosome will *not* be identical to their allelic partners on the homologue. Thus each crossover represents a chance to swap slightly different versions of hereditary instructions for particular traits.

We will look at the mechanism of crossing over in later chapters. For now, it is enough to remember this: *Crossing over leads to genetic recombination, which in turn leads to variation in the traits of offspring.*

Metaphase I Alignments

Major shufflings of whole chromosomes begin during the transition from prophase I to metaphase I, which is the second stage of meiosis. Suppose the shufflings are proceeding at this moment in one of your germ cells. By now, crossovers have made genetic mosaics of the chromosomes, but put this aside in order to simplify tracking. Just call the twenty-three chromosomes you inherited from your mother the *maternal* chromosomes and their twenty-three homologues from your father the *paternal* chromosomes.

Kinetochore microtubules have already oriented one chromosome of each pair toward one spindle pole and its homologue toward the other pole (refer to Section 10.3). Now they are moving all the chromosomes, which soon will become positioned at the spindle's equator.

Have all maternal chromosomes become attached to one spindle pole and all paternal chromosomes to the other pole? Maybe, but probably not. Remember, the initial contacts between kinetochore microtubules and the chromosomes are random. Because of the random grabs, the eventual positioning of maternal or paternal chromosomes at the spindle's equator at metaphase I follows no particular pattern. Now carry this thought one step further. *Either one* of each pair of homologous chromosomes can end up at either pole of the spindle after they move apart at anaphase I.

Think about the possibilities when you are tracking merely three pairs of homologues. As you can see from Figure 10.6, by metaphase I, three pairs of homologues may be arranged in any one of four possible positions. Here, eight combinations (2^3) of maternal and paternal chromosomes are possible for forthcoming gametes.

Of course, a human germ cell has twenty-three pairs of homologous chromosomes, not just three. So a grand total of 2^{23}—or *8,388,608*—combinations of maternal and paternal chromosomes is possible every time a human germ cell gives rise to sperm or eggs!

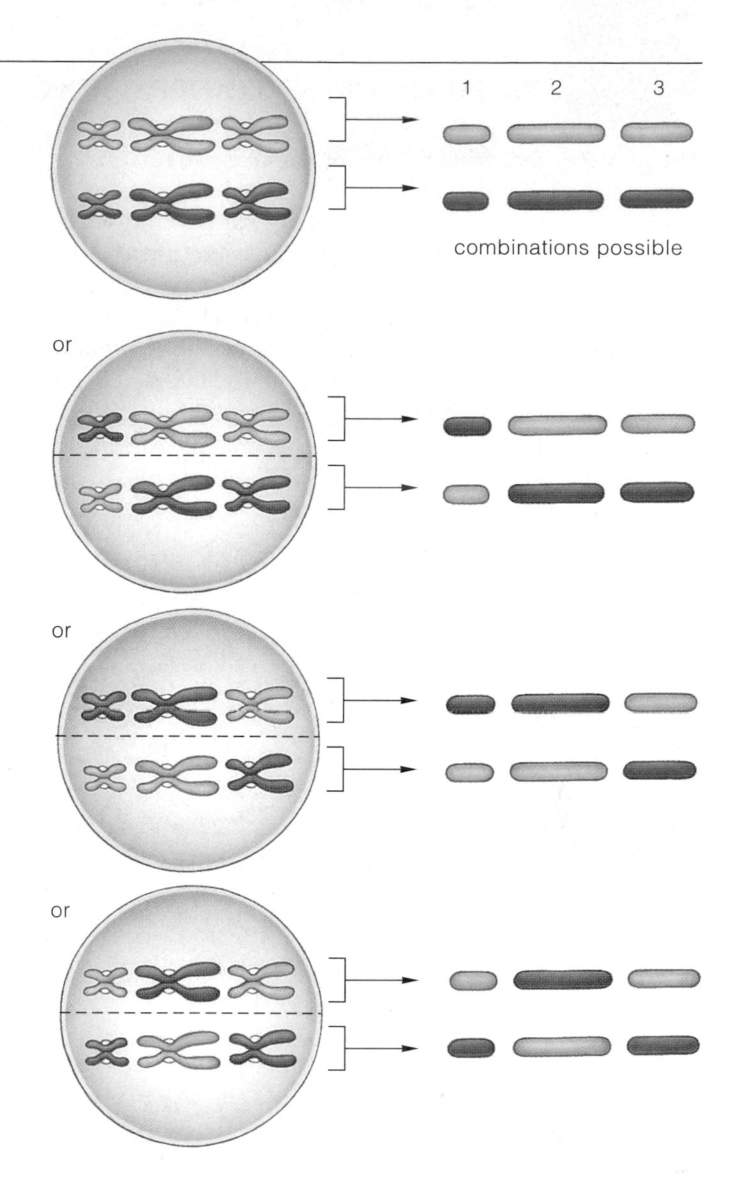

Figure 10.6 Possible outcomes of the random alignment of only three pairs of homologous chromosomes at metaphase I. Three types of chromosomes are labeled 1, 2, and 3. Maternal chromosomes are *purple;* paternal ones are *blue.* With merely four possible alignments, eight combinations of maternal and paternal chromosomes are possible in forthcoming gametes.

Moreover, in each sperm or egg, many hundreds of alleles inherited from the mother might not "say" the exact same thing about hundreds of different traits as the alleles inherited from the father. Are you beginning to get an idea of why such fascinating combinations of traits show up even in the same family?

Crossing over is an interaction between a pair of homologous chromosomes. It breaks up old combinations of alleles and puts new ones together during prophase I of meiosis.

The random attachment and subsequent positioning of each pair of maternal and paternal chromosomes at metaphase I lead to different combinations of maternal and paternal traits in each generation of offspring.

10.5 FROM GAMETES TO OFFSPRING

The gametes that form following meiosis are not all the same in their details. For example, human sperm have one tail, opossum sperm have two, and roundworm sperm have none. Crayfish sperm look like pinwheels. Most eggs are microscopic in size, yet an ostrich egg tucked inside its shell is as large as a baseball. From its appearance alone, you might not believe that a plant gamete is even remotely like an animal's.

Later chapters contain details of how gametes form in the life cycles of representative organisms, including humans. Figure 10.7 and the following points may help you keep the details in perspective.

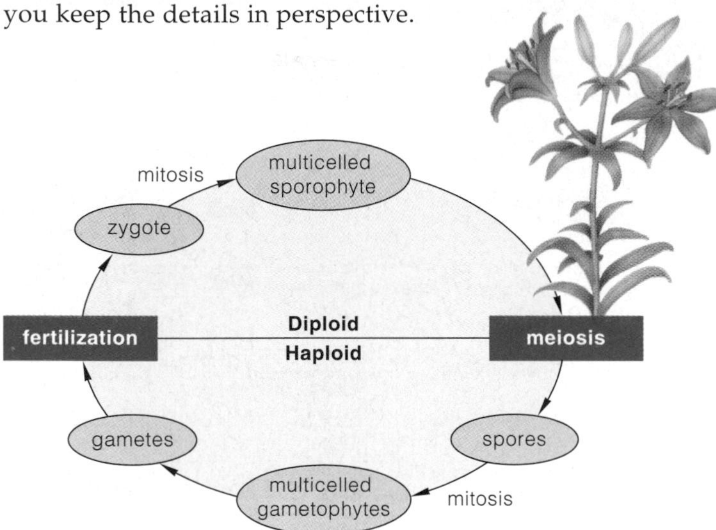

a Generalized life cycle for most kinds of plants

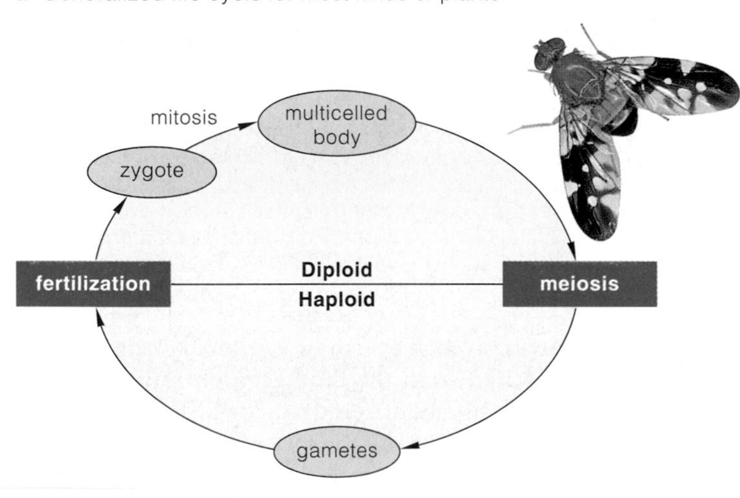

b Generalized life cycle for animals

Figure 10.7 Generalized life cycles for (**a**) most plants and (**b**) animals. The zygote is the first cell that forms when the nuclei of two gametes fuse together at fertilization.

For plants, a sporophyte (spore-producing body) develops, by way of mitotic cell divisions, from the zygote. After meiosis, gametophytes (gamete-producing bodies) form. A lily plant is a sporophyte. Gametophytes form in parts of its flowers.

Chapters 22 through 26, 31, and 44 include specific examples of life cycles of representative organisms.

Gamete Formation in Plants

For pine trees, apple trees, roses, dandelions, corn, and nearly all other familiar plants, certain events intervene between the times of meiosis and gamete formation. Among other things, spores form.

Spores are haploid resting cells, often walled, that are good at resisting drought, cold, and other adverse environmental conditions. When favorable conditions return, spores germinate (resume growth) and develop into a haploid body or structure that produces gametes. So *gamete*-producing bodies and *spore*-producing bodies develop during the life cycle of most kinds of plants. Figure 10.7*a* is a generalized diagram of these events.

Gamete Formation in Animals

In male animals, gametes form by a process known as spermatogenesis. Inside the male reproductive system, a diploid germ cell grows in size. It becomes a primary spermatocyte, a large immature cell that enters meiosis and cytoplasmic divisions. Four haploid cells result and develop into spermatids (Figure 10.8). These immature cells change in form and develop a tail, thus becoming a **sperm**, a common type of mature male gamete.

In female animals, gametes form by a process that is called oogenesis. In human females, for example, a diploid germ cell develops into an **oocyte**, or immature egg. Unlike a sperm cell, an oocyte accumulates many cytoplasmic components. Also, as Figure 10.9 indicates, its daughter cells differ in size and function. When the oocyte undergoes cytoplasmic division after meiosis I, one cell—the secondary oocyte—gets nearly all of the cytoplasm. The other cell, called the first polar body, is quite small. Some time afterward, both cells undergo meiosis II and then cytoplasmic division. One daughter cell of the secondary oocyte develops into a second polar body. The other gets most of the cytoplasm and develops into the gamete. A mature female gamete is called an ovum (plural, ova) or, more commonly, an **egg**.

Thus, one egg and three polar bodies have formed. Polar bodies don't have much in the way of nutrients or metabolic machinery and don't function as gametes. In time, they degenerate. But polar body formation allows the egg to end up with a suitable (haploid) number of chromosomes. Also, by getting most of the cytoplasm, the egg receives enough start-up machinery to support the new individual right after fertilization.

More Shufflings at Fertilization

The chromosome number characteristic of the parents is restored at **fertilization**, the time when a male gamete unites with a female gamete and their haploid nuclei

Further reading: Student Guide to InfoTrac on web site →

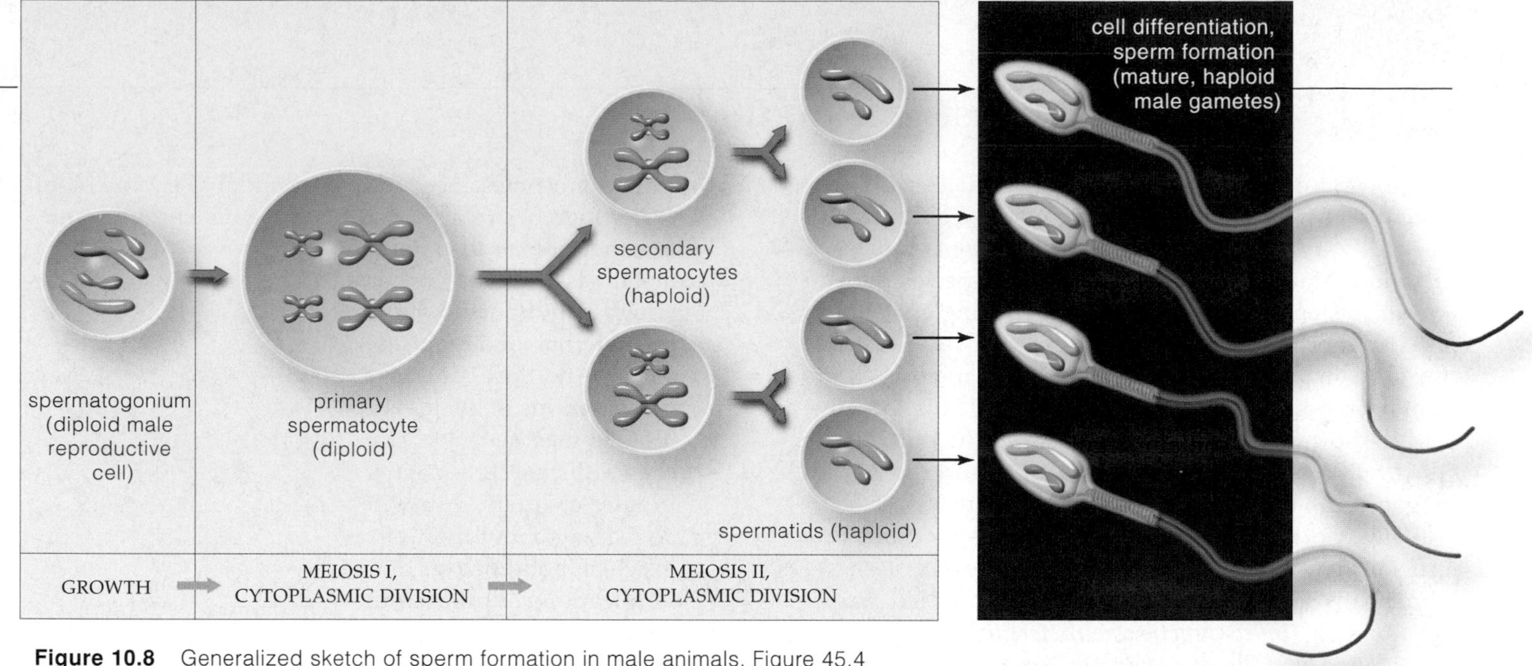

Figure 10.8 Generalized sketch of sperm formation in male animals. Figure 45.4 shows a specific example (how sperm form in human males).

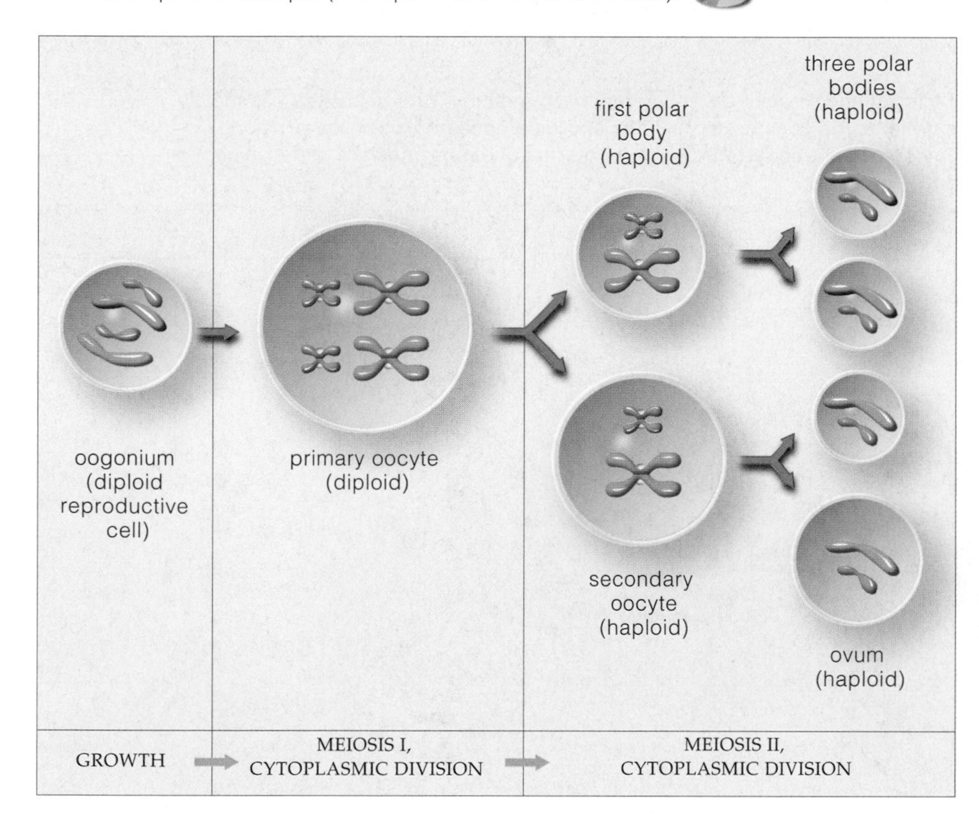

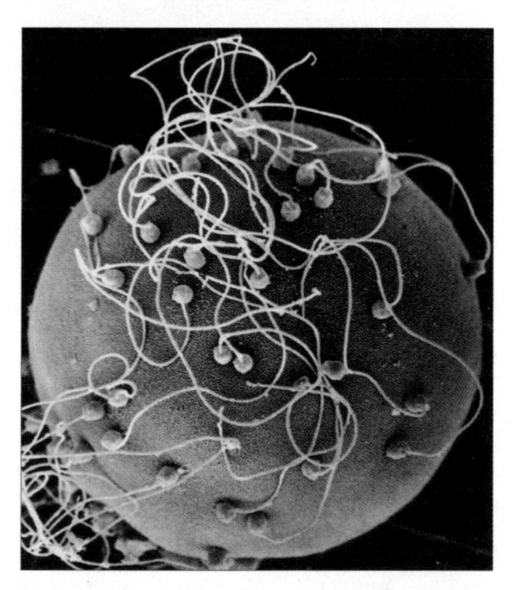

Figure 10.9 Egg formation in female animals. In animals, eggs are far larger than sperm, as the micrograph of sea urchin gametes (*above*) indicates. In addition, the three polar bodies are much smaller than an egg. Figure 45.7 shows a specific example (how eggs form in human females).

fuse. If meiosis did not precede it, fertilization would double the chromosome number each new generation. Such changes in chromosome number would disrupt the hereditary instructions encoded in chromosomes, usually for the worse. Why? Those instructions operate as a complex, fine-tuned package in each individual.

Fertilization contributes to variation in offspring. Reflect on the possibilities for humans alone. During prophase I, an average of two or three crossovers takes place in each human chromosome. Even without the crossovers, the random positioning of pairs of paternal and maternal chromosomes at metaphase I results in one of millions of possible chromosome combinations in each gamete. And of all the male and female gametes that are produced, which two actually get together is a matter of chance. The sheer number of combinations that can exist at fertilization is staggering!

Cumulatively, crossing over, the distribution of random mixes of homologous chromosomes into gametes, and fertilization contribute to variation in the traits of offspring.

10.6 MEIOSIS AND MITOSIS COMPARED

In this unit our focus has been on two nuclear division mechanisms. Single-celled eukaryotic species reproduce asexually by way of mitosis, followed by cytoplasmic division. Many multicelled eukaryotic species depend on mitosis and cytoplasmic division during episodes of asexual reproduction in their life cycle. All depend on it for growth and tissue repair. By contrast, meiosis occurs only in reproductive cells, such as germ cells that give rise to the gametes used in sexual reproduction. Figure 10.10 summarizes the basic similarities and differences between the two nuclear division mechanisms.

The end results of the two mechanisms differ in a crucial way. *Mitotic cell division only produces clones*—genetically identical copies of a parent cell. *But meiotic cell division, in conjunction with fertilization, promotes variation in traits among offspring*. First, crossing over at prophase I of meiosis puts new combinations of alleles in chromosomes. Second, the random assignment of either member of a pair of homologous chromosomes to either pole of the spindle at metaphase I affects gametes, which end up with mixes of maternal and paternal alleles. And third, different combinations of alleles are brought together simply by chance during fertilization. In later chapters, you will be reading about the ways in which both meiosis and fertilization contributed to the truly stunning diversity and evolution of sexually reproducing organisms.

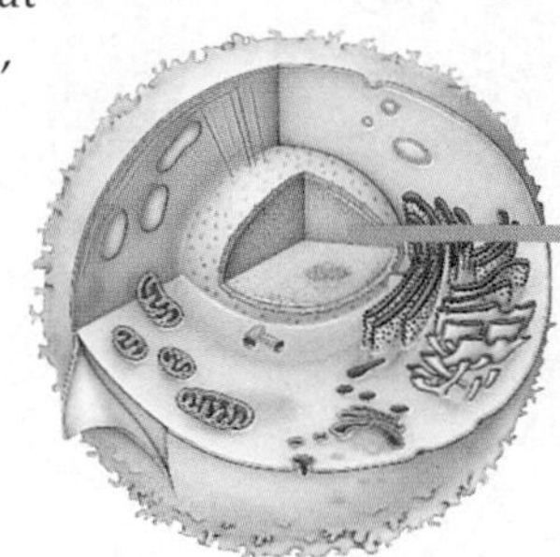

A *somatic cell* with a diploid chromosome number ($2n$) is at interphase. Before mitotic division begins, its DNA is replicated (all chromosomes are duplicated).

Figure 10.10 Summary of mitosis and meiosis. Both diagrams use a diploid ($2n$) animal cell as the example. They are arranged to help you compare similarities and differences between the division mechanisms. Maternal chromosomes are coded *purple*, and paternal chromosomes are coded *blue*.

MEIOSIS I

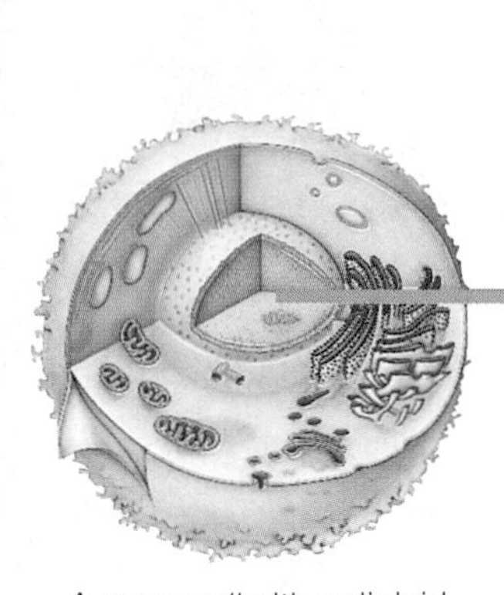

A *germ cell* with a diploid chromosome number ($2n$) is at interphase. Before mitotic division begins, its DNA is replicated (all chromosomes are duplicated).

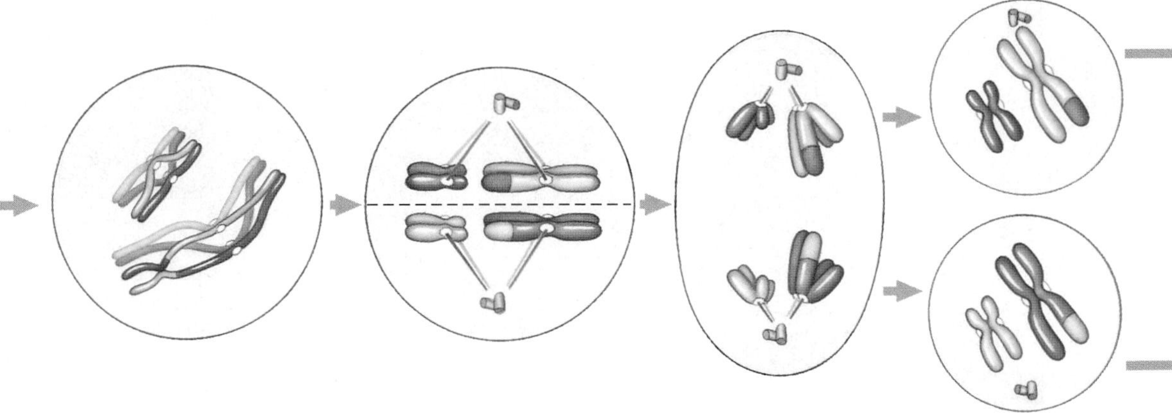

PROPHASE I

Each duplicated chromosome (consisting of two sister chromatids) condenses to threadlike form, then rodlike form. *Crossing over* occurs. Each chromosome unzips from its homologue. Each gets attached to the spindle in transition to metaphase.

METAPHASE I

All chromosomes are now positioned at the spindle's equator.

ANAPHASE I

Each chromosome is separated from its homologue. They are moved to opposite poles of the spindle.

TELOPHASE I

When the cytoplasm divides, there are two cells. Each has a haploid (n) number of chromosomes, but these are still in the duplicated state.

Further reading: Student Guide to InfoTrac on web site →

MITOSIS

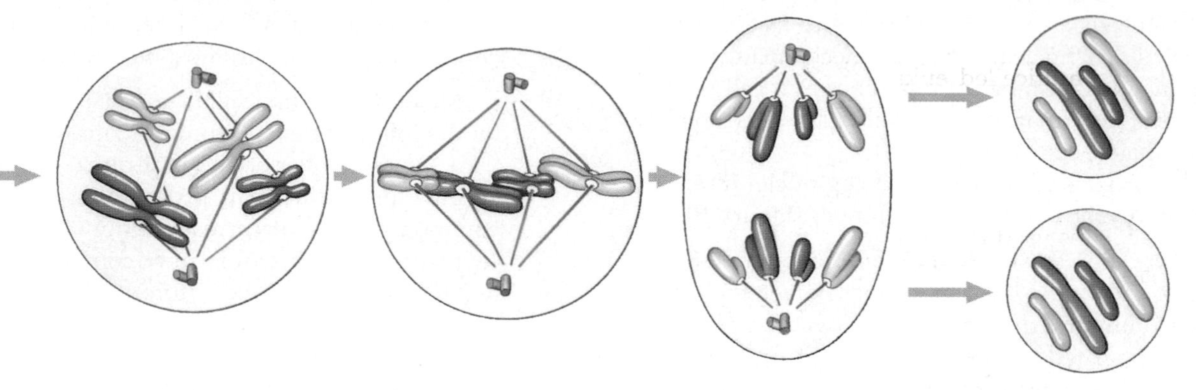

PROPHASE

Each duplicated chromosome (consisting of two sister chromatids) condenses from threadlike form to rodlike form. Each gets attached to the spindle during the transition to metaphase.

METAPHASE

All chromosomes are now positioned at the spindle's equator.

ANAPHASE

Sister chromatids of each chromosome are separated from each other. These new, daughter chromosomes are moved to opposite poles of the spindle.

TELOPHASE

When the cytoplasm divides, there are two cells. Each is diploid (2*n*)—*it has the same chromosome number as the parent cell.*

MEIOSIS II

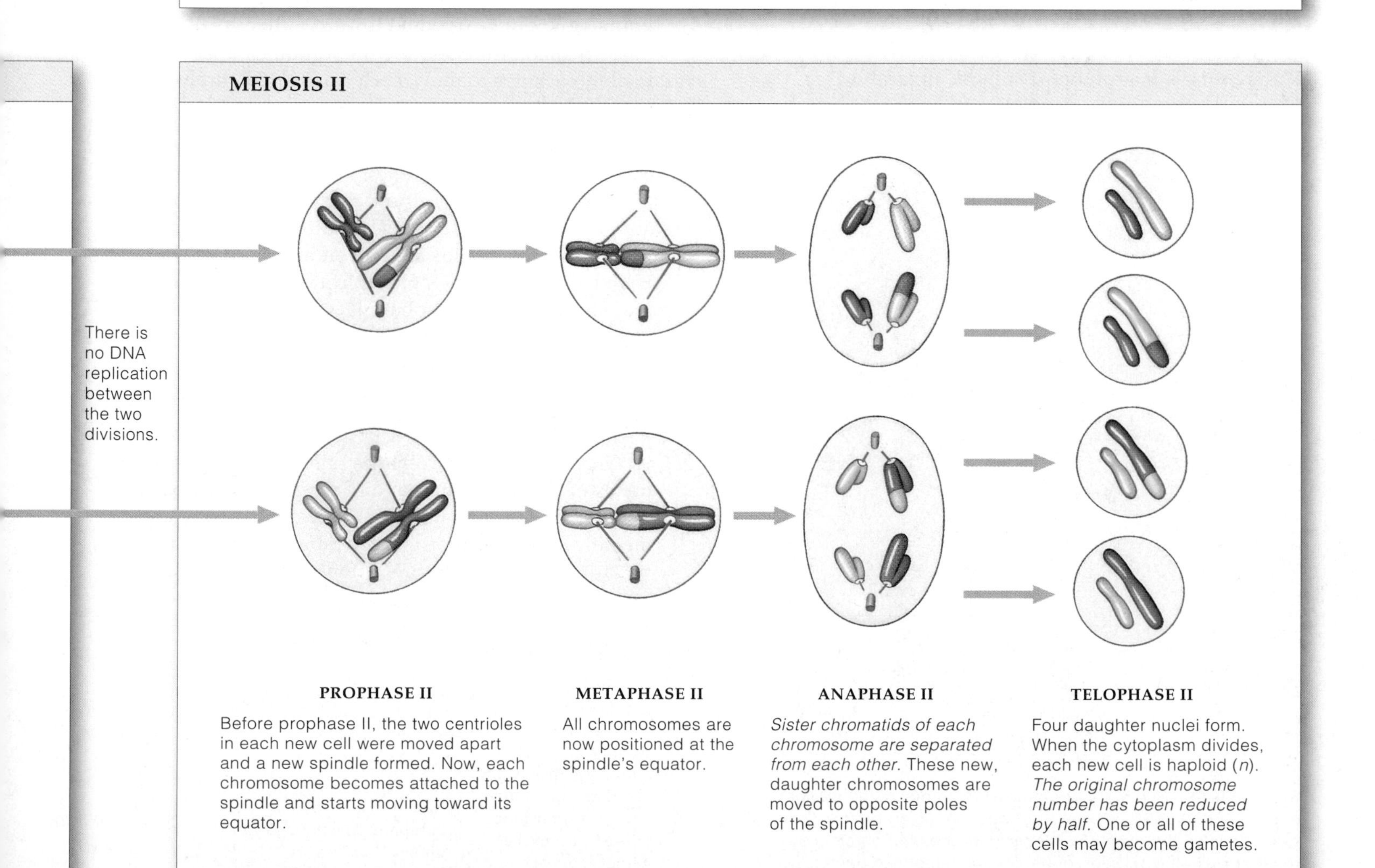

There is no DNA replication between the two divisions.

PROPHASE II

Before prophase II, the two centrioles in each new cell were moved apart and a new spindle formed. Now, each chromosome becomes attached to the spindle and starts moving toward its equator.

METAPHASE II

All chromosomes are now positioned at the spindle's equator.

ANAPHASE II

Sister chromatids of each chromosome are separated from each other. These new, daughter chromosomes are moved to opposite poles of the spindle.

TELOPHASE II

Four daughter nuclei form. When the cytoplasm divides, each new cell is haploid (*n*). *The original chromosome number has been reduced by half.* One or all of these cells may become gametes.

10.7 SUMMARY

1. The life cycle of each sexually reproducing species includes meiosis, gamete formation, and fertilization.

a. Meiosis, a nuclear division mechanism, reduces the chromosome number of a parent germ cell by half. It precedes the formation of haploid gametes (typically, sperm in males, eggs in females).

b. At fertilization, a sperm and egg nuclei fuse. This event restores the chromosome number (Figure 10.11).

2. A germ cell with a diploid chromosome number ($2n$) has *two* of each type of chromosome characteristic of its species. Commonly, one of each pair of chromosomes is maternal (inherited from a female parent) and the other is paternal (from a male parent).

3. Each pair of maternal and paternal chromosomes shows homology, meaning the two are alike. Generally the two have the same length, same shape, and same sequence of genes. And they interact during meiosis.

4. Chromosomes become duplicated during interphase. Each consists of two DNA molecules that will remain attached, as sister chromatids, during mitosis.

5. Meiosis consists of two consecutive divisions, which both require a microtubular spindle apparatus.

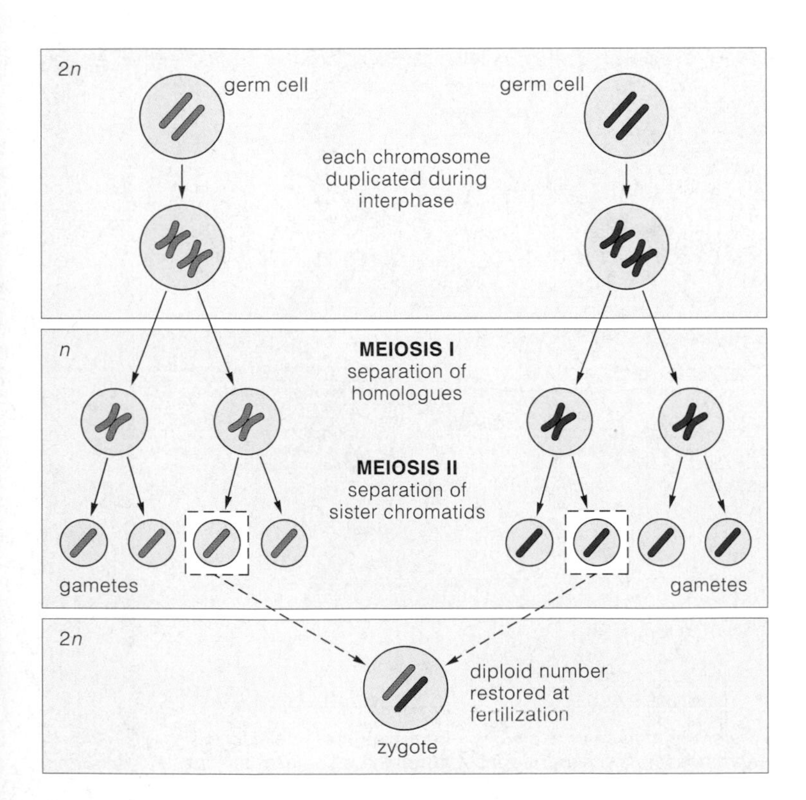

Figure 10.11 Summary of changes in chromosome number at different stages of sexual reproduction, using diploid ($2n$) germ cells as the example. Meiosis reduces the chromosome number by half (n). Union of haploid nuclei of two gametes at fertilization restores the diploid number.

a. In meiosis I, kinetochores (of chromosomes) and motor proteins (projecting from microtubules) interact to move each duplicated chromosome away from its partner, the homologous chromosome.

b. In meiosis II, similar interactions move the sister chromatids of each chromosome away from each other.

6. Meiosis I, the first nuclear division, is characterized by the following events and outcomes:

a. Crossing over occurs in prophase I. Two nonsister chromatids of each pair of homologous chromosomes commonly break at corresponding sites and exchange segments. And this puts new combinations of alleles together. Alleles (slightly different molecular forms of the same gene) specify different forms of the same trait.

b. Different combinations of alleles lead to variation in the details of a given trait among offspring.

c. Also in prophase I, a microtubular spindle forms outside the nucleus, and the nuclear envelope starts to break up. In cells with duplicated pairs of centrioles, one pair starts moving to the opposite spindle pole.

d. All the pairs of homologous chromosomes have become positioned at the spindle equator at metaphase I. For each pair, either the maternal chromosome or its homologue can be oriented toward either pole.

e. In anaphase I, spindle microtubules interact with the kinetochores to move each duplicated chromosome away from its homologue, to opposite spindle poles.

7. Meiosis II (second nuclear division) is characterized by these events and outcomes:

a. At metaphase II, all the duplicated chromosomes are positioned at the spindle equator.

b. Sister chromatids are moved apart in anaphase II. Each is now a separate, unduplicated chromosome.

c. By the end of telophase II, four nuclei that have a haploid chromosome number (n) have been formed.

8. When the cytoplasm divides, there are four haploid cells. One or all of these may function as gametes (or as plant spores that give rise to gamete-producing bodies).

9. Crossing over, the chance allocation of different mixes of pairs of maternal and paternal chromosomes to different gametes, and the chance of any two gametes meeting at fertilization all contribute to the immense variation in details of traits among offspring.

Review Questions

1. The diploid chromosome numbers for the somatic cells of a few organisms are listed at right. How many chromosomes will end up in the gametes of each organism? *10.2*

Fruit fly, *Drosophila melanogaster*	8
Garden pea, *Pisum sativum*	14
Corn, *Zea mays*	20
Frog, *Rana pipiens*	26
Earthworm, *Lumbricus terrestris*	36
Human, *Homo sapiens*	46
Chimpanzee, *Pan troglodytes*	48
Amoeba, *Amoeba*	50
Horsetail, *Equisetum*	216

2. A diploid germ cell has four pairs of homologous chromosomes, designated AA, BB, CC, and DD. How would the chromosomes of the gametes be designated? *10.2*

3. Look at the chromosomes in the germ cell in the diagram at the right. Is this cell at anaphase I or anaphase II? *10.3, 10.6*

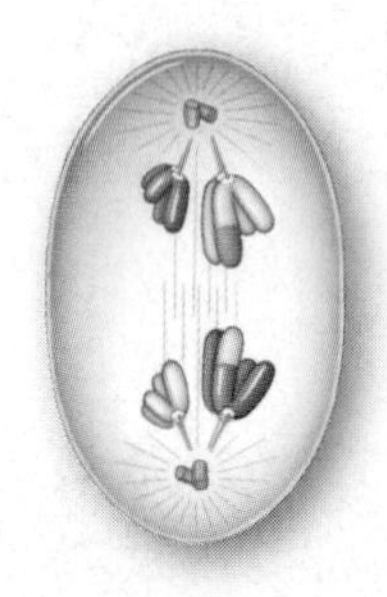

The cell is at anaphase ___ rather than anaphase ___. I know this because:
___
___
___
___

4. Define meiosis and describe its main stages. In what key respects is meiosis *not* like mitosis? *10.3, 10.6*

5. Actor Michael Douglas (Figure 10.12*a*) inherited a gene from each parent that influences the chin dimple trait. One form of the gene called for a dimple and the other didn't, but one is all it takes for this particular trait. Figure 10.12*b* shows what the chin of Mr. Douglas might have looked like if he had inherited two ordinary forms of the gene instead. What is the name for the alternative forms of the same gene? *10.1, 10.4*

6. Outline the main steps by which gametes form in plants. Do the same for gamete formation in animals. *10.5*

7. Genetically speaking, what is the key difference between the outcomes of sexual and asexual reproduction? *10.1, 10.6*

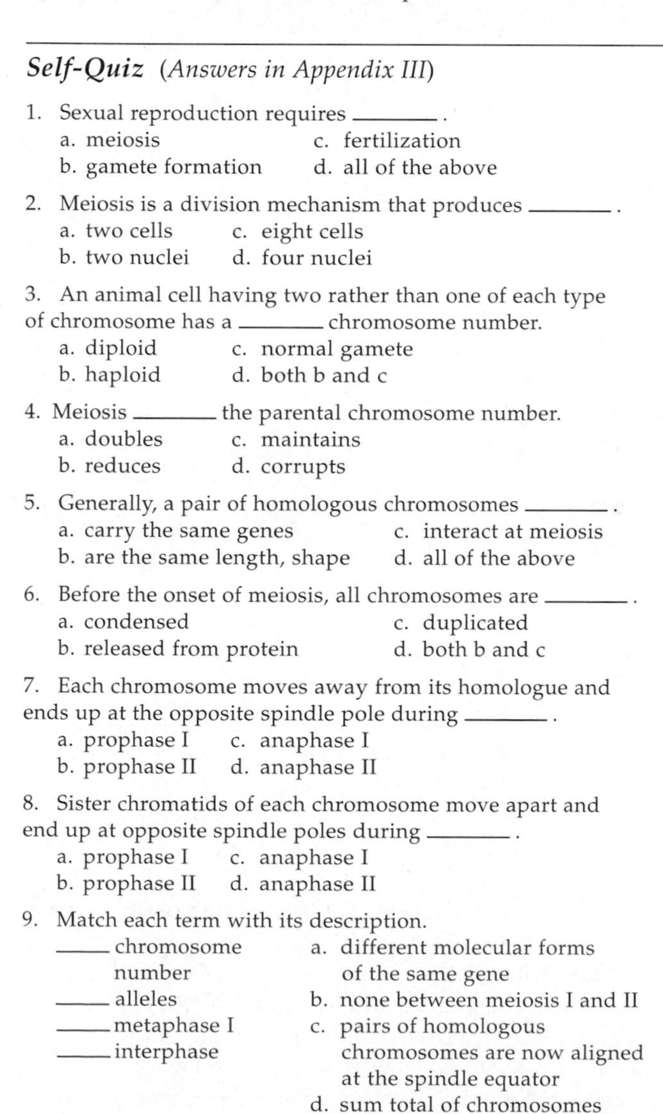

Self-Quiz (*Answers in Appendix III*)

1. Sexual reproduction requires ______.
   a. meiosis c. fertilization
   b. gamete formation d. all of the above

2. Meiosis is a division mechanism that produces ______.
   a. two cells c. eight cells
   b. two nuclei d. four nuclei

3. An animal cell having two rather than one of each type of chromosome has a ______ chromosome number.
   a. diploid c. normal gamete
   b. haploid d. both b and c

4. Meiosis ______ the parental chromosome number.
   a. doubles c. maintains
   b. reduces d. corrupts

5. Generally, a pair of homologous chromosomes ______.
   a. carry the same genes c. interact at meiosis
   b. are the same length, shape d. all of the above

6. Before the onset of meiosis, all chromosomes are ______.
   a. condensed c. duplicated
   b. released from protein d. both b and c

7. Each chromosome moves away from its homologue and ends up at the opposite spindle pole during ______.
   a. prophase I c. anaphase I
   b. prophase II d. anaphase II

8. Sister chromatids of each chromosome move apart and end up at opposite spindle poles during ______.
   a. prophase I c. anaphase I
   b. prophase II d. anaphase II

9. Match each term with its description.
   ___ chromosome number a. different molecular forms of the same gene
   ___ alleles b. none between meiosis I and II
   ___ metaphase I c. pairs of homologous chromosomes are now aligned at the spindle equator
   ___ interphase d. sum total of chromosomes in all cells of a given type

Figure 10.12 Example of the chin dimple trait (actually a fissure in the chin surface).

Critical Thinking

1. Assume you can measure the amount of DNA in a primary oocyte, then in a primary spermatocyte, which gives you a mass *m*. What mass of DNA would you expect to find in each mature gamete (egg and sperm) that forms after meiosis? What mass of DNA would you expect to find (1) in an egg fertilized by one of the sperm and (2) in that egg after the first DNA duplication?

2. Adam has a pair of alleles that influence whether a person is right- or left-handed. One allele says "left," and its partner says "right." Visualize one of his germ cells, in which chromosomes are being duplicated prior to meiosis. Visualize what happens to the chromosomes during anaphase I and II. (It might help to use index cards as models of the sister chromatids of each chromosome.) What fraction of Adam's sperm will carry the gene for right-handedness? For left-handedness?

3. Adam also has one allele for long eyelashes, and a partner allele (on the homologous chromosome) for short eyelashes. What fraction of his sperm will have these gene combinations:
   right-handed, long eyelashes *left-handed, long eyelashes*
   right-handed, short eyelashes *left-handed, short eyelashes*

Selected Key Terms

allele *10.1*
asexual reproduction *10.1*
chromosome number *10.2*
crossing over *10.4*
diploid number *10.2*
egg (ovum) *10.5*
fertilization *10.5*
gamete *CI*
gene *10.1*
germ cell *CI*
haploid number *10.2*
homologous chromosome *10.2*
meiosis *10.2*
oocyte *10.5*
sexual reproduction *10.1*
sister chromatid *10.2*
sperm *10.5*
spore *10.5*

Readings *See also www.infotrac-college.com*

Klug, W., and M. Cummings. 1994. *Concepts of Genetics.* Fourth edition. New York: Macmillan.

Wolfe, S. 1995. *Introduction to Molecular and Cellular Biology.* Belmont, California: Wadsworth.

11 OBSERVABLE PATTERNS OF INHERITANCE

A Smorgasbord of Ears and Other Traits

Basketball ace Charles Barkley has them. So does actor Tom Cruise. Actress Joan Chen doesn't, and neither did a monk named Gregor Mendel. To see how you fit in with these folks, use a mirror to check out your ears. Is the fleshy lobe at the base of each ear attached to the side of your head? If so, you and Barkley and Cruise have something in common. Or is the fleshy lobe not attached, so that you can flap it back and forth? If so, you are like Chen and Mendel (Figure 11.1).

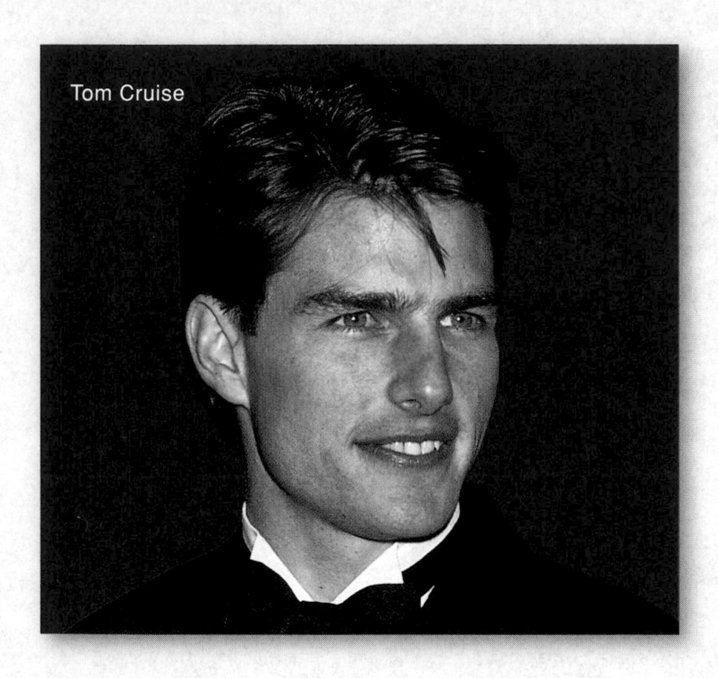

Figure 11.1 Attached and detached earlobes of a few representative humans. This sampling provides observable evidence of a trait governed by a certain gene, which exists in different molecular forms in the human population. Do you have one or the other version of the trait? It depends on which molecular forms of the gene you inherited from your mother and father. As Gregor Mendel perceived, such easily observable traits can be used to identify patterns of inheritance from one generation to the next.

Whether a person is born with attached or detached earlobes depends on a single kind of gene. That gene comes in slightly different molecular forms—alleles. Only one form has information about detached lobes. The information is put to use while a human body is developing inside the mother. It calls for a death signal, which is sent to all the cells positioned between the newly forming lobes and the head. Without the signal, the cells don't die and earlobes don't detach.

We all have genes for thousands of traits, including earlobes, cheeks, lashes, and eyeballs. Most traits vary in their details from one person to the next. Remember, we inherit pairs of genes, on pairs of chromosomes. In some pairs, one allele has strong effects and overwhelms the other allele's contribution. The outgunned allele is said to be recessive to the dominant one. If you have detached earlobes, dimpled cheeks, long lashes, or large eyeballs, you have at least one and quite possibly two dominant alleles that affect the trait in a particular way.

When both alleles of a pair are recessive, nothing masks their effect on a trait. You get *attached* earlobes with one pair of recessive alleles (and *flat* feet with another, a *straight* nose with another, and so on).

How did we discover such remarkable things about our genes? It started with Mendel. By analyzing garden

pea plants generation after generation, Mendel found indirect but *observable* evidence of how parents bestow units of hereditary information—genes—on offspring. This chapter focuses on both the methods and the results of Mendel's experiments. They remain a classic example of how a scientific approach can pry open important secrets about the natural world. And to this day, they serve as the foundation for modern genetics.

KEY CONCEPTS

1. Genes are units of information about heritable traits. Alleles, which are slightly different molecular forms of a gene, specify different versions of the same trait.

2. Each gene has a particular location on a particular chromosome of a species. Humans, pea plants, and other organisms with a diploid chromosome number inherit *pairs* of genes, at equivalent locations on pairs of homologous chromosomes.

3. When the two members of each pair of homologous chromosomes are moved apart from each other during meiosis, their pairs of genes are moved apart, also, and end up in different gametes. Gregor Mendel found indirect evidence of this gene segregation when he crossbred pea plants showing different versions of the same trait, such as purple or white flowers.

4. Each pair of homologous chromosomes in a germ cell is sorted out for distribution into one gamete or another independently of how the other pairs of homologous chromosomes are assorted. Mendel discovered indirect evidence of this when he tracked many plants having observable differences in two traits, such as flower color and plant height.

5. The contrasting forms of traits that Mendel happened to study were specified by nonidentical alleles. One allele was dominant, in that its effect on a trait masked the effect of a recessive allele paired with it.

6. Not all traits have such clearly dominant or recessive forms. One allele of a pair may be fully or partially dominant over its partner or codominant with it. Two or more gene pairs often influence the same trait, and some single genes influence many traits. Besides this, environmental factors induce further variation in traits.

11.1 MENDEL'S INSIGHT INTO INHERITANCE PATTERNS

More than a century ago, people wondered about the basis of inheritance. It was common knowledge that sperm and eggs both transmit information about traits to offspring. But few suspected that the information is organized in units (genes). Instead, the idea was that a father's blob of information "blended" with the mother's blob, like cream into coffee, at fertilization.

However, carried to its logical conclusion, blending would slowly dilute a population's shared pool of hereditary information until there was only a single version left of each trait. If that were so, why did, say, freckles keep showing up among the children of nonfreckled parents through the generations? Why weren't all the descendants of a herd of white stallions and black mares gray? The blending theory could scarcely explain the obvious variation in traits that people could observe with their own eyes. Nevertheless, few disputed the theory.

Charles Darwin was among the scholarly dissidents. According to the key premise of his theory of natural selection, individuals of a population show variation in heritable traits. Through the generations, the variations that improve the chance of surviving and reproducing show up with greater frequency than those that do not. Less advantageous variations may persist among fewer individuals, or they may even disappear. It is not that separate versions of a given trait are "blended out" of the population. Rather, *each version of a trait may persist in a population, at frequencies that can change over time.*

Just before Darwin presented his theory, someone was gathering evidence that eventually would support his key premise. A monk, Gregor Mendel, had already guessed that sperm and eggs carry distinct "units" of information about heritable traits. By carefully analyzing traits of pea plants generation after generation, Mendel found indirect but *observable* evidence of how parents transmit genes to offspring.

Mendel's Experimental Approach

Mendel spent most of his adult life in a monastery in Brno, a city near Vienna that has since become part of the Czech Republic. However, Mendel was not a man of narrow interests who accidentally stumbled onto principles of great import. The monastery of St. Thomas was close to European capitals that were the centers of scientific inquiry.

Having been raised on a farm, Mendel was aware of agricultural principles and their applications. He kept abreast of the breeding experiments and developments described in the available literature. He was a member of the regional agricultural society. He also won several awards for developing improved varieties of vegetables and fruits. Shortly after entering the monastery, he spent two years studying mathematics, physics, and botany at the University of Vienna. Few scholars of his time had combined talents in plant breeding and mathematics.

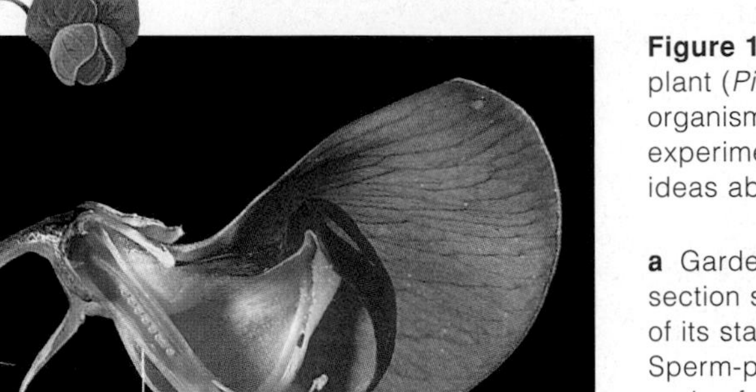

Figure 11.2 Garden pea plant (*Pisum sativum*), the organism Mendel chose for experimental tests of his ideas about inheritance.

a Garden pea flower. The section shows the location of its stamens and carpel. Sperm-producing pollen grains form in stamens. Eggs develop, fertilization takes place, and seeds mature inside the carpel.

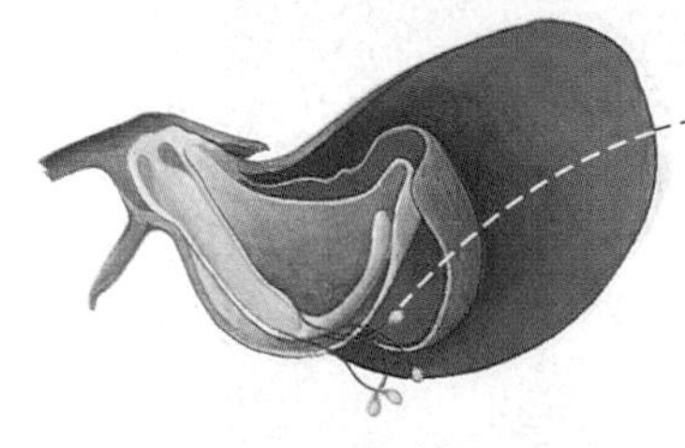

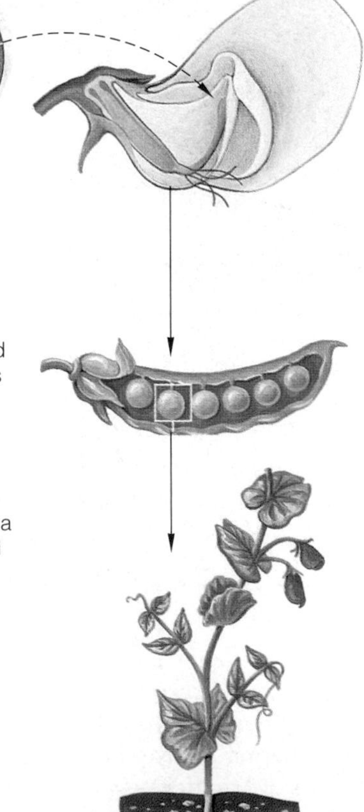

b Pollen from a garden pea plant that breeds true for purple flowers is brushed onto a floral bud of a plant that breeds true for white flowers. The white flower had its stamens snipped off. This is one way to assure cross-fertilization of plants.

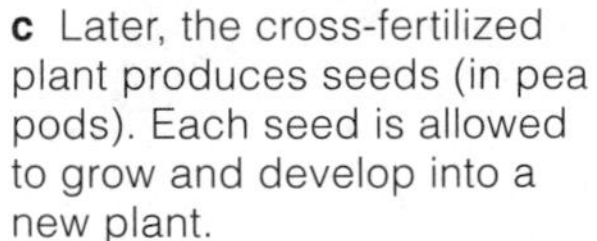

c Later, the cross-fertilized plant produces seeds (in pea pods). Each seed is allowed to grow and develop into a new plant.

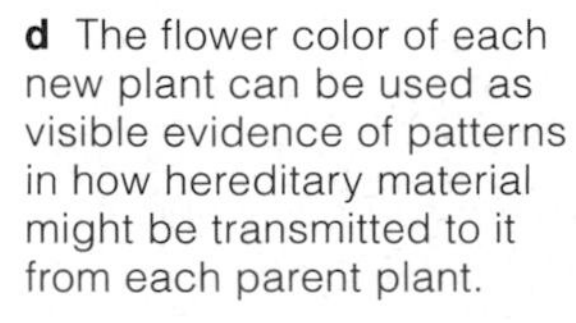

d The flower color of each new plant can be used as visible evidence of patterns in how hereditary material might be transmitted to it from each parent plant.

Further reading: Student Guide to InfoTrac on web site

Shortly after his university training, Mendel began experimenting with the garden pea plant, *Pisum sativum* (Figure 11.2). This plant is self-fertilizing. Male as well as female gametes (call them sperm and eggs) develop in different parts of the same flower, where fertilization takes place. Nearly all the plants breed true for certain traits. In other words, successive generations are just like their parents in one or more traits, as when all of the offspring grown from seeds of self-fertilized, white-flowered parent plants have white flowers.

As Mendel knew, we can cross-fertilize pea plants by transferring pollen from one plant's flower to the flower of another plant. For his experimental studies, he could open flower buds of a plant that bred true for a trait—say, white flowers—and snip out the stamens. (Stamens bear pollen grains in which sperm develop.) Then he could brush buds without stamens with pollen from a plant that bred true for a *different* version of the same trait: purple flowers. As Mendel hypothesized, he could use such clearly observable differences to track a trait through many generations. If there were patterns to the trait's inheritance, *then those patterns might tell him something about heredity itself.*

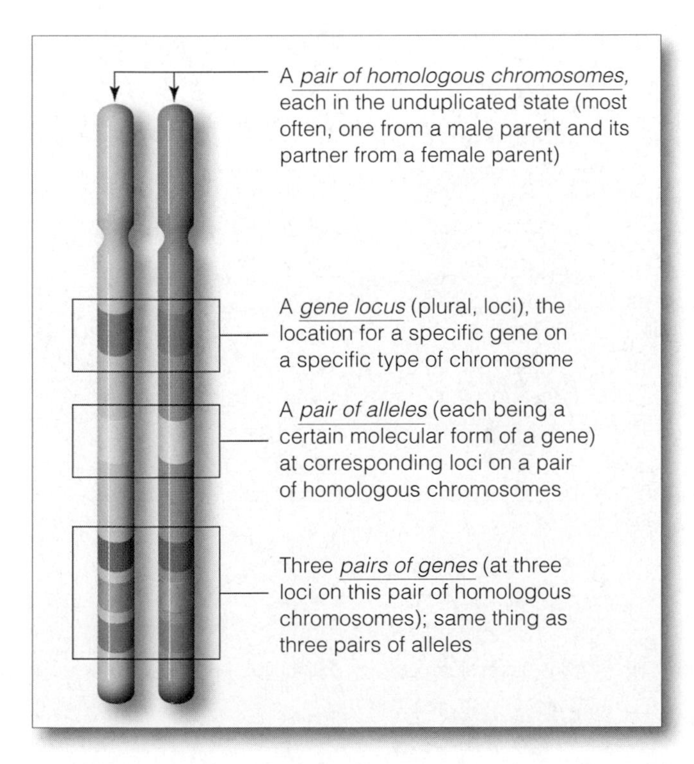

Figure 11.3 A few genetic terms illustrated. Diploid organisms have pairs of genes, on pairs of homologous chromosomes. For example, you inherited one chromosome of each pair from your mother and the other, homologous chromosome from your father.

Most genes can have slightly different molecular forms, called alleles. Different alleles specify different versions of the same trait. An allele at one location on a chromosome may or may not be identical to its partner on the homologous chromosome.

Some Terms Used in Genetics

Having read the chapter on meiosis, you already have insight into the mechanisms of sexual reproduction, which is more than Mendel had. Neither he nor anyone else of his era knew about chromosomes. So he could not have known that a chromosome number is reduced by half in gametes, then restored when gametes meet at fertilization. Even so, Mendel sensed what was going on. As we follow his thinking, let's simplify the story by substituting a few of the modern terms used in studies of inheritance (see also Figure 11.3):

1. **Genes** are units of information about specific traits, and they are passed from parents to offspring. Each gene has a specific location (locus) on a chromosome.

2. Cells with a diploid chromosome number ($2n$) have pairs of genes, on pairs of homologous chromosomes.

3. Mutation can alter a gene's molecular structure. The alteration may change the gene's information about a trait (as when the gene for flower color specifies purple and a mutated version specifies white). All the different molecular forms of the same gene are called **alleles**.

4. When offspring of genetic crosses inherit a pair of *identical* alleles for a trait, generation after generation, they are a **true-breeding lineage**. By contrast, when offspring of a genetic cross inherit a pair of *nonidentical* alleles for a trait, they are **hybrid offspring**.

5. When both alleles of a pair are identical, this is a *homozygous* condition. When the two are not identical, this is a *heterozygous* condition.

6. An allele is said to be *dominant* when its effect on a trait masks that of any *recessive* allele paired with it. We use capital letters for dominant alleles and lowercase letters for recessive ones. *A* and *a* are examples.

7. Putting this all together, a **homozygous dominant** individual has a pair of dominant alleles (*AA*) for a trait under study. A **homozygous recessive** individual has a pair of recessive alleles (*aa*). And a **heterozygous** individual has a pair of nonidentical alleles (*Aa*).

8. Two terms help keep the distinction clear between genes and the traits they specify. **Genotype** refers to the particular genes an individual carries. **Phenotype** refers to an individual's observable traits.

9. When tracking the inheritance of traits through generations of offspring, these abbreviations apply:

P	parental generation
F_1	first-generation offspring
F_2	second-generation offspring

11.2 MENDEL'S THEORY OF SEGREGATION

Mendel had an idea that in every generation, a plant inherits two "units" (genes) of information about a trait, one from each parent. To test this idea, he performed what is now known as **monohybrid crosses**. These are experimental crosses between two parents that breed true (are homozygous) for different versions of a single trait. The F_1 offspring are hybrids; each inherits a pair of nonidentical alleles (is heterozygous) for that trait.

Predicting Outcomes of Monohybrid Crosses

Mendel tracked many individual traits through two generations. For instance, in one series of experiments, he crossed true-breeding, purple-flowered plants and true-breeding, white-flowered ones. All plants grown from the seeds that resulted from this cross had purple flowers. Mendel allowed these plants to self-fertilize. Some plants grown from the seeds had white flowers!

If Mendel's hypothesis were correct—if each plant had inherited two units of information about flower color—then the unit for "purple" had to be dominant, because it had masked the unit for "white" in F_1 plants.

Let's rephrase his thinking. Germ cells of pea plants are diploid, with pairs of homologous chromosomes. Assume one parent is homozygous dominant (AA) and the other is homozygous recessive (aa) for flower color. Following meiosis, a sperm or egg carries one allele for flower color (Figure 11.4). Thus, when a sperm fertilizes an egg, only one outcome is possible: $A + a = Aa$.

Before continuing, you should know Mendel crossed hundreds of plants and tracked thousands of offspring. Besides this, he counted and recorded the number of plants showing dominance or recessiveness. As you can see from Figure 11.5, an intriguing ratio emerged. On average, three of every four F_2 plants had the dominant phenotype, and one had the recessive phenotype.

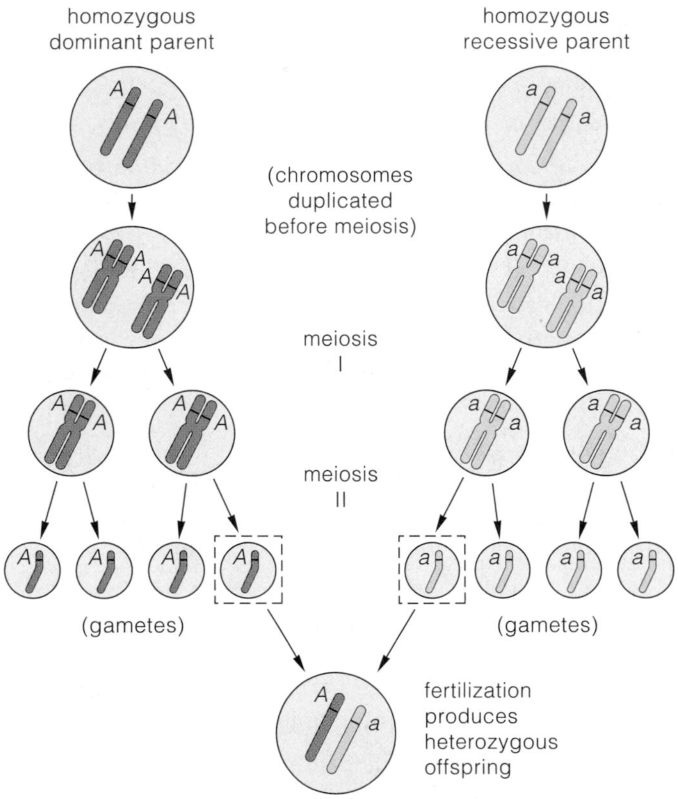

Figure 11.4 Monohybrid cross, showing how one gene of a pair segregates from the other gene. Two parents that breed true for two versions of a trait produce only heterozygous offspring.

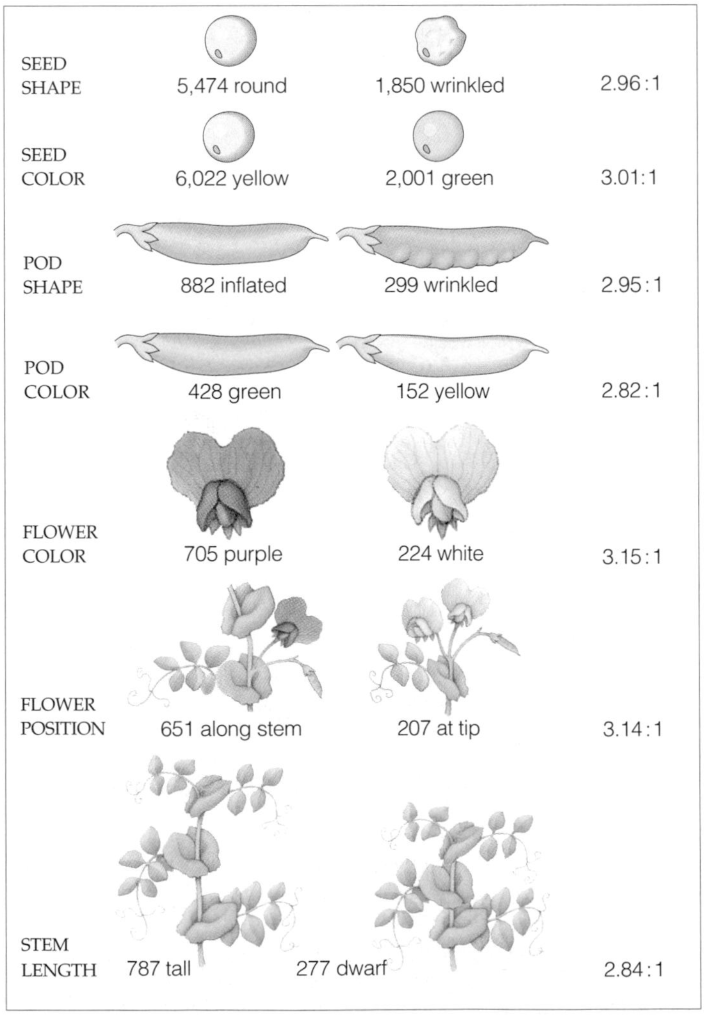

Trait Studied	Dominant Form	Recessive Form	F_2 Dominant-to-Recessive Ratio
SEED SHAPE	5,474 round	1,850 wrinkled	2.96 : 1
SEED COLOR	6,022 yellow	2,001 green	3.01 : 1
POD SHAPE	882 inflated	299 wrinkled	2.95 : 1
POD COLOR	428 green	152 yellow	2.82 : 1
FLOWER COLOR	705 purple	224 white	3.15 : 1
FLOWER POSITION	651 along stem	207 at tip	3.14 : 1
STEM LENGTH	787 tall	277 dwarf	2.84 : 1

Average ratio for all traits studied: **3 : 1**

Figure 11.5 *Right*: Numerical results from Mendel's monohybrid cross experiments with the garden pea plant (*P. sativum*). The numbers are his counts of the F_2 plants that carried dominant or recessive hereditary "units" (alleles) for the trait. On average, the dominant-to-recessive ratio was 3:1.

Further reading: Student Guide to InfoTrac on web site

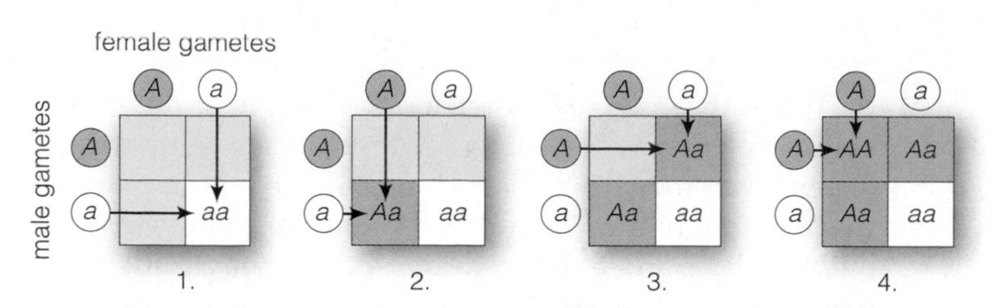

Figure 11.6 Punnett-square method of predicting the probable outcome of a genetic cross. Circles represent gametes. *Italic* letters on gametes represent dominant or recessive alleles. The squares show the different genotypes possible among offspring. In this case, gametes are from a self-fertilizing heterozygous (*Aa*) plant.

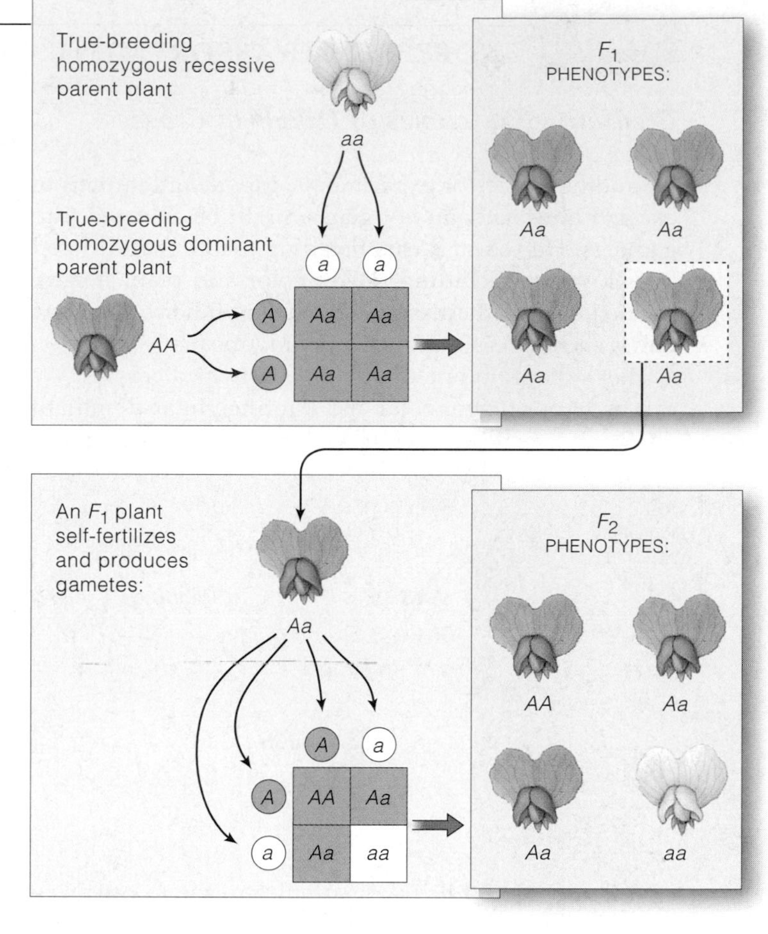

Figure 11.7 *Right*: Results from one of Mendel's monohybrid crosses. On average, the dominant-to-recessive ratio among the second-generation (F_2) plants was 3:1.

To Mendel, the ratio suggested that fertilization is a chance event, with a number of possible outcomes. And he had an understanding of probability, which applies to chance events *and therefore could help him predict the possible outcomes of genetic crosses.* **Probability** simply means this: The chance that each outcome of a given event will occur is proportional to the number of ways in which that event can be reached.

The **Punnett-square method**, explained in Figure 11.6 and applied in Figure 11.7, may help you visualize the possibilities. As you can see, if half of a plant's sperm (or eggs) were *a* and half were *A*, then four outcomes would be possible each time a sperm fertilized an egg:

POSSIBLE EVENT:	PROBABLE OUTCOME:
sperm *A* meets egg *A*	1/4 *AA* offspring
sperm *A* meets egg *a*	1/4 *Aa*
sperm *a* meets egg *A*	1/4 *Aa*
sperm *a* meets egg *a*	1/4 *aa*

By this prediction, an F_2 plant has three chances in four of getting at least one dominant allele (purple flowers). It has one chance in four of getting two recessive alleles (white flowers). That is a probable phenotypic ratio of three purple to one white, or 3:1.

Mendel's observed ratios were not *exactly* 3:1. You can see this for yourself by looking at the numerical results listed in Figure 11.5. Why did Mendel put aside the deviations? To understand why, flip a coin several times. As we all know, a coin is just as likely to end up heads as tails. But often a coin ends up heads, or tails, several times in a row. So if you flip the coin only a few times, the observed ratio might differ greatly from the predicted ratio of 1:1. Flip the coin many, many times, and you are more likely to come close to the predicted ratio. Mendel understood the rules of probability—and performed a large number of crosses. Almost certainly, this kept him from being confused by minor deviations from the predicted results of the experimental crosses.

Testcrosses

By running **testcrosses**, Mendel gained support for his prediction. In this type of experimental test, an organism shows dominance for a specified trait but its genotype is unknown, so it is crossed to a known homozygous recessive individual. Test results may reveal whether the organism is homozygous dominant or heterozygous.

Regarding the monohybrid crosses just described, Mendel tested his prediction that the purple-flowered F_1 offspring were heterozygous by crossing them with true-breeding, white-flowered plants. If they were all homozygous dominant, then all the F_2 offspring would show the dominant form of the trait. If heterozygous, there would be about as many dominant as recessive plants. Sure enough, when old enough to flower, about half the F_2 plants had purple flowers (*Aa*) and half had white (*aa*). Can you construct two Punnett squares that show the possible outcomes of this testcross?

The results from Mendel's monohybrid crosses and testcrosses became the basis of a theory of **segregation**, which we state here in modern terms:

MENDEL'S THEORY OF SEGREGATION. **Diploid cells have pairs of genes, on pairs of homologous chromosomes. The two genes of each pair are separated from each other during meiosis, so they end up in different gametes.**

11.3 INDEPENDENT ASSORTMENT

Predicting Outcomes of Dihybrid Crosses

By another series of experiments, Mendel attempted to explain how *two* pairs of genes might be assorted into gametes. He selected true-breeding plants that differed in two traits, including flower color and plant height. In such **dihybrid crosses,** F_1 offspring inherit two gene pairs, each consisting of two nonidentical alleles.

Let's diagram one of Mendel's dihybrid crosses. We can use *A* for flower color and *B* for height as dominant alleles and, as their recessive counterparts, *a* and *b*:

TRUE-BREEDING PARENTS: purple flowers, tall **AABB** x white flowers, dwarf **aabb**

GAMETES: AB AB ab ab

F_1 HYBRID OFFSPRING: **AaBb**

As Mendel would have predicted, the F_1 offspring from this cross are all purple-flowered and tall (*AaBb*).

When those F_1 plants reproduce, how will the two gene pairs be assorted into gametes? The answer partly depends on the chromosomal locations of those pairs. Assume one pair of homologous chromosomes carries the *Aa* alleles and a different pair carries the *Bb* alleles. Next, think of how all chromosomes become positioned at the spindle equator during metaphase I of meiosis (Figures 10.4 and 11.8). The chromosome with the *A* allele might be positioned to move to either spindle pole (then into one of four gametes). The same is true of its homologue. And the same is true of the chromosomes with the *B* and *b* alleles. Thus, after meiosis and gamete formation, four combinations of alleles are possible in the sperm or eggs: 1/4 *AB*, 1/4 *Ab*, 1/4 *aB*, and 1/4 *ab*.

Given the alternative alignments of chromosomes at metaphase I, several allelic combinations are possible at fertilization. Simple multiplication (four kinds of sperm times four kinds of eggs) tells us sixteen combinations of gametes are possible in the F_2 offspring of a dihybrid cross. Use the Punnett-square method to diagram the probabilities (Figure 11.9). Now add up all the possible phenotypes and you get 9/16 tall purple-flowered, 3/16 dwarf purple-flowered, 3/16 tall white-flowered, and 1/16 dwarf white-flowered plants. That is a probable phenotypic ratio of 9:3:3:1. Results from one dihybrid cross that Mendel described were close to this ratio.

Figure 11.8 Independent assortment. This example tracks two pairs of homologous chromosomes. An allele at one locus on a chromosome may or may not be identical with its partner allele on the homologous chromosome. At meiosis, either chromosome of a pair may become attached to either pole of the spindle. Thus, in this case, two different lineups are possible at metaphase I.

Nucleus of a diploid (2*n*) reproductive cell with only two pairs of homologous chromosomes

OR

Possible alignments of the homologous chromosomes at metaphase I of meiosis, as shown by two diagrams:

The resulting alignments of chromosomes at metaphase II:

The combinations of alleles possible in the forthcoming gametes:

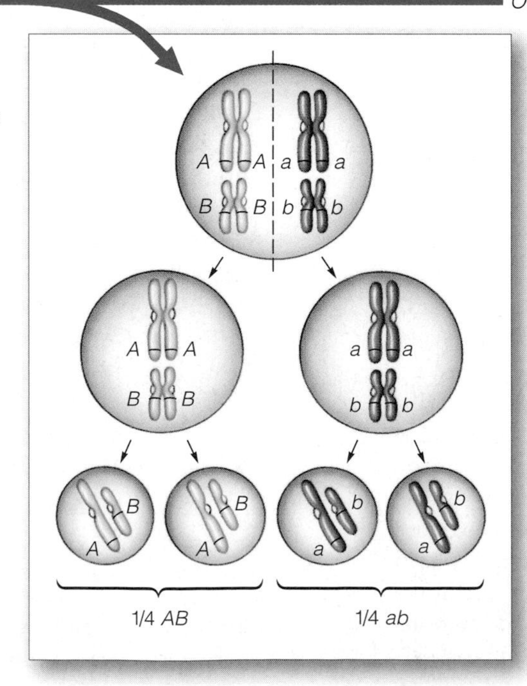

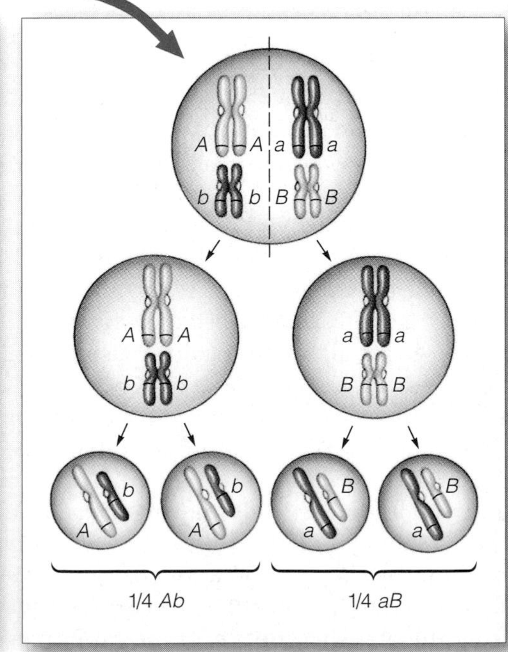

Further reading: Student Guide to InfoTrac on web site

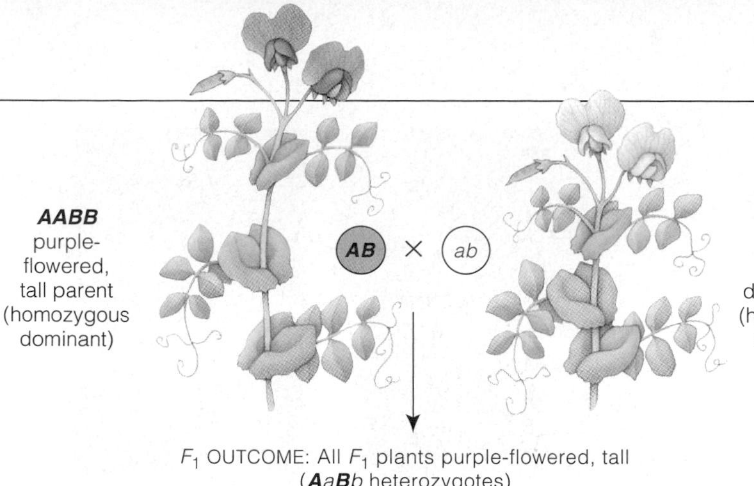

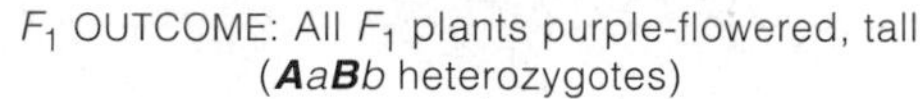

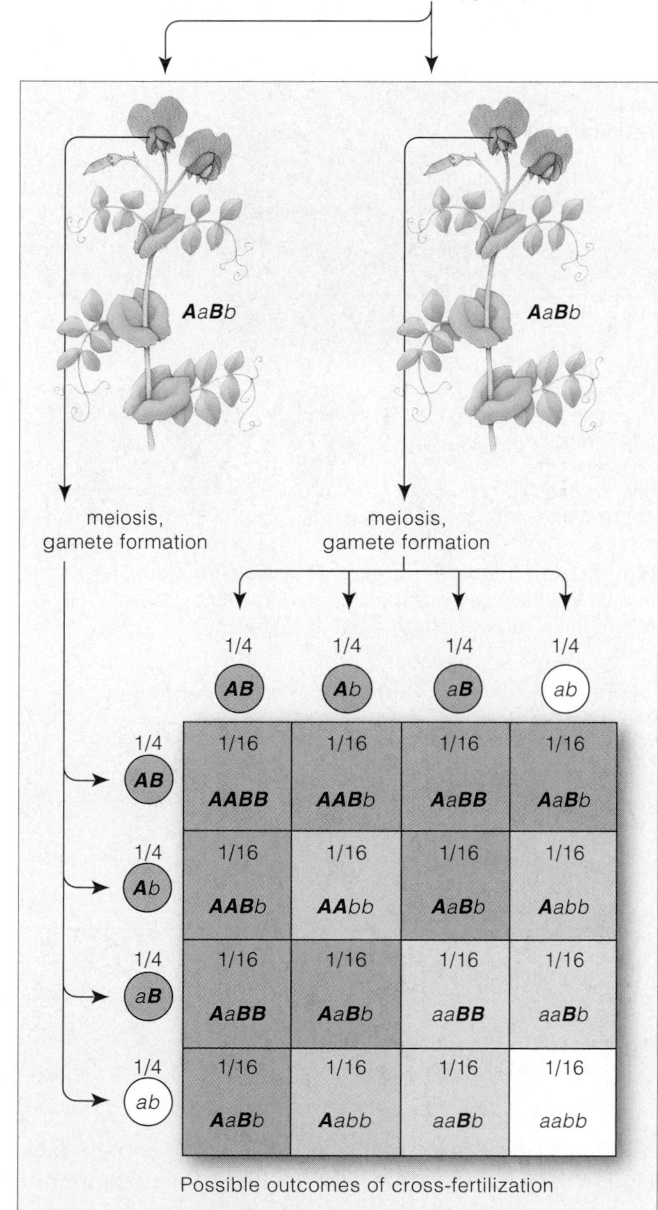

Figure 11.9 Results from Mendel's dihybrid cross between parent plants that bred true for different versions of two traits: flower color and plant height. *A* and *a* represent dominant and recessive alleles for flower color. *B* and *b* represent dominant and recessive alleles for plant height. As the Punnett square indicates, the probabilities of certain combinations of phenotypes among the F_2 offspring occur in a 9:3:3:1 ratio, on the average.

The Theory in Modern Form

Mendel could do no more than analyze the numerical results from his dihybrid crosses, because he didn't know that seven pairs of homologous chromosomes carry the pea plant's "units" of inheritance. It just seemed to him that the two units for the first trait he was tracking had been assorted into gametes independently of the two units for the other trait. In time his interpretation became known as the theory of **independent assortment**, which we state here in modern terms: By the end of meiosis, each pair of homologous chromosomes—and the genes they carry—have been sorted for shipment into gametes independently of how the other pairs were sorted out.

Independent assortment and hybrid crossing lead to stupendous genetic variation. In a monohybrid cross involving only a single gene pair, three genotypes are possible: *AA*, *Aa*, and *aa*. We can represent this as 3^n, where n is the number of gene pairs. When we consider more gene pairs, the number of possible combinations increases dramatically. Even if parents differ in merely ten pairs of genes, almost 60,000 genotypes are possible among their offspring. If they differ in twenty pairs of genes, the number approaches 3.5 billion!

In 1865 Mendel presented his ideas to the Brünn Natural History Society. His ideas had little impact. The next year his paper was published, and apparently it was read by few and understood by no one. In 1871 he became abbot of the monastery, and his experiments gradually gave way to administrative tasks. He died in 1884, never to know his experiments would become the starting point for the development of modern genetics.

Today, Mendel's theory of segregation still stands. Hereditary material is indeed organized in units (genes) that retain their identity and are segregated from each other for distribution into different gametes. But the theory of independent assortment does not apply to all gene combinations, as you will see in the next chapter.

MENDEL'S THEORY OF INDEPENDENT ASSORTMENT. **By the end of meiosis, genes on pairs of homologous chromosomes have been sorted out for distribution into one gamete or another independently of gene pairs of other chromosomes.**

11.4 DOMINANCE RELATIONS

For the most part, Mendel studied traits having clearly dominant or recessive forms. As the remaining sections of this chapter will make clear, however, the expression of other traits is not as straightforward.

Incomplete Dominance

In **incomplete dominance**, one allele of a pair isn't fully dominant over its partner, so a heterozygous phenotype *somewhere in between* the two homozygous phenotypes emerges. Cross a true-breeding red snapdragon and a true-breeding white one. All F_1 offspring will have pink flowers. Cross two F_1 plants and expect red, pink, and white snapdragons in a predictable ratio (Figure 11.10). What causes this inheritance pattern? Red snapdragons have two alleles that allow them to make an abundance of red pigment molecules. White snapdragons have two mutant alleles that render them pigment-free. The pink ones are heterozygous. Their one red allele can specify enough pigment to make flowers pink but not red.

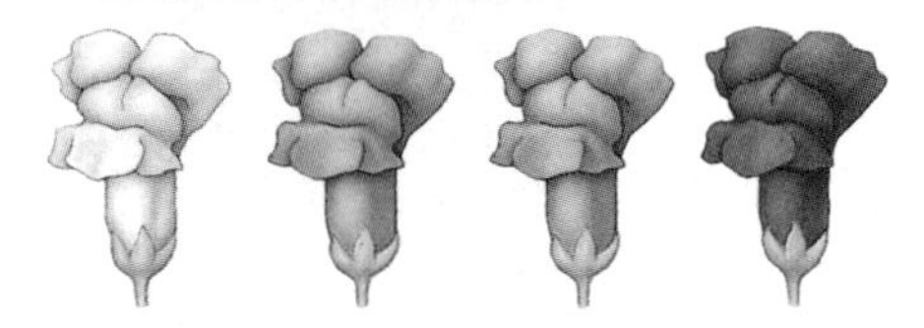

Figure 11.10 Visible evidence of incomplete dominance in heterozygous (pink) snapdragons, in which an allele for red pigment is paired with a "white" allele.

ABO Blood Types: A Case of Codominance

In **codominance**, a pair of nonidentical alleles specify two phenotypes, which are both expressed at the same time in heterozygotes. As an example, think about one of the glycolipids projecting from the plasma membrane of your red blood cells. It helps give red blood cells their unique identity, although it comes in slightly different molecular forms. A method of analysis known as *ABO blood typing* reveals which form a person has.

In humans, an enzyme dictates the glycolipid's final structure. The gene for that enzyme has three alleles. Two alleles, I^A and I^B, are codominant when paired. The third, i, is recessive; a pairing with either I^A or I^B masks its effect. Together, they are a **multiple allele system**, which we define as the presence of three or more alleles of a gene among individuals of a population.

Before each glycolipid molecule became positioned at the cell surface, it was modified in the cytomembrane system (Section 4.5). First an oligosaccharide chain was attached to a lipid molecule, then a sugar was attached to the end of that chain. Alleles I^A and I^B specify two slightly different versions of the enzyme catalyzing that final step. The two attach different sugars, which give the glycolipid its identity—either A or B.

Which alleles do you have? With either I^AI^A or I^Ai, you have type A blood. With I^BI^B or I^Bi, your blood is type B. With codominant alleles I^AI^B, it is AB—meaning you have both versions of the sugar-attaching enzyme. But if you are homozygous recessive (ii), the molecules never did get an additional sugar attached to them. Your blood type is neither A nor B; that is what type "O" means. Figure 11.11 summarizes the possibilities.

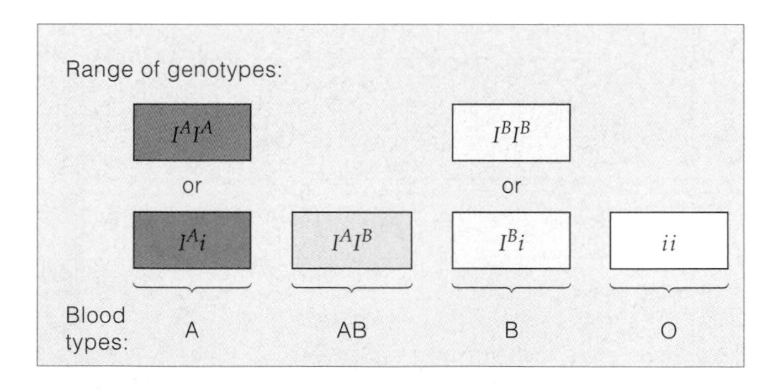

Figure 11.11 Allelic combinations that are related to ABO blood typing.

During *transfusions*, the blood of two people mixes. Unless their blood is compatible, the recipient's immune system perceives the red blood cells from the donor as "nonself." It will act against those cells and may cause death (Section 39.4).

One allele may be fully dominant, incompletely dominant, or codominant with its partner on the homologous chromosome.

Further reading: Student Guide to InfoTrac on web site

11.5 MULTIPLE EFFECTS OF SINGLE GENES

Expression of the alleles at just a single location on a chromosome may have positive or negative effects on two or more traits. This phenotypic outcome of a single gene's activity is known as **pleiotropy** (after the Greek *pleio–*, meaning more, and *–tropic*, meaning to change).

The genetic disorder *sickle-cell anemia* is a classic example of how the alleles at a single locus can have pleiotropic effects. The disorder arises from a mutated gene for beta-globin, one of two kinds of polypeptide chains in the hemoglobin molecule. Hemoglobin, recall, is the oxygen-transporting protein in red blood cells. We designate the mutant allele as Hb^S instead of Hb^A. Heterozygotes (Hb^A/Hb^S) usually show few symptoms of the disorder. Their red blood cells are able to produce enough normal hemoglobin molecules to compensate for the abnormal ones. In homozygotes (Hb^S/Hb^S), the red blood cells can only produce abnormal hemoglobin. This one abnormality may have drastic repercussions throughout the body, for it disrupts the concentration of oxygen in the bloodstream.

Humans, like most organisms, depend on the intake of oxygen for aerobic respiration. Oxygen in air flows into the lungs, then diffuses into the blood. There it binds to hemoglobin, which transports it through arteries, arterioles, and then small-diameter, thin-walled capillaries threading past all living cells in the body. A steep oxygen concentration gradient exists between blood and the fluid within the surrounding tissues, so most of the oxygen diffuses into the tissues and on into cells. With this extensive cellular uptake, the concentration of oxygen in blood declines. The decline is most pronounced at high altitudes and during strenuous activity.

In the red blood cells of people who bear the mutant gene, abnormal hemoglobin molecules stick together in rod-shaped arrangements. The rods distort cells into a sickle shape, as in Figure 11.12*b*. (A sickle is a farm tool with a long, crescent-shaped blade.) The distorted cells rupture easily, and their remnants clog and rupture capillaries. When these oxygen transporters are rapidly destroyed, the cells in affected tissues become starved for oxygen. Also, their clumping effect leads to local failures in the capacity of the circulatory system to deliver oxygen and to carry away carbon dioxide and other metabolic wastes.

Over time, ongoing expression of the mutant gene may damage tissues and organs throughout the body. Figure 11.12*c* tracks how the successive alterations in phenotype may proceed. In chapters to come, you will read about other aspects of this genetic disorder.

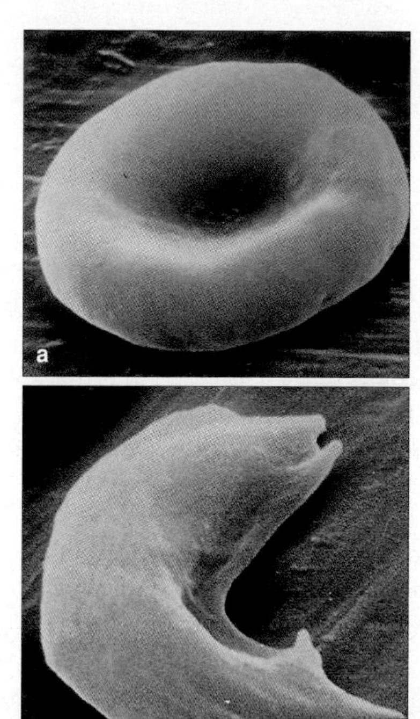

Figure 11.12 (**a**) Red blood cell from a person affected by the genetic disorder sickle-cell anemia. This scanning electron micrograph shows the surface appearance of the affected individual's red blood cells when the blood is adequately oxygenated. (**b**) This scanning electron micrograph shows the sickle shape that red blood cells assume when the concentration of oxygen in blood is low. (**c**) The diagram tracks the wide range of symptoms characteristic of an individual who is homozygous recessive for the disorder.

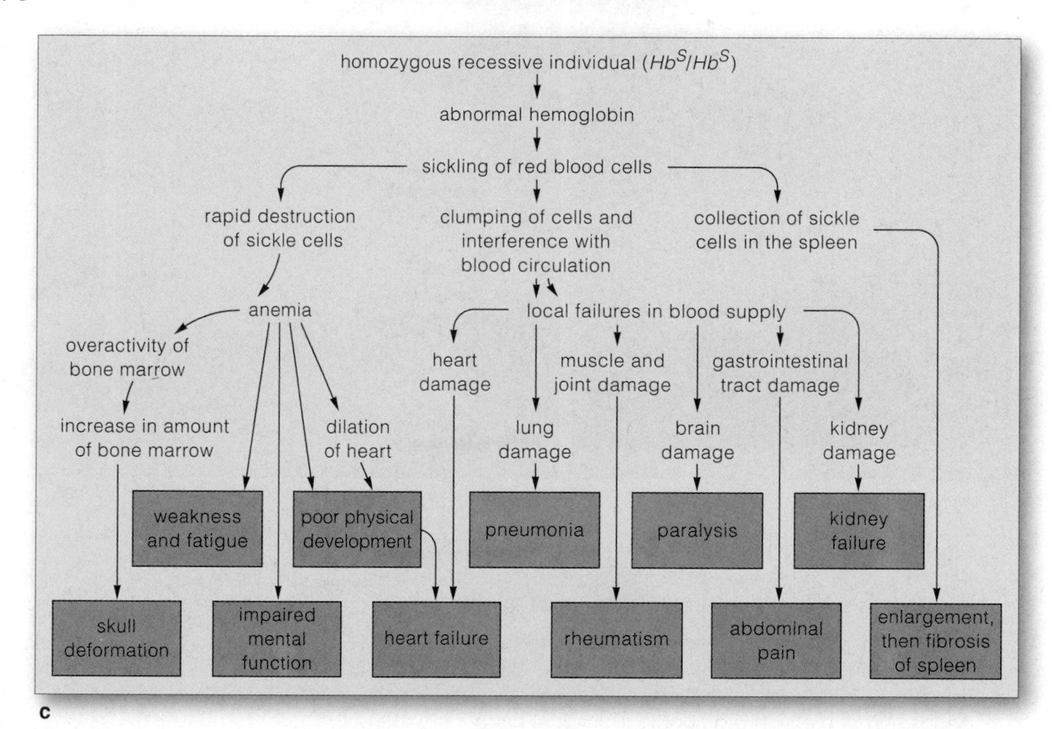

The alleles at a single gene location may have positive or negative effects on two or more traits.

The effects may not be simultaneous. Rather, they may have repercussions over time. The gene may lead to an alteration in one trait. That change may alter another trait, and so on.

11.6 INTERACTIONS BETWEEN GENE PAIRS

Often a trait results from interactions among products of two or more gene pairs. For example, two alleles of a gene may mask expression of another gene's alleles, so some expected phenotypes may not appear at all. Such interactions between the product of pairs of genes are called **epistasis** (meaning the act of stopping).

Hair Color in Mammals

Epistasis is common among the gene pairs responsible for skin or fur color in mammals. Consider the black, brown, or yellow fur of Labrador retrievers (Figure 11.13). The different colors arise from variations in the amount and distribution of melanin, a brownish black pigment. A variety of enzymes and other products of many gene pairs affect different steps in the production of melanin and its deposition in certain body regions.

The alleles of one gene specify an enzyme required to produce melanin. Expression of allele *B* (black) has a more pronounced effect and is dominant to *b* (brown). Alleles of a different gene control the extent to which molecules of melanin will be deposited in a retriever's hairs. Allele *E* permits full deposition. Two recessive alleles (*ee*) reduce deposition, and fur will be yellow.

In some individuals, those two gene pairs are not able to interact, owing to a certain allelic combination at still another gene locus. There, a gene (*C*) calls for tyrosinase, the first of several enzymes in a melanin-producing pathway. An individual bearing one or two dominant alleles (*CC* or *Cc*) can make the functional enzyme. An individual bearing two recessive alleles (*cc*) cannot. When the biosynthetic pathway for melanin production gets blocked, then *albinism*—the absence of melanin—is the resulting phenotype (Figure 11.14).

a BLACK LABRADOR

b YELLOW LABRADOR

c CHOCOLATE LABRADOR

Figure 11.13 The heritable basis of coat color among Labrador retrievers. The trait arises through interactions among the alleles of two pairs of genes.

One kind of gene is involved in melanin production. Allele *B* (black) of this gene is dominant to allele *b* (brown). A different kind of gene influences the deposition of melanin pigment in individual hairs. Allele *E* of this gene promotes deposition, but a pairing of recessive alleles (*ee*) of the gene blocks deposition, and a yellow coat results.

F_1 offspring of a dihybrid cross produce F_2 offspring in a 9:3:4 ratio, as the Punnett-square diagram at *right* indicates.

The yellow Labrador in photograph (**b**) probably has genotype *BBee*, because it can produce melanin but cannot deposit pigment in hairs. After studying the photograph, can you say why?

HOMOZYGOUS PARENTS: *BBEE* × *bbee*

F_1 PUPPIES: *BbEe*

ALLELIC COMBINATIONS POSSIBLE AMONG F_2 PUPPIES:

	BE	*Be*	*bE*	*be*
BE	*BBEE*	*BBEe*	*BbEE*	*BbEe*
Be	*BBEe*	*BBee*	*BbEe*	*Bbee*
bE	*BbEE*	*BbEe*	*bbEE*	*bbEe*
be	*BbEe*	*Bbee*	*bbEe*	*bbee*

RESULTING PHENOTYPES:

- 9/16 or 9 black
- 3/16 or 3 brown
- 4/16 or 4 yellow

Further reading: Student Guide to InfoTrac on web site

Figure 11.14 A rare albino rattlesnake. Like other animals that cannot produce melanin, it has pink eyes and its body surface is white, overall. (Eyes look pink because the absence of melanin from a tissue layer in the eyeball allows red light to be reflected from blood vessels in the eyes.) In birds and mammals, surface coloration results from pigments in feathers, fur, or skin. In fishes, amphibians, and reptiles, color-bearing cells give skin its surface coloration. Some of the cells contain melanin pigments or yellow-to-red pigments. Others contain crystals that reflect light and alter the effect of other pigments present.

The mutation affecting melanin production in the snake shown here had no effect on the production of yellow-to-red pigments and of light-reflecting crystals. Therefore, the snake's skin appears to be iridescent yellow as well as white.

Figure 11.15 Interaction between two genes that affect the same trait in domestic chicken breeds. The initial cross is between a Wyandotte (with a rose comb, **b**, on the crest of its head) and brahma (pea comb, **c**). With complete dominance at the locus for pea comb and at the locus for rose comb, products of the two gene pairs interact and give rise to a walnut comb (**a**). With full recessiveness at both gene loci, products interact and give rise to a single comb (**d**).

Comb Shape in Poultry

In some cases, interaction between two gene pairs results in a phenotype that neither pair can produce alone. The geneticists W. Bateson and R. Punnett identified two interacting gene pairs (*R* and *P*) that affect comb shape in chickens. Allelic combinations of *rr* at one gene locus and *pp* at the other locus result in the least common phenotype, the single comb. The presence of dominant alleles (*R*, *P*, or both) results in varied phenotypes.

Take a look at Figure 11.15. The diagram shows the combinations of alleles that interact to specify the rose, pea, and walnut combs shown in the photographs.

Genes often interact, as when alleles of one gene mask the expression of another gene, and when some expected phenotypes may not appear at all.

11.7 HOW CAN WE EXPLAIN LESS PREDICTABLE VARIATIONS?

Regarding the Unexpected Phenotype

As Mendel demonstrated, the phenotypic effects of one or two pairs of certain genes show up in predictable ratios when you track them from one generation to the next. Besides this, certain interactions among two or more gene pairs can produce phenotypes in predictable ratios, as the example of Labrador coat color clearly demonstrated in Section 11.6. However, even when you decide to track a single gene through the generations, you may discover that the resulting phenotypes are not quite what you had expected.

Consider *camptodactyly*—a rare genetic abnormality that affects both the shape and the movement of fingers. Certain people who carry the mutant allele for this heritable trait develop immobile, bent fingers on both hands. Other people develop immobile, bent fingers on the left or right hand only. Others who carry the mutant allele develop fingers that are not affected either way.

What is the source of such confounding variation? Recall that most organic compounds are synthesized by a series of metabolic steps, *and different enzymes—each a gene product—regulate different steps.* Maybe one gene has mutated in one of a number of different ways. Maybe the product of another gene blocks the pathway or causes it to run nonstop or not long enough. Or maybe poor nutrition or another factor that is variable in the individual's environment affects a crucial enzyme in the pathway. These are the sorts of variable factors that commonly introduce far less predictable variations in the phenotypes resulting from gene expression.

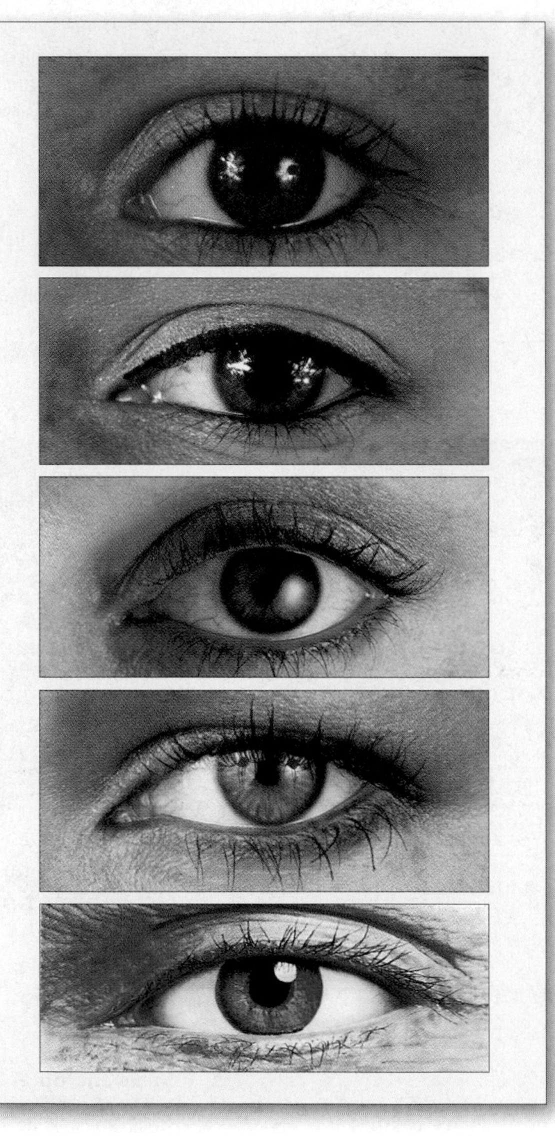

Figure 11.16 Samples from a range of continuous variation in human eye color. Different pairs of genes interact to produce and deposit melanin. Among other things, this pigment helps color the eye's iris. Different combinations of alleles result in small differences in eye color. Therefore, the frequency distribution for the eye-color trait appears to be continuous over a range from black to light blue.

Continuous Variation in Populations

Generally, the individuals of a population display a range of small differences in most traits. This characteristic of populations, which is known as **continuous variation**, is primarily an outcome of the number of genes affecting a trait and the number of environmental factors that influence their expression. In most cases, the greater the number of genes and environmental factors, the more continuous will be the expected distribution of all the versions of that trait.

Think about your own eye color. The colored part is the iris, a doughnut-shaped, pigmented structure just beneath the cornea. As is true of all humans, the color of your iris is the cumulative outcome of a number of gene products. Some products take part in the stepwise production and distribution of melanin, the same light-absorbing pigment that influences coat color in mammals. Dark eyes that seem to be almost black have abundant deposits of melanin molecules in the iris—so much so that most of the light that enters gets absorbed. Dark brown eyes don't have as many deposits of melanin, and some light that is not absorbed is reflected out from the iris. Light brown or hazel eyes have even less (Figure 11.16).

Green, gray, or blue eyes do not contain green, gray, or blue pigments. In these cases, the iris does incorporate different amounts of melanin, but not very much of it. As a result, many or most of the blue wavelengths of light that do enter the eye are reflected out.

How might you describe the continuous variation of some trait within a group, such as the college students in Figure 11.17*a*? They range from very short to very tall, with average heights much more common than either extreme. You can start out by dividing the full range of the different phenotypes into measurable categories. Next, count all the individual students in each category. This will give you the relative frequencies of all phenotypes distributed across the range of measurable values.

The bar chart in Figure 11.17*c* plots the proportion of students in each category against the range of the measured phenotypes. The vertical bars that are the shortest

Further reading: Student Guide to InfoTrac on web site →

a Students organized according to height, as an example of continuous variation

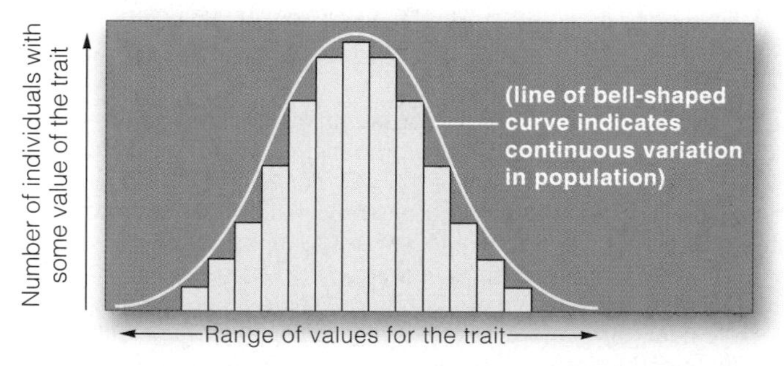

b Idealized bell-shaped curve for a population that displays continuous variation in some trait

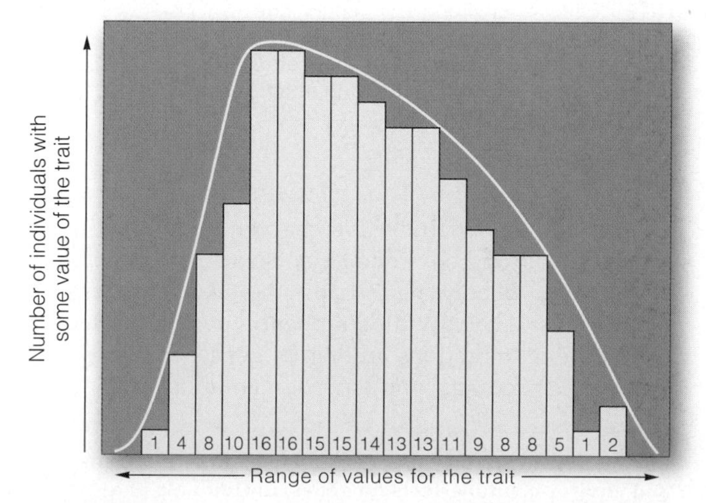

c Specific bell-shaped curve corresponding to the distribution of the trait (height) illustrated by the photograph in (**a**)

Figure 11.17 Continuous variation in body height, a trait that is one of the characteristics of the human population.

(**a**) Suppose you want to find the frequency distribution for height in a group of 169 biology students at Brigham Young University. You decide on how finely the range of possible heights should be divided. After this, you measure each student and assign her or him to the appropriate category. Finally, you divide the number in each category by the total number of all students in all categories.

(**b**) Commonly, a bar graph is used to depict continuous variation in a population. In such graphs, the proportion of individuals in each category is plotted against the range of measured phenotypes. The curved line above this particular set of bars is an idealized example of the kind of "bell-shaped" curve that emerges for populations showing continuous variation in a trait. The bell-shaped curve in (**c**) is a specific example of this type of diagram.

represent categories with the least number of students. The bar that is tallest represents the category with the greatest number of students. Finally, draw a graph line around all of the bars and you end up with a "bell-shaped" curve. Such curves are typical of populations that show continuous variation in a trait.

Enzymes and other products of genes regulate each step of most metabolic pathways. Mutations, gene interactions, and environmental conditions may affect one or more of the steps, and this leads to variations in phenotypes.

For most traits, the individuals of a population display continuous variation—that is, a range of small differences.

The greater the number of genes and environmental factors that can influence a trait, the more continuous will be the expected distribution of all the versions of that trait.

11.8 EXAMPLES OF ENVIRONMENTAL EFFECTS ON PHENOTYPE

We have mentioned, in passing, that environmental conditions often contribute to variable gene expression among individuals of a population. Before leaving this chapter, consider just two examples of environmentally induced variations in phenotype.

Possibly you have observed a Himalayan rabbit and probably a Siamese cat. Both of these furry mammals carry an allele that specifies a heat-sensitive version of one of the enzymes necessary for melanin production. At the surface of warm body regions, the enzyme is less active. Fur growing there is lighter in color than fur in cooler regions, which include the ears and other body parts that project away from the main body mass. The experiment shown in Figure 11.18 provided observable evidence of an environmental effect on gene expression.

The environment also influences genes that govern phenotypes of plants. You may have observed the color variation in the floral clusters of *Hydrangea macrophylla*, a species widely favored in home gardens (Figure 11.19). In this type of plant, the action of genes responsible for floral color produces different phenotypes, depending on the acidity of the soil in which the plant is growing.

Figure 11.18 Observable effect of differences in environmental conditions on gene expression in animals. A Himalayan rabbit normally has black hair only on its long ears, nose, tail, and lower leg limbs. For one experiment, a patch of a rabbit's white fur was plucked clean, then an icepack was secured over the hairless patch. Where the colder temperature had been maintained, the hairs that grew back were black.

Himalayan rabbits are homozygous for the *ch* allele of the gene for tyrosinase, an enzyme required to produce melanin. The allele specifies a heat-sensitive version of the enzyme, which is able to function only when the air temperature is below about 33°C. When cells that give rise to hairs grow under warmer conditions, they cannot produce melanin and hairs appear light. This happens in body regions that are massive enough to conserve a fair amount of metabolic heat. Ears and other slender extremities are cooler because they tend to lose metabolic heat more rapidly.

Figure 11.19 Effect of environmental conditions on gene expression in a favorite garden plant (*Hydrangea macrophylla*). Even plants that carry the same alleles have floral colors ranging from pink to blue, depending on the acidity of the soil in which they happen to be growing.

And so we conclude this chapter, which has dealt with heritable and environmental factors that give rise to variations in phenotype. What is the take-home lesson? Simply this: An individual's phenotype is an outcome of complex interactions among its genes, enzymes and other gene products, and environmental factors.

Owing to gene mutations, cumulative gene interactions, and environmental effects on genes, individuals of a population show degrees of variation for many traits.

Further reading: Student Guide to InfoTrac on web site

11.9 SUMMARY

1. A gene is a unit of information about a heritable trait. Alleles of a gene are different molecular versions of that information. Through experimental crosses with pea plants, Mendel gathered indirect evidence that diploid organisms have two genes for each trait and that genes retain their identity when transmitted to offspring.

2. *Monohybrid* crosses are experimental crosses between two individuals that are homozygous (true-breeding) for different versions of a trait; offspring are heterozygous (they inherit a pair of nonidentical alleles) for the trait. Mendel's monohybrid crosses with garden pea plants provided indirect evidence that some forms of a gene may be dominant over other, recessive forms.

3. A homozygous dominant individual has inherited two dominant alleles (*AA*) for the trait being studied. A homozygous recessive has two recessive alleles (*aa*), and a heterozygote has two nonidentical alleles (*Aa*).

4. In Mendel's monohybrid crosses ($AA \times aa$), all of the F_1 offspring were *Aa*. Then crosses between F_1 plants resulted in these combinations of alleles in F_2 offspring:

	A	*a*
A	*AA*	*Aa*
a	*Aa*	*aa*

AA (dominant)
Aa (dominant)
Aa (dominant)
aa (recessive)
} the expected phenotypic ratio of 3:1

5. Results from Mendel's monohybrid crosses led to the formulation of a theory of segregation. In modern terms, diploid organisms have pairs of genes, on pairs of homologous chromosomes. The two genes of each pair segregate from each other at meiosis, so that each gamete formed ends up with one or the other gene.

6. *Dihybrid* crosses are experimental crosses between two individuals that breed true (are homozygous) for different versions of two traits; so the F_1 offspring are heterozygous (inherit two nonidentical alleles) for both traits. The phenotypes of the F_2 offspring from Mendel's dihybrid crosses were close to a 9:3:3:1 ratio:

9 dominant for both traits
3 dominant for *A*, recessive for *b*
3 dominant for *B*, recessive for *a*
1 recessive for both traits

7. Mendel's dihybrid crosses led to the formulation of a theory of independent assortment. In modern terms, by the end of meiosis, the gene pairs of two homologous chromosomes have been sorted out for distribution into one gamete or another, independently of how the gene pairs of other chromosomes were sorted out.

8. Four factors commonly influence gene expression:

a. Degrees of dominance may occur between some pairs of genes.

b. The products of pairs of genes may interact to influence the same trait.

c. One gene may have positive or negative effects on two or more traits, a condition called pleiotropy.

d. Environmental conditions to which an individual is subjected may affect gene expression.

Review Questions

1. Distinguish between these terms: *11.1*
   a. gene and allele
   b. dominant allele and recessive allele
   c. homozygote and heterozygote
   d. genotype and phenotype
2. Define a true-breeding lineage. What is a hybrid? *11.1*
3. Distinguish between monohybrid and dihybrid crosses. What is a testcross, and why is it useful in genetic analysis? *11.2, 11.3*
4. Do segregation and independent assortment proceed during mitosis, meiosis, or both? *11.2, 11.3*
5. What do the vertical and horizontal arrows of the diagram at right represent? What do the bars and the curved line represent? *11.7*

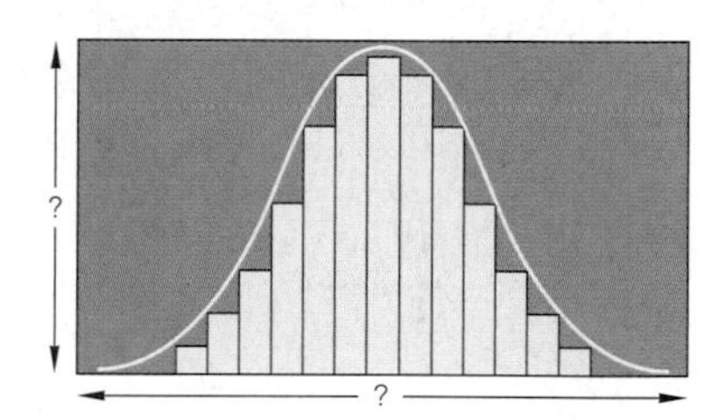

Self-Quiz (*Answers in Appendix III*)

1. Alleles are ______.
   a. different molecular forms of a gene
   b. different phenotypes
   c. self-fertilizing, true-breeding homozygotes
2. A heterozygote has a ______ for the trait being studied.
   a. pair of identical alleles
   b. pair of nonidentical alleles
   c. haploid condition, in genetic terms
   d. a and c
3. The observable traits of an organism are its ______.
   a. phenotype c. genotype
   b. sociobiology d. pedigree
4. F_1 offspring of the monohybrid cross $AA \times aa$ are ______.
   a. all *AA* c. all *Aa*
   b. all *aa* d. 1/2 *AA* and 1/2 *aa*
5. Second-generation offspring from a cross are the ______.
   a. F_1 generation c. hybrid generation
   b. F_2 generation d. none of the above
6. Assuming complete dominance will occur, the offspring of the cross $Aa \times Aa$ will show a phenotypic ratio of ______.
   a. 3:1 c. 1:2:1
   b. 9:1 d. 9:3:3:1
7. Crosses between F_1 individuals resulting from the cross $AABB \times aabb$ lead to F_2 phenotypic ratios close to ______.
   a. 1:2:1 c. 1:1:1:1
   b. 3:1 d. 9:3:3:1
8. Match each example with the most suitable description.
   ______ dihybrid cross a. *bb*
   ______ monohybrid cross b. $AaBb \times AaBb$
   ______ homozygous condition c. *Aa*
   ______ heterozygous condition d. $Aa \times Aa$

Critical Thinking—Genetics Problems
(*Answers in Appendix IV*)

1. One gene has alleles *A* and *a*. Another has alleles *B* and *b*. For each genotype listed, what type(s) of gametes will be produced? (Assume independent assortment occurs before gametes form.)
   a. *AABB* c. *Aabb*
   b. *AaBB* d. *AaBb*

2. Still referring to Problem 1, what will be the genotypes of the offspring from the following matings? Indicate the frequencies of each genotype among them.
   a. $AABB \times aaBB$ c. $AaBb \times aabb$
   b. $AaBB \times AABb$ d. $AaBb \times AaBb$

3. In one experiment, Mendel crossed a pea plant that bred true for green pods with one that bred true for yellow pods. All the F_1 plants had green pods. Which form of the trait (green or yellow pods) is recessive? Explain how you arrived at your conclusion.

4. Return to Problem 1, and assume you now study a third gene having alleles *C* and *c*. For each genotype listed, what type(s) of gametes will be produced?
   a. *AABB CC* c. *Aa BB Cc*
   b. *Aa BB cc* d. *Aa Bb Cc*

5. Mendel crossed a true-breeding tall, purple-flowered pea plant with a true-breeding dwarf, white-flowered plant. All F_1 plants were tall and had purple flowers. If an F_1 plant self-fertilizes, then what is the probability that a randomly selected F_2 offspring will be heterozygous for the genes specifying height and flower color?

6. At a gene location on a human chromosome, a dominant allele controls *tongue rolling*, an ability to curl up the two sides of the tongue (Figure 11.20). People who are homozygous for a recessive allele at that locus can't roll the tongue. At a different gene locus, a dominant allele controls whether the earlobes will be attached or detached (refer to Figure 11.1). These two pairs of genes assort independently. Suppose a tongue-rolling, detached-earlobed woman marries a man who has attached earlobes and cannot roll his tongue. Their first child has the phenotype of the father. Given this outcome,
   a. What are the genotypes of the mother, father, and child?
   b. What is the probability that a second child of theirs will have detached earlobes and won't be a tongue roller?

7. Bill and his wife, Marie, are hoping to have children. Both have notably flat feet and long eyelashes, and they tend to sneeze a lot (hence the name of the *achoo syndrome*). Dominant alleles give rise to these traits: *A* (foot arch), *E* (eyelash length), and *S* (chronic sneezing). Bill is heterozygous for all three dominant alleles and Marie is homozygous.
   a. What is Bill's genotype? What is Marie's genotype?
   b. Marie becomes pregnant four times. What is the probability each child will have the dominant phenotypes for all three traits?
   c. What is the probability that each child will have short lashes, high arches, and no chronic tendency to sneeze?

8. *DNA fingerprinting* is a method of identifying individuals by locating unique base sequences in their DNA molecules (Section 16.3). Before researchers refined the method, attorneys often relied on the ABO blood-typing system to settle disputes over paternity. Suppose that you, as a geneticist, are asked to testify during a paternity case in which the mother has type A blood, the child has type O blood, and the alleged father has type B blood. How would you respond to the following statements?
   a. Attorney of the alleged father: "The mother's blood is type A, so the child's type O blood must have come from the father. Because my client has type B blood, he could not be the father."
   b. Mother's attorney: "Because further tests prove this man is heterozygous, he must be the father."

9. Suppose you identify a new gene in mice. One of its alleles specifies white fur color. A second allele specifies brown fur color. You want to determine whether the relationship between the two alleles is one of simple dominance or incomplete dominance. What sorts of genetic crosses would give you the answer? On what types of observations would you base your conclusions?

10. Your sister moves away and gives you her purebred Labrador retriever, a female named Dandelion. Suppose you decide to breed Dandelion and sell puppies to help pay for your college tuition. Then you discover that two of her four brothers and sisters show *hip dysplasia*, a heritable disorder arising from a number of gene interactions. If Dandelion mates with a male Labrador known to be free of the harmful genes, can you guarantee to a buyer that puppies will not develop the disorder? Explain your answer.

11. A dominant allele *W* confers black fur on guinea pigs. If a guinea pig is homozygous recessive (*ww*), it has white fur. Fred would like to know whether his pet black-furred guinea pig is homozygous dominant (*WW*) or heterozygous (*Ww*). How might he determine his pet's genotype?

12. Red-flowering snapdragons are homozygous for allele R^1. White-flowering snapdragons are homozygous for a different allele (R^2). Heterozygous plants (R^1R^2) bear pink flowers. What phenotypes should appear among F_1 offspring of the crosses listed? What are the expected proportions for each phenotype?
   a. $R^1R^1 \times R^1R^2$ c. $R^1R^2 \times R^1R^2$
   b. $R^1R^1 \times R^2R^2$ d. $R^1R^2 \times R^2R^2$

Notice, in Problem 12, that in cases of incomplete dominance it is inappropriate to refer to either allele of a pair as dominant or recessive. When the phenotype of a heterozygous individual is halfway between those of the two homozygotes, then there is no dominance. Such alleles are usually designated by superscript numerals, as shown here, rather than by uppercase letters for dominance and lowercase letters for recessiveness.

Figure 11.20 A student at San Diego State University exhibiting the tongue-rolling trait for the benefit of a tongue-roll-challenged student.

13. Two pairs of genes affect comb type in chickens (Figure 11.15). When both genes are recessive, a chicken has a single comb. A dominant allele of one gene, *P*, gives rise to a pea comb. Yet a dominant allele of the other (*R*) gives rise to a rose comb. An epistatic interaction occurs when a chicken has at least one of both dominants, *P*— *R* —, which gives rise to a walnut comb. Predict the F_1 ratios resulting from a cross between two walnut-combed chickens that are heterozygous for both genes (*PpRr*).

14. As described in Section 11.5, a single mutant allele gives rise to an abnormal form of hemoglobin (Hb^S instead of Hb^A). Homozygotes (Hb^SHb^S) develop sickle-cell anemia. But the heterozygotes (Hb^AHb^S) show few outward symptoms.

Suppose a woman's mother is homozygous for the Hb^A allele and her father is homozygous for the Hb^S allele. She marries a male who is heterozygous for the allele, and they plan to have children. For *each* of her pregnancies, state the probability that this couple will have a child who is:

a. homozygous for the Hb^S allele
b. homozygous for the Hb^A allele
c. heterozygous Hb^AHb^S

15. Certain dominant alleles are so vital for normal development that an individual who is homozygous recessive for a mutant recessive form of the allele cannot survive. Such recessive, *lethal alleles* can be perpetuated by heterozygotes.

Consider the Manx allele (M^L) in cats. Homozygous cats (M^LM^L) die when they are still embryos inside the mother cat. In heterozygotes (M^LM), the spine develops abnormally, and the cats end up with no tail whatsoever (Figure 11.21).

Two M^LM cats mate. Among their *surviving* progeny, what is the probability that any one kitten will be heterozygous?

16. A recessive allele *a* is responsible for *albinism*, an inability to produce or deposit melanin in tissues. Humans and some other organisms can have this phenotype (Figure 11.22). In each of the following cases, what are the possible genotypes of the father, of the mother, and of their children?

a. Both parents have normal phenotypes; some of their children are albino and others are unaffected.
b. Both parents are albino and have only albino children.
c. The woman is unaffected, the man is albino, and they have one albino child and three unaffected children.

Figure 11.21 Manx cat.

Figure 11.22 An albino male in India.

17. Kernel color in wheat plants is determined by two pairs of genes. Alleles of one pair show incomplete dominance over alleles of the other pair.

For the gene pair at one locus on the chromosome, allele A^1 imparts one dose of red color to the kernel, whereas allele A^2 does not. At the second locus, allele B^1 gives one dose of red color to the kernel, whereas allele B^2 does not. One kernel with genotype $A^1A^1B^1B^1$ is dark red. A different kernel with genotype $A^2A^2B^2B^2$ is white. All other genotypes have kernel colors in between the two extremes.

a. Suppose you cross a plant grown from a dark red kernel with a plant grown from a white kernel. What genotypes and what phenotypes would you expect among the offspring?

b. If a plant with genotype $A^1A^1B^1B^2$ self-fertilizes, what genotypes and what phenotypes would be expected among the offspring? In what proportions?

Selected Key Terms

allele *11.1*
codominance *11.4*
continuous variation *11.7*
dihybrid cross *11.3*
epistasis *11.6*
F_1 *11.1*
F_2 *11.1*
gene *11.1*
genotype *11.1*
heterozygous *11.1*
homozygous dominant *11.1*
homozygous recessive *11.1*
hybrid offspring *11.1*
incomplete dominance *11.4*
independent assortment *11.3*
monohybrid cross *11.2*
multiple allele system *11.4*
phenotype *11.1*
pleiotropy *11.5*
probability *11.2*
Punnett-square method *11.2*
segregation *11.2*
testcross *11.2*
true-breeding lineage *11.1*

Readings *See also www.infotrac-college.com*

Fairbanks, D. J., and W. R. Andersen. 1999. *Genetics: The Continuity of Life*. Monterey, California: Brooks-Cole.

Orel, V. 1996. *Gregor Mendel: The First Geneticist*. New York: Oxford University Press.

12 CHROMOSOMES AND HUMAN GENETICS

The Philadelphia Story

Positioned at strategic locations in chromosomes are genes that work together to bring about orderly cell growth and division. Some specify enzymes and other proteins that perform these tasks, and neighboring genes control when, whether, and how fast the tasks proceed. If something disturbs the neighborhood, cell growth and division can spiral out of control and lead to cancer.

The first abnormal chromosome to be associated with cancer was named the *Philadelphia chromosome* after the city in which it was discovered. It shows up in cells of humans affected by a type of leukemia.

Leukemias arise when stem cells in bone marrow overproduce the white blood cells in charge of the body's housekeeping and defense. (All stem cells are unspecialized and retain the capacity for mitotic cell division. Their descendants also divide, and a portion become specialized cell types.) Leukemic cells infiltrate bone marrow and often crowd out the stem cells that give rise to red blood cells and platelets. Without red blood cells, anemia develops. Without platelets, the body starts to bleed internally. Leukemic cells also infiltrate the blood and organs, including the lymph nodes, spleen, and liver, and skew their functioning.

Untreated patients with acute leukemia often die within weeks or months. Those with chronic leukemia live longer but often die after infections overwhelm their body's weakened defenses. The Philadelphia chromosome is associated with a chronic leukemia.

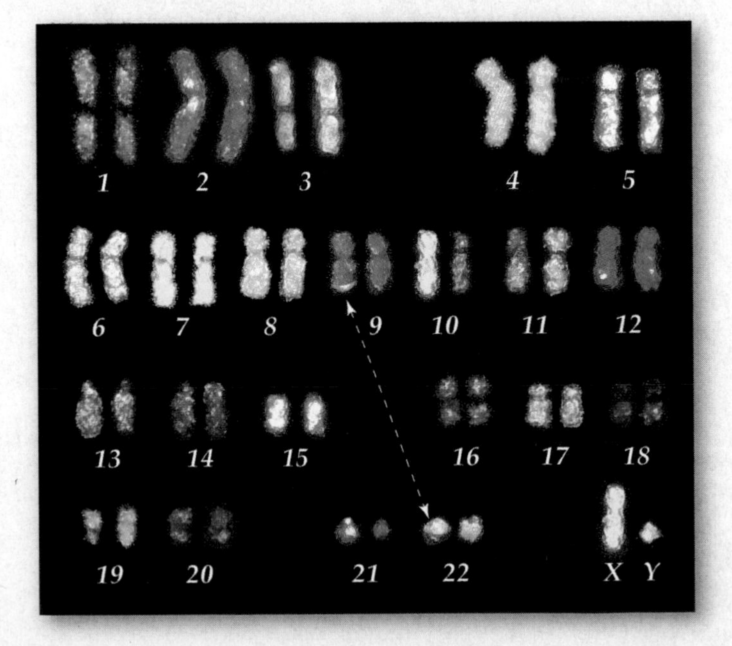

No one knew about the Philadelphia chromosome until microscopists learned how to identify its physical appearance. Chromosomes, recall, are most highly condensed at metaphase of mitosis. In that state, their size, length, and centromere location are easiest to identify. The chromosomes of many species also have distinct patterns of bands when stained a certain way. A preparation of metaphase chromosomes based on their defining features is a **karyotype**. Section 12.2 shows you how to construct a karyotype diagram from a photograph of metaphase chromosomes, just like microscopists do. The idea is to line up all the chromosomes, from largest to smallest, and position the sex chromosomes last in the lineup.

As it turned out, the Philadelphia chromosome is physically longer than its normal counterpart, human chromosome 9. The extra length is actually a piece of chromosome 22! What happened?

By chance, both chromosomes broke inside a stem cell in bone marrow. Then, in an instance of reciprocal translocation, each broken piece became reattached to the wrong chromosome—and a gene located at the end of chromosome 9 fused with a gene in chromosome 22. The altered gene region specifies an abnormal protein. In some way, that protein stimulates the unrestrained division of white blood cells.

A Philadelphia chromosome is not easy to identify in standard karyotypes. Things should change with *spectral* karyotyping, a newer research and diagnostic tool that artificially colors human chromosomes in an unambiguous way. Figure 12.1 is an example.

You began this unit of the book by looking at cell division, the starting point of inheritance. You thought about how chromosomes—and the genes they carry—are shuffled during meiosis and at fertilization. In this chapter you will delve more deeply into patterns of chromosomal inheritance. As the Philadelphia story suggests, the described methods of analysis are not remote from the world of your interests. An inherited collection of bits of information in chromosomal DNA gives rise to traits that, for better or worse, define each organism, young and old alike (Figure 12.2).

Figure 12.1 Pinpointing a killer—how a reciprocal translocation between human chromosome 9 and chromosome 22 might appear with the help of a new imaging technique called spectral karyotyping. One outcome of this translocation, the Philadelphia chromosome, is a risk factor in chronic myelogenous leukemia, a type of cancer.

Figure 12.2 The DNA icon for an Internet site, *Rare Genetic Diseases in Children*, which allows interested individuals to seek information about specific genetic diseases and disorders. At this writing, the Philadelphia chromosome is on the site's topic board.

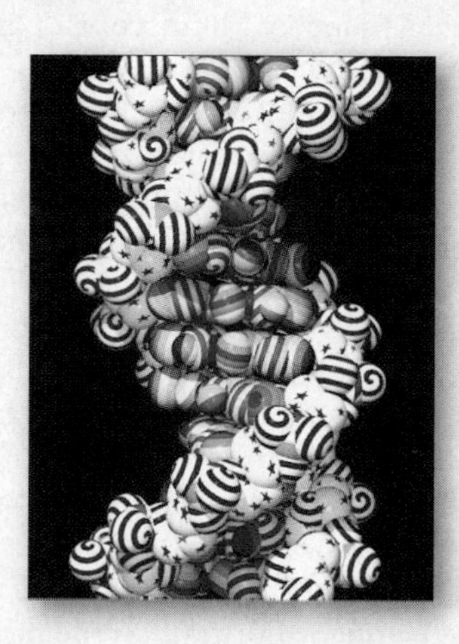

Insights into the chromosomal basis of inheritance started to accumulate after the rediscovery of Gregor Mendel's work. In 1884 Mendel had just passed away, and his paper on pea plants had been gathering dust in a hundred libraries for nearly two decades. Then improvements in the resolving power of microscopes rekindled efforts to find the hereditary material in cells. By 1882, Walther Flemming had seen threadlike bodies—chromosomes—inside the nuclear region of dividing cells. By 1884, a question was taking shape: Could chromosomes be the hereditary material?

Microscopists soon realized each gamete has half the number of chromosomes of a fertilized egg. In 1887, August Weismann hypothesized that a special division process must reduce the chromosome number by half before gametes form. Sure enough, in that same year meiosis was discovered. Weismann began to promote another hypothesis: If a chromosome number is halved during meiosis and restored at fertilization, then the cell's hereditary material must be half paternal and half maternal in origin. This view of heredity was hotly debated, and it prompted a flurry of experimental crosses—just like the ones Mendel had carried out.

Finally, in 1900, researchers came across Mendel's paper while checking literature related to their own genetic crosses. To their surprise, their experimental results confirmed what Mendel's results had already suggested: Diploid cells have two units (genes) for each heritable trait, and the two units are segregated from each other before gametes form.

In the decades that followed, researchers learned a great deal more about chromosomes. Turn now to a few high points of their work, which will be our starting point for exploring human inheritance.

KEY CONCEPTS

1. Cells of many organisms contain pairs of homologous chromosomes, which interact during meiosis. Typically, one chromosome of each pair is maternal in origin, and its homologue is paternal in origin.

2. One gene follows another in sequence along the length of a chromosome. Each kind of gene has its own position, or locus, in that sequence. Different molecular forms of the gene (alleles) may occupy the locus.

3. The combination of alleles in a chromosome does not necessarily remain intact throughout meiosis and gamete formation. During an event called crossing over, some alleles in the sequence swap places with their partner on the homologous chromosome. And alleles that swap places may or may not be identical.

4. Allelic recombinations contribute to variations in the phenotypes of offspring.

5. A chromosome may change structurally, as when a segment of it is deleted, duplicated, inverted, or moved to a new location. Also, the chromosome number of an individual's cells may change as a result of an improper separation of duplicated chromosomes during meiosis or mitosis.

6. Chromosome structure and the parental chromosome number rarely change. When changes do occur, they often give rise to genetic abnormalities or disorders.

12.1 THE CHROMOSOMAL BASIS OF INHERITANCE—AN OVERVIEW

Genes and Their Chromosome Locations

Earlier chapters provided you with a general sense of the structure of chromosomes and of what happens to them during meiosis. To refresh your memory and get a glimpse of where you are going from here with your reading, take a moment to study the following list:

1. **Genes** are units of information about heritable traits. The genes of eukaryotic species are distributed among a number of chromosomes, and each has its own location—a gene locus—in a particular type of chromosome.

2. A cell with a diploid chromosome number ($2n$) has inherited pairs of **homologous chromosomes**. All but one pair of homologous chromosomes are identical in their length, shape, and gene sequence. The single exception is a pairing of nonidentical sex chromosomes, such as X with Y. During meiosis, the two members of a pair of homologous chromosomes interact and then segregate from each other.

3. Compare the gene at one locus on a chromosome with its partner on a homologous chromosome, and you find that their molecular form may be the same or slightly different. Although many different forms of that gene may occur among members of a population, each diploid cell has only a pair of them.

4. The different molecular forms of a gene that are possible at a given locus are called **alleles**. They arise through mutation.

5. A *wild-type* allele is the most common form of a gene, either in a natural population or in standard, laboratory-bred strains of a species. Any other, less common form of a gene is called a *mutant* allele.

6. Genes on the same chromosome are physically linked together. The farther apart two linked genes are, the more vulnerable they are to **crossing over**, an event by which homologous chromosomes exchange corresponding segments. Crossing over results in **genetic recombination**: nonparental combinations of alleles in gametes, then in offspring.

7. Independent assortment (random alignment of each pair of homologous chromosomes during metaphase I of meiosis) results in nonparental combinations of alleles in gametes, then in offspring.

8. Abnormal occurrences during meiosis or mitosis occasionally change the structure of chromosomes and the parental chromosome number.

Autosomes and Sex Chromosomes

In all but one case, a pair of homologous chromosomes are exactly like each other in their length, shape, and gene sequence. Microscopists discovered the exception during the late 1800s. For many organisms, a distinctive chromosome occurs in female *or* male individuals, but not both. For example, a diploid cell of a human male has one **X chromosome** and one **Y chromosome** (XY), and that of a human female has two X chromosomes (XX). This inheritance pattern is common among many organisms, including all mammals and fruit flies.

A different pattern prevails for butterflies, moths, some birds, and certain fishes. For them, inheriting two identical sex chromosomes results in a male. Inheriting two nonidentical sex chromosomes results in a female.

Figure 12.1 shows examples of the human X and Y chromosomes. Notice that they are physically different; one is much shorter than the other. Besides this, they do not carry the same genes. Despite the differences, the two can synapse (become zippered together briefly) in a small region along their length, and this allows them to function as homologues during meiosis.

The human X and Y chromosomes fall in the more general category of **sex chromosomes**. The term refers to distinctive types of chromosomes which, in certain combinations, determine a new individual's sex—that is, whether a male or a female will develop. All other chromosomes in an individual's cells are the same in both sexes; they are called **autosomes**.

Karyotype Analysis

Today, microscopists can routinely analyze the physical appearance of sex chromosomes and autosomes taken from a cell at metaphase of mitosis. A preparation of metaphase chromosomes can be sorted out by their defining features and used to construct a karyotype diagram. You read about a new method of karyotyping in this chapter's introduction. You can use Section 12.2 to walk through a simple procedure for constructing your own karyotype diagrams.

Diploid cells have pairs of genes, on pairs of homologous chromosomes. At each gene locus, the alleles (alternative forms of a gene) may be identical or nonidentical.

As a result of crossing over and other events during meiosis, new combinations of alleles and parental chromosomes end up in offspring. Abnormal events during meiosis or mitosis also can change the structure and number of chromosomes.

Autosomes are the pairs of chromosomes that are the same in males and females of a species. One other pair, the sex chromosomes, govern the sex of a new individual.

Further reading: Student Guide to InfoTrac on web site →

12.2 KARYOTYPING MADE EASY

Karyotype diagrams help answer questions about an individual's chromosomes. Chromosomes are the most condensed and easiest to identify in cells that are going through metaphase of mitosis. Technicians don't count on finding a cell that happens to be dividing in the body when they go looking for it. Instead, they culture cells **in vitro** (literally, "in glass"). They put a sample of cells, usually from blood, into a glass container. The container holds a solution that stimulates cell growth and mitotic cell divisions.

Dividing cells can be arrested at metaphase by adding colchicine to a culture medium. As mentioned in Section 9.5, technicians and researchers use colchicine to block spindle formation. (Colchicine is an extract of autumn crocus and other plants of the genus *Colchicum*.) When a spindle can't form, the sister chromatids of duplicated chromosomes can't move to opposite spindle poles during nuclear division. As you can imagine, by using suitable colchicine concentrations and exposure times, technicians can stockpile metaphase cells, and this increases the chance of finding candidates for karyotype diagrams.

Following colchicine treatment, the culture medium is transferred to the tubes of a centrifuge, a motor-driven spinning device (Figure 12.3*a*). The cells have greater mass and density than the solution bathing them. As a result, the spinning force moves them farthest from the center of rotation, to the bottom of the attached tubes. Separation in response to a spinning force is called centrifugation.

Afterward, the cells are transferred to a saline solution. When immersed in this hypotonic fluid, they swell (by osmosis) and move apart. The metaphase chromosomes thus move apart also. The cells are ready to be mounted on a microscope slide, fixed as by air-drying, and stained.

Chromosomes take up some stains uniformly along their length, and this allows identification of chromosome size and shape. With other staining procedures, horizontal bands show up along the length of the chromosomes of certain species. If researchers direct a ray of ultraviolet light at the chromosomes, the bands will fluoresce. You can see an example of this in Figure 10.3.

In the final steps of karyotype preparation, the chromosomes are photographed through the microscope, and the image is enlarged. Next, the photograph itself is cut apart, one chromosome at a time, and individual cutouts are arranged according to size, shape, and length of arms. Then all the pairs of homologous chromosomes are horizontally aligned by centromeres. Figure 12.3*f* shows an example of a karyotype diagram prepared in this manner.

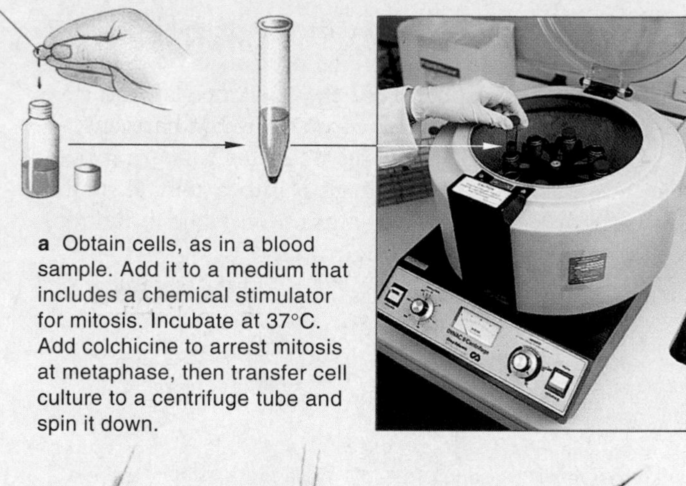

a Obtain cells, as in a blood sample. Add it to a medium that includes a chemical stimulator for mitosis. Incubate at 37°C. Add colchicine to arrest mitosis at metaphase, then transfer cell culture to a centrifuge tube and spin it down.

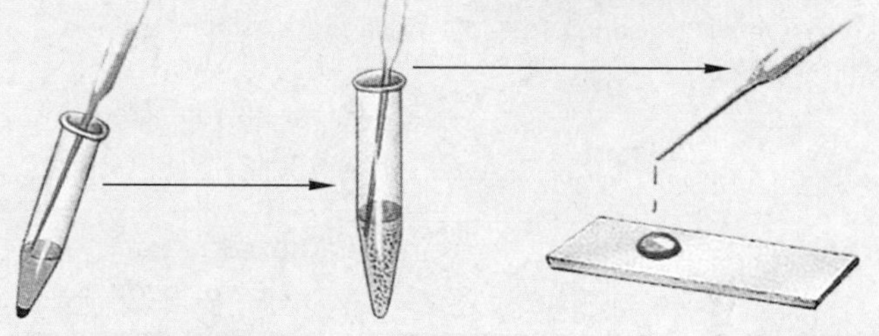

b Cells are now forced to bottom of the tube. Draw off culture medium. Add a dilute saline solution to the tube, then a fixative.

c Prepare and stain the cells for microscopy.

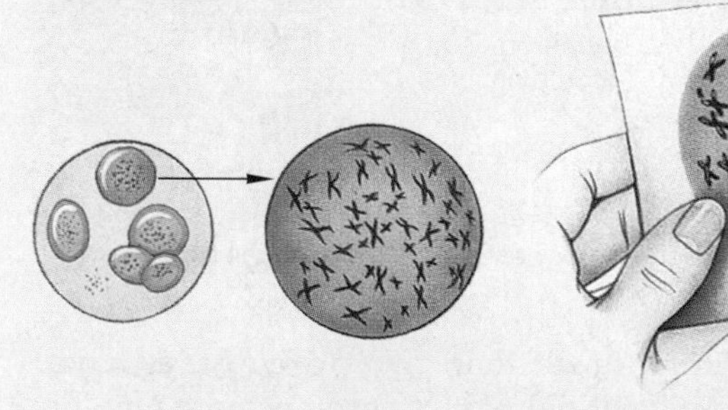

d Place cells on microscope slide. Observe.

e Photograph one cell through the microscope. Enlarge the image of its chromosomes. Cut the image apart. Arrange chromosomes as a set.

1 2 3 4 5 6 7 8 9 10 11 12
13 14 15 16 17 18 19 20 21 22 XX (or XY)

f

Figure 12.3 (**a**–**e**) Karyotype preparation. (**f**) Human karyotype. Human somatic cells have 22 pairs of autosomes and 1 pair of sex chromosomes (XX or XY). That is a diploid number of 46. These are metaphase chromosomes; each is in the duplicated state, with two sister chromatids joined at the centromere.

12.3 SEX DETERMINATION IN HUMANS

Analyzing human cells, as by karyotyping, has yielded evidence that each normal egg produced by a female has one X chromosome. Half the sperm cells produced by a male carry an X chromosome and half carry a Y.

If an X-bearing sperm fertilizes an X-bearing egg, the new individual will develop into a female. If the sperm happens to carry a Y chromosome, the individual will develop into a male (Figure 12.4).

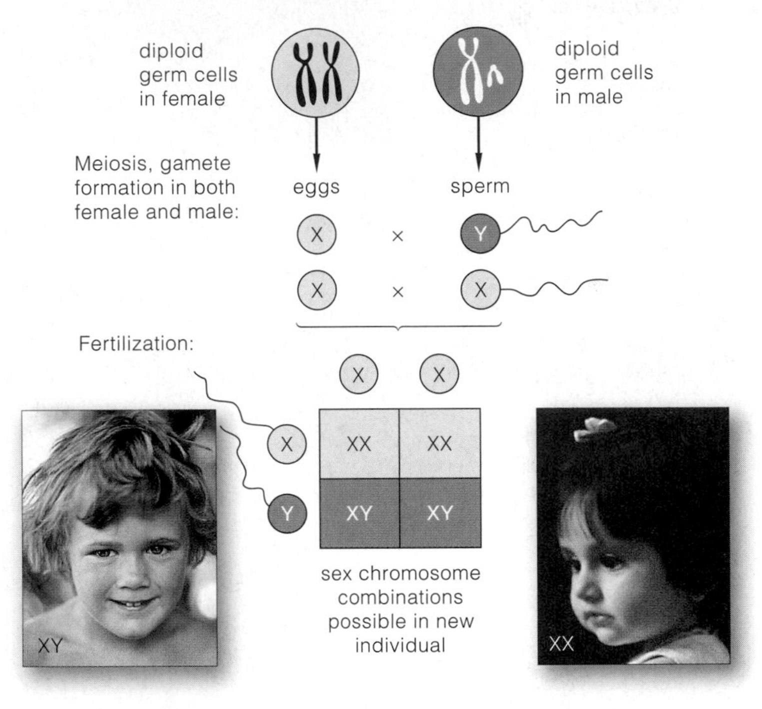

Figure 12.4 Pattern of sex determination in humans. 

At this writing, fewer than two dozen genes have been identified on the human Y chromosome. One of them is the master gene for male sex determination. Its expression leads to the formation of testes, the primary male reproductive organs (Figure 12.5). In that gene's absence, ovaries form automatically. Ovaries are the primary female reproductive organs. Testes and ovaries both produce important sex hormones that influence the development of particular sexual traits.

The human X chromosome carries more than 2,300 genes. Like other chromosomes, it carries some genes associated with sexual traits, such as the distribution of body fat and hair. However, most of its genes deal with *nonsexual* traits, such as blood-clotting functions. These genes can be expressed in males as well as in females (males, remember, also carry one X chromosome).

A certain gene on the human Y chromosome dictates that a new individual will develop into a male. In the absence of the Y chromosome (and the gene), a female develops.

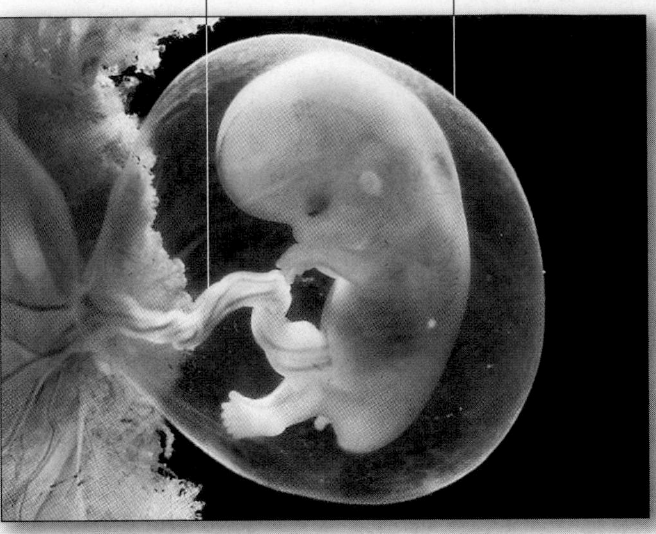

a A human embryo, eight weeks old and about an inch long. The mass at left is part of the placenta, an organ that forms from maternal and embryonic tissues.

Figure 12.5 Boys, girls, and the Y chromosome.

For about the first four weeks of its existence, a human embryo has neither male nor female traits, even though it normally carries either XY or XX chromosomes. Soon enough, however, ducts and other internal structures that can develop either way start forming.

(**a**–**c**) In an XX embryo, the ovaries (primary female reproductive organs) start to form automatically—*in the absence of a Y chromosome*. By contrast, in an XY embryo, the testes (primary male reproductive organs) start to form during the next four to six weeks. Apparently, a gene region on the Y chromosome governs a fork in the developmental road that can lead to maleness.

The newly forming testes start to produce testosterone and other sex hormones. These hormones are crucial for the development of a male reproductive system. By contrast, in an XX embryo, newly forming ovaries start to produce different kinds of sex hormones, particularly estrogens. Estrogens are crucial for the development of a female reproductive system.

The master gene for male sex determination is named *SRY* (short for the *S*ex-determining *R*egion of the *Y* chromosome). The same gene has been identified in DNA from male humans, chimpanzees, mice, rabbits, pigs, horses, cattle, and tigers, among others. None of the females tested had the gene. Tests with mice indicate that the gene region becomes active about the time that testes are starting to develop.

Further reading: Student Guide to InfoTrac on web site →

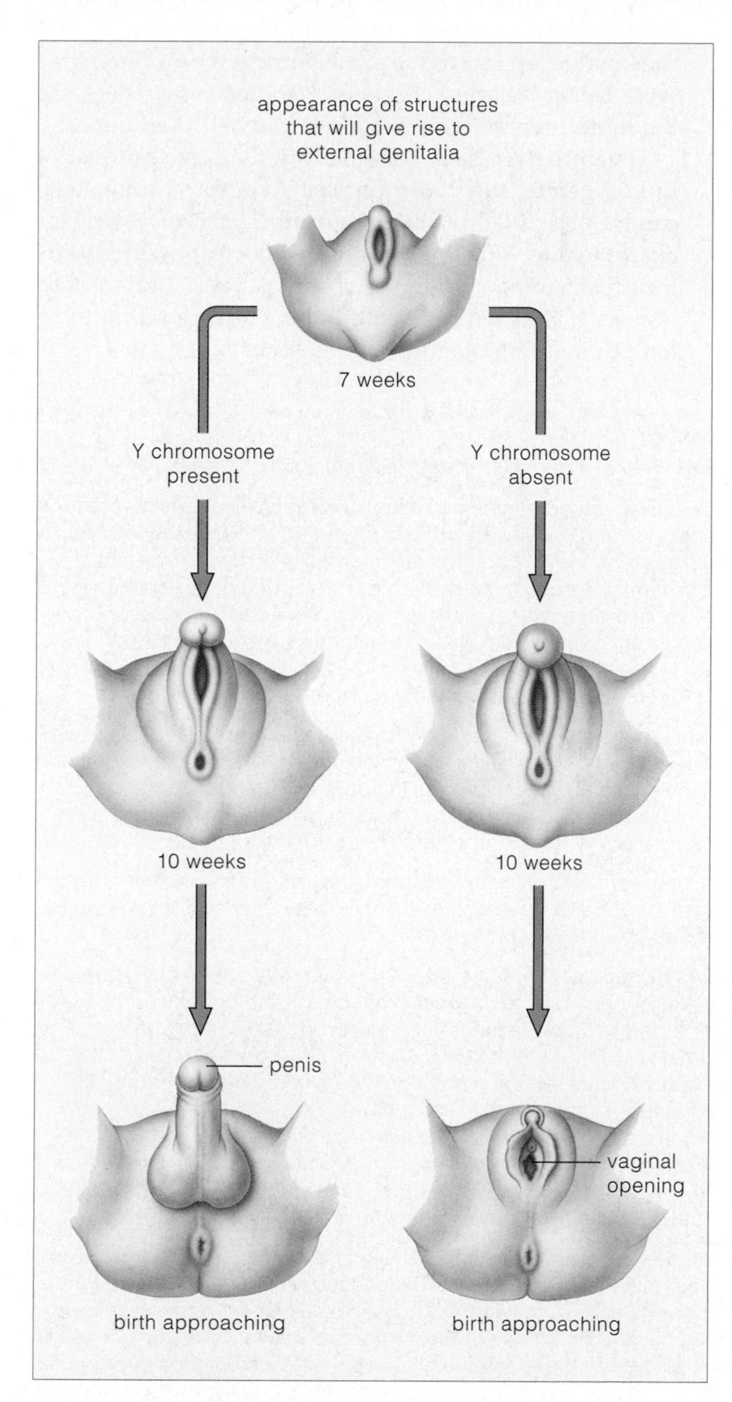

b External appearance of the newly forming reproductive organs in human embryos.

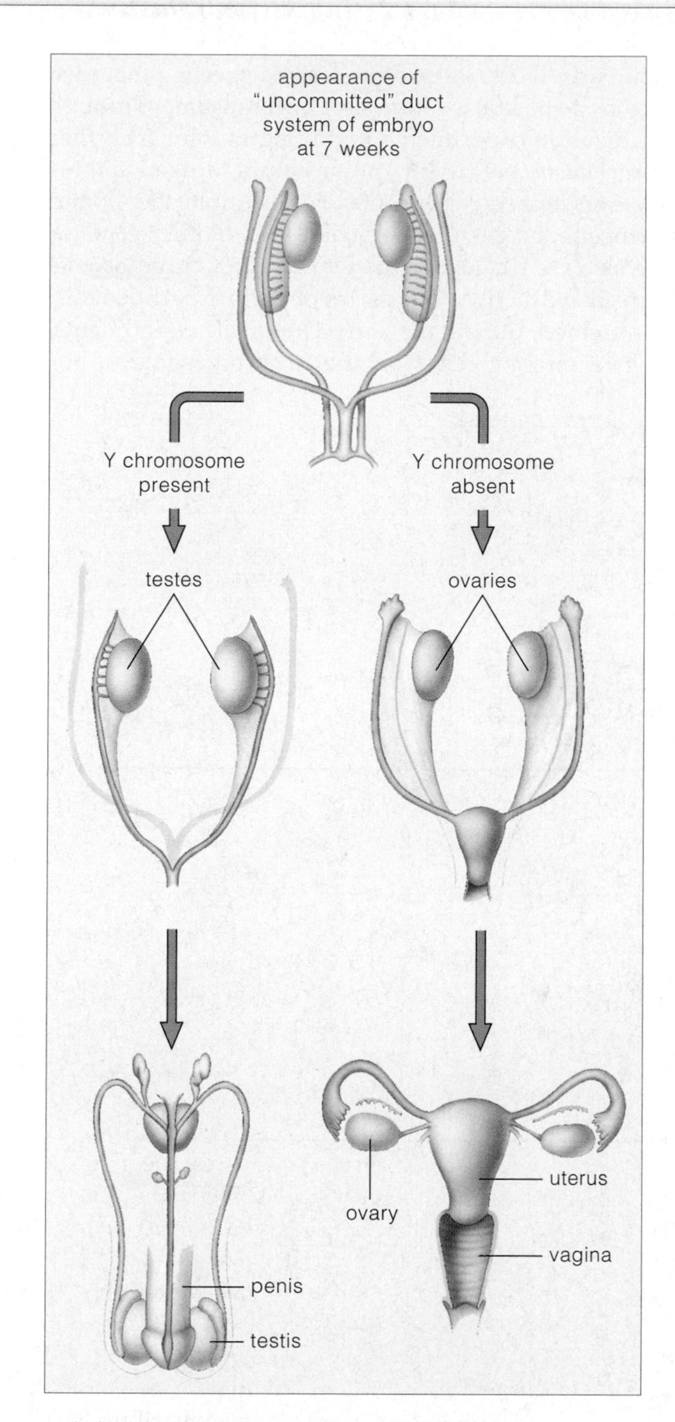

c A duct system that forms early on in the human embryo. The same internal system may develop into the male primary reproductive organs *or* the female primary reproductive organs. The presence or absence of the *SRY* gene in the embryo is the major factor that determines which developmental pathway will be followed.

The *SRY* gene resembles DNA regions that specify regulatory proteins. As described in Section 15.1, such proteins bind with certain parts of DNA and thereby turn genes on and off. It seems that the *SRY* gene product regulates a cascade of reactions that are necessary for male sex determination.

12.4 EARLY QUESTIONS ABOUT GENE LOCATIONS

Linked Genes—Clues to Inheritance Patterns

By the early 1900s, researchers were suspecting that each gene has a specific location on a chromosome. Through hybridization experiments involving mutant fruit flies (*Drosophila melanogaster*), Thomas Hunt Morgan and his coworkers helped confirm this. For example, they found evidence that a gene for eye color and another gene for wing size are located on the *Drosophila* X chromosome. Figure 12.6 describes one series of their experiments.

It seemed, during the early *Drosophila* experiments, that two mutant genes on the X chromosome (*w* for white eyes and *m* for miniature wings) were "linked." That is, they traveled together during meiosis, so that they ended up inside the same gamete. For a time, they were called "sex-linked genes." Today, researchers use the more precise terms **X-linked** and **Y-linked genes**.

Eventually, researchers identified a large number of linked genes, and those on each type of chromosome came to be called a **linkage group**. *D. melanogaster*, for example, has four linkage groups, corresponding to its four pairs of homologous chromosomes. Indian corn (*Zea mays*) has ten linkage groups, corresponding to its ten pairs of homologous chromosomes.

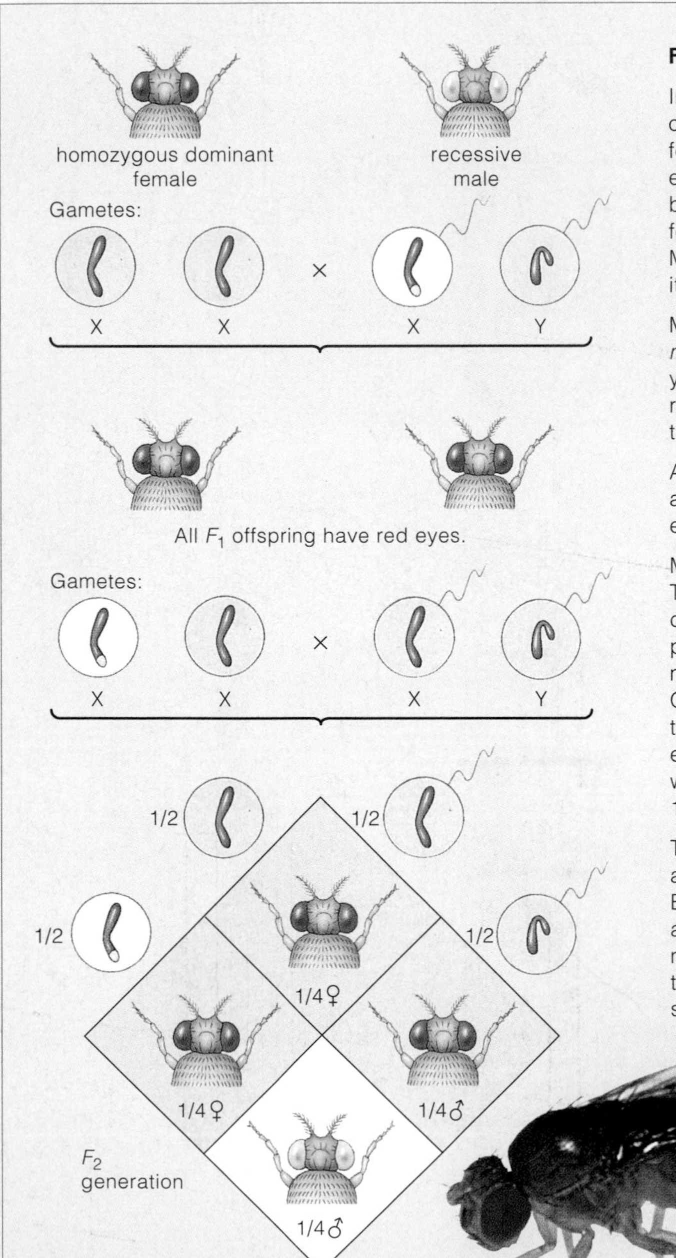

Figure 12.6 X-linked genes as clues to inheritance patterns.

In the early 1900s, the embryologist Thomas Morgan was studying patterns of inheritance. He and his coworkers discovered an apparent genetic basis for connections between sex determination and some nonsexual traits. For example, human males and females both have blood-clotting factors. Yet blood-clotting disorders (hemophilias) show up most often in males, not females, of a family lineage. Sex-specific outcomes were not like anything Mendel saw in his hybrid crosses of pea plants. With respect to phenotype, it made no difference which parent plant carried a recessive allele.

Morgan studied eye color and other nonsexual traits of fruit flies (*Drosophila melanogaster*). The flies can live in bottles on cornmeal, molasses, agar, and yeast. A female lays hundreds of eggs in a few days, and the offspring can reproduce in less than two weeks. So Morgan could track hereditary traits through nearly thirty generations of thousands of flies in a year's time.

At first, all flies were wild type for eye color; they had brick-red eyes. Then, as a result of an apparent mutation in a gene controlling eye color, a white-eyed male appeared in one of the bottles.

Morgan established true-breeding strains of white-eyed males and females. Then he performed paired, **reciprocal crosses**. (In the first of such paired crosses, one parent displays the trait of interest. In the second, the other parent displays it.) For the first cross, Morgan allowed white-eyed males to mate with homozygous red-eyed females. All F_1 offspring had red eyes. Of the F_2 offspring, however, only some of the males had white eyes. In the second cross, white-eyed females were mated with true-breeding red-eyed males. Half of the F_1 offspring were red-eyed females and half were white-eyed males. And of the F_2 offspring, 1/4 were red-eyed females, 1/4 white-eyed females, 1/4 red-eyed males, and 1/4 white-eyed males!

The seemingly odd results implied a relationship between an eye-color gene and sex determination. Probably the gene locus was on a sex chromosome. But which one? Because females (XX) could be white-eyed, the recessive allele would have to be on one of their X chromosomes. Suppose white-eyed males (XY) also carry the recessive allele on their X chromosome. Suppose there is no corresponding eye-color allele on the Y chromosome. If that were so, then the males would have white eyes, for they have no dominant allele to mask the effect of the recessive one.

The diagram at left illustrates the expected results when Morgan's idea of an X-linked gene is combined with Mendel's concept of segregation. By proposing that one particular gene is located on an X chromosome but not on the Y, Morgan was able to explain the outcome of his reciprocal crosses. His experimental results matched the predicted outcomes.

Further reading: Student Guide to InfoTrac on web site

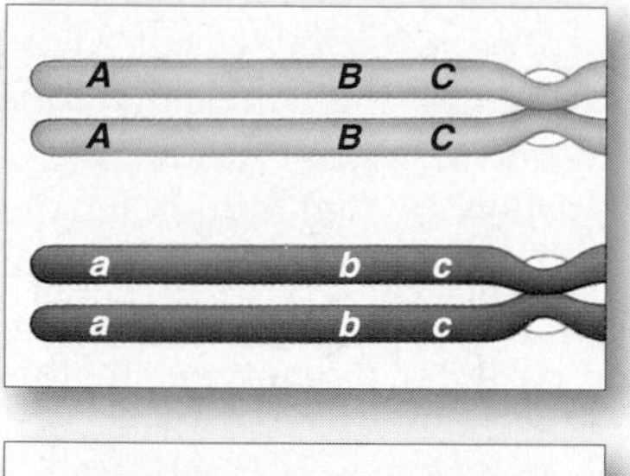

a One pair of homologous chromosomes in the duplicated state (each consists of two sister chromatids). One is shaded *blue*, the other *purple*. Three different genes are shown. Alleles at all three gene loci are nonidentical (*A* with *a*, *B* with *b*, and *C* with *c*).

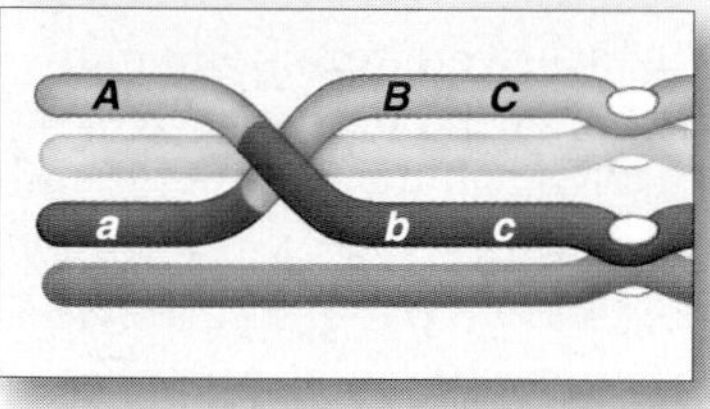

b In prophase I of meiosis, two nonsister chromatids exchange segments. The exchange represents one crossover event.

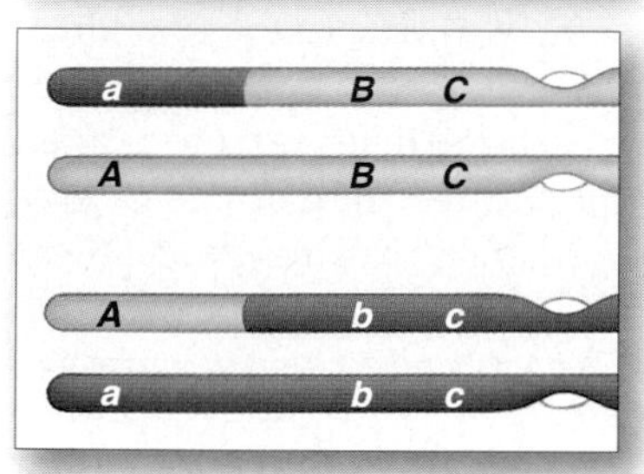

c This is the outcome of the crossover: genetic recombination between nonsister chromatids (which are shown after meiosis, as unduplicated, separate chromosomes).

Figure 12.7 Review of crossing over. As explained earlier in Section 10.4, this event occurs in prophase I of meiosis.

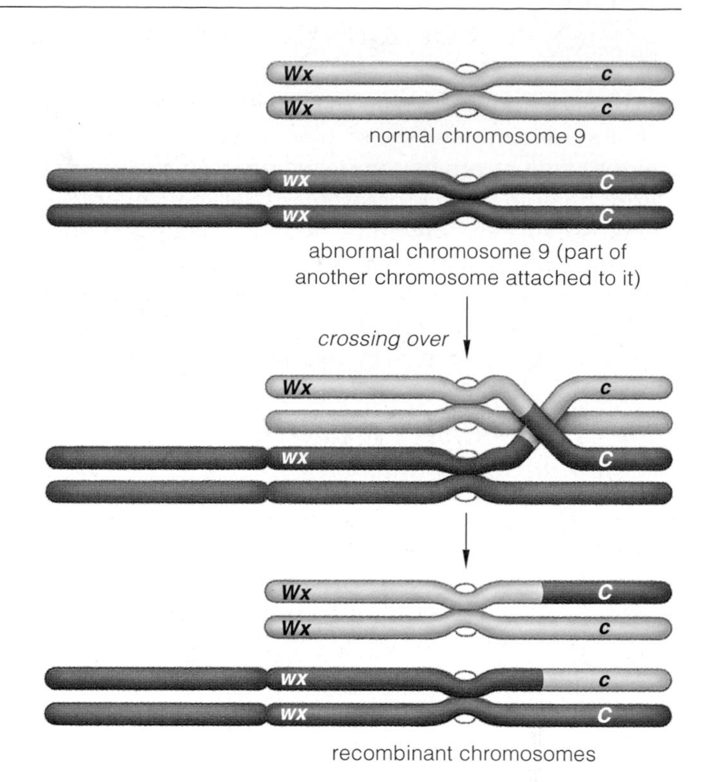

Figure 12.8 Harriet Creighton and Barbara McClintock's correlation of a cytological difference with a genetic difference in chromosome 9 from a strain of Indian corn (*Zea mays*).

Crossing Over and Genetic Recombination

If linked genes always stayed together through meiosis, then a dihybrid cross between true-breeding parents should always have a predictable outcome. Specifically, the most frequent phenotypes among the F_2 offspring should be those of the original parents. (Here you may wish to review Section 10.4 and Figure 12.7.) However, a number of puzzling results from the early *Drosophila* experiments did not match this expectation.

For example, Morgan crossed a *Drosophila* female that was recessive for white eyes and miniature wings with a wild-type male (red eyes and long wings). As expected, all F_1 males had white eyes and miniature wings, and all females were wild type. After crossing F_1 flies, Morgan analyzed 2,441 of the F_2 offspring. A significant number had white eyes and long wings or red eyes and miniature wings! As you will see in the next section, 900 (or 36.9 percent) were recombinants. As Morgan hypothesized, physical exchanges must have occurred between X chromosomes during meiosis.

It was not until 1931 that two genetic researchers, Harriet Creighton and Barbara McClintock, discovered evidence of such exchanges. They were experimenting with two *Z. mays* chromosomes that differed physically in a way that could be distinguished with a microscope. Such distinguishable features are **cytological markers** for the genes being studied.

The researchers used corn that was heterozygous for two genes on chromosome 9. Two alleles at one gene locus specified colored (*C*) or colorless (*c*) seeds. One allele at the other gene locus specified the synthesis of two forms of starch (*Wx*); its partner specified only one (*wx*). In one experiment, one of the chromosomes 9 had genotype *c Wx* and a normal appearance (Figure 12.8). Another chromosome 9 had genotype *C wx*, and it was longer; an abnormal event caused a piece of a different chromosome to become attached to one of its ends.

During meiosis, crossing over sometimes occurred between the two gene loci on a pair of the cytologically different chromosomes. The outcome was two kinds of genetic recombinants: *CWx* and *cwx*. As Creighton and McClintock realized, when genetic recombination had occurred, cytological features of the chromosomes also had changed. When *wx* ended up with *c* instead of *C*, so did the extra length. They observed no such correlation in F_1 generations that retained the parental genotype.

Genes at different loci on the same chromosome belong to the same linkage group and do not assort independently during meiosis.

Crossing over between homologous chromosomes disrupts gene linkages and results in nonparental combinations of genes in chromosomes. Correlations between specific genes and cytological markers provide evidence of recombination.

12.5 RECOMBINATION PATTERNS AND CHROMOSOME MAPPING

As we now know, crossing over is not a rare event. In fact, for humans and most other eukaryotic species, meiosis cannot even be completed properly unless each pair of homologous chromosomes takes part in at least one crossover. The preceding section briefly described experimental evidence of this remarkable event. Let us now consider a few specific examples of the ways in which crossing over disrupts gene linkages and what researchers do with this information.

How Close Is Close? A Question of Recombination Frequencies

Figure 12.9 shows Morgan's cross between a wild-type *Drosophila* male and a female that was recessive for white eyes and for miniature wings. Again, the two parental genotypes showed up equally among the F_1 offspring (50 percent white-eyed, miniature-winged males and 50 percent wild-type females). Of the 2,441 F_2 offspring analyzed, 36.9 percent were recombinants.

Morgan's group also extended their investigations to other X-linked genes. In one series of experiments, a true-breeding white-eyed, yellow-bodied mutant female was crossed with a wild-type male (with red eyes and a gray body). As expected, 50 percent of the F_1 offspring showed one or the other parental phenotype. In this case, however, only 129 of 2,205 of the F_2 offspring that were analyzed—or 1.3 percent—were recombinants.

From the results of many such experimental crosses, it appeared that the genes controlling eye color and wing size were not as "tightly linked" as those for eye color and body color. Also, the two parental genotypes were represented in approximately equal numbers of the F_2 offspring, and the same was true of the two kinds of recombinant genotypes. What was going on here? As Morgan hypothesized, certain alleles tend to remain together during meiosis more often than others *because they are positioned closer together on the same chromosome.*

Imagine any two genes at two different locations on the same chromosome. *The probability that a crossover*

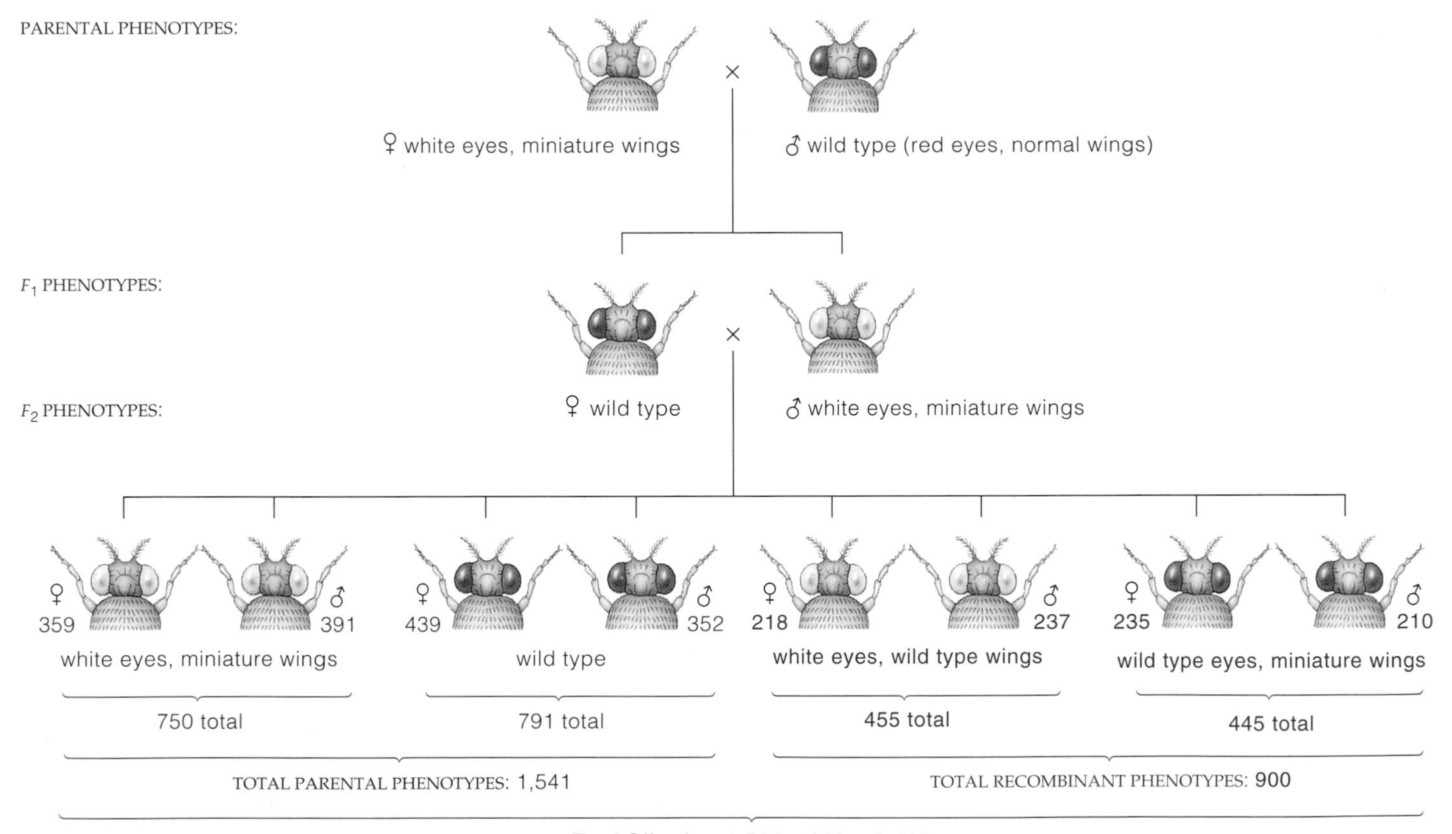

Figure 12.9 Experimental crosses with *Drosophila melanogaster* that provided indirect evidence of how crossing over can disrupt linkage groups. In this example, Morgan's group tracked a mutant gene for white eyes and a different mutant gene for miniature wings. As you probably know, the symbol ♀ signifies female and the symbol ♂ signifies male.

Further reading: Student Guide to InfoTrac on web site

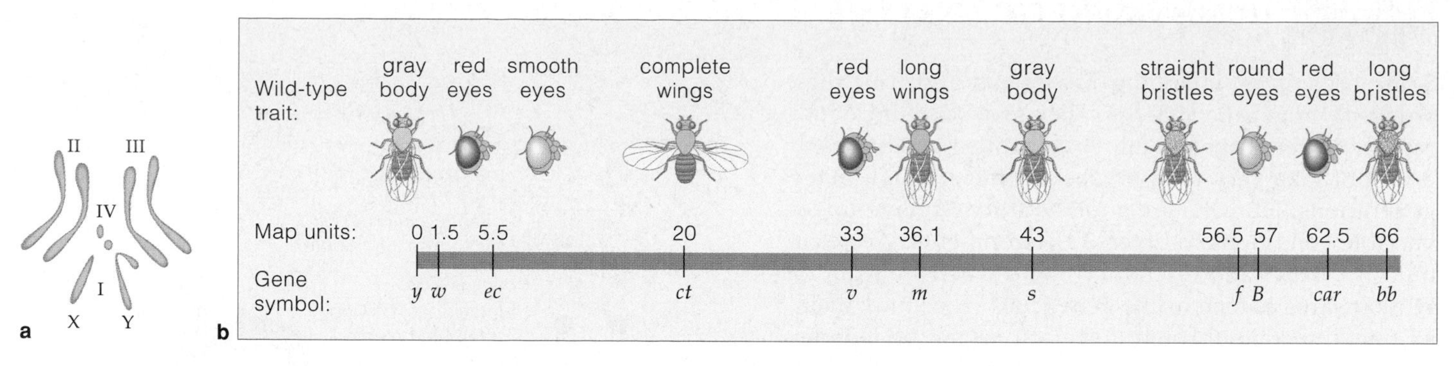

Figure 12.10 (**a**) *Drosophila melanogaster* chromosomes. (**b**) A linkage map for a number of genes on chromosome I (here, the X chromosome). As you can see, more than one gene can influence the same trait, such as eye color.

will disrupt their linkage is proportional to the distance that separates the two loci. Suppose genes *A* and *B* are twice as far apart as two other genes, *C* and *D*:

We would expect crossing over to disrupt the linkage between *A* and *B* much more often.

Two genes are very closely linked when the distance between their loci is small; their allelic combinations nearly always end up in the same gamete. Linkage is more vulnerable to crossover when the distance between the two loci is greater. When two gene loci are very far apart, crossing over disrupts linkage so often that those genes assort independently of each other into gametes.

Linkage Mapping

After using testcrosses to analyze crossover patterns, Alfred Sturtevant, one of Morgan's students, proposed that the percentage of recombinants in gametes might be a quantitative measure of the relative positions of genes along a chromosome. This investigative approach is now called **linkage mapping**.

The relative linear distance between any two linked genes on a genetic map is expressed in map units. One map unit corresponds to a crossover frequency of 1 percent, twenty map units correspond to a crossover frequency of 20 percent, and so on. For instance, take a look at Figure 12.10*b*, which is a linkage map for one of the *D. melanogaster* chromosomes. The amount of gene recombination to be expected between "smooth eyes" and "long wings" would be 30.6 percent (5.5 map units subtracted from 36.1 map units).

Linkage maps do not show *actual* physical distances between gene loci. The most accurate approximations have been calculated only for closely linked genes. Why? As map distance increases, multiple crossovers occur and skew recombination frequencies. Even so, the *map distance* between two gene loci as estimated by genetic crosses generally correlates with the *physical distance*, or length of DNA between them, as calculated by other methods. (Exceptions to this generalization occur in the centromere region.)

Of the several thousand known genes in the four types of *Drosophila* chromosomes, the positions of about a thousand have been mapped. What about the 50,000 to 100,000 genes on the twenty-three types of human chromosomes? Unlike fruit flies, humans do not lend themselves to experimental crosses. Nevertheless, some tight gene linkages have been identified by tracking the resulting phenotypes, one generation after another, in certain families.

For example, recessive alleles at two gene loci on the X chromosome cause *color blindness* and *hemophilia* (to be described shortly). One female carried both alleles, but she herself was symptom-free. So was her father, so she must have inherited a normal X chromosome from him (remember, males have only one X and one Y). The X chromosome that she inherited from her mother must have carried both mutant alleles. The female gave birth to six sons. Three sons developed color blindness and hemophilia; two were unaffected. Here was phenotypic evidence that recombination had not occurred. But her sixth son started life as a fertilized, recombinant egg; he was color-blind only. Thus, for the two mutant alleles in this family, the recombination frequency is 1/6 (or 0.167 percent).

Many affected families would have to be examined to get a good estimate of the map distance between the two gene loci. Generally speaking, these genes do not have high recombination frequencies because they are very closely linked at one end of the X chromosome.

The farther apart two genes are on a chromosome, the greater will be the frequency of crossing over and therefore of genetic recombination between them.

Linkage mapping is a method of measuring the relative linear distances between genes on the same chromosome. Such maps roughly correspond to actual physical distances, or lengths of DNA, as calculated by other methods.

12.6 HUMAN GENETIC ANALYSIS

Some organisms, including pea plants and fruit flies, are ideal for genetic analysis. They grow and reproduce rapidly in small spaces, under controlled conditions. It does not take very long to track a trait through many generations. Humans are another story. We live under variable conditions in diverse environments. We select our own mates and reproduce if and when we want to. Humans live as long as the geneticists who study them, so tracking traits through generations can be tedious. Most human families are so small, there are not enough offspring for easy inferences about inheritance.

Constructing Pedigrees

To get around the problems associated with analyzing human inheritance, geneticists put together pedigrees. A **pedigree** is a chart that shows genetic connections among individuals. When genetic researchers construct them, they use standardized methods and definitions, as well as standardized symbols to represent individuals, as shown in Figure 12.11*a*.

When they analyze pedigrees, geneticists rely on their knowledge of probability and Mendelian inheritance patterns, which may yield clues to the genetic basis for a trait. For example, clues might suggest that an allele responsible for a disorder is dominant or recessive or that it is located on a particular autosome or sex chromosome.

Gathering a great many family pedigrees increases the numerical base for analysis. When any trait follows a simple Mendelian inheritance pattern, a geneticist has greater confidence in predicting the probability of its occurrence among children of prospective parents. We will return to this topic later.

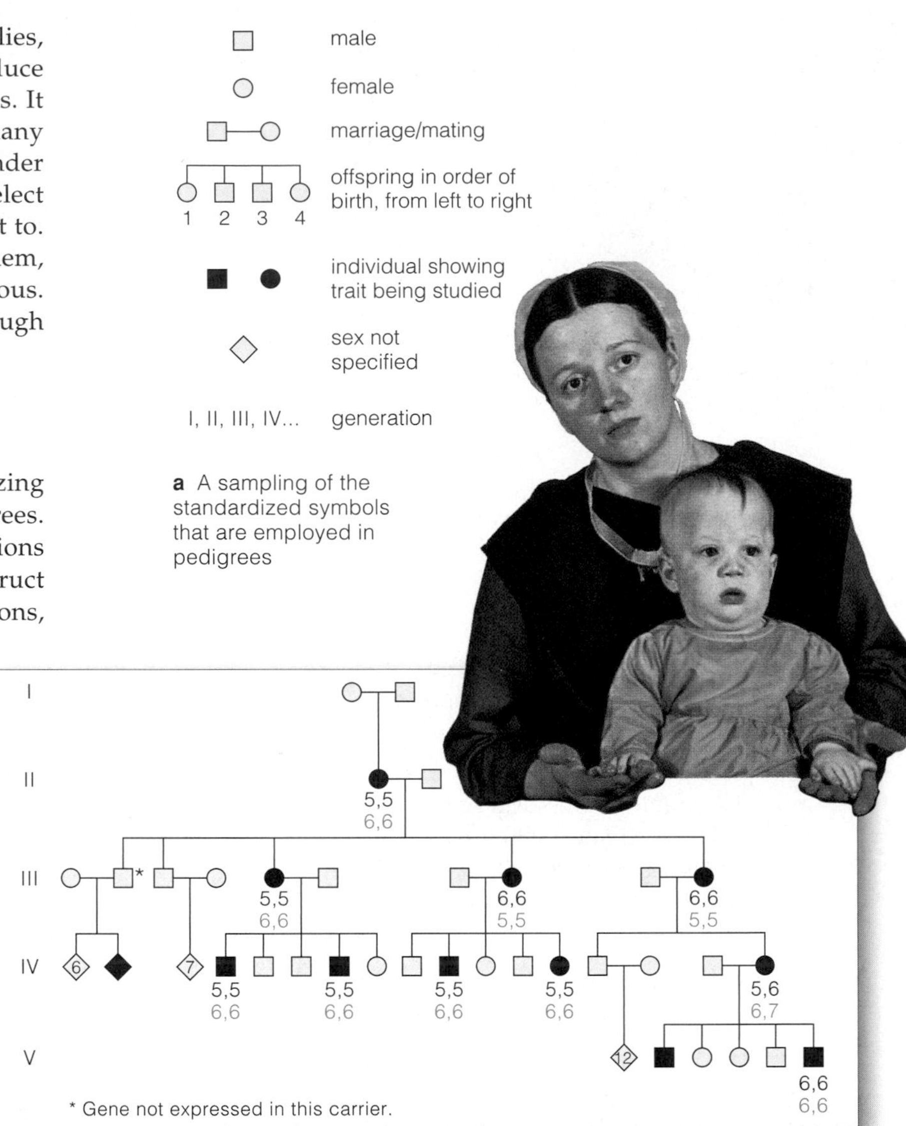

a A sampling of the standardized symbols that are employed in pedigrees

b Pedigree for a family in which polydactyly recurs

c Nancy Wexler and a pedigree she constructed to track Huntington disorder

Figure 12.11 (**a**) Some standardized symbols used in constructing pedigree diagrams. (**b**) Example of a pedigree for *polydactyly*, a condition in which an individual has extra fingers, extra toes, or both. Expression of the gene for this trait can vary from one individual to the next. *Black* numerals signify the number of fingers on each hand where data were available; *blue* numerals signify the number of toes on each foot.

(**c**) From human genetic researcher Nancy Wexler, a pedigree for *Huntington disorder*, by which the human nervous system progressively degenerates. Wexler and her team constructed an extended family tree for nearly 10,000 people in Venezuela. As analysis of affected and unaffected individuals revealed, a dominant allele on human chromosome 4 is the genetic culprit. Wexler has a special interest in Huntington disorder; she herself has a 50 percent chance of developing it.

Further reading: Student Guide to InfoTrac on web site →

Table 12.1 Examples of Human Genetic Disorders and Genetic Abnormalities

Disorder or Abnormality*	Main Consequences
AUTOSOMAL RECESSIVE INHERITANCE	
Albinism *11.6; 11.9 CT*	Absence of pigmentation
Blue offspring *18.12 CT*	Bright blue skin coloration
Cystic fibrosis *4.12 CT*	Excessive glandular secretions leading to tissue, organ damage
Ellis–van Creveld syndrome *18.11*	Extra fingers, toes, short limbs
Galactosemia *12.7*	Brain, liver, eye damage
Phenylketonuria (PKU) *12.11*	Mental retardation
Sickle-cell anemia *11.5; 14.1; 14.5; 18.6*	Adverse pleiotropic effects on organs throughout body
AUTOSOMAL DOMINANT INHERITANCE	
Achondroplasia *12.7*	One form of dwarfism
Achoo syndrome *11.9 CT*	Chronic sneezing
Camptodactyly *11.7*	Rigid, bent little fingers
Familial cholesterolemia *16 CI*	High cholesterol levels in blood; eventually clogged arteries
Huntington disorder *12.6; 12.7*	Nervous system degenerates progressively, irreversibly
Marfan syndrome *12.12 CT*	Abnormal or no connective tissue
Polydactyly *12.6*	Extra fingers, toes, or both
Progeria *12.8*	Drastic premature aging
Neurofibromatosis *14.7*	Tumors of nervous system, skin
X-LINKED RECESSIVE INHERITANCE	
Color blindness *12.7*	Inability to distinguish among some or all colors
Fragile X syndrome *12.7*	Mental retardation
Hemophilia *12.7; 12.12 CT*	Impaired blood-clotting ability
Muscular dystrophies *12.12 CT; 15.7 CT*	Progressive loss of muscle function
Testicular feminization syndrome *37.2*	XY individual but having some female traits; sterility
X-linked anhidrotic dysplasia *15.4*	Mosaic skin (patches with or without sweat glands); other effects
CHANGES IN CHROMOSOME NUMBER	
Down syndrome *12.10; 45.4*	Mental retardation; heart defects
Klinefelter syndrome *12.10*	Sterility; retardation
Turner syndrome *12.10*	Sterility; abnormal ovaries, abnormal sexual traits
XYY condition *12.10*	Mild retardation or free of symptoms
CHANGES IN CHROMOSOME STRUCTURE	
Chronic myelogenous leukemia *12 CI*	Overproduction of white blood cells in bone marrow; organ malfunctions
Cri-du-chat syndrome *12.9*	Mental retardation; abnormally shaped larynx

* *Italic* numbers indicate sections in which a disorder is described. *CI* signifies *C*hapter *I*ntroduction. *CT* signifies an end-of-chapter *C*ritical *T*hinking question.

Regarding Human Genetic Disorders

Table 12.1 is a list of some heritable traits that have been studied in detail. A few are abnormalities, or deviations from the average condition. Said another way, a **genetic abnormality** is nothing more than a rare, uncommon version of a trait, as when a person is born with six toes on each foot instead of five. Whether an individual or society at large views an abnormal trait as disfiguring or merely interesting is subjective. As the classic novel *The Hunchback of Notre Dame* suggests, there is nothing inherently life-threatening or even ugly about it.

By comparison, a **genetic disorder** is an inherited condition that sooner or later will cause mild to severe medical problems. A **syndrome** is a recognized set of symptoms that characterize a given disorder.

Because alleles underlying severe genetic disorders put people at great risk, they are rare in populations. Why, then, don't they disappear entirely? There are two reasons. First, rare mutations introduce new copies of the alleles into the population. Second, in heterozygotes, the harmful allele is paired with a normal one that may cover its functions, so it still can be passed to offspring.

You may hear someone refer to a genetic disorder as a disease, but the terms are not always interchangeable. A disease also is an abnormal alteration in the way the body functions, and it, too, is characterized by a set of symptoms. But **disease** is illness caused by infectious, dietary, or environmental factors, not by inheritance of mutant genes. If an individual's previously workable genes get altered in a way that disrupts body functions, the illness might be called a *genetic* disease.

With these qualifications in mind, we turn next to examples of inheritance in the human population. As you will see, genetic analyses of family pedigrees have often revealed simple Mendelian inheritance patterns for certain traits. Researchers have traced many of the traits to dominant or recessive alleles on an autosome or X chromosome. They have traced others to changes in the structure or number of chromosomes.

For many genes, pedigree analysis might reveal simple Mendelian inheritance patterns that will allow inferences about the probability of their transmission to children.

A genetic abnormality is a rare or less common version of an inherited trait. A genetic disorder is an inherited condition that results in mild to severe medical problems.

12.7

INHERITANCE PATTERNS

Autosomal Recessive Inheritance

For some traits, inheritance patterns reveal two clues that point to a recessive allele on an autosome. *First,* if both parents are heterozygous, any child of theirs will have a 50 percent chance of being heterozygous and a 25 percent chance of being homozygous recessive, as Figure 12.12 indicates. *Second,* if the parents are both homozygous recessive, any child of theirs will be, also.

On average, 1 in 100,000 newborns is homozygous for a recessive allele that causes *galactosemia*. It cannot produce functional molecules of an enzyme that stops a product of lactose breakdown from accumulating to toxic levels. Lactose normally is converted to glucose and galactose, then to glucose–1–phosphate (which is broken down by glycolysis or converted to glycogen). In affected persons, the full conversion is blocked:

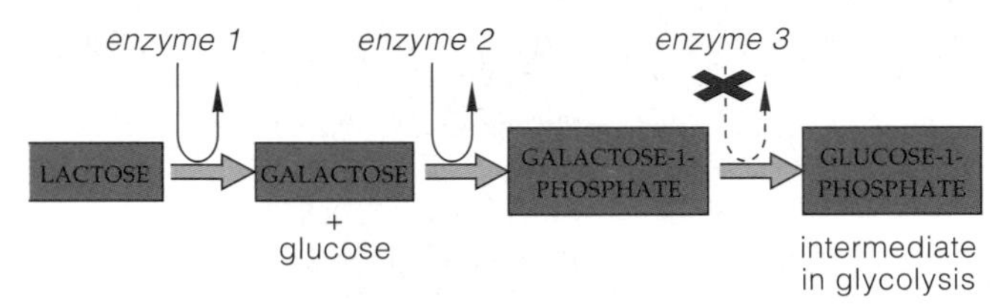

A high blood level of galactose damages the eyes, liver, and brain. This telling symptom of the disorder can be detected in urine samples. Malnutrition, diarrhea, and vomiting are other symptoms. Untreated galactosemics often die in childhood. If affected individuals are quickly placed on a restricted diet that excludes dairy products, however, they can grow up symptom-free.

Autosomal Dominant Inheritance

Two clues of a different sort indicate that an autosomal dominant allele is responsible for a trait. *First,* the trait typically appears in each generation, for the allele is usually expressed, even in heterozygotes. *Second,* if one parent is heterozygous and the other is homozygous recessive, there is a 50 percent chance that any child of theirs will be heterozygous (Figure 12.13*a*).

A few dominant alleles persist in populations even though they cause severe genetic disorders. Some persist by spontaneous mutations. For others, expression of the dominant allele may not interfere with reproduction or affected people reproduce before symptoms are severe.

For example, *Huntington disorder* is characterized by progressive involuntary movements and deterioration of the nervous system, and eventual death. Symptoms may not even start to show up until an affected person is past age thirty. Most people have already reproduced by then. Affected individuals usually die during their forties or fifties, perhaps before they realize that they have transmitted the mutant allele to their children.

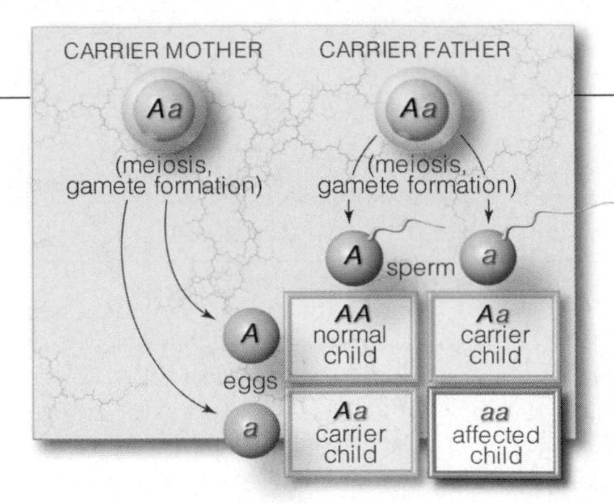

Figure 12.12 A pattern for autosomal recessive inheritance. In this case, both parents are heterozygous carriers of the recessive allele (shown in *red*).

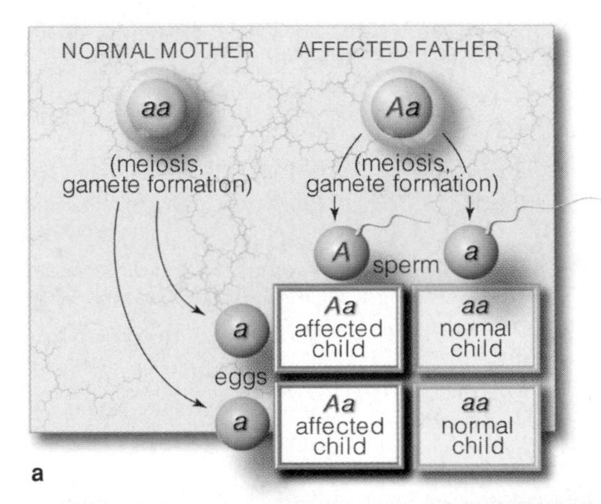

Figure 12.13 (**a**) A pattern for autosomal dominant inheritance. This dominant allele (*red*) is fully expressed in the carriers. (**b**) Infanta Margarita Teresa of the Spanish court and her maids, including the achondroplasic woman at the far right.

As another example, *achondroplasia* affects about 1 in 10,000 people. Commonly, the homozygous dominant condition leads to stillbirth, but heterozygotes are able to reproduce. While they are young, cartilage components of their skeleton form improperly. At maturity, affected people have abnormally short arms and legs relative to other body parts (Figure 12.13*b*). Adult achondroplasics are less than 4 feet, 4 inches tall. Often the dominant allele has no other phenotypic effects.

Further reading: Student Guide to InfoTrac on web site

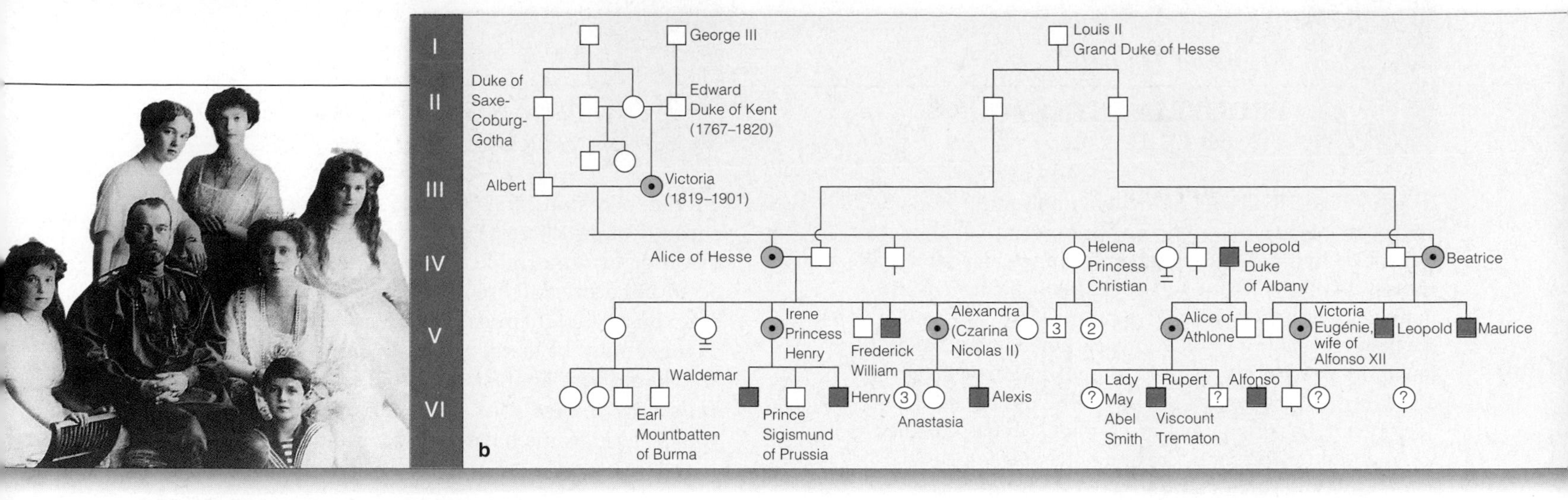

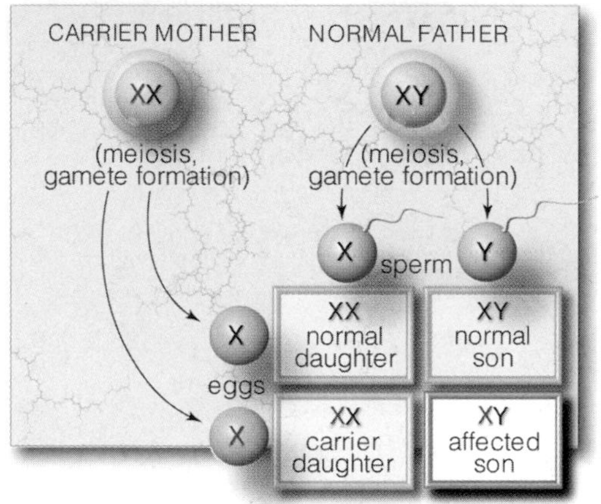

Figure 12.14 (**a**) One pattern for X-linked inheritance. In this example, assume the mother carries the recessive allele on one of her X chromosomes (*red*).

(**b**) *Above:* Partial pedigree for Queen Victoria's descendants, showing carriers and affected males who carried the X allele for hemophilia A. Of the Russian royal family members in the photograph, the mother was a carrier and Crown Prince Alexis was hemophilic. He was a focus of political intrigue that helped trigger the Russian Revolution of 1917.

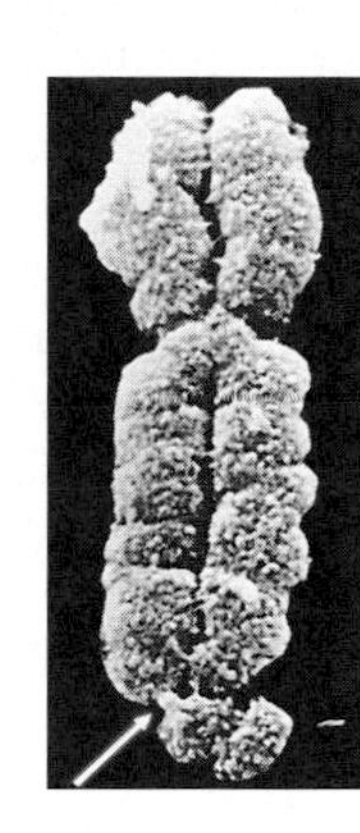

Figure 12.15 Scanning electron micrograph of the fragile X chromosome from a cultured cell. The arrow points to the fragile site.

X-Linked Recessive Inheritance

Distinctive clues often show up when a recessive allele on an X chromosome causes a genetic disorder. *First,* males show the recessive phenotype more often than females. The recessive allele can be masked in females, who may inherit a dominant allele on their other X chromosome. The allele is not masked in males, who have only one X chromosome (Figure 12.14*a*). *Second,* a son cannot inherit the recessive allele from his father. A daughter can. If a daughter is heterozygous, there is a 50 percent chance each son of hers will inherit the allele.

Color blindness, an inability to distinguish among some or all colors, is a common X-linked recessive trait. For instance, in red–green color blindess, the individual lacks some or all of the sensory receptors that normally respond to visible light of red or green wavelengths.

Hemophilia A, a blood-clotting disorder, is a case of X-linked recessive inheritance. In most people, a blood-clotting mechanism quickly stops bleeding from minor injuries. Clot formation requires the products of several genes, some of which are on the X chromosome. If any of the X-linked genes is mutated in a male, the absence of its functional product prolongs bleeding. About 1 in 7,000 males inherits the mutant allele for hemophilia A. Clotting is close to normal in heterozygous females.

The frequency of hemophilia A was unusually high in royal families of nineteenth-century Europe. Queen Victoria of England was a carrier (Figure 12.14*b*). At one time, the recessive allele was present in eighteen of her sixty-nine descendants.

Fragile X syndrome, an X-linked recessive disorder that causes mental retardation, affects 1 in 1,500 males in the United States. In cultured cells, an X chromosome carrying the mutant allele is constricted near the end of its long arm (Figure 12.15). This constriction is called a fragile site because the end of the chromosome tends to break away. The term is misleading, for the end breaks away only in cultured cells, not in cells in the body.

A mutant gene, not breakage, causes the syndrome. It specifies a protein required for normal development of brain cells. Within that gene, a segment of DNA is repeated several times. Certain mutations result in the addition of many more repeats in the DNA; hence their name, *expansion* mutations. The outcome is a mutant allele that cannot function properly. Because that allele is recessive, males who inherit it and females who are homozygous for it have fragile X syndrome. Expansion mutations are now known to be the cause of Huntington disorder and some other genetic disorders.

Genetic analyses of family pedigrees have revealed simple Mendelian inheritance patterns for certain traits, as well as for many genetic disorders that arise from expression of specific alleles on an autosome or X chromosome.

Focus on Health

12.8 PROGERIA—TOO YOUNG TO BE OLD

Imagine being ten years old with a mind trapped inside a body that is rapidly getting a bit more shriveled, more frail—*old*—with each passing day. You are just barely tall enough to peer over the top of the kitchen counter, and you weigh less than thirty-five pounds. Already you are bald and have a crinkled nose. Possibly you have a few more years to live. Could you, like Mickey Hayes and Fransie Geringer, still laugh with your friends?

Of every 8 million newborn humans, one is destined to grow old far too soon. On one of its autosomes, that rare individual carries a mutant gene that gives rise to *Hutchinson–Gilford progeria syndrome*. Through hundreds, thousands, then many billions of DNA replications and mitotic cell divisions, terrible information encoded in that gene was systematically distributed to every cell in the growing embryo, and later in the newborn. Its legacy will be accelerated aging and a greatly reduced life span. The photograph of Mickie Hayes and Fransie Geringer in Figure 12.16 shows some of the symptoms.

The mutation causes gross disruptions in interactions among genes that bring about the body's growth and development. Observable symptoms start before age two. Skin that should be plump and resilient starts to thin. Skeletal muscles weaken. Tissues in limb bones that should lengthen and grow stronger start to soften. Hair loss is pronounced; extremely premature baldness is inevitable. There are no documented cases of progeria running in families, so it seems likely the gene mutates spontaneously, at random. It appears to be dominant over its normal partner on the homologous chromosome.

Most progeriacs can expect to die in their early teens as a result of strokes or heart attacks. These final insults are brought on by a hardening of the walls of arteries, a condition typical of advanced age. When Mickey Hayes turned eighteen, he was the oldest living progeriac. Fransie was seventeen when he died.

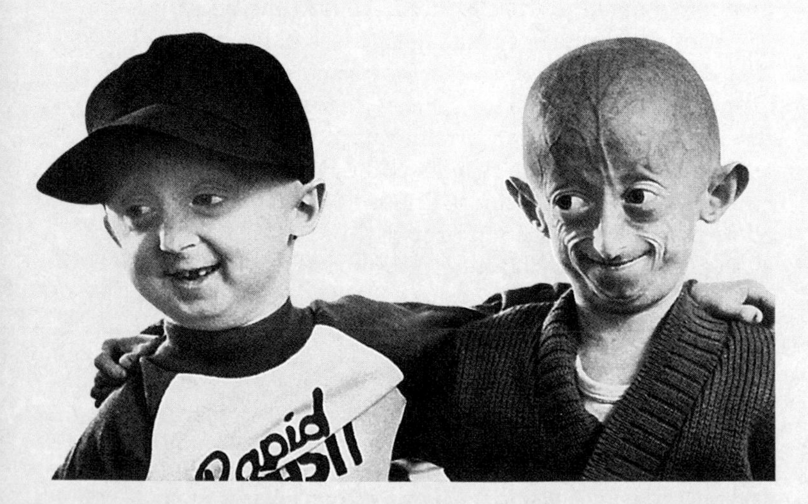

Figure 12.16 Two boys who met at a gathering of progeriacs at Disneyland, California, when they were not yet ten years old.

12.9 CHANGES IN CHROMOSOME STRUCTURE

On rare occasions, the physical structure of one or more chromosomes changes, and the outcome is a genetic disorder or abnormality. Such changes spontaneously occur in nature and are induced in research laboratories by exposure to chemicals or by irradiation. Either way, changes may be detected by microscopic examination and karyotype analysis of cells at mitosis or meiosis. Let's now review four kinds of structural change. As you will see, some have serious or lethal consequences.

Major Categories of Structural Change

DUPLICATION Even normal chromosomes have gene sequences that are repeated several to many hundreds or thousands of times. These are **duplications**:

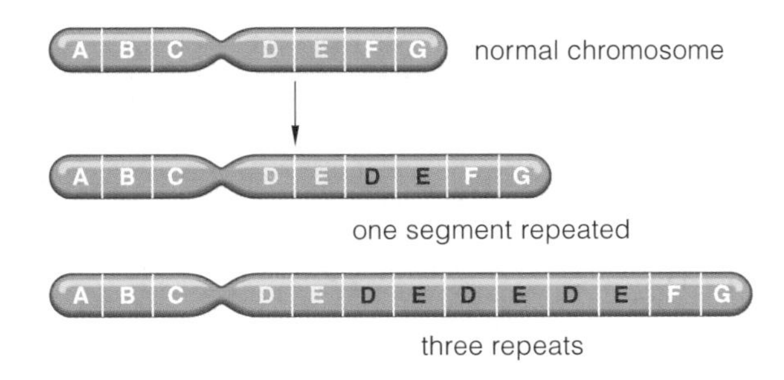

INVERSION With an **inversion**, a linear stretch of DNA within the chromosome becomes oriented in the reverse direction, with no molecular loss:

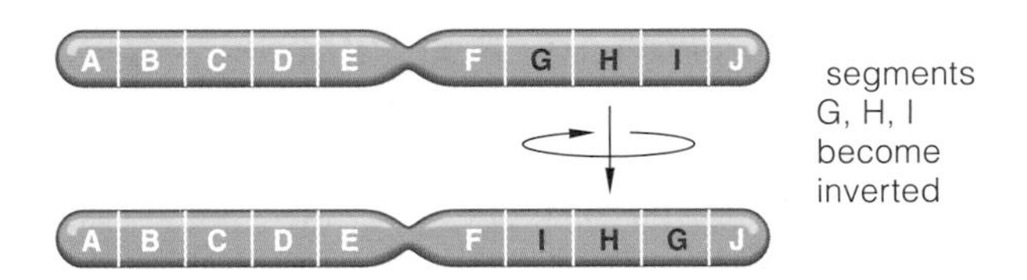

TRANSLOCATION As the chapter introduction showed, the Philadelphia chromosome that has been linked to a form of leukemia results from **translocation**, whereby a broken part of a chromosome becomes attached to a *non*homologous chromosome. Most translocations are reciprocal (both chromosomes exchange broken parts):

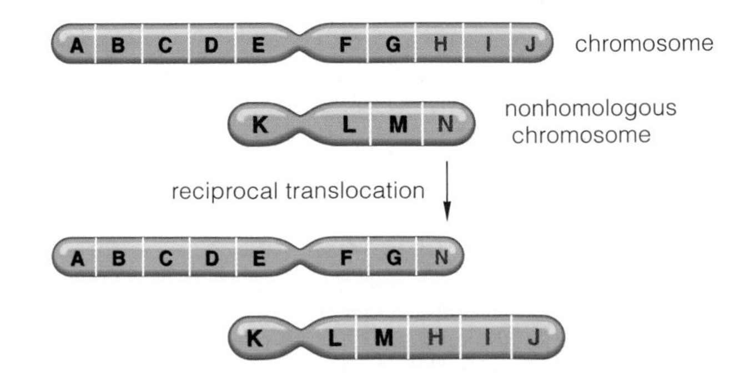

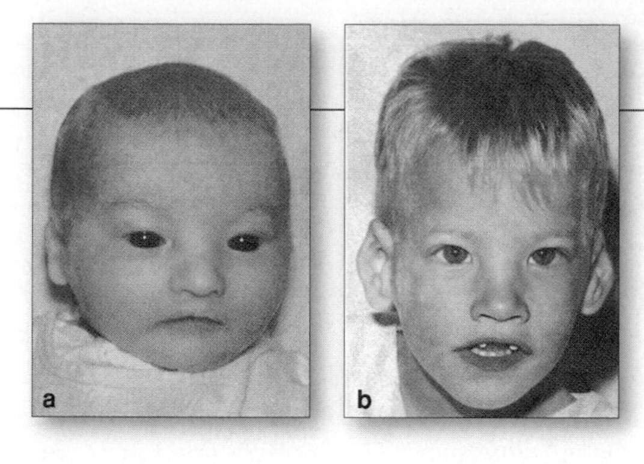

Figure 12.17 (**a**) A male infant who later developed cri-du-chat syndrome. The ears are positioned low on the side of the head relative to the eyes. (**b**) The same boy, four years later. By this age, affected humans no longer make mewing sounds typical of the syndrome.

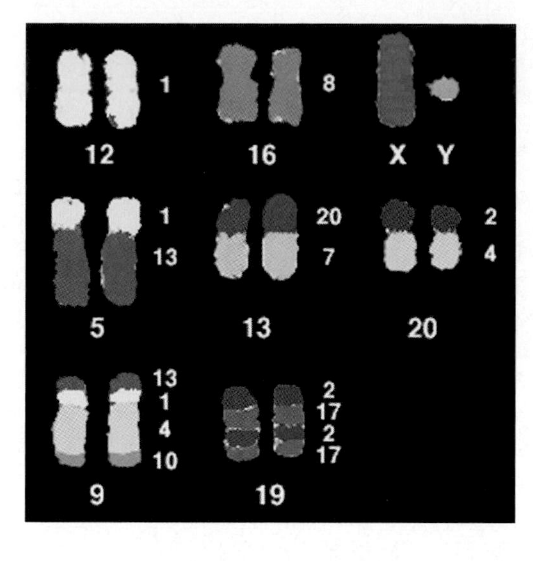

Figure 12.18 Spectral karyotype of duplicated chromosomes from an ape. Colors identify which parts are structurally identical in human chromosomes.

DELETION Viral attacks, irradiation (ionizing radiation especially), chemical assaults, or other environmental factors may trigger a **deletion**, the loss of some segment of a chromosome:

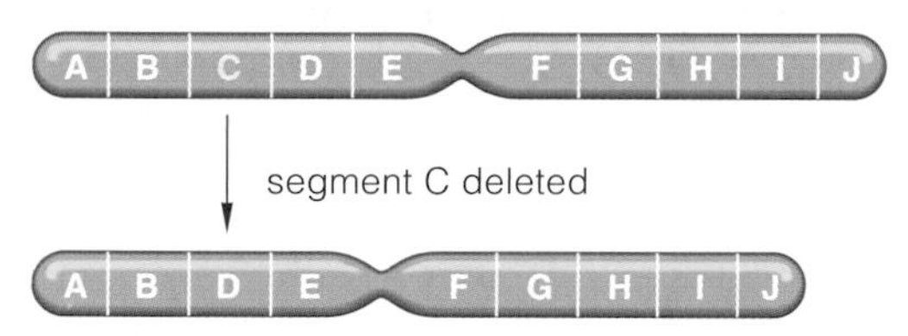

Don't species with diploid cells have an advantage after a deletion occurs? That is, wouldn't genes on the homologous chromosome cover the loss? Maybe, but the sword of chance cuts both ways. If the remaining segment happens to carry a harmful recessive allele, nothing will mask or compensate for *its* effects.

Most deletions are lethal or cause serious disorders in mammals, for they disrupt normal gene interactions that underlie a program of growth, development, and maintenance activities. For example, one deletion from human chromosome 5 results in mental retardation and the development of an abnormally shaped larynx. When affected infants cry, they produce sounds rather like a cat's meow. Hence *cri-du-chat* (cat-cry), the name of this disorder. Figure 12.17 shows an affected child.

Does Chromosome Structure Evolve?

Changes in chromosome structure tend to be selected against rather than conserved over evolutionary time. Even so, for many species, we have interesting signs of changes that occurred in the past. To give one example, many duplications have done their bearers no harm. Although duplications are relatively rare events, they have accumulated over millions, even billions, of years and are now built into the DNA of all species.

Biologists speculate that duplicates of some gene sequences with neutral effects could have an adaptive advantage. In effect, a copy could free up a gene for chance mutations that might turn out to be useful, and the normal gene would still issue the required product. One or more duplicated gene sequences could become slightly modified, and then the products of those genes could function in slightly different or new ways.

Several kinds of duplicated, modified genes seem to have had pivotal roles in evolution. Consider the gene regions for the polypeptide chains of hemoglobin, as shown in Section 3.5. In humans and other primates, gene regions for the polypeptide chains of hemoglobin contain multiple sequences that are remarkably similar. The sequences specify whole families of chains, which have slight structural differences. Each of the resulting hemoglobin molecules transports oxygen with slightly different efficiencies, depending on cellular conditions.

Certain duplications, inversions, and translocations probably helped put primate ancestors of humans on a unique evolutionary road. They apparently contributed to divergences that led to the modern apes and humans. Of our twenty-three pairs of chromosomes, eighteen are nearly identical to their counterparts in chimpanzees and gorillas. The other five pairs differ at inverted and translocated regions. You can observe such similarities for yourself by comparing chromosomes of a human with those of a gibbon, one of the apes (Figure 12.18).

On rare occasions, a segment of a chromosome may get lost, inverted, moved to a new location, or duplicated.

Most chromosome changes are harmful or lethal, especially when they alter gene interactions that underlie growth, development, and maintenance activities.

Over evolutionary time, many changes have been conserved; they confer adaptive advantages or have had neutral effects.

12.10 CHANGES IN CHROMOSOME NUMBER

Occasionally, abnormal events occur before or during cell division, then gametes and new individuals end up with the wrong chromosome number. The consequences range from minor to lethal physical changes.

Categories and Mechanisms of Change

With **aneuploidy**, individuals usually have one extra or one less chromosome. This condition is a major cause of human reproductive failure. Possibly it affects about one-half of all fertilized eggs. As autopsies reveal, most *miscarriages* (spontaneous aborting of embryos before pregnancy reaches full term) are aneuploids.

With **polyploidy**, individuals have three or more of each type of chromosome. About one-half of all species of flowering plants are polyploid (Section 19.4). Often, researchers can induce polyploidy in undifferentiated plant cells by exposing them to colchicine (Section 12.2). Some species of insects, fishes, and other animals are polyploids. However, polyploidy is lethal for humans. All but about 1 percent of human polyploids die before birth, and the rare newborns die soon after birth.

Chromosome numbers can change during mitotic or meiotic cell divisions. Suppose a cell cycle proceeds through DNA duplication and mitosis but is arrested before the cytoplasm divides. The cell is now *tetra*ploid, with four of each type of chromosome. Suppose one or more pairs of chromosomes fail to separate in mitosis or meiosis, an event called **nondisjunction**. Some or all of the forthcoming cells will have too many or too few chromosomes, as in the Figure 12.19 example.

The chromosome number also may get changed at fertilization. Visualize a normal gamete that unites by chance with an $n + 1$ gamete (one extra chromosome). The new individual will be "trisomic" ($2n + 1$); it will have three of one type of chromosome and two of every other type. What if an $n - 1$ gamete unites with a normal n gamete? Then, the new individual will turn out to be "monosomic" ($2n - 1$).

Change in the Number of Autosomes

Nearly all changes in the number of autosomes arise by nondisjunction during gamete formation. Let's look at one of the most common of the resulting disorders. A trisomic 21 newborn, with three chromosomes 21, will develop *Down syndrome*. The symptoms vary greatly, but most affected individuals show moderate to severe mental impairment. About 40 percent develop heart defects. Abnormal development of the skeleton means older children have shortened body parts, loose joints, and poorly aligned bones of the hips, fingers, and toes. Muscles and muscle reflexes are weaker than normal, and speech and other motor skills develop slowly. With special training, trisomic 21 individuals often take part in normal activities; Figure 12.20 gives examples. As a group, they tend to be cheerful and affectionate, and they derive great pleasure from socializing.

Down syndrome is one of many disorders that can be detected before birth. Before detection procedures were widespread, about 1 in 700 newborns of all ethnic groups was trisomic 21. The number is now closer to 1 in 1,100, for some women elect to terminate pregnancy when the condition is detected (Section 12.11). The risk is greater if pregnant women are more than thirty-five years old, as indicated by the diagram in Figure 12.20*b*.

Change in the Number of Sex Chromosomes

Most sex chromosome abnormalities arise as a result of nondisjunction during meiosis and gamete formation. Let's look at a few phenotypic outcomes.

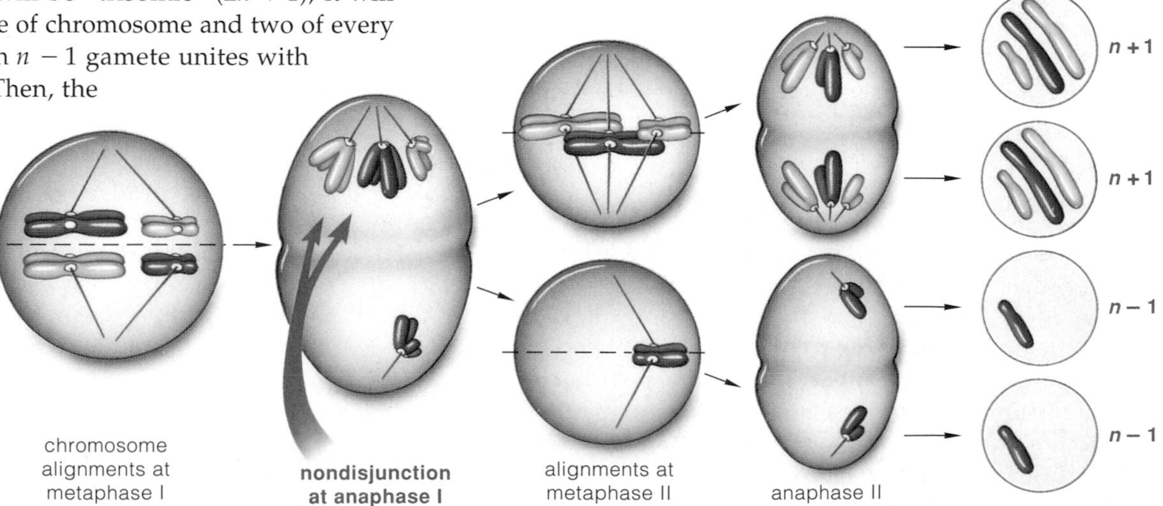

Figure 12.19 Example of nondisjunction. Of two pairs of homologous chromosomes shown, one pair fails to separate at anaphase I of meiosis. The chromosome number changes in the gametes. (Make a sketch of nondisjunction at anaphase II. What will the chromosome numbers be in gametes?)

Further reading: Student Guide to InfoTrac on web site →

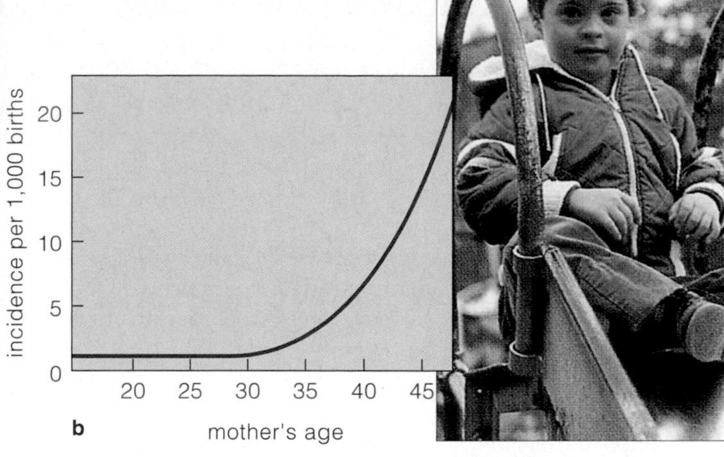

Figure 12.20 Down syndrome. (**a**) Karyotype revealing a trisomic 21 condition (*arrows*). (**b**) Relationship between the frequency of Down syndrome and mother's age at the time of childbirth. Results are from a study of 1,119 affected children who were born in Victoria, Australia, between 1942 and 1957. The young lady above was a lively participant in the Special Olympics, held annually in San Mateo, California.

TURNER SYNDROME Inheriting one X chromosome with no corresponding X or Y chromosome gives rise to *Turner syndrome*, which affects 1 in 2,500 to 10,000 or so newborn girls. Nondisjunction affecting sperm accounts for 75 percent of the cases. We see fewer people with Turner syndrome compared to people having other sex chromosome abnormalities. The likely reason is that at least 98 percent of all X0 zygotes spontaneously abort early in pregnancy. Approximately 20 percent of all the spontaneously aborted embryos in which chromosome abnormalities were detected have been X0.

Despite the near-lethality, X0 survivors are not as disadvantaged as other aneuploids. They grow up well proportioned, albeit short—4 feet, 8 inches tall, on the average. Generally, their behavior is normal during childhood. But most Turner females are infertile. They do not have functional ovaries and so cannot produce eggs or sex hormones. Without sex hormones, breast enlargement and the development of other secondary sexual traits are reduced. Possibly as a result of their arrested sexual development and small size, X0 females often become passive and are easily intimidated by peers during their teens. Some patients have benefited from hormone therapy and corrective surgery.

KLINEFELTER SYNDROME Of every 500 to 2,000 liveborn males, one has inherited two X chromosomes and one Y chromosome. This XXY condition results mainly from nondisjunction in the mother (about 67 percent of the time, compared to 33 percent in the father). Symptoms of the resulting *Klinefelter syndrome* develop after the onset of puberty. XXY males are taller than average and are sterile or nearly so. Their testes usually are much smaller than average; the penis and scrotum are not. Often facial hair is sparse, and breasts are somewhat enlarged. Injections of the hormone testosterone can reverse the feminized traits but not the low fertility. Some XXY males display mild mental impairment, but many fall in the normal range for intelligence. Except for their low fertility, many affected individuals show no outward symptoms at all.

XYY CONDITION About 1 in 1,000 males has one X and two Y chromosomes. *XYY males* tend to be taller than average. Some may be mildly retarded, but most are phenotypically normal. At one time, XYY males were thought to be genetically predisposed to become criminals. The erroneous conclusion was based on small numbers of cases in narrowly selected groups, such as prison inmates. Investigators often knew who the XYY males were, and this may have biased their evaluations. There were no **double-blind studies**, by which different investigators gather data independently of one another and then match them up only after both sets of data are completed. In this case, the same investigators gathered the karyotypes and the personal histories. Fanning the stereotype was a sensationalized report in 1968 that a mass-murderer of young nurses was XYY. He wasn't.

In 1976, a Danish geneticist issued a report on a large-scale study based on records of 4,139 tall males, twenty-six years old, who had reported to their draft board. Besides giving results of physical examinations and intelligence testing, the records provided clues to social and economic status, educational history, and any criminal convictions. Only twelve of the males were XYY, which left more than 4,000 for the control group. The only finding of significance was that tall, mentally impaired males who engage in criminal activity are just more likely to get caught—irrespective of karyotype.

Most changes in chromosome number arise as an outcome of nondisjunction during meiosis and gamete formation.

12.11 PROSPECTS IN HUMAN GENETICS

With the arrival of their newborn, parents typically ask, "Is our baby normal?" Quite naturally, they want their baby to be free of genetic disorders, and most of the time it is. But what are the options when it is not?

We do not approach heritable disorders and diseases the same way. We attack diseases with antibiotics, surgery, and other weapons. But how do we attack a heritable "enemy" that can be transmitted to offspring? Do we institute regional, national, or global programs to identify people who might be carrying harmful alleles? Do we tell them they are "defective" and run a risk of bestowing a disorder on their children? Who decides which alleles are "harmful"? Should society bear the cost of treating genetic disorders before and after birth? If so, should society also have a say in whether an affected embryo will be born at all, or whether it should be aborted? An **abortion** is the expulsion of a pre-term embryo or fetus from the uterus.

Questions such as these are only the tip of an ethical iceberg. And we do not have anwers that are universally acceptable throughout our society.

PHENOTYPIC TREATMENTS Often, the symptoms of genetic disorders can be either minimized or suppressed by (1) exerting dietary controls, (2) making adjustments to specific environmental conditions, and (3) intervening surgically or by way of hormone replacement therapy.

For example, dietary control works in *phenylketonuria*, or PKU. A certain gene specifies an enzyme that converts one amino acid to another—phenylalanine to tyrosine. If an individual is homozygous recessive for a mutated form of the gene, the first of these amino acids accumulates inside the body. If excess amounts are diverted into other pathways, then phenylpyruvate and other compounds may form. High levels of phenylpyruvate in the blood can impair the functioning of the brain. When affected people restrict their intake of phenylalanine, they are not required to dispose of excess amounts, so they can lead normal lives. Among other things, they can avoid soft drinks and other food products that are sweetened with aspartame, a compound that contains phenylalanine.

Environmental adjustments help counter or minimize the symptoms of some disorders, as when albinos avoid exposure to direct sunlight. Surgical reconstructions also minimize many physical problems. For example, surgeons close up a form of *cleft lip* in which a vertical fissure cuts through the lip and extends into the roof of the mouth.

GENETIC SCREENING Through large-scale screening programs in the general population, affected persons or carriers of a harmful allele often can be detected early enough to start preventive measures before symptoms develop. For example, most hospitals in the United States routinely screen newborns for PKU, so today it is less common to see people with symptoms of this disorder.

GENETIC COUNSELING If a first child or close relative has a severe heritable problem, prospective parents may worry about their next child. They may request help in evaluating their options from a qualified professional counseler. *Genetic counseling* often includes diagnosis of parental genotypes, detailed pedigrees, and genetic testing for hundreds of known metabolic disorders. Geneticists, too, may be contacted to help predict risks for genetic disorders. During genetic counseling, the prospective parents must be reminded that the same risk usually applies to each pregnancy.

PRENATAL DIAGNOSIS Methods of *prenatal diagnosis* can be used to determine the sex as well as more than a hundred genetic conditions of the embryo or fetus. (*Prenatal* means "before birth." The term *embryo* applies until eight weeks after fertilization, after which the term *fetus* is appropriate.)

For example, suppose a woman who is forty-five years old becomes pregnant. She worries that her forthcoming child may develop Down syndrome. She might request prenatal diagnosis by *amniocentesis* (Figure 12.21). With this diagnostic procedure, a clinician withdraws a tiny sample of the fluid inside the amnion, the membranous sac surrounding the fetus. Some cells that the fetus has sloughed

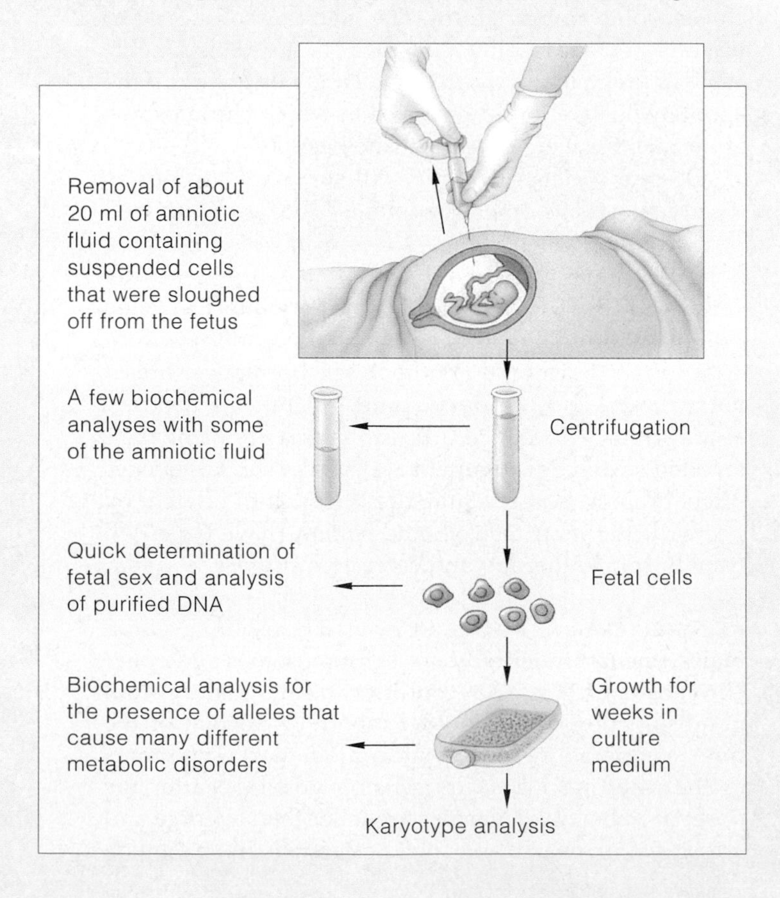

Figure 12.21 Steps in amniocentesis, a prenatal diagnostic tool.

Further reading: Student Guide to InfoTrac on web site

off are suspended in the sample. The cells are cultured and then analyzed.

Chorionic villi sampling (CVS) is a different diagnostic procedure. A clinician withdraws cells from the chorion, a fluid-filled, membranous sac that surrounds the amnion. CVS can be perfomed weeks before amniocentesis. It can yield results as early as the ninth week of pregnancy.

Direct visualization of the developing fetus is possible with *fetoscopy*. A fiber-optic device, an endoscope, uses pulsed sound waves to scan the uterus and visually locate particular parts of the fetus, umbilical cord, or placenta (Figure 12.22). Fetoscopy has been used to diagnose blood cell disorders, such as sickle-cell anemia and hemophilia.

All three procedures may accidentally cause infection or puncture the fetus. Occasionally the punctured amnion does not reseal itself quickly, and an excessive amount of amniotic fluid may leak out and cause problems for the fetus. A mother-to-be who requests amniocentesis runs a 1 to 2 percent greater risk of miscarriage. For CVS, she runs a 0.3 percent risk that her future child will have missing or underdeveloped fingers and toes. Fetoscopy increases the risk of a miscarriage by 2 to 10 percent.

Parents-to-be probably should seek counseling from their doctor to help them weigh the risks and benefits of such procedures as applied to their own circumstances. They may wish to (1) ask about the small overall risk of 3 percent that any child will have some kind of birth defect, (2) ask about the severity of genetic disorder that a child might be at risk of developing, and (3) consider how old the woman is at the time of pregnancy.

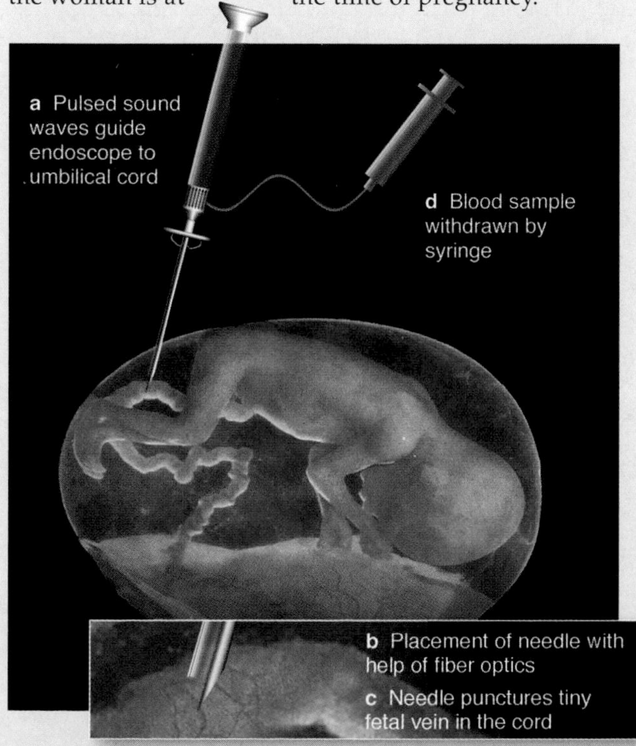

Figure 12.22 Fetoscopy for prenatal diagnosis.

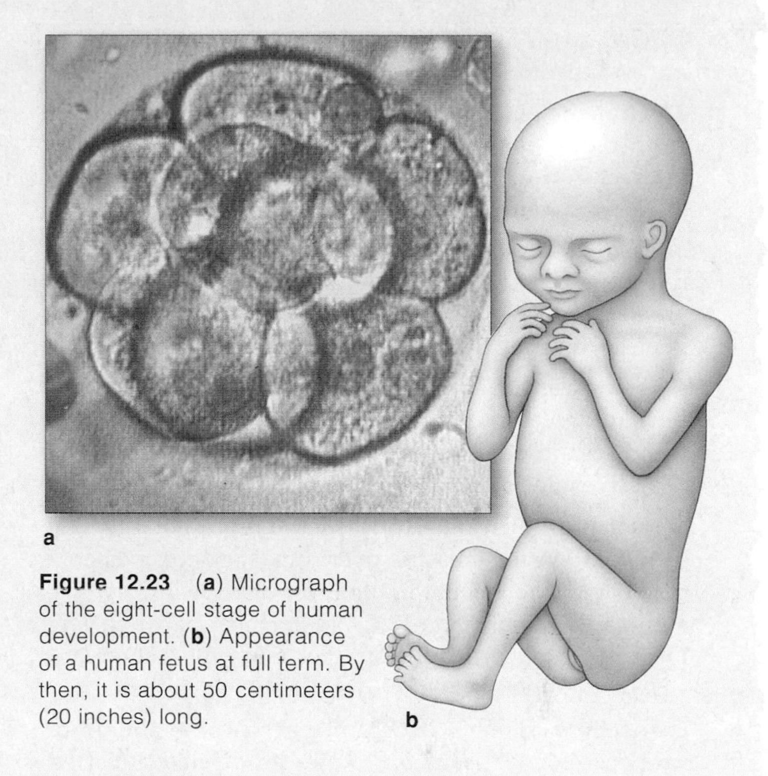

Figure 12.23 (**a**) Micrograph of the eight-cell stage of human development. (**b**) Appearance of a human fetus at full term. By then, it is about 50 centimeters (20 inches) long.

REGARDING ABORTION What happens when prenatal diagnosis does reveal a serious problem? Do prospective parents opt for induced abortion? We can only say here that they must weigh their awareness of the severity of the genetic disorder against ethical and religious beliefs. Worse, they must play out their personal tragedy on a larger stage, dominated by a nationwide battle between fiercely vocal "pro-life" and "pro-choice" factions. We return to this volatile issue in Section 45.13.

PREIMPLANTATION DIAGNOSIS Another procedure, *preimplantation diagnosis*, relies on **in-vitro fertilization**. In vitro, recall, means "in glass." Sperm and eggs taken from prospective parents are quickly transferred to an enriched medium in a glass petri dish. One or more eggs may become fertilized. In two days, mitotic cell divisions may convert a fertilized egg into a ball of eight cells, such as the one shown in Figure 12.23*a*.

According to one view, the tiny, free-floating ball is a *pre*-pregnancy stage. Like the unfertilized eggs discarded monthly from a woman, the ball is not attached to the uterus. All cells in the ball have the same genes and are not yet committed to giving rise to specialized cells of a heart, lungs, and other organs. Doctors take one of the undifferentiated cells and analyze its genes for suspected disorders. If the cell has no detectable genetic defects, the ball is inserted into the uterus.

Some couples who are at risk of passing on muscular dystrophy, cystic fibrosis, and other disorders have opted for the procedure. Many *"test-tube" babies* have been born in good health and are free of the mutant alleles.

12.12 SUMMARY

1. Genes, the units of instruction for heritable traits, are arranged one after the other along chromosomes.

2. Human somatic cells are diploid ($2n$). They contain twenty-three pairs of homologous chromosomes that interact during meiosis. Each pair has the same length, shape, and gene sequence (except for the XY pairing).

3. Human females have two X chromosomes. Males have one X paired with one Y. All other chromosomes are autosomes (the same in both females and males). A gene on the Y chromosome determines sex.

4. Pedigrees, or charts of genetic connections through lines of descent, often provide clues to inheritance of a trait. Certain patterns are characteristic of dominant or recessive alleles on autosomes or the X chromosome.

5. Genes on the same chromosome represent a linkage group. However, crossing over (breakage and exchange of segments between homologues) disrupts linkages, as summarized in Section 12.4. The farther apart two gene loci are along the length of a chromosome, the greater will be the frequency of crossovers between them.

6. A chromosome's structure may become altered on rare occasions. A segment might be deleted, inverted, moved to a new location (translocated), or duplicated.

7. The chromosome number can change. Gametes and offspring might get one more or one less chromosome than their parents (aneuploidy). They might also get three or more of each type of chromosome (polyploidy). Nondisjunction at meiosis (prior to gamete formation) accounts for most of these chromosome abnormalities.

8. Crossing over adds to potentially adaptive variation in traits in a population. By contrast, most changes in chromosome number or structure are harmful or lethal. Over evolutionary time, however, changes occasionally establish themselves in chromosomes of the species.

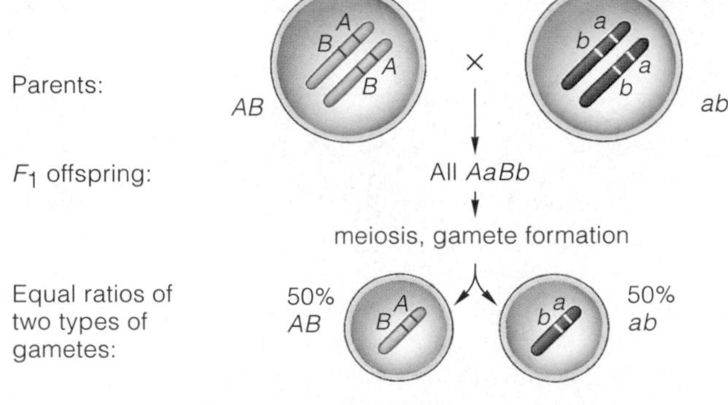

a Full linkage between two genes (no crossovers): half of the gametes have one parental genotype and half have the other.

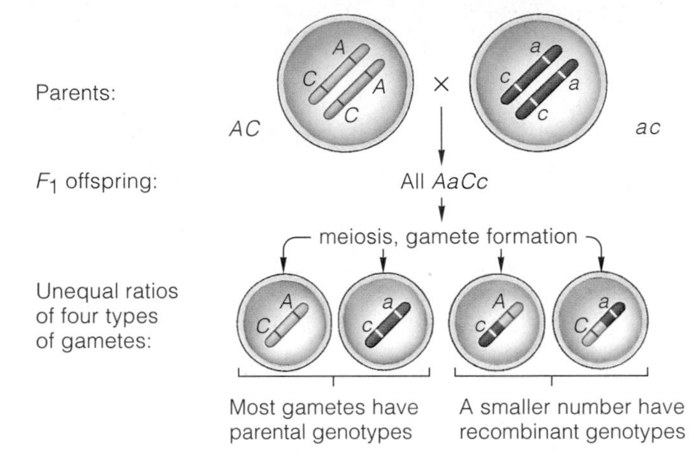

b Incomplete linkage; crossing over affected the outcome.

Figure 12.24 Summary examples of how (**a**) complete and (**b**) incomplete gene linkage can affect a dihybrid cross.

Review Questions

1. What is a gene? What are alleles? *12.1*
2. Distinguish between: *CI, 12.1*
   a. homologous and nonhomologous chromosomes
   b. sex chromosomes and autosomes
   c. karyotype and karyotype diagram
3. Define genetic recombination, and describe how crossing over can bring it about. *12.1, 12.4*
4. Define pedigree, genetic abnormality, and genetic disorder. *12.6*
5. Contrast a typical pattern of autosomal recessive inheritance with that of autosomal dominant inheritance. *12.7*
6. Describe two clues that often show up when a recessive allele on an X chromosome causes a genetic disorder. *12.7*
7. Distinguish among a chromosomal deletion, duplication, inversion, and translocation. *12.9*
8. Define aneuploidy and polyploidy. Make a sketch of an example of nondisjunction. *12.10*

Self-Quiz (*Answers in Appendix III*)

1. ______ segregate during ______ .
   a. Homologues; mitosis
   b. Genes on nonhomologous chromosomes; meiosis
   c. Homologues; meiosis
   d. Genes on one chromosome; mitosis
2. The probability of a crossover occurring between two genes on the same chromosome is ______ .
   a. unrelated to the distance between them
   b. increased if they are closer together on the chromosome
   c. increased if they are farther apart on the chromosome
3. Chromosome structure can be altered by ______ .
   a. deletions c. inversions e. all of the above
   b. duplications d. translocations
4. Nondisjunction can be caused by ______ .
   a. crossing over in mitosis
   b. segregation in meiosis
   c. failure of chromosomes to separate during meiosis
   d. multiple independent assortments
5. A gamete affected by nondisjunction would have ______ .
   a. a change from the normal chromosome number
   b. one extra or one missing chromosome
   c. the potential for a genetic disorder
   d. all of the above
6. Genetic disorders can be caused by ______ .
   a. altered chromosome number c. mutation
   b. altered chromosome structure d. all of the above

Figure 12.25 Mutant fruit fly with vestigial wings

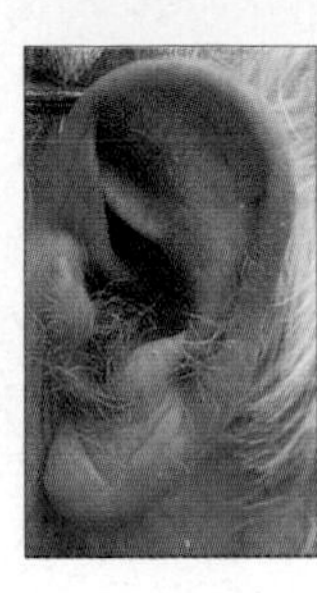

Figure 12.26 Hairy pinnae

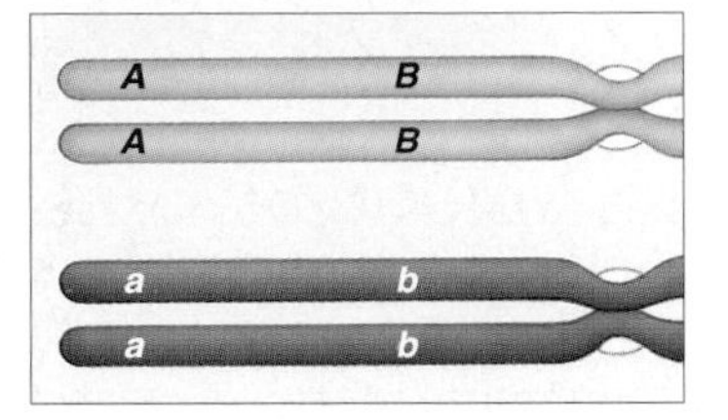

Figure 12.28 Two of your linked genes, on a pair of homologous chromosomes.

7. Is this statement true or false: Independent assortment, crossing over, and changes in chromosome structure and number contribute to phenotypic variation in a population.

8. Match the chromosome terms appropriately.

___ crossing over	a. number and defining features of an individual's metaphase chromosomes
___ deletion	b. chromosome segment moves to a nonhomologous chromosome
___ nondisjunction	c. disrupts gene linkages at meiosis
___ translocation	d. causes gametes to have abnormal chromosome numbers
___ karyotype	e. loss of a chromosome segment

Critical Thinking—Genetics Problems

(Answers in Appendix IV)

1. Human females are XX and males are XY.
   a. Does a male inherit the X from his mother or father?
   b. With respect to X-linked alleles, how many different types of gametes can a male produce?
   c. If a female is homozygous for an X-linked allele, how many types of gametes can she produce with respect to that allele?
   d. If a female is heterozygous for an X-linked allele, how many types of gametes can she produce with respect to that allele?

2. Expression of a dominant allele at a gene locus governing wing length in *D. melanogaster* results in long wings. Figure 12.25 shows how homozygosity for a recessive allele results in formation of vestigial (short) wings. You cross a homozygous dominant, long-winged fly with a homozygous recessive, vestigial-winged fly. Then you ask a technician to expose the fertilized eggs to a level of x-rays known to induce mutation and deletions. Later, when the irradiated eggs develop into adults, most flies are heterozygous and have long wings. A few have vestigial wings. What might explain these results?

3. Suppose expression of one allele of a Y-linked gene results in nonhairy ears in males. Expression of another allele results in rather long hairs, a condition called *hairy pinnae* (Figure 12.26).
   a. Why would you *not* expect females to have hairy pinnae?
   b. Any son of a hairy-eared male will also be hairy-eared, but no daughter will be. Explain why.

4. *Marfan syndrome* is a genetic disorder with pleiotropic effects. Expression of a mutant allele leads to the absence or abnormal formation of a connective tissue (fibrillin) in many organs. Often affected persons are tall and thin, with double-jointed fingers. Arms, fingers, and lower limbs are disproportionately long. The spine is abnormally curved. Eye lenses are easy to dislocate. Thin tissue flaps in the heart that act like valves to direct blood flow may billow the wrong way as the heart contracts. The heart

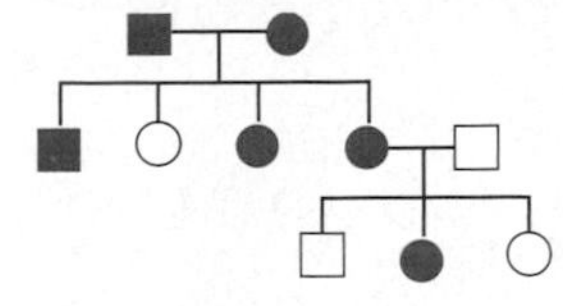

Figure 12.27 Identify the mystery pedigree.

beats irregularly. The aorta, the main artery carrying blood away from the heart, is fragile. Its diameter widens; its wall may tear.

An estimated 40,000 people in the United States are affected, including a few prominent athletes. Spontaneous mutation gives rise to the allele. Genetic analysis shows it follows a pattern of autosomal dominant inheritance. What is the chance any child will inherit the allele if one parent is heterozygous for it?

5. In the Figure 12.27 pedigree, does the phenotype indicated by *red* circles and squares follow a Mendelian inheritance pattern that is autosomal dominant, autosomal recessive, or X-linked?

6. One kind of *muscular dystrophy*, a genetic disorder, is due to a recessive X-linked allele. Usually, symptoms start in childhood. Over time, a slowly progressing loss of muscle function leads to death, usually by age twenty or so. Unlike color blindness, this disorder is nearly always restricted to males. Suggest why.

7. Suppose you carry two linked genes with alleles *Aa* and *Bb*, respectively, as in Figure 12.28. If the crossover frequency between these two genes is zero, what genotypes would be expected among the gametes you produce, and with what frequencies?

8. Say you have alleles for lefthandedness and straight hair on one chromosome and alleles for righthandedness and curly hair on the homologous chromosome. If the two loci for the alleles are very close together along the chromosome, how likely is it that crossing over will occur between them? If they are distant from each other, is a crossover more or less likely to occur?

9. Individuals showing *Down syndrome* usually have an extra chromosome 21, so their body cells contain 47 chromosomes.
   a. At which stages of meiosis I and II could a mistake occur that could result in the altered chromosome number?
   b. In a few cases, 46 chromosomes are present, including two normal-appearing chromosomes 21 and a longer-than-normal chromosome 14. Explain how this situation can arise.

10. In the human population, mutation of two different genes on the X chromosome causes two types of X-linked *hemophilia* (types A and B). In a few known cases, a woman is heterozygous for both mutant alleles (one on each of her two X chromosomes). All her sons should have either hemophilia A or B. Yet, on very rare occasions, such a woman gives birth to a son who does not have hemophilia, and his one X chromosome does not have either mutant allele. Explain how such an X chromosome could arise.

Selected Key Terms

abortion *12.11*
allele *12.1*
aneuploidy *12.10*
autosome *12.1*
crossing over *12.1*
cytological marker *12.4*
deletion *12.9*
disease *12.6*
double-blind study *12.10*
duplication *12.9*
gene *12.1*
genetic abnormality *12.6*
genetic disorder *12.6*
genetic recombination *12.1*
homologous chromosome *12.1*
inversion *12.9*
in vitro *12.2*
in-vitro fertilization *12.11*
karyotype *CI*
linkage group *12.4*
linkage mapping *12.5*
nondisjunction *12.10*
pedigree *12.6*
polyploidy *12.10*
reciprocal cross *12.4*
sex chromosome *12.1*
syndrome *12.6*
translocation *12.9*
X chromosome *12.1*
X-linked gene *12.4*
Y chromosome *12.1*
Y-linked gene *12.4*

Readings

Fairbanks, D., and W. R. Andersen. 1999. *Genetics: The Continuity of Life.* Monterey, California: Brooks-Cole.

13

DNA STRUCTURE AND FUNCTION

Cardboard Atoms and Bent-Wire Bonds

One might have wondered, in the spring of 1868, why Johann Friedrich Miescher was collecting cells from the pus of open wounds and, later, from the sperm of a fish. Miescher, a physician, wanted to identify the chemical composition of the nucleus. These particular cells have very little cytoplasm, which makes it easier to isolate the nuclear material for analysis.

Miescher finally succeeded in isolating an organic compound with the properties of an acid. Unlike other substances in cells, it incorporated a notable amount of phosphorus. Miescher called the substance nuclein. He had discovered what came to be known many years later as **deoxyribonucleic acid**, or **DNA**.

The discovery did not cause even a ripple through the scientific community. At the time, no one really knew much about the physical basis of inheritance—that is, *which chemical substance encodes the instructions for reproducing parental traits in offspring*. Few even suspected that the cell nucleus might hold the answer. For a time, researchers generally believed hereditary instructions had to be encoded in the structure of some unknown class of proteins. After all, heritable traits are spectacularly diverse. Surely the molecules encoding information about those traits were structurally diverse also. Proteins are put together from potentially limitless combinations of twenty different amino acids, so the thinking was that they could function as the sentences (genes) in each cell's book of inheritance.

Figure 13.1 James Watson and Francis Crick posing in 1953 by their newly unveiled structural model of DNA.

By the early 1950s, however, the results of many ingenious experiments clearly indicated that DNA was the substance of inheritance. Moreover, in 1951, Linus Pauling did something that no one had done before. Through his training in biochemistry, a talent for model building, and a few great educated guesses, he deduced the three-dimensional structure of the protein collagen. Pauling's discovery was truly electrifying. If someone could pry open the secrets of proteins, then why not assume the same might be done for DNA? And once the structural details of the DNA molecule were understood, wouldn't they provide clues to its biological functions? *Who would go down in history as having discovered the very secrets of inheritance?*

Scientists around the world started scrambling after that ultimate prize. Among them were James Watson, a young postdoctoral student from Indiana University, and Francis Crick, an unflappably exuberant researcher working at Cambridge University. Exactly how could DNA, a molecule that consisted of only four kinds of subunits, hold genetic information? Watson and Crick spent long hours arguing over everything they had read about the size, shape, and bonding requirements of the subunits of DNA. They fiddled with cardboard cutouts of the subunits. They even badgered chemists to help them identify any potential bonds they might have overlooked. Then they assembled models from bits of metal, held together with wire "bonds" bent at seemingly suitable angles.

In 1953, Watson and Crick put together a model that fit all the pertinent biochemical rules and all the facts about DNA that they had gleaned from other sources. They had discovered the structure of DNA (Figures 13.1 and 13.2). And the breathtaking simplicity of that structure enabled them to solve another long-standing riddle—*how the world of life can show such unity at the molecular level and yet show such spectacular diversity at the level of whole organisms.*

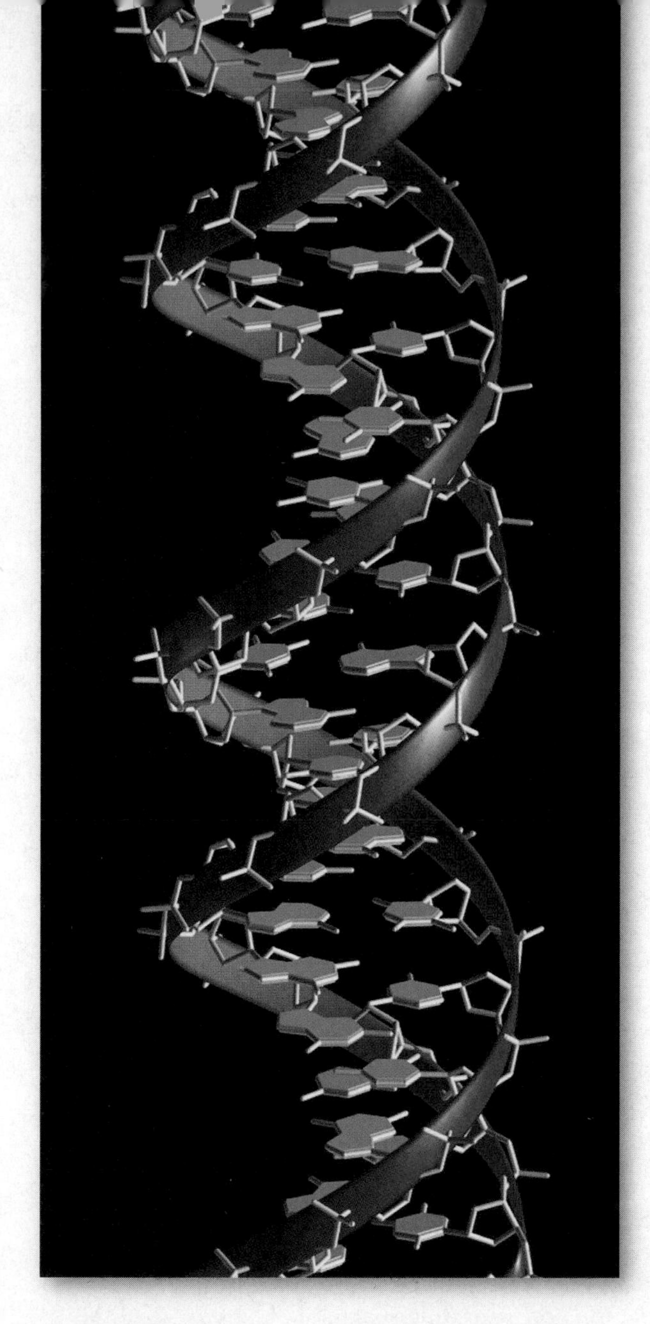

Figure 13.2 A recent computer-generated model of DNA, corresponding to the prototype that Watson and Crick put together decades ago.

With this chapter, we turn to investigations that led to our current understanding of DNA. The story is more than a march through details of its structure and function. *It also is revealing of how ideas are generated in science*. On the one hand, having a shot at fame and fortune quickens the pulse of men and women in any profession, and scientists are no exception. On the other hand, science proceeds as a community effort, with individuals sharing not only what they can explain but also what they do not understand. Even when an experiment fails to produce the anticipated results, it might turn up information that others can use or lead to questions that others can answer. Unexpected results, too, might be clues to something important about the natural world.

KEY CONCEPTS

1. In all living cells, DNA molecules are the storehouses of information about heritable traits.

2. In a DNA molecule, two strands of nucleotides twist together, like a spiral stairway. Each strand consists of four kinds of nucleotides that are the same except for one component—a nitrogen-containing base. The four bases are adenine, guanine, thymine, and cytosine.

3. Great numbers of nucleotides are arranged one after another in each strand of the DNA molecule. In at least some regions, the order in which one kind of nucleotide follows another is unique for each species. Hereditary information is encoded in the particular sequence of nucleotide bases.

4. Hydrogen bonds connect the bases of one strand of the DNA molecule to bases of the other strand. As a rule, adenine pairs (hydrogen-bonds) with thymine, and guanine with cytosine.

5. Before a cell divides, its DNA is replicated with the assistance of enzymes and other proteins. Each double-stranded DNA molecule starts unwinding. As it does so, a new, complementary strand is assembled bit by bit on the exposed bases of each parent strand, according to the base-pairing rule stated above.

13.1 DISCOVERY OF DNA FUNCTION

Early and Puzzling Clues

The year was 1928. Frederick Griffith, an army medical officer, was attempting to develop a vaccine against *Streptococcus pneumoniae*, a bacterium that is one cause of the lung disease pneumonia. (When introduced into the body, vaccines mobilize internal defenses against a real attack. Many vaccines are preparations of either killed or weakened bacterial cells.) Griffith never did develop a vaccine. But his work unexpectedly opened a door to the molecular world of heredity.

Griffith isolated and cultured two different strains of the bacterium. He noticed that colonies of one strain had a rough surface appearance, but those of the other strain appeared smooth. He designated the two strains *R* and *S* and used them in a series of four experiments:

1. Laboratory mice were injected with live R cells. The mice did not develop pneumonia, as indicated in Figure 13.3. *The R strain was harmless.*

2. Other mice were injected with live S cells. The mice died. Blood samples taken from them teemed with live S cells. *The S strain was pathogenic* (disease-causing).

3. S cells were killed by exposure to high temperature. Mice injected with these cells did not die.

4. Live R cells were mixed with heat-killed S cells and injected into mice. The mice died—and blood samples from them teemed with *live* S cells!

What was going on in the fourth experiment? Maybe heat-killed S cells in the mixture weren't really dead. But if that were true, then mice injected with heat-killed S cells alone (experiment 3) would have died. Maybe harmless R cells in the mixture had mutated into a killer form. But if that were true, then mice injected with the R cells alone (experiment 1) would have died.

The simplest explanation was as follows: *Heat killed the S cells but did not destroy their hereditary material—including the part that specified "how to cause infection."* Somehow, that material had been transferred from the dead S cells to living R cells, which put it to use.

Further experiments showed the harmless cells had indeed picked up information on causing infections and were permanently transformed into pathogens. After a few hundreds of generations, descendants of those transformed bacterial cells were still infectious!

The unexpected results of Griffith's experiments intrigued Oswald Avery and his fellow biochemists. In time they were even able to transform harmless bacterial cells with *extracts* of killed pathogenic cells. Finally in 1944, after rigorous chemical analyses, they felt confident in reporting that the hereditary substance in their extracts probably was DNA—not proteins, as was then widely believed. To give experimental evidence for their conclusion, they reported that they had added certain protein-digesting enzymes to some extracts, but cells exposed to those extracts were transformed anyway. To other extracts, they had added an enzyme that digests DNA but not proteins. Doing so blocked hereditary transformation.

genetic material
viral coat
sheath
base plate
tail fiber
a

Despite these impressive experimental results, many biochemists refused to give up on the proteins. Avery's findings, they said, probably applied only to bacteria.

Confirmation of DNA Function

By the early 1950s molecular detectives, including Max Delbrück, Alfred Hershey, Martha Chase, and Salvador Luria, were using viruses as experimental subjects. The viruses they had selected, called **bacteriophages**, infect *Escherichia coli* and other bacteria.

Viruses are biochemically simple infectious agents. Although they are not alive, they do contain hereditary information about building more new virus particles. At some point after a virus has infected a host cell, viral enzymes take over that cell's metabolic machinery—which starts churning out substances that are necessary to construct new virus particles.

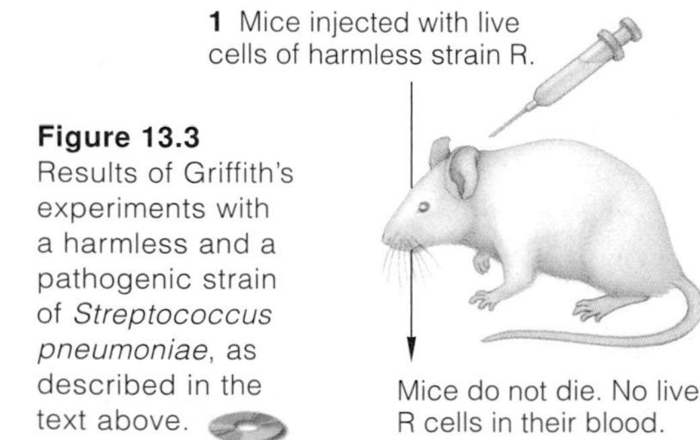

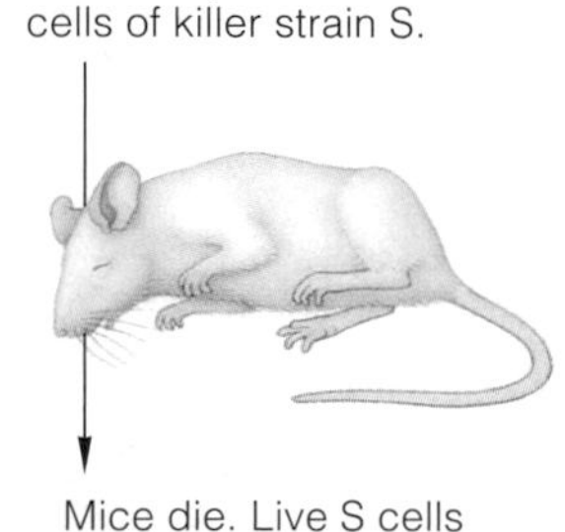

3 Mice injected with heat-killed S cells.
Mice do not die. No live S cells in their blood.

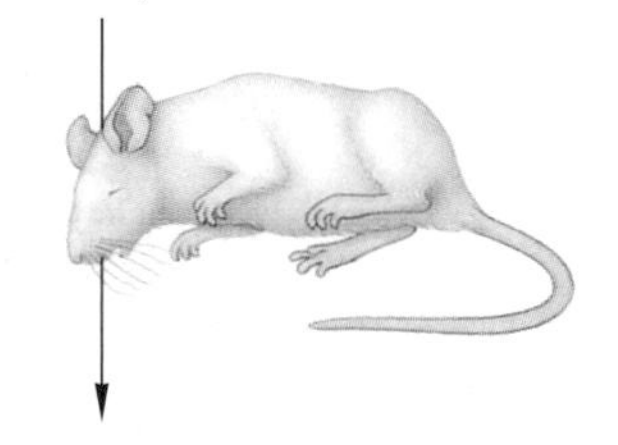

Figure 13.3 Results of Griffith's experiments with a harmless and a pathogenic strain of *Streptococcus pneumoniae*, as described in the text above.

Further reading: Student Guide to InfoTrac on web site

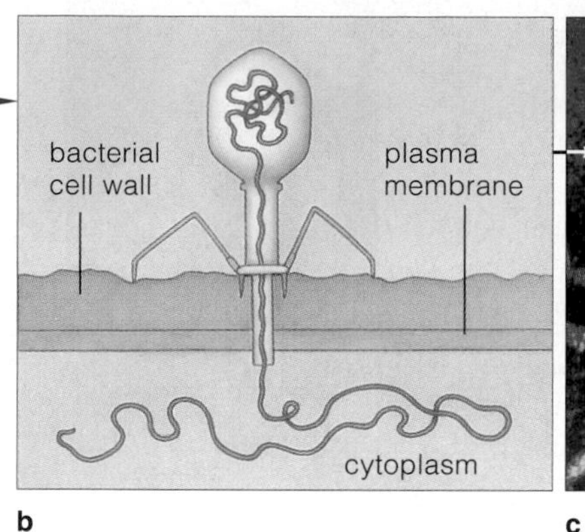

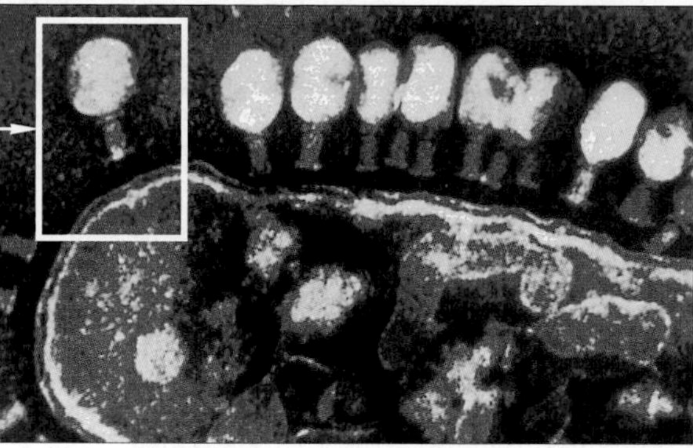

Figure 13.4 (**a**,**b**) *Far left*: Structural organization of a T4 bacteriophage. The diagram shows the genetic material of this type of virus being injected into the cytoplasm of a host cell. For this bacteriophage, it is DNA (the *blue*, threadlike strand). (**c**) Electron micrograph of T4 virus particles infecting a bacterium (*Escherichia coli*), which has just become an unwilling host.

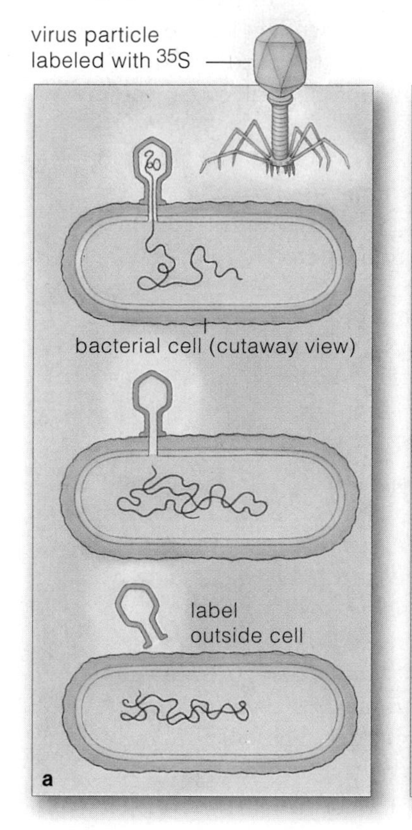

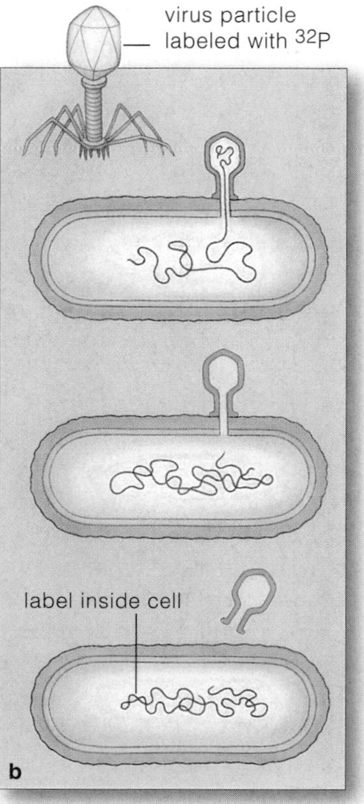

Figure 13.5 Examples of the landmark experiments that pointed to DNA as the substance of heredity. In the 1940s, Alfred Hershey and his colleague, Martha Chase, were studying the biochemical basis of inheritance. They were aware that certain bacteriophages consist of proteins and DNA. *Did the proteins, DNA, or both contain viral genetic information?*

To find a possible answer, Hershey and Chase started with two known biochemical facts to design two experiments: First, the proteins of bacteriophages incorporate sulfur (S) but not phosphorus (P). Second, the DNA of bacteriophages contains phosphorus but not sulfur.

(**a**) In one experiment, some bacterial cells were grown on a culture medium that included a radioisotope of sulfur, ^{35}S. When the bacterial cells synthesized proteins, they had to take up that radioisotope—which the researchers used as a tracer. (Here you may wish to review Section 2.2.)

After the cells were labeled with the tracer, bacteriophages were allowed to infect them. As the infection ran its course, viral proteins were synthesized inside the host cells. These proteins also became labeled with ^{35}S. And so did the new generation of virus particles.

Next, the labeled bacteriophages were allowed to infect a new batch of unlabeled bacteria that were suspended in a fluid culture medium. Afterward, Hershey and Chase whirred the fluid in a kitchen blender. Whirring dislodged the viral protein coats from the cells. The particles became suspended in the fluid medium. Chemical analysis revealed the presence of labeled protein in the fluid. But there was no evidence of labeled protein *inside* the bacterial cells.

(**b**) For the second experiment, Hershey and Chase cultured more bacterial cells. The phosphorus that was available to them for synthesizing DNA included the radioisotope ^{32}P. Later, bacteriophages were allowed to infect the cells.

As predicted, the viral DNA synthesized inside infected cells became labeled, as did the new generation of virus particles. Next, the labeled particles were allowed to infect bacteria that were suspended in a fluid medium. Then they were dislodged from the host cells.

Analysis showed the labeled viral DNA was not in the fluid. DNA stayed *inside* host cells, where its hereditary instructions had to be used to make more virus particles. Here was evidence that DNA is the genetic material of this type of virus.

By 1952, researchers knew that some bacteriophages consist only of DNA and a protein coat. Also, electron micrographs revealed that the main part of the viruses remains *outside* the cells they are infecting (Figure 13.4). Possibly, such viruses were injecting genetic material alone *into* host cells. If that were true, was the material DNA, protein, or both? Through many experiments, researchers accumulated strong evidence that DNA, not proteins, serves as the molecule of inheritance. Figure 13.5 describes two of these landmark experiments.

Information for producing the heritable traits of single-celled and multicelled organisms is encoded in DNA.

13.2

DNA STRUCTURE

What Are the Components of DNA?

Long before the bacteriophage studies were under way, biochemists knew that DNA contains only four types of nucleotides that are the building blocks of nucleic acids. Each **nucleotide** consists of a five-carbon sugar (which, in DNA, is deoxyribose), a phosphate group, and one of the following nitrogen-containing bases:

adenine	**guanine**	**thymine**	**cytosine**
(A)	(G)	(T)	(C)

As you can see from Figure 13.6, all four types of nucleotides in DNA have their component parts joined together in much the same way. However, T and C are pyrimidines, which are single-ring structures. A and G are purines, which are larger, bulkier molecules; they have double-ring structures.

By 1949, Erwin Chargaff, a biochemist, had shared two crucial insights into the composition of DNA with the scientific community. First, the amount of adenine relative to guanine differs from one species to the next. Second, the amount of adenine in DNA always equals that of thymine, and the amount of guanine always equals that of cytosine. We may show this as:

$$A = T \quad \text{and} \quad G = C$$

The proportions of those four kinds of nucleotides relative to each other were tantalizing clues. In some way, the proportions almost certainly were related to the arrangement of the nucleotides in a DNA molecule.

The first convincing evidence of that arrangement emerged from Maurice Wilkins's research laboratory in England. Rosalind Franklin, one of Wilkins's colleagues, had obtained especially good **x-ray diffraction images** of DNA fibers. (Maybe for the reasons sketched out in Section 13.3, Franklin's contribution has only recently been acknowledged.) X-ray diffraction images can be made by directing a beam of x-rays at a molecule. The molecule scatters the beam in patterns that are captured on film. The pattern consists only of dots and streaks; it alone does not reveal molecular structure. However, researchers use photographic images of those patterns to calculate the positions of the molecule's atoms.

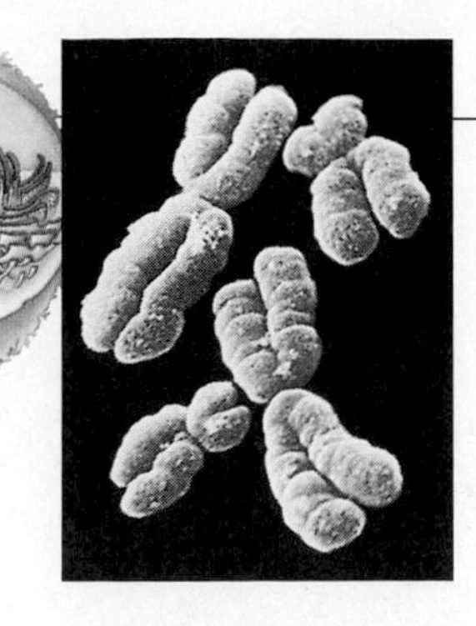

All chromosomes in a cell contain DNA. What does DNA contain? Four kinds of nucleotides, A, G, C, and T. Here are the structural formulas for those nucleotides:

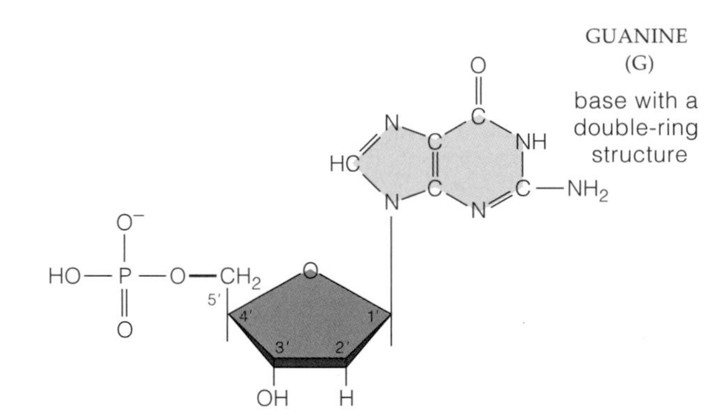

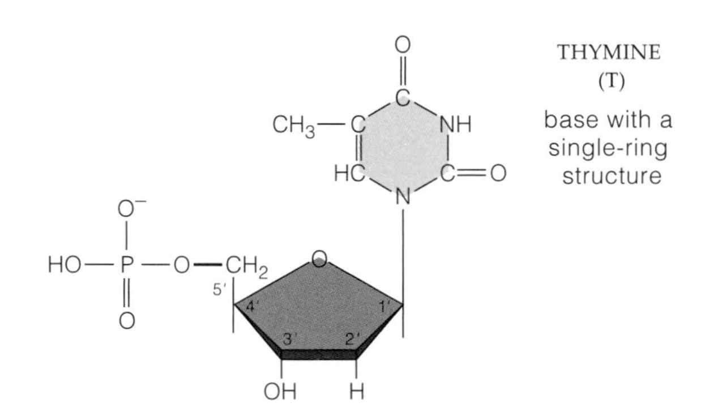

THYMINE
(T)
base with a single-ring structure

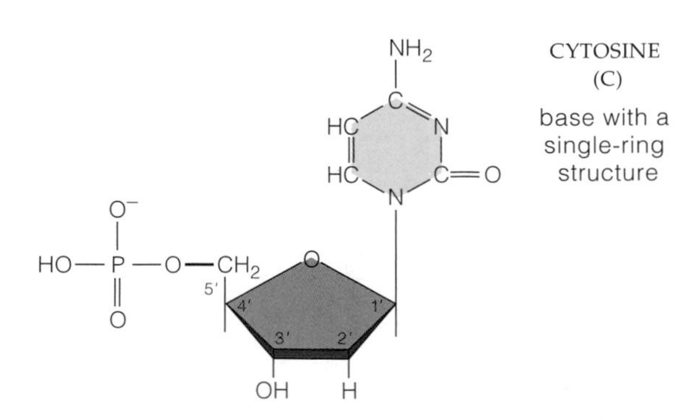

Figure 13.6 Four kinds of nucleotides that serve as building blocks for DNA. The small numerals on the structural formulas identify the carbon atoms to which other parts of the molecule are attached.

As you can see, each nucleotide in a DNA molecule has a five-carbon sugar (shaded *red*), which has a phosphate group attached to the fifth carbon atom of its carbon ring structure. A nucleotide also has one of four kinds of nitrogen-containing bases (*blue*), which is attached to the first carbon atom. The four kinds of nucleotides in DNA differ only in which base they have: adenine, guanine, thymine, or cytosine.

Further reading: Student Guide to InfoTrac on web site

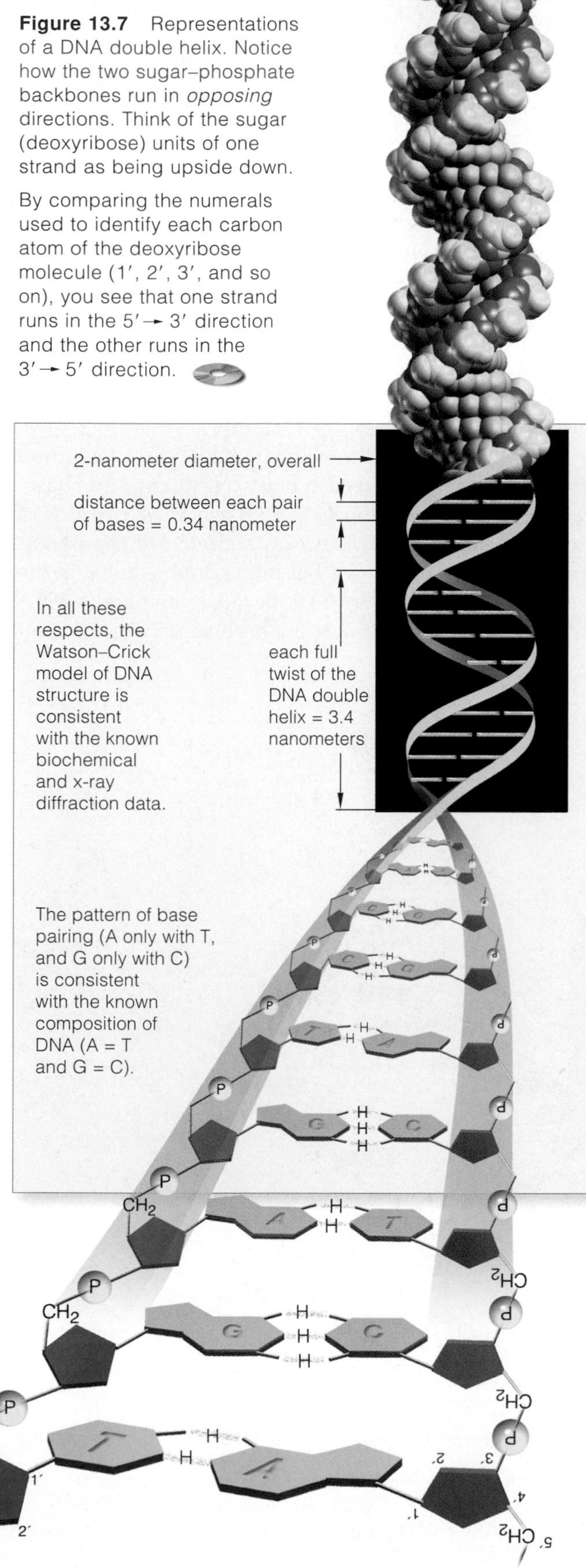

Figure 13.7 Representations of a DNA double helix. Notice how the two sugar–phosphate backbones run in *opposing* directions. Think of the sugar (deoxyribose) units of one strand as being upside down.

By comparing the numerals used to identify each carbon atom of the deoxyribose molecule (1′, 2′, 3′, and so on), you see that one strand runs in the 5′→3′ direction and the other runs in the 3′→5′ direction.

DNA does not readily lend itself to x-ray diffraction. However, researchers could rapidly spin a suspension of DNA molecules, spool them onto a rod, and gently pull them into gossamer fibers, like cotton candy. If the atoms in DNA were arranged in a regular order, x-rays directed at a fiber should scatter in a regular pattern that could be captured on film.

Calculations based on Franklin's images strongly indicated that the DNA molecule had to be long and thin, with a 2-nanometer diameter. Some molecular configuration was being repeated every 0.34 nanometer along its length, and another one every 3.4 nanometers.

Could the sequence of nucleotide bases be twisting, like a circular stairway? Certainly Pauling thought so. After all, he discovered the helical shape of collagen. He and everybody else—including Wilkins, Watson, and Crick—were thinking "helix." Watson later wrote, "We thought, why not try it on DNA? We were worried that *Pauling* would say, why not try it on DNA? Certainly he was a very clever man. He was a hero of mine. But we beat him at his own game. I still can't figure out why."

Pauling, it turned out, made a big chemical mistake. His model had hydrogen bonds at phosphate groups holding DNA's structure together. That does happen in highly acidic solutions. It doesn't happen in cells.

Patterns of Base Pairing

As Watson and Crick perceived, DNA consists of *two* strands of nucleotides, held together at their bases by hydrogen bonds. The bonds form when the two strands run in opposing directions and twist together into a double helix (Figure 13.7). Two kinds of base pairings form along the length of the molecule: A—T and G—C. This bonding pattern permits variation in the order of bases in any given strand. For example, in even a tiny stretch of DNA from a rose, gorilla, human, or any other organism, the sequence might be:

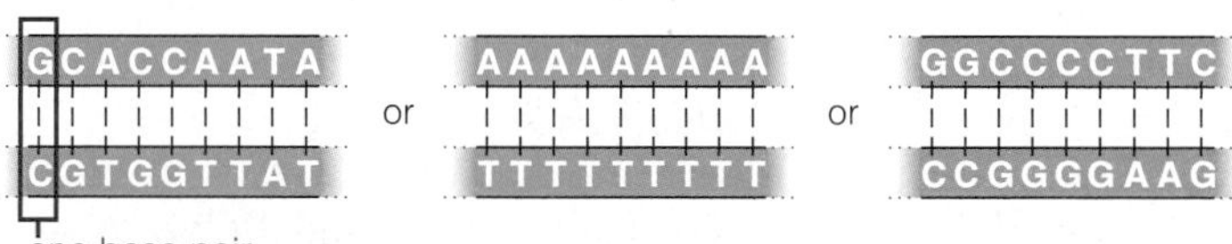

In fact, even though all DNA molecules show the same bonding pattern, each species has unique base sequences in its DNA. *This molecular constancy and variation among species is the foundation for the unity and diversity of life.*

The pattern of base pairing between the two strands in DNA is constant for all species—A with T, and G with C. However, the DNA molecules of each species show unique differences in the sequence of base pairs along their length.

Focus on Bioethics

13.3 ROSALIND'S STORY

In 1951, Rosalind Franklin arrived at King's Laboratory of Cambridge University with impressive credentials. Earlier, in Paris, she had refined existing procedures for x-ray diffraction while studying the structure of coal. She also had devised a new mathematical approach to interpreting x-ray diffraction images and had built three-dimensional models of molecules, as Pauling had done. Now she had been asked to run an x-ray crystallography laboratory, which she would create with state-of-the-art equipment. Her assignment? Investigate the structure of DNA.

No one bothered to tell her that, down the hall, Maurice Wilkins was already working on the puzzle. Even the graduate student assigned to assist her failed to mention it. And no one bothered to tell Wilkins about Franklin's assignment, so he assumed she was just a technician hired to do his x-ray crystallography work because he could not do it himself. And so began a poisonous clash. To Franklin, Wilkins seemed inexplicably prickly; to Wilkins, Franklin displayed an appalling lack of deference that technicians usually show to researchers.

Figure 13.8 Rosalind Franklin.

Wilkins had a prized cache of crystalline fibers of DNA, each having parallel arrays of hundreds of millions of DNA molecules—and these he gave to his "technician."

Five months later, Franklin gave a talk on what she had learned so far. DNA, she said, might have two, three, or four parallel chains twisted into a helix, with phosphate groups projecting outward. She had measured DNA's density and assigned DNA fibers to 1 of 230 categories of crystals, based on the symmetry of their parallel chains.

With his background in crystallography, Crick would have recognized the significance of that symmetry *if* he had been present. (To wit, *paired* chains running in opposite directions would look the same even if flipped 180°. Two paired chains? No. DNA's density ruled that out. But *one pair* of chains? Yes!) Watson was in the audience, but he didn't have a clue to what Franklin was talking about.

Later, Franklin created an outstanding x-ray diffraction image of wet DNA fibers that fairly screamed *Helix!* She also worked out DNA's length and diameter. But she had been working with dry fibers for so long, she did not dwell on her new data. Wilkins, however, did. In 1953, he let Watson see Franklin's exceptional x-ray diffraction image and reminded him of what she had reported fourteen months earlier. And when Watson and Crick finally did focus on her data, they had the final bits of information necessary to start building a DNA model—one with two helically twisted chains running in opposing directions.

Not until ten years after Franklin's untimely death in 1958 did Watson acknowledge her pivotal discoveries.

13.4 DNA REPLICATION AND REPAIR

How Is a DNA Molecule Duplicated?

The discovery of DNA structure was a turning point in studies of inheritance. Until then, no one could explain **DNA replication**, or how the molecule of inheritance is duplicated before the cell divides. Once Watson and Crick had assembled their model, Crick understood at once how this might be done.

As he knew, enzymes can easily break the hydrogen bonds between the two nucleotide strands of a DNA molecule. When these enzymes and other proteins act on the molecule, one strand can unwind from the other, thereby exposing stretches of nucleotide bases. Cells have stockpiles of free nucleotides, and these can pair with the exposed bases.

Each parent strand remains intact, and a companion strand is assembled on each one according to this base-pairing rule: A to T, and G to C. As soon as a stretch of a new, partner strand forms on a stretch of the parent strand, the two twist together into a double helix, in the manner shown in Figure 13.9. Because the parent DNA strand is conserved during the replication process, half

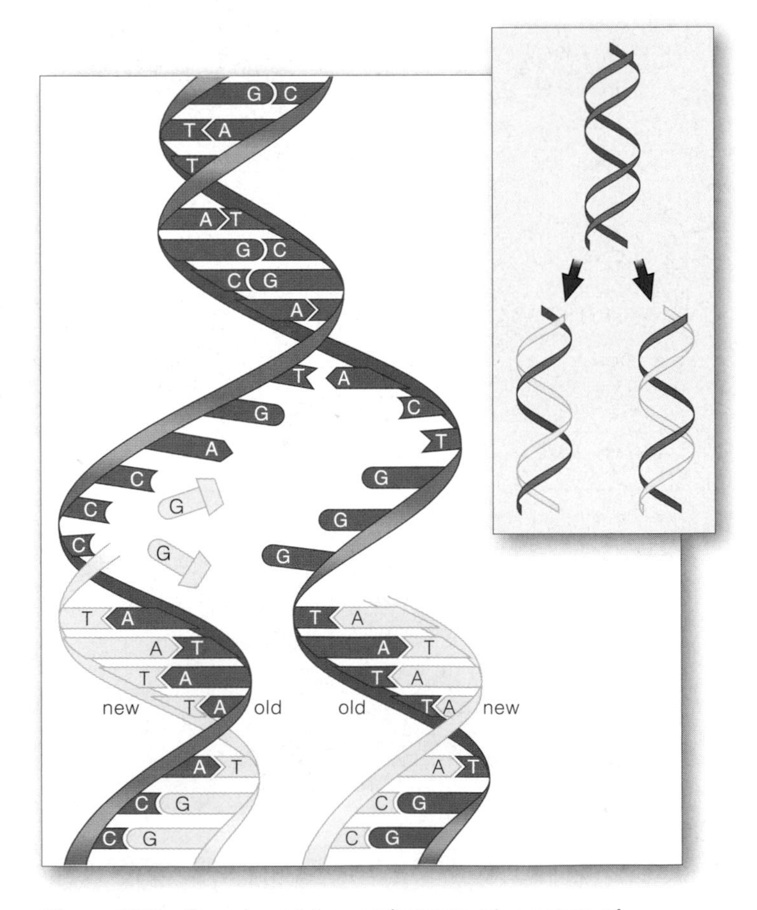

Figure 13.9 Overview of the semiconservative nature of DNA replication. The original two-stranded DNA molecule is shown in *blue*. Each parent strand remains intact, and a new strand (*yellow*) is assembled on each one.

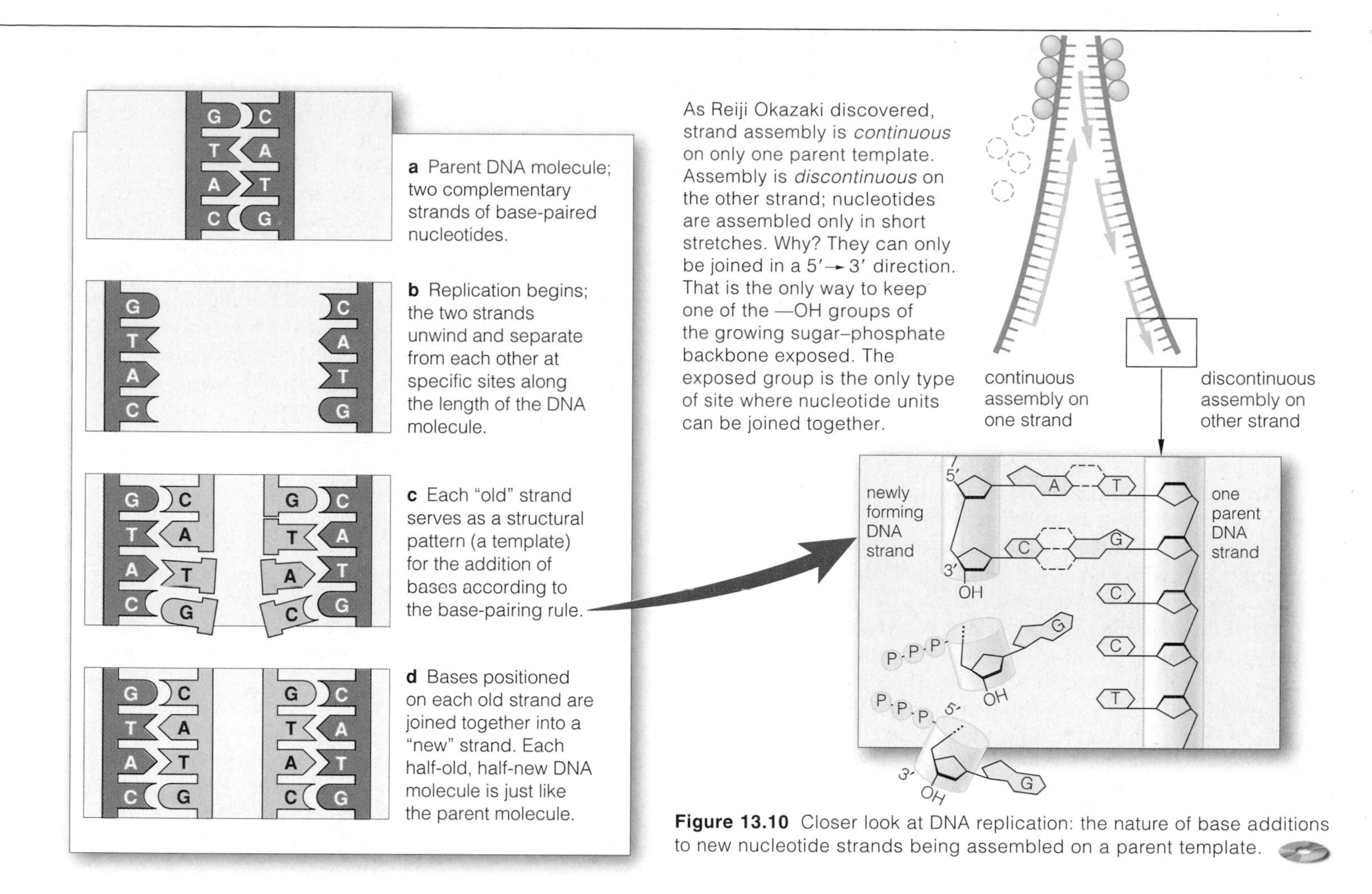

Figure 13.10 Closer look at DNA replication: the nature of base additions to new nucleotide strands being assembled on a parent template.

of every double-stranded DNA molecule is "old" and half is "new" (Figure 13.9). That is why biologists refer to the process as *semiconservative* replication.

DNA replication uses a team of molecular workers. In response to cellular signals, the replication enzymes become active along the length of the DNA molecule. Together with other proteins, some enzymes unwind the strands in both directions and prevent them from rewinding. Enzyme action jump-starts the unwinding but is not required to unzip hydrogen bonds between the strands; hydrogen bonds are individually weak.

Now enzymes called **DNA polymerases** can attach short stretches of free nucleotides to unwound portions of the parent template (Figure 13.10). Free nucleotides brought up for strand assembly supply the energy that drives the replication process. Each has three phosphate groups attached. DNA polymerase splits away two of the groups, and some energy released by the reaction is used to attach the nucleotide to a growing strand.

DNA ligases fill in tiny gaps between the new short stretches to form a continuous strand. Then enzymes wind the template strand and complementary strand together to form a DNA double helix.

As you will read in Section 16.2, some replication enzymes are used in recombinant DNA technology.

Monitoring and Fixing the DNA

DNA polymerases, DNA ligases, and other enzymes also engage in **DNA repair**. By this process, enzymes excise and repair altered parts of the base sequence in one strand of a double helix. Suppose a few bases are lost from one strand during replication. DNA polymerases can "read" the complementary sequence on the other strand. With assistance from other repair enzymes, they restore the original sequence. And remember crossing over, whereby nonsister chromatids of homologous chromosomes break and exchange segments at meiosis? The same enzymes that bring about that recombination also repair double-strand breaks. They initiate strand exchange with homologous DNA. Section 14.5 gives you an idea of what happens when mutation or some other factor compromises the excision–repair function.

DNA is replicated prior to cell division. Enzymes unwind its two strands. Each strand remains intact throughout the process—it is conserved—and enzymes assemble a new, complementary strand on each one.

Enzymes involved in replication also repair the DNA where base-pairing errors have crept into the nucleotide sequence.

13.5 DOLLY, DAISIES, AND DNA

Imagine the possibility of making a genetically identical copy—a clone—of yourself. Is the image that far-fetched? Consider this: Researchers have been cloning complex animals for more than a decade. For example, some use in vitro fertilization methods to grow cattle embryos in petri dishes. After cleavage is under way in a new embryo, they split the tiny cluster of embryonic cells, and the two tinier clusters continue to develop as two identical-twin cattle embryos. These are implanted in surrogate mothers, and they are born as cloned calves. Figure 13.11*a* shows cloned Holsteins that are prized for their milk production.

A researcher who clones farm animals derived from embryonic cells has to wait for the clones to grow up to see if they display a desired trait. Using a differentiated cell from a prized adult would be faster, for the prized genotype would already be demonstrated and could be maintained indefinitely. However, it seemed impossible to trick a differentiated cell into using its DNA in a new way to become the first cell of an embryo.

What does "differentiated" mean? When an embryo first grows from a fertilized egg, all of its cells have the same DNA and are pretty much alike. Then different embryonic cells start using different parts of their DNA. Their unique selections commit them to being liver cells, heart cells, brain cells, and other specialists in structure, composition, and function (Section 15.3).

Then, in 1997 in Scotland, a research group led by Ian Wilmut coaxed a differentiated cell from a sheep udder into becoming an "uncommitted" first cell of an embryo. Thus a lamb named Dolly was born a clone of her mother without help from a father. Wilmut's group had managed to transfer nuclei from differentiated cells into enucleated unfertilized eggs, which they had removed earlier from a pregnant ewe. (*Enucleated* means that the nucleus was surgically removed from a cell with a microscopically small needle.) After hundreds of attempts, one egg proceeded to grow, by mitotic cell divisions, into a whole sheep. Dolly developed into a healthy adult and gave birth to a lamb of her own (Figure 13.11*b*).

In science, extraordinary claims call for extraordinary proof—in this case, successful repeats of the experiment. Resounding success came in 1998. Ryozo Yanagimachi and his coworkers at the University of Hawaii cloned three generations of mice (*Mus musculus*). They quickly transferred nuclei from mature cumulus cells from an ovary into unfertilized, enucleated eggs. (Cumulus cells provide nutritional support to neighboring eggs.) Shortly afterward, they chemically activated the eggs as a way to jump-start development into fully formed mice. The resting period may have helped the cells adjust to the unconventional mode of fertilization.

By the year's end, experimenters in Japan gave us a new way of looking at Daisy the Cow. Yukio Tsunoda and his coworkers slipped nuclei from a cow's cumulus cells and oviduct cells into enucleated eggs. Of 240 transfers, 67 embryos formed. Ten of the survivors were transferred into cows that served as surrogate mothers. Eight cloned calves were born. Only four survived, but the results suggest that cloning cows from differentiated cells may become at least as efficient as producing them by in vitro fertilization.

In short, individuals of three genera of mammals have now been cloned. More importantly, the efficiency of cloning techniques is increasing at a startling pace.

All of which raises some far-out questions. Is human cloning next? For example, will a human Daisy be able to reproduce copies of herself without involving men? Will men be able to pay to have their DNA inserted into an enucleated cell, have the cell implanted in a surrogate mother, and make a little repeat of themselves?

Recently Lee Silver, a molecular biologist at Princeton University, said he knows of two qualified specialists in human fertility who are interested in cloning humans. He suggested that anyone who thinks the technology will move slowly is being naive. Ethically speaking, should such an attempt be made at all? At this writing, scientists and nonscientists alike are actively debating the issue.

Figure 13.11 (**a**) A clone of Holsteins resulting from in vitro fertilization. (**b**) Dolly, a cloned sheep. She started her life as a differentiated cell that was extracted from an adult ewe, then induced to undergo mitotic cell divisions. She is shown with her first lamb. She clearly is able to breed normally and reproduce the old-fashioned way. However, her DNA may be deteriorating (aging) faster than it should.

Further reading: Student Guide to InfoTrac on web site

13.6 SUMMARY

1. The hereditary information of cells and multicelled organisms is encoded in DNA (deoxyribonucleic acid).

2. DNA consists of nucleotide subunits. Each of these has a five-carbon sugar (deoxyribose), one phosphate group, and one of four kinds of nitrogen-containing bases (adenine, thymine, guanine, or cytosine).

3. A DNA molecule consists of two nucleotide strands twisted together into a double helix. The bases of one strand pair (hydrogen-bond) with bases of the other.

4. The bases of the two strands in a DNA double helix pair in constant fashion. Adenine pairs with thymine (A to T), and guanine with cytosine (G to C). *Which* base pair follows the next (A–T, T–A, G–C, or C–G) varies along the length of the strands.

5. Overall, the DNA of one species includes a number of unique stretches of base pairs that set it apart from the DNA of all other species.

6. During DNA replication, enzymes unwind the two strands of a double helix and assemble a new strand of complementary sequence on each parent strand. Two double-stranded molecules result. One strand of each molecule is "old" (it is conserved); the other is "new."

7. Some of the enzymes involved in DNA replication also repair DNA where base-pairing errors have been introduced into the nucleotide sequence.

Review Questions

1. Name the three molecular parts of a nucleotide in DNA. Also name the four different bases in these nucleotides. *13.2*

2. What kind of bond joins two DNA strands in a double helix? Which nucleotide base-pairs with adenine? With guanine? *13.2*

3. Explain how DNA molecules can show both constancy and variation from one species to the next. *13.2*

Self-Quiz (*Answers in Appendix III*)

1. Which is *not* a nucleotide base in DNA?
   a. adenine c. uracil e. cytosine
   b. guanine d. thymine

2. What are the base-pairing rules for DNA?
   a. A–G, T–C c. A–U, C–G
   b. A–C, T–G d. A–T, G–C

3. A DNA strand having the sequence C–G–A–T–T–G would be complementary to the sequence ______.
   a. C–G–A–T–T–G c. T–A–G–C–C–T
   b. G–C–T–A–A–G d. G–C–T–A–A–C

4. One species' DNA differs from others in its ______.
   a. sugars c. base sequence
   b. phosphate groups d. all of the above

5. When DNA replication begins, ______.
   a. the two DNA strands unwind from each other
   b. the two DNA strands condense for base transfers
   c. two DNA molecules bond
   d. old strands move to find new strands

6. DNA replication requires ______.
   a. free nucleotides c. many enzymes
   b. new hydrogen bonds d. all of the above

7. Match the DNA terms appropriately.
   ____ DNA polymerase a. two nucleotide strands that are twisted together
   ____ constancy in base pairing b. A with T, G with C
   ____ replication c. hereditary material duplicated
   ____ DNA double helix d. replication enzyme

Critical Thinking

1. Chargaff's data suggested that adenine pairs with thymine, and guanine pairs with cytosine. What other data available to Watson and Crick suggested that adenine-guanine and cytosine-thymine pairs normally do not form?

2. One of Matthew Meselson and Frank Stahl's experiments supported the semiconservative model of DNA replication. The researchers made "heavy" DNA by growing *Escherichia coli* in a medium enriched with ^{15}N, a heavy isotope of nitrogen. They prepared "light" DNA by growing *E. coli* in the presence of ^{14}N, the more common isotope. An available technique helped them identify which replicated molecules were heavy, light, or hybrid (one heavy strand, one light). Use two pencils of two different colors, one for heavy strands and one for light strands. Starting with a DNA molecule having two heavy strands, sketch the daughter molecules that would form after one replication in a ^{14}N-containing medium. Now sketch the four DNA molecules that would result if these daughter molecules were replicated a second time in the ^{14}N medium.

3. Mutations (permanent changes in base sequences of genes) are the original source of genetic variation. This variation is the raw material of evolution. Yet how can both statements be true, given that cells have efficient mechanisms to repair DNA before mutations can become established?

4. As indicated in Section 4.11, a pathogenic strain of *E. coli* has acquired an ability to produce a dangerous toxin that has caused medical problems and fatalities. This is especially the case for young children who have ingested undercooked, contaminated beef. Develop a hypothesis to explain how a normally harmless bacterium such as *E. coli* can become a pathogen.

5. In October 1999, scientists announced the amazing discovery of a woolly mammoth that had been frozen in glacial ice for the past 20,000 years. Its soft organs are intact. They plan to carefully use hair dryers to thaw it. They want to minimize disruption of its cells, and its DNA—which they plan to use to clone a woolly mammoth. Consider Section 13.5, then speculate on the pros and cons of cloning an extinct animal.

Selected Key Terms

adenine (A) *13.2*
bacteriophage *13.1*
cytosine (C) *13.2*
deoxyribonucleic acid (DNA) *CI*
DNA ligase *13.4*
DNA polymerase *13.4*
DNA repair *13.4*
DNA replication *13.4*
guanine (G) *13.2*
nucleotide *13.2*
thymine (T) *13.2*
x-ray diffraction image *13.2*

Readings *See also www.infotrac-college.com*

Watson, J. 1978. *The Double Helix*. New York: Atheneum. Highly personal view of scientists and their methods, interwoven into an account of how DNA structure was discovered.

Wolfe, S. 1995. *Introduction to Molecular and Cellular Biology*. Belmont, California: Wadsworth. Comprehensive and accessible.

14 FROM DNA TO PROTEINS

Beyond Byssus

Picture a mussel, of the sort shown in Figure 14.1. Hard-shelled but soft of body, it is using its muscular foot to probe a wave-scoured rock. At any moment, pounding waves can whack the mussel into the water, hurl it repeatedly against the rock with shell-shattering force, and so offer up a gooey lunch for gulls.

By chance, the mussel's foot comes across a crevice in the rock. The foot moves, broomlike, and sweeps the crevice clean. It presses down, forcing air out from underneath it, then arches up. The result is a vacuum-sealed chamber, rather like the one that forms when a plumber's rubber plunger is being squished down and up to unclog a drain. Into this vacuum chamber the mussel spews a fluid, consisting of keratin and other proteins, which bubbles into a sticky foam. Now, by curling its foot into a small tubular shape and pumping the foam through it, the mussel produces sticky threads about as wide as a human whisker. As a final touch, it varnishes the threads with another type of protein and ends up with an adhesive called byssus, which anchors the mussel to the rock.

Byssus is the world's premier underwater adhesive. Nothing that humans have manufactured even comes close. (Sooner or later, water chemically degrades or deforms synthetic adhesives.) Byssus truly fascinates biochemists, dentists, and surgeons looking for better ways to do tissue grafts and to rejoin severed nerves. Genetic engineers insert mussel DNA into yeast cells,

Figure 14.1 Mussels (*Mytilus californianus*) busily demonstrating the importance of proteins for survival. When mussels come across a suitable anchoring site, they use their muscular foot rather like a plumber's plunger to create a vacuum chamber. In this chamber they manufacture the world's best underwater adhesive from a mix of proteins. The adhesive anchors them to substrates in their wave-swept habitat.

which go on to reproduce in large numbers and serve as "factories" for translating mussel genes into useful quantities of proteins. This exciting work, like the mussel's own byssus-building efforts, starts with one of life's universal precepts: *Every protein is synthesized in accordance with instructions contained in DNA.*

You are about to trace the steps leading from DNA to proteins. Many enzymes are players in this pathway. So is another kind of nucleic acid besides DNA. The same steps produce *all* proteins, from mussel-inspired adhesives to the keratin in your hair and fingernails to the insect-digesting enzymes of a Venus flytrap.

Start out by thinking of each cell's DNA as a book of protein-building instructions. The alphabet used to create the book is simple enough: A, T, G, and C (for the nucleotide bases adenine, thymine, guanine, and cytosine). How do you get from that alphabet to a protein? The answer starts with DNA's structure.

DNA, recall, is a double-stranded molecule. Which kind of nucleotide base follows the next along the length of a strand—that is, the **base sequence**—differs from one kind of organism to the next. As you read in the preceding chapter, before a cell divides, its DNA is replicated as the two strands unwind entirely from each other. However, at other times in a cell's life, the two strands unwind only in certain regions to expose particular base sequences—genes. Most of the genes contain instructions for building proteins.

It takes two steps, **transcription** and **translation**, to carry out a gene's protein-building instructions. In eukaryotic cells, transcription proceeds in the nucleus. At this step, a selected base sequence in DNA serves as a structural pattern—a template—for assembling a strand of **ribonucleic acid** (RNA) from the cell's pool of free nucleotides. Afterward, the RNA moves into the cytoplasm, where translation proceeds. At this second step, RNA directs the assembly of amino acids into polypeptide chains. The newly formed chains become folded into the three-dimensional shapes of proteins.

In short, DNA guides the synthesis of RNA, then RNA guides the synthesis of proteins:

DNA —*transcription*→ RNA —*translation*→ PROTEIN

The newly synthesized proteins will play structural and functional roles in cells. And some even will have roles in synthesizing more DNA, RNA, and proteins.

KEY CONCEPTS

1. Life cannot exist without enzymes and other proteins. Proteins consist of polypeptide chains, which consist of amino acids. The sequence of amino acids corresponds to a gene, which is a sequence of nucleotide bases in a DNA molecule.

2. The path leading from genes to proteins consists of two steps, called transcription and translation.

3. In transcription, the double-stranded DNA molecule is unwound at a gene region, then an RNA molecule is assembled on the exposed bases of one of the strands.

4. In translation, a certain type of RNA directs the linkage of one amino acid after another, in the sequence required to produce the specified polypeptide chain.

5. With few exceptions, the genetic "code words" by which DNA's instructions are translated into proteins are the same in all species of organisms.

6. A mutation is a permanent alteration in a gene's base sequence. Mutations are the original source of genetic variation in populations.

7. Mutations cause changes in protein structure, protein function, or both. The changes may lead to small or large differences in traits among individuals of a population.

Focus on Science

14.1 CONNECTING GENES WITH PROTEINS

GARROD'S HYPOTHESIS Early in the 1900s Archibald Garrod, a physician, was puzzling over certain illnesses. They appeared to be heritable, for they kept recurring in the same families. Those illnesses also appeared to be metabolic disorders. In each case, blood or urine samples from affected patients contained abnormally high levels of a substance known to be produced at a certain step in a metabolic pathway.

Most likely, the enzyme that operated at the *next* step in the metabolic pathway was defective. Something was preventing it from chemically recognizing or interacting properly with the substance that is supposed to be its substrate. If Garrod's reasoning were correct, then the metabolic pathway would be blocked from that step onward. Figure 14.2 illustrates this concept.

Garrod's hypothesis could explain why molecules of a particular substance were accumulating in excess amounts in the body fluids of affected individuals. He suspected that his patients differed from unaffected individuals in a key aspect of their metabolism: Each one had inherited a single metabolic defect. Garrod concluded that specific "units" of inheritance (genes) are expressed through the synthesis of specific enzymes of metabolic pathways.

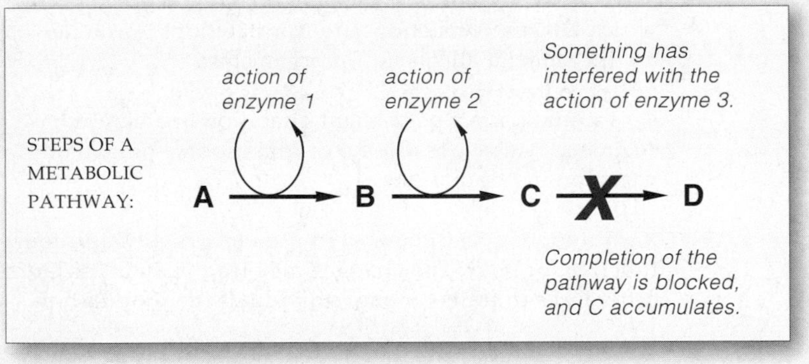

STEPS OF A METABOLIC PATHWAY:

A → B → C ✗→ D

action of enzyme 1

action of enzyme 2

Something has interfered with the action of enzyme 3.

Completion of the pathway is blocked, and C accumulates.

Figure 14.2 Example of how a defective enzyme can block completion of a metabolic pathway.

BEADLE, TATUM, AND A BREAD MOLD Thirty-three years later two researchers, George Beadle and Edward Tatum, were experimenting with the red bread mold (*Neurospora crassa*). This fungus is a common spoiler of baked goods (Figure 14.3). But it also lends itself to genetic experiments and has become an organism of choice in the laboratory. *N. crassa* can be grown easily on an inexpensive culture medium that contains only sucrose, mineral salts, and biotin, which is one of the B vitamins. The fungal cells can synthesize all of the other nutrients they require, including other vitamins.

Suppose a fungal enzyme that takes part in a synthesis pathway is defective as a result of a gene mutation. The researchers suspected that this had happened in some of the strains of *N. crassa* that they were studying. One strain grew only when it was supplied with vitamin B_6, another with vitamin B_1, and so on.

Chemical analysis of cell extracts revealed a different defective enzyme in each mutant strain. In other words, *each inherited mutation corresponded to a defective enzyme*. Here was evidence favoring the "one-gene, one-enzyme" hypothesis.

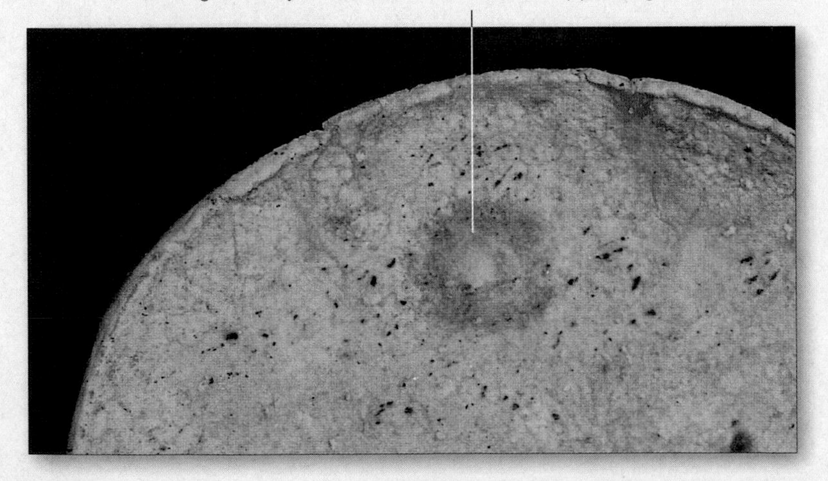

Figure 14.3 Colony of the red bread mold (*Neurospora crassa*) and other fungal species on a stale tortilla.

CLUES FROM GEL ELECTROPHORESIS Later on, the one-gene, one-enzyme hypothesis was refined as a result of investigations into the genetic basis of *sickle-cell anemia*. This heritable disorder arises from the presence of an abnormal version of a protein, hemoglobin, in the red blood cells of affected individuals. The abnormal hemoglobin is designated HbS instead of HbA (Section 11.5).

In 1949, the biochemists Linus Pauling and Harvey Itano subjected molecules of HbS and HbA to **gel electrophoresis**. This laboratory procedure uses an electric field to move molecules through a viscous gel and separate them according to their size, shape, and net surface charge. Often the gel is sandwiched between glass or plastic plates to form a viscous slab. The two ends of the slab are suspended in two salt solutions that are connected by electrodes to a power source (Figure 14.4). When voltage is applied to the apparatus, the molecules present in the gel migrate through the electric field according to their individual charge, and they move away from one another in the gel. Later on, the molecules can be pinpointed by staining the gel after a predetermined period of electrophoresis.

Pauling and Itano carefully layered a mixture of HbS and HbA molecules on the gel at the top of the slab. As the molecules gradually migrated down through the gel, they separated into distinct bands. The band that moved fastest carried the greatest

Further reading: Student Guide to InfoTrac on web site →

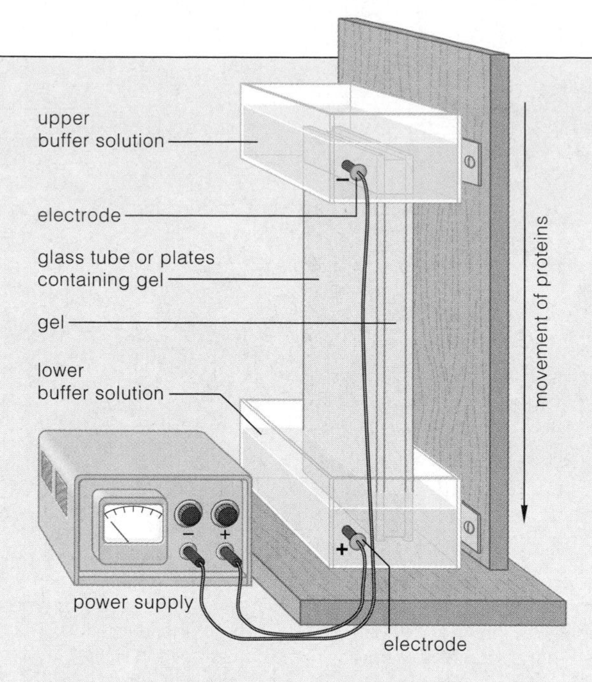

Figure 14.4 Diagram of one type of apparatus that is used for gel electrophoresis studies.

surface charge, and this turned out to be composed of HbA molecules. HbS molecules moved more slowly.

ONE GENE, ONE POLYPEPTIDE Later, Vernon Ingram pinpointed the biochemical difference between HbS and HbA. Hemoglobin, recall, has four polypeptide chains (Figure 14.5). Two of the chains are designated alpha and the other two, beta. An abnormal HbS chain arises from a gene mutation that affects protein synthesis. The mutation causes valine instead of glutamate to be added as the sixth amino acid of the beta chain (Figure 14.5*c*). Whereas glutamate carries an overall negative charge, valine has no net charge. That is the reason why HbS chains behaved differently in the electrophoresis studies.

Because of that one mutation, HbS hemoglobin has a "sticky" (hydrophobic) patch. In blood capillaries, where oxygen concentrations are at their lowest, hemoglobin molecules interact at the sticky patches. They aggregate into rods and distort red blood cells, and the consequences adversely affect organs throughout the body (Section 11.5).

The discovery of the genetic difference between alpha and beta chains of hemoglobin suggested that *two* genes code for hemoglobin—one for each kind of polypeptide chain. More importantly, it provided further evidence that genes code for proteins in general, not just for enzymes.

And so a more precise hypothesis emerged: *The amino acid sequences of polypeptide chains—the structural units of proteins—are encoded in genes.*

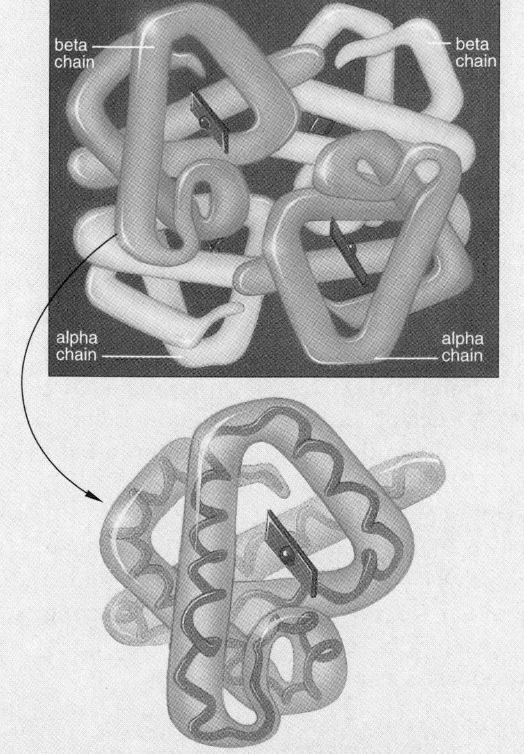

a *Above*: Arrangement of the four polypeptide chains and heme groups of a hemoglobin molecule. *Below*: Closer view of one of the beta chains.

VALINE HISTIDINE LEUCINE PROLINE THREONINE GLUTAMATE GLUTAMATE

b Normal sequence of amino acids at the start of the beta chain that is characteristic of HbA molecules.

VALINE HISTIDINE LEUCINE PROLINE THREONINE VALINE GLUTAMATE

c The single amino acid substitution (*yellow*) that results in the abnormal beta chain that is characteristic of HbS molecules.

Figure 14.5 A single amino acid substitution that starts with a gene mutation and ends with the symptoms of sickle-cell anemia, which are described in Section 11.5.

14.2 HOW IS DNA TRANSCRIBED INTO RNA?

The Three Classes of RNA

Before turning to the details of protein synthesis, let's clarify one point. The chapter introduction might have left you with the impression that synthesis of proteins requires only one class of RNA molecules. Actually, it requires three. Transcription of most genes produces **messenger RNA**, or **mRNA**—the only class of RNA that carries *protein-building* instructions. Transcription of some other genes produces **ribosomal RNA**, or **rRNA**, a major component of ribosomes. Ribosomes, recall, are the structural units upon which polypeptide chains are assembled. Transcription of still other genes produces **transfer RNA**, or **tRNA**, which delivers amino acids one by one to a ribosome in the order specified by mRNA.

The Nature of Transcription

An RNA molecule is almost but not quite like a single strand of DNA. RNA, too, consists of only four types of nucleotides. Each nucleotide has a five-carbon sugar, ribose (not DNA's deoxyribose), a phosphate group, and a base. Three types of bases—adenine, cytosine, and guanine—are the same in RNA and DNA. But in RNA, the fourth type of base is **uracil**, not thymine (Figure 14.6). Like thymine, uracil can pair with adenine. This means a new RNA strand can be put together on a DNA region according to base-pairing rules (Figure 14.7).

Transcription resembles DNA replication in another respect. Enzymes add nucleotides to a growing RNA strand one at a time, in the 5′ → 3′ direction. Here you might wish to refer to the simple explanation of strand assembly in Section 13.4 (Figure 13.10.)

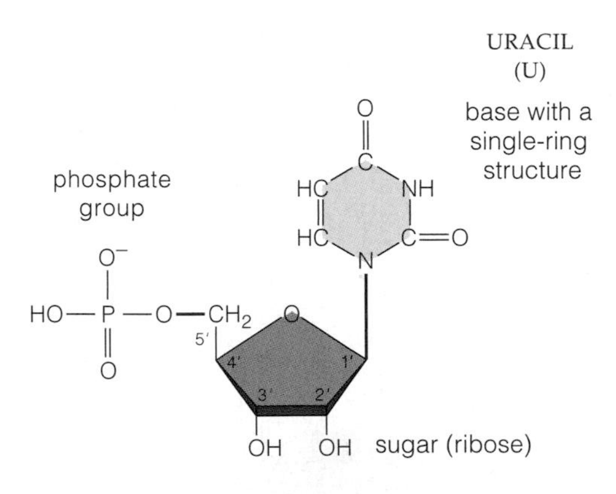

Figure 14.6 Structural formula for one of the four types of RNA nucleotides. The three others have a different base (adenine, guanine, or cytosine instead of the uracil shown here). Compare Section 13.2, which shows DNA's four nucleotides. Notice that the sugars of DNA and RNA differ at one group only (*yellow*).

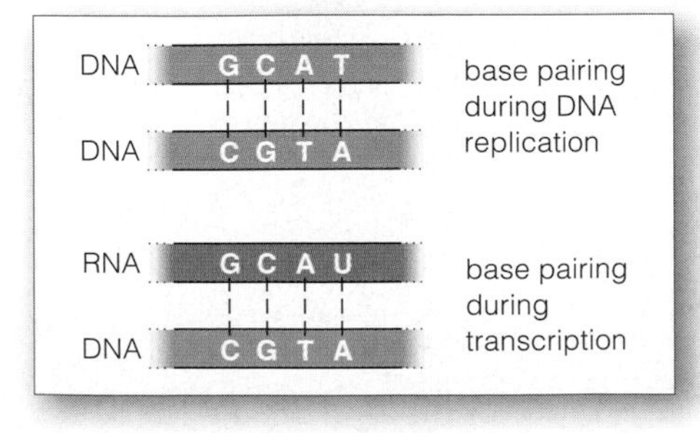

Figure 14.7 An example of base pairing of RNA with DNA during transcription, as compared to base pairing during DNA replication.

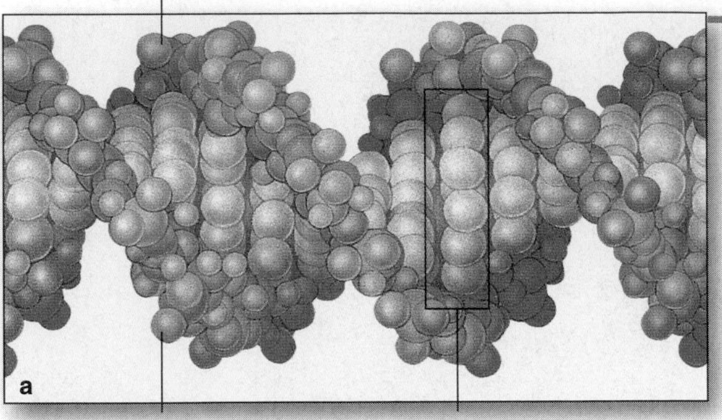

Figure 14.8 The process of gene transcription, by which an RNA molecule is assembled on a DNA template. The sketch in (**a**) shows a gene region in part of a DNA double helix. In this region, the base sequence of one of the two nucleotide strands (not both) is about to be transcribed into an RNA molecule, in the manner shown in (**b**) through (**e**).

Transcription *differs* from DNA replication in three key respects. First, only a selected stretch of one DNA strand, not the whole molecule, serves as the template. Second, instead of DNA polymerase, a different enzyme, **RNA polymerase**, catalyzes the addition of nucleotides to the 3′ end of a growing strand of RNA. Third, transcription results in a single, free strand of RNA nucleotides, not in a double helix.

Transcription is initiated at a **promoter**, a base sequence in DNA that signals the start of a gene. Proteins position an RNA polymerase on DNA and thereby help it bind to the promoter. The enzyme moves along the DNA strand, joining nucleotides one after another (Figure 14.8). When that enzyme reaches a particular base, however, the new RNA molecule is released as a free transcript.

Finishing Touches on mRNA Transcripts

In eukaryotic cells alone, a new mRNA molecule is unfinished. That "pre-mRNA" must be modified before its protein-building instructions can be put to use. Just as a dressmaker might snip off some threads or add bows on a dress before it leaves the shop, so do eukaryotic cells tailor their pre-mRNA.

For example, enzymes attach a cap to the 5′ end of a pre-mRNA molecule. The cap is a nucleotide

Further reading: Student Guide to InfoTrac on web site →

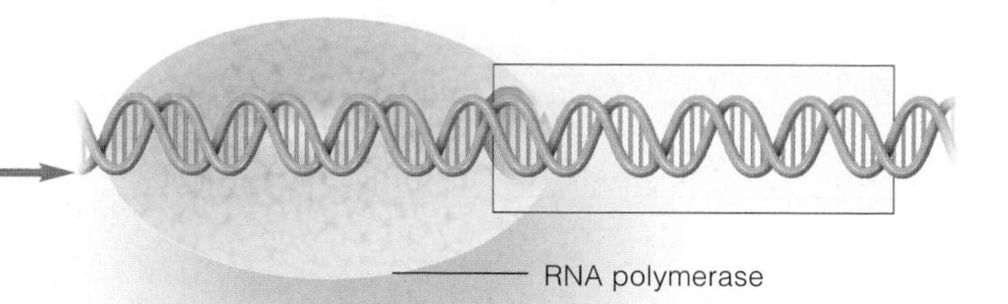

b A molecule of RNA polymerase binds with a promoter region in the DNA. It will recognize the base sequence located downstream from that site as a template for linking together the nucleotides adenine, cytosine, guanine, and uracil into a strand of RNA.

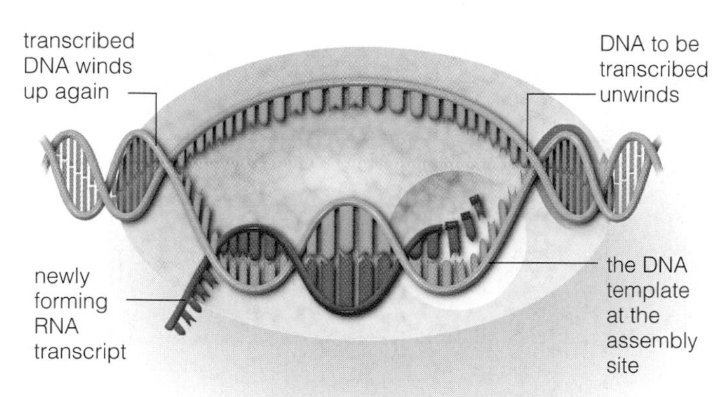

c All through transcription, the DNA double helix is unwound in front of the RNA polymerase. Short lengths of the newly forming RNA strand briefly wind up with its DNA template strand. New stretches of RNA unwind from the template (and the two DNA strands wind up again).

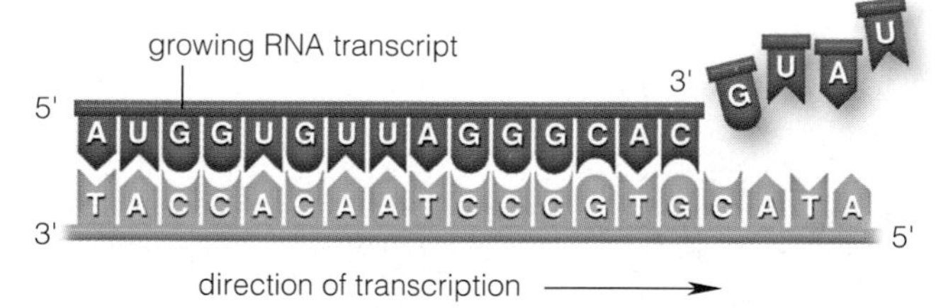

d What happened at the assembly site? RNA polymerase catalyzed the base-pairing of RNA nucleotides, one after another, with exposed bases on the DNA template strand.

5' A U G G U G U U A G G G C A C G U A U 3'

e At the end of the gene region, the last stretch of the new mRNA transcript is unwound and released from the DNA.

that has a methyl group and phosphate groups bonded with it. Also, enzymes attach a tail of about 100 to 200 adenine-containing nucleotides to the 3′ end of most pre-mRNA transcripts. Hence the name, "poly-A tail" (for multiple adenine units). The tail gets wound up with proteins. Later, in the cytoplasm, the cap will assist the binding of mRNA to a ribosome. Also, enzymes will

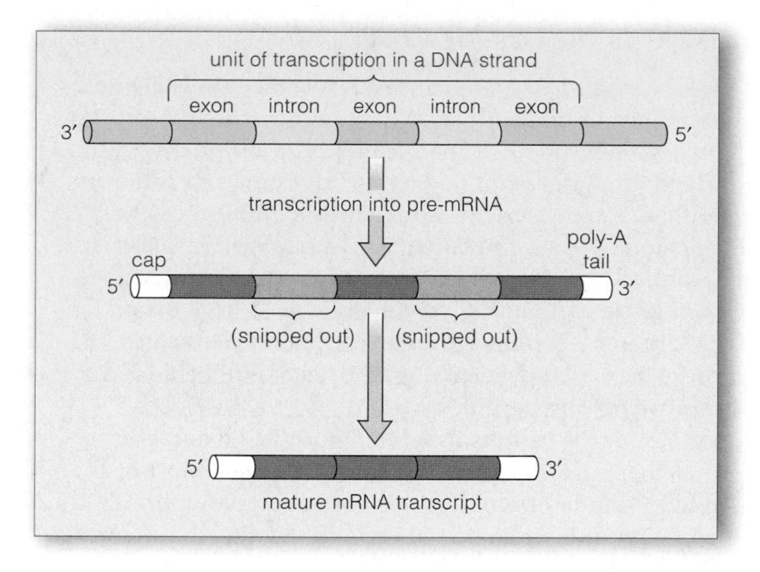

Figure 14.9 Transcription and modification of new mRNA in the nucleus of eukaryotic cells. Its cap is a nucleotide with functional groups attached. Its tail is a string of adenine nucleotides.

slowly destroy the wound-up tail from the tip on back. Such tails "pace" the access of enzymes to mRNA and thereby dictate how long an mRNA molecule will last. Apparently, they help keep protein-building messages intact for as long as the cell requires them.

Besides these alterations, the pre-mRNA itself gets modified. Most eukaryotic genes contain one or more **introns**, base sequences that must be removed before a pre-mRNA molecule can be translated. The introns intervene between **exons**, the parts that remain in the mRNA when it gets translated into protein. As Figure 14.9 shows, the introns are transcribed right along with the exons, but they are snipped out before the mRNA leaves the nucleus in mature form.

It could be that some introns are evolutionary junk, the leftovers of past mutations that led nowhere. Yet many introns are sites where instructions for building a particular protein can be snipped apart and spliced back together in various ways. The alternative splicing allows different cells in your body to use the same gene to make different versions of a pre-mRNA transcript, and therefore different versions of the resulting protein. We return to this topic in Section 15.3.

During gene transcription, a sequence of exposed bases in one of the two strands of a DNA molecule serves as the template for assembling a single strand of RNA. The assembly follows these base-pairing rules: adenine only with uracil, and cytosine only with guanine.

Before leaving the nucleus, each new mRNA transcript, or pre-mRNA, undergoes modification into final form.

14.3 DECIPHERING THE mRNA TRANSCRIPTS

What Is the Genetic Code?

Like a strand of DNA, an mRNA molecule is a linear sequence of nucleotides. What are the protein-building "words" encoded in that sequence? Gobind Khorana, Marshall Nirenberg, and other investigators came up with the answer. They showed that ribosomes "read" nucleotide bases *three at a time*, as triplets. Base triplets in an mRNA strand were given this name: **codons**.

Figure 14.10 will give you an idea of how the order of different codons in an mRNA strand dictates the order in which particular amino acids will be added to a growing polypeptide chain.

Count the codons listed in Figure 14.11, and you see that there are sixty-four kinds. Notice how most of the twenty kinds of amino acids correspond to more than one codon. Glutamate corresponds to the code words GAA *and* GAG, for example. Also notice how AUG has dual functions. It codes for the amino acid methionine, and it also is an initiation codon, a START signal for translating an mRNA transcript at a ribosome. That is, "three-bases-at-a-time" selections start at a particular AUG in the transcript's nucleotide sequence. Codons UAA, UAG, and UGA do not correspond to amino acids. They serve as STOP signals, which prevent further addition of amino acids to a new polypeptide chain.

The set of sixty-four different codons is the **genetic code.** It is the basis of protein synthesis in all organisms.

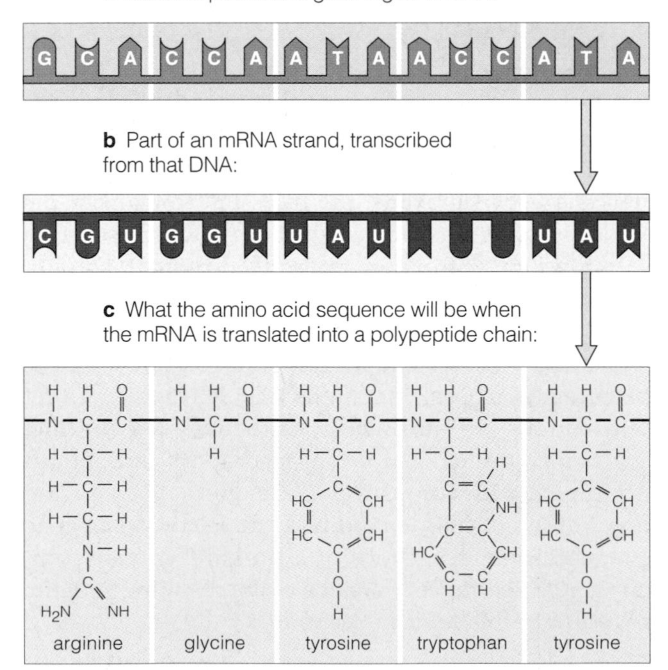

Figure 14.10 The steps from genes to proteins. (**a**) This diagram represents a region of a DNA double helix that was unwound during transcription. (**b**) The exposed bases on one DNA strand served as a template for assembling an mRNA strand. In the new mRNA transcript, every three nucleotide bases equaled one codon. Each codon calls for one amino acid in a polypeptide chain. (**c**) Referring to Figure 14.11, can you fill in the blank codon for tryptophan in the chain?

Amino acids that correspond to base triplets:

FIRST BASE	SECOND BASE OF A CODON: U	C	A	G	THIRD BASE
U	phenylalanine	serine	tyrosine	cysteine	U
U	phenylalanine	serine	tyrosine	cysteine	C
U	leucine	serine	STOP	STOP	A
U	leucine	serine	STOP	tryptophan	G
C	leucine	proline	histidine	arginine	U
C	leucine	proline	histidine	arginine	C
C	leucine	proline	glutamine	arginine	A
C	leucine	proline	glutamine	arginine	G
A	isoleucine	threonine	asparagine	serine	U
A	isoleucine	threonine	asparagine	serine	C
A	isoleucine	threonine	lysine	arginine	A
A	methionine (or START)	threonine	lysine	arginine	G
G	valine	alanine	aspartate	glycine	U
G	valine	alanine	aspartate	glycine	C
G	valine	alanine	glutamate	glycine	A
G	valine	alanine	glutamate	glycine	G

Figure 14.11 The genetic code. The codons in mRNA are nucleotide bases, "read" in blocks of three. Sixty-one of the base triplets correspond to specific amino acids. Three others serve as signals that stop translation. The left column of the diagram shows the first of the three nucleotides in each codon in mRNA. The middle columns show the second nucleotide. The right column shows the third. Reading from left to right, for instance, the triplet UGG corresponds to tryptophan. Both UUU and UUC correspond to phenylalanine.

Structure and Function of tRNA and rRNA

In a cell's cytoplasm are pools of free amino acids and free tRNA molecules. The tRNAs each have a molecular "hook," an attachment site for amino acids. They also have an **anticodon**, a nucleotide triplet that can base-pair with a codon (Figure 14.12). When tRNAs bind to

Further reading: Student Guide to InfoTrac on web site

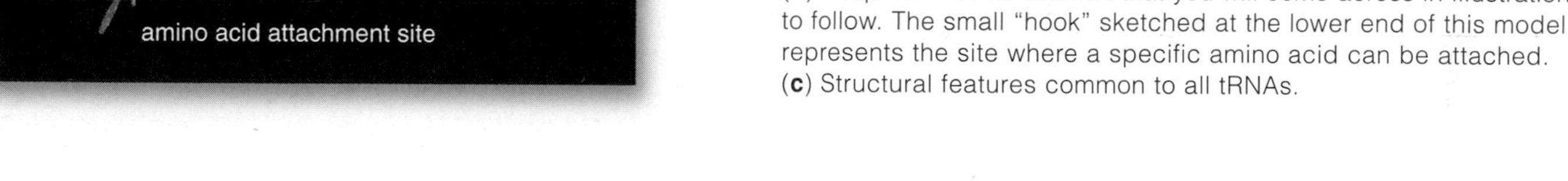

Figure 14.12 (**a**) Computer-generated, three-dimensional model of one type of tRNA molecule.

(**b**) Simplified model of tRNA that you will come across in illustrations to follow. The small "hook" sketched at the lower end of this model represents the site where a specific amino acid can be attached. (**c**) Structural features common to all tRNAs.

a A small ribosomal subunit. (The *red* arrow points to a platform for chain assembly on the surface)

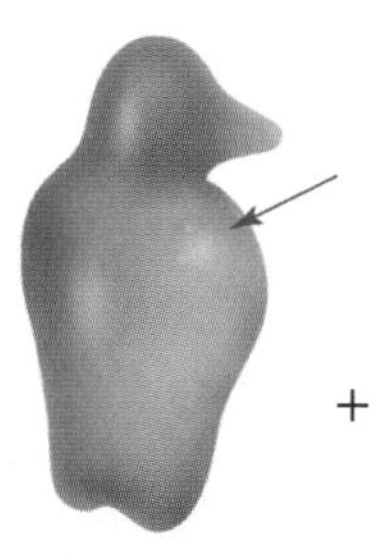

+

b A large ribosomal subunit, with a tunnel through parts of its interior

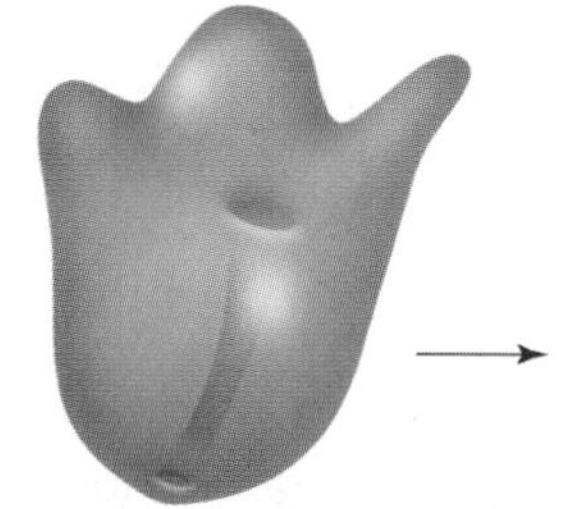

→

c Side view of an intact ribosome, showing how the platform and tunnel are aligned.

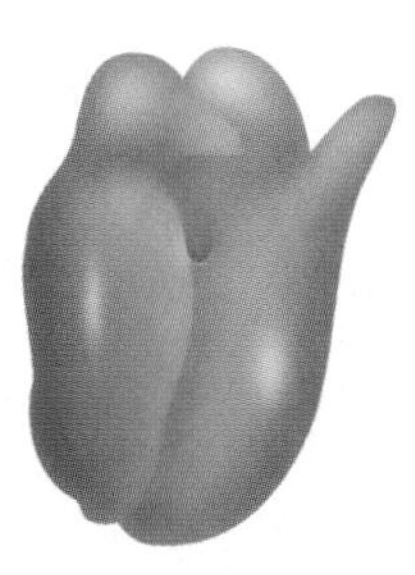

Figure 14.13 Model of eukaryotic ribosomes. Polypeptide chains are assembled on the platform of the small ribosomal subunit. Newly forming chains may move through the tunnel of the large ribosomal subunit.

the codons, they automatically position their attached amino acids in the order specified by mRNA.

A cell has a cytoplasmic pool of sixty-four kinds of codons, but it is able to utilize fewer kinds of tRNAs. How do tRNAs match up with more than one type of codon? According to base-pairing rules, adenine must pair with uracil, and cytosine with guanine. For codon–anticodon interactions, however, the rules loosen up for the first and third bases. To give one example, AUU, AUC, and AUA specify isoleucine. All three of these codons can pair with a single type of tRNA that carries isoleucine. Such freedom in codon–anticodon pairing at a base is known as the "wobble effect."

Even before anticodons interact with the codons of an mRNA strand, that strand must bind to specific parts of the surface of ribosomes. As shown in Figure 14.13, each ribosome has two subunits. These are assembled inside the nucleus from rRNA and protein components, some of which show enzyme activity.

At some point the subunits are shipped separately to the cytoplasm. There, intact, functional ribosomes are put together, each from two subunits—but only when messages encoded in mRNA are to be translated.

The nucleotide sequence of both DNA and mRNA encodes protein-building instructions. The genetic code is a set of sixty-four base triplets, which are nucleotide bases read in blocks of three. A codon is a base triplet in mRNA.

Different combinations of codons specify the amino acid sequence of different polypeptide chains, start to finish.

mRNAs are the only molecules that carry protein-building instructions from DNA into the cytoplasm.

tRNAs deliver amino acids to ribosomes and base-pair with codons in the order specified by mRNA. Their action translates mRNA into a sequence of amino acids.

rRNAs are components of ribosomes, the structures upon which amino acids are assembled into polypeptide chains.

14.4 HOW IS mRNA TRANSLATED?

Stages of Translation

The protein-building code built into mRNA transcripts of DNA becomes translated at intact ribosomes in the cytoplasm. Translation proceeds through three stages: initiation, elongation, and termination.

During the stage called *initiation*, a particular tRNA that can start transcription and an mRNA transcript are both loaded onto a ribosome. First, the initiator tRNA binds with the small ribosomal subunit. AUG, the start codon for the transcript, matches up with this tRNA's anticodon. At the same time, the AUG binds with the small subunit. Second, a large ribosomal subunit binds with the small subunit. When joined together this way, the three form an initiation complex (Figure 14.14*b*). Now the next stage can begin.

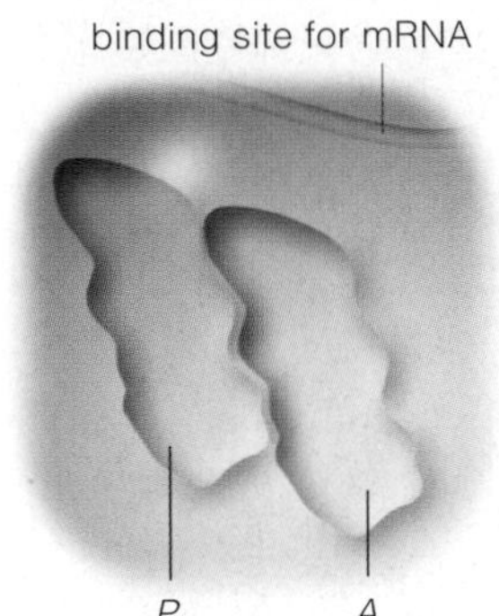

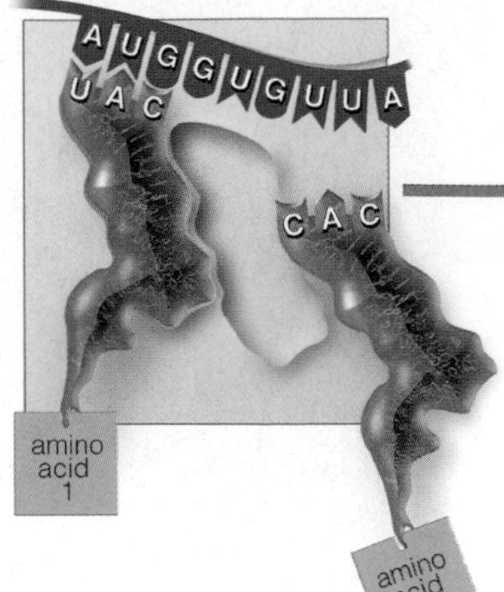

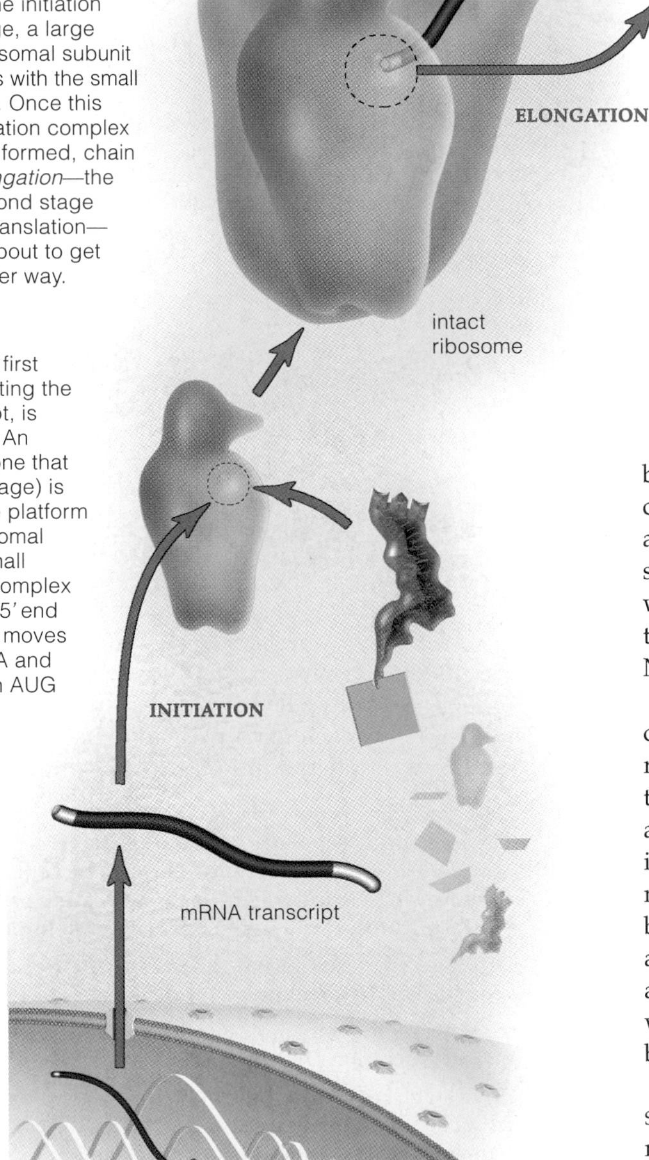

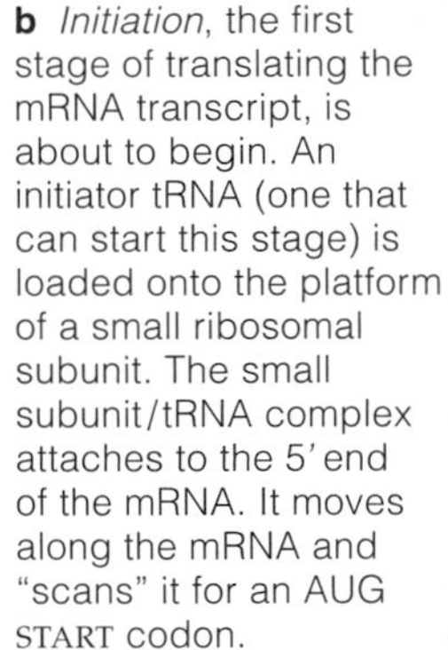

a A mature mRNA transcript leaves the nucleus by passing through pores across the nuclear envelope. Thus it enters the cytoplasm, which contains pools of many free amino acids, tRNAs, and ribosomal subunits.

b *Initiation*, the first stage of translating the mRNA transcript, is about to begin. An initiator tRNA (one that can start this stage) is loaded onto the platform of a small ribosomal subunit. The small subunit/tRNA complex attaches to the 5′ end of the mRNA. It moves along the mRNA and "scans" it for an AUG START codon.

c As the final step of the initiation stage, a large ribosomal subunit joins with the small one. Once this initiation complex has formed, chain *elongation*—the second stage of translation—is about to get under way.

d This close-up of the small ribosomal subunit's platform shows the relative positions of binding sites for an mRNA transcript and for tRNAs that deliver amino acids to the intact ribosome.

e The initiator tRNA has become positioned in the first tRNA binding site (designated *P*) on the ribosome platform. Its anticodon matches up with the START codon (AUG) of the mRNA, which also has become positioned in *its* binding site. Another tRNA is about to move into the platform's second tRNA binding site (designated *A*). It is one that can bind with the codon following the START codon.

Figure 14.14 Translation, the second step of protein synthesis.

In the *elongation* stage of translation, a polypeptide chain is assembled as the mRNA passes between two ribosomal subunits, like a thread being moved through the eye of a needle. Some components of the ribosomes are enzymes. They join individual amino acids together in a sequence dictated by the sequence of codons in the mRNA molecule. Figure 14.14*f–i* shows how a peptide bond forms between the most recently attached amino acid of a growing polypeptide chain and the next amino acid delivered to the intact ribosome. Here you might wish to refer to the description and sketch of peptide bond formation in Section 3.4 (Figure 3.15).

During the last stage of translation, *termination*, a STOP codon in the mRNA moves onto the platform, and no tRNA has a corresponding anticodon. Now proteins called release factors bind to the ribosome. They trigger enzyme activity that detaches the mRNA *and* the chain from the ribosome (Figure 14.14*j–l*).

Further reading: Student Guide to InfoTrac on web site

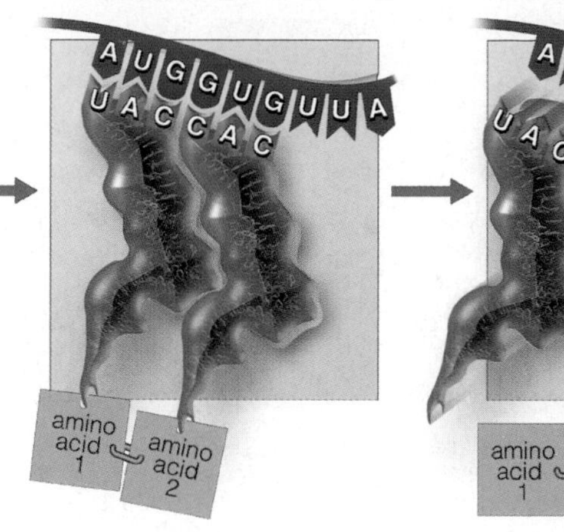

f Enzyme action breaks the bond between the initiator tRNA and the amino acid hooked to it. At the same time, enzyme action catalyzes the formation of a peptide bond between that amino acid and the one hooked to the second tRNA. Then the initiator tRNA is released from the ribosome.

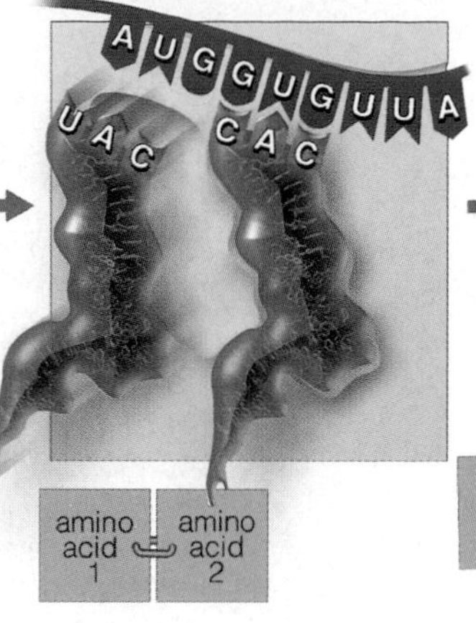

g Now the first amino acid is attached only to the second one—which is still hooked to the second tRNA. This tRNA is about to move into the ribosomal platform's *P* site and slide the mRNA along with it by one codon. This will align the third codon in the *A* site.

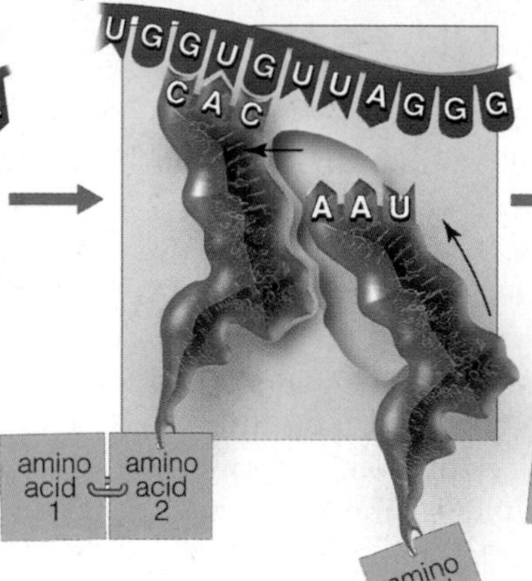

h A third tRNA is about to move into the vacated *A* site. Its anticodon is able to base-pair with the third codon of the mRNA transcript. Enzymes will now catalyze the formation of a peptide bond between amino acids 2 and 3.

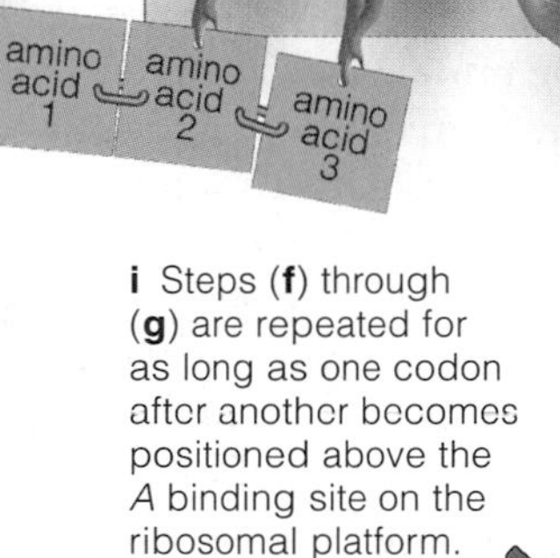

i Steps (**f**) through (**g**) are repeated for as long as one codon after another becomes positioned above the *A* binding site on the ribosomal platform.

What Happens to the New Polypeptides?

Unfertilized eggs and other cells that will be called upon to rapidly synthesize many copies of different proteins usually stockpile mRNA transcripts in their cytoplasm. In cells that are rapidly using or secreting proteins, you often observe polysomes. Each polysome is a cluster of many ribosomes translating one mRNA transcript at the same time. The transcript threads through all of them, one after another, like the thread of a pearl necklace.

After new polypeptide chains are synthesized, many join the cytoplasmic pool of free proteins. Many others enter the ribosome-studded, flattened sacs of rough ER, which is part of the cytomembrane system (Section 4.4). There they take on final form before they are shipped to their ultimate destinations inside or outside the cell.

Translation is initiated when a small ribosomal subunit and an initiator tRNA arrive at an mRNA's START codon and a large ribosomal subunit binds to them.

tRNAs deliver amino acids to the ribosome in the order dictated by the sequence of mRNA codons, to which the tRNA anticodons base-pair. A polypeptide chain lengthens as peptide bonds form between the amino acids.

Translation ends when a STOP codon triggers events that cause the chain and mRNA to detach from the ribosome.

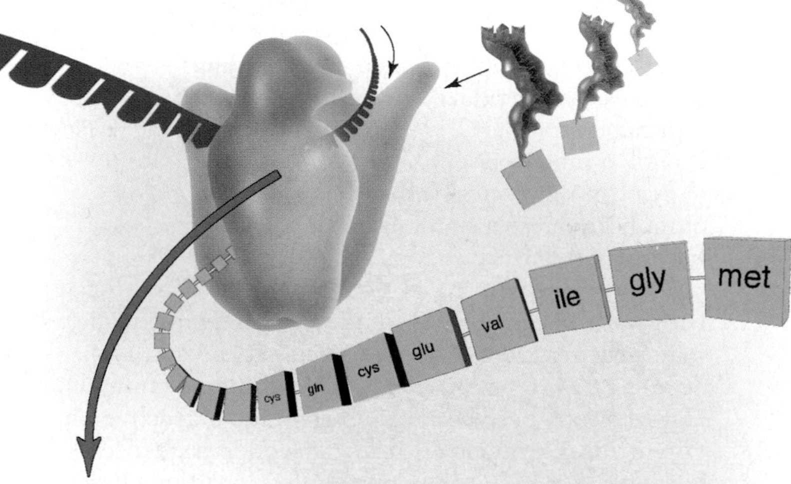

TERMINATION

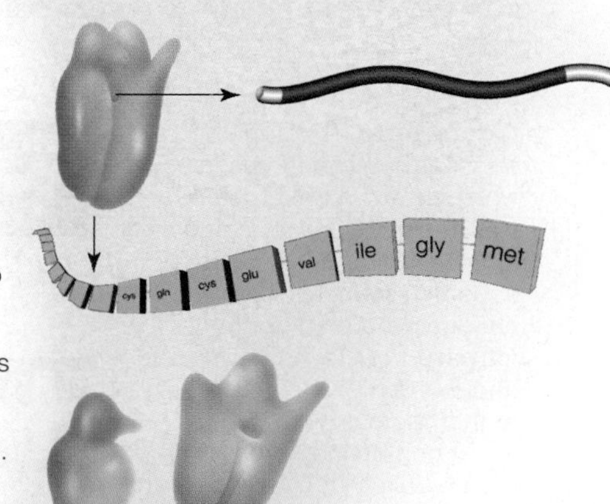

j A STOP codon moves onto the ribosomal assembly platform. It is the signal to release the mRNA transcript from the ribosome.

k The newly formed polypeptide chain also is released from the ribosome. It is free to join the pool of proteins in the cytoplasm or to enter rough ER of the cytomembrane system.

l The two ribosomal subunits separate.

14.5 DO MUTATIONS AFFECT PROTEIN SYNTHESIS?

Whenever a cell puts its genetic code into action, it is making precisely those proteins that it requires for its structure and functions. If something changes a gene's code words, the resulting protein may change, also. If the protein is central to cell architecture or metabolism, we can expect the outcome to be an abnormal cell.

Gene sequences do change. Sometimes one base gets substituted for another in the nucleotide sequence. At other times, an extra base is inserted into the sequence or a base is lost from it. Such small-scale changes in the nucleotide sequence of genes in the DNA molecule are **gene mutations**. There is some leeway here; remember, more than one codon may specify the same amino acid. For example, if a mutation were to change UCU to UCC, it probably would not have dire effects, for both codons specify serine. More often, however, mutations give rise to proteins with altered or lost functions.

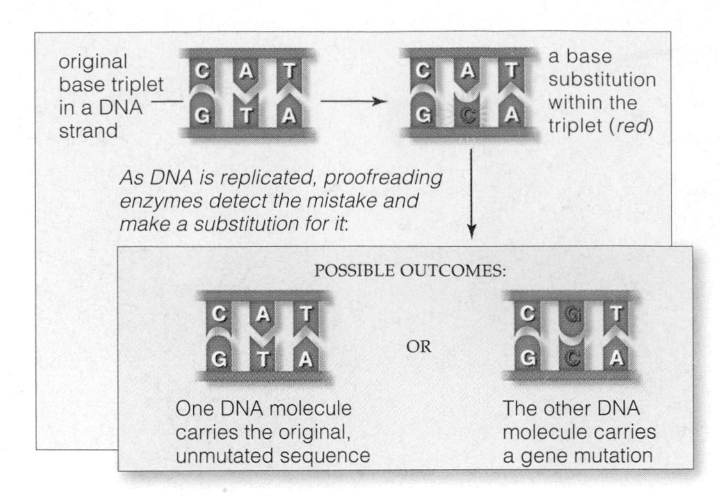

a Example of a base-pair substitution

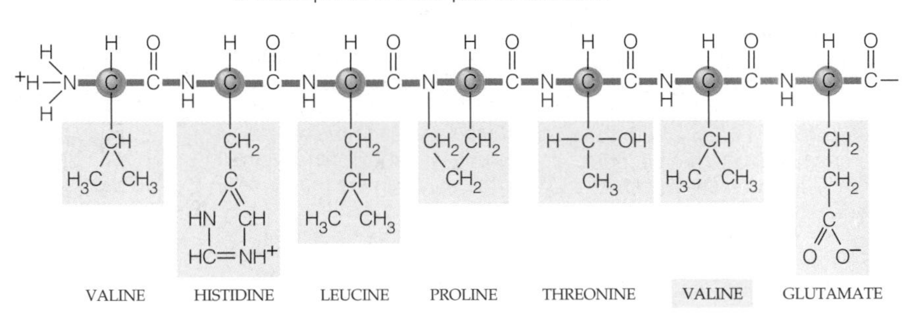

b Outcome of the base-pair substitution

Figure 14.15 Common types of mutations. (**a**) One example of a base-pair substitution. (**b**) Remember Figure 14.5? The base-pair substitution shown here is the type of molecular change that caused a different amino acid (valine, not glutamate) to be substituted in beta chains of hemoglobin.

Common Gene Mutations and Their Sources

Figure 14.15 gives an example of a common gene mutation. Here, one base (adenine) is wrongly paired with another base (cytosine) when DNA is being replicated. Proofreading enzymes can recognize an error in a newly replicated strand of DNA and fix it. If they do not, a mutation will become established in one DNA molecule during the next round of replication. As a result of this mutation, a **base-pair substitution**, one amino acid can replace another during protein synthesis. This is what happens, recall, in people who carry Hb^S, a mutant allele that causes sickle-cell anemia (Sections 11.5 and 14.1).

Figure 14.16 has an example of a different mutation. Here, an *extra* base became inserted into a gene region. Polymerases, remember, read nucleotide sequences in blocks of three. This insertion shifted the "three-bases-at-a-time" reading frame by one base; hence the name, *frameshift* mutation. The affected gene has a different message, and an altered version of the protein will be synthesized. Frameshift mutations fall within broader categories of gene mutation known as **insertions** and **deletions**. In such cases, one to several base pairs are inserted into a DNA molecule or deleted from it.

As a final example, Barbara McClintock discovered that mutations may result when transposable elements,

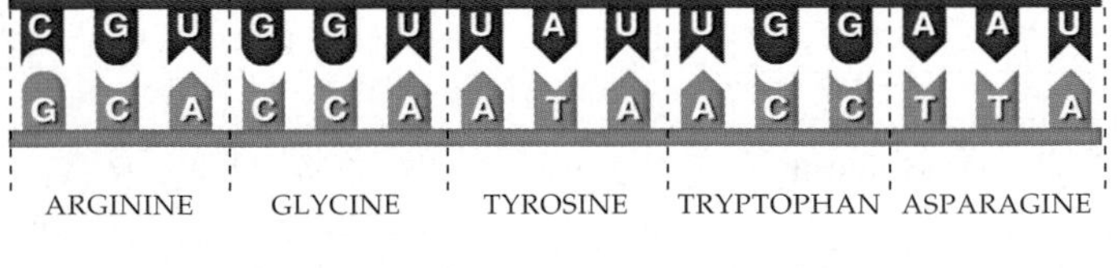

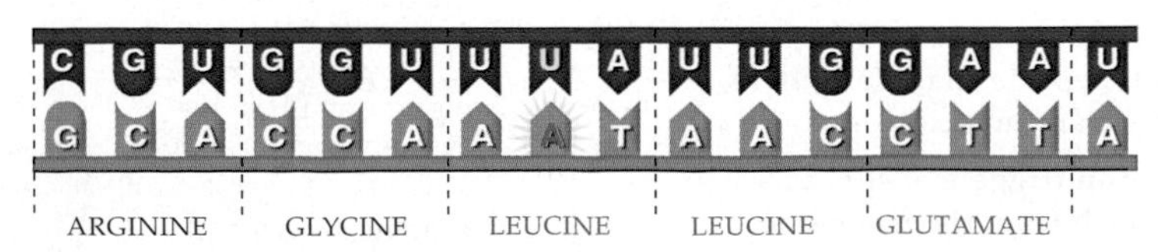

Figure 14.16 Example of an insertion, a mutation in which an extra base gets inserted into a gene region of DNA. This insertion has caused a *frameshift*; it has changed the reading frame for base triplets in the DNA and in the mRNA transcript of that region. As a result, the wrong amino acids will be called up when the mRNA transcript becomes translated into protein.

Further reading: Student Guide to InfoTrac on web site

Figure 14.17 Barbara McClintock, who won a Nobel Prize for her meticulous research and eventual insight that some DNA segments can slip into and out of different locations in DNA molecules. We call these segments transposons. In her hands is an ear of Indian corn (*Zea mays*). The curiously nonuniform coloration of its kernels sent her on the road to discovery.

Each corn kernel is a seed, which has the potential to grow into a new corn plant. All of its cells have the same pigment-coding genes. However, some of the kernels are colorless or spottily colored. In the ancestor of the plant from which this ear of corn was plucked, a gene in a germ cell left its position in a DNA molecule, invaded another DNA molecule, and shut down a pigment-encoding gene.

The plant inherited the mutation. As cell divisions proceeded in the growing plant, none of the mutated cell's descendants was able to synthesize pigment molecules. Each gave rise to colorless kernel tissue. Later, in some cells, the movable DNA slipped out of the pigment-encoding gene. All the descendants of *those* cells produced pigment—and colored kernel tissue.

or **transposons**, are on the move. These DNA segments move spontaneously from one location to another in the same DNA molecule or to a different one. Often they inactivate genes into which they become inserted. Their unpredictability can cause interesting variations in traits. Figure 14.17 and *Critical Thinking* question 5 at the end of this chapter provide two examples.

Causes of Gene Mutations

Many mutations arise spontaneously as DNA is being replicated. This should not come as a surprise, given the swift pace of replication and the huge pools of free nucleotides concentrated near the growing DNA strands. Proofreading and repair enzymes detect and fix most of them. However, a low number of mistakes do slip past those enzymes with predictable frequency.

Each gene has a characteristic **mutation rate**, which is the probability it will mutate spontaneously during a specified interval, such as each DNA replication cycle. (This is not the same as mutation *frequency*, the number of times a gene mutation has occurred in a population, as in 1 million gametes that produced 500,000 people.)

Not all mutations are spontaneous. Many result after exposure to mutagens (mutation-causing agents in the environment). Ultraviolet radiation, especially the 260-nanometer wavelength in sunlight, is mutagenic. This is the wavelength DNA absorbs most strongly, and it may induce crosslinks to form between two pyrimidine neighbors on the same DNA strand. Skin cancers are one outcome. Other mutagens are gamma rays and x-rays. They can ionize water and other molecules around the DNA, so free radicals form. Free radicals are molecular fragments having an unpaired electron—and they can attack DNA's structure. The next section takes a closer look at the effects of such **ionizing radiation**.

Natural and synthetic chemicals in the environment can accelerate the rate of spontaneous mutations. For example, substances called **alkylating agents** transfer methyl or ethyl groups to reactive sites on the bases or phosphate groups of DNA. At an alkylated site, DNA becomes more susceptible to base-pair disruptions that invite mutation. Many **carcinogens**, which are cancer-causing agents, operate by alkylating the DNA.

The Proof Is in the Protein

Spontaneous mutations are rare in terms of a human life. (The rate for eukaryotes ranges between 10^{-4} and 10^{-6} per gene per generation.) If one arises in a somatic cell, any good or bad consequences will not endure, for it cannot be passed on to offspring. If the mutation arises in a germ cell or gamete, however, it may enter the evolutionary arena. The same can happen with a mutation in an asexually reproducing organism or cell.

In all such cases, nature's test is this: *A protein that is specified by a heritable mutation may have harmful, neutral, or beneficial effects on an individual's ability to function in the prevailing environment.* As you will read in the next unit of the book, the outcomes of gene mutations can have powerful evolutionary consequences.

A gene mutation is an alteration in one to several bases in the nucleotide sequence of DNA. The most common are base-pair substitutions, base insertions, and base deletions.

Each gene has a spontaneous and characteristic mutation rate, which may be accelerated by exposure to harmful radiation and certain chemicals in the environment.

A protein specified by a mutated gene may have harmful, neutral, or beneficial effects on the ability of an individual to function in the prevailing environment.

Focus on Health

14.6 MUTAGENIC RADIATION

Each day, radiation bombards the human body. However, only two categories of radiation are mutagenic; they can cause mutations.

Protons, neutrons, x-rays, and gamma rays are among the forms of *ionizing* radiation. Their wavelengths are highly energetic, and they can damage DNA that takes a direct hit. More commonly, their effects are indirect. When such high-energy rays strike molecules—including water molecules that bathe the cell's organic compounds—they often strip its atoms of an electron. In this way the atom becomes an ion (hence the name, ionizing radiation). The unpaired electron left behind is highly reactive, and its molecular owner has become a free radical.

When free radicals collide with other molecules, they can create new free radicals. When they collide with the phosphate bonds in DNA, they can break them. A break in one strand of double-stranded DNA doesn't amount to much, because repair enzymes easily patch things up. When both strands break, repair enzymes often fail to restore the original base sequence. It is not the ionizing radiation itself that directly causes mutation. Rather, its agents are the free radicals that it induces to form.

Ionizing radiation is potentially quite dangerous. It can deeply penetrate living tissue, in which case a trail of free radicals marks its passage. Think about this when your doctor or dentist recommends an x-ray examination. X-rays are especially useful for diagnosing medical and dental problems. Because they pass more easily through some tissues than others, they can be used to produce an image of the body's interior on photographic film. The patient's exposure time is usually brief, and the dosage is extremely low to minimize the risk of mutations and radiation-induced disease.

But ionizing radiation has a cumulative mutagenic effect. Repeated exposure to low levels over many years can cause problems. That is why x-ray technicians use lead aprons and shielding booths. With respect to long-term exposure, there may be no safe level of radiation.

Unlike ionizing radiation, *nonionizing* radiation is not energetic enough to invite the formation of free radicals. Rather, it simply excites electrons enough to boost them to a higher energy level. Nonionizing radiation can only penetrate single-celled organisms (bacteria and protistans) and the outermost layers of living cells of multicelled organisms.

Ultraviolet (UV) light is the most significant form of mutagenic nonionizing radiation. The nucleotides of DNA easily absorb it. Maximum absorption—and maximum mutagenesis—occurs at a wavelength of 254 nanometers. Cytosine and thymine are particularly vulnerable to excitation; they may be converted to lesions with altered base-pairing properties. The introduction to Chapter 15 describes one such lesion—a pyrimidine dimer. If left unrepaired, such lesions can cause mutations in a newly synthesized DNA strand during replication.

14.7 SUMMARY

1. Cells, and multicelled organisms, cannot stay alive without enzymes and other proteins. A protein consists of one or more polypeptide chains, each of which is composed of a linear sequence of amino acids.

a. An amino acid sequence of a polypeptide chain corresponds to a gene region in a DNA molecule. Each gene is a sequence of nucleotide bases in one of the DNA molecule's two strands. The bases are A, T, G, and C (adenine, thymine, guanine, and cytosine).

b. For most genes, that sequence corresponds to a linear sequence of specific amino acids for a particular polypeptide chain. (Some genes specify tRNA or rRNA, not the mRNA that is translated into proteins.)

c. Understanding of the connection between genes and proteins started with studies of gene mutations that affected enzymes known to catalyze steps in metabolic pathways. Comparisons between normal and abnormal proteins, hemoglobin especially, led to the hypothesis that the amino acid sequence of polypeptide chains is encoded in genes.

2. The biochemical path from genes to proteins has two steps, called transcription and translation:

DNA —*transcription*→ RNA —*translation*→ PROTEIN

a. During transcription, the double-stranded DNA is unwound at a gene region. Enzymes use its exposed bases as a template, or a structural pattern, to assemble a strand of ribonucleic acid (RNA) from the cell's pool of free nucleotides.

b. During translation, three classes of RNAs interact in the synthesis of polypeptide chains, which later will twist, fold, and often become modified into the final, three-dimensional shape of the protein.

c. Figure 14.18 is a visual summary of this flow of genetic information from DNA to proteins, as it occurs in eukaryotic cells. The DNA is transcribed into RNA in the nucleus, but RNA is translated in the cytoplasm. Prokaryotic cells (bacteria) lack a nucleus; transcription and translation proceed in their cytoplasm.

3. Here are the key points concerning transcription:

a. When RNA is transcribed from exposed bases of DNA, base-pairing rules govern its assembly. Guanine pairs with cytosine, as in DNA replication, but uracil (not thymine) pairs with adenine in RNA:

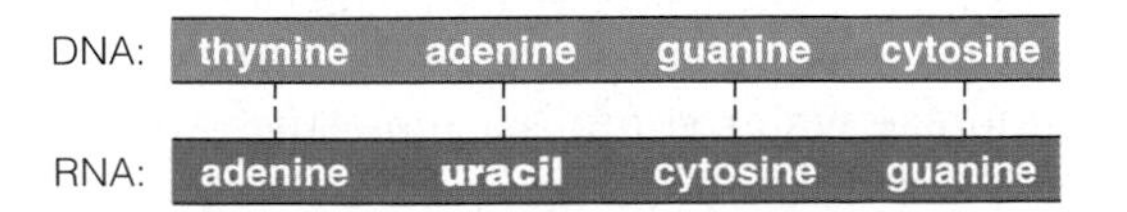

b. Different gene regions in DNA serve as templates for assembling different RNA molecules.

Further reading: Student Guide to InfoTrac on web site →

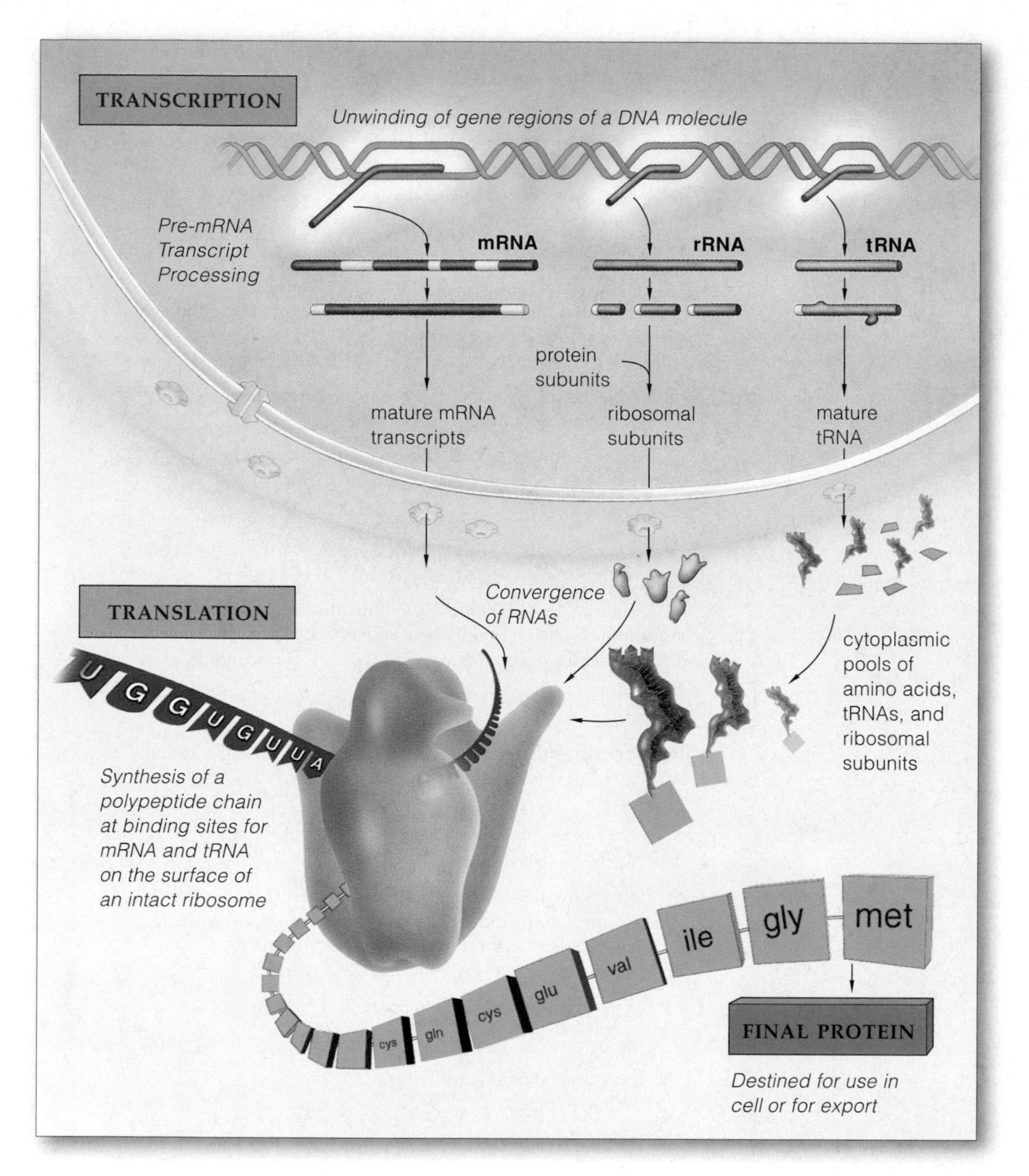

Figure 14.18 Visual summary of protein synthesis—transcription and translation—as it proceeds in eukaryotic cells. All three classes of RNA are assembled in the nucleus and shipped to the cytoplasm. In prokaryotic cells, transcription as well as translation proceeds in the cytoplasm.

c. Messenger RNA (mRNA) is the only class of RNA that carries protein-building instructions.

d. A ribosome has two subunits made of ribosomal RNA (rRNA) and other factors. Each is a physical site where polypeptide chains are assembled. Its subunits are built in the nucleus (or, in prokaryotic cells, in the cytoplasm). They meet up only in the cytoplasm.

e. Transfer RNA (tRNA) is the vehicle of translation; it will latch on to free amino acids in the cytoplasm and deliver them to ribosomes. It will do so in a sequence that corresponds to the sequential message of mRNA.

f. RNA transcripts of eukaryotic cells are processed into final form before being shipped from the nucleus. For example, mRNA's noncoding portions (introns) are excised, so only its coding portions (exons) get spliced together. Only mature mRNA transcripts are translated into protein.

4. Here are the key points about translation:

a. mRNA interacts with many tRNAs and ribosomes in ways that assemble amino acids in a sequence that produces a specific kind of polypeptide chain.

b. Translation is based on the genetic code: a set of sixty-four base triplets. A triplet is a series of nucleotide bases that ribosomal proteins "read" in blocks of three.

c. One base triplet in mRNA is called a codon. An anticodon is a complementary triplet in tRNA. Some combination of different codons specifies what the amino acid sequence of a polypeptide chain will be, start to finish.

5. Translation proceeds through the following three stages:

a. Initiation stage. One small ribosomal subunit, one initiator tRNA, and one mRNA transcript converge. The small subunit binds reversibly with a large ribosomal subunit.

b. Chain elongation stage. A number of tRNAs deliver many amino acids to the ribosome. Their anticodons base-pair with codons in mRNA. The amino acids become joined one after another by peptide bonds to form a new polypeptide chain.

c. Chain termination. A STOP codon in mRNA causes the chain and the mRNA to detach from the ribosome.

6. Gene mutations are potentially heritable, small-scale alterations in the nucleotide sequence of DNA.

a. Many gene mutations arise spontaneously during DNA replication. Others arise after the DNA is exposed to mutagens, such as ultraviolet radiation and other mutation-causing agents in the environment.

b. A base-pair substitution (one base replaces a base of a different kind) changes only one codon, but this may cause an amino acid substitution that alters protein function. Insertions of one or more bases into a gene or deletions from it can shift the reading frame to specify different amino acids. Transposons (movable elements) may inactivate genes into which they become inserted.

c. Each gene has a characteristic mutation rate: the probability that it will spontaneously mutate in some specified time interval, such as a DNA replication cycle.

7. A protein specified by a mutant gene might have harmful, neutral, or beneficial effects on the individual. The outcome depends on prevailing conditions in the internal and external environments. Somatic mutations affect individuals only. Mutations in reproductive cells are heritable and can enter the evolutionary arena.

Review Questions

1. Are polypeptide chains assembled on DNA? If so, state how. If not, state how and where they are assembled. *CI, 14.1*

2. Define gene transcription and translation, the two stages of events by which polypeptide chains of proteins are synthesized. Both stages proceed in the cytoplasm of prokaryotic cells. Where does each stage proceed in eukaryotic cells? *CI*

3. Briefly state how gel electrophoresis, a common laboratory procedure, works. Explain how it provided a clue that small differences in normal and abnormal versions of the same protein may lead to big differences in how the proteins function. *14.1*

4. Name the three classes of RNA and briefly describe their functions. *14.2, 14.3*

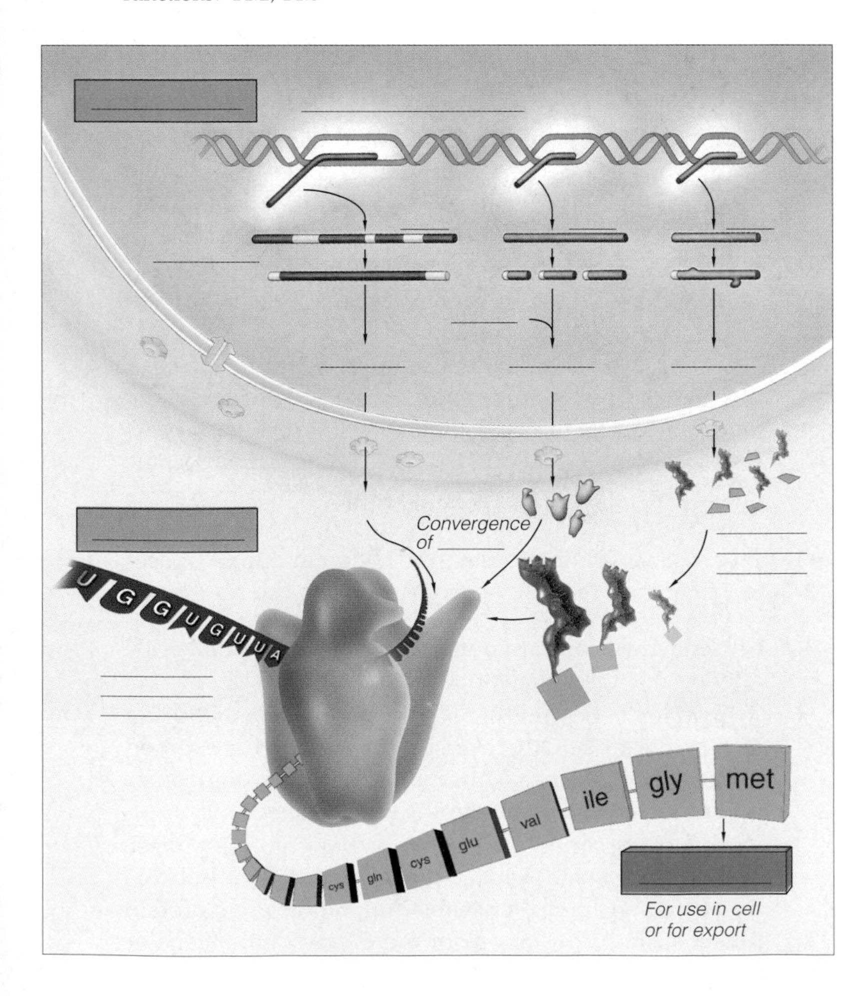

5. In what key respect does the sequence of nucleotide bases in RNA differ from those in DNA? *14.2*

6. How does the process of gene transcription resemble DNA replication? How does it differ from DNA replication? *14.2*

7. Pre-mRNA transcripts of eukaryotic cells contain introns and exons. Are the introns or exons snipped out before the transcript leaves the nucleus? *14.2*

8. Distinguish between codon and anticodon. *14.3*

9. Cells use the set of sixty-four codons in the genetic code to build polypeptide chains from twenty kinds of amino acids. Are some amino acids specified by different codons? If so, in what respect do the codons differ? *14.3*

10. Name the three stages of translation and briefly describe the key events of each one. *14.4*

11. Review Figure 14.18 on the preceding page. Then, on your own, fill in the blanks of the diagram at lower left.

12. Define gene mutation. *14.5*

13. Do all mutations arise spontaneously? Do environmental agents trigger change in each case? *14.5*

14. Define and state the possible outcomes of the following mutations: a base-pair substitution, a base insertion, and an insertion of a transposon at a new DNA location. *14.5*

15. Define and explain the difference between mutation rate and mutation frequency. *14.5*

16. What determines whether an altered product of a mutation will have helpful, neutral, or harmful effects? *14.5*

Self-Quiz (*Answers in Appendix III*)

1. Garrod, then Beadle and Tatum, deduced that the individual, heritable mutations they investigated corresponded to ______ .
   a. metabolic disorders
   b. defective enzymes
   c. defective tRNAs
   d. both a and b

2. DNA's many genes are transcribed into different ______ .
   a. proteins
   b. mRNAs only
   c. mRNAs, tRNAs, and rRNAs
   d. All are correct.

3. An RNA molecule is ______ .
   a. a double helix
   b. usually single-stranded
   c. always double-stranded
   d. usually double-stranded

4. An mRNA molecule is produced by ______ .
   a. replication
   b. duplication
   c. transcription
   d. translation

5. Each codon calls for a specific ______ .
   a. protein
   b. polypeptide
   c. amino acid
   d. carbohydrate

6. Referring to Figure 14.11, use the genetic code to translate the mRNA sequence UAUCGCACCUCAGGAGACUAG. Notice that the first codon in the frame is UAU. Which of the following three amino acid sequences is being specified?

   a. TYR—ARG—THR—SER—GLY—ASP—

   b. TYR—ARG—THR—SER—GLY

   c. TYR—ARG—TYR—SER—GLY—ASP—

7. Anticodons pair with ______.
   a. mRNA codons c. tRNA anticodons
   b. DNA codons d. amino acids

8. Match the terms with the suitable description.

____ alkylating agent	a. part remaining in mRNA transcript
____ chain elongation	b. base triplet coding for amino acid
____ exon	c. second stage of translation
____ genetic code	d. base triplet that pairs with codon
____ anticodon	e. one environmental agent that induces mutation in DNA
____ intron	f. set of sixty-four codons for mRNA
____ codon	g. part removed from a pre-mRNA transcript

Critical Thinking

1. Sandra discovered a tRNA with a mutation in DNA that encodes the anticodon 3′-AAU instead of 3′-AUU. In cells with the mutated tRNA, what will be the effect on protein synthesis?

2. A DNA polymerase made an error during the replication of an important gene region of DNA. None of the DNA repair enzymes detected or repaired the damage. A portion of the DNA strand with the error is shown here:

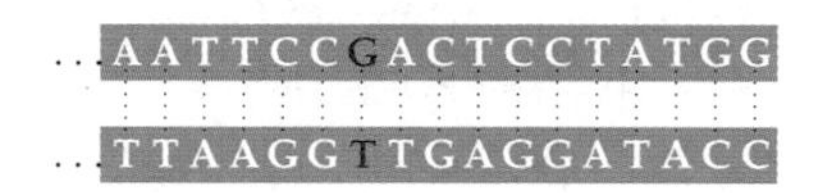

After the DNA molecule is replicated and two daughter cells have formed, one cell is carrying a mutation and the other cell is normal. Develop a hypothesis to explain this observation.

3. In the bacterium *E. coli*, the end product (E) of the following metabolic pathway is absolutely essential for life:

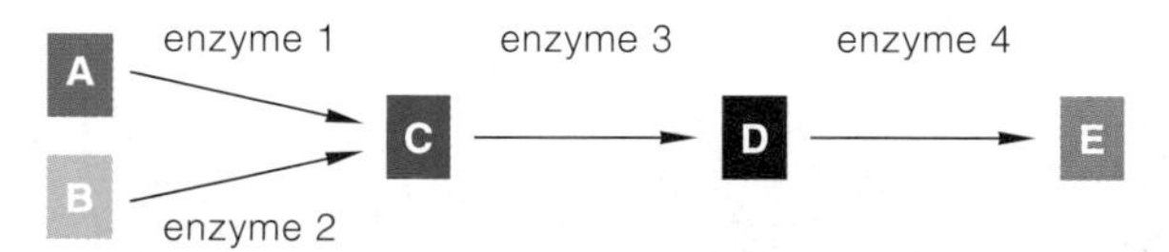

Ginetta, a geneticist, is attempting to isolate mutations in the genes for the four enzymes of this pathway. She has been able to isolate mutations in the genes for enzymes 1 and 2. In each case, the mutant *E. coli* cells synthesize a reduced amount of the pathway's end product E. However, Ginetta has not been able to isolate *E. coli* cells that have mutations in the genes for enzymes 3 and 4. Develop a hypothesis to explain why.

4. List all possible codons in the genetic code that could be changed into a STOP codon by a single nucleotide substitution.

5. *Neurofibromatosis* is a human autosomal dominant disorder caused by mutations in the *NF1* gene. It is characterized by soft, fibrous tumors in the peripheral nervous system and skin as well as abnormalities in muscles, bones, and internal organs (Figure 14.19).

Because the gene is dominant, an affected child usually has an affected parent. Yet in 1991, scientists reported on a boy who had neurofibromatosis yet whose parents did not. When they examined both copies of his *NF1* gene, they found the copy he had inherited from his father contained a transposon, a DNA segment that had moved to a new location. Neither the father nor the mother had a transposon in any of the copies of their own *NF1* genes. Speculate on the cause of neurofibromatosis in the human body and how it may have arisen.

6. The genome of a species is all the DNA in a haploid number of its chromosomes. The human genome has about 3.2 billion base pairs. But half of that may consist of transposons, which some researchers call *parasitic DNA*.

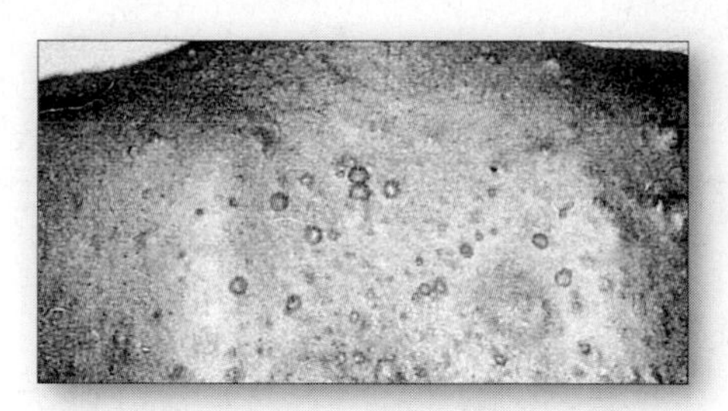

Figure 14.19 Soft skin tumors on a person with neurofibromatosis, a type of genetic disorder.

Some transposons contain instructions for synthesizing an enzyme that cuts the transposon out of the DNA and inserts it into a new location. *Retrotransposons* are more complex. These gene segments contain instructions for a reverse transcriptase, a type of enzyme that uses RNA as a template to make DNA. That DNA is a movable duplicate of the retrotransposon. Its mode of replication is uncannily similar to that of retroviruses (Section 22.8). Many biologists suspect that the first retroviruses were retrotransposons that left one species for a new one, maybe after inserting themselves into the DNA of a virus.

Either way, parasitic DNA can be duplicated hundreds of times and passed on through generations. As it moves in and out of a genome, it can switch genes on or off, maybe at unsuitable times or in the wrong cells. John McDonald of the University of Georgia argues that their destabilizing effects represent threats to survival. Over evolutionary time, natural selection may have worked at the molecular level to silence these genetic parasites.

Think about how eukaryotic chromosomes have many more genes and greater structural complexity than bacterial DNA. In the distant past, did some chromosomal proteins start to interact to isolate parasitic DNA from vital genes? Was this the impetus for the evolution of controls over complex genomes?

Or think about how vertebrate DNA is highly methylated. Methyl groups (alkylating agents) tend to shut down genes, and they are pervasive in vertebrate genomes. Did methylation first evolve as a way to control invading transposons? Was that control later extended to control vital genes?

If you find these ideas provocative, you may opt to continue your reading on this subject. Start with Ayala Ochert's highly readable article "Transposons" (*Discover*, December 1999).

Selected Key Terms

alkylating agent *14.5*
anticodon *14.3*
base sequence *CI*
base-pair substitution *14.5*
carcinogen *14.5*
codon *14.3*
deletion (of base) *14.5*
exon *14.2*
gel electrophoresis *14.1*
gene mutation *14.5*
genetic code *14.3*
insertion (of base) *14.5*
intron *14.2*
ionizing radiation *14.5*
mRNA (messenger RNA) *14.2*
mutation rate *14.5*
promoter (RNA) *14.2*
ribonucleic acid (RNA) *CI*
RNA polymerase *14.2*
rRNA (ribosomal RNA) *14.2*
transcription *CI*
translation *CI*
transposon *14.5*
tRNA (transfer RNA) *14.2*
uracil *14.2*

Readings *See also www.infotrac-college.com*

Crick, F. 1988. *What Mad Pursuit: A Personal View of Scientific Discovery*. New York: HarperCollins. Crick's autobiography.

Fairbanks, F., and W. R. Andersen. 1999. *Genetics: The Continuity of Life*. Monterey, California: Brooks/Cole.

15

CONTROLS OVER GENES

When DNA Can't Be Fixed

1992 was an unforgettable year for Laurie Campbell. She finally turned eighteen. In that same year she just happened to notice a suspiciously black mole on her skin. It had an odd lumpiness about it, a ragged border, and an encrusted surface. No fool, Laurie quickly made an appointment with her family doctor, who quickly ordered a biopsy. The mole turned out to be a *malignant melanoma*, the deadliest form of skin cancer.

Laurie was lucky. She detected a cancer in its earliest stage, before it could spread through her body. Now she regularly checks out the appearance of other moles on her skin. She is aware of having become a statistic—one of 500,000 people in the United States who develop skin cancer in any given year and one of the 23,000 with malignant melanoma. She knows now that, of 7,500 who will die each year from skin cancer, 5,600 have malignant melanoma.

Laurie is smart. She plotted out the position of every mole on her body. Once a month, that body map is her guide for a quick but thorough self-examination. Figure 15.1 shows examples of what she looks for. Laurie also schedules a medical examination every six months.

Changes in DNA trigger skin cancer. The ultraviolet wavelengths in rays from the sun, tanning lamps, and other sources of nonionizing radiation can cause the bad molecular changes. For example, they can promote covalent bonding between two adjacent thymine bases in a nucleotide strand. The two nucleotides to which the bases belong become an abnormal, bulky structure—a thymine dimer—within the DNA. Normally, at least seven gene products interact as a DNA repair mechanism to remove such bulky lesions. But mutation in one or more of the genes can disrupt the repair machinery. Thymine dimers can accumulate in skin cells and trigger cancers and other lesions.

thymine dimer

The risk of skin cancer is greater for some than others. At one extreme are people affected by the genetic disorder *xeroderma pigmentosum*. Even brief exposure to sunlight puts them at risk of developing skin tumors and dying early from cancer.

You are at risk if you have moles that are chronically irritated, as by shaving or abrasive clothing. You are at risk if your skin, including the lip surface, is chronically chapped, cracked, or sore. You are at risk if your family has a history of cancer or if you have had radiation therapy. And, like Laurie, you are at risk if you burn

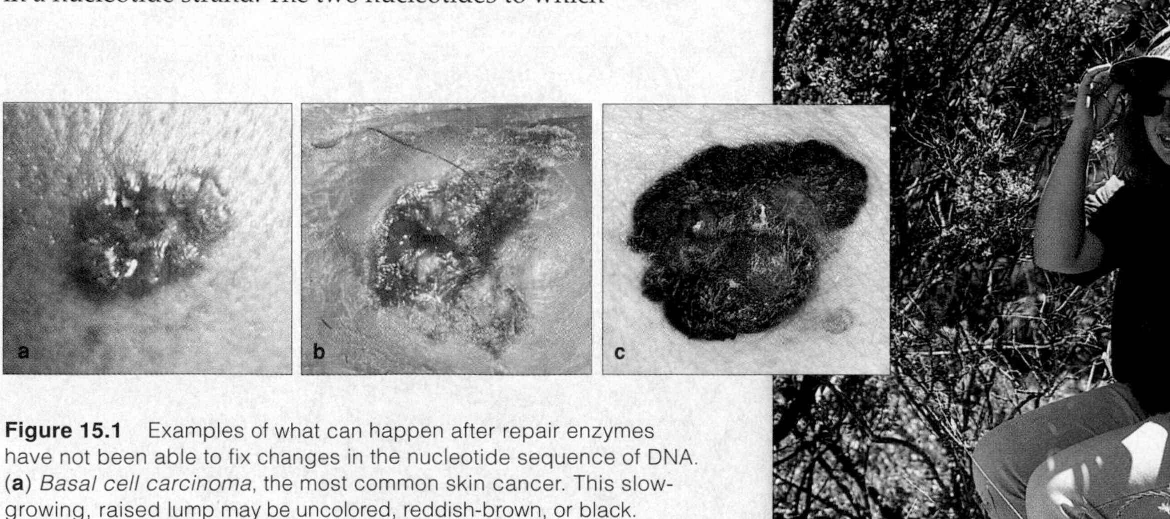

Figure 15.1 Examples of what can happen after repair enzymes have not been able to fix changes in the nucleotide sequence of DNA. (**a**) *Basal cell carcinoma*, the most common skin cancer. This slow-growing, raised lump may be uncolored, reddish-brown, or black. (**b**) *Squamous cell carcinoma*, the second most common skin cancer. The pink growths, firm to the touch, grow rapidly under the surface of skin exposed to the sun. (**c**) *Malignant melanoma*, which spreads most rapidly. The malignant cells form very dark, encrusted lumps. They may itch like an insect bite or bleed easily. (**d**) Laurie Campbell avoiding the sun—and melanoma.

Figure 15.2 A patrol of white blood cells in an encounter with a body cell that has undergone cancerous transformation.

easily. Damaged DNA is the reality for all of us in spite of the ill-advised, socially promoted allure of tanning and staying out unprotected under the sun.

By definition, all **cancers** are malignant forms of tumors, which are tissue masses that arise through mutations in the genes that govern cell growth and division. Tumor cells don't respond to normal controls; they go on dividing as long as conditions for growth remain favorable. Cells of common skin moles and other *benign* tumors grow in an unprogrammed way. But they grow slowly, and they still have the surface recognition proteins that hold them together in their home tissue. Most benign tumors are left alone unless they become overly large or irritating.

In a *malignant* tumor, the abnormal cells grow and divide more rapidly, with destructive physical and metabolic effects on surrounding tissues. The cells are grossly disfigured (Figure 15.2). They cannot construct a normal cytoskeleton or plasma membrane, and they cannot synthesize normal recognition proteins. Also, malignant cells can break loose from their home tissue, enter lymph or blood vessels, travel along, then slip out and invade other parts of the body where they do not belong. And there they may start growing as new tumors. **Metastasis** (meh-TAH-stu-SIS) is the name for the process of abnormal cell migration and tissue invasion.

This chapter is your invitation to learn about the controls that govern when and how fast genes will be transcribed and translated, and whether gene products will be activated or shut down. By starting out with cancerous transformations, it invites you to reflect on how lucky we are when proper gene controls are in place and cells are operating as they should. Each year in the developed countries alone, 15 to 20 percent of all deaths result from cancer. It is not just a human problem. Researchers also have observed cancer in most of the animal species they have studied to date.

KEY CONCEPTS

1. In cells, a variety of controls govern when, how, and to what extent genes are expressed. The control elements operate in response to changing chemical conditions and to reception of external signals.

2. Control is exerted by way of regulatory proteins and other molecules that operate before, during, or after gene transcription. Different controls interact with DNA, with RNA that has been transcribed from the DNA, or with gene products—that is, with the resulting polypeptide chains or final proteins.

3. Prokaryotic cells depend on rapid control over short-term shifts in nutrient availability and other aspects of their surrounding environment. They commonly rely on regulatory proteins that help make quick adjustments in rates of gene transcription.

4. All eukaryotic cells depend on controls over short-term shifts in diet and levels of activity. In complex multicelled species, they also depend on controls over an intricate, long-term program of growth and development.

5. Controls over eukaryotic cells come into play when new cells contact one another in developing tissues. They also come into play when those cells start interacting with their neighbors by way of hormones and other signaling molecules.

6. Although all cells of a multicelled organism inherit the same genes, different cell types activate or suppress many of those genes in different ways. The controlled, selective use of genes leads to the synthesis of the proteins that give each type of cell its distinctive structure, function, and products.

15.1 OVERVIEW OF GENE CONTROL

At this very moment, bacteria are feeding on nutrients in your gut. Red blood cells are binding, transporting, or giving up oxygen, and great numbers of epithelial cells in your skin are busily synthesizing the protein keratin. Like cells everywhere, they are functioning by virtue of the protein products of genes.

Cells are selective about which gene products they make or require. Different kinds express certain genes just once, only at certain times, all the time, or not at all. *Which genes are being expressed depends on the type of cell, its moment-by-moment adjustments to changing chemical conditions, which external signals it happens to be receiving —and its built-in control systems.*

For example, availability of nutrients shifts rapidly and often for enteric bacteria (those living in intestines). Like other prokaryotes, such bacteria rapidly transcribe genes for nutrient-digesting enzymes. They make many copies of those enzymes whenever nutrients are moving through the intestines. They cut back on synthesis of the enzymes when nutrients are scarce. By comparison, your cells do not encounter drastic shifts in the solute concentrations and composition of the fluid that bathes them, and few exhibit rapid shifts in transcription.

The "systems" that control the expression of genes consist of molecules. For instance, **regulatory proteins** influence transcription, translation, and gene products. As you will see, such molecules exert their effects by interacting with DNA, RNA, new polypeptide chains, or final proteins (such as enzymes). Some components are activated or inhibited by signaling molecules, such as hormones. Others operate in response to changing concentrations of substances outside or inside the cell.

Negative control systems block a particular activity in a cell, and **positive control systems** promote it. For instance, one regulatory protein inhibits transcription of a gene when it binds with DNA, but the action of a different regulatory protein enhances its transcription. Usually, regulatory proteins do not act alone. To give an example, some types bind to DNA when required to do so, then release their grip on the binding site when a different control element interacts with them.

Summing up, these are the main concepts to keep in mind as you read through the rest of this chapter:

Cells exert selective control over when, how, and to what extent each of their genes is expressed.

The expression of a given gene depends on the type of cell and its functions, on chemical conditions, and on signals from the outside environment.

Many regulatory proteins and other molecules exert control over gene expression through their interactions with DNA or RNA. Others exert control through their interaction with gene products—new polypeptide chains or final proteins.

15.2 CONTROL IN BACTERIAL CELLS

Let's first consider some examples of gene control in prokaryotic cells—that is, bacteria. When nutrients are plentiful and other environmental conditions also favor growth, bacteria tend to grow and divide indefinitely. Gene controls promote the rapid synthesis of enzymes that have roles in nutrient digestion and other growth-related activities. Transcription is fast, and translation is initiated even before mRNA transcripts are finished. Remember, bacteria have no nucleus; nothing separates their DNA from ribosomes in the cytoplasm.

When a nutrient-degrading pathway utilizes several enzymes, all genes for those enzymes are transcribed, often as a continuous mRNA molecule. The genes are not transcribed when conditions turn unfavorable. The rest of this section gives two cases of how controls can adjust transcription rates downward or upward.

Negative Control of Transcription

The enteric bacterium *Escherichia coli* inhabits the gut of all mammals. It survives on glucose, lactose (a sugar in milk), and other ingested nutrients. Like other adult mammals, you probably do not drink milk around the clock. When you do, however, *E. coli* cells in your gut rapidly transcribe three genes for enzymes that have roles in breakdown reactions that begin with lactose.

A promoter precedes the three genes, which are next to one another. A **promoter**, recall, is a base sequence that signals the start of a gene. A different sequence, an **operator**, intervenes between a promoter and bacterial genes. It is a binding site for a **repressor**, a regulatory protein that can block transcription (Figure 15.3). An arrangement in which a promoter and operator service more than one gene is an **operon**. Elsewhere in *E. coli* DNA, a different gene codes for this repressor, which can bind with the operator *or* with a lactose molecule.

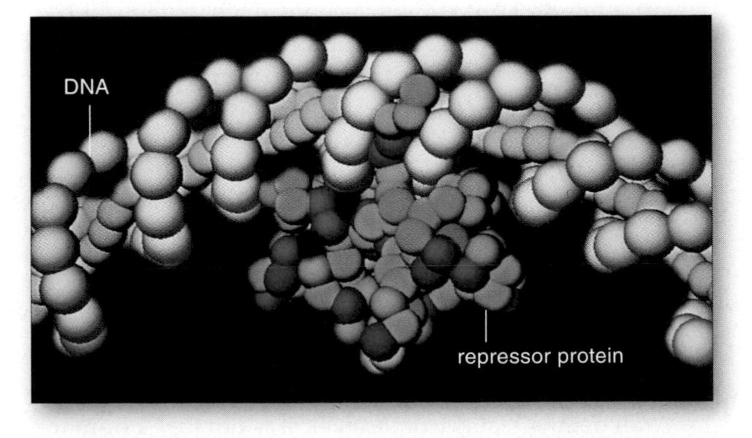

Figure 15.3 Computer model of one type of repressor protein binding to an operator at a site in a bacterial DNA molecule.

Further reading: Student Guide to InfoTrac on web site →

a A repressor protein exerts negative control over three genes of the lactose operon by binding to the operator and inhibiting transcription.

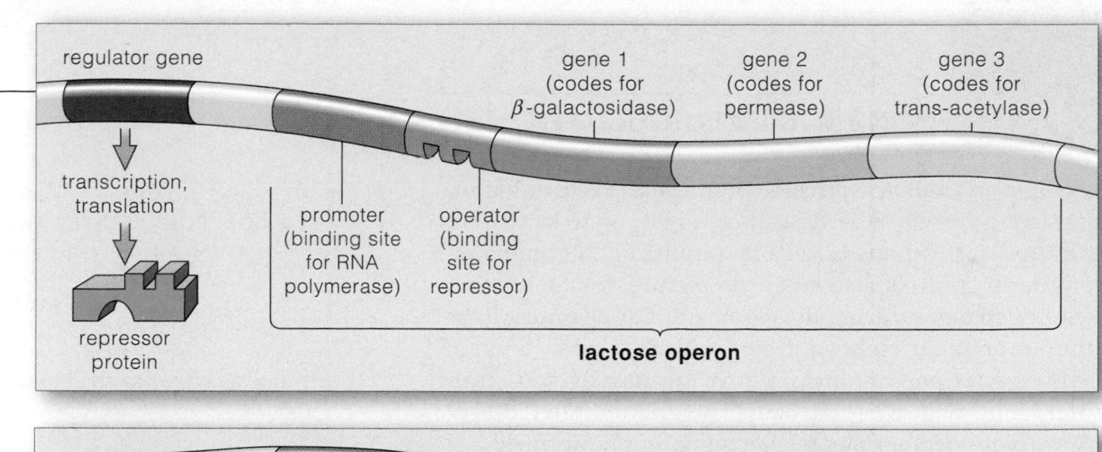

b When the concentration of lactose is low, the repressor is free to block transcription. Being bulky, it overlaps the promoter and prevents binding by RNA polymerase. The enzymes (not needed) are not produced.

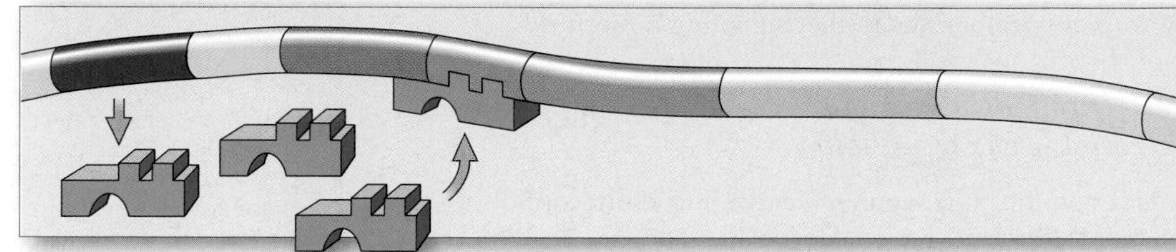

c At high concentration, lactose is an inducer of transcription. It binds to and distorts the shape of the repressor—which now cannot bind to the operator. The promoter is exposed and the genes can be transcribed.

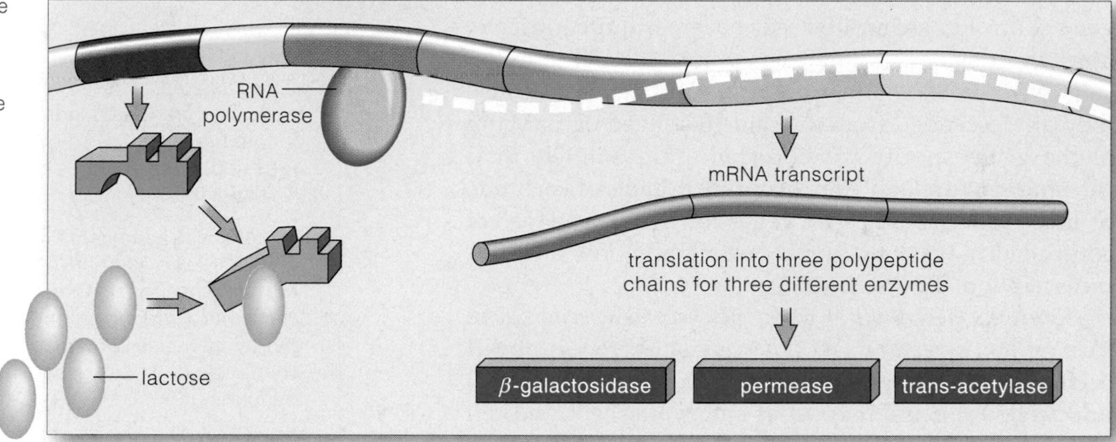

Figure 15.4 Negative control of the lactose operon. The operon's first gene codes for an enzyme that splits lactose, a disaccharide, into two subunits (glucose and galactose). The second enzyme codes for an enzyme that transports lactose into cells. The third enzyme functions in metabolizing certain sugars.

When the lactose concentration is low, a repressor binds with the operator, as in Figure 15.4. Being a large molecule, it overlaps the promoter, so transcription is blocked. Thus, *lactose-degrading enzymes are not built if they are not required.* When many lactose molecules are around, chances are greater that one of them will bind with the repressor. Binding alters the repressor's shape, so it cannot bind with the operator. RNA polymerase is free to transcribe the genes. Therefore, *lactose-degrading enzymes are synthesized only when required.*

Positive Control of Transcription

E. coli cells pay far more attention to glucose than to lactose. They transcribe genes for glucose breakdown faster, and continuously. Even if lactose is present, the lactose operon isn't used much *unless glucose is absent.* At such times, CAP, a regulatory protein, acts on the operon. CAP is one of the **activator proteins**, which are part of positive control systems. To understand how it works, you have to know the lactose operon's promoter is not good at binding RNA polymerase. It does better when CAP adheres to it first. But CAP will not do this until it is activated by a small molecule called cAMP.

Among other things, cAMP forms from ATP, which *E. coli* makes by breaking down glucose (in glycolysis). Not much cAMP is available in the cell when glucose is plentiful and glycolysis is running full bore. When such conditions prevail, CAP does not get primed to adhere to the promoter, so transcription of the lactose operon genes slows almost to a standstill.

Suppose glucose is scarce but lactose is available. cAMP accumulates, and CAP–cAMP complexes form. Now the lactose operon genes are transcribed rapidly, and the cell uses lactose as an alternative energy source.

Prokaryotic cells, which must respond rapidly to changing conditions, commonly rely on a small number of regulatory proteins that exert rapid, on-off control of transcription.

15.3 CONTROL IN EUKARYOTIC CELLS

Like bacteria, eukaryotic cells control short-term shifts in diet and in levels of activity. If cells happen to be one of hundreds or trillions of cells in a multicelled organism, long-term controls also enter the picture, for their gene activity changes during development. For example, as a human or plant embryo grows and develops, its new cells contact one another in growing tissues and start interacting with their particular neighbors. They do so by way of hormones and other signaling molecules.

Cell Differentiation and Selective Gene Expression

Later on in the book, you will be reading about controls over development, especially in Chapters 32, 44, and 45. For now, simply become familiar with the idea that gene control in eukaryotic cells becomes quite intricate. Start by considering this basic point: All of the living cells in your body inherited the same genes (because they all descended from the same fertilized egg). Many of the genes specify proteins that are absolutely vital for the structure and everyday functioning of each one of those cells. That is why genes for those proteins are controlled in ways that promote ongoing, low levels of transcription.

Even so, *nearly all of your cells became specialized in composition, structure, and function*. This process of **cell differentiation** proceeds during the development of all multicelled species. It arises as embryonic cells and cell lineages descended from them activate and suppress a small fraction of their genes in unique, selective ways. As examples, only immature red blood cells use the genes for making hemoglobin. Only certain white blood cells use the genes for making weapons called antibodies.

Controls Before and After Transcription

Many of the genes that govern the housekeeping tasks in all eukaryotic cells are under positive controls that promote continuous, low levels of transcription. In the case of many others, we see fluctuating adjustments of transcription. Why? In every multicelled organism, the internal environment (tissue fluids and blood) affords much more stable operating conditions for the body's individual cells. Most often, gene transcription rates rise or fall by small degrees in response to slight shifts in concentrations of signaling molecules, nutrients, and products in that environment.

As the examples in Figure 15.5*a* indicate, some gene sequences are repeatedly duplicated or rearranged in genetically programmed fashion prior to transcription. Besides this, programmed chemical modifications often shut down many genes. So does the orderly packaging of DNA by histones and other chromosomal proteins, as you may realize after reflecting on the organization of eukaryotic chromosomes (Section 9.5).

Controls also come into play after genes have been transcribed. As Figure 15.5*b–d* indicates, many controls govern RNA transcript processing, transport of mature RNAs from the nucleus, and rates of translation in the cytoplasm. Other controls deal with the modification of new polypeptide chains. Others deal with activating, inhibiting, and breaking down existing proteins. As one example, consider enzymes, the proteins that catalyze

a CONTROLS RELATED TO TRANSCRIPTION. At any given time, most genes of a multicelled organism are shut down, either permanently or temporarily. The genes necessary for a cell's everyday tasks are under positive controls that promote ongoing, low levels of transcription. With the help of these controls, a cell is assured of having enough enzymes and other proteins to carry out its most basic functions. Transcription of many other genes often shifts only slightly. In this case, controls work to assure chemical responsiveness even when concentrations of specific substances rise or fall only slightly.

Also, even before some genes are transcribed, a portion of their base sequences may be amplified, rearranged, or chemically modified in temporary or reversible ways. These changes are not mutations. They are heritable, programmed events that affect the manner in which particular genes will be expressed, if at all.

1. *Gene amplification.* Some cells that require enormous numbers of certain molecules briefly increase the number of the required genes. Sometimes multiple rounds of DNA replication produce hundreds or thousands of gene copies prior to transcription! This happens in immature amphibian eggs and glandular cells of some insect larvae (Sections 15.4 and 15.5).

2. *DNA rearrangements.* In a few cell types, many base sequences are alternative "choices" for parts of a gene. Prior to transcription, they are snipped out of the DNA. Combinations of the snippets are spliced together as the gene's final base sequence. For instance, this happens when B lymphocytes, a special class of white blood cells, are forming (Section 40.9). Different cells transcribe and translate their uniquely rearranged DNA into staggeringly diverse versions of protein weapons called antibodies, which act specifically against one of staggeringly diverse kinds of pathogens and other foreign agents in the body.

3. *Chemical modification.* Numerous histones and other proteins interact with eukaryotic DNA in highly organized fashion. The DNA–protein packaging, as well as chemical modifications to particular sequences, influences gene activity. Usually, only a small fraction of a cell's genes are available for transcription. A dramatic shutdown occurs in female mammals. As Section 15.4 describes, this event is called X chromosome inactivation.

Figure 15.5 Examples of the levels of control over gene expression in eukaryotes.

Further reading: Student Guide to InfoTrac on web site

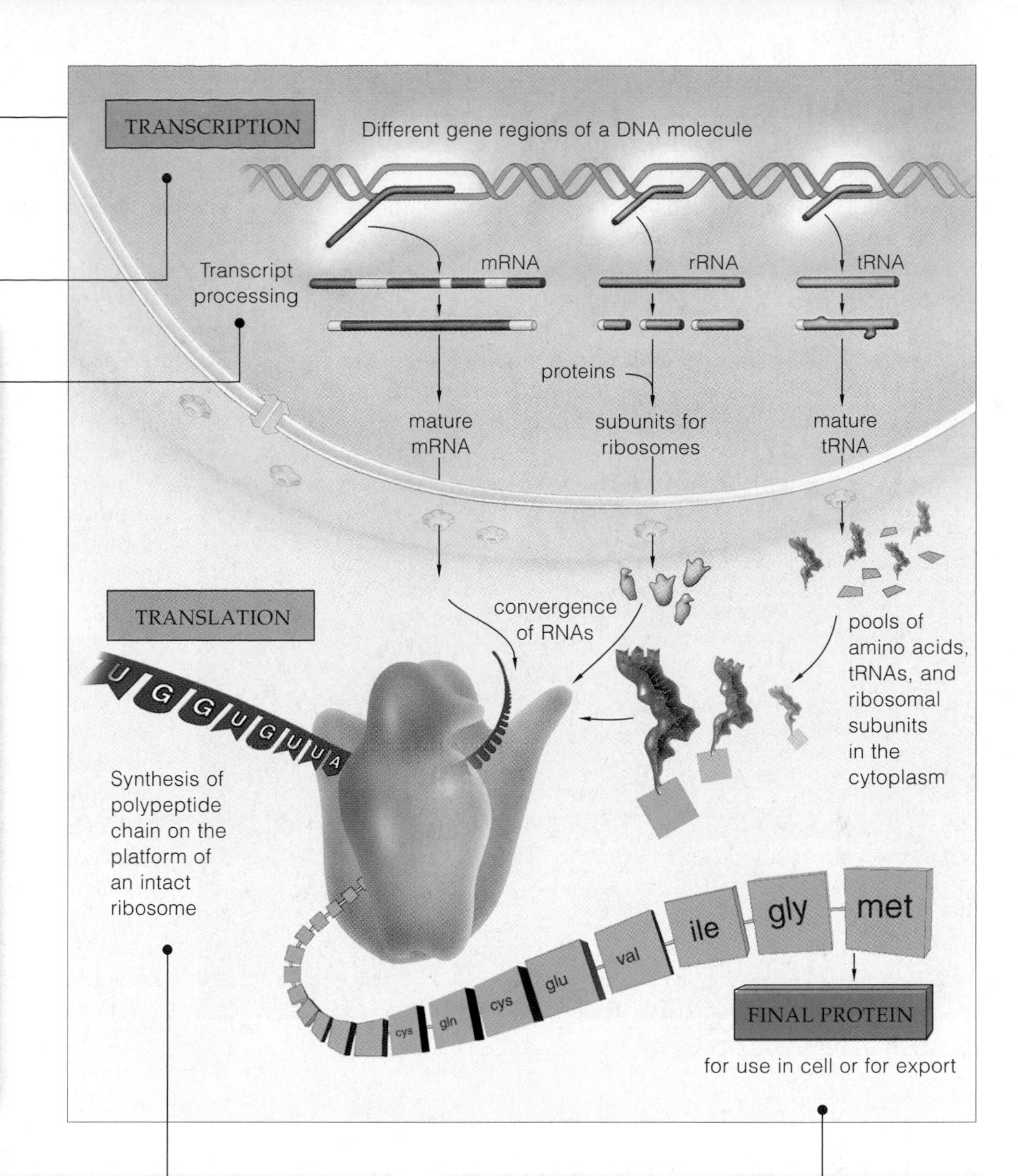

b TRANSCRIPT PROCESSING CONTROLS. As you read earlier in Section 14.2, pre-mRNA transcripts undergo modifications before they leave the nucleus.

For example, introns can be cut out and exons can be spliced together in more than one way. That is, a transcript from a single gene may undergo *alternative splicing*.

Pre-mRNA transcribed from a gene that specifies a contractile protein, troponin-1, is like this. Enzymes excise different portions of the pre-mRNA transcript in different cells. When the exons are spliced together, the final protein-building message is slightly different from one cell type to the next. The new proteins are all very similar, but each is unique in a certain region of its amino acid sequence. The proteins function in slightly different ways, which may account for the subtle variations we observe in the functioning of different types of muscles in the body.

c CONTROLS OVER TRANSLATION. Diverse controls govern when, how rapidly, and how often a given mRNA transcript will be translated. Sections 37.2, and 44.3 provide elegant examples.

The stability of a transcript influences the number of protein molecules that can be produced from it. Enzymes destroy transcripts from the poly-A tail on up (Section 14.2). The tail's length and its attached proteins affect the pace of degradation. Also, after leaving the nucleus, some transcripts are temporarily or permanently inactivated. For example, in unfertilized eggs, many transcripts are inactivated and stored in the cytoplasm. These "masked messengers" will not be available for translation until after fertilization, when great numbers of protein molecules will be required for the early cell divisions of the new individual.

d CONTROLS FOLLOWING TRANSLATION. Before they can become fully functional, many polypeptide chains must pass through the cytomembrane system (Section 4.5). There they undergo modification, as when enzymes attach specific oligosaccharides or phosphate groups to them.

Also, a variety of control mechanisms govern the activation, inhibition, and stability of the enzymes and other molecules that are involved in protein synthesis. Allosteric control of tryptophan synthesis, as described in Section 6.7, is an example.

nearly all metabolic reactions. In addition to selectively transcribing and translating genes for enzymes, diverse control systems activate and inhibit the molecules of enzymes that have already been synthesized in the cell.

Just imagine the coordination necessary to govern which of the cell's thousands of types of enzymes are to be stockpiled, deployed, or degraded in a specified interval. *That coordination governs all short-term and long-term aspects of cell structure and function.*

In multicelled species, gene controls guide the moment-by-moment activities of cells. They also guide intricate, long-term patterns of bodily growth and development.

All cells of complex organisms inherit the same genes, yet most become specialized in composition, structure, and function. This process of cell differentiation arises when different populations of cells activate and suppress some of their genes in highly selective, unique ways.

EVIDENCE OF GENE CONTROL

By some estimates, the cells of a multicelled organism rarely use more than 5 to 10 percent of their genes at a given time. One way or another, control mechanisms are keeping most of the genes repressed. Besides this, *which* genes are being expressed varies, depending on the stage of growth and development the organism is passing through. The following examples only hint at the kinds of controls operating at different stages.

Transcription in Lampbrush Chromosomes

Amphibian life cycles are such that the unfertilized eggs of females grow extremely fast while they are maturing. Growth requires numerous copies of enzymes and other proteins. Each germ cell that gives rise to eggs stockpiles large numbers of RNAs and ribosomes, all of which will participate in rapid protein synthesis.

At prophase I of meiosis, a protein scaffold forms between pairs of duplicated, homologous chromosomes. The sister chromatids of each pair decondense, and the DNA loops outward from the scaffold. The loops are profuse and stiffened with a coat of new RNA transcripts and proteins. The overall appearance is bristly, hence the name **lampbrush chromosomes** (Figure 15.6).

What is going on here? Histones and other proteins that structurally organize the DNA have loosened their grip, so the genes specifying RNA molecules are now accessible. Histones, recall, are part of nucleosomes, the unit of organization in eukaryotic chromosomes. As Section 9.5 describes, a nucleosome is a stretch of DNA looped around a core of histone molecules. Figure 15.7*b* shows a model based on photomicrographs that suggest how nucleosome packaging might be loosened up during transcription.

The chromosomes are packaged further into distinct domains. This higher order packaging is still poorly understood, but it is known to have a crucial role in controlling transcription. There are about 10,000 loops in many amphibian eggs. Each corresponds to a specific DNA sequence. Yet most of the DNA remains condensed in regions called chromomeres, which for the most part are *not* transcribed. Controls keep some parts of the chromosomes extended for transcription even as they keep other parts condensed and thereby inactive. The domains are large, and they are precise.

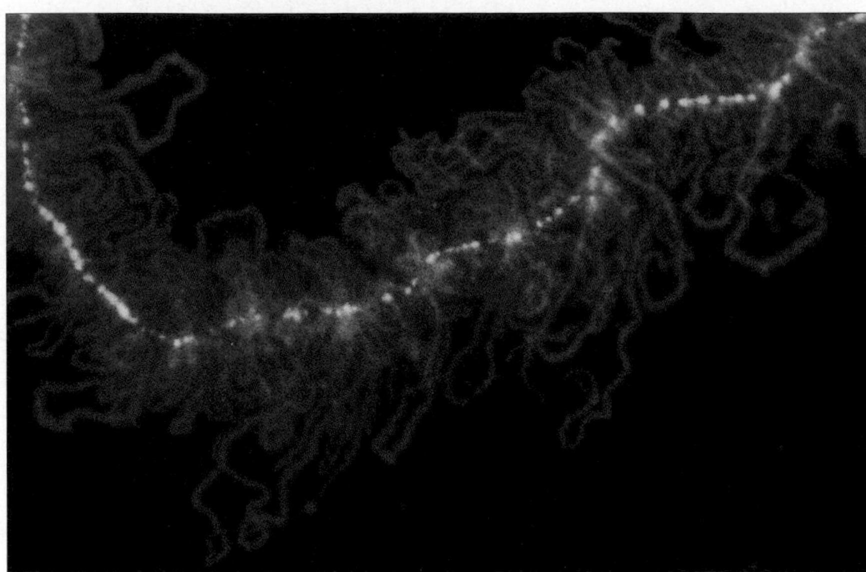

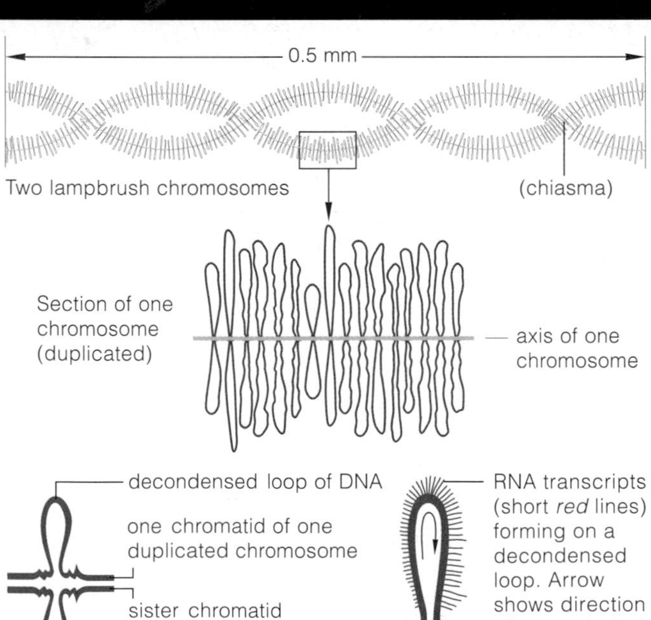

Figure 15.6 Micrograph of a lampbrush chromosome from a germ cell of a newt (*Notophthalmus viridiscens*). During prophase I, gene regions of the duplicated chromosomes decondensed and looped outward. A *red* fluorescent dye labeled a component of RNA–protein complexes, the presence of which is evidence of gene activity. (A *white* fluorescent dye labeled the DNA making up the axis of each duplicated chromosome's structural framework.)

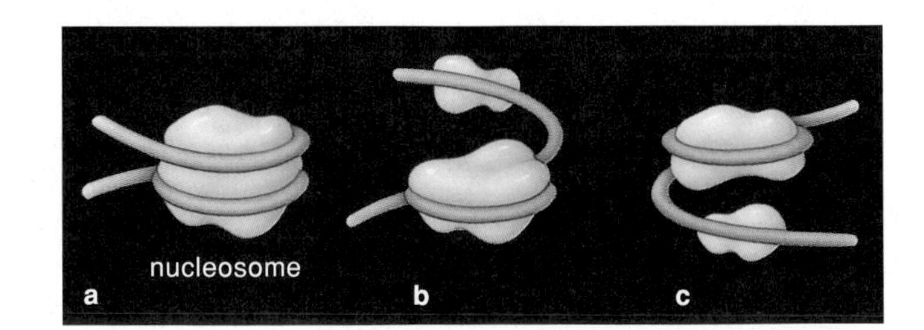

Figure 15.7 Model of the kinds of changes in the structural organization of chromosomal DNA that might promote gene transcription. The tight DNA–histone packing in nucleosomes (**a**) might loosen up at gene sequences that are about to be transcribed in the manner shown in (**b**–**c**). Eukaryotic genes are known to be transcribed primarily in regions where the packing of chromosomal DNA is most relaxed.

Further reading: Student Guide to InfoTrac on web site →

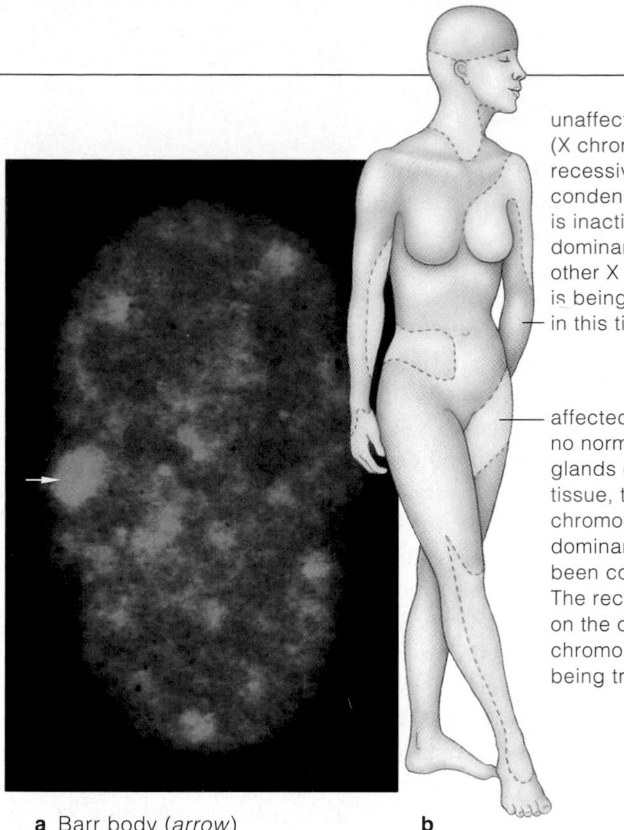

Figure 15.8 (**a**) Micrograph of an inactivated X chromosome, called a Barr body, as it appears in a human female's somatic cell during interphase. The X chromosome is not condensed this way in a human male's cells. (**b**) Anhidrotic ectodermal dysplasia, a mosaic pattern of gene expression. The condition arises as a result of random X chromosome inactivation.

Figure 15.9 Why is this female calico cat "calico"? In her cells, one X chromosome carries a dominant allele for the brownish-black pigment melanin. The allele on her other X chromosome specifies yellow fur. At an early stage of the cat's embryonic development, one of the two X chromosomes was inactivated at random in each cell that had formed by then.

In all descendants of those cells, the same chromosome also became inactivated, which left them with only one functional allele for the coat-color trait. We see patches of different colors, depending on which allele was inactivated in cells that formed a given tissue region. (The white patches result from a gene interaction involving the "spotting gene," which blocks melanin synthesis entirely.)

X Chromosome Inactivation

As you know, proteins impart structural organization to eukaryotic DNA. In female mammals, the structure of one of the two X chromosomes in each body cell is chemically modified in a way that completely closes off access to its genes.

A mammalian embryo destined to become female inherits one X chromosome from the mother and another from the father. One of the two condenses in each cell and is inactivated. Condensation is part of the program of development, but the outcome is random. *One or the other chromosome may be inactivated.* We observe such a condensed X chromosome in the interphase nucleus; it shows up as a dense spot (Figure 15.8*a*). The condensed X chromosome was named a **Barr body** (after Murray Barr, its discoverer).

Whenever a maternal (or paternal) X chromosome is inactivated in a cell, it also becomes inactivated in all descendants of that cell. Thus, when development ends, the female is a "mosaic" for those X chromosomes. *She bears patches of tissue where the genes of the maternal X chromosome are expressed—and patches of tissues in which the genes of the paternal X chromosome are expressed.* Any pair of alleles on those two chromosomes may or may not be identical. That is why tissues such as skin may have different features in different body regions. Mary Lyon discovered this **mosaic tissue effect**, which arises from random X chromosome inactivation.

We see the effect in human females who have one X chromosome with a recessive allele that blocks sweat gland formation. The absence of sweat glands in such heterozygotes is one symptom of *anhidrotic ectodermal dysplasia*. In skin patches without the glands, only genes on the X chromosome with the mutant allele are being transcribed; the X chromosome with the dominant allele is condensed. In patches with the glands, the dominant allele is active (Figure 15.8*b*). The same effect shows up in female calico cats, which are heterozygous for black and yellow coat-color alleles on their X chromosomes. The coat color in a given body region depends on which X chromosome's genes are transcribed (Figure 15.9).

In a typical cell of multicelled organisms, controls repress all but about 5 to 10 percent of the genes at a given time.

The remaining fraction of genes is selectively expressed at different stages of growth and development. Lampbrush chromosome formation, X chromosome inactivation, and other cellular events provide evidence of this.

15.5 EXAMPLES OF SIGNALING MECHANISMS

The story that opened this chapter introduced the idea that a great variety of signals influence gene activity. The following examples from animals and plants will give you a sense of their effects at the molecular level.

Hormonal Signals

Hormones, a major category of signaling molecules, can stimulate or inhibit gene activity in target cells. Any cell with receptors for a given hormone is a target. Animal cells secrete hormones into tissue fluid. Most hormone molecules are picked up by the bloodstream, which distributes them to cells some distance away. In plants, hormones do not travel far from cells that secrete them.

Certain hormones bind to membrane receptors at the surface of target cells. Others enter target cells, bind to regulatory proteins, and so help initiate transcription. In many cases, a hormone molecule must first combine briefly with an enhancer. **Enhancers** are base sequences that serve as binding sites for suitable activator proteins. They may or may not be adjacent to a promoter in the same DNA molecule. When some distance separates the two, a loop forms in the DNA, and looping unites the enhancer with the promoter. RNA polymerase can bind avidly to the bound complex at the base of the loop.

Consider the effect of **ecdysone**, a hormone with key roles in many insect life cycles. Immature stages called larvae grow rapidly during the cycles. They continually feed on organic matter, such as decaying leaves, and require copious amounts of saliva to prepare food for digestion. DNA in their salivary gland cells has been replicated repeatedly. The copies of DNA molecules have remained together in parallel array, forming what is known as a **polytene chromosome**. When ecdysone binds to a receptor on the gland cells, its signal triggers very rapid transcription of multiple copies of genes in the DNA. The gene regions affected by this hormonal signal puff out during transcription, as in Figure 15.10. Afterward, translation of mRNA transcripts made from the genes produces the protein components of saliva.

In vertebrates, certain hormones have widespread effects on gene expression because many types of cells have receptors for them. As one example, the pituitary gland secretes somatotropin, or growth hormone. This hormonal signal stimulates synthesis of all the proteins required for cell division and, ultimately, the body's growth. Most cells have receptors for somatotropin.

Other vertebrate hormones signal only certain cells at certain times. Prolactin, a secretion from the pituitary gland, is like this. A few days after a female mammal gives birth, prolactin can be detected in her blood. This hormone activates genes in mammary gland cells that have receptors for it. Those genes have responsibility for milk production. Liver cells and heart cells have the same genes but do not have the receptors necessary to respond to signals from prolactin.

Explaining hormonal control of gene activity is like explaining a full symphony orchestra to someone who has never seen one or heard it perform. Many separate parts must be defined before their interactions can be understood! We will return to this topic later on. As you will especially see in Chapter 32, 37, 44, and 45, some of the most elegant examples of hormonal controls are drawn from studies of plant and animal reproduction and development.

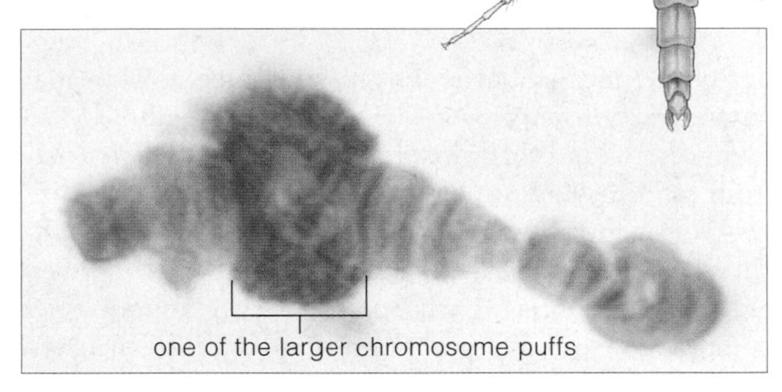

Figure 15.10 Visible evidence of transcription in a polytene chromosome from a larva of a midge (*Chironomus*). Midges are flies, one to ten millimeters long. The short-lived, winged adults often congregate in great swarms, which helps them find mates in a hurry. They might seem to be pests merely because of their large numbers. They also look somewhat like mosquitoes, but they don't bite.

Most midge larvae develop in aquatic habitats. To sustain their rapid growth, they feed continually, as on decaying organic material. They require a lot of saliva, and they must continuously transcribe genes for saliva's protein components. Those genes have undergone amplification. Ecdysone, a hormone, serves as a regulatory protein that helps promote their transcription. Midge chromosomes loosen and puff out in regions where the genes are being transcribed in response to the hormonal signal. The puffs become large and appear quite diffuse when transcription is most intense. Staining techniques reveal banding patterns in the chromosomes, as evident in the micrograph.

Sunlight as a Signal

Plant a few seeds from a corn or bean plant in a pot that contains moist, nutrient-rich soil. Next, let the seeds germinate, but keep them in total darkness. After eight days have passed, they will develop into seedlings that are spindly and notably pale, owing to the absence of chlorophyll (Figure 15.11). Now expose the seedlings to a single burst of dim light from a flashlight. Within ten minutes, they will start converting stockpiles of certain molecules to the activated forms of chlorophylls, the

Further reading: Student Guide to InfoTrac on web site →

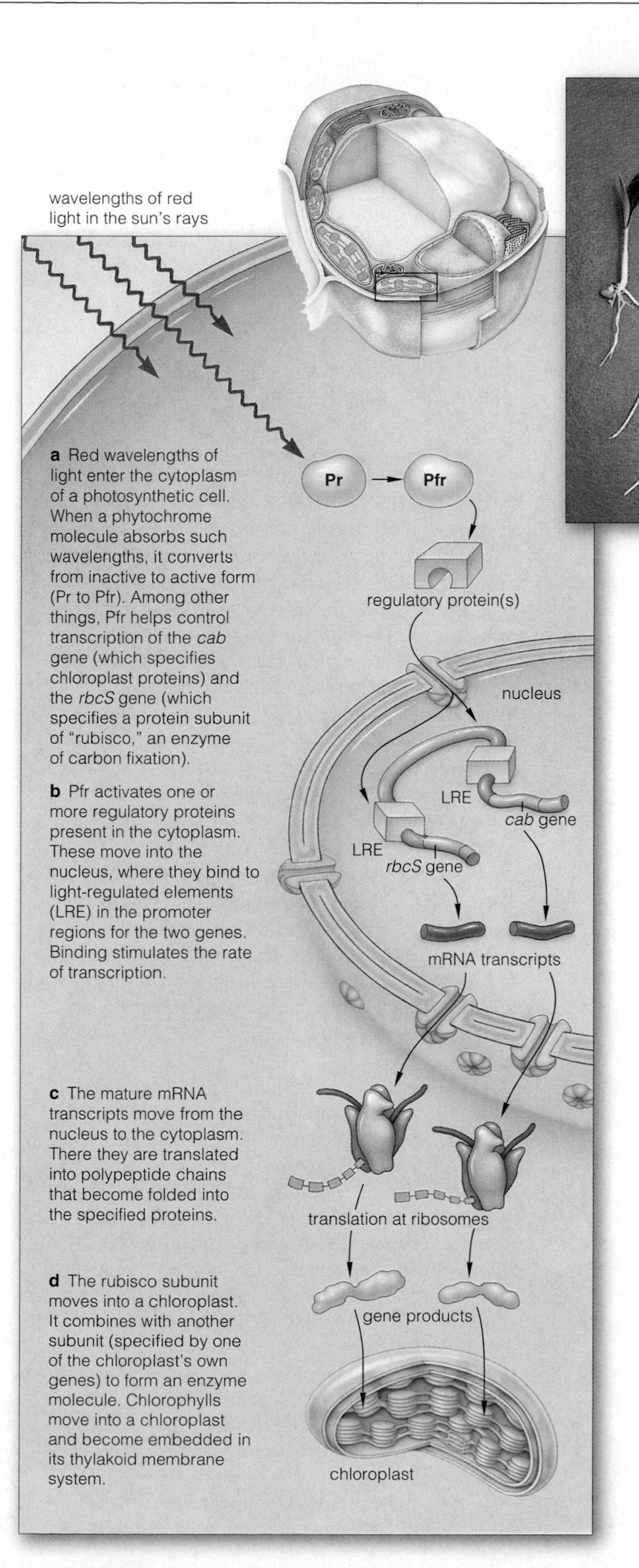

Figure 15.11 Sunlight as a signal for gene activity. The photograph shows the effect of an absence of light on corn seedlings. The two seedlings at *left* were the control group; they were grown inside a sunlit greenhouse. The two seedlings positioned next to them were grown in total darkness for eight days. The dark-grown plants were not able to convert their stockpiled precursors of chlorophyll molecules to active form, and they never did green up. The diagram illustrates one model for the mechanism by which phytochrome might help to control gene transcription in plants. Red wavelengths can convert the phytochrome molecule from inactive form (here designated Pr) to active form (Pfr). In this form, the phytochrome can serve as a regulator of transcription.

light-trapping pigment molecules that donate electrons for the reactions of photosynthesis.

Phytochrome, a blue-green pigment, is a signaling molecule that helps plants adapt over the short term to changes in light conditions. Section 32.4 takes a close look at this molecule. For now, simply be aware that it alternates between active and inactive forms. At sunset, at night, or in shade, far-red wavelengths predominate. At such times, the phytochrome in cells is inactive. It is activated at sunrise, when red wavelengths dominate the sky. Also, the amount of red or far-red wavelengths that a plant intercepts varies from day to night and as seasons change. Such variations act as control signals over phytochrome activity. That activity influences the transcription of certain genes at certain times of day and year. The genes specify a number of enzymes and other proteins that help seeds to germinate, stems to lengthen and then branch, and leaves to grow. Gene products also help flowers, fruits, and seeds to form.

Elaine Tobin and her coworkers at the University of California, Los Angeles, performed experiments that provided evidence in favor of phytochrome control. Using dark-grown seedlings of duckweed (*Lemna*), they discovered a marked increase in the number of certain mRNA transcripts after a one-minute exposure to red light. Exposure had enhanced transcription of the genes for proteins that bind chlorophylls and for rubisco, an enzyme that mediates carbon fixation (Sections 7.6 and 7.7). In the absence of the proteins, chloroplasts do not develop properly and they don't turn green.

Hormones and other signaling molecules, including diverse kinds that respond to environmental stimuli, have profound influence on gene expression.

LOST CONTROLS AND CANCER

THE CELL CYCLE REVISITED Every second, millions of cells in your skin, bone marrow, gut lining, liver, and elsewhere divide and replace their worn-out, dead, and dying predecessors. They do not divide willy-nilly. Controls govern the expression of genes that specify enzymes and other proteins required for cell growth, DNA replication, chromosome movements, and division of the cytoplasm. Diverse regulatory proteins control the synthesis and use of those gene products. And they control when the division machinery is put to rest.

Controls over the cell cycle are extensive. At various interrelated points, gene activity can be stepped up or slowed down, and gene products can advance, delay, or block the cycle. Some genes directly regulate passage through the cycle and are its primary controllers. Their products include **protein kinases**, a class of enzymes that attach phosphate groups to proteins. They operate at the boundary between G1 and S of the cell cycle, and during the transition from G2 to mitosis (Section 9.2).

Other genes encode proteins that modify the activity of the primary control genes or their products. Among them are genes for **growth factors**: transcriptional signals sent by one cell to trigger growth in other cells. For example, when epidermal growth factor (EGF) binds to its receptor on target cells, it triggers tyrosine kinase activity—a signal for mitosis. Other genes specify the receptors for growth factors.

Certain gene products exert indirect control over the cell cycle. For example, some are enzymes that help maintain DNA replication. Others take part in stockpiling weapons to be released when the cell is scheduled to die.

If genes specifying any of these products become mutated, the cell may be deprived of crucial enzymes or proteins. Cancer may follow.

CHARACTERISTICS OF CANCER At the least, four features characterize all cancer cells. First, *their plasma membrane and cytoplasm change profoundly.* The membrane becomes more permeable; its proteins are lost or altered, and abnormal ones form. The cytoskeleton becomes disorganized, shrinks, or both. Enzyme action shifts, as in amplified reliance on glycolysis. Second, *cancer cells grow and divide abnormally.* Controls that prevent overcrowding in tissues are lost. Cell populations reach high densities. Proteins trigger abnormal increases in the number of small blood vessels that can service the growing cell mass. Third, *cancer cells have a weakened capacity for adhesion.* Recognition proteins are altered or lost, so cells cannot stay anchored in proper tissues. Thus they can break away and establish colonies in distant tissues (Figure 15.12). Fourth, *cancer cells are lethal.* Unless eradicated, they will put the individual on a painful road to death.

Figure 15.12 Evidence of loss of gene control: (**a**) A benign tumor, a mass of cells with an outwardly normal appearance that are enclosed in a capsule of connective tissue. (**b**) A malignant tumor, a disorganized collection of cancer cells. (**c**) Steps in metastasis.

ONCOGENES Any gene having the potential to induce cancerous transformation is an **oncogene**. Oncogenes were first identified in retroviruses (a class of RNA viruses). They are mutated forms of normal genes that specify the growth factors and other proteins that are necessary for normal cell functioning. Think of the normal forms as *proto*-oncogenes.

One category of oncogenes specifies proteins that induce cell proliferation. To give examples, *erbB* specifies the receptor for epidermal growth factor, *src* specifies tyrosine kinase, and *myc* specifies a factor necessary for transcription.

Another category specifies proteins that are *anti*-oncogenes; they inhibit cell proliferation. The product of one such oncogene, *p53*, prevents some cancers. To date, mutation in *p53* has been implicated in 90 percent of the reported cases of small-cell lung cancer and more than 50 percent of breast and colon cancer.

A third category of oncogenes suppresses or triggers programmed cell death. One of these, *bcl-21*, helps keep normal body cells from dying before their time.

Oncogene expression is highly controlled in all normal cells. Typically, mitosis is induced

Further reading: Student Guide to InfoTrac on web site

Figure 15.13 An artist's representation of weapons of cell death being unleashed inside a cell.

when one cell type secretes a growth factor, which acts on the cells bearing the receptor for that factor. Mitosis is suppressed when one cell type secretes anti-oncogenes. Uncontrolled activation of any oncogene may spark tumor formation. Inactivation of any anti-oncogene releases one of the brakes on tumor formation.

Oncogenes can form following certain insertions of viral DNA into cellular DNA. They may also form after carcinogens (cancer-inducing agents) change the DNA. As you read earlier, ultraviolet radiation and forms of nonionizing radiation (such as x-rays) are carcinogens. So are many natural and synthetic compounds, including asbestos and certain components of tobacco smoke.

Remember those chromosome alterations and gene mutations? Some cancers arise when base substitutions or deletions alter a proto-oncogene or one of the controls over its transcription. Others arise by translocations, as described in the introduction to Chapter 12 and Section 12.9. Destabilizing transposons also cause some cancers.

HERE'S TO SUICIDAL CELLS! We conclude with a case study of a gene and its cancerous transformation.

The first cell of a new multicelled individual contains marching orders that will guide its descendants along a program of growth, development, and reproduction, then on to death. As part of that program, many cells heed calls to self-destruct when they complete a prescribed function. If they become altered in ways that might pose a threat to the body as a whole, as by infection or cancer, they can execute themselves. **Apoptosis** (app-oh-TOE-sis) is the name for this form of cell death. It starts with molecular signals that activate and unleash lethal weapons of self-destruction, which were stockpiled earlier within the cell.

Protein-cleaving enzymes called **ICE-like proteases** are such weapons.Think of them as folded pocketknives or lethal Ninja weapons. When popped open, they chop apart structural proteins, including the building blocks of cytoskeletal elements and nucleosomes that organize the DNA (Figure 15.13).

A body cell in the act of suicide shrinks away from its neighbors. Its cytoplasm seems to roil, and its surface repeatedly bubbles outward and inward. No longer are its chromosomes extended through the nucleoplasm; they are bunched up near the nuclear envelope. The nucleus, then the cell, breaks apart. Phagocytic white blood cells patrol tissues. If they encounter suicidal cells or remnants of them, the patrolling cells will swiftly engulf them.

The timing of cell death is predictable in some cells, such as pigment-packed keratinocytes. These form the densely packed sheets of dead cells that are continually sloughed off and replaced at the skin surface. They have a three-week life span, more or less. Keratinocytes and other body cells, even the kinds that are supposed to last for a lifetime, can be induced to die ahead of schedule. All it takes is sensitivity to signals that can activate ICE-like proteases and other enzymes of death.

The knives remain sheathed in cancer cells, which are supposed to—but don't—commit suicide on cue. Maybe they have lost normal receptors that would allow contact with their signaling neighbors. Maybe they are receiving abnormal signals about when to grow, divide, or cease dividing. In some cancers, for instance, the anti-oncogene *bcl-2* either has been tampered with or it has been shut down. Apoptosis cannot occur.

As you might deduce from all of this, cancer is a multistep process, involving more than one oncogene. Researchers have much to learn. But they have already identified many of the genes in the process. Through their work, it may be possible to diagnose carriers early and frequently. Cancer that can be detected early enough might still be curable, as by surgery.

15.7 SUMMARY

1. Cells are equipped with controls that govern gene expression; that is, which gene products appear, when, and in what amounts. When control mechanisms come into play depends on cell type, on prevailing chemical conditions, and on signals from other cell types that can change a target cell's activities.

2. Regulatory proteins, enzymes, hormones, and other molecules are components of control systems. Various kinds interact with one another, with DNA and RNA, and with products of genes. Control systems operate before, during, and after transcription and translation.

3. In all cell types, two of the most common categories of control systems operate to inhibit or enhance gene transcription. Their effects are reversible.

a. With negative control systems, a regulatory protein binds at a specific DNA sequence and prevents one or more genes from being transcribed.

b. With positive control systems, a regulatory protein binds to DNA and promotes transcription.

4. Unlike eukaryotic cells, most bacteria (prokaryotic cells) do not require many genes to live, grow, and reproduce. Most of their control systems affect rates of transcription. Control of operons (groupings of related bacterial genes and control elements) is an example.

5. Promoters are binding sites in DNA that signal the start of a gene. In all cells, activator proteins bind next to promoters, and the complex promotes transcription. In bacterial cells only, operators serve as binding sites for regulatory proteins that can inhibit transcription.

6. Eukaryotic cells use more complex gene controls. Gene activity must change rapidly in response to short-term shifts in the surroundings, as in prokaryotic cells. However, for all multicelled species, it also must be intricately adjusted during growth and development, when great numbers of cells grow and divide, make physical contact with one another in diverse tissues, and interact chemically over the long term.

7. Being descended from the same cell, all of the cells in a multicelled organism inherit the same assortment of genes. However, different types of cells activate and suppress some fraction of the genes in different ways. This behavior is called selective gene expression.

8. Cell differentiation is one outcome of selective gene expression. By this process, different lineages of cells in multicelled organisms become specialized in function, appearance, and composition.

9. Genes associated with the cell cycle are extensively controlled. Gene activity can be stepped up or slowed. Gene products can advance, delay, or block the cycle. Products of the primary control genes include protein receptors for control signals. Other gene products, such as diverse growth factors and protein kinases, regulate primary control genes or their products. Cancer may follow mutation of any of these genes.

10. Cancer involves loss of normal controls over the cell cycle and mechanisms of programmed cell death.

a. In cancer cells, the structure and function of both the plasma membrane and the cytoplasm are severely compromised. These cells grow and divide abnormally. They have a weakened capacity for adhesion to their home tissue. Unless eradicated, they are lethal.

b. Growth factors, enzymes, and other products of certain genes normally induce cell proliferation, inhibit cell proliferation, or suppress or trigger programmed cell death (apoptosis). Oncogenes are mutated forms of those essential genes, and they have the potential to induce cancer.

Review Questions

1. Briefly describe the general characteristics of normal cells. Then explain the difference between a benign tumor and a malignant tumor. *CI, 15.6*

2. In what fundamental way do negative and positive controls of transcription differ? Is the effect of one or the other form of control (or both) reversible? *15.2*

3. Distinguish between: *15.2*
   a. repressor protein and activator protein
   b. promoter and operator
   c. repressor and enhancer

4. Describe one type of control of transcription for the lactose operon in *E. coli*, a prokaryotic cell. *15.2*

5. A plant, fungus, or animal consists of diverse cell types. How might the diversity arise, given that body cells in each of these organisms inherit the same set of genetic instructions? As part of your answer, define cell differentiation and the general way that selective gene expression brings it about. *15.3*

6. Using the diagram on the facing page, define at least three types of gene controls in eukaryotic cells and indicate the levels at which they take effect. *15.3*

7. What is a Barr body? Does it appear in the cells of males, females, or both? Explain your answer. *15.4*

Self-Quiz (*Answers in Appendix III*)

1. The expression of a given gene depends on the ______.
   a. type of cell and its functions
   b. chemical conditions
   c. environmental signals
   d. all of the above

2. Regulatory proteins interact with ______.
   a. DNA
   b. RNA
   c. gene products
   d. all of the above

3. In prokaryotic cells but not eukaryotic cells, a(n) ______ precedes the genes of an operon.
   a. lactose molecule
   b. promoter
   c. operator
   d. both b and c

4. A base sequence signaling the start of a gene is a(n) ______.
   a. promoter
   b. operator
   c. enhancer
   d. activator protein

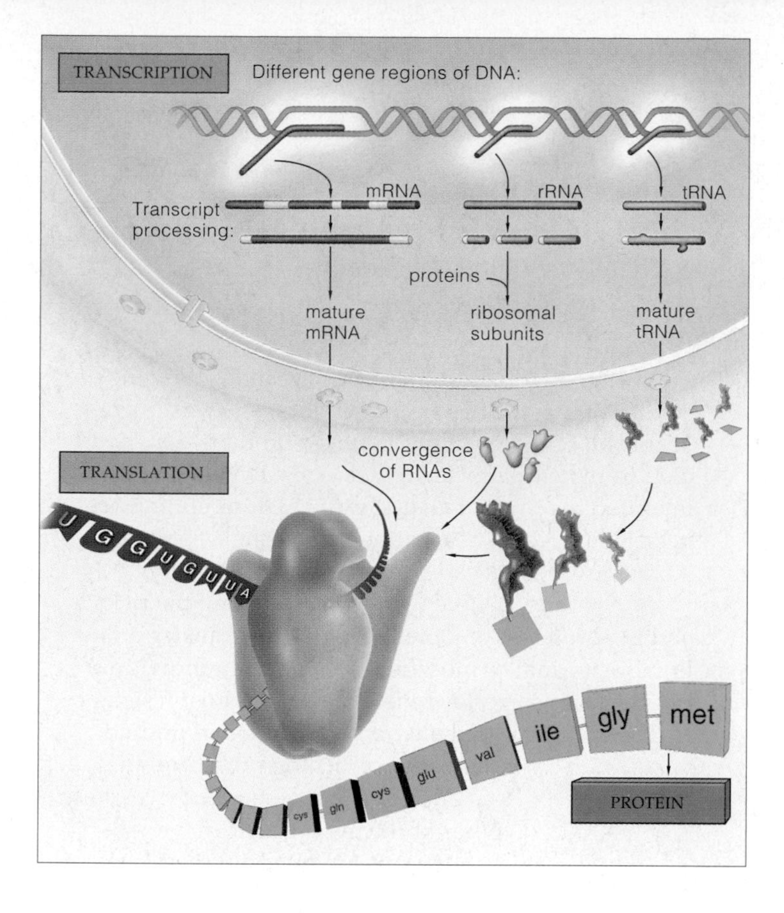

5. An operon most typically governs ______ .
   a. bacterial genes
   b. a eukaryotic gene
   c. genes of all types
   d. DNA replication

6. Eukaryotic genes guide ______ .
   a. rapid short-term activities
   b. overall growth
   c. development
   d. all of the above

7. Cell differentiation ______ .
   a. occurs in all complex multicelled organisms
   b. requires different genes in different cells
   c. involves selective gene expression
   d. both a and c
   e. all of the above

8. X chromosome inactivation may result in a ______ for some traits.
   a. male phenotype
   b. uniform tissue effect
   c. mosaic tissue effect
   d. patterned tissue effect

9. Hormones interact with ______ .
   a. membrane receptors
   b. regulatory proteins
   c. enhancers
   d. all of the above

10. Apoptosis is ______ .
   a. cell division after severe tissue damage
   b. programmed cell death by suicide
   c. a popping sound in mutated toes

11. ICE-like proteases are ______ .
   a. structural proteins
   b. lethal weapons
   c. environmental signals
   d. low-temperature enzymes

12. Match the terms with their most suitable descriptions.
   ______ phytochrome
   ______ Barr body
   ______ oncogene
   ______ repressor
   ______ lampbrush chromosome

   a. inhibits gene transcription
   b. gene that induces cancerous transformation
   c. intense, controlled transcription in selected domains of DNA
   d. helps plants adapt to daily and seasonal changes in light
   e. inactivated X chromosome

Critical Thinking

1. Geraldo isolated a strain of *E. coli* in which a mutation has affected the capacity of CAP to bind to a region of the lactose operon, as it would do normally. State how the mutation affects transcription of the lactose operon when cells of this bacterial strain are subjected to these conditions:
   a. Lactose and glucose are available.
   b. Lactose is available but glucose is not.
   c. Both lactose and glucose are absent.

2. *Duchenne muscular dystrophy*, a genetic disorder, affects boys almost exclusively. Early in childhood, muscles begin to atrophy (waste away) in affected individuals, who typically die in their teens or early twenties as a result of respiratory failure. Muscle biopsies of women who carry a gene associated with the disorder reveal some regions of atrophied muscle tissue. Yet the muscle tissue adjacent to these regions was normal or even larger and more chemically active, as if to compensate for the weakness of the adjoining region. How can you explain these observations?

3. Unlike most rodents, guinea pigs are already well developed at the time of birth. Within a few days, they are able to eat grass, vegetables, and other plant material. Suppose a breeder decides to separate the baby guinea pigs from their mothers after three weeks. He wants to keep the males and females in different cages, but it is quite difficult to identify the sex of guinea pigs when they are so young. Suggest a simple test that the breeder can perform to identify their sex.

4. Individuals affected by *pituitary dwarfism* cannot synthesize somatotropin (also called growth hormone). Children with this genetic abnormality will not grow to normal height. Develop a hypothesis to explain why therapy that is based on somatotropin injections may be effective.

5. Transcription is generally controlled by the binding of a protein to a DNA sequence "upstream" from a gene, rather than by modification of RNA polymerase. Develop a hypothesis to explain why this is so.

6. The closer a mammalian species is to humans in its genetic makeup, the more useful information it may yield in laboratory studies of the mechanisms of cancer. Do you support the use of any mammal for cancer research? Why or why not?

Selected Key Terms

activator protein *15.2*
apoptosis *15.6*
Barr body *15.4*
cancer *CI*
cell differentiation *15.3*
ecdysone *15.5*
enhancer *15.5*
growth factor *15.6*
hormone *15.5*
ICE-like protease *15.6*
lampbrush chromosome *15.4*
metastasis *CI*
mosaic tissue effect *15.4*
negative control system *15.1*
oncogene *15.6*
operator *15.2*
operon *15.2*
phytochrome *15.5*
polytene chromosome *15.5*
positive control system *15.1*
promoter *15.2*
protein kinase *15.6*
regulatory protein *15.1*
repressor *15.2*

Readings *See also www.infotrac-college.com*

Duke, R., D. Ojcius, and J. Ding-E Young. December 1996. "Cell Suicide in Health and Disease." *Scientific American*, 80–87.

Murray, A., and M. Kirschner. March 1991. "What Controls the Cell Cycle?" *Scientific American*, 264(3): 56–63.

Tijan, R. February 1995. "Molecular Machines That Control Genes." *Scientific American*, 54–61.

16

RECOMBINANT DNA AND GENETIC ENGINEERING

Mom, Dad, and Clogged Arteries

Butter! Bacon! Eggs! Ice cream! Cheesecake! Possibly you think of such foods as enticing, off-limits, or both. After all, who among us doesn't know about animal fats and the dreaded cholesterol?

Soon after you feast on those fatty foods, cholesterol enters the bloodstream. Cholesterol is important. It is a structural component of animal cell membranes, and without membranes, there would be no cells. Cells also remodel cholesterol into a variety of molecules, such as the vitamin D necessary for the development of good bones and teeth. Normally, however, the liver itself synthesizes enough cholesterol for your cells.

Some proteins circulating in blood combine with cholesterol and other substances to form lipoprotein particles. The high-density lipoproteins, or *HDLs*, transport cholesterol to the liver, where it can be metabolized. The low-density lipoproteins (*LDLs*) normally end up in cells that store or use cholesterol.

Sometimes too many LDLs form. The excess infiltrates the elastic walls of arteries, where it contributes to the formation of atherosclerotic plaques (Figure 16.1). These abnormal masses interfere with blood flow and narrow the arterial diameter. If the plaques clog one of the tiny coronary arteries that deliver blood to the heart, chest pains or a heart attack may result.

How well you handle dietary cholesterol depends on genes you inherited from your parents. One gene specifies a protein that serves as the cell's receptor for LDLs. Inherit two "good" alleles of that gene, and your blood level of cholesterol will tend to remain so low that your arteries may never clog up, even with a high-fat diet. Inherit two copies of a certain mutated allele, however, and you are destined to develop a rare genetic disorder called *familial cholesterolemia*. Levels of cholesterol become abnormally high. Many affected people die of heart attacks in childhood or their teens.

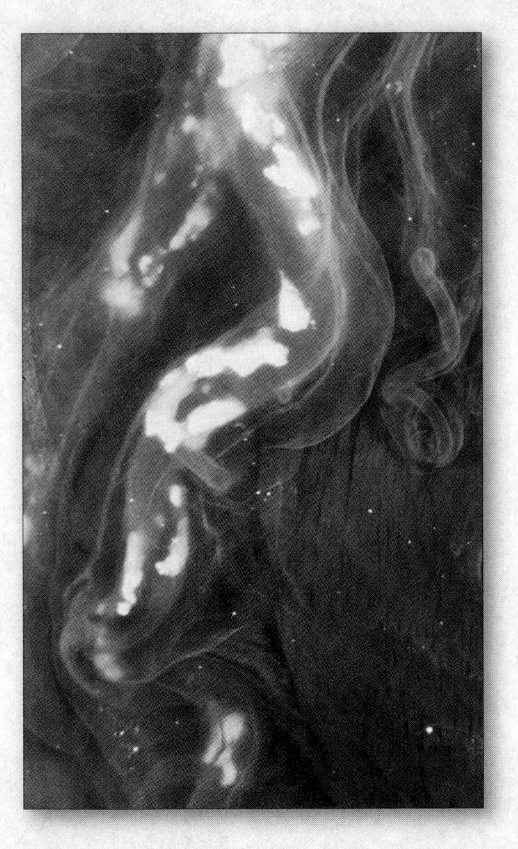

Figure 16.1 Life-threatening plaques (*bright yellow-white*) in one of the coronary arteries, small vessels that deliver blood to the heart. Abnormally high levels of cholesterol contribute to plaque formation.

In 1992 a woman from Quebec, Canada, became a milestone in the history of genetics. She was thirty years old. Like two of her younger brothers who had died from heart attacks in their early twenties, she inherited the mutant allele for the LDL receptor. She herself survived a heart attack when she was sixteen. At twenty-six, she had coronary bypass surgery.

At the time, people were hotly debating the risks and the promises of **gene therapy**—the transfer of one or more normal or modified genes into an individual's body cells to correct a genetic defect or boost resistance to disease. Even so, the woman opted for an untried, physically wrenching procedure that was designed to give her body working copies of the good gene.

Surgeons removed about 15 percent of the woman's liver. Researchers put cells from it in a nutrient-rich medium to promote cell growth and division. *And they spliced the functional allele for the LDL receptor into the genetic material of a harmless virus.* That modified virus served roughly the same function as a hypodermic needle. It was allowed to infect the cultured liver cells and insert copies of the good gene into them.

Later, about a billion modified cells were infused into the woman's portal vein, the main blood vessel that leads directly to the liver. There, some cells took up residence and started to make the missing cholesterol receptor. Two years after that, a fraction of the woman's liver cells were behaving normally and sponging up cholesterol from the blood. Her blood levels of LDLs had declined nearly 20 percent. Also, scans of her arteries showed no evidence of the progressive clogging that had nearly killed her.

Her cholesterol levels remain more than twice as high as normal, and it is too soon to know whether the gene therapy will prolong her life. Yet the intervention does provide strong evidence that gene therapy may be able to help some patients.

As you might gather from this pioneering clinical application, recombinant DNA technology has great potential for medicine. It also has staggering potential

Figure 16.2 One outcome of genetic manipulation by way of artificial selection practices: a large kernel from a modern strain of corn next to the tiny kernels of an ancestral species, which was recovered from a prehistoric cave in Mexico.

for agriculture and industry. Think of it! For thousands of years, we humans have been changing genetically based traits of species. By artificial selection practices, we produced new crop plants and breeds of cattle, cats, dogs, and birds from wild ancestral stocks. We were selective agents for meatier turkeys, sweeter oranges, seedless watermelons, flamboyant ornamental roses, and big juicy corn kernels (Figure 16.2). We produced splendid hybrid organisms, including the mule (horse × donkey) and the plants that produce the tangelo (tangerine × grapefruit).

Of course, we have to remember we are newcomers on the evolutionary stage. During the 3.8 billion years before we even made our entrance, nature conducted uncountable numbers of genetic experiments by way of mutation, crossing over, and other events that introduce changes in genetic messages. Those countless changes are the source of life's rich diversity.

But the striking thing about human-directed changes is that the pace has picked up. Today, researchers use **recombinant DNA technology** to analyze genes. They cut and recombine DNA from different species, then insert it into bacterial, yeast, or mammalian cells that replicate their DNA and divide at a rapid pace. The cells copy the foreign DNA as if it were their own and churn out useful quantities of recombinant DNA molecules. The technology also is the basis of **genetic engineering**, by which genes are isolated, modified, and inserted back into the same organism or into a different one. The protein products of those modified genes might cover the functions of their missing or malfunctioning counterparts.

The new technology does not come without risks. With this chapter, we consider its basic aspects. At the chapter's end, we also address some ecological, social, and ethical questions related to its application.

KEY CONCEPTS

1. Genetic experiments have been proceeding in nature for billions of years, through gene mutations, crossing over, recombination, and other natural events.

2. Humans are now purposefully bringing about genetic changes by way of recombinant DNA technology. Such manipulations are called genetic engineering.

3. With this technology, researchers isolate, cut, and splice together gene regions from different species, then greatly amplify the number of copies of the genes that interest them. The genes, and in some cases the proteins they specify, are produced in quantities that are large enough to use for research and for practical applications.

4. Three activities are at the heart of recombinant DNA technology. First, procedures based on specific types of enzymes are used to cut DNA molecules into fragments. Second, the fragments are inserted into cloning tools, such as plasmids. Third, the fragments containing the genes of interest are identified, then copied rapidly and repeatedly.

5. Genetic engineering involves isolating, modifying, and inserting genes back into the same organism or into a different one. The goal is to beneficially modify traits that the genes influence. Human gene therapy, which focuses on controlling or curing genetic disorders, is an example.

6. The new technology raises social, legal, ecological, and ethical questions regarding its benefits and risks.

16.1 A TOOLKIT FOR MAKING RECOMBINANT DNA

Restriction Enzymes

In the 1950s, the scientific community was agog over the discovery of DNA's structure. The excitement gave way to frustration, for no one could figure out the sequence of nucleotides—the order of genes and gene regions along a chromosome. Robert Holley and his colleagues managed to sequence a relatively small tRNA by using digestive enzymes that could break the molecule, step by step, into fragments small enough to characterize chemically. But DNA? DNA molecules are much longer than RNA. No one knew how to cut one into fragments long enough to have unique, and therefore analyzable, sequences. Digestive enzymes can cut DNA, but they chop away in no particular order. So there was no telling how the fragments had been arranged in the molecule.

Then, by accident, Hamilton Smith discovered that *Haemophilus influenzae*, a bacterium, swiftly responds to a bacteriophage attack by chopping the foreign DNA. Wonderfully, extracts from *H. influenzae* cells held an enzyme that restricts its activity to a specific kind of site in DNA. It was suitably named a **restriction enzyme**.

In time, several hundred strains of bacteria offered up a toolkit of restriction enzymes that recognize and cut specific sequences of four to eight bases in DNA. Table 16.1 lists a few. They are among many that make *staggered* cuts, which leave single-stranded "tails" on the end of a DNA fragment. Tails made by *Taq*I are two bases long (CG). *Eco*RI makes tails four bases long (AATT).

How many cuts do restriction enzymes make? That depends partly on the molecule. To give one example, the eight-base sequence recognized by *Not*I is rare in the DNA of mammals. Usually the resulting fragments are tens of thousands of base pairs long. That is long enough to be useful for studying the organization of a **genome**, which is all the DNA in a haploid number of chromosomes for each species. For the human species, the genome is about 3.2 billion base pairs long.

Table 16.1 Examples of Restriction Enzymes

Bacterial Source	Enzyme's Abbreviation	Its Specific Cut
Thermus aquaticus	*Taq*I	5′ T|C G A 3′ 3′ A G C|T 5′
Escherichia coli	*Eco*RI	5′ G|A A T T C 3′ 3′ C T T A A|G 5′
Nocardia otitidus caviarum	*Not*I	5′ G C|G G C C G C 3′ 3′ C G C C G G|C G 5′

Modification Enzymes

DNA fragments with staggered cuts have sticky ends. "Sticky" means a restriction fragment's single-stranded tail has a capacity to base-pair with a complementary tail of any other DNA fragment or molecule cut by the same restriction enzyme. *Mix together DNA fragments cut by the same restriction enzyme, and the sticky ends of any two fragments having complementary base sequences will base-pair and form a recombinant DNA molecule:*

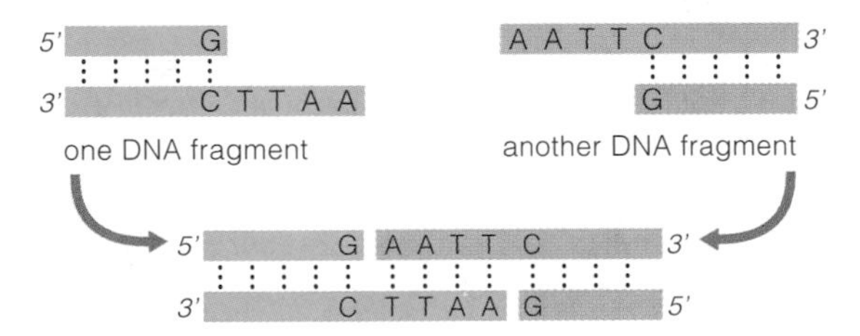

Notice the nicks where such fragments base-pair. **DNA ligase**, a modification enzyme, can seal these nicks:

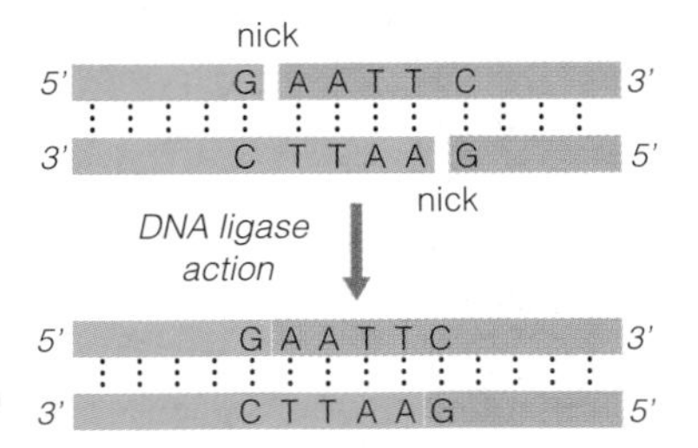

Cloning Vectors for Amplifying DNA

Restriction and modification enzymes made it possible to insert foreign DNA into bacterial cells that have the capacity to replicate it. As you know, a bacterial cell has a single chromosome, a circular DNA molecule with all the genes that it requires to grow and reproduce. Many species of bacteria also may inherit plasmids. As Figure 16.3 shows, a **plasmid** is a small circle of extra DNA. It carries a few genes, and it gets replicated, also. Bacterial cells usually survive without plasmids, although some

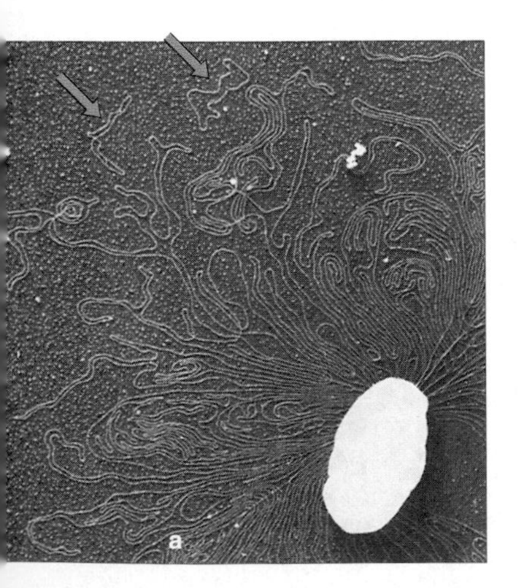

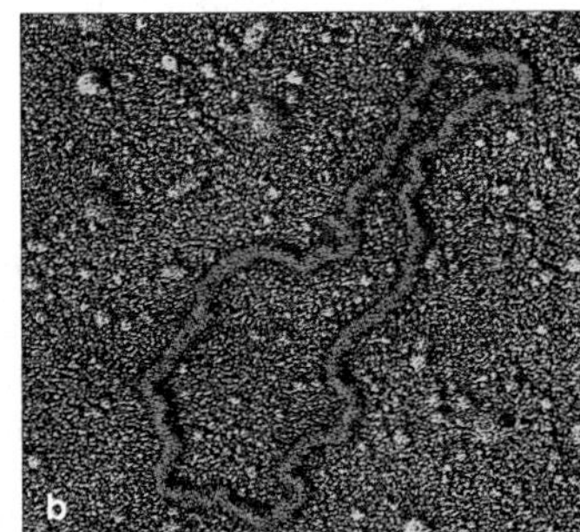

Figure 16.3 (**a**) Plasmids (*blue* arrows) released from a ruptured *Escherichia coli* cell. (**b**) One plasmid at higher magnification.

Further reading: Student Guide to InfoTrac on web site

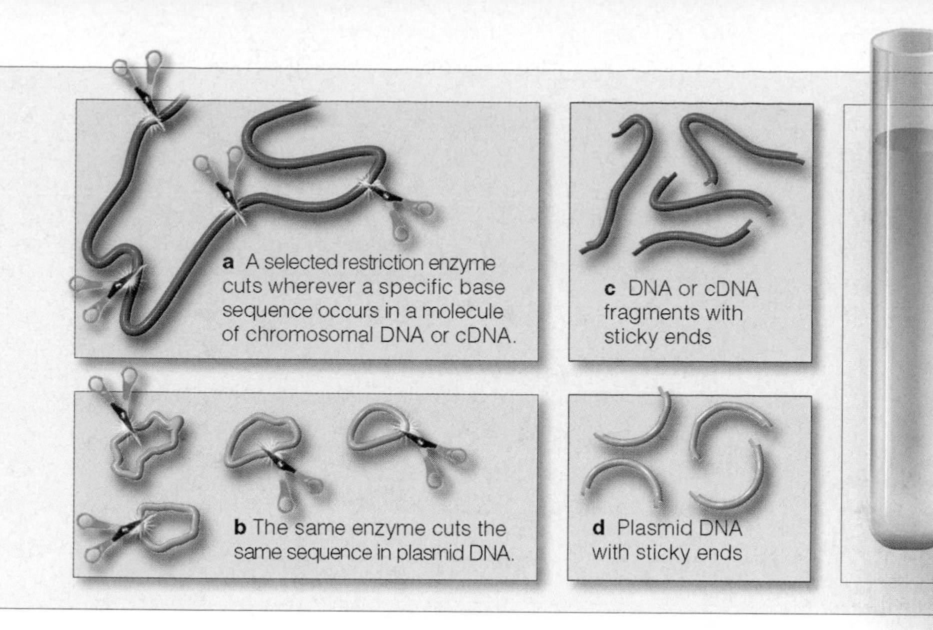

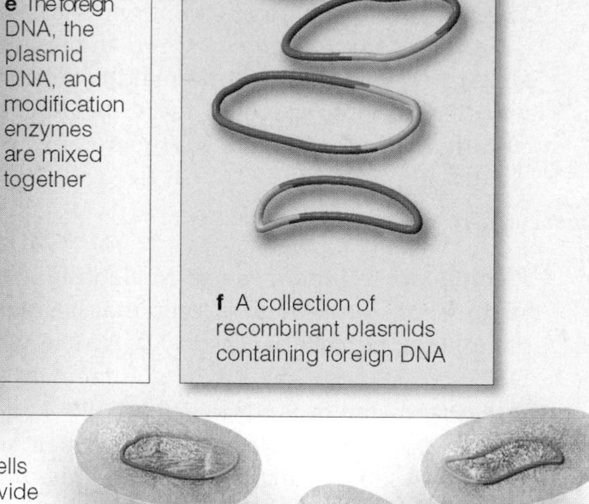

g Host cells able to divide rapidly take up recombinant plasmids

Figure 16.4 (**a**–**f**) Formation of recombinant DNA—in this case, a collection of either chromosomal DNA fragments or cDNA sealed into bacterial plasmids. (**g**) Recombinant plasmids are inserted into host cells that can rapidly amplify the foreign DNA of interest.

plasmid genes do offer special benefits (as when they confer resistance to antibiotics).

Under favorable conditions, bacteria divide rapidly and often—every thirty minutes, for some species. In short order, huge populations of genetically identical cells form. Before every division, each cell's replication enzymes duplicate the bacterial chromosome. They also replicate the plasmid, sometimes repeatedly, so a cell can end up with many identical copies of foreign DNA.

In research laboratories, such foreign DNA can be inserted into a plasmid for replication. The outcome is called a **DNA clone**, because bacterial cells make many identical, "cloned" copies of it.

A modified plasmid that can accept foreign DNA is called a **cloning vector**. It serves as a taxi for delivering foreign DNA into a bacterium, yeast, or some other cell that starts a "cloning factory"—a population of rapidly dividing descendants, all with identical copies of the foreign DNA (Figure 16.4).

Reverse Transcriptase To Make cDNA

Researchers analyze genes to unlock secrets about gene products and how they are put to use. However, even when a host cell takes up a gene, it might not be able to synthesize the protein. For example, most eukaryotic genes incorporate noncoding sequences (introns). New mRNA transcripts of those genes cannot be translated until introns are snipped out and coding regions (exons) spliced together. Bacterial cells can't remove introns, so they often can't translate human genes into proteins.

Researchers get around this problem with **cDNA**, a DNA strand "copied" from a mature mRNA transcript. **Reverse transcriptase**, an enzyme from RNA viruses, catalyzes transcription in reverse. That is, it assembles a complementary DNA strand on an mRNA transcript (Figure 16.5). Other enzymes remove the RNA from the hybrid mRNA–cDNA molecule, make a complementary DNA strand, and thus make double-stranded cDNA.

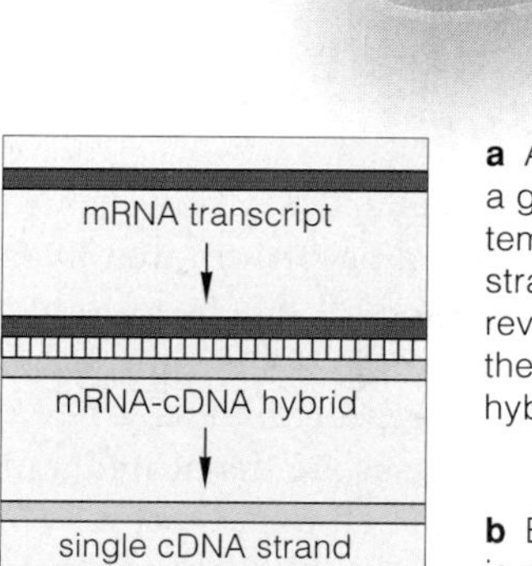

a A mature mRNA transcript of a gene of interest is used as a template to assemble a single strand of DNA. The enzyme reverse transcriptase catalyzes the assembly. An mRNA–cDNA hybrid molecule is the result.

b By enzyme action, the mRNA is removed and a second strand of DNA is assembled on the DNA strand remaining. The outcome is double-stranded cDNA, "copied" from an mRNA template.

Figure 16.5 Formation of cDNA from an mRNA transcript.

Double-stranded cDNA can be further modified, as by attaching signals for transcription and translation. The modified cDNA can be inserted into a plasmid for amplification. Then the recombinant plasmids can be inserted into bacterial cells, which may use the cDNA instructions for synthesizing a protein of interest.

Restriction enzymes and modification enzymes can cut apart chromosomal DNA or cDNA and splice it into plasmids and other cloning vectors. Then the recombinant plasmids can be inserted into bacteria or other rapidly dividing cells that can make multiple, identical quantities of the foreign DNA.

16.2 PCR—A FASTER WAY TO AMPLIFY DNA

The polymerase chain reaction, widely known as **PCR**, is another way to amplify fragments of chromosomal DNA or cDNA. These copy-making reactions proceed with astounding speed in test tubes, not in bacterial cloning factories. Primers get them going.

What Are Primers?

Primers are synthetic, short nucleotide sequences, ten to thirty or so nucleotides long, that base-pair with any complementary sequences in DNA. The workhorses of DNA replication—the DNA polymerases—chemically recognize primers as START tags. Following a computer program, machines synthesize a primer one step at a time. How do researchers decide on the order of those nucleotides? First, they must identify short nucleotide sequences located just before and just after the DNA region from a cell that interests them. Then they build primers that have the complementary sequences.

What Are the Reaction Steps?

For PCR, the enzyme of choice is a DNA polymerase extracted from *Thermus aquaticus*, a bacterium that lives in superheated water of hot springs. It has been found even in water heaters. The enzyme is not destroyed at the elevated temperatures required to unwind a DNA double helix. Most DNA polymerases are denatured and permanently lose their activity at such temperatures.

Researchers mix together primers, the polymerase, cellular DNA from an organism, and free nucleotides. Next, they expose the mixture to precise temperature cycles. During each temperature cycle, the two strands of all the DNA molecules in the mixture unwind from each other.

Primers become positioned on exposed nucleotides at targeted sites according to base-pairing rules (Figure 16.6). Each round of reactions doubles the number of DNA molecules amplified from the target site. Thus, if there are 10 such molecules in the test tube, there soon will be 20, then 40, 80, 160, 320, 640, 1,280, and so on. Thus a targeted region from a single DNA molecule can be rapidly amplified to billions of molecules.

In short, *PCR amplifies samples that contain even tiny amounts of DNA*. At this writing, it is the amplification procedure of choice in thousands of laboratories all over the world. As you will see in the next section, such samples can be obtained even from a single hair follicle or drop of blood left at the scene of a crime.

PCR is a method of amplifying chromosomal DNA or cDNA inside test tubes. Compared to cloning methods, PCR is far more rapid.

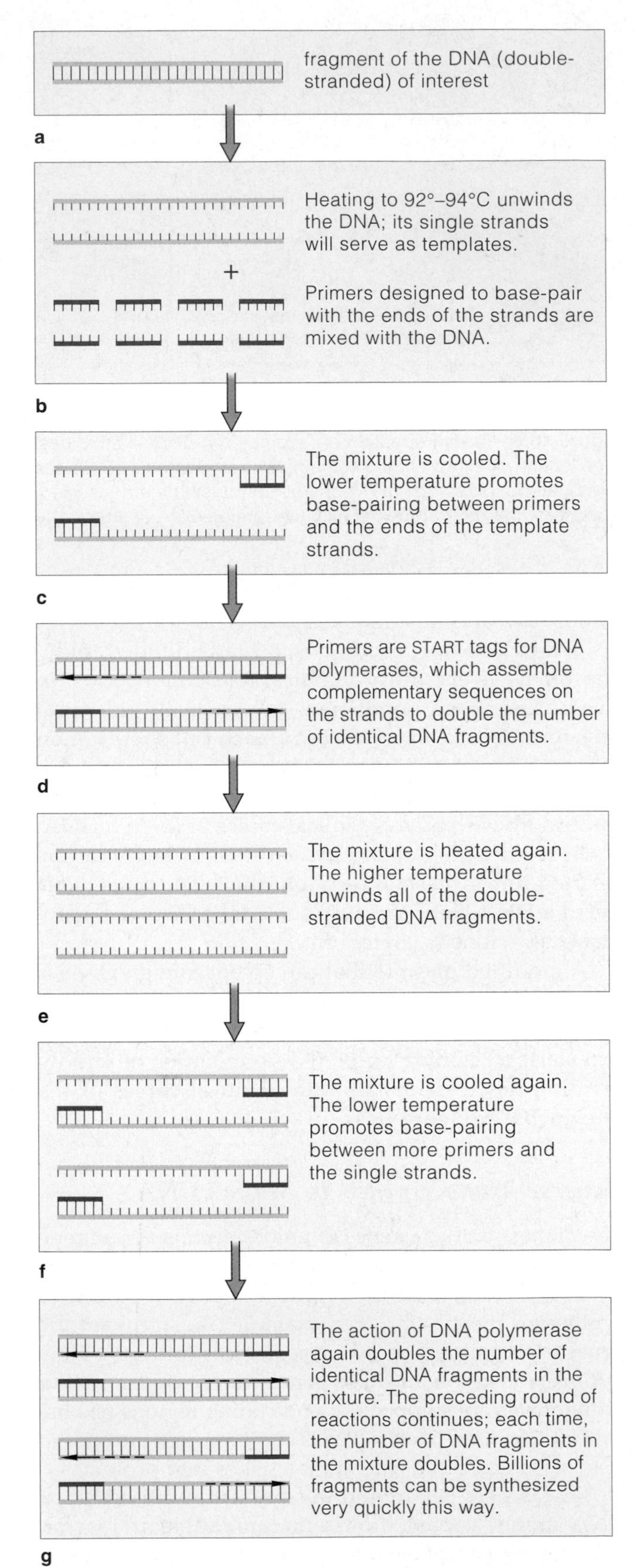

Figure 16.6 The polymerase chain reaction (PCR).

Further reading: Student Guide to InfoTrac on web site

DNA FINGERPRINTS

Except for identical twins, no two people have exactly the same sequence of bases in their DNA. By detecting the differences in DNA sequences, scientists can distinguish one person from another. As you know, each human has a unique set of fingerprints, a marker of his or her identity. Like all other sexually reproducing species, each human also has a **DNA fingerprint**, which is a unique array of DNA fragments that were inherited from each parent in a Mendelian pattern. DNA fingerprints are so accurate that even full siblings are readily distinguished from one another.

More than 99 percent of the DNA is exactly the same in all humans. But DNA fingerprinting focuses only on the part that tends to differ from one person to the next. Throughout the human genome are **tandem repeats**—short regions of repeated DNA—that differ substantially among people. For example, the five bases TTTTC are repeated anywhere from four to fifteen times in tandem in different people, and three bases (CGG) are repeated five to fifty times in tandem.

The number of repeats may increase or decrease during DNA replication. Mutation rates in tandem repeats are much higher than the rates at most other sites in DNA. Because such mutations occurred over many generations in many different family lineages, each person is usually heterozygous for a repeat number at any given tandem-repeat locus. By examining many tandem-repeat sites, researchers found out that each person carries a unique combination of repeat numbers.

Researchers detect differences at tandem-repeat sites with **gel electrophoresis**, a laboratory technique that uses an electric field to force molecules through a viscous gel (Section 14.1, Figure 14.4). In this case, it separates DNA fragments according to their length. Size alone dictates how far each fragment moves through the gel, so tandem repeats of different sizes migrate at different rates.

A gel is immersed in a buffered solution, then DNA fragments from individuals are added to the gel. When an electric current is applied to the solution, one end of the gel takes on a negative charge, and the other end a positive charge. DNA molecules carry a negative charge (because of the negatively charged phosphate groups), so they migrate through the gel toward the positively charged pole. They do so at different rates, and so they separate into bands according to length. The smaller the fragment, the farther it will migrate. After a set time, researchers can identify fragments of different lengths by staining the gel or by specifically highlighting fragments that contain tandem repeats.

Figure 16.7 shows a series of tandem-repeat DNA fragments that were separated by gel electrophoresis. They are DNA fingerprints from seven individuals and from blood collected at a crime scene. Notice how much their DNA fingerprints differ. Can you identify which one of the patterns exactly matches the pattern from the crime scene? DNA fingerprints help forensic scientists identify criminals and victims, and exclude innocent suspects. A few drops of blood, semen, or cells from a hair follicle at a crime scene or on a suspect's clothing often yield enough to do the trick. Such an analysis even confirmed that human bones exhumed from a shallow pit in Siberia belonged to five members of the Russian imperial family, all shot to death in secrecy in 1918. In addition, because DNA fragments that make up a DNA fingerprint are inherited, scientists use them to resolve questions of paternity.

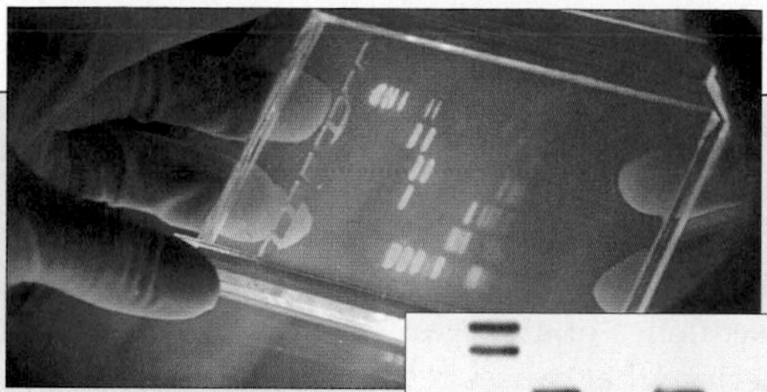

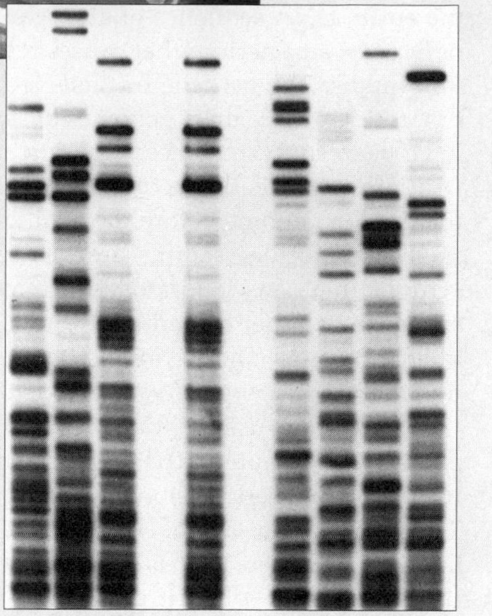

Figure 16.7 *Right:* Comparison of DNA fingerprints from a bloodstain discovered at a crime scene and from blood samples of seven suspects (the circled numbers). Given what you have learned about gel electrophoresis, point out which of the seven is a match. *Above:* The gel shown in this photograph had been stained earlier with a substance that can make DNA fragments fluoresce when placed under ultraviolet light.

When DNA fingerprinting was first used as evidence in court, attorneys challenged conclusions based on it. But DNA fingerprinting is now firmly established as an accurate and unambiguous way to identify individuals and paternity. It is widely used to convict the guilty and exonerate innocent suspects.

For example, in 1998, an eleven-year-old girl was raped, stabbed, and strangled to death in Elisabethfehn, Germany. In the largest mass screening to date, police requested blood samples from 16,400 males in and around the vicinity. A single DNA fingerprint from a thirty-year-old mechanic matched the evidence, and he confessed to the crime.

The variation in tandem repeats also can be detected as restriction fragment length polymorphisms, or RFLPs (DNA fragments of different sizes that had been cleaved by restriction enzymes). In the case of tandem repeats, a restriction enzyme cleaves the DNA flanking the repeat. Alternatively, researchers might use PCR to amplify the tandem repeat region. Either way, differences in the size of the fragments, which reveal genetic differences, can be detected with electrophoresis.

16.4 HOW IS DNA SEQUENCED?

In 1995, researchers accomplished a feat that was little more than a dream a few decades ago. They determined the entire DNA sequence of an organism—*Haemophilus influenzae*, a bacterium that causes respiratory infections in humans. The genome of this species has 1.8 million nucleotides. Since then, genomes of several other species have been fully sequenced. By current projections, we will know the sequence of all 3.2 billion nucleotides of the human genome before the year 2002.

The molecular sleuths are using **automated DNA sequencing**. This laboratory technique can reveal the sequence of either cloned DNA or PCR-amplified DNA in a few hours. The technique has all but replaced the more laborious, expensive methods that preceded it.

Researchers use the four standard nucleotides (T, C, A, and G) for automated DNA sequencing. They also use four modified versions, which we can represent as T*, C*, A*, and G*. Each modified version has been labeled. Attached to it is a molecule that fluoresces a particular color when it passes through a laser beam. Each time a modified nucleotide becomes incorporated in a growing DNA strand, it blocks DNA synthesis.

Before the reactions, researchers mix the eight kinds of nucleotides together. To that mixture they add the millions of copies of the DNA to be sequenced, a type of primer, and DNA polymerase. Then they separate the DNA into single strands, and the reactions begin.

The primer binds with its complementary sequence on one of the strands. DNA polymerase synthesizes a new DNA strand right behind the primer. One by one, it adds nucleotides in the order dictated by the exposed sequence. Each time, one of the standard nucleotides *or* one of the modified versions may be attached.

Suppose that DNA polymerase encounters a T in a template strand. It will catalyze the base-pairing of either A or A* to it. If A is added to the new strand, replication will continue. If A* is added, replication will stop; the modified nucleotide will *block* addition of more nucleotides to that DNA strand. The same thing happens at each nucleotide in a template strand. If a standard nucleotide is added, replication will continue; if a modified version is added, replication stops.

Remember, the starting mixture contained millions of identical copies of the DNA sequence. Because either a standard or modified nucleotide could be added at each exposed base, the new strands ended at different locations in the sequence. The mixture now contains millions of copies of tagged fragments having different lengths. These can be separated by length into *sets* of fragments. And each set corresponds to just one of the nucleotides in the entire base sequence.

The automated DNA sequencer is a machine that separates the sets of fragments by gel electrophoresis. The set having the shortest fragments migrates fastest through the gel and peels away from it first. The last set to peel away has the longest fragments. Because its fragments have a modified nucleotide at the 3′ end, the set fluoresces a particular color as it passes through a laser beam (Figure 16.8*a*). The automated sequencer detects the color and thus determines which nucleotide is on the end of the fragments in each set. It assembles the information from all the nucleotides in the sample and reveals the entire DNA sequence.

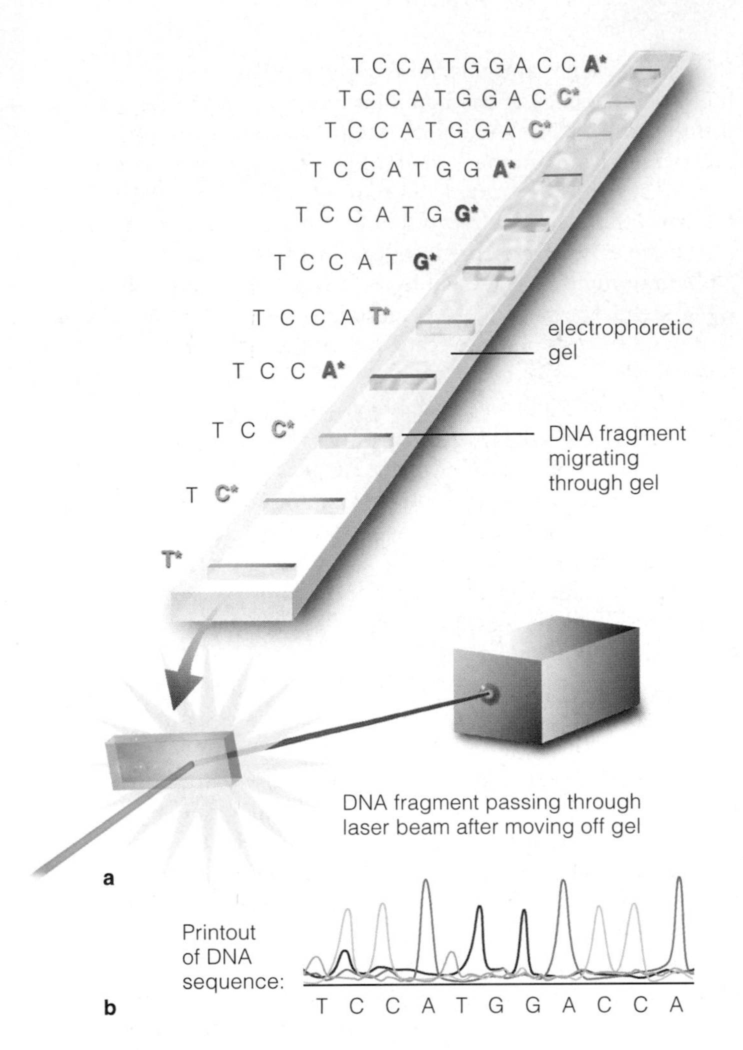

Figure 16.8 Automated DNA sequencing. (**a**) DNA fragments from an organism's genome become labeled at their end with a modified nucleotide that fluoresces a certain color. (**b**) Printout of the DNA sequence used in this example. Each peak indicates absorbance by a particular labeled nucleotide.

Figure 16.8*b* shows the printout from an automated DNA sequencer. Each peak in the tracing represents the detection of a particular color as the sets of fragments reached the end of the gel.

With automated DNA sequencing, the order of nucleotides in a cloned or amplified DNA fragment can be determined.

Further reading: Student Guide to InfoTrac on web site

16.5 FROM HAYSTACKS TO NEEDLES—ISOLATING GENES OF INTEREST

Any genome consists of thousands of genes. There are about 4,000 in the *E. coli* genome, for example, and over 100,000 in the human genome. What if you wanted to learn about or modify the structure of any one of those genes? First you would have to isolate that gene from all others in the genome.

There are several ways to do this. If part of the gene sequence is already known, you can design primers to amplify the whole gene or part of it with PCR. Often, though, researchers must isolate and clone a gene. First they make a **gene library**. This is a mixed collection of bacteria that contain different cloned DNA fragments, one of which is the gene of interest. A *genomic* library contains cloned DNA fragments from an entire genome. A *cDNA* library contains DNA derived from mRNA. A cDNA library is usually the most useful, for it is free of introns. Even then, the gene of interest is like a needle in a haystack. How can you isolate it from all the others in the library? One way is to use a nucleic acid probe.

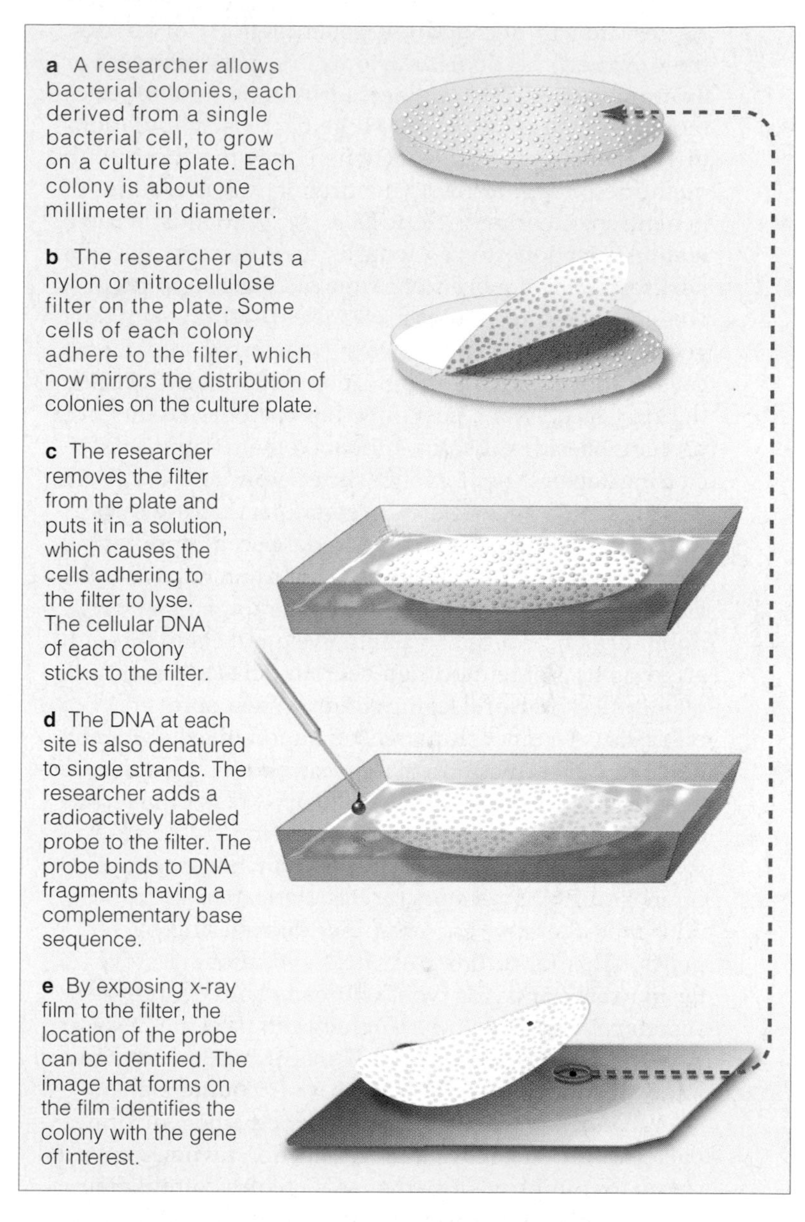

Figure 16.9 Use of a probe to identify bacterial colonies that have taken up a gene library.

What Are Probes?

A **probe** is a very short stretch of DNA labeled with a radioisotope so that it can be distinguished from other DNA molecules in a given sample. Part of the probe must be able to base-pair with some portion of the gene of interest. Any base-pairing that takes place between sequences of DNA (or RNA) from different sources is called **nucleic acid hybridization**.

How do you acquire a suitable probe? Sometimes part of the gene or a closely related gene has already been cloned, in which case you can use it as a probe. If the gene's structure is a mystery, you still may be able to work backward from the amino acid sequence of its protein product, assuming the protein is available. By using the genetic code as a guide (Section 14.3), maybe you can build a DNA probe that is more or less similar to the gene of interest.

Screening for Genes

Once you have a gene library and a suitable probe, you are ready to hunt down the gene. Figure 16.9 shows the steps of one isolation method. The first step is to take bacterial cells of the library and spread them apart on the surface of a gelled growth medium in a petri plate. When spread out sufficiently, individual cells undergo division. Each cell starts a colony of genetically identical cells. The bacterial colonies appear as hundreds of tiny white spots on the surface of a culture medium.

After colonies appear, you lay a nylon filter on top of the colonies. Some cells stick to the filter at locations that mirror the locations of the original colonies. You use solutions to rupture the cells, and the DNA so released sticks to the filter. You make the DNA unwind to single strands and then add the probes. The probes hybridize only with DNA from the colony that took up the gene of interest. If you expose the probe-hybridized DNA to x-ray film, the pattern formed by the radioactivity will identify that colony. With this information you can now culture cells from that one colony, which will replicate only the cloned gene.

Probes may be used to identify one particular gene among many in gene libraries. Bacterial colonies that have taken up the library can be cultured to isolate the gene.

16.6 USING THE GENETIC SCRIPTS

As researchers decoded the genetic scripts of species, they opened the door to astonishing possibilities. For example, genetically engineered bacteria now produce medically valued proteins. Huge bacterial populations thrive in stainless steel vats, where they produce useful quantities of the desired gene products. Beneficiaries of genetic engineering include *diabetics*, who must receive insulin injections for as long as they live. Insulin is a pancreatic hormone. At one time, medical supplies of it were extracted from pigs and cattle. Later on, synthetic genes for human insulin were transferred into *E. coli* cells. Those cells gave rise to populations that became the first large-scale, cost-effective bacterial factory for proteins. Besides insulin, human somatotropin, blood-clotting factors, hemoglobin, interferon, and a variety of drugs and vaccines are also manufactured this way.

Genetically engineered bacteria also hold potential for industry and for cleaning up environmental messes. For instance, as you know, many microorganisms break down organic wastes and help cycle nutrients through ecosystems. Some modified bacteria break down crude oil into less harmful compounds. When sprayed onto oil spills, as from a shipwrecked supertanker, they may help avert an environmental disaster. Other bacteria are genetically engineered to sponge up excess phosphates or heavy metals from the environment.

In addition, bacterial species that house plasmids offer astonishing benefits for basic research, agriculture, and gene therapy. For example, deciphering bacterial genes helps us reconstruct life's evolutionary story. In Sections 20.4 and 21.6, you will read about evolutionary secrets revealed when researchers compare the DNA or RNA of different organisms. The next two sections give a few more examples.

What about the "bad" bunch—the pathogenic fungi, bacteria, and viruses? Natural selection favors mutated genes that improve a pathogen's chances of evading a host organism's natural defenses. Mutation is frequent among pathogens that reproduce rapidly, so designing new antibiotics and other defenses against new gene products is a constant challenge. But if we learn about the genes, we may have advance warning of the "plan of attack." The story of HIV, a rapidly mutating virus that causes AIDS, gives insight into the magnitude of the problem (Section 40.11).

Knowing about genes allows us to genetically engineer beneficial microorganisms for uses in medicine, industry, agriculture, and environmental remediation.

Knowing about genes can help us reconstruct evolutionary histories of species.

Knowing about genes can help us devise counterattacks against rapidly mutating pathogens.

16.7 DESIGNER PLANTS

Regenerating Plants From Cultured Cells

Many years ago, Frederick Steward and his coworkers cultured cells from carrot plants and induced some of them to develop into small embryos. As you will read in Section 31.7, some embryos grew into whole plants. Today, researchers routinely regenerate crop plants and many other plant species from cultured cells. They use various methods to pinpoint a gene in a culture that contains, say, millions of cells. Suppose the researcher includes a toxic product of a pathogen in the culture medium. If only a few cells carry a gene that confers resistance to the toxin, they will be the only ones left.

Once whole plants are regenerated from preselected cultured cells, researchers can hybridize them with other varieties. The idea is to transfer desired genes to a plant lineage to improve herbicide resistance, pest resistance, and other traits that can benefit crops and gardens.

To protect the food supply for most of the human population, botanists are combing the world for seeds of the wild ancestors of potatoes, corn, and other crop plants. They send their prizes—seeds that carry genes of a plant's lineage—to seed banks. These safe storage facilities preserve the genetic diversity. The supply is vulnerable; farmers focus mainly on genetically similar varieties of high-yield crop plants. Genetic uniformity makes our food crops dangerously vulnerable to many kinds of pathogenic fungi, viruses, and bacteria.

To give an example, a new fungal strain of *Southern corn leaf blight* destroyed much of the 1970 corn crop in the United States. This widespread epidemic happened because all of the plants were genetically similar; they all carried the same gene that conferred susceptibility to the disease. Since then, seed companies have become more attentive to offering corn with genetic diversity. To introduce genetic diversity into existing varieties, plant breeders must tap into the seeds in seed storage.

The search for beneficial genes is a race against time. Farmers throughout the world want high yields, which many of the genetically uniform varieties provide. They discard older, more diverse varieties. By doing so, they also discard the genetic diversity that is so essential for the future of our food supply.

Figure 16.10 Crown gall tumor on a woody plant, an abnormal tissue growth triggered by a gene carried by the Ti plasmid.

Further reading: Student Guide to InfoTrac on web site →

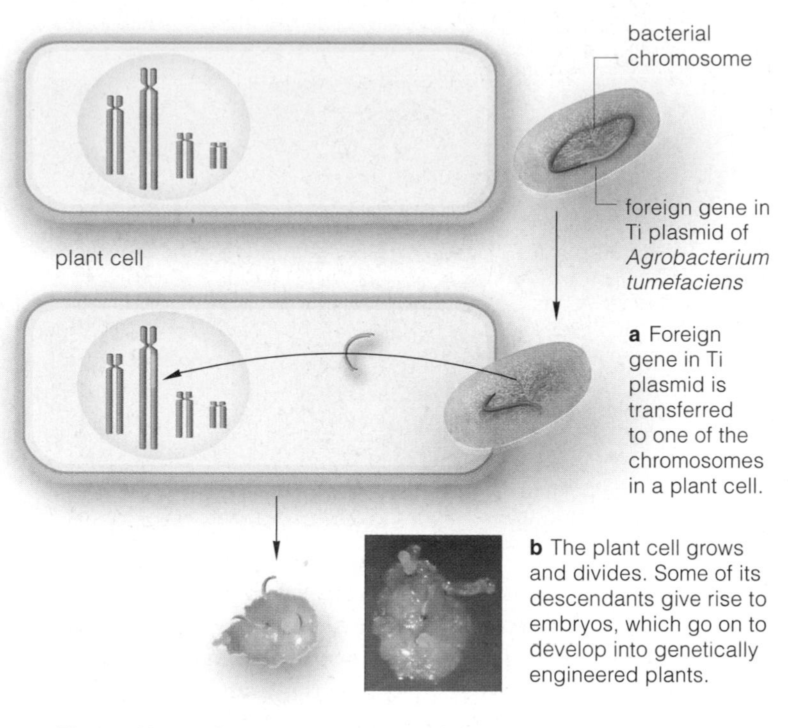

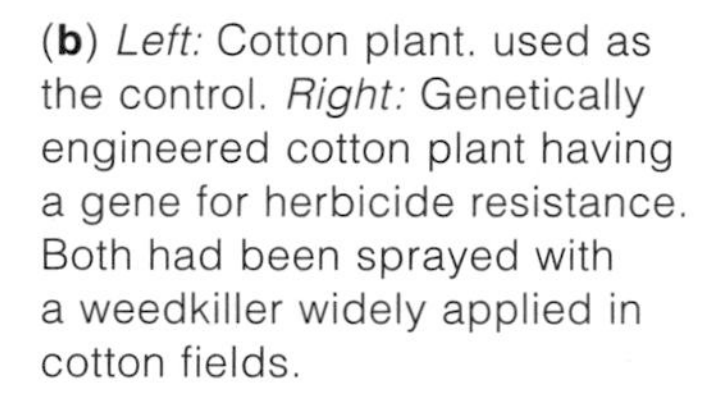

Figure 16.11 Gene transfer from the bacterium *Agrobacterium tumefasciens* to a plant cell, by way of the Ti plasmid gene.

Figure 16.13 (**a**) Control plant (*left*) and three genetically engineered aspen seedlings. Vincent Chiang and coworkers suppressed a regulatory gene of a lignin biosynthetic pathway. The modified plants synthesized normal lignin, but not as much. Lignin production decreased by as much as 45 percent—yet cellulose production increased 15 percent. Root, stem, and leaf growth were greatly enhanced. Plant structure did not suffer. Wood harvested from such trees might make it easier to manufacture paper and some clean-burning fuels, such as ethanol. (Lignin, a tough polymer, reinforces secondary cell walls of plants. It must be chemically extracted from the wood used to make paper.)

(**b**) *Left:* Cotton plant. used as the control. *Right:* Genetically engineered cotton plant having a gene for herbicide resistance. Both had been sprayed with a weedkiller widely applied in cotton fields.

How Are Genes Transferred Into Plants?

A simple example gives insight into how genes can be transferred into a plant. Genetic engineers often insert new or modified genes into the Ti (*T*umor-*I*nducing) plasmid from *Agrobacterium tumefaciens*. This bacterium infects many species of flowering plants. Some plasmid genes invade a plant's DNA, then induce the formation of abnormal tissue masses known as crown gall tumors (Figure 16.10). Before introducing the Ti plasmid into plant cells, researchers first remove the tumor-inducing genes and insert a desired gene into it. They place plant cells in a culture of the modified bacteria. When some cells take up the gene, whole plants may be regenerated from the cellular descendants, as in Figure 16.11. The expression of foreign genes in plants can sometimes be quite dramatic, as shown in Figure 16.12, at left.

In nature, *A. tumefaciens* infects the plants called dicots,

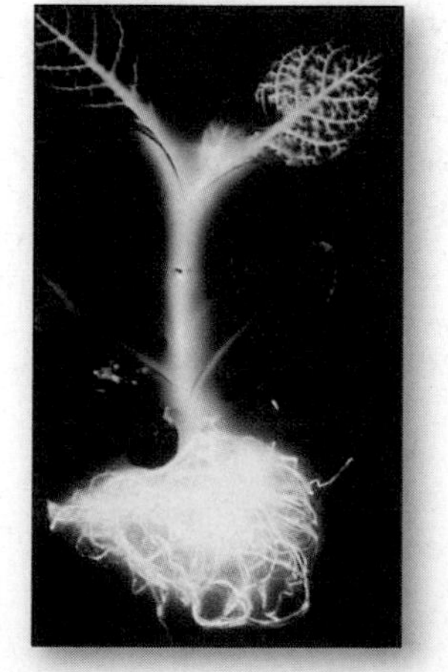

Figure 16.12 A modified plant that glows in the dark owing to a gene transfer. A firefly gene was inserted into the plant's DNA and is being expressed. As described in the Chapter 6 introduction, the gene's product is luciferase, an enzyme required for bioluminescence.

which include beans, peas, potatoes and other essential crops. Geneticists are working to modify the bacterium so it also can infect monocots that are vital food crops, including wheat, corn, and rice. In some cases, they use electric shocks or chemicals to deliver modified genes into plant cells. As another delivery trick, they blast microscopic particles coated with DNA into the cells.

Despite many obstacles, improved varieties of crop plants have been engineered or are in the works. For example, genetically engineered cotton plants display resistance to an herbicide (Figure 16.13). Farmers spray the herbicide in a field of the modified cotton plants and kill the weeds, and the plants remain unaffected.

Also on the horizon are engineered plants that can serve as factories for pharmaceuticals. A few years ago, genetically engineered tobacco plants that make human hemoglobin and other proteins were planted in a field in North Carolina. Ecologists later found no trace of foreign genes or proteins in the soil or in other plants or animals in the vicinity. As a final example, mustard plant cells grown in a Stanford University laboratory have synthesized biodegradable plastic beads that may be used for manufacturing plastics.

Genetically diverse plant species are essential to protecting our vulnerable food supply. Genetic engineers can design plants with new beneficial traits.

16.8 GENE TRANSFERS IN ANIMALS

Supermice and Biotech Barnyards

The first mammals enlisted for experiments in genetic engineering were laboratory mice. Consider an example of this work. R. Hammer, R. Palmiter, and R. Brinster corrected a hormone deficiency that leads to dwarfism in mice. Insufficient levels of somatotropin (or growth hormone) cause the abnormality. The researchers used a microneedle to inject the gene for rat somatotropin into fertilized mouse eggs, which they implanted in an adult female. The gene was successfully integrated into mouse DNA, and the eggs developed into mice. Young mice in which that foreign gene was expressed were 1–1/2 times larger than the dwarf littermates. In other experiments, researchers transferred the gene for human somatotropin into a mouse embryo. There, it became integrated into mouse DNA. The genetically modified embryo developed into a "supermouse" (Figure 16.14).

Today, human genes are being inserted into mouse embryos as part of ongoing research into the molecular basis of genetic disorders, such as Alzheimer's disease. In addition to microneedles, microscopic laser beams are used to penetrate the plasma membrane of cultured cells. The new methods have varying degrees of success.

Animals of "biotech barnyards" are competing with bacterial factories as genetically engineered sources of proteins. For example, goats produce CFTR protein (for treating cystic fibrosis) and TPA (which may diminish the severity of heart attacks). Cattle may soon produce human collagen for repairing cartilage, bone, and skin.

In 1997 researchers reported the first case of cloning a mammal from an adult cell. They extracted a nucleus from a ewe's mammary gland cell and inserted it into an egg (from another ewe) from which the nucleus had been removed. Signals from the egg cytoplasm sparked the development of an embryo, which was implanted into a surrogate mother. As described in Section 13.5, the result was the clone *Dolly*.

Although animals have been routinely cloned from embryonic tissue for years, cloning from adult animal cells has advantages. An adult animal already displays a version of some trait that may be sought after, and cloning might maintain its genotype indefinitely.

The experiment that produced Dolly opened up the possibility of making genetically engineered clones of sheep, cattle, and other farm animals from adults with proven genetic potential. (It also raised the jarring idea that human clones might be just around the corner. The United States promptly imposed a ban on the federal funding of research into human cloning.)

One year later, Steve Stice and his colleagues used similar cloning procedures and went further down the road to designer cattle. Starting with a cell culture from cattle, they induced the growth and development of six genetically identical calves. More importantly, they also engineered specific, targeted alterations in the cloning cell lineage. Their goal is to genetically modify cattle for resistance to bovine spongiform encephalopathy. (You may know this condition as mad cow disease. Humans who eat infected meat are vulnerable to it; Section 22.9.) Their technique also is being used to insert the gene for human serum albumin into the chromosomes of dairy cows. This protein is medically useful because it helps control blood pressure. At present, it must be separated from large quantities of donated human blood. Imagine how much easier it would be to obtain large quantities of the protein from bountiful supplies of milk.

Figure 16.14 Evidence of a successful gene transfer. Two ten-week-old mouse littermates. *Left:* This one weighs 29 grams. *Right:* This one weighs 44 grams. It grew from a fertilized egg into which a gene for human somatotropin had been inserted.

Mapping and Using the Human Genome

In the early 1980s, biologists were in an uproar over a costly proposal before the National Institutes of Health (NIH) to map the entire human genome. Many argued that benefits would be incalculable—not only for pure research, but also for efforts to cure genetic disorders and combat diseases. To others, it was a boondoggle; such mapping could not be done, and it would divert NIH funding from worthy endeavors. But then Leroy Hood, a molecular biologist, invented automated DNA sequencing (Section 16.4), and the race was on.

By 1990, the Human Genome Initiative was under way. The international effort came with a projected price tag of 3 billion dollars—about 92 cents for each base pair in the heritable script for human life. For the first phase, biologists were to pinpoint the location of the stretches of nucleotides that were already sequenced. Once those "reference markers" were assigned to the chromosomes, biologists could fill in the long spans between them.

By 1991, not even 2,000 genes had been sequenced. But the pace quickened after Craig Ventner realized that even a tiny bit of cDNA could be used as a molecular hook to drag the whole sequence of its parent gene out

Further reading: Student Guide to InfoTrac on web site →

of a cDNA library. He named the hooks ESTs (*E*xpressed *S*equence *T*ags). By tapping commercial sources, Ventner and his colleague, Mark Adams, put together a cDNA library that had been produced from various neurons, the communication cells of nervous systems. Within a few months, they had sequenced 2,000 more genes. By 1994, about 35,000 genes had been decoded.

Later, Ventner and Hamilton Smith used a software program, the TIGR Assembler, to work out the first full genome of an organism—the bacterium *H. influenzae* you read about in Section 16.1. Through application of advanced technologies, the Human Genome Initiative will be completed ahead of schedule.

Even when the whole human genome is sequenced, it will not be easy to manipulate it to advantage. Put yourself in the position of some genetic researcher. You have modified a human gene and you have to figure out how to get it into a host cell of a particular tissue. That's not all. You have to be sure the modified gene gets inserted at a suitable site—say, at or very near the locus for its abnormal counterpart in a chromosome. You must also make sure the cell will synthesize the specified protein at suitable times, in suitable amounts.

Today, most experimenters employ stripped-down viruses as vectors to put genes into cultured human cells, which may incorporate the foreign genes into their own DNA. However, the viral genetic material can undergo rearrangements, deletions, and other alterations, which can shut down or disrupt gene expression. Synthetic, streamlined versions of human chromosomes might be developed that will be recognized by a cell's machinery for DNA replication and protein synthesis.

In some cases of gene therapy, simply inserting a number of genetically modified cells into a tissue does the individual good, even if the cells make as little as 10 to 15 percent of the required protein. The introduction to this chapter is a case in point. So far, however, no one can predict where new genes will end up. The danger is that a gene insertion may disrupt the function of some other gene, including the genes controlling cell growth and cell division. One-for-one gene swaps by way of a process called homologous recombination are possible. But Oliver Smithies, one of the best at this, currently puts genes right where they are supposed to go, once every 100,000 or so tries. The body may be somewhat forgiving of a small load of cells bearing near-misses. But mistakes in germ cells—which give rise to all cells of a new individual—probably would have unexpected and disastrous effects on the body's development.

Although the technical details are still being worked out, modified genes are being transferred into cells of humans and other mammals during experimental and clinical trials.

Focus on Bioethics

16.9 WHO GETS ENHANCED?

This chapter opened with a historic, inspiring case, the first proof that gene therapy may help save human lives. It closes with questions that invite you to consider some social and ethical issues surrounding the application of recombinant DNA technology to the human genome.

To most of us, human gene therapy to correct genetic abnormalities seems like a socially acceptable goal. Let's take this idea one step further. Is it also socially desirable or acceptable to change certain genes of a normal human individual (or sperm or egg) to alter or enhance traits?

The idea of selecting desirable human traits is called *eugenic engineering*. Yet who decides which forms of a trait are most "desirable"? For example, would it be okay to engineer taller or blue-eyed or fair-skinned boys and girls? Would it be okay to engineer "superhumans" with amazing strength or intelligence? Are there some people narcissistic enough to commission a clone of themselves, one with a few genetically engineered "enhancements"?

Recently, more than 40 percent of Americans surveyed say gene therapy would be okay to make their offspring smarter or better looking. A poll of British parents found 18 percent willing to use genetic enhancement to prevent children from being aggressive and 10 percent to keep them from growing up to be homosexual.

There are those who say the DNA of any organism must never be altered. Put aside the fact that nature itself alters DNA much of the time and has done so for nearly all of life's history. The concern is that *we* do not have the wisdom to bring about beneficial changes without causing great harm to ourselves or to the environment. To be sure, when it comes to altering human genes, one is reminded of our very human tendency to leap before we look. And yet, when it comes to restricting genetic modifications of any sort, one also is reminded of another old saying: "If God had wanted us to fly, he would have given us wings." Something about the human experience did give us a capacity to imagine wings of our own making—and that capacity carried us to the frontiers of space.

Where are we going from here? To gain perspective on the question, spend some time reading the history of the human species. It is a history of survival in the face of challenges, threats, bumblings, and sometimes disasters on a grand scale. It is also a story of our connectedness with the environment and with one another.

The basic questions confronting you are these: Should we be more cautious, believing the risk takers may go too far? And what do we as a species stand to lose if risks are not taken? At this writing, a young man with a genetic disorder died after gene therapy that used an adenovirus as a vector. Yet two *SCID-X1* infants with a mutant gene that disabled their immune system may be cured. A viral vector put copies of the nonmutated gene into stem cells taken from their bone marrow. Months after the modified cells were reinfused into their bone marrow, both "bubble boys" left their isolation tents. Their immune system is working normally; they are healthy and living at home.

16.10 SAFETY ISSUES

Many years have passed since foreign DNA was first transferred into a plasmid. That gene transfer ignited a debate that will continue well into the next century. The issue is this: *Do potential benefits of gene modifications and gene transfers outweigh potential dangers?*

Genetically engineered bacteria are "designed" so they cannot survive except in the laboratory. As added precautions, "fail-safe" genes are built into the foreign DNA in case they escape. These genes are silent unless the captives are exposed to environmental conditions—whereupon the genes get activated, with lethal results for their owner. Say the package includes a *hok* gene next to a promoter of the lactose operon (Section 15.2). Sugars are plentiful in the environment. If they were to activate the *hok* gene, the protein product of that gene will destroy membrane function and the wayward cell.

What about a worst-case scenario? Remember how retroviruses are now being used to insert genes into cultured cells? If they escape from laboratory isolation and infect organisms, what might be the consequences?

And what about genetically engineered plants and animals released into the environment? For example, Steven Lindlow thought about how frost destroys many crops. Knowing that a surface protein of a bacterium promotes formation of ice crystals, he excised the "ice-forming" gene from bacterial cells. As he hypothesized, spraying "ice-minus bacteria" on strawberry plants in an isolated field prior to a frost would help plants resist freezing. He actually had deleted a *harmful* gene from a species, yet a bitter legal battle ensued. The courts ruled in his favor. His coworkers sprayed a strawberry patch; nothing bad happened.

Then there was the potato plant designed to kill the insects that attack it. It also was too toxic for people to eat. Or think of how crop plants compete poorly with weeds for nutrients. Many have been designed to resist weedkillers so that farmers can spray for weeds and not worry about killing their crops. Herbicides do have toxic effects on more than their targets (Section 3.6). If crop plants offer herbicide resistance, will farmers be less or more apt to spread them about?

And what if engineered plants or animals transfer modified genes to organisms in the wild? Think of how the advantage in acquiring resources would tilt toward vigorous weeds blessed with herbicide-resistant genes. Such possibilities are why standards for rigorous and extended safety tests must be in place *before* modified organisms enter the environment.

For more on safety issues, turn to *Critical Thinking* question 8 at the chapter's end.

Rigorous safety tests are carried out before genetically modified organisms are released into the environment.

16.11 SUMMARY

1. Uncountable numbers of gene mutations and other forms of genetic "experiments" have been proceeding in nature for at least 3 billion years.

2. Through artificial selection practices, humans have manipulated the genetic character of many species for many thousands of years. Currently, recombinant DNA technology has enormously expanded our capacity to genetically modify organisms.

3. A genome is all the DNA in the haploid chromosome number for a species. By recombinant DNA technology, a genome can be cut into fragments, then the fragments can be amplified (copied over and over again) to make useful quantities that permit analysis of the nucleotide sequence of the genome or a specific portion of it.

4. Researchers work with chromosomal DNA or with cDNA (a DNA strand transcribed by the enzyme reverse transcriptase from a mature mRNA transcript). The first choice is better for questions about DNA regions that control gene expression or that contain introns. cDNA is a better choice for questions about the amino acid sequence of a protein. Either way, researchers use restriction enzymes, which cut the DNA into fragments.

5. The use of plasmids or other cloning vectors is one way to amplify DNA fragments. Many bacteria contain plasmids, which are small circles of DNA with a few genes in addition to those of the bacterial chromosome.

a. Certain restriction enzymes make staggered cuts that leave the fragments with single-stranded tails. Such tails base-pair with complementary tails of any other DNA cut by the same enzyme, such as plasmid DNA.

b. DNA ligase, a modification enzyme, seals base-pairing sites between plasmid DNA and foreign DNA. A plasmid modified to accept foreign DNA is a cloning vector; it can deliver foreign DNA into a bacterium or some other cell that can start a population of rapidly dividing descendant cells. All of the descendants have identical copies of the foreign DNA. Collectively, all of the identical copies are a DNA clone.

6. Currently, the polymerase chain reaction (PCR) is the fastest way to amplify fragments of chromosomal DNA or cDNA. Bacterial factories are not needed; the reactions occur in test tubes. The reactions use a supply of nucleotide building blocks and primers: synthetic, short nucleotide sequences that will base-pair with any complementary DNA sequence and that enzymes (DNA polymerases) recognize to start replication. Each round of replication doubles the number of fragments.

7. For sexually reproducing species, no two individuals have exactly the same DNA base sequence (except for identical twins). When restriction enzymes are used to cut an individual's DNA, the result is a unique array of restriction fragments called a DNA fingerprint.

Further reading: Student Guide to InfoTrac on web site →

a. The DNA in all humans is more than 99 percent identical except at tandem repeats (short stretches of repeated base sequences, such as TTTC). The number and combination of tandem repeats is unique in each individual and can be detected by gel electrophoresis.

b. DNA fingerprinting has uses in forensic science and in resolving paternity suits.

8. Automated DNA sequencing can rapidly reveal the base sequence of cloned DNA or PCR-amplified DNA fragments. The method labels each fragment with one of four modifed nucleotides, then separates them according to length by gel electrophoresis. Each label fluoresces a certain color under a laser beam. A machine reads each as it peels off the gel and assembles the whole sequence.

9. A gene library is a mixed collection of bacterial cells that took up different cloned DNA or cDNA fragments. A gene may be isolated from the library with use of a probe, a very short stretch of radioactively labeled DNA that is known or suspected to be similar to or identical with part of that gene and can base-pair with it. A base pairing between nucleotide sequences from different sources is called nucleic acid hybridization.

10. In genetic engineering, specific genes are modified and inserted into the same organism or a different one. In gene therapy, copies of normal or modified genes are inserted into individuals in order to correct a genetic defect or boost resistance to disease.

11. Generally, recombinant DNA technology and genetic engineering have enormous potential for research and applications in medicine and agriculture, in the home and industry. As with any new technology, however, the potential benefits must be weighed against the potential risks, including ecological and social repercussions.

Review Questions

1. Distinguish between recombinant DNA technology and genetic engineering. *CI*
2. Distinguish these terms from one another: *16.1*
   a. chromosomal (genomic) DNA and cDNA
   b. cloning vector and DNA clone
3. Define PCR. Can fragments of chromosomal DNA, cDNA, or both be amplified by PCR? *16.2*
4. Define DNA fingerprint. Briefly describe which portions of the DNA are used in DNA fingerprinting. *16.3*
5. Outline the steps of automated DNA sequencing. *16.4*
6. Define cDNA library, then briefly explain how a gene can be isolated from it. Define probe and nucleic acid hybridization as part of your answer. *16.5*
7. Give three examples of applications that can be derived from knowledge of an organism's genome. *CI, 16.6–16.8*
8. Name one of the ways in which modified genes have been inserted into mammalian cells. *16.8*
9. Define gene therapy. Once the human genome has been fully sequenced, why will it be difficult to manipulate its genes to advantage? *CI, 16.8*

Self-Quiz (*Answers in Appendix III*)

1. ______ is the transfer of normal genes into body cells to correct a genetic defect.
   a. Reverse transcription
   b. Nucleic acid hybridization
   c. Gene mutation
   d. Gene therapy

2. DNA fragments result when ______ cut DNA molecules at specific sites.
   a. DNA polymerases
   b. DNA probes
   c. restriction enzymes
   d. RFLPs

3. ______ are small circles of bacterial DNA that are separate from the circular bacterial chromosome.

4. Foreign DNA that was inserted into a plasmid and then replicated many times in a population of bacteria is a ______.
   a. DNA clone
   b. gene library
   c. DNA probe
   d. gene map

5. By reverse transcription, ______ is assembled on ______.
   a. mRNA; DNA
   b. cDNA; mRNA
   c. DNA; enzymes
   d. DNA; agar

6. PCR stands for ______.
   a. polymerase chain reaction
   b. polyploid chromosome restrictions
   c. polygraphed criminal rating
   d. politically correct research

7. By gel electrophoresis, fragments of a gene library can be separated according to ______.
   a. shape
   b. length
   c. species

8. Automated DNA sequencing relies on ______.
   a. supplies of standard and labeled nucleotides
   b. primers and DNA polymerases
   c. gel electrophoresis and a laser beam
   d. all of the above

9. Match the terms with the most suitable description.

______ DNA fingerprint	a. selecting "desirable" traits
______ Ti plasmid	b. deciphering 3.2 billion base pairs of 23 human chromosomes
______ nature's genetic experiments	c. used in some gene transfers
______ nucleic acid hybridization	d. unique array of DNA fragments inherited in Mendelian pattern from each of two parents
______ Human Genome Initiative	e. base pairing of nucleotide sequences from different DNA or RNA sources
______ eugenic engineering	f. mutations, crossovers

Critical Thinking

1. In the following diagram, which restriction enzyme made the cuts (indicated in *red*) in part of a DNA molecule from two different organisms?

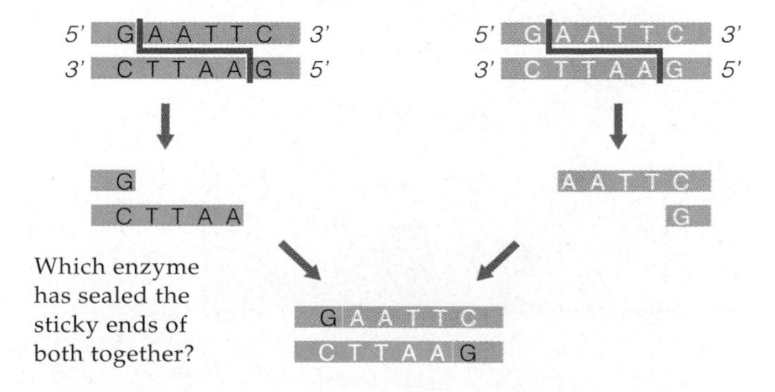

2. Ryan, a forensic scientist, obtained a very small DNA sample from a crime scene. In order to examine the sample by DNA fingerprinting, he must amplify the sample by the polymerase chain reaction. He estimates there are 50,000 copies of the DNA in his sample. Calculate the number of copies he will have after fifteen cycles of PCR.

3. A game warden in Africa confiscated eight ivory tusks from elephants. Some tissue is still attached to the tusks. Now he must determine whether the tusks were taken illegally from northern populations of endangered elephants or from other populations of elephants to the south that can be hunted legally. How can he use DNA fingerprinting to find the answer?

4. The Human Genome Initiative is not yet completed, yet knowledge about a number of the newly discovered genes is already being used to detect genetic disorders. Ask yourself: What will be done with genetic information about individuals? Will insurance companies and potential employers request it? At this writing, many women have already refused to take advantage of genetic screening for a gene associated with the development of breast cancer. Should medical records about people participating in genetic research and genetic clinical services be made available to other individuals? If not, how could such information be protected?

5. Lunardi's Market put out a bin of tomatoes having splendid vine-ripened redness, flavor, and texture. The sign posted above the bin identified them as genetically engineered produce. Most shoppers selected unmodified tomatoes in the adjacent bin, even though those tomatoes were pale pink, mealy-textured, and tasteless. Which ones would you pick? Why?

6. Avoiding genetically engineered food is probably impossible in the United States. At least 45 percent of its cotton, 38 percent of its soybean, and 25 percent of its corn crops are engineered to withstand weedkillers or make their own pesticides. For years, the corn and soybeans have found their way into breakfast cereals, tofu, soy sauce, vegetable oils, beer, soft drinks, and other food products. They are fed to farm animals.

By contrast, public resistance to genetically engineered food is high in Europe, especially Britain. Many people, including the Prince of Wales, speak out against what the tabloids call "Frankenfood." Protesters routinely vandalize crops (Figure 16.15). Worries abound that such foods may be more toxic, have lower nutritional value, and promote natural selection for antibiotic resistance. Some people worry that designer plants will cross-pollinate with wild plants to produce "superweeds."

Biotechnologists envision a new Green Revolution. They argue that designer plants can hold down food production costs, reduce dependence on pesticides and herbicides, enhance crop yields, and offer improved flavor, nutritional value, even salt tolerance and drought tolerance. Yet the chorus of critics in Europe may provoke a trade war with the United States. The issue is not small potatoes, so to speak. In 1998, the value of agricultural exports reached about 50 billion dollars. Flattery, threats, and bullying are rampant on both sides of the Atlantic. Restrictions on genetic engineering will profoundly impact United States agriculture, and inevitably the impact will trickle down to what you eat and how much you pay for it.

All of which invites you to read up on scientific research on this issue and form your own opinions. The alternatives are being swayed either by catchy, frightful phrases (Frankenfood, for example) or by biased reports from groups (such as chemical manufacturers) with their own agendas.

Possibly start with Christopher Bond's article, "Politics, Misinformation, and Biotechnology" in the 18 February 2000 *Science* (287: 1201). Dodd argues that biotechnology attempts to solve real-world problems of sickness, hunger, and dwindling resources; and that . . ."hysteria and unworkable propositions advanced by those who can afford to take their next meal for granted have little currency among those who are hungry."

7. What if it were possible to create life in test tubes? This is the question behind attempts to model and eventually create *minimal organisms*, which we define as living cells having the smallest set of genes necessary to survive and reproduce.

As recent experiments by Craig Ventner and Claire Fraser revealed, *Mycoplasma genitalium*, a bacterium with 517 genes (and 2,209 transposons) might be a candidate. By disabling its genes one at a time in the laboratory, they discovered that it may have only 265–350 essential protein-coding genes.

What if those genes were to be synthesized one at a time and inserted into an engineered cell consisting only of a plasma membrane and cytoplasm? Would the cell come to life? The possibility that it might prompted Ventner and Fraser to seek advise from a panel of bioethicists and theologians. As Arthur Caplan, a bioethicist at the University of Pennsylvania reported, no one on the panel objected to synthetic life research. They felt that much good might come of it, provided scientists didn't claim to have found the "secret of life." The 10 December 1999 issue of *Science* includes an essay from the panel and an article on *M. genitalium* research. Read both, then write down your thoughts about creating life in a test tube.

Figure 16.15 Activists ripping some genetically modified crop plants from a field in Great Britain.

Selected Key Terms

automated DNA sequencing *16.4*
cDNA *16.1*
cloning vector *16.1*
DNA clone *16.1*
DNA fingerprint *16.3*
DNA ligase *16.1*
gel electrophoresis *16.3*
gene library *16.5*
gene therapy *CI*
genetic engineering *CI*
genome *16.1*
nucleic acid hybridization *16.5*
PCR *16.2*
plasmid *16.1*
primer *16.2*
probe (nucleic acid) *16.5*
recombinant DNA technology *CI*
restriction enzyme *16.1*
reverse transcriptase *16.1*
tandem repeats *16.3*

Readings *See also www.infotrac-college.com*

Cho, M. et al. 10 December 1999. "Ethical Considerations in Synthesizing a Minimal Genome." *Science* 286:2087–2090.

Fairbanks, D., and W. R. Andersen. 1999. *Genetics: The Continuity of Life*. Monterey, California: Brooks-Cole.

Joyce, G. December 1992. "Directed Molecular Evolution." *Scientific American* 267(6): 90–97.

Watson, J. D. et al., 1992. *Recombinant DNA*. Second edition. New York: Scientific American Books.

APPENDIX I. CLASSIFICATION SCHEME

The classification scheme that follows is a composite of several that microbiologists, botanists, and zoologists use. Major groupings are agreed upon, more or less. There is not always agreement on what to name a particular grouping or where it may fit within the overall hierarchy. There are several reasons for the lack of total consensus.

First, the fossil record varies in its quality and in its completeness. Therefore, the phylogenetic relationship of one group to others is sometimes open to interpretation. Comparative studies at the molecular level are firming up the picture, but this work is still under way.

Second, ever since the time of Linnaeus, classification schemes have been based on the perceived morphological similarities and differences among organisms. Although some original interpretations are now open to question, we are so used to thinking about organisms in certain ways that reclassification often proceeds slowly.

For example, birds and reptiles traditionally have been placed in separate classes (Reptilia and Aves), yet there are many compelling arguments for grouping the lizards and snakes in one class and the crocodilians, dinosaurs, and birds in a separate class. Some favor six kingdoms but others favor three domains: archaebacteria, eubacteria, and eukaryota (alternatively, archaea, bacteria, and eukarya).

Third, researchers in microbiology, mycology, botany, zoology, and the other fields of biological inquiry have inherited a wealth of literature, based on classification schemes that were developed over time in each of those fields. Many see no good reason to give up the established terminology and thereby disrupt access to the past.

For instance, many microbiologists and botanists use *division*, and zoologists *phylum*, for taxa that are equivalent in the hierarchy of classification. Also, opinions are still polarized with respect to the kingdom Protista, certain members of which could just as easily be grouped with plants, fungi, or animals. Indeed, the term protozoan is a holdover from an earlier scheme in which some single-celled organisms were ranked as simple animals.

Given the problems, why do we even bother imposing artificial frameworks on the history of life? We do this for the same reason that a writer might decide to break up the history of civilization into several volumes, a number of chapters, and many paragraphs. Both efforts are attempts to impart obvious structure to what might otherwise be an overwhelming body of knowledge and to enhance the retrieval of information from it.

Finally, bear in mind that we include this classification scheme primarily for your reference purposes. Besides being open to revision, it also is by no means complete. It does not include the most recently discovered species, as from the mid-ocean. Many existing and extinct organisms of the so-called lesser phyla are not represented here. Our strategy is to focus mainly on organisms mentioned in the text. A few examples of organisms also are listed.

SUPERKINGDOM PROKARYOTA. Prokaryotes. Almost all microscopic species. DNA organized at nucleoid (a region of cytoplasm), not inside a membrane-bound nucleus. All are bacteria, either single cells or simple associations of cells. Autotrophs and heterotrophs. Table A on the following page lists representative types. Reproduce by prokaryotic fission, sometimes by budding and by bacterial conjugation.

The authoritative reference in bacteriology, *Bergey's Manual of Systematic Bacteriology,* calls this "a time of taxonomic transition." It groups bacteria mostly by numerical taxonomy (Section 22.6), not on phylogeny. The scheme presented here reflects strong evidence of evolutionary relationships for at least some bacterial groupings.

KINGDOM EUBACTERIA. Gram-negative, gram-positive forms. Peptidoglycan present in cell wall. Photosynthetic autotrophs, chemosynthetic autotrophs, and heterotrophs.

PHYLUM GRACILICUTES. Typical Gram-negative, thin wall. Autotrophs (photosynthetic and chemosynthetic) and heterotrophs. *Anabaena* and other cyanobacteria. *Escherichia, Pseudomonas, Neisseria, Myxococcus.*

PHYLUM FIRMICUTES. Typical Gram-positive, thick wall. Heterotrophs. *Bacillus, Staphylococcus, Streptococcus, Clostridium, Actinomycetes.*

PHYLUM TENERICUTES. Gram-negative, wall absent. Heterotrophs (saprobes, pathogens). *Mycoplasma.*

KINGDOM ARCHAEBACTERIA. Methanogens, extreme halophiles, extreme thermophiles. Evolutionarily closer to eukaryotic cells than to eubacteria. Strict anaerobes. Distinctive cell wall, membrane lipids, ribosomes, RNA sequences. *Methanobacterium, Halobacterium, Sulfolobus.*

SUPERKINGDOM EUKARYOTA. Eukaryotes. Both single-celled and multicelled species. Cells start out life with a nucleus (encloses the DNA) and usually other membrane-bound organelles. Chromosomes have many histones and other proteins attached.

KINGDOM PROTISTA. Diverse single-celled, colonial, and multicelled eukaryotic species. Existing species are unlike bacteria in characteristics and are most like the earliest, structurally simple eukaryotes. Autotrophs, heterotrophs, or both (Table 23.3). Reproduce sexually and asexually (by meiosis, mitosis, or both). Many related evolutionarily to plants, fungi, and possibly animals.

PHYLUM CHYTRIDIOMYCOTA. Chytrids. Heterotrophs; saprobic decomposers or parasites. *Chytridium.*

PHYLUM OOMYCOTA. Water molds. Heterotrophs. Decomposers, some parasites. *Saprolegnia, Phytophthora, Plasmopara.*

PHYLUM ACRASIOMYCOTA. Cellular slime molds. Heterotrophs with free-living, phagocytic amoeboid cells and spore-bearing stages. *Dictyostelium.*

PHYLUM MYXOMYCOTA. Plasmodial slime molds. Heterotrophs with free-living, phagocytic amoeboid cells and spore-bearing stages. Aggregate into streaming mass of cells that discard plasma membranes. *Physarum.*

Table A Representative Eubacteria and Archaebacteria Grouped on the Basis of Numerical Taxonomy

Some Major Groups	Main Habitats	Characteristics	Representatives
EUBACTERIA			
Photoautotrophs:			
Cyanobacteria, green sulfur bacteria, and purple sulfur bacteria	Mostly lakes, ponds; some marine, terrestrial habitats	Photosynthetic; use sunlight energy, carbon dioxide; cyanobacteria use oxygen-producing noncyclic pathway; some also use cyclic route	*Anabaena, Nostoc, Rhodopseudomonas, Chloroflexus*
Photoheterotrophs:			
Purple nonsulfur and green nonsulfur bacteria	Anaerobic, organically rich muddy soils, and sediments of aquatic habitats	Use sunlight energy; organic compounds as electron donors; some purple nonsulfur may also grow chemotrophically	*Rhodospirillum, Chlorobium*
Chemoautotrophs:			
Nitrifying, sulfur-oxidizing, and iron-oxidizing bacteria	Soil; freshwater, marine habitats	Use carbon dioxide, inorganic compounds as electron donors; influence crop yields, cycling of nutrients in ecosystems	*Nitrosomonas, Nitrobacter, Thiobacillus*
Chemoheterotrophs:			
Spirochetes	Aquatic habitats; parasites of animals	Helically coiled, motile; free-living and parasitic species; some major pathogens	*Spirochaeta, Treponema*
Gram-negative aerobic rods and cocci	Soil, aquatic habitats; parasites of animals, plants	Some major pathogens; some fix nitrogen (e.g., *Rhizobium*)	*Pseudomonas, Neisseria, Rhizobium, Agrobacterium*
Gram-negative facultative anaerobic rods	Soil, plants, animal gut	Many major pathogens; one bioluminescent (*Photobacterium*)	*Salmonella, Escherichia, Proteus, Photobacterium*
Rickettsias and chlamydias	Host cells of animals	Intracellular parasites; many pathogens	*Rickettsia, Chlamydia*
Myxobacteria	Decaying organic material; bark of living trees	Gliding, rod-shaped; aggregation and collective migration of cells	*Myxococcus*
Gram-positive cocci	Soil; skin and mucous membranes of animals	Some major pathogens	*Staphylococcus, Streptococcus*
Endospore-forming rods and cocci	Soil; animal gut	Some major pathogens	*Bacillus, Clostridium*
Gram-positive nonsporulating rods	Fermenting plant, animal material; gut, vaginal tract	Some important in dairy industry, others major contaminators of milk, cheese	*Lactobacillus, Listeria*
Actinomycetes	Soil; some aquatic habitats	Include anaerobes and strict aerobes; major producers of antibiotics	*Actinomyces, Streptomyces*
ARCHAEBACTERIA (ARCHAEA)			
Methanogens	Anaerobic sediments of lakes, swamps; animal gut	Chemosynthetic; methane producers; used in sewage treatment facilities	*Methanobacterium*
Extreme halophiles	Brines (extremely salty water)	Heterotrophic; also, unique photosynthetic pigments (bacteriorhodopsin) form in some	*Halobacterium*
Extreme thermophiles	Acidic soil, hot springs, hydrothermal vents	Heterotrophic or chemosynthetic; use inorganic substances as electron donors	*Sulfolobus, Thermoplasma*

PHYLUM SARCODINA. Amoeboid protozoans. Heterotrophs, free-living or endosymbiotic, some pathogens. Soft-or shelled bodies, locomotion by pseudopods. The rhizopods (naked amoebas, foraminiferans), *Amoeba proteus*, *Entomoeba*. Also actinopods (radiolarians, heliozoans).

PHYLUM CILIOPHORA. Ciliated protozoans. Heterotrophs, predators or symbionts, some parasitic. All have cilia. Free-living, sessile, or motile. *Paramecium, Didinium,* hypotrichs.

PHYLUM MASTIGOPHORA. Animal-like flagellated protozoans. Heterotrophs, free-living, many internal parasites. All with one to several flagella. *Trypanosoma, Trichomonas, Giardia.*

APICOMPLEXA. Heterotrophs, sporozoite-forming parasites. Complex structures at head end. Most familiar members called sporozoans. *Cryptosporidium, Plasmodium, Toxoplasma.*

PHYLUM EUGLENOPHYTA. Euglenoids. Mostly heterotrophs, some autotrophs (photosynthetic), some switch depending on environmental conditions. Most with one short, one long flagellum; red, green, or colorless. *Euglena, Peranema.*

PHYLUM PYRRHOPHYTA. Dinoflagellates. Photosynthetic, mostly, but some heterotrophs. *Fiesteria, Gymnodinium breve.*

PHYLUM CHRYSOPHYTA. Golden algae, yellow-green algae, diatoms. Photosynthetic. Some flagellated, others not. *Mischococcus, Synura, Vaucheria.*

PHYLUM RHODOPHYTA. Red algae. Mostly photosynthetic, some parasitic. Nearly all marine, some in freshwater habitats. *Porphyra. Bonnemaisonia, Euchema.*

PHYLUM PHAEOPHYTA. Brown algae. Photosynthetic, nearly all in temperate or marine waters. *Macrocystis, Fucus, Sargassum, Ectocarpus, Postelsia.*

PHYLUM CHLOROPHYTA. Green algae. Mostly photosynthetic, some parasitic. Most freshwater, some marine or terrestrial. *Chlamydomonas, Spirogyra, Ulva, Volvox, Codium, Halimeda.*

KINGDOM FUNGI. Nearly all multicelled eukaryotic species. Heterotrophs, mostly saprobic decomposers, some parasites. Nutrition based upon extracellular digestion of organic matter and absorption of nutrients by individual cells. Multicelled species form absorptive mycelia within substrates and structures that produce asexual spores (and sometimes sexual spores).

PHYLUM ZYGOMYCOTA. Zygomycetes. Zygosporangia (zygote inside thick wall) formed by sexual reproduction. Bread molds, related forms. *Rhizopus, Philobolus.*

PHYLUM ASCOMYCOTA. Ascomycetes. Sac fungi. Sac-shaped cells form sexual spores (ascospores). Most yeasts and molds, morels, truffles. *Saccharomycetes, Morchella, Neurospora, Sarcoscypha, Claviceps, Ophiostoma, Candida, Aspergillus, Penicillium.*

PHYLUM BASIDIOMYCOTA. Basidiomycetes. Club fungi. Most diverse group. Produce basidiospores inside club-shaped structures. Mushrooms, shelf fungi, stinkhorns. *Agaricus, Amanita, Craterellus, Gymnophilus, Puccinia, Ustilago.*

IMPERFECT FUNGI. Sexual spores absent or undetected. The group has no formal taxonomic status. If better understood, a given species might be grouped with sac fungi or club fungi. *Arthobotrys, Histoplasma, Microsporum, Verticillium.*

LICHENS. Mutualistic interactions between fungal species and a cyanobacterium, green alga, or both. *Lobaria, Usnea, Cladonia.*

KINGDOM PLANTAE. Multicelled eukaryotes. Nearly all photosynthetic autotrophs with chlorophylls *a* and *b*. Some parasitic. Nonvascular and vascular species, generally with well-developed root and shoot systems. Nearly all adapted in form and function to survive dry conditions on land; a few in aquatic habitats. Sexual reproduction predominant; also asexual reproduction by vegetative propagation and other mechanisms.

PHYLUM RHYNIOPHYTA. Earliest known vascular plants; muddy habitats. Extinct. *Cooksonia, Rhynia.*

PHYLUM PROGYMNOSPERMOPHYTA. Progymnosperms. Ancestral to early seed-bearing plants; extinct. *Archaeopteris.*

PHYLUM PTERIDOSPERMOPHYTA. Seed ferns. Fernlike gymnosperms; extinct. *Medullosa.*

PHYLUM CHAROPHYTA. Stoneworts.

PHYLUM BRYOPHYTA. Bryophytes: mosses, liverworts, hornworts. Seedless, nonvascular, haploid dominance. *Marchantia, Polytrichum, Sphagnum.*

PHYLUM PSILOPHYTA. Whisk ferns. Seedless, vascular. No obvious roots, leaves on sporophyte. *Psilotum.*

PHYLUM LYCOPHYTA. Lycophytes, club mosses. Seedless, vascular. Leaves, branching rhizomes, vascularized roots and stems. *Lepidodendron* (extinct), *Lycopodium, Selaginella.*

PHYLUM SPHENOPHYTA. Horsetails. Seedless, vascular. Some sporophyte stems photosynthetic, others nonphotosynthetic, spore-producing. *Calamites* (extinct), *Equisetum.*

PHYLUM PTEROPHYTA. Ferns. Largest group of seedless vascular plants (12,000 species), mainly tropical, temperate habitats. *Pteris, Trichomanes, Cyathea* (tree ferns), *Polystichum.*

PHYLUM CYCADOPHYTA. Cycads. Gymnosperm group (vascular, bears "naked" seeds). Tropical, subtropical. Palm-shaped leaves, simple cones on male and female plants. *Zamia.*

PHYLUM GINKGOPHYTA. Ginkgo (maidenhair tree). Type of gymnosperm. Seeds with fleshy outer layer. *Ginkgo.*

PHYLUM GNETOPHYTA. Gnetophytes. Only gymnosperms with vessels in xylem and double fertilization (but endosperm does not form). *Ephedra, Welwitchia.*

PHYLUM CONIFEROPHYTA. Conifers. Most common and familiar gymnosperms. Generally cone-bearing species with needle-like or scale-like leaves.
- Family Pinaceae. Pines, firs, spruces, hemlock, larches, Douglas firs, true cedars. *Abies, Cedrus, Pinus, Pseudotsuga.*
- Family Cupressaceae. Junipers, cypresses. *Cupressus, Juniperus.*
- Family Taxodiaceae. Bald cypress, redwoods, bigtree, dawn redwood. *Metasequoia, Sequoia, Sequoiadendron, Taxodium.*
- Family Taxaceae. Yews. *Taxus.*

PHYLUM ANTHOPHYTA. Angiosperms (the flowering plants). Largest, most diverse group of vascular seed-bearing plants. Only organisms that produce flowers, fruits.

Class Dicotyledonae. Dicotyledons (dicots). Some families of some representative orders are listed:
- Family Nymphaeaceae. Water lilies.
- Family Papaveraceae. Poppies.
- Family Brassicaceae. Mustards, cabbages, radishes.
- Family Malvaceae. Mallows, cotton, okra, hibiscus.
- Family Solanaceae. Potatoes, eggplant, petunias.
- Family Salicaceae. Willows, poplars.
- Family Rosaceae. Roses, apples, almonds, strawberries.
- Family Fabaceae. Peas, beans, lupines, mesquite.
- Family Cactaceae. Cacti.
- Family Euphorbiaceae. Spurges, poinsettia.
- Family Cucurbitaceae. Gourds, melons, cucumbers, squashes.
- Family Apiaceae. Parsleys, carrots, poison hemlock.
- Family Aceraceae. Maples.
- Family Asteraceae. Composites. Chrysanthemums, sunflowers, lettuces, dandelions.

Class Monocotyledonae. Monocotyledons (monocots). Some families of several different orders are listed:
- Family Liliaceae. Lilies, hyacinths, tulips, onions, garlic.
- Family Iridaceae. Irises, gladioli, crocuses.
- Family Orchidaceae. Orchids.
- Family Arecaceae. Date palms, coconut palms.
- Family Cyperaceae. Sedges.
- Family Poaceae. Grasses, bamboos, corn, wheat, sugarcane.
- Family Bromeliaceae. Bromeliads, pineapples, Spanish moss.

KINGDOM ANIMALIA. Multicelled eukaryotes, nearly all with tissues, organs, and organ systems and motility during at least part of their life cycle. Heterotrophs, predators (herbivores, carnivores, omnivores), parasites, detritivores. Reproduce sexually and, in many species, asexually. Embryonic development proceeds through a series of continuous stages.

PHYLUM PLACOZOA. Marine. Simplest known animal. Two cell layers, no mouth, no organs. *Trichoplax.*

PHYLUM MESOZOA. Ciliated, wormlike parasites, about the same level of complexity as *Trichoplax.*

PHYLUM PORIFERA. Sponges. No symmetry, tissues. *Euplectella.*

PHYLUM CNIDARIA. Radial symmetry, tissues, nematocysts.
- Class Hydrozoa. Hydrozoans. *Hydra, Obelia, Physalia, Prya.*
- Class Scyphozoa. Jellyfishes. *Aurelia.*
- Class Anthozoa. Sea anemones, corals. *Telesto.*

PHYLUM CTENOPHORA. Comb jellies. Modified radial symmetry.

PHYLUM PLATYHELMINTHES. Flatworms. Bilateral, cephalized; simplest animals with organ systems. Saclike gut.
- Class Turbellaria. Triclads (planarians), polyclads. *Dugesia.*
- Class Trematoda. Flukes. *Clonorchis, Schistosoma.*
- Class Cestoda. Tapeworms. *Diphyllobothrium, Taenia.*

PHYLUM NEMERTEA. Ribbon worms. *Tubulanus.*

PHYLUM NEMATODA. Roundworms. *Ascaris, Caenorhabditis elegans, Necator (hookworms), Trichinella.*

PHYLUM ROTIFERA. Rotifers. *Asplancha, Philodina.*

PHYLUM MOLLUSCA. Mollusks.
- Class Polyplacophora. Chitons. *Cryptochiton, Tonicella.*

Class Gastropoda. Snails (periwinkles, whelks, limpets, abalones, cowries, conches, nudibranchs, tree snails, garden snails), sea slugs, land slugs. *Aplysia, Ariolimax, Cypraea, Haliotis, Helix, Liguus, Limax, Littorina, Patella.*

Class Bivalvia. Clams, mussels, scallops, cockles, oysters, shipworms. *Ensis, Chlamys, Mytelus, Patinopectin.*

Class Cephalopoda. Squids, octopuses, cuttlefish, nautiluses. *Dosidiscus, Loligo, Nautilus, Octopus, Sepia.*

PHYLUM BRYOZOA. Bryozoans (moss animals).

PHYLUM BRACHIOPODA. Lampshells.

PHYLUM ANNELIDA. Segmented worms.

Class Polychaeta. Mostly marine worms. *Eunice, Neanthes.*

Class Oligochaeta. Mostly freshwater and terrestrial worms, but many marine. *Lumbricus* (earthworms), *Tubifex.*

Class Hirudinea. Leeches. *Hirudo, Placobdella.*

PHYLUM TARDIGRADA. Water bears.

PHYLUM ONYCHOPHORA. Onychophorans. *Peripatus.*

PHYLUM ARTHROPODA.

Subphylum Trilobita. Trilobites; extinct.

Subphylum Chelicerata. Chelicerates. Horseshoe crabs, spiders, scorpions, ticks, mites.

Subphylum Crustacea. Shrimps, crayfishes, lobsters, crabs, barnacles, copepods, isopods (sowbugs).

Subphylum Uniramia.
Superclass Myriapoda. Centipedes, millipedes.
Superclass Insecta.
Order Ephemeroptera. Mayflies.
Order Odonata. Dragonflies, damselflies.
Order Orthoptera. Grasshoppers, crickets, katydids.
Order Dermaptera. Earwigs.
Order Blattodea. Cockroaches.
Order Mantodea. Mantids.
Order Isoptera. Termites.
Order Mallophaga. Biting lice.
Order Anoplura. Sucking lice.
Order Homoptera. Cicadas, aphids, leafhoppers, spittlebugs.
Order Hemiptera. Bugs.
Order Coleoptera. Beetles.
Order Diptera. Flies.
Order Mecoptera. Scorpion flies. *Harpobittacus.*
Order Siphonaptera. Fleas.
Order Lepidoptera. Butterflies, moths.
Order Hymenoptera. Wasps, bees, ants.

PHYLUM ECHINODERMATA. Echinoderms.
Class Asteroidea. Sea stars. *Asterias.*
Class Ophiuroidea. Brittle stars.
Class Echinoidea. Sea urchins, heart urchins, sand dollars.
Class Holothuroidea. Sea cucumbers.
Class Crinoidea. Feather stars, sea lilies.
Class Concentricycloidea. Sea daisies.

PHYLUM HEMICHORDATA. Acorn worms.

PHYLUM CHORDATA. Chordates.

Subphylum Urochordata. Tunicates, related forms.

Subphylum Cephalochordata. Lancelets.

Subphylum Vertebrata. Vertebrates.
Class Agnatha. Jawless vertebrates (lampreys, hagfishes).
Class Placodermi. Jawed, heavily armored fishes; extinct.
Class Chondrichthyes. Cartilaginous fishes (sharks, rays, skates, chimaeras).
Class Osteichthyes. Bony fishes.
Subclass Dipnoi. Lungfishes.
Subclass Crossopterygii. Coelacanths, related forms.
Subclass Actinopterygii. Ray-finned fishes.
Order Acipenseriformes. Sturgeons, paddlefishes.
Order Salmoniformes. Salmon, trout.
Order Atheriniformes. Killifishes, guppies.
Order Gasterosteiformes. Seahorses.
Order Perciformes. Perches, wrasses, barracudas, tunas, freshwater bass, mackerels.
Order Lophiiformes. Angler fishes.
Class Amphibia. Mostly tetrapods; embryo in amnion.
Order Caudata. Salamanders.
Order Anura. Frogs, toads.
Order Apoda. Apodans (caecilians).
Class Reptilia. Skin with scales, embryo enclosed in amnion.
Subclass Anapsida. Turtles, tortoises.
Subclass Lepidosaura. *Sphenodon,* lizards, snakes.
Subclass Archosaura. Dinosaurs (extinct), crocodiles, alligators.
Class Aves. Birds. (In some of the more recent schemes, dinosaurs, crocodilians, and birds are grouped in the same category.)
Order Struthioniformes. Ostriches.
Order Sphenisciformes. Penguins.
Order Procellariiformes. Albatrosses, petrels.
Order Ciconiiformes. Herons, bitterns, storks, flamingoes.
Order Anseriformes. Swans, geese, ducks.
Order Falconiformes. Eagles, hawks, vultures, falcons.
Order Galliformes. Ptarmigan, turkeys, domestic fowl.
Order Columbiformes. Pigeons, doves.
Order Strigiformes. Owls.
Order Apodiformes. Swifts, hummingbirds.
Order Passeriformes. Sparrows, jays, finches, crows, robins, starlings, wrens.
Class Mammalia. Skin with hair; young nourished by milk-secreting glands of adult.
Subclass Prototheria. Egg-laying mammals (duckbilled platypus, spiny anteaters).
Subclass Metatheria. Pouched mammals or marsupials (opossums, kangaroos, wombats, Tasmanian devil).
Subclass Eutheria. Placental mammals.
Order Insectivora. Tree shrews, moles, hedgehogs.
Order Scandentia. Insectivorous tree shrews.
Order Chiroptera. Bats.
Order Primates.
Suborder Strepsirhini (prosimians). Lemurs, lorises.
Suborder Haplorhini (tarsioids and anthropoids).
Infraorder Tarsiiformes. Tarsiers.
Infraorder Platyrrhini (New World monkeys).
Family Cebidae. Spider monkeys, howler monkeys, capuchin.
Infraorder Catarrhini (Old World monkeys and hominoids).
Superfamily Cercopithecoidea. Baboons, macaques, langurs.
Superfamily Hominoidea. Apes and humans.
Family Hylobatidae. Gibbon.
Family Pongidae. Chimpanzees, gorillas, orangutans.
Family Hominidae. Existing and extinct human species (*Homo*) and australopiths.
Order Lagomorpha. Rabbits, hares, pikas.
Order Rodentia. Most gnawing animals (squirrels, rats, mice, guinea pigs, porcupines, beavers, etc.).
Order Cetacea. Whales, porpoises.
Order Carnivora. Carnivores.
Suborder Feloidea. Cats, mongooses, hyenas.
Suborder Canoidea. Dogs, weasels, skunks, otters, raccoons, pandas, bears.
Order Pinnipedia. Seals, walruses, sea lions.
Order Proboscidea. Elephants; mammoths (extinct).
Order Sirenia. Sea cows (manatees, dugongs).
Order Perissodactyla. Odd-toed ungulates (horses, tapirs, rhinos).
Order Artiodactyla. Even-toed ungulates (camels, deer, bison, sheep, goats, antelopes, giraffes, etc.).
Order Edentata. Anteaters, tree sloths, armadillos.
Order Tubulidentata. African aardvarks.

APPENDIX II. UNITS OF MEASURE

Metric-English Conversions

Length

English		*Metric*
inch	=	2.54 centimeters
foot	=	0.30 meter
yard	=	0.91 meter
mile (5,280 feet)	=	1.61 kilometer
To convert	*multiply by*	*to obtain*
inches	2.54	centimeters
feet	30.00	centimeters
centimeters	0.39	inches
millimeters	0.039	inches

Weight

English		*Metric*
grain	=	64.80 milligrams
ounce	=	28.35 grams
pound	=	453.60 grams
ton (short) (2,000 pounds)	=	0.91 metric ton
To convert	*multiply by*	*to obtain*
ounces	28.3	grams
pounds	453.6	grams
pounds	0.45	kilograms
grams	0.035	ounces
kilograms	2.2	pounds

Volume

English		*Metric*
cubic inch	=	16.39 cubic centimeters
cubic foot	=	0.03 cubic meter
cubic yard	=	0.765 cubic meters
ounce	=	0.03 liter
pint	=	0.47 liter
quart	=	0.95 liter
gallon	=	3.79 liters
To convert	*multiply by*	*to obtain*
fluid ounces	30.00	milliliters
quart	0.95	liters
milliliters	0.03	fluid ounces
liters	1.06	quarts

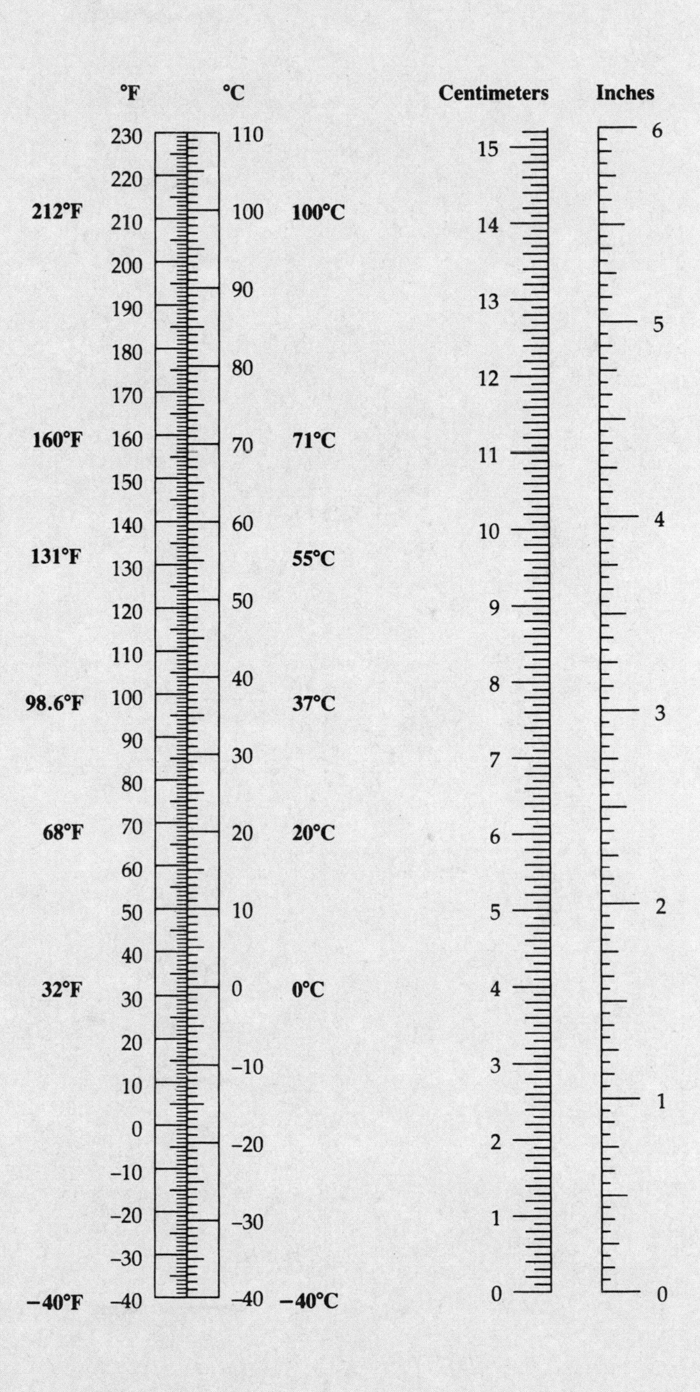

APPENDIX III. ANSWERS TO SELF-QUIZZES *Italicized numbers refer to relevant text pages*

1
1. **Metabolism** *5*
2. **Homeostasis** *5*
3. **Cell** *6*
4. **Adaptive** *10*
5. **Mutations** *10*
6. d *4*
7. d *4, 10*
8. d *10–11*
9. d *13*
10. c *13, 16*
11. a *15*
12. c *10*
    e *11*
    d *13*
    b *12*
    a *12*

2
1. a *24*
2. c *26–27*
3. f *28–29*
4. e *29*
5. f *30–31*
6. acid, base *30*
7. c *20*
    a *31*
    b *24*
    d *29*

3
1. d *36*
2. e *37*
3. f *38*
4. b *40*
5. d *42, 47*
6. d *45*
7. d *47–48*
8. c *42–43*
    e *47*
    b *41*
    d *48–49*
    a *38*

4
1. c *54–55*
2. d *58*
3. d *58, 72*
4. False (many cells have a wall) *58, 72, 73*
5. c *74*
6. e *66*
    d *67*
    a *54, 58, 64*
    b *64*
    c *64*

5
1. c *80–81*
2. b *80*
3. a 81
4. a (e.g., receptors, adhesion proteins not part of internal cell membranes) *79, 81*
5. a *89*
6. centrifuge *82*
7. d *84*
8. b *87*

6
1. c *97*
2. d *98–99*
3. d *100*
4. d *101*
5. b *103*
6. a *102*
7. e *104, 106*
8. a *104–105*
9. a *107*
10. c *107*
    g *103*
    a *103*
    d *103*
    e *103*
    b *104*
    f *103*

7
1. Carbon dioxide; sunlight *112*
2. d *114, 115i*
3. b *114, 115i*
4. e *114, 115i*
5. d *120*
6. c *114, 115i*
7. c *123*
8. c *123*
9. c *123*
    d *123*
    e *121*
    b *120–121*
    f *120*
    a *118*

8
1. d *132*
2. c *134*
3. b *132*
4. c *132–133*
5. d *140*
6. c *140*
7. b *140–141*
8. d *142–143*
9. b *134*
    c *140*
    a *136–137*
    d *138*

9
1. a *150*
2. b *150*
3. c *150*
4. c *151*
5. a *150, 153*
6. d *150*
7. b *151*
8. d *152–153*
    b *152*
    c *153*
    a *153*

10
1. d *162*
2. d *163, 165*
3. a *162*
4. b *163*
5. d *162*
6. c *164*
7. c *164*
8. d *165*
9. d *162*
    a *162*
    c *164*
    b *163*

11
1. a *177*
2. b *177*
3. a *177*
4. c *178–179*
5. b *177*
6. a *179*
7. d *180*
8. b *180–181*
    d *178–179*
    a *177*
    c *177*

12
1. c *194*
2. c *199*
3. e *206–207*
4. c *208*
5. d *208*
6. d *203*
7. True *194*
8. c *199*
    e *207*
    d *208*
    b *206*
    a *192*

13
1. c *218*
2. d *218*
3. d *219*
4. c *219*
5. a *220*
6. d *220–221*
7. d *221*
    b *220*
    c *220*
    a *221*

14
1. d *226–227*
2. c *228*
3. b *228*
4. c *228–229*
5. c *230*
6. a *230*
7. a *230–231*
8. e *235*
    c *232–233*
    a *229*
    f *230*
    d *230–231*
    g *229*
    b *230*

15
1. d *242*
2. d *242*
3. d *242*
4. a *242*
5. a *242*
6. d *244*
7. d *244*
8. c *247*
9. d *248*
10. b *251*
11. b *251*
12. d *249*
    e *247*
    b *250*
    a *242*
    c *246*

16
1. d *254*
2. c *256*
3. Plasmids *256*
4. a *257*
5. b *257*
6. a *258*
7. b *259*
8. d *260*
9. d *259*
    c *263*
    f *255*
    e *261*
    b *264*
    a *265*

APPENDIX IV. ANSWERS TO GENETICS PROBLEMS

CHAPTER 11

1. a. *AB*

   b. *AB*, *aB*

   c. *Ab*, *ab*

   d. *AB*, *Ab*, *aB*, *ab*

2. a. All of the offspring will be *AaBB*.

   b. 1/4 *AABB* (25% each genotype)
   1/4 *AABb*
   1/4 *AaBB*
   1/4 *AaBb*

   c. 1/4 *AaBb* (25% each genotype)
   1/4 *Aabb*
   1/4 *aaBb*
   1/4 *aabb*

   d. 1/16 *AABB* (6.25%)
   1/8 *AaBB* (12.5%)
   1/16 *aaBB* (6.25%)
   1/8 *AABb* (12.5%)
   1/4 *AaBb* (25%)
   1/8 *aaBb* (12.5%)
   1/16 *AAbb* (6.25%)
   1/8 *Aabb* (12.5%)
   1/16 *aabb* (6.25%)

3. Yellow is recessive. Because F_1 plants have a green phenotype and must be heterozygous, green must be dominant over the recessive yellow.

4. a. *ABC*

   b. *ABc*, *aBc*

   c. *ABC*, *aBc*, *ABc*, *aBc*

   d. *ABC*, *aBC*, *AbC*, *abC*,
   ABc, *aBc*, *Abc*, *abc*

5. Because all F_1 plants of this dihybrid cross had to be heterozygous for both genes, then 1/4 (25%) of the F_2 plants will be heterozygous for both genes.

6. a. The mother must be heterozygous for both genes. Both the father and their first child are homozygous recessive for both genes.

   b. The probability that their second child will not be a tongue roller and will have detached earlobes is 1/4 (25%).

7. a. Bill *AaEeSs*, and Marie *AAEESS*

   b. There is a 100% probability that all their children will have Bill's phenotype (flat feet, long eyelashes, and tendency to sneeze).

   c. Zero probability; no child of theirs will have high arches or short eyelashes, and none will be sneezy.

8. a. The mother must be heterozygous I^Ai. The male with type B blood could have fathered the child if he were heterozygous I^Bi .

   b. Genotype alone cannot prove the accused male is the father. Even if he happens to be heterozygous, *any* male who carries the *i* allele could be the father, including those heterozygous for type A blood (I^Ai) or type B blood (I^Bi) and those with type O blood (*ii*).

9. A mating between a mouse from a true-breeding, white-furred strain and a mouse from a true-breeding, brown-furred strain would provide you with the most direct evidence. Because true-breeding strains typically are homozygous for a trait being studied, all F_1 offspring from this mating should be heterozygous. Record the phenotype of each F_1 mouse, then let them mate with one another. Assuming only one gene locus is involved, these are possible outcomes for the F_2 offspring:

   a. All F_1 mice are brown, and their F_2 offspring segregate 3 brown : 1 white. *Conclusion*: Brown is dominant to white.

   b. All F_1 mice are white, and their F_2 offspring segregate 3 white : 1 brown. *Conclusion*: White is dominant to brown.

   c. All F_1 mice are tan, and the F_2 offspring segregate 1 brown : 2 tan : 1 white. *Conclusion*: The alleles at this locus show incomplete dominance.

10. You cannot guarantee that the puppies won't develop the disorder without more information about Dandelion's genotype. You could do so only if she is a heterozygous carrier, if the male is free of the alleles, and if the alleles are recessive.

11. Fred could use a testcross to find out if his pet's genotype is *WW* or *Ww*. He can let his black guinea pig mate with a white guinea pig having the genotype *ww*.
If any F_1 offspring are white, then his pet's genotype is *Ww*. If the two parents are allowed to mate repeatedly and all the offspring of the matings are black, there is a high probability that his pet is *WW*. (If, say, ten offspring are all black, then the probability that the male is *WW* is about 99.9 percent. The greater the number of offspring, the more confident Fred can be of his conclusion.)

12. a. 1/2 red, 1/2 pink

    b. All pink

    c. 1/4 red, 1/2 pink, 1/4 white

    d. 1/2 pink, 1/2 white

13. 9/16 walnut comb
3/16 rose comb
3/16 pea comb
1/16 single comb

14. Because both parents are heterozygotes (Hb^AHb^S), the following are the probabilities for each child:

a. 1/4 Hb^SHb^S

b. 1/4 Hb^AHb^A

c. 1/2 Hb^AHb^S

15. A mating of two M^LM cats yields:

1/4 homozygous dominant (MM)

1/2 heterozygous (M^LM)

1/4 homozygous recessive (M^LM^L)

Because M^LM^L is lethal, the probability that any one kitten among the survivors will be heterozygous is 2/3.

16. a. Both parents must be heterozygotes (*Aa*). Their children may be albino (*aa*) or unaffected (*AA* or *Aa*).

b. All are *aa*.

c. The albino father must be *aa*. They have an albino child, so the mother must be *Aa*. (If she were *AA*, they could not have an albino child.) The albino child is *aa*. The three unaffected children are *Aa*. There is a 50% chance that any child of theirs will be albino. The observed 3:1 ratio is not surprising, given the small number of offspring.

17. a. All of the offspring will have medium-red color corresponding to the genotype $A^1A^2B^1B^2$.

b. All possible genotypes could appear in the following proportions:

1/16	$A^1A^2B^1B^1$	dark red
1/8	$A^1A^1B^1B^2$	medium-dark red
1/16	$A^1A^1B^2B^2$	medium red
1/8	$A^1A^2B^1B^1$	medium-dark red
1/4	$A^1A^2B^1B^2$	medium red
1/8	$A^1A^2B^2B^2$	light red
1/16	$A^2A^2B^1B^1$	medium red
1/8	$A^2A^2B^1B^2$	light red
1/16	$A^2A^2B^2B^2$	white

CHAPTER 12

1. a. Human males (XY) inherit their X chromosome only from their mother.

b. In males, an X-linked allele will only be found on his one X chromosome. Males can produce two kinds of gametes: one kind with a Y chromosome free of the gene, and the other kind with an X chromosome bearing the X-linked allele.

c. One. Each gamete of a woman who is homozygous for an X-linked allele will have an X chromosome that carries the allele.

d. Two. If a female is heterozygous for an X linked allele, half of the gametes that she produces will contain one of the alleles and the other half will contain the other allele.

2. All of the offspring should be heterozygous for the gene and all should have long wings. However, because some have vestigial wings, the dominant allele might have mutated because of radiation.

3. a. Females normally do not have a Y chromosome, hence they would not have a Y-linked gene.

b. All sons inherit their Y chromosome from their father, but a daughter can only inherit his X chromosome. So the allele for hairy pinnae can be transmitted only to his sons, not to his daughters.

4. Because Marfan syndrome is a case of autosomal dominant inheritance and because one parent bears the allele, the probability of any child inheriting the mutant allele is 50%.

5. Because the phenotype appeared in every generation shown in the diagram, this must be a pattern of autosomal dominant inheritance.

6. A daughter could develop this type of muscular dystrophy only if she were to inherit two X-linked recessive alleles—one from her father and one from her mother. However, if a male bears the allele on his X chromosome, it will be expressed, then he will develop the disorder, and most likely he will not father children because of his early death.

7. If no crossover occurs between the two genes, then half the chromosomes will carry alleles *AB* and half will carry alleles *ab*.

8. If the alleles are close together, it is highly unlikely that a crossover will separate them from each other during meiosis. The greater the distance between the two loci, the greater the probability that a crossover will separate them from each other.

9. a. Nondisjunction could occur in anaphase I or anaphase II of meiosis.

b. As a result of a translocation, chromosome 21 (which is small) may become attached to the end of chromosome 14. Even though the chromosome number of the new individual would be 46, its somatic cells would contain the translocated chromosome 21, in addition to two normal chromosomes 21.

10. In the mother, a crossover between the two genes at meiosis generates an X chromosome that carries neither mutant allele.

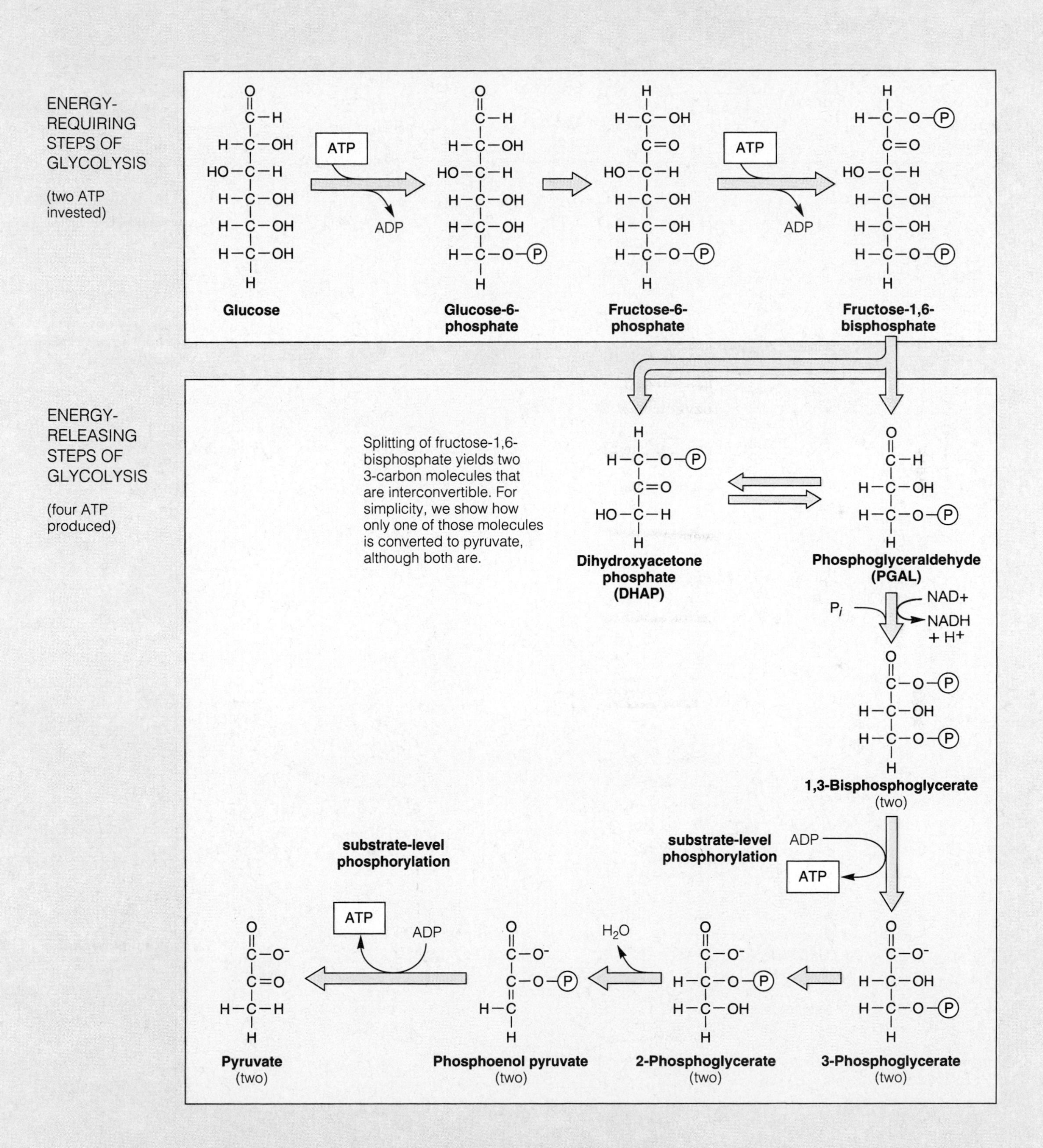

Figure A Glycolysis, ending with two 3-carbon pyruvate molecules for each 6-carbon glucose entering the reactions. The *net* energy yield is two ATP molecules (two invested, four produced).

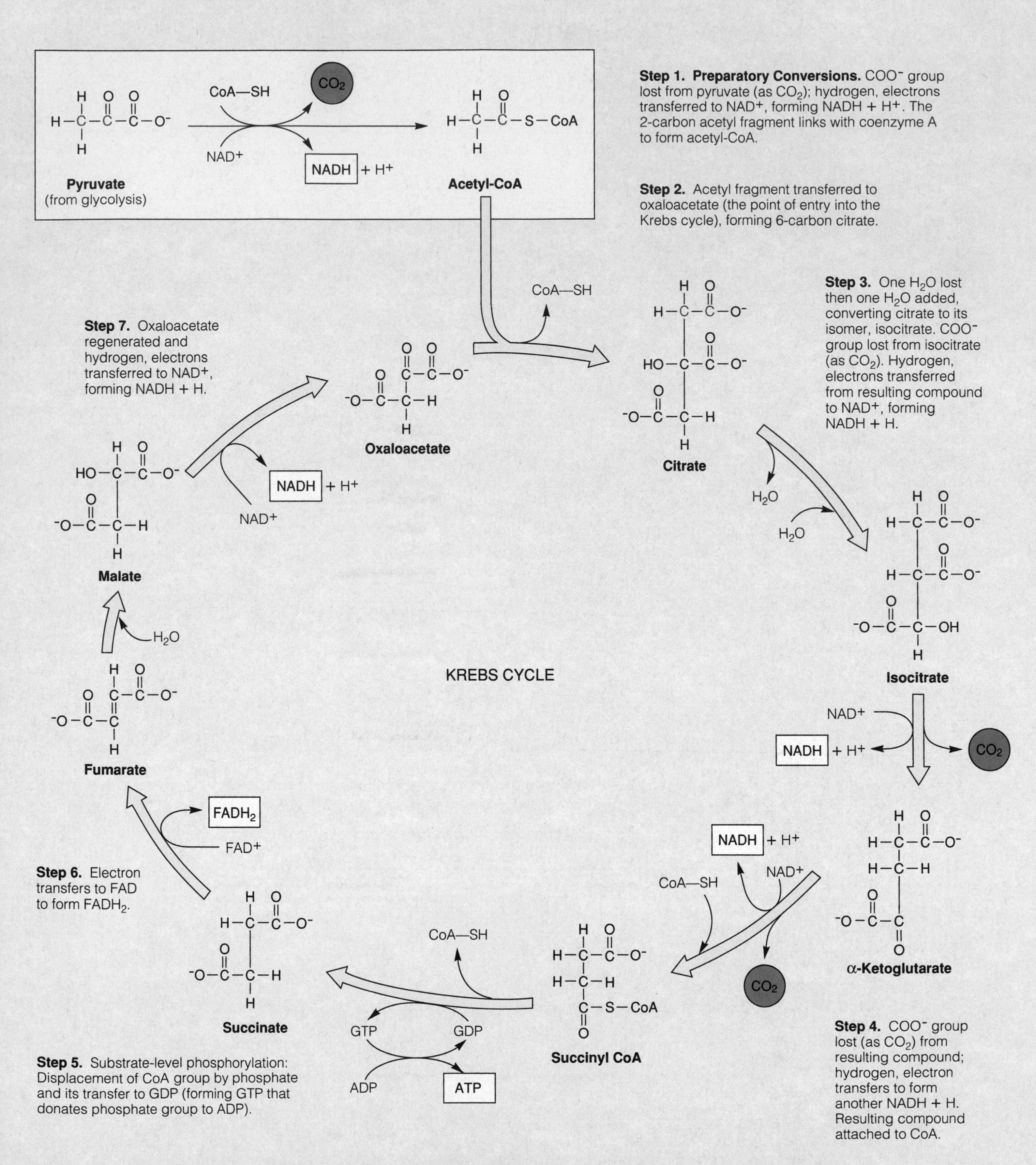

Figure B Krebs cycle (also known as the citric acid cycle). *Red* identifies the carbon atoms entering the cyclic pathway by way of acetyl-CoA.

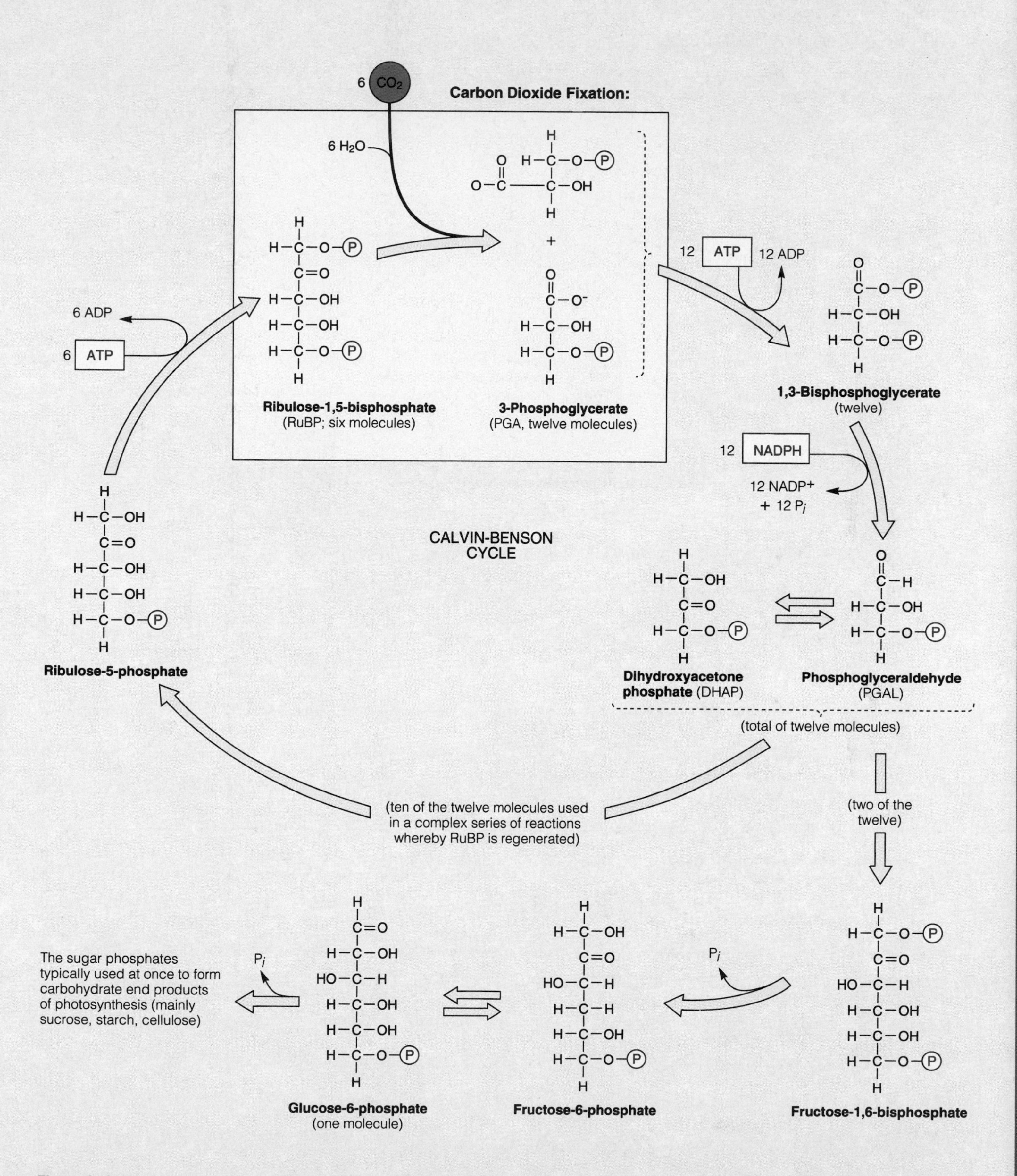

Figure C Calvin–Benson cycle of the light-independent reactions of photosynthesis.

APPENDIX VI. PERIODIC TABLE OF THE ELEMENTS

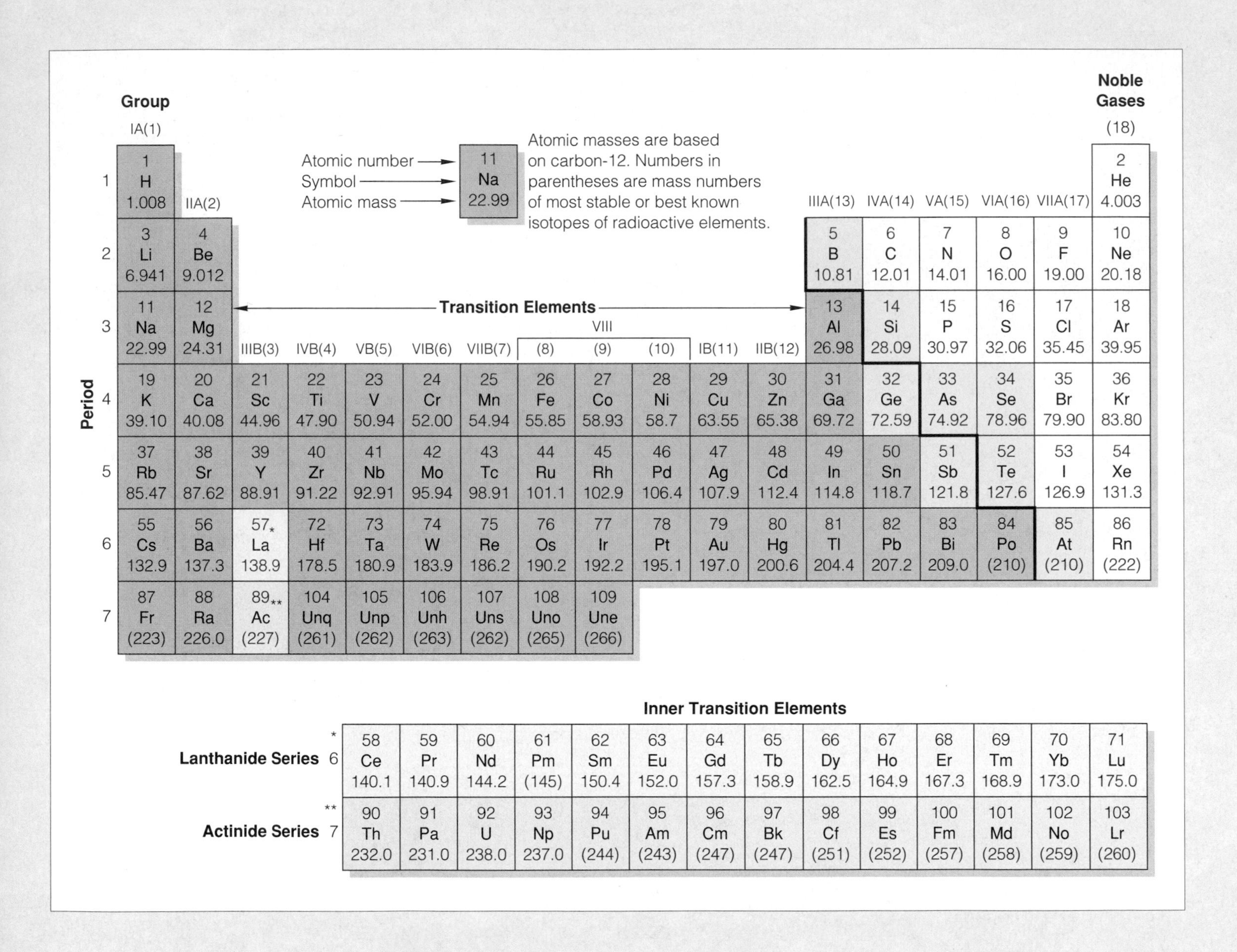

Atomic number / Symbol / Atomic mass: 11 / Na / 22.99

Atomic masses are based on carbon-12. Numbers in parentheses are mass numbers of most stable or best known isotopes of radioactive elements.

Period	IA(1)	IIA(2)	IIIB(3)	IVB(4)	VB(5)	VIB(6)	VIIB(7)	VIII (8)	VIII (9)	VIII (10)	IB(11)	IIB(12)	IIIA(13)	IVA(14)	VA(15)	VIA(16)	VIIA(17)	Noble Gases (18)
1	1 H 1.008																	2 He 4.003
2	3 Li 6.941	4 Be 9.012											5 B 10.81	6 C 12.01	7 N 14.01	8 O 16.00	9 F 19.00	10 Ne 20.18
3	11 Na 22.99	12 Mg 24.31											13 Al 26.98	14 Si 28.09	15 P 30.97	16 S 32.06	17 Cl 35.45	18 Ar 39.95
4	19 K 39.10	20 Ca 40.08	21 Sc 44.96	22 Ti 47.90	23 V 50.94	24 Cr 52.00	25 Mn 54.94	26 Fe 55.85	27 Co 58.93	28 Ni 58.7	29 Cu 63.55	30 Zn 65.38	31 Ga 69.72	32 Ge 72.59	33 As 74.92	34 Se 78.96	35 Br 79.90	36 Kr 83.80
5	37 Rb 85.47	38 Sr 87.62	39 Y 88.91	40 Zr 91.22	41 Nb 92.91	42 Mo 95.94	43 Tc 98.91	44 Ru 101.1	45 Rh 102.9	46 Pd 106.4	47 Ag 107.9	48 Cd 112.4	49 In 114.8	50 Sn 118.7	51 Sb 121.8	52 Te 127.6	53 I 126.9	54 Xe 131.3
6	55 Cs 132.9	56 Ba 137.3	57* La 138.9	72 Hf 178.5	73 Ta 180.9	74 W 183.9	75 Re 186.2	76 Os 190.2	77 Ir 192.2	78 Pt 195.1	79 Au 197.0	80 Hg 200.6	81 Tl 204.4	82 Pb 207.2	83 Bi 209.0	84 Po (210)	85 At (210)	86 Rn (222)
7	87 Fr (223)	88 Ra 226.0	89** Ac (227)	104 Unq (261)	105 Unp (262)	106 Unh (263)	107 Uns (262)	108 Uno (265)	109 Une (266)									

Transition Elements: groups IIIB(3) through IIB(12).

Inner Transition Elements

* **Lanthanide Series** 6	58 Ce 140.1	59 Pr 140.9	60 Nd 144.2	61 Pm (145)	62 Sm 150.4	63 Eu 152.0	64 Gd 157.3	65 Tb 158.9	66 Dy 162.5	67 Ho 164.9	68 Er 167.3	69 Tm 168.9	70 Yb 173.0	71 Lu 175.0
** **Actinide Series** 7	90 Th 232.0	91 Pa 231.0	92 U 238.0	93 Np 237.0	94 Pu (244)	95 Am (243)	96 Cm (247)	97 Bk (247)	98 Cf (251)	99 Es (252)	100 Fm (257)	101 Md (258)	102 No (259)	103 Lr (260)

GLOSSARY

ABO blood typing Method of using the presence of proteins A, B, or both at surface of red blood cells to characterize an individual's blood. O signifies absence of both proteins.

abortion Premature, spontaneous or induced expulsion of the embryo or fetus from uterus. Spontaneous abortion also called miscarriage.

absorption Uptake of water and solutes from the environment by cell or multicelled organism; e.g., movement of nutrients, fluid, and ions across gut lining and into internal environment.

acetyl-CoA (uh-SEED-ul) Coenzyme A with a two-carbon fragment from pyruvate attached. In the second stage of aerobic respiration, it transfers the fragment to oxaloacetate for the Krebs cycle.

acidity Of a solution, an excess of hydrogen ions relative to hydroxyl ions.

acid [L. *acidus*, sour] Any dissolved substance that donates hydrogen ions to other solutes or to water molecules.

actin (AK-tin) Cytoskeletal protein; subunit of microfilaments.

action potential Abrupt, brief reversal in the resting membrane potential of a neuron and other excitable cells.

activation energy For each type of reaction, the minimum amount of collision energy that will drive reactant molecules to an activated state, from which the reaction will proceed spontaneously.

active site Crevice in the surface of an enzyme molecule where a specific reaction is catalyzed (made to proceed more rapidly than it would spontaneously).

active transport Pumping of a specific solute across a cell membrane, through the interior of a transport protein, against its concentration gradient. Requires an energy boost, as from ATP.

adaptation [L. *adaptare*, to fit] Of evolution, being adapted (or becoming more adapted) to a set of environmental conditions. Of a sensory neuron, a decrease or cessation in the frequency of action potentials when a stimulus is maintained at constant strength.

adaptive radiation Macroevolutionary pattern; a burst of genetic divergences from a lineage that gives rise to many new species, each adapted to using a novel resource or a new (or recently vacated) habitat.

adaptive trait Any aspect of form, function, or behavior that helps the individual survive and reproduce under prevailing conditions.

adenine (AH-de-neen) A purine; a nitrogen-containing base in certain nucleotides.

ADP Adenosine diphosphate (ah-DEN-uh-seen die-FOSS-fate). Nucleotide coenzyme; typically accepts inorganic (unbound) phosphate or a phosphate group, thus becoming ATP.

aerobic respiration (air-OH-bik) [Gk. *aer*, air, + *bios*, life] Main ATP-forming pathway; proceeds from glycolysis through Krebs cycle and then electron transport phosphorylation. Final electron acceptor is oxygen. Typical net energy yield: 36 ATP per glucose molecule.

agglutination (ah-glue-tin-AY-shun) Forced clumping together of nonself markers when antibodies circulating in blood chemically recognize them. Clumping makes bearers of those markers more easily destroyed by phagocytes. Potential problem in recipients of transfused blood of a different type.

aging Of any multicelled organism showing extensive cell differentiation, a gradual and expected deterioration of the body over time.

AIDS Acquired immunodeficiency syndrome. A set of chronic disorders following infection by the human immunodeficiency virus (HIV), which destroys cells of the immune system.

alcohol Organic compound that includes one or more hydroxyl groups (—OH); it dissolves readily in water. Sugars are examples.

alcoholic fermentation One of the anaerobic ATP-forming pathways. The pyruvate from glycolysis is degraded to acetaldehyde, which accepts electrons from NADH to form ethanol; NAD^+ needed for the reactions is regenerated. Net yield: two ATP.

allele (uh-LEEL) For a given gene locus, one of two or more slightly different molecular forms of a gene that arise through mutation and that code for different versions of the same trait.

allele frequency For a given locus, the relative abundance of each kind of allele among all the individuals of a population.

amino acid (uh-MEE-no) An organic molecule with a hydrogen atom, an amino group, an acid group, and an R group, all covalently bonded to a carbon atom. Twenty kinds are the subunits of polypeptide chains.

ammonification (uh-moan-ih-fih-KAY-shun) Process by which some soil bacteria and fungi break down nitrogenous wastes and organic remains; part of nitrogen cycle.

amnion (AM-nee-on) An extraembryonic membrane; the boundary layer of a fluid-filled sac (amniotic cavity) in which the embryos of some vertebrate embryos grow and develop, move freely, and are protected from sudden impacts and temperature shifts.

amphibian Only type of vertebrate making the transition from water to land (evolutionarily and in their embryonic development). Existing gropus are salamanders, frogs, toads, and caecilians.

anaerobic electron transport ATP-forming pathway in which electrons stripped from an organic compound move through transport systems in the bacterial plasma membrane; an inorganic compound in the environment often serves as the final electron acceptor. Variable but always small net energy yield.

anaerobic pathway (an-uh-ROW-bik) [Gk. *an*, without, + *aer*, air] Metabolic pathway in which a substance other than oxygen is final acceptor of electrons stripped from substrates.

analogous structures (ann-AL-uh-gus) [Gk. *analogos*, similar to one another] Body parts that once differed in evolutionarily distant lineages but converged in their structure and function in response to similar environmental pressures.

anaphase (AN-uh-faze) Of mitosis, stage when sister chromatids of each chromosome move apart to opposite spindle poles. In anaphase I of meiosis, each duplicated chromosome and its homologue move to opposite spindle poles. In anaphase II of meiosis, sister chromatids of each chromosome move to opposite poles.

aneuploidy (AN-yoo-ploy-dee) Having one extra or one less chromosome relative to the parental chromosome number.

animal Multicelled heterotroph that feeds on other organisms, is motile for at least part of life cycle, develops through embryonic stages, has tissues (except for *Trichoplax* and sponges), and most often organs and organ systems.

Animalia Kingdom of animals.

antibiotic Metabolic product of soil microbes that kills bacterial competitors for nutrients.

antibody [Gk. *anti*, against] Antigen-binding receptor. Only B cells make antibodies, then position them at their surface or secrete them.

anticodon Sequence of three nucleotide bases in a tRNA molecule that can base-pair with a codon in an mRNA molecule.

antigen (AN-tih-jen) [Gk. *anti*, against, + *genos*, race, kind] Any molecular configuration that certain lymphocytes recognize as nonself and that triggers an immune response.

antigen-presenting cell Cell that processes and bears antigen fragments, bound to MHC molecules, at its surface. The antigen–MHC complexes promote immune responses.

apoptosis (APP-oh-TOE-sis) Programmed cell death. Molecular signals activate weapons of self-destruction in body cells that finished their prescribed functions or became altered, as by infection or cancerous transformation.

Archaebacteria Kingdom of prokaryotes more like eukaryotic cells than eubacteria; includes methanogens, halophiles, and thermophiles.

artificial selection Selection of traits among individuals of a population in an artificial environment, under contrived conditions.

asexual reproduction Any of a number of modes of reproduction by which offspring arise from a single parent and inherit the genes of that parent only.

atom Smallest particle unique to an element; has one or more positively charged protons, electrons, and (except for hydrogen), neutrons.

atomic number Number of protons in the nucleus of each atom of an element.

ATP Adenosine triphosphate (ah-DEN-uh-seen try-FOSS-fate). Energy-carrying nucleotide with adenine, ribose, and three phosphate groups. Phosphate-group transfers from ATP drive most energy-requiring metabolic reactions.

australopith (OHSS-trah-low-pith) [L. *australis*, southern, + Gk. *pithekos*, ape] One of earliest known hominids; a primate that may be on or near evolutionary road to modern humans.

autoimmune response Misdirected immune response in which lymphocytes mount an attack against normal body cells.

autonomic nervous system (auto-NOM-ik) All nerves from central nervous system to the smooth muscle, cardiac muscle, and glands of viscera (soft internal organs and structures).

autosome Type of chromosome that is the same in males and in females of the species.

autotroph (AH-toe-trofe) [Gk. *autos*, self, + *trophos*, feeder] Organism that makes its own organic compounds using an environmental energy source (e.g., sunlight) along with carbon dioxide as its carbon source.

B cell B lymphocyte; only cell that produces antibodies. Key player in immune responses.

bacterial conjugation The transfer of plasmid DNA from one bacterial cell to another.

bacteriophage (bak-TEER-ee-oh-fahj) [Gk. *baktērion*, small staff, rod, + *phagein*, to eat] Category of viruses that infect bacterial cells.

Barr body In body cells of female mammals, one of either of the two X chromosomes that was condensed to inactivate its genes.

basal body A centriole which, after giving rise to microtubules of a flagellum or cilium, remains attached to its base in the cytoplasm.

base Any substance that accepts hydrogen ions when dissolved in water.

base sequence Particular order in which one nucleotide base follows the next in a strand of DNA or RNA. The order is unique in at least some regions for each species.

behavior Response to external and internal stimuli based on sensory, neural, endocrine, and effector components. Has a genetic basis, can evolve, and can be modified by learning.

bilateral symmetry Body plan in which left and right halves generally are mirror images.

binary fission Asexual reproductive mode of protozoans and some other animals. The body divides into two parts of the same or different sizes. *Compare* prokaryotic fission.

biological species concept Defines a species as one or more populations of individuals that are interbreeding under natural conditions, that are producing fertile offspring, and that are isolated reproductively from other such populations. Applies to sexually reproducing species only.

bioluminescence Any organism flashing with fluorescent light by way of an ATP-driven reaction involving enzymes (luciferases).

biosphere [Gk. *bios*, life, + *sphaira*, globe] All regions of the Earth's waters, crust, and atmosphere in which organisms live.

biosynthetic pathway Metabolic pathway by which organic compounds are synthesized.

blood Fluid connective tissue of water, solutes, and formed elements (blood cells and platelets). Blood transports substances to and from cells, and helps maintain internal environment.

bone Vertebrate organ with mineral-hardened connective tissue (bone tissue); helps move body, protects other organs, stores minerals. Some (e.g., breastbone) produce blood cells.

brain Of most nervous systems, integrating center that receives and processes sensory input and issues coordinated commands for responses by muscles and glands.

buffer system A weak acid and the base that forms when it dissolves in water. The two work as a pair to counter slight shifts in pH.

bulk flow In response to a pressure gradient, a movement of more than one kind of molecule in the same direction in the same medium, as in blood, sap, or air.

C3 plant Plant that uses three-carbon PGA as the first intermediate for carbon fixation during the second stage of photosynthesis.

C4 plant Plant that uses oxaloacetate (a four-carbon compound) as the first intermediate for carbon fixation during the second stage of photosynthesis. CO_2 is fixed twice, in two cell types. Carbon dioxide accumulates in leaf and helps counter photorespiration.

Calvin–Benson cycle Cyclic, light-independent reactions; "synthesis" part of photosynthesis. Uses ATP and NADPH from light-dependent reactions. RuBP or some compound to which carbon has been affixed is rearranged and regenerated, and a sugar phosphate forms.

CAM plant Type of plant that conserves water by opening stomata only at night, when it fixes carbon dioxide by means of a C4 pathway.

cancer Malignant tumor; mass of cells that have grossly altered plasma membrane and cytoplasm, grow and divide abnormally, and adhere weakly to home tissue (which leads to metastasis). Lethal unless eradicated.

carbohydrate [L. *carbo*, charcoal, + *hydro*, water] Molecule of carbon, hydrogen, and oxygen mostly in a 1:2:1 ratio. Carbohydrates are structural materials, energy stores, and transportable energy forms. Monosaccharide, oligosaccharide, or polysaccharide.

carbon cycle An atmospheric cycle. Carbon moves from its largest reservoirs (sediments, rocks, and the ocean), through the atmosphere (mostly as CO_2), through food webs, and back to the reservoirs.

carbon dioxide fixation First step of light-independent reactions. Enzyme action affixes carbon (from CO_2) to RuBP or to some other compound for entry into the Calvin–Benson cycle.

carcinogen (kar-SIN-uh-jen) Any substance or agent that can trigger cancer.

carotenoid (kare-OTT-en-oyd) An accessory pigment. Different kinds absorb blue-violet and blue-green wavelengths, the energy of which is transferred to chlorophylls. They reflect yellow, orange, and red wavelengths.

cDNA DNA molecule copied from a mature mRNA transcript by reverse transcription.

cell [L. *cella*, small room] Smallest living unit; organized unit with a capacity to survive and reproduce on its own, given DNA instructions, energy sources, and raw materials.

cell cycle Events by which a cell increases in mass, roughly doubles its cytoplasmic components, duplicates its DNA, then divides its nucleus and cytoplasm. Extends from the time a cell forms until it completes division.

cell differentiation Key development process. Different cell lineages become specialized in their composition, structure, and function by activating and suppressing some fraction of the genome in different ways.

cell junction Site that joins cells physically, functionally, or both (e.g., tight junction in animals; plasmodesma in plants).

cell plate Disklike structure that forms in a plant cell after nuclear division; becomes a crosswall, with new plasma membrane on both surfaces, that divides the cytoplasm.

cell theory Theory stating that all organisms consist of one or more cells, the cell is the smallest unit with a capacity for independent life, and all cells arise from preexisting cells.

cell wall Of most bacteria, many protistan and fungal cells, and plant cells, the outermost, semirigid, permeable structure that helps the cell retain its shape and resist rupturing when the internal fluid pressure increases.

central nervous system Brain and spinal cord.

central vacuole Large, fluid-filled organelle of living, mature plant cell. Stores amino acids, sugars, ions, and toxic wastes. As it enlarges during growth, it forces the primary cell wall to expand and cell surface area to increase.

centriole (SEN-tree-ohl) Structure that gives rise to microtubules of cilia and flagella.

centromere (SEN-troh-meer) [Gk. *kentron*, center, + *meros*, a part] Constricted portion of each chromosome; location of a kinetochore to which spindle microtubules become attached.

channel protein Transport protein that serves as an open or gated channel where ions and other solutes move across a cell membrane.

chemical bond A union between the electron structures of two or more atoms or ions.

chemical energy Potential energy of molecules.

chemoautotroph (KEE-moe-AH-toe-trofe) Type of bacterium that can synthesize its own organic compounds using only carbon dioxide as the carbon source and an inorganic substance as the energy source.

chemoreceptor Sensory receptor that detects chemical energy (ions or molecules dissolved in the fluid bathing it).

chlorophyll (KLOR-uh-fill) [Gk. *chloros*, green, + *phyllon*, leaf] Main photosynthetic pigment; absorbs violet-to-blue and red wavelengths but transmits green.

chloroplast (KLOR-uh-plast) The organelle of photosynthesis in plants and many protistans.

chordate Animal with a notochord, dorsal hollow nerve cord, pharynx, and gill slits in pharynx wall during at least part of life cycle.

chromatid (CROW-mah-tid) Of a duplicated eukaryotic chromosome, one of two DNA molecules (with associated proteins) attached at centromere. Mitosis or meiosis separates them; each becomes a separate chromosome.

chromatin A cell's collection of DNA and all of the proteins associated with it.

chromosome (CROW-moe-some) [Gk. *chroma*, color, + *soma*, body] Of eukaryotic cells, a DNA molecule, duplicated or unduplicated, with a profusion of associated proteins. Of prokaryotic cells (bacteria), a circular DNA molecule with few, if any, proteins attached.

chromosome number All chromosomes in a given type of cell. *See* haploidy; diploidy.

cilium (SILL-ee-um), plural **cilia** Short motile or sensory structure projecting from surface of certain eukaryotic cells; its core is a 9 + 2 array of microtubules.

circulatory system Organ system that moves substances to and from cells, and often helps stabilize body temperature and pH. Typically consists of a heart, blood vessels, and blood.

classification scheme A way of organizing and retrieving information about species.

cleavage Early stage of animal development. Mitotic cell divisions divide a fertilized egg into many smaller, nucleated cells; original volume of egg cytoplasm does not increase.

cleavage furrow Ringlike depression defining cleavage plane for dividing animal cells.

cleavage reaction Enzyme action that splits a molecule in two or more parts; hydrolysis is an example.

cloning vector Plasmid that has been modified in the laboratory to accept foreign DNA.

coal Nonrenewable energy source that formed over 280 million years ago from submerged, undecayed, and compacted plant remains.

codominance In heterozygotes, simultaneous expression of a pair of nonidentical alleles that specify different phenotypes.

codon mRNA base triplet; its linear sequence corresponds to a linear sequence of amino acids in a polypeptide chain. Of 64 codons, 61 code for amino acids; 3 of these are also START signals for translation, 1 is a STOP signal for translation.

coenzyme Nucleotide that acts as an enzyme helper; it accepts electrons and hydrogen atoms stripped from substrates at a reaction site and transfers them to another reaction site.

cofactor Metal ion or coenzyme; it helps an enzyme catalyze a reaction or it transfers electrons, atoms, or functional groups to a different substrate.

cohesion Capacity to resist rupturing when placed under tension (stretched).

community All populations in a habitat. Also, a group of organisms with similar life-styles in a habitat, such as a bird community.

comparative morphology [Gk. *morph*, form] Scientific study of comparable body parts of adults or embryonic stages of major lineages.

competitive exclusion Theory that two or more species that require identical resources cannot coexist indefinitely.

compound Substance consisting of two or more elements in unvarying proportions.

concentration gradient Between two adjoining regions, a difference in the number of molecules (or ions) of a substance. All molecules are in constant motion. As they collide, they careen outward to an adjoining region where they are less concentrated. Barring other forces, all substances diffuse down such a gradient.

condensation reaction Two molecules become covalently bonded into a larger molecule, and water often forms as a by-product.

connective tissue proper Animal tissue with a characteristic proportion of fibroblasts and other cells, fibers (e.g., collagen, elastin), and a ground substance consisting of modified polysaccharides.

conservation biology International efforts to identify all species (particularly in marine and land ecosystems most vulnerable to habitat destruction and high extinction rates) and to design long-term management programs based on ecological and evolutionary principles.

consumer [L. *consumere*, to take completely] A heterotroph that feeds on cells or tissues of other organisms for carbon and energy (e.g., herbivores, carnivores, and parasites).

continuous variation Of a population, a more or less continuous range of small differences in a given trait among its individuals.

contractile vacuole (kun-TRAK-till VAK-you-ohl) [L. *contractus*, to draw together] Organelle that takes up excess water in cell body and contracts to expel water through a pore.

control group A group used as a standard for comparison with an experimental group. Ideally, a control group is identical with the experimental group in all respects except for the one variable being studied.

covalent bond (koe-VAY-lunt) [L. *con*, together, + *valere*, to be strong] Sharing of one or more electrons between atoms or groups of atoms. If electrons shared equally, bond is nonpolar. If shared unequally, it is polar (slightly positive at one end, slightly negative at the other).

crossing over At prophase I of meiosis, an interaction in which nonsister chromatids (of a pair of homologous chromosomes) break at corresponding sites and exchange segments; genetic recombination is the result.

culture Sum of behavior patterns of a social group, passed between generations by way of learning and symbolic behavior, especially language.

cyclic AMP (SIK-lik) Short for cyclic adenosine monophosphate. A nucleotide having roles in intercellular communication.

cyclic pathway of ATP formation Most ancient photosynthetic pathway, at plasma membrane of some bacteria and at chloroplast's thylakoid membrane. Photosystems in the membrane give up electrons to transport systems, which return them to photosystems. Electron flow across the membrane sets up H^+ gradients that drive ATP formation at nearby sites.

cyst Of many microorganisms, a resting stage with thick outer layers that typically forms under adverse conditions. Of skin, abnormal, fluid-filled sac without an external opening.

cytochrome (SIGH-toe-krome) [Gk. *kytos*, hollow vessel, + *chrōma*, color] Iron-containing protein molecule of electron transport systems, as used in photosynthesis, aerobic respiration.

cytokinesis (SIGH-toe-kih-NEE-sis) [Gk. *kinesis*, motion] Cytoplasmic division; splitting of a parent cell into daughter cells.

cytology Study of cell structure and function.

cytomembrane system System of organelles that modify, package, and distribute newly formed proteins and lipids. Endoplasmic reticulum, Golgi bodies, lysosomes, and various vesicles are its components.

cytoplasm (SIGH-toe-plaz-um) All cell parts, particles, and semifluid substances between the plasma membrane and the nucleus (or nucleoid, in bacteria).

cytoplasmic localization Parceling of a portion of maternal messages in the egg cytoplasm to each blastomere that forms during cleavage.

cytosine (SIGH-toe-seen) Pyrimidine; one of the nitrogen-containing bases in nucleotides.

cytoskeleton Dynamic internal framework of eukaryotic cells. Its microtubules and other components structurally support the cell and organize and move its internal parts. It helps free-living cells move in their environment.

decomposer [partly L. *dis-*, to pieces, + *companere*, arrange] Fungal or bacterial heterotroph. Obtains carbon and energy from remains, products, or wastes of organisms. Collectively, decomposers help cycle nutrients to producers in ecosystems.

deductive logic Pattern of thinking by which an individual makes inferences about specific consequences or specific predictions that must follow from a hypothesis.

degradative pathway A stepwise series of metabolic reactions that break down organic compounds to products of lower energy.

deletion At the cytological level, loss of a segment from a chromosome. At the molecular level, loss of one to several base pairs from a DNA molecule.

denaturation (deh-NAY-chur-AY-shun) Loss of a molecule's three-dimensional shape as weak bonds (e.g., hydrogen bonds) are disrupted.

dermis Skin layer beneath the epidermis that consists primarily of dense connective tissue.

detrital food web Network of food chains in which energy flows mainly from plants through arrays of detritivores and decomposers.

development Of multicelled species, emergence of specialized, morphologically distinct body parts according to a genetic program.

diffusion Net movement of like molecules or ions down their concentration gradient. In the absence of other forces, they collide constantly and randomly owing to their inherent energy. It occurs most frequently where they are most crowded, so the net movement is outward from regions of higher to lower concentrations.

digestive system Body sac or tube having one or two openings and often specialized regions for ingesting, digesting, and absorbing food, then eliminating undigested residues.

dihybrid cross Experimental cross between true-breeding parents that differ in two traits. The hybrid F_1 offspring inherit two gene pairs, each with two nonidentical alleles.

diploidy (DIP-loyd-ee) Presence of two of each type of chromosome (pairs of homologous chromosomes) in a cell nucleus at interphase. *Compare* haploidy.

disaccharide (die-SAK-uh-ride) [Gk. *di*, two, + *sakcharon*, sugar] A common oligosaccharide; two covalently bonded sugar monomers.

disease Outcome of infection when defenses against a pathogen cannot be mobilized fast enough, and the pathogen's activities interfere with normal body functions.

diversity of life Sum total of all variations in form, function, and behavior that accumulated in different lineages. Such variations generally are adaptive to prevailing conditions or were adaptive to conditions in the past.

DNA Deoxyribonucleic acid (dee-OX-ee-RYE-bow-new-CLAY-ik). For cells and many viruses, a nucleic acid that is the molecule of inheritance. Hydrogen bonds hold its two helically twisted nucleotide strands together. DNA's nucleotide sequence encodes instructions for synthesizing proteins, hence new individuals of a species.

DNA clone Many identical copies of foreign DNA that was inserted into plasmids and later replicated repeatedly after being taken up by population of host cells (typically, bacteria).

DNA–DNA hybridization *See* nucleic acid hybridization.

DNA fingerprint DNA fragments, inherited in a Mendelian pattern from each parent, that give each individual a unique identity. For humans, the most informative fragments are from tandem repeats (short regions of repeated DNA) that differ greatly among individuals.

DNA ligase (LYE-gaze) Enzyme that seals new base-pairings during DNA replication; also used in recombinant DNA technology.

DNA polymerase (poe-LIM-uh-raze) Enzyme of replication and repair that assembles a new strand of DNA on a parent DNA strand; it uses exposed nucleotide bases as the template.

DNA repair Process that restores the original base sequence when part of a DNA molecule gets altered. DNA polymerases, DNA ligases, and other enzymes execute the repair.

DNA replication Process by which molecules of DNA are duplicated for later distribution to daughter nuclei. Completed before mitosis and meiosis in eukaryotic cells and during prokaryotic fission in bacterial cells.

dominant allele Of diploid cells, an allele that masks phenotypic effect of any recessive allele paired with it.

double fertilization Of flowering plants only, fusion of a sperm nucleus with an egg nucleus (a zygote forms), and fusion of another sperm nucleus with nuclei of endosperm mother cell, which gives rise to a nutritive tissue.

drug addiction Chemical dependence on a drug, which assumes an "essential" biochemical role in the body following habituation and tolerance.

duplication Gene sequence repeated several to many hundreds or thousands of times. Even normal chromosomes have such sequences.

ecdysone Hormone that has major influence over early development of many insects.

ecology [Gk. *oikos*, home, + *logos*, reason] Scientific study of how organisms interact with one another and with their physical and chemical environment.

ecosystem Array of organisms and their physical environment, all interacting by a flow of energy and a cycling of materials.

ectoderm [Gk. *ecto*, outside, + *derma*, skin] The first-formed, outermost primary tissue layer of animal embryos; gives rise to nervous system tissues and integument's outer layer.

effector Muscle (or gland) that responds to signals from an integrator (e.g., a brain) by producing movement (or chemical change) to adjust the body to changing conditions.

effector cell Differentiated cell of a lymphocyte subpopulation that immediately engages and destroys an antigen-bearing agent during an immune response.

egg Mature female gamete; an ovum.

electromagnetic spectrum All wavelengths from radiant energy less than 10^{-5} nm long to radio waves more than 10 km long.

electron Negatively charged unit of matter, with particulate and wavelike properties, that occupies one of the orbitals around the atomic nucleus. Atoms gain, lose, or share electrons.

electron transfer A molecule donates one or more electrons to another molecule.

electron transport phosphorylation (FOSS-for-ih-LAY-shun) Last stage of aerobic respiration, when electrons from reaction intermediates flow through a membrane transport system that gives them up to oxygen. The flow sets up an electrochemical gradient that drives ATP formation at other sites in the membrane.

electron transport system Organized array of membrane-bound enzymes and cofactors that accept and donate electrons in series. It sets up an electrochemical gradient that makes H^+ flow across the membrane. The flow energy drives ATP formation at other reaction sites.

element Substance that cannot be degraded by ordinary means into a substance having different properties.

emerging pathogen Deadly pathogen, either a newly mutated strain of an existing type or one that evolved long ago and is now exploiting an increased presence of human hosts.

emulsification In chyme, a suspension of fat droplets coated with bile salts.

end product Substance present at the end of a metabolic pathway.

endergonic reaction (en-dur-GONE-ik) Chemical reaction having a net gain in energy.

endocrine gland Ductless gland that secretes hormones, which later enter the bloodstream.

endocrine system Integrative system of cells, tissues, and organs, functionally linked to the nervous system, that exerts control by way of its hormones and other chemical secretions.

endocytosis (EN-doe-sigh-TOE-sis) Cell uptake of substances when part of plasma membrane forms a vesicle around them. Three routes are receptor-mediated endocytosis, bulk transport of extracellular fluid, and phagocytosis.

endoplasmic reticulum or **ER** (EN-doe-PLAZ-mik reh-TIK-yoo-lum) Organelle that starts at nucleus and curves through cytoplasm. New polypeptide chains acquire side chains inside rough ER (with ribosomes on its cytoplasmic side); smooth ER (with no ribosomes) is a site of lipid synthesis.

endoskeleton [Gk. *endon*, within, + *sklēros*, hard] Of chordates, an internal framework of cartilage, bone, or both that works with skeletal muscle to support and move body, and to maintain posture.

energy Capacity to do work.

energy carrier Molecule that delivers energy from one metabolic reaction site to another. ATP is the most common energy carrier.

energy flow pyramid Pyramidal diagram of an ecosystem's trophic structure. It depicts the energy losses at each transfer to another trophic level.

enhancer A short DNA base sequence that is a binding site for an activator protein.

entropy (EN-trow-pee) Measure of the degree of disorder in a system (how much energy has become so disorganized, usually as dissipating heat, that it is no longer available to do work). A system must receive and use energy to stay organized; it tends toward entropy when its energy outputs exceed energy inputs.

enzyme (EN-zime) A protein or one of a few RNAs that greatly speed (catalyze) reactions between substances, most often at functional groups.

epistasis (eh-PISS-tah-sis) Interaction among the products of two or more gene pairs.

epithelium (EP-ih-THEE-lee-um) Tissue that covers the animal body's external surfaces and lines its internal cavities and tubes. It has one free surface and one resting on a basement membrane that is next to a connective tissue.

equilibrium, dynamic [Gk. *aequus*, equal, + *libra*, balance] Point at which a reaction runs forward as fast as in reverse, so no net change in reactant or product concentrations.

essential amino acid Any amino acid that an organism cannot synthesize for itself and must obtain from a food source.

essential fatty acid Any fatty acid that an organism cannot synthesize for itself and must obtain from a food source.

Eubacteria Kingdom of all prokaryotic cells except archaebacteria.

eukaryotic cell (yoo-CARE-EE-oh-tic) [Gk. *eu*, good, + *karyon*, kernel] Cell having a nucleus and other membrane-bound organelles.

evaporation [L. *e-*, out, + *vapor*, steam] The conversion of a substance from liquid state to gaseous state under input of heat energy.

evolution, biological [L. *evolutio*, unrolling] Genetic change in a line of descent over time, brought about by microevolutionary events (gene mutation, natural selection, genetic drift, and gene flow).

excretion Removal of excess water, solutes, and wastes, and some harmful substances from body by way of a urinary system or glands.

exergonic reaction (EX-ur-GONE-ik) Chemical reaction that shows a net loss in energy.

exocrine gland (EK-suh-krin) [Gk. *es*, out of, + *krinein*, to separate] Glandular structure that secretes products, usually through ducts or tubes, to a free epithelial surface.

exocytosis (EK-so-sigh-TOE-sis) Release of the contents of a vesicle at the cell surface, where the vesicle's membrane fuses with and becomes part of the plasma membrane.

exon One of the base sequences of an mRNA transcript that will become translated.

exoskeleton [Gk. *sklēros*, hard, stiff] External skeleton, as in arthropods.

exotic species Species that left its established home range, deliberately or accidentally, and successfully took up residence elsewhere.

experiment Test that simplifies observation in nature or in the laboratory by manipulating and controlling conditions under which the observations are made.

extinction Irrevocable loss of a species.

extracellular fluid Of most animals, all fluid not inside cells; plasma (the liquid portion of blood) plus interstitial fluid (the liquid that occupies spaces between cells and tissues).

extracellular matrix The ground substance, fibrous proteins, and other materials between cells of animal tissues (e.g., cartilage).

FAD Flavin adenine dinucleotide; a type of nucleotide coenzyme that transfers electrons and unbound protons (H^+) from one reaction site to another. At such times it is abbreviated $FADH_2$.

fat Lipid with a glycerol head and one, two, or three fatty acid tails. Tryglycerides have three. The carbon backbone of unsaturated tails has single covalent bonds; that of saturated tails has one or more double bonds.

fatty acid Molecule with a backbone of as many as thirty-six carbon atoms, a carboxyl group ($-COO^-$) at one end, and hydrogen atoms at most or all of the other bonding sites.

feedback inhibition Mechanism by which a cellular change resulting from some activity shuts down the activity that brought it about.

fermentation [L. *fermentum*, yeast] Anaerobic pathway of ATP formation that starts with glycolysis and ends with transfer of electrons to a breakdown product or an intermediate. NAD^+ is regenerated. Net energy yield: two ATP per glucose molecule.

fertilization [L. *fertilis*, to carry, to bear] The Fusion of a sperm nucleus with the nucleus of an egg, which thus becomes a zygote.

fever Any core temperature higher than the set point in the hypothalamic region that functions as the body's thermostat.

first law of thermodynamics [Gk. *therme*, heat, + *dynamikos*, powerful] Law of nature that the total amount of energy in the universe remains constant. Energy cannot be created from nothing and existing energy cannot be destroyed.

fitness Increase in adaptation to environment brought about by genetic change.

fixation Loss of all but one kind of allele at a gene locus; all individuals in a population are homozygous for it.

flagellum (fluh-JELL-um), plural **flagella** Motile structure of many free-living eukaryotic cells. Its core has a 9 + 2 array of microtubules.

fluid mosaic model Idea that cell membranes consist of a lipid bilayer and proteins. The lipids impart basic structure, impermeability to water-soluble molecules, and (by packing variations and movements) fluidity. Diverse proteins span the bilayer or associate with one of its surfaces and perform most membrane functions (e.g., transport, signal reception).

fossil Recognizable, physical evidence of an organism that lived in the distant past.

fossil fuel Coal, petroleum, or natural gas; a nonrenewable energy source that formed long ago from remains of swamp forests.

fossilization How fossils form. An organism or traces of it become buried in sediments or volcanic ash. Water and dissolved inorganic compounds infiltrate remains. Accumulating sediments exert pressure above the burial site. Over time, the pressure and chemical changes transform the remains to stony hardness.

free radical Any highly reactive molecular fragment having an unpaired electron.

functional group An atom or a group of atoms that is covalently bonded to the carbon backbone of an organic compound and that influences its chemical behavior.

functional-group transfer Enzyme-mediated event in which a molecule donates one or more functional groups to another molecule.

Fungi Kingdom of fungi which, as a group, are major decomposers. Also includes diverse pathogens and parasites.

gamete (GAM-eet) [Gk. *gametēs*, husband, and *gametē*, wife] Haploid cell, formed by meiotic cell division of a germ cell; required for sexual reproduction. Eggs and sperm are examples.

gamete formation Formation of sex cells (e.g., sperm and eggs); occurs in reproductive tissues or organs in most eukaryotic species.

gametophyte (gam-EET-oh-fite) [Gk. *phyton*, plant] Haploid gamete-producing body that forms during plant life cycles.

gel electrophoresis Laboratory technique used to distinguish different molecules in a given sample. Application of an electric field forces molecules to migrate through a viscous gel and distance themselves from one another on the basis of length, size, or electric charge.

gene [short for German *pangan*, after Gk. *pan*, all, + *genes*, to be born] Unit of information about a heritable trait, passed from parents to offspring. Each gene has a specific location on a chromosome (e.g., its locus).

gene flow Microevolutionary process; alleles enter and leave a population as an outcome of immigration and emigration, respectively.

gene frequency Abundance of a given allele relative to other alleles at same locus in a population.

gene library Mixed collection of bacteria that house many different cloned DNA fragments.

gene locus A gene's chromosomal location.

gene pair Two alleles at the same gene locus on a pair of homologous chromosomes.

gene pool All genotypes in a population.

gene therapy Generally, the transfer of one or more normal genes into an organism to correct or lessen adverse effects of a genetic disorder.

genetic code [After L. *genesis*, to be born] The correspondence between nucleotide triplets in DNA (then mRNA) and specific sequences of amino acids in resulting polypeptide chains; the basic language of protein synthesis in cells.

genetic disease Illness in which expression of one or more genes increases susceptibility to an infection or weakens immune response to it.

genetic disorder Inherited condition that causes mild to severe medical problems.

genetic divergence Gradual accumulation of differences in gene pools of populations or subpopulations of a species after a geographic barrier arises and separates them; mutation, natural selection, and genetic drift thereafter are operating independently in each one.

genetic drift Change in allele frequencies over the generations, as brought about by chance alone. Population size influences its effect on genetic and phenotypic diversity, because small populations are more vulnerable to losing alleles entirely.

genetic engineering Altering the information content of DNa molecules with recombinant DNA technology.

genetic equilibrium In theory, a state in which a population is not evolving. These conditions must be met: no mutation, the population very large in size and isolated from others of same species, and no natural selection (all members reproduce equally by random mating).

genetic recombination Result of any process that can incorporate new genetic information into a chromosome or DNA fragment. As one example, allelic combinations in chromosomes emerging from meiosis usually differ from the parental combinations (say, *Ab* and *aB* compared to parental types *AB* and *ab*). Also, nonreciprocal gene transfers in nature or the laboratory result in genetic recombination.

genome All of the DNA in a haploid number of chromosomes for a given species.

genotype (JEEN-oh-type) Genetic constitution of an individual; a single gene pair or the sum total of an individual's genes.

genus, plural **genera** (JEEN-US, JEN-er-ah) [L. *genus*, race or origin] A grouping of all species perceived to be more closely related to one another in their morphology, ecology, and evolutionary history than to other species at the same taxonomic level.

germ cell Animal cell of a lineage set aside for sexual reproduction; gives rise to gametes.

gland Secretory cell or structure derived from epithelium and often connected to it.

glyceride (GLISS-er-eyed) Molecule with one, two, or three fatty acid tails attached to a glycerol backbone; one of the fats or oils.

glycerol (GLISS-er-ohl) [Gk. *glykys*, sweet, + L. *oleum*, oil] Three-carbon compound with three hydroxyl groups; component of fats and oils.

glycocalyx Sticky mesh of polysaccharides, polypeptides, or both around the cell wall of many bacteria.

glycogen (GLY-kuh-jen) A highly branched polysaccharide made of glucose monomers; the main storage carbohydrate in animals.

glycolysis (gly-CALL-ih-sis) [Gk. *glykys*, sweet, + *lysis*, breaking apart] Breakdown of glucose or another organic compound to two pyruvate molecules. First stage of aerobic respiration, fermentation, and anaerobic electron transport. Oxygen has no role in glycolysis, which occurs in the cytoplasm of all cells. Two NADH form. Net yield: two ATP per glucose molecule.

glycoprotein Protein with linear or branched oligosaccharides covalently bonded to it. Nearly all surface proteins of animal cells and many proteins circulating in blood are examples.

Golgi body (GOHL-gee) Organelle of lipid assembly, polypeptide chain modification, and packaging of both in vesicles for export or for transport to locations in cytoplasm.

gonad (GO-nad) Primary reproductive organ in which animal gametes are produced.

green alga Type of protistan evolutionarily, structurally, and biochemically most similar to plants (e.g., nearly all are photoautotrophs with starch grains and chlorophylls *a* and *b* in chloroplasts; and some have cell walls of cellulose, pectin, and other polysaccharides typical of plants).

ground substance Intercellular material in some animal tissues; made of cell secretions and other noncellular components.

growth Of multicelled species, increases in the number, size, and volume of cells. Of bacteria, increase in the number of cells in a population.

guanine Nitrogen-containing base in one of the four nucleotide monomers in DNA or RNA.

half-life The time it takes for half of a given quantity of any radioisotope to decay into a different, and less unstable, daughter isotope.

halophile Archaebacterium of saline habitats.

haploidy (HAP-loyd-ee) Presence of only half of the parental number of chromosomes in a spore or gamete, as brought about by meiosis.

Hardy–Weinberg rule Statement that allele frequencies stay the same over the generations when there is no mutation, the population is infinitely large and is isolated from other populations of the same species, mating is random, and all individuals are reproducing equally and randomly (no natural selection).

heart Muscular pump; its contractions keep blood circulating through the animal body.

heat Thermal energy; a form of kinetic energy.

helper T cell T lymphocyte with central roles in both antibody-mediated and cell-mediated immune responses. When activated, it makes and secretes chemicals that induce responsive T and B cells to undergo rapid divisions into populations of effector and memory cells.

hemoglobin (HEEM-oh-glow-bin) [Gk. *haima*, blood, + L. *globus*, ball] Iron-containing, oxygen-transporting protein of red blood cells.

hermaphrodite Individual having both male and female gonads.

heterotroph (HET-er-oh-trofe) [Gk. *heteros*, other, + *trophos*, feeder] Organism unable to make its own organic compounds; feeds on autotrophs, other heterotrophs, organic wastes.

heterozygous condition (HET-er-oh-ZYE-guss) [Gk. *zygoun*, join together] For a given trait, having a pair of nonidentical alleles at a gene locus (that is, on a pair of homologous chromosomes).

higher taxon (plural, **taxa**) One of ever more inclusive groupings that reflect relationships among species. Family, order, class, phylum, and kingdom are examples.

histone Type of protein intimately associated with eukaryotic DNA and largely responsible for organization of eukaryotic chromosomes.

homeostasis (HOE-me-oh-STAY-sis) [Gk. *homo*, same, + *stasis*, standing] State in which physical and chemical aspects of internal environment (blood, interstitial fluid) are being maintained within ranges suitable for cell activities.

homeotic gene A master gene governing the development of specific body parts.

hominid [L. *homo*, man] All species on or near evolutionary road leading to modern humans.

hominoid Apes, humans, and recent ancestors.

homologous chromosome (huh-MOLL-uh-gus) [Gk. *homologia*, correspondence] Of cells with a diploid chromosome number, one of a pair of chromosomes that are identical in size, shape, and gene sequence, and that interact at meiosis. Nonidentical sex chromosomes (e.g., X and Y) also interact as homologues during meiosis.

homologous structures The same body parts that became modified differently, in different lines of descent from a common ancestor.

homology Similarity in one or more body parts in different species; attributable to descent from a common ancestor.

homozygous condition (HOE-moe-ZYE-guss) For a specified trait, having a pair of identical alleles at a gene locus (on a pair of homologous chromosomes).

homozygous dominant condition Having a pair of dominant alleles at a gene locus (on a pair of homologous chromosomes).

homozygous recessive condition Having a pair of recessive alleles at a gene locus (on a pair of homologous chromosomes).

hormone [Gk. *hormon*, stir up, set in motion] Signaling molecule secreted by one cell that stimulates or inhibits activities of any other cell having receptors for it. Animal hormones are picked up by bloodstream, which delivers them to cells some distance away. In plants, hormones do not travel far from source cells.

hybrid offspring Of a genetic cross, offspring with a pair of nonidentical alleles for a trait.

hydrogen bond Weak interaction between a small, highly electronegative atom of a molecule and a neighboring hydrogen atom already taking part in a polar covalent bond.

hydrogen ion Free (unbound) proton; that is, a hydrogen atom that lost its electron and now bears a positive charge (H^+).

hydrolysis (high-DRAWL-ih-sis) [L. *hydro*, water, + Gk. *lysis*, loosening] Cleavage reaction that breaks covalent bonds and splits a molecule into two or more parts. Commonly, H^+ and OH^- (derived from a water molecule) become attached to the exposed bonding sites.

hydrophilic substance [Gk. *philos*, loving] A polar substance that dissolves easily in water. Sugars are examples.

hydrophobic substance [Gk. *phobos*, dreading] A nonpolar substance; it strongly resists being dissolved in water. Oil is an example.

hydrosphere Collectively, all of the Earth's liquid or frozen water.

hydrostatic pressure Pressure exerted by a volume of fluid against a wall, membrane, or some other structure that encloses the fluid.

hypertonic solution A fluid having a greater concentration of solutes relative to another fluid.

hypothesis In science, a possible explanation of a phenomenon, one that has the potential to be proved false by experimental tests.

hypotonic solution A fluid that has a lower concentration of solutes relative to another fluid.

immune response Events by which B cells and T cells recognize antigen andgive rise to antigen-sensitized populations of effector cells and memory cells.

immunoglobulin (Ig) One of five classes of antibodies, each with antigen-binding sites as well as other sites with specialized functions.

in vitro fertilization Conception outside the body ("in glass" petri dishes or test tubes).

inbreeding Nonrandom mating among close relatives that share many identical alleles; a form of genetic drift in a small group of relatives that are preferentially interbreeding.

incomplete dominance Condition in which one allele of a pair is not fully dominant over the other; a heterozygous phenotype in between both homozygous phenotypes emerges.

independent assortment theory Mendelian theory that by the end of meiosis, each pair of homologous chromosomes (and linked genes on each one) are sorted out for shipment to gametes independently of how all the other pairs were sorted. Later modified to account for the disruptive effect of crossing over on linkages.

indirect selection theory Idea that altruistic individuals can pass on their genes indirectly by helping relatives survive and reproduce.

induced-fit model Idea that a substrate alters the shape of an enzyme's active site when bound to it, causing a more precise molecular fit between the two that promotes reactivity.

inductive logic Pattern of thinking by which an individual derives a general statement from specific observations.

infection Invasion and multiplication of a pathogen in a host. Disease follows if defenses are not mobilized fast enough; the pathogen's activities interfere with normal body functions.

inflammation, acute Rapid response to tissue injury by phagocytes and diverse proteins (e.g., histamine, complement, clotting factors). Signs include localized redness, heat, swelling, pain.

inheritance The transmission, from parents to offspring, of genes that specify structures and functions characteristic of the species.

inhibitor Substance able to bind with a specific molecule and interfere with its functioning.

integument Of animals, protective body cover (e.g., skin). Of seed-bearing plants, one or more layers around an ovule; becomes a seed coat.

intermediate Substance that forms between the start and end of a metabolic pathway.

intermediate filament One of the ropelike cytoskeletal elements that impart mechanical strength to animal cells and tissues.

interphase Of a cell cycle, interval between nuclear divisions when a cell increases in mass and roughly doubles the number of its cytoplasmic components. It also duplicates its chromosomes (replicates its DNA) during interphase, but *not* between meiosis I and II.

interstitial fluid (IN-ter-STISH-ul) [L. *interstitus*, to stand in the middle of something] The portion of extracellular fluid that occupies the spaces between animal cells and tissues.

intron One of the noncoding portions of a pre-mRNA transcript. All introns are excised before translation.

inversion A linear stretch of DNA within a chromosome that has become oriented in the reverse direction, with no molecular loss.

invertebrate Any animal without a backbone. Of the 2 million named species in the animal kingdom, all but 50,000 are invertebrates.

ion, negatively charged (EYE-on) An atom or a molecule that acquired an overall negative charge by gaining one or more electrons.

ion, positively charged An atom or a molecule that acquired an overall positive charge by losing one or more electrons.

ionic bond Two ions being held together by the attraction of their opposite charge.

isotonic solution A fluid having the same solute concentration as a fluid against which it is being compared.

isotope (EYE-so-tope) Of an element, an atom with more or fewer neutrons than the atoms having the most common number.

joint Area of contact or near-contact between bones.

juvenile Of some animals, a post-embryonic stage that changes only in size and proportion to become the adult (no metamorphosis).

karyotype (CARE-ee-oh-type) For an individual or a species, a preparation of metaphase chromosomes sorted by length, centromere location, and other defining features.

key innovation A structural or functional modification to the body that, by chance, gives a lineage the opportunity to exploit the environment in more efficient or novel ways.

keystone species A species that dominates a community and dictates its structure.

kilocalorie 1,000 calories of heat energy (the amount required to raise the temperature of 1 kilogram of water by 1°C). Used as the unit of measure for the caloric content of foods.

kinase Enzyme that catalyzes phosphate-group transfers (e.g., a protein kinase).

kinetic energy Energy of motion.

kinetochore Cluster of proteins and DNA at the centromere region of a chromosome; spindle microtubules become attached to it at mitosis or meiosis. One is present on each chromatid of a duplicated chromosome.

Krebs cycle Cyclic pathway in mitochondria only; together with a few preparatory steps, the stage of aerobic respiration in which pyruvate is broken down to carbon dioxide and water. Coenzymes accept electrons and unbound protons (H^+) from intermediates and deliver them to next stage; two ATP form.

lactate fermentation An anaerobic pathway of ATP formation. Pyruvate from glycolysis is converted to three-carbon lactate, and NAD^+ is regenerated. Net energy yield: two ATP.

learned behavior Lasting modification in behavior as a result of experience or practice.

lethal mutation Mutation with drastic effects on phenotype; usually causes death.

life cycle Recurring pattern of genetically programmed events from the time individuals are produced until they themselves reproduce.

light-dependent reactions The first stage of photosynthesis. Sunlight energy is trapped and converted to chemical energy of ATP, NADPH, or both, depending on the pathway.

light-independent reactions Second stage of photosynthesis. ATP makes phosphate-group transfers required to build sugar phosphates. NADPH delivers electrons and hydrogen atoms for the synthesis reactions, which also require carbon from carbon dioxide. Sugar phosphates enter other reactions by which starch, cellulose, and other end products are assembled.

lineage (LIN-ee-age) Line of descent.

linkage group All the genes on a chromosome.

lipid Mainly a greasy or oily hydrocarbon. Lipid molecules strongly resist dissolving in water but quickly dissolve in nonpolar substances. Some types serve as the main reservoirs of stored energy in all cells; others are structural materials (as in membranes) and cell products (e.g., surface coatings).

lipid bilayer Structural basis of cell membranes. Two layers of mostly phospholipid molecules. Hydrophobic tails of the lipids are sandwiched between the hydrophilic heads, and heads are dissolved in intracellular or extracellular fluid.

lipoprotein Molecule that forms when proteins circulating in blood combine with cholesterol, triglycerides, and phospholipids absorbed from the small intestine.

logic Thought patterns by which an individual draws a conclusion that does not contradict evidence used to support that conclusion.

lysis [Gk. *lysis*, a loosening] Gross damage to a plasma membrane, cell wall, or both that lets the cytoplasm to leak out; causes cell death.

lysogenic pathway Latent period that extends many viral replication cycles. Viral genes get integrated into host chromosome and may stay inactivated through many host cell divisions but eventually are replicated in host progeny.

lysosome (LYE-so-sohm) Important organelle of intracellular digestion.

lysozyme Infection-fighting enzyme present in mucous membranes (e.g., of mouth, vagina).

lytic pathway Of viruses, a rapid replication pathway that ends with lysis of host cell.

macroevolution Large-scale patterns, trends, and rates of change among higher taxa.

macrophage Phagocytic white blood cell; roles in nonspecific defenses and immune responses. One of the key antigen-presenting cells.

mass number Sum of all protons and neutrons in an atom's nucleus.

meiosis (my-OH-sis) [Gk. *meioun*, to diminish] Two-stage nuclear division process that halves the chromosome number of a parental germ cell (to haploid number). Each daughter nucleus receives one of each type of chromosome. Basis of gamete formation. Also, basis of formation of spores that give rise to gamete-producing bodies (gametophytes).

memory cell B or T cell that forms during an immune response but that does not act at once; it enters a resting phase, from which it is released for a secondary immune response.

mesoderm (MEH-zoe-derm) [Gk. *mesos*, middle, + *derm*, skin] Primary tissue layer important in evolution of all large, complex animals; gives rise to many internal organs and part of the integument.

messenger RNA (mRNA) A single strand of ribonucleotides transcribed from DNA, then translated into a polypeptide chain. The only RNA encoding protein-building instructions.

metabolic pathway (MEH-tuh-BALL-ik) Orderly sequence of enzyme-mediated reactions by which cells maintain, increase, or decrease the concentrations of particular substances.

metabolism (meh-TAB-oh-lizm) [Gk. *meta*, change] All the controlled, enzyme-mediated chemical reactions by which cells acquire and use energy to synthesize, store, degrade, and eliminate substances in ways that contribute to growth, survival, and reproduction.

metaphase Of meiosis I, a stage when all pairs of homologous chromosomes have become positioned at the spindle equator. Of mitosis or meiosis II, a stage when all the duplicated chromosomes have become positioned at the equator of the microtubular spindle.

methanogen Anaerobic archaebacterium that produces methane gas as by-product.

MHC marker Self-marker protein. Some are on all body cells of the individual; others are unique to macrophages and lymphocytes.

microfilament [Gk. *mikros*, small, + L. *filum*, thread] Cytoskeletal element. Each consists of two thin, twisted polypeptide chains; it has roles in cell movement and in producing and maintaining cell shapes.

micrograph Photograph of an image that came into view with the aid of a microscope.

microorganism Organism, usually single celled, too small to be observed without a microscope.

microspore Walled haploid spore; becomes a pollen grain in gymnosperms and angiosperms.

microtubular spindle Bipolar array of many microtubules; forms during nuclear division and moves chromosomes apart in controlled ways.

microtubule (my-crow-TUBE-yool) Cylindrical, hollow cytoskeletal element that consists of tubulin subunits; roles in cell shape, growth, and motion (e.g., key skeletal element of cilia, flagella, spindle apparatus).

microtubule organizing center MTOC; mass of substances in cytoplasm of eukaryotic cells; number, type, and location dictate orientation and organization of cell's microtubules.

mimicry (MIM-ik-ree) Close resemblance in form, behavior, or both between one species (the mimic) and another (its model). Serves in deception, as when an orchid mimics a female insect and so attracts males that pollinate it.

mineral Any element or inorganic compound that formed by natural geologic processes and is required for normal cell functioning.

mitochondrion (MY-toe-KON-dree-on) Double-membrane organelle of ATP formation. Only site of aerobic respiration's second and third stages. May have endosymbiotic origins.

mitosis (my-TOE-sis) [Gk. *mitos*, thread] Type of nuclear division that maintains the parental chromosome number for daughter cells. The basis of growth in size, tissue repair, and often asexual reproduction for eukaryotes.

mixture Two or more elements intermingled in proportions that can and usually do vary.

model Theoretical, detailed description or analogy that helps people visualize something that has not yet been directly observed.

molecular clock Model used to calculate the time of origin of one lineage or species relative to others. The underlying assumption is that neutral mutations accumulate in a lineage at predictable rates that can be measured as a series of ticks back through time.

molecule A unit of matter in which chemical bonds hold together two or more atoms of the same or different elements.

Monera In earlier classification schemes, a prokaryotic kingdom that encompasses both archaebacteria and eubacteria.

monohybrid cross Experimental cross between two parents that are homozygous for different versions of the same trait (e.g., *AA* and *aa*). F_1 offspring are heterozygous; each inherits a pair of nonidentical alleles (*Aa*) for the trait.

monomer Small molecule used as a subunit of polymers, such as sugar monomers of starch.

monosaccharide (MON-oh-SAK-ah-ride) [Gk. *monos*, alone, single, + *sakcharon*, sugar] One of the simple carbohydrates; a single sugar monomer. Glucose is an example.

monosomy Presence of a chromosome that has no homologue in a diploid cell.

morphogenesis (MORE-foe-JEN-ih-sis) [Gk. *morphe*, form, + *genesis*, origin] Inherited program of orderly changes in size, shape, and proportions of an animal embryo, leading to specialized tissues and early organs.

morphological convergence Macroevolutionary pattern. In response to similar environmental pressures over time, evolutionarily distant lineages evolve in similar ways and end up being alike in appearance, functions, or both.

morphological divergence Macroevolutionary pattern. Genetically diverging lineages slowly undergo change from the body form of their common ancestor.

multicelled organism Organism composed of many cells with coordinated metabolic activity; most show extensive cell differentiation into tissues, organs, and organ systems.

multiple allele system Three or more slightly different molecular forms of a gene that occur among individuals of a population.

mutagen (MEW-tuh-jen) Any environmental agent, such as a virus or ultraviolet radiation, that can alter DNA's molecular structure.

mutation [L. *mutatus*, a change, + *-ion*, an act, a result, or a process] Heritable change in the molecular structure of DNA. Original source of all new alleles and, ultimately, the diversity of life.

mutation frequency Of a population, the number of times that a mutation at a particular locus has arisen.

mutation rate Of a gene locus, the probability that a spontaneous mutation will occur during or between DNA replication cycles.

myosin (MY-uh-sin) Motor protein, often bound to microtubules; key roles in cell movements.

NAD^+ Nicotinamide adenine dinucleotide. A nucleotide coenzyme; abbreviated NADH when carrying electrons and H^+ to a reaction site.

$NADP^+$ Nicotinamide adenine dinucleotide phosphate. A phosphorylated nucleotide coenzyme; abbreviated $NADPH_2$ when it is carrying electrons and H^+ to a reaction site.

natural killer cell Cytotoxic lymphocyte that reconnoiters for tumor cells and virus-infected cells, then touch-kills them.

natural selection Microevolutionary process; the outcome of differences in survival and reproduction among individuals that vary in details of heritable traits. Over generations, it typically leads to increased fitness.

nerve Cordlike bundle of the axons of sensory neurons, motor neurons, or both sheathed in connective tissue.

nervous system Integrative organ system with nerve cells interacting in signal-conducting and information-processing pathways. Detects and processes stimuli, and elicits responses from effectors (e.g., muscles and glands).

neutral mutation Mutation that has little or no effect on phenotype. Natural selection cannot change its frequency in a population because it does not affect survival or reproduction.

neutron Unit of matter, one or more of which occupies the atomic nucleus and has mass but no electric charge.

niche (NITCH) [L. *nidas*, nest] Sum total of all activities and relationships in which individuals of a species engage as they secure and use the resources required to survive and reproduce.

noncyclic pathway of ATP formation (non-SIK-lik) [L. *non*, not, + Gk. *kylos*, circle] Light-dependent reactions of photosynthesis that requires photolysis, two photosystems, and two transport chains. Water molecules are split. They release electrons and hydrogen that are used in ATP and NADPH formation, plus oxygen as a by-product (the basis of Earth's oxygen-rich atmosphere).

nondisjunction Failure of sister chromatids or failure of a pair of homologous chromosomes to separate at meiosis or mitosis. As a result, daughter cells end up with too many or too few chromosomes.

nucleic acid (new-CLAY-ik) Single- or double-stranded chain of four kinds of nucleotides joined at phosphate groups. Nucleic acids differ in their base sequences. DNA and RNA are examples.

nucleic acid hybridization Any base-pairing between sequences of DNA or RNA from different sources.

nucleoid (NEW-KLEE-oid) Portion of bacterial cell interior in which the DNA is physically organized but not enclosed by a membrane.

nucleolus (new-KLEE-oh-lus) [L. *nucleolus*, little kernel] In the nucleus of a nondividing cell, an assembly site for the protein and RNA subunits that will later form ribosomes in the cytoplasm.

nucleosome (NEW-klee-oh-sohm) A stretch of eukaryotic DNA looped twice around a spool of histone molecules; one of many units that give condensed chromosomes their structure.

nucleotide (NEW-klee-oh-tide) Small organic compound consisting of a five-carbon sugar (deoxyribose), a nitrogen-containing base, and a phosphate group. The structural unit of adenosine phosphates, nucleotide coenzymes, and nucleic acids.

nucleotide coenzyme Protein that assists an enzyme by delivering electrons and hydrogen atoms released at a reaction site to another reaction site.

nucleus (NEW-klee-us) [L. *nucleus*, a kernel] Of atoms, a central core of one or more protons and (in all but hydrogen atoms) neutrons. In a eukaryotic cell, the organelle that physically separates DNA from cytoplasmic machinery.

numerical taxonomy Study of the degree of relatedness between an unidentified organism and a known group through comparisons of traits. Used to classify prokaryotes, which are poorly represented in the fossil record.

nutrient Element with a direct or indirect role in metabolism that no other element fulfills.

nutrition Processes by which food is selectively ingested, digested, absorbed, and converted to the body's own organic compounds.

obesity Excess of fat in adipose tissue; caloric intake has exceeded the body's energy output.

oligosaccharide (oh-LIG-oh-SAC-uh-RID) Short-chain carbohydrate of two or more covalently bonded sugar monomers. Disaccharides (two monomers) are examples.

oncogene (ON-koe-jeen) Any gene having the potential to induce cancerous transformation.

oocyte Immature egg of all animals and some protistans.

oogenesis (oo-oh-JEN-uh-sis) Process by which a germ cell develops into a mature oocyte.

operator Very short base sequence between a promoter and bacterial genes; a binding site for a repressor that can block transcription.

operon Promoter–operator sequence that services more than one bacterial gene; part of a control mechanism that adjusts transcription rates upward or downward.

orbital One of the volumes of space around the atomic nucleus in which one or at most two electrons are likely to be at any instant.

organ Body structure having definite form and function that consists of more than one tissue.

organ formation Developmental stage in which primary tissue layers give rise to cell lineages unique in structure and function. Descendants of those lineages give rise to all the different tissues and organs of the adult.

organ system Two or more organs that are interacting chemically, physically, or both in a common task.

organelle (or-GUN-ell) Membrane-bound sac or compartment in the cytoplasm having one or more specialized metabolic functions. Most eukaryotic cells have a profusion of them.

organic compound Molecule of one or more elements covalently bonded to some number of carbon atoms.

osmosis (oss-MOE-sis) [Gk. *osmos*, pushing] The diffusion of water in response to water concentration gradient between two regions that are separated by a selectively permeable membrane. The greater the number of ions and molecules dissolved in a solution, the lower its water concentration.

osmotic pressure Force that operates after hydrostatic pressure develops in a cell or in an enclosed body region; the amount of force that stops further increases in fluid volume by countering the inward diffusion of water.

oxaloacetate (ox-AL-oh-ASS-ih-tate) A four-carbon compound with roles in metabolism (e.g., the point of entry into the Krebs cycle).

oxidation–reduction reaction An electron transfer between atoms or molecules. Often an unbound proton (H^+) is transferred at the same time.

parasite [Gk. *para*, alongside, + *sitos*, food] Organism that lives in or on a host organism for at least part of its life cycle. It feeds on specific tissues and usually does not kill its host outright.

parasitism Symbiotic interaction in which one species (a parasite) benefits and the other (its host) is harmed. The parasite lives inside or on a host and feeds on its cells or tissues.

parasitoid Type of insect larva that grows and develops in a host organism (usually another insect), consumes its soft tissues, and kills it.

parasympathetic nerve (PARE-uh-SIM-pu-THET-ik) An autonomic nerve. Signals carried by such nerves tend to slow overall body activities and divert energy to basic tasks, and to help make small adustments in internal organ activity by acting continually in opposition to sympathetic nerve signals.

parthenogenesis (par-THEN-oh-GEN-uh-sis) An unfertilized egg giving rise to an embryo.

passive transport Process by which a transport protein that spans a cell membrane passively permits a solute to diffuse through its interior. Also called facilitated diffusion.

PCR Polymerase chain reaction. A method of enormously amplifying the quantity of DNA fragments cut by restriction enzymes.

pedigree Diagram of the genetic connections among related individuals through successive generations; uses standardized symbols.

penis Male copulatory organ by which sperm is deposited in a female reproductive tract.

peroxisome Vesicle in which fatty acids and amino acids are first digested to hydrogen peroxide, then converted to harmless products.

PGA Phosphoglycerate (FOSS-foe-GLISS-er-ate) Intermediate of glycolysis and of the Calvin–Benson cycle.

PGAL Phosphoglyceraldehyde. Intermediate of glycolysis and of the Calvin-Benson cycle.

pH scale Measure of the concentration of free hydrogen ions (H^+) in blood, water, and other solutions. pH 0 is the most acidic, 14 the most basic, and 7, neutral.

phagocyte (FAG-uh-sight) [Gk. *phagein*, to eat, + *kytos*, hollow vessel] Cell that captures prey by phagocytosis (e.g., amoebas); also cells that use same process for defense and day-to-day tissue housekeeping (e.g., macrophages).

phagocytosis (FAG-uh-sigh-TOE-sis) [Gk. *phagein*, to eat, + *kytos*, hollow vessel] Engulfment of foreign cells or particles by way of pseudopod formation and endocytosis.

phenotype (FEE-no-type) [Gk. *phainein*, to show, + *typos*, image] Observable trait or traits of an individual that arise from gene interactions and gene–environment interactions.

phospholipid Organic compound that has a glycerol backbone, two fatty acid tails, and a hydrophilic head of two polar groups (one being phosphate). Phospholipids are the main structural material of cell membranes.

phosphorylation (FOSS-for-ih-LAY-shun) A common means of activating molecules for a reaction. An enzyme either attaches inorganic phosphate to a molecule or mediates a transfer of a phosphate group from one molecule to another (as when ATP phosphorylates glucose).

photoautotroph Photosynthetic autotroph; any organism that synthesizes its own organic compounds using carbon dioxide as the source of carbon atoms and sunlight as the energy source. Nearly all plants, some protistans, and a few bacteria are photoautotrophs.

photolysis (foe-TALL-ih-sis) [Gk. *photos*, light, +-*lysis*, breaking apart] Reaction sequence in which photon energy splits water molecules. The released electrons and hydrogen take part in the noncyclic pathway of photosynthesis, and the oxygen is released as a by-product.

photoperiodism Biological response to change in relative lengths of daylight and darkness.

photoreceptor Light-sensitive sensory cell.

photosynthesis Trapping of sunlight energy, followed by its conversion to chemical energy (ATP, NADPH, or both) and then synthesis of sugar phosphates, which become converted into sucrose, cellulose, starch, and other end products. The main pathway by which energy and carbon enter the web of life.

photosystem One of many clusters of light-trapping pigment molecules embedded in photosynthetic membranes. A chlorophyll of the system gives up electrons necessary for the light-dependent reactions of photosynthesis.

phycobilin (FIE-koe-BY-lin) Type of accessory pigment that extends the functional range of chlorophyll in photosynthesis. Abundant in red algae and in cyanobacteria especially.

phylogeny Evolutionary relationships among species, starting with an ancestral form and including branches leading to descendants.

phytochrome A light-sensitive pigment. Its controlled activation and inactivation take part in plant hormone activities that govern leaf expansion, stem branching, stem lengthening and often seed germination and flowering.

pigment Any light-absorbing molecule.

pioneer species Any opportunistic colonizer of barren or disturbed habitats. Adapted for rapid growth and dispersal.

plankton [Gk. *planktos*, wandering] Of aquatic habitats, a community of suspended or weakly swimming organisms, mostly microscopic.

plant Generally, a multicelled photoautotroph with well-developed root and shoot systems; photosynthetic cells that include starch grains as well as chlorophylls *a* and *b* ; and cellulose, pectin, and other polysaccharides in cell walls.

Plantae Kingdom of plants.

plasma (PLAZ-muh) Liquid portion of blood; mainly water in which ions, proteins, sugars, gases, and other substances are dissolved.

plasma membrane Outermost cell membrane; structural and functional boundary between cytoplasm and the fluid outside the cell.

plasmid Of many bacteria, a small, circular molecule of extra DNA that carries only a few genes and that is replicated independently of the bacterial chromosome.

plasmolysis Osmotically induced shrinkage of a cell's cytoplasm.

pleiotropy (PLEE-oh-troe-pee) [Gk. *pleon*, more, + *trope*, direction] Positive or negative effects on two or more traits owing to expression of alleles at a single gene locus. Effects may or may not emerge at the same time.

polar body One of four cells that form by the meiotic cell division of an oocyte but that does not become the ovum.

pollutant Natural or synthetic substance with which an ecosystem has no prior evolutionary experience, in terms of kinds or amounts; it accumulates to disruptive or harmful levels.

polymer (POH-lih-mur) [Gk. *polus*, many, + *meris*, part] Large molecule consisting of three to millions of monomers of the same or different kinds.

polymerase (puh-LIM-ur-aze) Enzyme that catalyzes a polymerization reaction (e.g., the DNA polymerase of DNA replication/repair).

polymorphism (poly-MORE-fizz-um) [Gk. *polus*, many, + *morphe*, form] The persistence of two or more qualitatively different forms of a trait (morphs) in a population.

polypeptide chain Organic compound with a sequence of three or more amino acids. Peptide bonds between them result in a regular pattern of nitrogen atoms in the carbon backbone: —N—C—C—N—C—C— . Every protein consists of one or more polypeptide chains.

polyploidy (POL-ee-PLOYD-ee) Having three or more of each type of chromosome in the nucleus of cells at interphase.

polysaccharide [Gk. *polus*, many, + *sakcharon*, sugar] Straight or branched chain of many covalently linked sugar units of the same or different kinds. In nature, the most common polysaccharides are cellulose, starch, and glycogen.

polysome A number of ribosomes translating the same mRNA molecule at the same time, one after the other, during protein synthesis.

population All individuals of the same species occupying the same area.

population density Count of individuals of a population occupying a specified area or specified volume of a habitat.

population distribution Dispersal pattern for individuals of a population through a habitat.

population size The number of individuals that make up the gene pool of a population.

positive feedback mechanism A homeostatic control mechanism. It sets in motion a chain of events that intensifies change from an original condition; after a limited time, intensification reverses the change.

potential energy Capacity of any stationary object to do work owing to its position in space or to the arrangement of its parts (e.g., a cat in a frozen posture, about to spring at a mouse).

predation Ecological interaction in which a predator feeds on a prey organism.

predator [L. *prehendere*, to grasp, seize] A heterotroph that feeds on other living organisms (its prey), that lives neither in or nor on them (as parasites do), and that may or may not end up killing them.

prediction Statement about what you should observe in nature if you were to go looking for a particular phenomenon; the if–then process.

primary immune response Defensive actions by white blood cells and their secretions, as elicited by first-time recognition of antigen. Includes antibody- and cell-mediated responses.

primary wall A wall of polysaccharides, glycoproteins, and cellulose that is flexible and thin enough to allow new plant cells to divide or change shape during growth and development.

primate Mammalian lineage dating from the Eocene; includes prosimians, tarsioids, and anthropoids (monkeys, apes, and humans).

primer Short nucleotide sequence designed to base-pair with any complementary DNA sequence; later, DNA polymerases recognize it as a START tag for replication.

probability The chance that each outcome of a given event will occur is proportional to the number of ways the outcome can be reached.

probe Very short stretch of DNA designed to base-pair with part of a gene being studied and labeled with an isotope to distinguish it from DNA in the sample being investigated.

producer Autotroph (self-feeder); it nourishes itself using sources of energy and carbon from its physical environment. Photoautotrophs and chemoautotrophs are examples.

prokaryotic cell (pro-CARE-EE-oh-tic) [L. *pro*, before, + Gk. *karyon*, kernel] Archaebacterium or eubacterium; single-celled organism, most often walled; lacks the profusion of membrane-bound organelles observed in eukaryotic cells.

prokaryotic fission Cell division mechanism by which a bacterial cell reproduces.

promoter Short stretch of DNA to which RNA polymerase can bind and start transcription.

prophase Of mitosis, a stage when duplicated chromosomes start to condense, microtubules form a spindle, and the nuclear envelope starts to break up. Duplicated pairs of centrioles (if present) are moved to opposite spindle poles.

prophase I The first stage of meoisis I. Each duplicated chromosome starts to condense. It pairs with its homologue; nonsister chromatids usually undergo crossing over. Each becomes attached to microtubular spindle. One of the duplicated pairs of centrioles (if present) is moved to opposite spindle pole.

prophase II First stage of meiosis II. In each daughter cell, spindle microtubules attach to kinetochores of each chromosome and move them toward spindle's equator. One centriole pair (if present) is already at each spindle pole.

protein Organic compound composed of one or more polypeptide chains.

Protista Kingdom of protistans. Chytrids; water molds; slime molds; protozoans; sporozoans; euglenoids; chrysophytes; dinoflagellates; and red, brown, and green algae are major groups.

protistan (pro-TISS-tun) [Gk. *prōtistos*, primal, very first] Diverse species, ranging from single cells to giant kelps, that are photoautotrophs, heterotrophs, or both. Some are thought to be most like the earliest eukaryotic cells. All are unlike bacteria in having a nucleus, large ribosomes, mitochondria, ER, Golgi bodies, chromosomes with many proteins attached, and cytoskeletal microtubules.

proton Positively charged particle; one or more reside in nucleus of each atom. An unbound (free) proton is called a hydrogen ion (H^+).

proto-oncogene Gene sequence similar to an oncogene but coding for a protein that is used in normal cell functions. When mutated, it may trigger cancerous transformation.

protozoan Type of protistan that may resemble the single-celled heterotrophs that gave rise to animals. Amoeboid, animal-like, and ciliated protozoans are major categories.

Punnett-square method Construction of a diagram of a genetic cross that is a simple way to predict the probable outcomes.

purine Nucleotide base having a double ring structure (e.g., adenine or guanine).

pyrimidine (pih-RIM-ih-deen) Nucleotide base having a single ring structure (e.g., cytosine or thymine).

pyruvate (PIE-roo-vate) A small organic compound with a backbone of three carbon atoms. Two molecules form as end products of glycolysis.

radial symmetry Animal body plan having four or more roughly equivalent parts around a central axis (e.g., sea anemone).

radioisotope Unstable atom (uneven number of protons and neutrons). It spontaneously emits particles and energy; over a predictable time span, it decays into a different atom.

radiometric dating Method of measuring the proportions of (1) a radioisotope in a mineral trapped long ago in newly formed rock and (2) a daughter isotope that formed from it by radioactive decay in the same rock. Used to assign absolute dates to fossil-containing rocks and to the geologic time scale.

rearrangement, molecular Conversion of one organic compound to another through changes in its internal bonds.

receptor, molecular Type of membrane protein that binds an extracellular substance (e.g., hormone).

receptor, sensory Sensory cell or specialized cell adjacent to it that can detect a stimulus.

recessive allele [L. *recedere*, to recede] In heterozygotes, an allele whose expression is fully or partially masked by expression of its partner. Fully expressed only in homozygous recessives.

recombinant chromosome Of eukaryotes, a chromosome that emerges from meiosis with a combination of alleles that differs from a parental combination of alleles.

recombinant DNA Any molecule of DNA that incorporates one or more nonparental nucleotide sequences. Outcome of microbial gene transfer in nature or recombinant DNA technology.

recombinant DNA technology Procedures by which DNA molecules from different species are isolated, cut up, and spliced together. A population of rapidly dividing bacterial cells or PCR is then used to amplify the recombinant molecules to useful quantities.

recombination Any enzyme-mediated reaction that inserts one DNA sequence into another. "Generalized" recombination uses any pair of homologous sequences between chromosomes as substrates, as during crossing over. Site-specific recombination uses only a short stretch of homology between viral and bacterial DNA. A different reaction can insert transposable elements at new, random sites in bacterial or eukaryotic genomes; no homology is required.

regulatory protein Component of mechanisms that control transcription, translation, and gene products by interacting with DNA, RNA, new polypeptide chains, or proteins (e.g., enzymes).

repressor Protein that binds with an operator on bacterial DNA to block transcription.

reproduction Any process by which a parental cell or organism produces offspring. Among eukaryotes, asexual modes (e.g., binary fission, budding, vegetative propagation) and sexual modes. Bacteria employ prokaryotic fission. Viruses do not reproduce themselves; host organisms execute their replication cycle.

reproductive isolating mechanism A heritable feature of body form, functioning, or behavior that prevents interbreeding between two or more genetically divergent populations.

reproductive success Production of viable offspring by the individual.

respiration [L. *respirare*, to breathe] Of all animals, exchange of environmental oxygen with carbon dioxide from cells (e.g., through integumentary exchange or a respiratory system).

restriction enzyme One of a class of bacterial enzymes that cut apart foreign DNA injected into the cell body, as by viruses. Important tool of recombinant DNA technology.

reverse transcription Synthesis of DNA on an RNA template by using reverse transcriptase, a viral enzyme. Basis of RNA virus replication cycle and of cDNA synthesis in laboratory.

RFLP Short for restriction fragment length polymorphism. DNA fragments of different sizes, cleaved by restriction enzymes, that reveal genetic differences among individuals.

Rh blood typing Method of characterizing red blood cells according to a self-marker protein at their surface. Rh^+ cells have it; Rh^- cells do not.

ribosomal RNA (rRNA) Type of RNA that combines with proteins to form ribosomes, on which polypeptide chains are assembled.

ribosome Structure composed of two subunits of rRNA and proteins. Has binding sites for mRNA and tRNAs, which interact to produce a polypeptide chain in translation stage of protein synthesis.

RNA Ribonucleic acid. Any of a class of single-stranded nucleic acids that function in transcribing and translating the genetic instructions encoded in DNA into proteins. A molecule of mRNA, rRNA, or tRNA.

rubisco RuBP carboxylase; an enzyme that catalyzes attachment of the carbon atom from CO_2 to RuBP and so starts the Calvin–Benson cycle of the light-independent reactions.

RuBP Short for ribulose bisphosphate. Organic compound that has a backbone of five carbon atoms that serves in carbon fixation and that is regenerated in the Calvin–Benson cycle in C3 plants.

salinization Salt buildup in soil by evaporation, poor drainage, and heavy irrigation.

salt Compound that releases ions other than H^+ and OH^- in solution.

second law of thermodynamics A law of nature stating that the spontaneous direction of energy flow is from forms organized to less organized forms. The total amount of energy in the universe is spontaneously flowing from forms of higher to lower quality; with each conversion, some energy becomes randomly dispersed in a form (heat, most often) not as readily available to do work.

second messenger Molecule within a cell that mediates a hormonal signal by initiating the cellular response to it.

secondary immune response Immune action against previously encountered antigen, more rapid and prolonged than a primary response owing to swift participation of memory cells.

secondary wall A wall on the inner surface of the primary wall of an older plant cell that stopped growing but needs structural support. Contains lignin in older cells of woody plants.

secretion A cell acting on its own or as part of glandular tissue releases a substance across its plasma membrane, to the surroundings.

segregation, theory of [L. *se-*, apart, + *grex*, herd] Mendelian theory. Sexually reproducing organisms inherit pairs of genes (on pairs of homologous chromosomes), the two genes of each pair are separated from each other at meiosis, and they end up in separate gametes.

selective gene expression Control of which gene products a cell makes or activates during a specified interval. Depends on the type of cell, its adjustments to changing chemical conditions, which external signals it is receiving, and its built-in control systems.

selective permeability Of a cell membrane, a capacity to let some substances but not others cross it at certain sites, at certain times. The capacity arises as an outcome of its lipid bilayer structure and its transport proteins.

semiconservative replication [Gk. *hēmi*, half, + L. *conservare*, to keep] Mechanism of DNA duplication. The DNA double helix unzips, and a complementary strand is assembled on exposed bases of each strand. Each conserved strand and its new partner wind up together to form a double helix, thus being a half-old, half-new molecule.

senescence (sen-ESS-cents) [L. *senescere*, to grow old] Processes leading to the natural death of an organism or to parts of it (e.g., leaves).

sensation Conscious awareness of a stimulus.

sensory neuron Type of neuron that detects a stimulus and relays information about it toward an integrating center (e.g., a brain).

sensory system The "front door" of a nervous system; it detects external and internal stimuli and relays information to integrating centers that issue commands for responses.

sex chromosome A chromosome with genes that influence primary sex determination (whether male or female gonads will develop in the new individual). Depending on the species, somatic cells have one or two sex chromosomes, of the same or different type. In mammals, females are XX and males XY.

sexual dimorphism Occurrence of female and male phenotypes among the individuals of a sexually reproducing species.

sexual reproduction Production of offspring by meiosis, gamete formation, and fertilization.

sexual selection A microevolutionary process. Natural selection favors a trait that gives the individual a competitive edge in attracting or keeping a mate, hence in reproductive success.

shell model Model of electron distribution in which all orbitals available to electrons of atoms occupy a nested series of shells.

sign stimulus Simple environmental cue that triggers a response to a stimulus that the nervous system is prewired to recognize.

sister chromatid Of a duplicated chromosome, one of two DNA molecules (and associated proteins) attached at the centromere until they are separated from each other during mitosis or meiosis. After separation, each is then called a chromosome in its own right.

six-kingdom classification scheme A recent phylogenetic scheme that groups all organisms into the kingdoms Eubacteria, Archaebacteria, Protista, Fungi, Plantae, and Animalia.

skeletal muscle An organ with hundreds to many thousands of muscle cells bundled inside a sheath of connective tissue, which extends past the muscle as tendons.

sodium–potassium pump Type of membrane transport protein that, when activated by ATP, selectively transports potassium ions across a membrane, against its concentration gradient, and passively allows sodium ions to cross it in the opposite direction.

solute (SOL-yoot) [L. *solvere*, to loosen] Any substance dissolved in a solution. Spheres of hydration around charged parts of its ions and molecules keep them dispersed.

solvent Any fluid (e.g., water) in which one or more substances are dissolved.

somatic cell (so-MAT-ik) [Gk. *somā*, body] Any body cell that is not a germ cell. (Germ cells are the forerunners of gametes.)

somatic nervous system Nerves leading from a central nervous system to skeletal muscles.

speciation (spee-see-AY-shun) The formation of a daughter species from a population or subpopulation of a parent species by way of microevolutionary processes. Routes vary in their details and in length of time before the required reproductive isolation is completed.

species (SPEE-sheez) [L. *species*, a kind] One kind of organism. Of sexually reproducing organisms, one or more natural populations in which individuals are interbreeding and are reproductively isolated from other such groups.

sperm [Gk. *sperma*, seed] Mature male gamete.

spermatogenesis (sper-MAT-oh-JEN-ih-sis) Formation of mature sperm from a germ cell.

sphere of hydration A clustering of water molecules around individual molecules or ions of a substance placed in water owing to positive and negative interactions among them.

spinal cord The part of the central nervous system in a canal inside the vertebral column; site of direct reflex connections between sensory and motor neurons; also has tracts to and from the brain.

spore Reproductive or resting structure of one or a few cells, often walled or coated and adapted for resisting adverse conditions, for dispersal, or both. May be nonsexual or sexual (formed by way of meiosis). Sporozoans, fungi, plants, and some bacteria form spores.

stabilizing selection Mode of natural selection by which intermediate phenotypes in the range of variation are favored and extremes at both ends are eliminated.

start codon Base triplet in mRNA that serves as the START signal for translation.

stem cell Self-perpetuating animal cell that stays unspecialized. Some of its daughter cells also are self-perpetuating; others differentiate into specialized cells (e.g., red blood cells that arise from stem cells in bone marrow).

steroid hormone Lipid-soluble hormone made from cholesterol that acts on a target cell's DNA by entering the nucleus alone or bound to intracellular receptor. Some act by binding to a receptor on a target's plasma membrane.

sterol (STAIR-all) Lipid with a rigid backbone of four fused carbon rings. Sterols differ in the number, position, and type of their functional groups. Cholesterol is one; it is a precursor of steroid hormones and occurs in animal cell membranes.

stimulus [L. *stimulus*, goad] A specific form of energy (e.g., pressure, light, and heat) that activates a sensory receptor able to detect it.

stop codon Base triplet in a strand of mRNA that serves as a STOP signal during translation; it blocks further additions of amino acids to a newly forming polypeptide chain.

strain One of two organisms with differences that are too minor to classify it as a separate species (e.g., *Escherichia coli* strain 018:K1:H).

stroma [Gk. *strōma*, bed] A semifluid matrix between the thylakoid membrane system and the two outer membranes of a chloroplast; a zone where sucrose, starch, cellulose, and other end products of photosynthesis are assembled.

substrate Reactant or precursor for a specific enzyme-mediated metabolic reaction.

substrate-level phosphorylation The direct, enzyme-mediated transfer of a phosphate group from a substrate to a molecule, as when an intermediate of glycolysis gives up a phosphate group to ADP to form ATP.

surface-to-volume ratio Mathematical relation in which volume increases with the cube of the diameter, but surface area increases only with the square. If a growing cell were simply to expand in diameter, its volume of cytoplasm would increase faster than the surface area of the plasma membrane required to service it. In general, this constraint keeps cells small, elongated, or with infoldings or outfoldings of its plasma membrane.

symbiosis (sim-by-OH-sis) [Gk. *sym*, together, + *bios*, life, mode of life] Individuals of one species live near, in, or on those of another species for at least part of life cycle (e.g., in commensalism, mutualism, and parasitism).

sympathetic nerve An autonomic nerve that deals mainly with increasing overall body activities at times of heightened awareness, excitement, or danger; also works continually in opposition with parasympathetic nerves to make minor adjustments in internal organ activities.

syndrome A set of symptoms that may not individually be a telling clue but collectively characterize a genetic disorder or disease.

T cell T lymphocyte; a type of white blood cell vital to immune responses (e.g., helper T cells and cytotoxic T cells).

target cell Any cell with molecular receptors that can bind with a particular hormone or some other signaling molecule.

taxonomy Field of biology that deals with identifying, naming, and classifying species.

telophase (TEE-low-faze) Of meiosis I, the stage when one member of each pair of homologous chromosomes has arrived at a spindle pole. Of mitosis and of meiosis II, the stage when chromosomes decondense into threadlike structures and two daughter nuclei form.

temperature A measure of the kinetic energy of ions or molecules in a specified region.

territory An area that an animal is defending against competitors for mates, food, water, living space, other resources.

test A means to determine the accuracy of a prediction, as by conducting experimental or observational tests and by developing models. Scientific tests are made under controlled conditions in nature or the laboratory.

testcross Experimental cross to determine whether an individual of unknown genotype that shows dominance for a trait is either homozygous dominant or heterozygous.

theory, scientific A testable explanation of a broad range of related phenomena, one that has been subjected to extensive experimental testing and can be used with a high degree of confidence. A scientific theory remains open to tests, revision, and tentative acceptance or rejection.

thermophile A type of archaebacterium that is adapted to unusually hot aquatic habitats, such as hot springs and hydrothermal vents.

thermoreceptor Sensory cell or specialized cell next to it that detects radiant energy (heat).

thylakoid Of chloroplasts, part of an internal membrane system folded repeatedly into a stack of disks. Such stacks (grana) have light-absorbing pigments and enzymes required to form ATP, NADPH, or both in photosynthesis. The stacks connect (by membranous channels) as a single functional compartment.

thymine Nitrogen-containing base; one of the nucleotides in DNA.

tissue Of multicelled organisms, a group of cells and intercellular substances that function together in one or more specialized tasks.

tonicity (TOE-niss-ih-TEE) Relative solute concentrations of two fluids (e.g., cytoplasmic fluid relative to extracellular fluid).

touch-killing Mechanism by which cytotoxic T cells directly release perforins and toxins onto a target cell and cause its destruction.

toxin A normal metabolic product of one species with chemical effects that can hurt or kill individuals of a different species.

trace element Any element that represents less than 0.01 percent of body weight.

tracer Substance with a radioisotope attached, like a shipping label, that researchers can track after delivering it into a cell, body, ecosystem, or some other system. Laboratory devices detect emissions from the tracer as it moves through a pathway or reaches a destination.

transcription [L. *trans*, across, + *scribere*, to write] First stage of protein synthesis. An RNA strand is assembled on exposed bases of one unwound strand of a DNA double helix. The transcript's base sequence is complementary to that of the DNA template.

transfer RNA (tRNA) An RNA that binds with and delivers amino acids to a ribosome and that pairs with an mRNA codon during the translation stage of protein synthesis.

translation Stage of protein synthesis when an mRNA's base sequence becomes converted to a sequence of particular amino acids in a new polypeptide chain. rRNA, tRNA, and mRNA interact to bring this about.

translocation Of cells, a stretch of DNA that moved to a new location in a chromosome or in a different chromosome, with no molecular loss. Of vascular plants, a process by which organic compounds are distributed through the phloem.

transport protein One of many kinds of membrane proteins involved in active or passive transport of water-soluble substances across the lipid bilayer of a cell membrane. Solutes on one side of the membrane pass through the protein's interior to the other side.

transposable element A stretch of DNA that can move at random from one location to another in the individual's genome. Often it inactivates the genes into which it becomes inserted and causes changes in phenotype.

triglyceride (neutral fat) A type of lipid that has three fatty acid tails attached to a glycerol backbone. Triglycerides are the body's most abundant lipids and its richest energy source.

trisomy (TRY-so-mee) The presence of three chromosomes of a given type in a cell rather than the two characteristic of a parental diploid chromosome number.

true breeding Of a sexually reproducing species, a lineage in which one version only of a trait shows up through the generations in all parents and their offspring.

turgor pressure (TUR-gore) [L. *turgere*, to swell] Internal fluid pressure on a cell wall when water moves into the cell by osmosis.

ultrafiltration Bulk flow of a small amount of protein-free plasma from a blood capillary when the outward-directed force of blood pressure is greater than the inward-directed osmotic force of interstitial fluid.

uniformity theory Early theory that the Earth's surface changes in gradual, uniformly repetitive ways (major floods, earthquakes, and other infrequent annual catastrophes were not considered unusual). Helped change Darwin's view of evolution. Has since been replaced by plate tectonics theory.

uracil (YUR-uh-sill) Nitrogen-containing base of a nucleotide in RNA but not DNA. Like thymine, uracil can base-pair with adenine.

urinary system Organ system that adjusts the volume and composition of blood, and thereby helps maintain extracellular fluid.

vaccination Immunization procedure against a specific pathogen.

vaccine An antigen-containing preparation, swallowed or injected, designed to increase immunity to certain diseases by inducing formation of armies of effector and memory B and T cells.

variable Of an experimental test, a specific aspect of an object or event that may differ over time and among individuals. A single variable is directly manipulated in an attempt to support or disprove a prediction; any other variables that might influence the results are identical (ideally) in both the experimental group and one or more control groups.

vertebrate Animal with a backbone.

vesicle (VESS-ih-kul) [L. *vesicula*, little bladder] One of a variety of small, membrane-bound sacs in the cytoplasm that function in the transport, storage, or digestion of substances.

vestigial (ves-TIDJ-ul) Applies to a small body part, tissue, or organ abnormally developed or degenerated and unable to function like its normal counterpart (e.g., vestigial wings of mutant fruit flies; "tail bones" of humans).

viroid An infectious particle consisting only of very short, tightly folded strands or circles of RNA. Viroids might have evolved from introns, which they resemble.

virus A noncellular infectious agent that is composed of DNA or RNA and a protein coat; it can become replicated only after its genetic material enters a host cell and subverts the host's metabolic machinery.

vitamin Any of more than a dozen organic substances that an organism requires in small amounts for metabolism but that it generally cannot synthesize for itself.

water potential Sum of two opposing forces (osmosis and turgor pressure) that can cause a directional movement of water into or out of an enclosed volume (e.g., a walled cell).

wavelength A wavelike form of energy in motion. The horizontal distance between the crests of every two successive waves.

wax Molecule having long-chain fatty acids packed together and linked to long-chain alcohols or to carbon rings. Waxes have a firm consistency and repel water.

wild-type allele Allele that occurs normally or with the greatest frequency at a given gene locus among individuals of a population.

X chromosome A type of sex chromosome. In mammals, an XX pairing causes an embryo to develop into a female; an XY pairing causes it to develop into a male.

X-linked gene Any gene that is located on an X chromosome.

X-linked recessive inheritance Recessive condition in which the responsible, mutated gene is located on the X chromosome.

Y chromosome Distinctive chromosome in males or females of many species, but not both (e.g., human males are XY and human females, XX).

Y-linked gene Gene on a Y chromosome.

yolk sac An extraembryonic membrane. In most shelled eggs, it holds nutritive yolk; in humans, part of the yolk sac becomes a site of blood cell formation and some cells give rise to forerunners of gametes.

zooplankton A community of suspended or weak-swimming heterotrophs of freshwater or marine habitats. Most of its species are microscopic; commonly, rotifers and copepods are among the most abundant.

zygote (ZYE-goat) The first cell of a new individual, formed by fusion of a sperm nucleus with egg nucleus at fertilization; a fertilized egg.

CREDITS AND ACKNOWLEDGMENTS

This page constitutes an extension of the copyright page. We have made every effort to trace the ownership of all copyrighted material and to secure permission from copyright holders. In the event of any question arising as to the use of any material, we will be pleased to make the necessary corrections in future printings. Thanks are due to the following authors, publishers, and agents for permission to use the material indicated.

ART BY LISA STARR **1.2; 2.7** (b); **2.10** (b); **2.17; 3.6** Art and Photograph; **3.19** (c); **4.12; 4.17** (above); **4.19; 5.11; 5.12; 6.5; 6.10; 6.11; 6.15; 6.16** (c); **6.19; 6.20; 6.21; 7.2; 7.3** (d, e); **7.5; 7.9; 7.10; p. 120; 7.11; 7.12; 7.14; 7.15; 8.4; 8.5** (b, c); **8.6; 8.7; 8.9; 8.10** (a); **9.9** (a, d); **9.10; 10.8; 10.9; 11.3; 12.12; 12.13; 12.14** (a); **12.22; 13.2; 13.7; 14.8; 14.12; 14.13; 14.14; 14.15; 14.16; 14.18; 15.5; 16.4; 16.5; 16.6; 16.8, 16.9; 16.11**

CHAPTER 1 **1.1** Frank Kaczmarek / **1.3** Jack de Coningh / **1.4** © Y. Arthrus-Bertrand/ Peter Arnold, Inc. / **p. 6** Jack de Coningh / **1.6** (a) Walt Anderson/Visuals Unlimited; (b) Gregory Dimijian/Photo Researchers; (c) Alan Weaving/Ardea London / **1.7** (a) R. Robinson/Visuals Unlimited; (b) Tony Brain/SPL/Photo Researchers; (c) M. Abbey/Visuals Unlimited; (d,f) Edward S. Ross; (e) Dennis Brokaw; (g,h) Pat & Tom Leeson/Photo Researchers / **1.8** J. A. Bishop, L. M. Cook / **1.9** Courtesy of Derrell Fowler, Tecumseh, OK / **1.11** Gary Head / **1.13** James Carmichael, Jr./ NHPA / **p. 19** James M. Bell/Photo Researchers /

CHAPTER 2 **2.1** Gary Head for Norman Terry, University of California, Berkeley / **2.2** Jack Carey / **2.4** Gary Head / **2.5** (a) Hank Morgan/Rainbow; (b) Art by Raychel Ciemma; (c) Harry T. Chugani, M.D., UCLA School of Medicine / **2.6** Micrograph Maris and Cramer / **2.10** Micrograph © Bruce Iverson / **2.11** Vandystadt/Photo Researchers / **2.13** Art by Raychel Ciemma; photograph © Kennan Ward/The Stock Market / **2.14** Copyright © Kennan Ward/The Stock Market / **2.15** H. Eisenbeiss/Frank Lane Picture Agency / **2.18** Michael Grecco/Picture Group /

CHAPTER 3 **3.1** Dave Schiefelbein / **p. 35** NASA / **3.3** Tim Davis/ Photo Researchers / **3.6** Photograph by Lisa Starr / **3.8** David Scharf/Peter Arnold, Inc. / **3.10** Clem Haagner/Ardea, London / **3.11** (c) micrograph, Lewis L. Lainey; (d) Larry Lefever/Grant Heilman; (e) Kenneth Lorenzen / **3.12** (a) Ron Davis/Shooting Star; (b) Frank Trapper/Sygma / **3.18** Art by Palay/Beaubois / **3.19** (a) CNRI/SPL/Photo Researchers; Art by R. Demarest / **3.20** (a) © Inga Spence/Tom Stack & Associates; (inset) Gary Head; (e) Photograph by Sue Hartzell / **3.22** (c) Art by Precision Graphics /

CHAPTER 4 **4.1** (a) Corbis/Bettmann; (b) Armed Forces Institute of Pathology; (c) National Library of Medicine; (e) The Francis A. Countway Library of Medicine / **4.3** (a, b, c) Art by Raychel Ciemma / **4.4** (a,c) Art by Raychel Ciemma; (b) George S. Ellmore / **4.5** (a) Leica Microsystems, Inc., Deerfield, IL.; (b, d) Art by Gary Head; (c) © George Musil/Visuals Unlimited / **4.6** Micrograph by Driscoll, Youngquist, and Baldeschwieler/Cal Tech/Science Source/Photo Researchers; Art by Raychel Ciemma / **4.7** (a–d) Jeremy Pickett-Heaps, School of Botany, University of Melbourne / **4.8** Art by Raychel Ciemma and ACG / **4.9** M.C. Ledbetter, Brookhaven National Laboratory; Art by Raychel Ciemma and ACG / **4.10** G.L. Decker; Art by Raychel Ciemma and ACG / **4.11** Stephen L. Wolfe / **4.12** (left) Don W. Fawcett/Visuals Unlimited; (right) A.C. Faberge, *Cell and Tissue Research*, 151: 403-415, 1974. / **4.13** Art by Raychel Ciemma and ACG / **4.14** (a,b) Micrographs, Don W. Fawcett/Visuals Unlimited; (above) Art by Raychel Ciemma / **4.15** Gary W. Grimes; Art by R. Demarest after a model by J. Kephart / **4.16** Keith R. Porter / **4.17** L. K. Shumway; (below right) Art by Palay/Beaubois / **4.18** J. Victor Small, Gottfried Rinnerthaler / **4.20** (a) © Lennart Nilsson; (b) CNRI/SPL/Photo Researchers / **4.21** Art by Precision Graphics after Stephen L. Wolfe, *Molecular and Cellular Biology*, Wadsworth, 1993 / **4.23** Ron Hoham, Dept. of Biology, Colgate University / **4.24** (a, c) Art by Raychel Ciemma; (b) Biophoto Associates/Photo Researchers / **4.25** (a) George S. Ellmore; Art by Raychel Ciemma; (b) Ed Reschke; Art by Joel Ito / **4.26** Art by Raycehl Ciemma / **4.27** (a) Art by Raychel Ciemma; (b) Courtesy of Dr. G. Cohen-Bazire; (c) K. G. Murti/Visuals Unlimited; (d) R. Calentine/Visuals Unlimited; (e)Gary Gaard and Arthur Kelman /

CHAPTER 5 **5.1** (a) Runk/Schoenberger/Grant Heilman; (b) Ingio Everson/Bruce Coleman Ltd. / **5.2** (a, b) Art by Precision Graphics; (c) Art by Raychel Ciemma / **5.3** Art by Raychel Ciemma / **5.4** Charles D. Winters/Photo Researchers; Art by Precision Graphics after Stephen L. Wolfe, *Molecular and Cellular Biology*, Wadsworth / **5.5** P. Pinto da Silva, D. Branton, *Journal of Cell Biology*, 45:98, by permission of The Rockefeller University Press; Art by Palay/Beaubois / **5.8** Art by Palay/Beaubois / **5.9** Art by Precision Graphics / **5.10** Art by Raychel Ciemma / **5.14** M. Sheetz, R. Painter, and S. Singer, *Journal of Cell Biology*, 70:193 (1976) by permission of The Rockefeller University Press / **5.17** Art by Leonard Morgan / **5.18** M. M. Perry and A. M. Gilbert / **5.19** M. Abbey/Visuals Unlimited / **5.22** Frieder Sauer/Bruce Coleman Ltd. /

CHAPTER 6 **6.1** (a, above) © Oxford Scientific Films/Animals Animals; (a, below) © Raymond Mendez/Animals Animals; (b) Keith V. Wood / **6.2** (a, b) C. Contag, *Molecular Microbiology*, November 1985, Vol. 18, No. 4, pp. 593-603. "Photonic Detection of Bacterial Pathogens in Living Hosts." Reprinted by permission of Blackwell Science / 6.3 Evan Cerasoli / **6.4** (above) NASA 97; (below) Manfred Kage/Peter Arnold, Inc. / **6.7** Art by Nadine Sokol; Frog photos NHPA/A.N.T. Photo Library / **6.10** Art by Raychel Ciemma and ACG after B. Alberts, et al, *Molecular Biology of the Cell*, Garland Publishing, 1983 / **6.16** (a, b) Thomas A. Steitz; (c) Art by Palay/Beaubois / **6.17** (b) Douglas Faulkner/Sally Faulkner Collection / **6.21** (c, d) © Gary Head /

CHAPTER 7 **7.1** Carolina Biological Supply Company; Art by Raychel Ciemma / **7.2** Photograph PhotoVault / **7.4** (b) Barker-Blakenship/FPG; (c) Art by Precision Graphics after Govindjee / **7.5** Art by Precision Graphics after Stephen L. Wolfe, *Molecular and Cellular Biology*, Wadsworth / **7.6** Barker-Blakenship/FPG / **7.7** Art by Precision Graphics / **7.8** Larry West/FPG / **7.13** E. R. Degginger / **7.15** Art by Precision Graphics / **7.16** (a) Grant Heilman, Inc.; Art by Raychel Ciemma (b) Dick Davis/Photo Researchers; (c, left) B.J. Miller, Fairfax, VA/BPS; (c, right) Martin Grosnick/Ardea London / **p. 125** Steve Chamberlain, Syracuse University / 7.17 NASA /

CHAPTER 8 **8.1** Stephen Dalton/Photo Researchers / **8.3** (a, both) Gary Head; (b) Paolo Fioratti / **8.4** (right) Art by \ Palay/Beaubois / **8.5** (a) Keith R. Porter; (b) Art by L. Calver / **8.7** Art by Raychel Ciemma / **8.10** (b) Adrian Warren/Ardea London; (c) David M. Phillips/Visuals Unlimited / **8.11** William Grenfell/Visuals Unlimited / **8.12** Gary Head / **p. 14** R. Llewellyn/Superstock, Inc. / **p. 147** © Lennart Nilsson /

CHAPTER 9 **9.1** (left, right) Chris Huss; (top,bottom) Tony Dawson / **9.2** C. J. Harrison et al, *Cytogenetics and Cell Genetics*, 35:21-27, © 1983, S. Karger, and A. G. Basel / **9.4** A. S. Bajer, University of Oregon / **9.5** Photos © Ed Reschke; Art by Raychel Ciemma / **9.6** (left) Micrograph © R. Calentine/Visuals Unlimited; (right) Micrograph B. A. Palevita and E. H. Newcomb, Univ. of Wisconsin/BPS/Tom Stack & Associates / **9.7** (d) Micrograph © D. M. Phillips/Visuals Unlimited / **9.8** (a–c, e) Lennart Nilsson from *A Child Is Born* © 1966, 1967 Dell Publishing Company, Inc.; (d) Lennart Nilsson from *Behold Man*, © 1974 by Albert Bonniers Forlag and Little, Brown & Company, Boston / **9.9** (a) C. J. Harrison et al *Cytogenetics and Cell Genetics*, 35: 21-27 © 1983 S. Karger, A. G. Basel; (b) B. Hamkalo; (c) O. L. Miller, Jr., Steve L. McKnight / **9.11** (a) Courtesy of The Family of Henrietta Lacks; (b) Dr. Pascal Madaule, France /

CHAPTER 10 **10.1** (a) Jane Burton/Bruce Coleman Ltd.; (b) Dan Kline/Visuals Unlimited / **10.3** CNRI/SPL/Photo Researchers / 10.4, 10.6 Art by Raychel Ciemma / **10.9** David M. Phillips/Visuals Unlimited / **10.10** Art by Raychel Ciemma / **10.12** © Richard Corman/Outline /

CHAPTER 11 **11.1** (left) Frank Trapper/Sygma; (center) Focus on Sports; (right, above) Moravian Museum, Brno; (right, below) Fabian/Sygma / **11.2** Jean M. Labat/Ardea, London; Art by Jennifer Wardrip / **11.5, 11.7** Art by Hans & Cassady, Inc. / **11.8** Art by Raychel Ciemma / **11.10** William E. Ferguson / **11.12** (a, b) Micrographs Stanley Flegler/Visuals Unlimited / **11.13** (a, b) Michael Stuckey/Comstock, Inc.; (c) © Russ Kinne/Comstock, Inc. / **11.14** David Hosking / **11.15** Tedd Somes / **11.16** (top to bottom) Frank Cezus; Frank Cezus; Michael Keller;Ted Beaudin; Stan Sholik/all FPG / **11.17** (a) Dan Fairbanks, Brigham Young University / **11.18** Jane Burton/Bruce Coleman Ltd.; Art by D. & V. Hennings / **11.19** (left) William E. Ferguson; (right) Eric Crichton/Bruce Coleman Ltd. / **11.20** Evan Cerasoli / **11.21** Leslie Faltheisek/Clacritter Manx / **11.22** Joe McDonald/Visuals Unlimited /

CHAPTER 12 **12.1** From "Multicolor Spectral Karyotyping of Human Chromosomes," by E. Schrock, T. Ried, et al, *Science*, 26, July 1966, 273:495. Used by permission of E. Schrock and T. Reid and the American Association for the Advancement of Science / **12.2** Paul A. Thiessen / **12.3** (a, f) Photographs Omikron/Photo Researchers / **12.4** (left) Gary Head; (right) Jack Carey and Lisa Starr / **12.5** (a) From Lennart Nilsson, *A Child Is Born*, © 1966, 1977 Dell Publishing Company, Inc.; (b) Art by R. Demarest after Patten, Carlson & others; (c) Redrawn by R. Demarest by permission from p. 126 of M. Cummings, *Human Heredity: Principles and Issues,* 3rd Ed. © 1994 by Brooks/Cole. All rights reserved / **12.6** Photograph Carolina Biological Supply Company / **12.9** Art by Raychel Ciemma / **12.11** (b) Photograph Dr. Victor A. McKusick; (c) Steve Uzzell / **12.13** (b) Giraudon/Art Resource, NY / **12.14** (a) After V. A. McKusick, *Human Genetics,* 2nd Ed., © 1969. Reprinted by permission of Prentice-Hall, Inc., Englewood Cliffs, NJ; (b) Photograph Corbis-Bettmann / **12.15** C. J. Harrison / **12.16** Eddie Adams/ AP Photo / **12.17** (a, b) Courtesy G. H. Valentine / 12.18 From "Multicolor Spectral Karyotyping of Human Chromosomes," by E. Schrock, T. Ried, et al, *Science,* 26, July 1966, 273:496. Used by permission of E. Schrock and T. Reid and the American Association for the Advancement of Science / **12.19** Art by Raychel Ciemma / **12.20** (a, left) Cytogenetics Laboratory, University of California, San Francisco; (a, right) Photograph courtesy of Peninsula Association for Retarded Children and Adults, San Mateo Special Olympics, Burlingame, CA; (b, left) After Collman and Stoller, *American Journal of Public Health,* 52, 1962; (b, right) Photograph used by permission of Carole Lafrate / **12.21** Art by Raychel Ciemma / **12.23** (a) Fran Heyl Associates © Jacques Cohen, computer enhanced by © Pix Elation; (b) Art by Raychel Ciemma / **12.25** Carolina Biological Supply Company / **12.27** Bonnie Kamin/Stuart Kenter Associates /

CHAPTER 13 **13.1** A C. Barrington Brown © 1968 J. D. Watson / **13.3** Art by Raychel Ciemma / **13.4** (c) Lee D. Simon/Science Source/Photo Researchers / **13.6** Micrograph Biophoto Associates/SPL/Photo Researchers / **13.8** Courtesy of the Cold Springs Harbor Laboratory Archives / **13.11** (b) PA News Photo Library /

CHAPTER 14 **14.1** (above) Dennis Hallinan/FPG; (below) © Bob Evan/Peter Arnold, Inc. / **14.3** Gary Head / **14.4** From Stephen L. Wolfe, *Molecular and Cellular Biology,* Wadsworth / **14.5** Art by Palay/Beaubois and Precision Graphics / **14.8** Art by Hans & Cassady, Inc. / **14.12** (a) 3-D Model of tRNA by David B. Goodin, Ph.D. / **14.17** (left) Nik Kleinberg; (right) Peter Starlinger / **14.18** Art by Raychel Ciemma / **14.19** Courtesy of the National Neurofibromatosis Foundation /

CHAPTER 15 **15.1** (a) Ken Greer/Visuals Unlimited; (b) Biophoto Associates/Science Source/Photo Researchers; (c) James Stevenson/SPL/Photo Researchers; (d) Gary Head / **15.2** Lennart Nilsson © Boehringer Ingelheim International GmbH / **15.3** Brian Matthews, University of Oregon / **15.4** Art by Palay/Beaubois and Hans & Cassady. / **15.5** Art by Raychel Ciemma / **15.6** Micrograph M. Roth and J. Gall / **15.7** Art by Palay/Beaubois / **15.8** (a) Dr. Karen Dyer Montgomery / **15.9** Jack Carey / **15.10** W. Beerman; Art by Raychel Ciemma / **15.11** Art by Precision Graphics; Photograph Frank B. Salisbury / **15.12** Art by Betsy Palay/Atemis / **15.13** Slim Films /

CHAPTER 16 **16.1** Lewis L. Lainey / **16.2** Science VU/Visuals Unlimited / **16.3** (a) Dr. Huntington Potter and Dr. David Dressler; (b) Stanley N. Cohen/ Science Source/Photo Researchers / **16.4** Art by Raychel Ciemma / **16.7** Cellmark Diagnostics, Abingdon, UK. / **16.10** Stephen L. Wolfe, *Molecular and Cellular Biology* / **16.11** Photograph Herve Chaumeton, Agence Nature / **16.12** Keith V. Wood / **16.13** (a) Dr. Vincent Chiang, School of Forestry and Wood Products, Michigan Technology University; (b) Courtesy Calgene LLC / **16.14** R. Brinster, R. E. Hammer, School of Veterinary Medicine, University of Pennsylvania / **16.15** © Adrian Arbib/ Still Pictures

INDEX

The letter i *designates illustration; the letter* t *designates table.*

D

E

G

Q

R